ANIMAL BEHAVIOUR

A Synthesis of Ethology and Comparative Psychology

ANIMAL BEHAVIOUR

A Synthesis of Ethology and Comparative Psychology

second edition

ROBERT A. HINDE
Sub-Department of Animal Behaviour
University of Cambridge, Madingley

McGRAW-HILL BOOK COMPANY
New York St. Louis San Francisco Düsseldorf
London Mexico Panama Sydney Toronto

ANIMAL BEHAVIOUR

A Synthesis of Ethology and Comparative Psychology

Library of Congress Catalog Card Number 66–16333

28927

1 2 3 4 5 6 7 8 9 0 MAMM 7 9 8 7 6 5 4 3 2 1 0

This book was set in Palatino by Brown Bros. Linotypers, Inc., and printed on permanent paper and bound by The Maple Press Company. The designer was Richard Paul Kluga, the drawings were done by Juan Barberis, and the cover was designed by Barbara Ellwood. The editors were James R. Young and Andrea Stryker-Rodda. Peter D. Guilmette supervised the production.

PREFACE TO THE SECOND EDITION

The preparation of a second edition only three years after the first seemed as though it would be a simple task. In practice, the new material which has become available and some necessary rethinking on old issues have demanded extensive rewriting of every chapter. To help the reader, I have expanded the discussion at a number of places. In addition, further section headings have been added, the table of contents and the index have been extended, and the illustrations have been increased in number and redrawn. The general organization of the book, however, remains the same, though the chapter on the integration of sequences of behaviour has been transferred to Section 3.

I would like to express my continuing gratitude to W. H. Thorpe for creating the department in which this work was done, for guidance over many issues, and for much helpful discussion. Among the many to whom I am grateful for comments arising from the first edition, there are some whom I would particularly like to mention. F. A. Beach, V. H. Denenberg, J. W. Kling, and L. de Ruiter sent me detailed comments on the whole book which were the greatest help in its revision. T. I. Thompson similarly made detailed and constructive criticisms on every chapter; it was fortunate for me that he was working at Madingley in 1967–1968 so that I was able to have many illuminating discussions with him. Joan Stevenson also read all parts of the book, and I was greatly helped by her comments. The new Chapter 25 is based largely on two papers written jointly with her, and she also gave me unstinting help on numerous other issues. Many other colleagues made detailed comments on parts of the book, and I profited much from their advice: here I would mention especially P. P. G. Bateson, W.

Baum, G. Horn, J. B. Hutchison, R. G. W. Prescott, Yvette Spencer-Booth, and H. P. Zeigler. I am especially grateful to Yvette Spencer-Booth, who was responsible for much of the data on rhesus monkeys cited in Chapter 26, and carried the greater burden of our joint research while I was engaged in this revision. Elizabeth Steel, who was largely responsible for many of the experiments on canaries cited in the text, also helped in a number of ways. Finally, Rachel Pollard-Urquhart attended to many detailed points in the final preparation of the manuscript. To these and many others I would like to express my gratitude.

ROBERT A. HINDE

PREFACE TO THE FIRST EDITION

The obsolescence of the old distinctions between the biological sciences is nowhere more evident than in the study of animal behaviour. The problems it raises not only provide a meeting ground for psychologists, zoologists, physiologists, anatomists, geneticists, ecologists and many others—they demand their cooperation. As a result of this, the subject of animal behaviour cannot be rigidly circumscribed: its students both contribute to and depend on the traditional disciplines of biology.

In this book I have attempted to survey the area where psychology, physiology, and ethology overlap. The term ethology is applied particularly to the work of students who, although differing widely in the problems they tackle, the level of analysis at which they work, the methods they use and the theoretical interpretations (if any) they adopt, share certain orienting attitudes which are, perhaps, more important as unifying factors than any particular theoretical scheme. Many of these attitudes are a consequence of the zoological training of the early ethologists. They felt, for instance, that the description and classification of behaviour is a necessary preliminary to its analysis; that the behaviour of an animal cannot be properly studied without some knowledge of the environment to which its species has become adapted in evolution; and that questions about the evolution and biological function of behaviour are, in priciple, as valid and important as those about its immediate causation.

In practice it is not profitable to make a rigid distinction between their interests and those of many physiologists and psychologists. In studying causation, ethologists have sought explanations in physiological terms; and there have long been comparative psychologists interested in the problems with which ethologists are concerned. More recently the mutual benefit to be gained from a marriage of the ethological approach with the detailed analyses made by those psychologists especially concerned with learning has become apparent. It is for these reasons that, although there will always, rightly, be some who continue in the traditions of classical comparative psychology and others who will pursue classical ethology, the term ethology is now often extended to cover the whole area of common interest.

Because the common ground involves, primarily, questions about the causation and development of behaviour in the individual, these have been my main concern: the evolution and function of behaviour, although the principal interest of the early ethologists such as Heinroth, Whitman, Craig, and Lorenz, are considered more briefly. While believing that the student of the causation of behaviour must work towards physiological explanations, limitations of space and my own competence have prohibited, with few exceptions, detailed discussion of neurophysiological mechanisms. At the other extreme, I have touched only briefly on social organization, dominance hierarchies, and other consequences of individual behaviour for the group. I have limited myself to those aspects of the study of learning for which ethologists' observations have a special relevance, and I have omitted any discussion of the genetics of behaviour. Where it seemed relevant, I have attempted to show how my subject matter is related to these other fields and to provide one or two recent key references for those who wish to follow them.

Within these limits I have drawn my material primarily from studies of vertebrate behaviour, to some extent from studies of arthropods and molluscs, but hardly at all from forms lower than these: the wider the range of species discussed, the greater the diversity of the mechanisms which underlie their behaviour and, therefore, the more superficial the generalizations that can be made.

To achieve a wholly satisfactory classification of the material has proved an impossible task: such traditional categories as motivation and perception, or maturation and learning, can be artificial, and constrain nature. Nevertheless, a book must be divided into chapters, so I have had to resort to frequent cross-references. The chapters are grouped into four sections. The first is concerned with general questions of method and theory. The immediate causation of behaviour is considered in Section 2, and its development in the individual in Section 3. Questions of the evolution and function of behaviour are discussed more briefly in Section 4.

It would have been impossible for me to have completed this book without the help of numerous friends and colleagues. I would like especially to thank W. H. Thorpe, Director of the Sub-department of Animal Behaviour at Cambridge; it has been my good fortune to work with him for the last fifteen years, and I am happy to have this opportunity to acknowledge my debt to him. He read the manuscript in full, and it has benefited greatly from his criticisms. The book was originally planned as a joint one with N. Tinbergen, but extraneous circumstances made a collaborative project impossible. He has nevertheless read part of the manuscript, and I am greatly indebted to him for his comments on it, and also for the many discussions we have had and for all I have learnt from him. Yvette Spencer-Booth read a complete draft of the manuscript. Her attention to points of logic and exposition and her hard-headed constructive criticism forced me to re-think many issues, and led to the elimination of many errors and inconsistencies: I am very grateful to her. In my attempts to bring together the studies of ethologists and comparative psychologists I have been helped by four American colleagues. Joan Stevenson read critically the whole manuscript, making many suggestions for its improvement on both major issues and points of detail, and introducing me to material with which I had previously been unfamiliar: I profited greatly from discussions with her. F. A. Beach, Frances Clayton, and J. S. Rosenblatt each read a number of chapters and, by their comments and suggestions, helped to reduce their shortcomings. I also received much help from others who read and commented on parts of the manuscript—P. Bateson, P. Bertelson, J. M. Cullen, G. Horn, J. S. Kennedy, Barbara Lade, H. W. Lissmann, C. J. Pennycuick, M. Richards, and a number of other colleagues. The final draft owed much to the care with which my wife amended earlier and clumsier versions. I am grateful to Yvette Spencer-Booth for a number of figures, and to Elizabeth Steel for some of the diagrams. Finally, I would like to acknowledge my debt to Moira Bruce for the care with which she looked after the preparation of the manuscript and helped in checking references and proofs.

ROBERT A. HINDE

CONTENTS

ANIMAL BEHAVIOUR

A Synthesis of Ethology and Comparative Psychology

section one

AIMS AND METHODS

1

INTRODUCTION

1.1 THE NETWORK OF EVENTS

The moment we start to observe behaviour, we begin to abstract. In the first instance this is necessary because the methods at our disposal cannot cope with unique events: if we could measure behavioural events sufficiently accurately, we should find that every one is unique. In practice, therefore, we have to group them into classes by abstracting properties which, to a degree of precision appropriate to the task in hand, recur in more than one event.

But there are other reasons why we have to abstract. One is that, while observing one aspect of behaviour, such as nest-building, we must select some activities but neglect others going on simultaneously. Yet the extent to which different but simultaneous activities interact with each other is an open problem: if we are interested in the nest-building of a canary we may be able to neglect respiration, but if the building of the bubble nest of the air-breathing fish *Betta splendens* is under study, we cannot.

In addition, we must confine our observations to a finite span of time. Yet each activity we study is related to events which immediately preceded it in time—electrical and chemical changes in nerve fibres, for instance, or activity in endocrine glands. These events may also be related to concurrent activities other than the one under observation. Furthermore, they occur in structures which have a developmental history extending further into the past.

Moving forward in time, the activity selected for study will have consequences reflected in changes in the organism, its environment, or both. Some of these consequences will influence the subsequent behaviour of the organism and some will not: in either case, some will be of no long-term significance, while others will affect its chances of surviving and reproducing in its environment. These significant consequences, through the mechanisms of inheritance and natural selection, may influence the behaviour of future members of the species.

All this is not merely a way of saying that behaviour is complicated. Only when we see that behaviour involves a complex nexus of events in time, and that its study involves both the selection of an interval of time and also selection from among the events occurring within that interval, can we even formulate the problems it poses. Conversely, the problems we choose will determine how we describe behaviour, what we select for analysis or measurement, and what we ignore or reject.

1.2 THE PROBLEMS

The study of behaviour involves the identification of relations between behaviour and events inside or outside the organism which precede, accompany, or follow it. It is convenient to divide the problems which arise into three groups which, though theoretically separable, are in practice interrelated.

First, the events selected for study must be related to other events or conditions which immediately precede them. This is usually called "causal analysis": provided we are sufficiently sophisticated to bear in mind that temporal correlations may be of more than one kind, we need not be chary of using the word "cause" in its everyday sense.

Causal analysis does not follow any set pattern, but characteristically it proceeds through a sequence of stages—a sequence which has not yet been completed for even the simplest examples of behaviour. As an essential preliminary, the behavioural events must be identified and classified (see pages 13–16). The relations of the behavioural events with each other, and with antecedent conditions, must then be specified: this leads to the identification of "causal factors" which may be either events in the real world, such as stimuli, or postulated variables which relate such events to the behaviour, or the behavioural events to each other (see page 14). For instance, if we find that a certain display posture of the great tit is consistently associated with certain other patterns of behaviour, such as striking another bird with the beak, we may classify them together (see pages 370–372), perhaps giving them the label "aggressive behaviour". If the display can be elicited by the bird's mirror image, we can be sure visual stimuli are important: their nature can then be analysed further (Chaps. 4–6). If the display is elicited more easily at certain times of day, or at certain seasons, than at others, internal factors must also be important (Chaps. 8–14). Such a level of description may be adequate for prediction and control of the behaviour, at any rate within a specified range of conditions, but complete understanding requires analysis at every level of complexity—not only the patterns of behaviour of the whole animal but also the functioning of sense organs and effectors, and of the nervous mechanisms which relate them. One route to this involves specification of the properties which the neuromuscular mechanisms must have to produce the observed behaviour, and then attempts to identify mechanisms with some of the requisite properties within the sense organ–nervous system–effector complex. From here analysis can proceed by similar steps to neuronal, subneuronal, and physicochemical levels of analysis.

The second group of problems is also concerned with causation but involves delving further into the past. As an animal develops, its behaviour changes: the course of ontogeny demands analysis and understanding. Here we are concerned with questions such as "How does behaviour change?" "To what extent do the changes depend on environmental influences?" "How do such influences produce their effects?" The behaviour we study in any one interval of time is influenced by the developmental processes which have given rise to the dynamic structure the animal has at that time, and influences its structure in the future: problems of ontogeny are thus closely linked to those of immediate causation. Indeed, the distinction between "developmental" changes and the changes which occur from moment to moment is arbitrary, though it is convenient to distinguish between long-term more or less irreversible changes and those which are seasonal or of shorter duration.

The processes studied in the analysis of development have as consequences the processes we studied in our analysis of immediate causation. But these too have their consequences. Indeed, since the interval of time over which our causal analysis is made must have a finite length, the earlier events within it may affect the later ones: the consequences of behaviour can become causes. Furthermore the immediate consequences of behaviour, through changes within the organism itself or in the environment, may influence the probability that the same behaviour will be repeated during later intervals of time. The problems of cause and immediate consequence are thus closely linked. But the consequences of an activity may have further, longer-term effects: they may, for instance, affect the organism's chances of survival or reproduction. Since individuals differ in their behaviour, natural selection can then operate. Here, therefore, lies a third group of interrelated problems about causation extending both further forward in time and much further backward into the past than those we have considered so far. In this case we ask questions such as: "What are the consequences of each pattern of behaviour, and which of them provide material for the action of natural selection?" "How did the behavioural differences between species arise?" "What have been the nature and course of the evolutionary changes in the behaviour of each species?" "How does behaviour affect the formation of new species?" Since questions about evolution refer to events in the past, an experimental approach is usually impossible. Nevertheless, the comparative study of existing species has permitted considerable progress (see Sec. 4).

1.3 THE LIMITS OF GENERALIZATIONS

These then are the problems with which we are concerned. What precisely can be achieved? In the case of causal analysis, at present we are beginning to understand a few types of behaviour in the repertoire of a handful of species at one level of analysis. About the behaviour of the great majority of species we know nothing. Except for certain limited aspects of behaviour, it would be presumptuous for us to think we could devise principles applicable to all organisms which were not mere truisms. Thus we must first limit our goals, attempting to elucidate unifying principles or generalizations applicable within a specified range. These may be relatively superficial and applicable to a wide range of phenomena or species, or else intensive and concerned only with one aspect of the behaviour of one animal. Most work lies somewhere in between these two extremes; but because generalizations made on the basis of one can be partially tested by information obtained with the other, both approaches are necessary and complementary.

The extent of the validity of any statement is, of course, inversely related to the degree of precision with which it is stated: every generalization must thus be accompanied by a qualification which defines its scope. For instance, the statement, "In the breeding season most birds defend territories from other individuals of the same species and sex" is true for most male passerine birds but not, for example, for the emperor penguin (*Aptenodytes forsteri*). The statement "courtship displays depend on conflicting tendencies to behave in incompatible ways" is useful descriptively for many birds and fishes but probably not for frogs and toads. The statement "chaffinch song attracts females but repels males" is true in the breeding season but not outside. More quantitative statements may be valid only for one strain of one species reared in a particular laboratory with one kind of preexperimental treatment. Generalizations about behaviour and statements about the range of their validity must go hand in hand.

1.4 THE COMPARATIVE APPROACH

At some stage it becomes necessary to make statements valid for more than one species: these must be based on comparative study of a number of species. But the comparative method has to be used cautiously: apparently close similarities between species may prove to be merely parallel evolutionary adaptations to a similar environment, and rest on quite different causal bases. Just as the wings of pterodactyls, birds, and bats evolved independently and have entailed different modifications to the primitive forelimb, so also does the dehusking of seeds by tits (*Parus* spp.) and cardueline finches involve quite different movements. In order to assess the validity and depth of generalizations, it is therefore useful to choose first, for comparison, species believed on other grounds to be closely related phylogenetically, and only then proceed to the more distantly related ones (Chaps. 27 and 28). Resemblances between the behaviour of distantly related species may turn out on analysis to be merely superficial (e.g., Schneirla, 1959).

Pursued in this way, a comparative approach does more than permit generalization. Even in the initial stage of selective description, knowledge of the behaviour of one species may be helpful in the study of a related one. The present writer would certainly never have noticed the occasional pivoting movements made by greenfinches (*Chloris chloris*) if he had not already been familiar with the highly ritualized and thus conspicuous pivoting display of the related goldfinch (*Carduelis carduelis*) (Hinde, 1955/1956). Conversely many complex behaviour patterns which have been elaborated after a long evolutionary history may be more easily understood if studied first in a relatively primitive and uncomplicated form. Such comparison is also the only method available for investigating the evolution of behaviour—the origin of many display postures would be quite obscure if it were not possible to observe them in a less ritualized form in related species (Lorenz, 1935, 1950, 1951).

The differences revealed by comparative study may be as interesting and important as the similarities: both will provide new theoretical insight into problems of causation, function, and evolution. Such differences must, however, be considered in relation to the biology of the species: for example, measures of so-called "intelligence" obtained by applying the same test to widely differing species are meaningless unless this is done (Baerends, 1941; Tinbergen, 1951).

1.5 LEVELS OF ANALYSIS—BEHAVIOUR AND PHYSIOLOGY

The material with which we start is usually at the behavioural level. If prediction of behaviour, given the antecedent conditions, was the sole aim, there might be no need to reduce to a physiological level: reference to underlying mechanisms could be unnecessary. But even if the complete prediction of behaviour were possible, we should still have advanced only one stage towards its full understanding: a further stage would be reached if the regularities in behaviour could be understood in terms of the physiological organization which they reflect. Thus hypotheses must be judged not only at the behavioural level, but also in terms of their compatibility with lower ones.

Indeed, a theoretical system which by its nature cannot be related to physiological data is unlikely to have a wide validity even at the behavioural level. To take a parallel from physics, Toulmin (1953) has pointed out that it is a "great virtue of a good model that it does suggest further questions, taking us beyond the phenomena from which we began, and tempts us to formulate hypotheses which turn out to be experimentally fertile". Thus the Greek antenna model of sight, and the models of thermal and gravitational phenomena as the effects of caloric and gravitational fluids, prompted unprofitable questions; on the other hand, "the model of light as a substance in motion is a good model, not only because it provides us with an easily intelligible interpretation of the diagrams of geometrical optics, but also because it . . . leads us to speculate about light-particles or lightwaves as the things which travel, or are propagated: these speculations have borne fruit". Of course, just as an explanation of shadow-casting in terms of the wave theory of light tends to be laborious, prediction of behaviour via all the physiological intermediaries would be hopelessly cumbersome. But a behavioral model should be at least compatible with, and, if possible, translatable into, a physiological one. Analysis down to a physiological level is therefore a desirable, but not at present usually attainable, aim.

"Physiologizing" is, however, more than a long-term aim. Physiology is a valuable source of information for the student of behaviour. A fertile psychology cannot afford to ignore physiological data that may suggest useful concepts: psychologists who deny themselves such data may be neglecting an important source of evidence on which to base their predictions of behaviour. While the student of behaviour in choosing his units for study must initially be guided by the regularity of their relations with other variables, such as stimuli and other antecedent conditions, he will be wise also to aim at an analysis of behaviour into units which the physiologist will be able to handle: that analysis can be guided by a consideration of present physiological knowledge (Hebb, 1955).

Yet the student who aims to pursue his analysis from the behavioural level to the physiological one is exposed to a special danger—concepts useful at one stage in the analysis may be misleading at another. A classic example of this—and we shall meet others later—is the concept of "drive", "urge", or "tendency", which is useful at an initial behavioural level of analysis but can become a handicap at a physiological one. The concepts used must be appropriate to the task in hand, and their limitations recognized.

1.6 USE OF MODELS

The crossing of the bridge between behaviour and physiology can often be facilitated by the use of models. Those used in this book are "functional schemata, constructed in order to illustrate definite causal relationships", and "are quite abstract, although the consequences they predict are concrete and experimentally verifiable" (von Holst, 1954). They serve merely to indicate the job which the physiological machinery must be doing and to provide the physiologist with an indication of what he must look for (see also Hinde, 1956a, 1960a).

1.7 SUMMARY

1. The study of behaviour involves selection from a complex nexus of events in time.
2. The study of behaviour involves three main kinds of problem—its immediate causation, its ontogeny, and its functional and evolutionary aspects.
3. In the present state of knowledge, it is advisable for the student of behaviour to pursue limited goals, attempting to elucidate principles applicable over a specified range. Superficial generalizations of wide validity and precise ones of limited scope are complementary, and both are necessary.
4. Comparative study is necessary before generalizations applicable to more than one species can be made. It is also essential for studies of the evolution of behaviour.
5. The analysis of behaviour can be aided by physiological data, and the student of behaviour should attempt to devise hypotheses compatible with the physiological information available.
6. The analysis of behaviour can be aided by the use of models which indicate the nature of the causal relationships involved.

2

THE DESCRIPTION AND CLASSIFICATION OF BEHAVIOUR

The importance of the initial descriptive phase in the study of behaviour has often been neglected, especially by those who have modelled their methods on those of classical physics. The earlier physicists dealt largely with everyday events—falling apples, floating boats, rainbows, melting ice—which were already familiar to them and to their colleagues. In biology, by contrast, the work of generations of taxonomists and systematists was necessary before progress could be made. The task of the taxonomist of species is to sort out the almost unlimited diversity of individuals in nature into easily recognizable groups, to describe those groups in terms of their significant characters, and to define the differences among them. Similarly, a complex stream of behaviour must be described and broken down into units suitable for study, which then must be classified into groups according to common properties. No two units will be absolutely identical in all of their properties, and so classification into groups depends upon abstracting common characters (MacCorquodale and Meehl, 1954).

2.1 METHODS OF DESCRIPTION

In practice, there are two methods for describing behaviour. One involves reference ultimately to the strength, degree, and patterning of muscular contractions (or glandular activity, or change in some other physiological property). The other involves reference not to these changes but to their consequences. This is not a distinction between types of behaviour, as the "molecular" versus "molar" distinction is often used by learning theorists, but between criteria for describing behaviour: it is of the utmost importance since the whole character of the subsequent analysis may be influenced by which type of description is chosen.

The first method is often limited to patterns of limb or body movement, since a complete description would be not only impracticable but also unnecessarily refined and cumbersome. Terms such as "knee jerk", "positive supporting reaction", and "sleeping posture" are of this type. Of course, the degree of precision required in description depends on the nature of the analysis undertaken.

In classifying on the basis of such descriptions, we are grouping together similar spatiotemporal patterns of muscular contraction, though the classification will depend only on selected characteristics, others being ignored. The choice of the characteristics employed should, of course, have an objective justification, but it may take cognisance of the fact that variability of motor patterns is not necessarily paralleled by a similar variability of mechanisms; just as, although pecking at a piece of food could involve an infinite variety of motor patterns, it involves localization in only three dimensions. In addition, the criteria may vary greatly in complexity. Thus the crouching of a bird could be characterized merely by a certain degree of bending in the tarsal joint. "Tail flicks" are more complicated, involving a temporal pattern of contractions in a number of different muscles. Courtship displays, to take an even more complicated case, may consist of the simultaneous or sequential occurrence of several such spatiotemporal patterns. In such cases the selection of criteria is further complicated by the disappearance of some components when the response is given at low intensity. The usual practice here is to describe low- and high-intensity forms and indicate the intermediate variations, a method roughly comparable to that used by taxonomists when the characteristics of a species change gradually over its area of distribution, but one that often leads to the description only of rare extreme cases.

The second method of describing behaviour may be called "description by consequence". "Picking up nest material", "pressing down the lever", or "approaching" do not refer to particular patterns of muscular contraction but cover all patterns which lead (or could lead, and there is room for ambiguity here) to the specified result. Descriptions like this are normally used when the behaviour involves orientation to objects in the environment, and when the motor patterns, though leading to a constant result, are themselves diverse. They thus tend to be used especially for "appetitive" or searching behaviour.

There are several advantages in this kind of description. First, a brief description may cover a multitude of motor patterns, each of which in itself is variable—for instance approaching covers hopping, walking, flying, and sidling: it therefore provides an essential shorthand. Often the number of smaller groups, characterized by physical description, into which it is possible to divide each behaviour class characterized by consequence is large. Sometimes, however, their number is surprisingly limited. For example the grey-lag goose (*Anser anser*) uses only one method of retrieving eggs which have rolled out of its nest—it puts its beak beyond them and tries to manoeuvre them back under its breast. In spite of the inefficiency of this technique, the bird never attempts to use its apparently more suitable webbed feet or wings (see Fig. 2.1).

Second, descriptions by consequence often permit units of behaviour to be described in terms of objectively defined changes in the environment, and thus in a manner relatively free from errors of observation. Whether a canary did or did not place a piece of grass in its nest pan, or whether a rat did or did not press a lever sufficiently to produce a microswitch contact, are matters which brook no argument.

A third advantage of descriptions by consequence is that they often call attention to essential features of the behaviour that may not appear in physical descriptions, such as orientation with respect to the environment, or responsiveness to external stimuli. For instance, the hammering movement used by great tits (*Parus major*) to open nuts is almost identical with the movement used in attacking rivals, being distinguished only by the speed of the movement: they differ, however, in the object at which they are directed (Blurton-Jones, 1968). Again, an animal's behaviour may

Fig. 2.1. Grey-lag goose retrieving an egg. (After Tinbergen, 1951.)

change radically (e.g., from one type of appetitive behaviour to another) with little change in the motor pattern; or, alternatively, a radical change in the patterns of muscular contraction (e.g., from flying to walking) may be, for some purposes, of little significance. Description by consequence is thus often essential for a full understanding of the behaviour.

Of course, description by consequence is not always applicable. For instance, the environmental consequences of the particular posture in which an animal sleeps are of little relevance. Furthermore, descriptions of this type have certain dangers. First, the manner in which categories of behaviour are described may have important effects on subsequent hypotheses; whether, for instance, we say "entering the goal box" or "escaping from the alley".

Second, as stressed by MacCorquodale and Meehl (1954), such descriptions are susceptible to overinterpretation. The statement "the rat learned to press the lever" can be interpreted as pure description, but can also be overinterpreted to imply reciprocal relations between the various behaviour classes by which the lever could be depressed, such that a decrease in frequency of one is followed by a compensatory increase in another. This may be so, but does not necessarily follow from the initial descriptive statement. As we have seen in the case of the egg-rolling response, the number of alternative response classes which can be used in bringing about a given consequence may be surprisingly limited.

Whichever type of description is used, it inevitably involves the rejection of some of the data and a selection (arbitrary, intuitive, or rational) of the features to be studied—as soon as we move away from the initial cinematographic record of the behaviour, we are neglecting information. Just because description and classification involve rejection of data and the selection of criteria, there are always the dangers that the essential data are rejected and that the units used are irrelevant, too small, or too large for the task in hand. For instance, descriptions of reproductive behaviour in terms of "maternal behaviour" or "sexual behaviour" are inadequate for a proper analysis of the stimuli controlling them (Tinbergen, 1951). Similarly, understanding of the apparently "insightful" tool-using behaviour of primates is greatly facilitated by examination of its components (Schiller, 1957).

On the other hand, insufficient rejection of data may lead to a hopelessly confusing mass of detail and divert attention from the essentials of the problem. Thus, any attempt to describe precisely the patterns of muscular contraction involved when a fighting bird strikes its adversary with its beak would be difficult because of the variations in the initial posture and relative positions of the combatants—it is more useful to describe the behaviour in terms of consequence.

The two types of description are thus not always alternative. Description in terms of muscular contraction is unsuitable in some cases because of the complexity of the data it provides and in others because the data cannot be subdivided into classes with an objective validity. In other cases, description by consequence is equally unsuitable. Sometimes—as in the threat and courtship postures of birds, which involve both a relatively stereotyped motor pattern and an orientation with respect to the environment—both must be used.

2.2 CLASSIFICATION OF BEHAVIOUR TYPES

The isolation of instances of behaviour allows us to associate them with each other by virtue of common characteristics and thus recognize them as examples of the "same" type of behaviour. These types may be defined either as involving movements whose form varies only within prescribed limits or as involving a common effect. Mere enumeration of such types is, however, almost valueless unless they are further classified into groups. Classification requires criteria, and, in principle, the number of criteria that could be chosen is infinite. They could, for example, depend on common features in the form of the movement, or on the oxygen consumption needed for its performance, or on the consequences the behaviour has. In practice we choose criteria suitable for the particular problem we are concerned with: three kinds which have been especially valuable must be mentioned.

1. Classification in Terms of Immediate Causation

The types of behaviour shown by an animal can be grouped together according to the causal factors on which they depend, activities which share causal factors being classified together. Thus, all activities whose frequency or intensity is significantly increased by male sex hormone can be grouped together as male sexual behaviour. All activities similarly influenced by the stimulus situation "rival male" can be described as "agonistic behaviour". This type of classification is essential for an understanding of the behaviour of the whole animal, and it works well in practice. There are, however, some difficulties. Two methods are used to assess whether patterns of behaviour share a causal factor. One is to administer it and see whether both types of behaviour increase in frequency. The other, useful when the factor itself has not been isolated, is to investigate the temporal correlations between the appearances of various activities: two activities which consistently appear together are likely to be related causally (see pages 370–380). But in either case, a positive result does not necessarily mean a direct effect of any factor on both patterns (page 606). Nor does the non-occurrence of a pattern when a factor is administered mean that the factor is without influence on it—additional ones may be necessary before the behaviour appears (see Hanson, 1955, for a discussion of the pitfalls involved in thinking in terms of simple causal chains).

2. Functional Classification

Activities can also be classified according to the adaptive (in an evolutionary sense) consequences they serve. "Threat", "courtship", "hunting" refer to categories of this type. Such terms correspond to words such as "legs" or "eyes" in morphology. A structure is classed as a leg because of its function, and there is no implication that the legs of, for example, arthropods and vertebrates are related either developmentally or phylogenetically. A similar structure which has a different function is given a different name; for instance, "maxilliped".

In practice, when any one species is being considered, a functional classification of behaviour very often corresponds to a causal one: considered from the point of view of evolution, this is not hard to understand —the mechanisms which underlie adaptive behaviour need be far less complex if functionally related activities share causal factors than if the various causal factors are distributed at random among the functional groups. For this reason functional-sounding terms, like "sexual behaviour" and "parental behaviour", are sometimes used to refer to categories which are also defined in causal terms. But although the functional classification is a good guide to the causal, this is not always the case. For instance, Beer (1963a) has shown that a causal classification of the activities shown by black-headed gulls (*Larus ridibundus*) early in the reproductive season does not coincide with a functional classification. In particular "sideways building" has more causal affinities with activities which contribute functionally to incubation than with others which contribute to nest-building. As another example, in many animals it is quite unjustifiable to talk about "reproductive behaviour" when implying a causal category, since sexual and parental behaviour depend on different causal factors. Causal analysis must precede causal classification if functional and causal categories are not to be confused. Furthermore, the same movement pattern may have different functions in different contexts. Thus Beach (1965) points out that to label penile erection in a week-old puppy as sexual behaviour just because the adult male has an erection while copulating is as absurd as labelling urination by the infant male as territorial marking. Even greater care is necessary in generalizing from species to species, for the same functional category (e.g., courtship) may rest on quite different causal bases in different animals (Tinbergen, 1952, 1953, 1963). Functional categories are often used to include behaviour patterns from unrelated species. "Territorial behaviour" and "migratory behaviour" are cases in point. Such a procedure is quite justifiable so long as the categories are used in discussions of function only. Here care is again necessary, for almost identical behaviour elements may have dissimilar functions in different species.

3. Historical Classification

Two quite distinct systems must be recognized:

(*a*) *Classification according to source.* Here patterns of behaviour believed to have a common historical origin are grouped together. This is much used in the study of the evolution of behaviour, especially the evolution of "fixed action patterns". The basic criterion for grouping together is then similarity in the pattern of muscular contraction, but, as in the systematics of organisms, due allowance must be made for possibilities of convergence, divergence, etc. This depends on the comparative study of closely related species (Chap. 27).

(*b*) *Classification according to method of acquisition*. Here behaviour classes are grouped according to the nature of changes which have occurred in their development or evolution; for instance, "learned", "ritualized".

THE INDEPENDENCE OF THE CLASSIFICATORY SYSTEMS

Although the classificatory systems discussed in the previous section are basically independent of each other, possibilities of confusion constantly arise. Just because, for reasons we have already seen, functional and causal categories often overlap, the temptation to confuse them may be severe. This is particularly dangerous when the functional categories are nothing more than descriptive generalizations. Reference to "self-preservative" as opposed to "reproductive" instincts, for instance, both confuses functional and causal categories and implies something causally in common between the members of each group—for example between feeding and preening, or nest-building and feeding young. Similarly, although some authors have grouped together activities such as scratching, stretching, and shaking as "comfort movements", Weidmann (1958) has shown that they occur independently of each other.

2.3 SUMMARY

1. Behaviour may be described and categorized in terms of spatiotemporal patterns of muscular contraction or in terms of the consequences of the behaviour.
2. Types of behaviour can be classified according to a theoretically unlimited number of systems. Three useful ones depend on common causal factors, common functional consequences, and common historical (ontogenetic or evolutionary) source or method of acquisition. Classificatory systems based on different types of criteria must be treated as independent.

section two

THE STUDY OF CAUSATION

3

THE CONTROL OF MOVEMENT

We have already seen that the movements shown by animals can be described in two ways: in terms of the spatiotemporal patterns of muscular contraction involved, or in terms of their consequences. Our main concern in this chapter is with the means by which the patterning of movements is controlled. Any movement involves a number of muscles contracting in particular relationships with each other: is the temporal patterning of their contractions determined from the start in the central nervous system, or is it adjusted by peripheral stimuli received as a consequence of earlier phases of the movement? When the movement is specified more appropriately in terms of its environmental consequences than in terms of a particular sequence of muscular contractions, environmental stimuli are likely to play a part in its control. When the movement itself is more stereotyped, however, the question is an open one, for even if environmental stimuli play no part, proprioceptive feedback may yet control the sequence and extent of muscular contractions.

There are, thus, three possible sources for the patterning of complex movements: environmental stimuli (and, in the case of movements directed towards the animal's own body, we must include here some stimuli from its own body surface), proprioceptive stimuli, and coordinating mechanisms within the central nervous system itself. In any particular case the pattern is likely to be derived from more than one of these sources, and our first task is to separate their effects.

3.1 FIXED ACTION PATTERNS

There are some movements whose form seems to be independent of environmental stimuli, even though they may be elicited by such stimuli in the first instance. A female canary uses a special weaving movement to push loose strands of material into the cup of its nest. If the bird is given neither a nest site nor nest material, this movement may be performed on the floor of the cage and without nest material; its form then differs little if at all from that when it is performed in its normal functional context, and so stimuli from the nest cup or nest material can play little part in its control. The soliciting posture of a female chaffinch, which includes flexion of the legs, withdrawal of the head, raising of the tail, and rapid vibration of the wings, is remarkably constant in form: although it involves most of

Fig. 3.1. Four intensity stages of the posture used by female chaffinches in soliciting for copulation. (From a drawing by Yvette Spencer-Booth.)

the muscles in the body, the relations among their contractions vary little, and they are thus presumably independent of immediate environmental control (Fig. 3.1).

Such movements are often referred to as fixed action patterns (FAPs). Each one, although it may consist of a quite complicated spatiotemporal pattern of muscular contractions, cannot be split into successive responses which depend on qualitatively different external stimuli (Lorenz, 1935, 1937; Tinbergen, 1942). Since such movements depend on external factors for their elicitation only, they vary in degree of completeness but not in the relationships among their parts. The soliciting posture of the female chaffinch may involve any degree of leg-flexion, tail-raising or wing-shivering, but a given degree of one is always combined with an approximately constant degree of the others. In some cases even such intensity differences are almost absent: the duration of the "head-throw" display of the golden-eye duck is remarkably constant, having a mean duration of 1.29 sec and a standard deviation of only $\pm$ 0.08 sec (Dane et al., 1959). Such constancy, we shall see later (Chaps. 16 and 27), is especially characteristic of "signal" movements used in intra-specific communication, though not all signal movements are as constant as this one.

Just because the diversity of nature does not fit readily into man-made pigeonholes, many concepts in biology prove useful at a particular level of analysis but yet must not be regarded as absolute. The category of fixed action patterns is such a one and requires some further comment. First, it is defined solely by reference to the absence of a role of external stimuli in controlling the form of movement. Although many of the FAPs studied are in fact rather constant in form, such constancy is not a defining characteristic. Indeed there are unfortunately very few studies in which the intra-individual variability in the form of a movement pattern has been assessed (but see Siegel et al., 1965; Hailman, 1967; Hutchison et al., 1968b). Furthermore, as discussed below, many movements are elicited by one group of stimuli but oriented with respect to the environment by another: constancy of form may be apparent only when the latter are held constant. But in any case stereotypy is a relative matter, and FAPs intergrade in variability with movements subject to more complex control (Barlow, 1968; Brown, 1968).

There is, however, a way in which constancy of temporal pattern is relevant. In some cases a single stimulus situation (e.g., a model of a rival) can elicit a number of responses (e.g., attack, threat, flight) which do not depend on qualitatively different aspects of the stimulus. If the order of which the units occur is variable, it is of course not useful to refer to the *group* of responses as an FAP, even though all are elicited by the same stimulus characters. It may, however, be useful to describe the individual responses in this way.

A further point about the category of fixed action patterns concerns their mode of development. Their form is often said to be "innate" or "species-characteristic". The difficulties which arise when the term "innate" is used in this way are discussed in Chap. 18. Many of the FAPs studied by ethologists are in fact characteristic of a particular systematic category, but this is not how FAPs are defined: they may equally well be characteristic of individuals.

Even if the defining characteristics of fixed action patterns are limited to the original one, namely that once elicited their form is independent of external stimuli, the distinction between movements which do and do not meet this criterion is not so clear cut as once appeared, as evidence presented later in this chapter shows. However, in spite of all these difficulties, and in spite of the fact that movements are often labelled as fixed action patterns intuitively, even though the nature of their stimulus dependence has not been analysed experimentally, the category has proved its value in the study of behaviour: as with all such categories, its value is relative to the task in hand.

Each animal has a repertoire of these fixed action patterns and only a limited ability for developing new ones. The limitation may be set by the availability of the effectors and the mechanics of the body: the vocal apparatus of a song bird cannot produce some sounds, and a hippopotamus just cannot walk on two legs. Beyond this, however, limitations are imposed by the nature of the nervous system. Thus, as we have seen, the grey-lag goose and many other ground-nesting birds have a special movement for retrieving eggs which roll out of the nest: they stand up, place the bill beyond the egg, and then roll the egg back into the nest, moving the bill from side to side to prevent the egg from slipping away (Fig. 2.1). The whole act is clumsy, for the egg frequently rolls away from the narrow ventral side of the bill, and it would seem much more efficient to use one of the wings or feet. Nevertheless, no bird has ever been known to develop a new movement for retrieving eggs (Lorenz and Tinbergen, 1939; Tinbergen, 1953).

As mentioned above, any particular movement of this type may be characteristic of a genus or higher systematic category, a species, or merely one individual. All passerine birds and their close relatives, as well as the plovers, scratch their heads by bringing a foot over the partly lowered wing of the same side, whereas nearly all other birds bring the foot under the wing (Wickler, 1961a, b); and the neonates of nearly all primate species use special movements of the head in searching for the nipple (e.g., Prechtl, 1958; Hinde et al., 1964). As we shall see later (Chap. 27), in such cases the pattern of movement can be used as an indicator of phylogenetic relationship in exactly the same way as a morphological structure. On the other hand, the lateral courtship display of the male chaffinch is given by that species alone, other finches having different displays, and the song of each kind of song bird is limited to that species alone. Finally, particular individuals may develop stereotyped movements of their own: these are especially evident in captive animals, which often develop tics as a result of the impoverished conditions of confinement (see Chaps. 19 and 22).

3.2 ORIENTATION OF FIXED ACTION PATTERNS WITH RESPECT TO THE ENVIRONMENT

By definition, the form of a fixed action pattern is independent of environmental control. Nevertheless, we frequently observe patterns of muscular contraction which, although recognizably similar from one occasion to the next, are yet modified by stimuli from the environment. The modifications usually serve to orient the movement, and it is often possible to separate the stimuli responsible for the orientation from those which elicit the fixed action pattern itself. Thus, in the egg-rolling response (see page 22) by the grey-lag goose, the side-to-side movements of the beak are continuously governed by stimuli from the irregular movements of the egg. If the egg is removed, or if it is replaced by a cylinder which rolls in a regular fashion, these lateral movements cease. On the other hand, the movement of the bill towards the breast, once elicited, is independent of further stimuli from the egg: if the egg is removed once the movement has started, it nevertheless goes through to completion in the normal way. It is thus only the movement of the bill towards the breast which, once elicited, is independent of further external stimuli and constitutes the fixed action pattern (Lorenz and Tinbergen, 1939).

Similarly the alarm call, given by chaffinches when a flying predator appears, elicits flying to cover in other members of the species. The direction of the flight, however, is determined by stimuli from suitable cover in the neighbourhood—the alarm call merely elicits it (Marler, 1956b). A comparable case has been analysed in some detail by Baerends (1957a). The alarm response of the young of the mouthbreeding cichlid *Tilapia mosambica* is elicited by the approach of a large object or by turbulence in the water. The direction in which they swim, however, is determined by stimuli normally provided by the female. Some experiments to determine the nature of these stimuli are illustrated in Fig. 3.2. The fry direct themselves towards the lower parts of a dummy object, and to dark patches on it, and push themselves repeatedly into holes on the surface. As a consequence most of them find their way into the female's mouth.

In these cases the fixed action pattern is related in a constant fashion to environmental stimuli, but many fixed action patterns can be oriented in two or more ways. The male guppy (*Lebistes reticulatus*) uses a "sigmoid posture" in courting the female. This movement appears at two points in courtship: first, when the male is "leading" the female away from the school, when his body is aligned with hers; and again in the "checking"

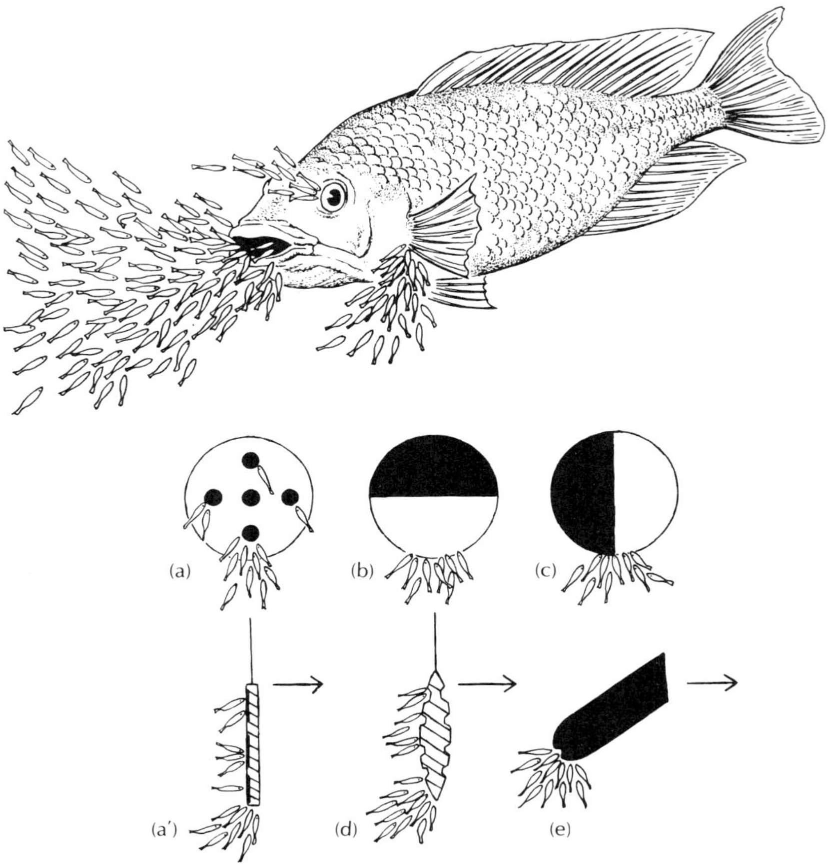

Fig. 3.2. Some model experiments on the stimulus directing the approach of young towards the female of the mouthbreeding Tilapia mosambica. (After Baerends, 1957*a*.)

Models *a*, *b*, and *c* are flat discs (cross section, *a′*), *d* is a disc with pits, *e* is a black-painted test tube with an opening in the bottom. The underside of the object is attractive (*a*, *b*, *c*, *d*, *e*) but not because it is the darkest part (*b*, *c*). In addition, black spots attract the young (*a*, *a′*), while hollows are found by constant pushing against the surface of the model (*d*, *e*).

phase, when the male orients his body at right angles to the female and just in front of her, and thereby stops further forward movement (Fig. 3.3). The fixed action pattern is the same in both cases, only the orientation differing (Baerends et al., 1955).

Of course, it is not possible to separate eliciting from orienting stimuli in all stereotyped movements. In some, an orientation component

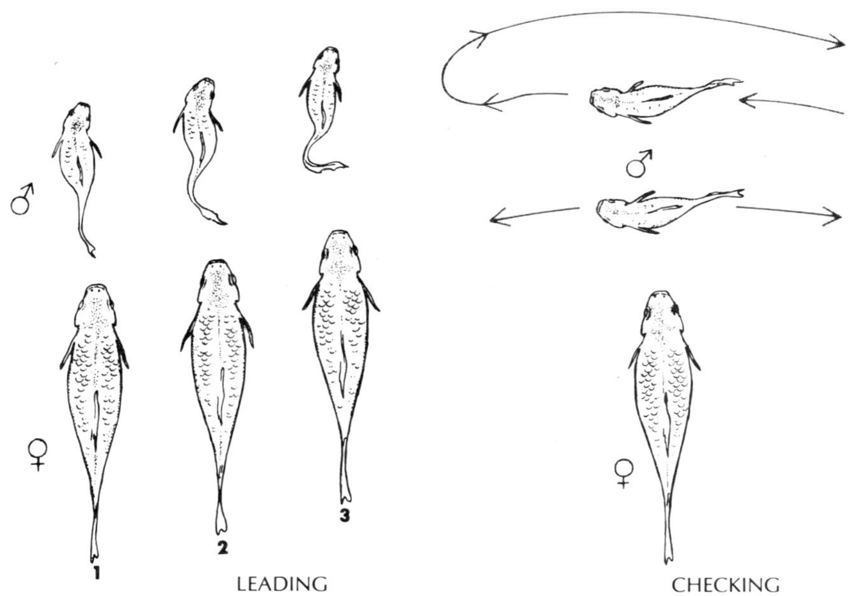

Fig. 3.3. The sigmoid posture of the male guppy (Lebistes reticulatus). (After Baerends et al., 1955.)

On the left the male's body is oriented away from the female: this position is used in leading the female. The transverse orientation, shown on the right, is used in checking her.

is entirely lacking: swallowing and ejaculation are examples. In others the stimuli which orient the movement are the same as those which elicit it. In others the stimuli are different, but their roles are too closely interwoven to be separated. We shall mention the problem of the integration of such movements again later in this chapter. First, however, we must consider the relative roles of central and peripheral factors in the control of some fixed action patterns.

3.3 COORDINATION OF MOVEMENT PATTERNS

Central Patterning

Some complicated movements can be elicited by a wide range of external stimuli or by crude electrical stimulation applied to parts of the central nervous system (e.g., Hess, 1943; Hüber, 1955; Sheer, 1961; von Holst and von St. Paul, 1963; Åkerman, 1966a, b). In such cases (see example in Fig. 12.3) the form of the movement cannot be dictated by either the eliciting stimuli or the mechanics and physiology of the musculature (cf. Wilson and Wyman, 1963). At the extremes, two possibilities are open. The sequence in which the muscles contract could be determined solely by the properties of central nervous mechanisms underlying the movement, so that, once initiated, the movement is independent of sensory stimuli, internal or external to the animal. Alternatively, the muscles contracting later in the sequence could be activated by feedback effects from those contracting earlier: in this case the form of the movement would be continually under sensory control. Even fixed action patterns, in which the sequence of muscular contractions, once elicited, is by definition independent of external control, could depend on proprioceptive feedback.

In either case the first phase of the response must be centrally determined: the alternative possibility that these stereotyped movements always begin with a random movement which is corrected proprioceptively can be dismissed (Horridge, 1961). The fundamental issue is, thus, whether or not their continuance depends on the afferent effects of their own earlier phases. In order to tackle this problem experimentally it is necessary to compare the form of the movement, or of the impulses in the motor nerves, under two conditions: when the sensory stimulation from the earlier phases of the movement is available to the central nervous system, and when it is not. For obvious reasons, rhythmic movements lend themselves especially to analyses of this sort: the question then becomes, "Does the rhythmic pattern persist in the absence of feedback from the periphery?"

One of the simplest and best analysed cases concerns the cardiac ganglion of decapod crustacea, which contains only nine neurones (Fig. 3.4). The electrical impulses in the efferent nerve consist of a regular sequence of intermittent patterned bursts, each containing large and small spikes separated by silent periods. In the intact animal the heart rate may be affected by a variety of tonic inputs (Larimer and Tindel, 1966), but the sequence of intermittent bursts continues even when the ganglion is isolated from phasic influences. Although occasional variants occur, the detailed sequence of the impulses which constitute each burst is remarkably constant—the order of the first few spikes often remaining the same for over 3,500 bursts. Since afferent influences are excluded, this patterning must be a consequence of the structure of the nervous mechanism itself.

The component cells, when isolated from synaptic input, usually either fire at a uniform low frequency or are quiescent: their integration depends on complex interrelations among the four smaller posterior cells, which act as pacemakers, and the five anterior followers (Maynard, 1955; Hagiwara, 1961; Hartline, 1967, and refs. cited).

A similar conclusion has been reached for the sound-producing mechanism of the cicada *Graptopsaltria nigrofuscata* (Hagiwara and Watanabe, 1956). This involves contractions in two main sound muscles, one on each

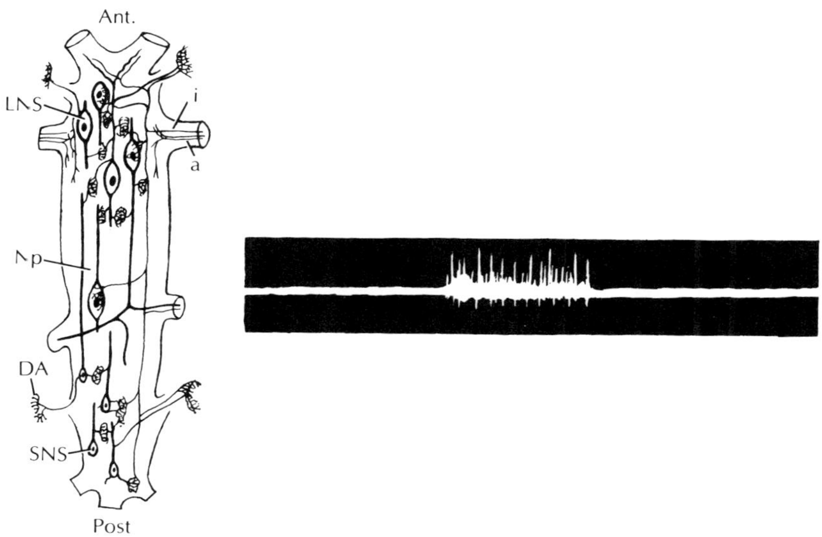

Fig. 3.4. Cardiac ganglion of the lobster and patterned burst from cardiac ganglion. LNS, large nerve cells; SNS, small nerve cells; DA, dendritic arborization; Np, neuropile; a, accessory nerve fibre. (From Hagiwara, 1961, after Maynard, 1955*a*.)

side of the body, which are each innervated by a single efferent nerve fibre from the mesothoracic ganglion. When the sensory nerves leading from the hair sensillae to the ganglion are stimulated, the electrical response of the sound muscles consists of a series of spikes. As the intensity of the stimulus is increased, the frequency of these spikes remains constant at about 100/sec, but the length of each series increases to 10 or more. If the electrical discharges of the two muscles are recorded simultaneously, it is found that each spike in one muscle occurs halfway between two spikes in the other. Thus, without feedback control, a simple sensory stimulus results in a pattern of alternate responses in the two muscles at 100/sec: this patterning must originate in the ganglion. Hagiwara and Watanabe have been able to suggest how this happens. Since the effect of an antidromic impulse on the motor neurone does not change the patterning of the reflex discharge, the motor rhythm must originate in the premotor neurone. Furthermore, internuncial neurones have been found which, on stimulation of the sensory nerves, gave repetitive spikes at 200/sec. It thus seems that the rhythm in the two muscles depends on a common pacemaker which fires at twice the frequency of the motor neurones and activates the two sides alternately. On this view the alternation between the two sides would be due to some form of mutual inhibition.

As always, of course, care must be taken in generalizing across species. Sound production in crickets, for instance, is more complex. The initiation and patterning of the song seems to be determined in the mushroom bodies of the brain as a consequence of impulses from the periphery. The thoracic ganglia, from which the stridulatory organs are innervated, can by themselves start only a basic form of the stridulatory rhythm: normally this is modulated by impulses from the brain. The role of afferent feedback from the stridulatory organs is not yet clear (Hüber, 1960, 1964).

As another example of central patterning, Horridge (1961) has studied the sequence of motor impulses from the cerebral ganglion of the bivalve mollusc *Mya*, which normally brings about retraction of the mantle and closure of the shell. Single impulses in a preganglionic (sensory) axon can induce a sequence of motor impulses from the ganglion in which up to ten motor axons can be individually identified, the patterning of the impulses being similar (but not identical) in successive repetitions. The sequence is unchanged when all other nerves of the cerebral ganglion are severed and when the motor nerve is cut beyond the recording electrodes: the sequence thus does not depend on proprioceptive feedback from the movement it causes. Traditionally, a patterned output of this type is ascribed to patterned connections between the internuncial neurones, but Horridge points out that the neuropile consists of a tangled mass of small fibres in which specific connections between cells seem to be absent. He therefore suggests that the internuncial and motor neurones respond differentially to a transmitter substance liberated in the neuropile by the preganglionic fibre.

In these cases it has been shown that the patterning in the output of a single ganglion can arise within the ganglion independently of any sort of feedback control. Complete central patterning of this type has not yet been established with certainty for any complex species-characteristic movement in vertebrates. Although Weiss (1941a, b) claimed that complete deafferentation did not affect the swimming of tadpoles, later work on other species (see pages 39–45) suggests that this study should be repeated. However, the evidence is highly suggestive in the case of reflex swallowing in mammals (Doty and Bosma, 1956). The initial phase of swallowing is a complicated activity in which motor neurones scattered from mesencephalic to third cervical levels are integrated to produce a coordinated pattern of contraction among about twenty different muscles. The sequence of contraction in some

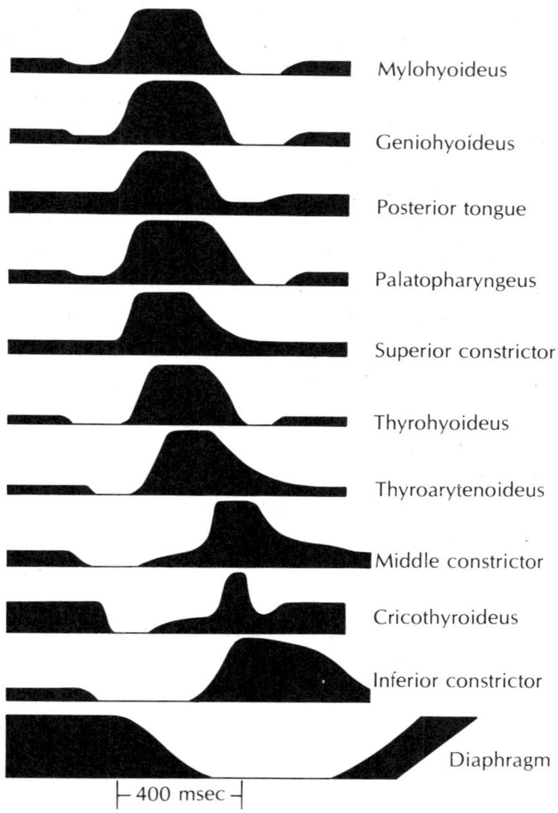

Fig. 3.5. Schematic summary of electromyographic activity in deglutition for unanaesthetized dog. Height of line indicates intensity of action in each muscle. (From Doty and Bosma, 1956.)

of these muscles in the dog is illustrated in Fig. 3.5. The temporal pattern, duration, and amplitude of the contractions are independent of the means used to elicit swallowing, whether this be electrical stimulation of the superior laryngeal nerve at frequencies ranging from 3 to 100/sec, stimulation of the pharynx with a cotton swab, or rapid injection of water into the mouth: the patterning thus arises within the mechanism of swallowing itself. Once swallowing is elicited, its coordination could be due to the patterning of internuncial neurones in the brainstem, to collateral linkages between motor neurones arranged in such a manner that the firing of one leads sequentially to the firing of another, or to the activation of the motor neurones which fire later in the sequence by afferents from muscles which contracted earlier. Sequential linkages between motor neurones are rendered unlikely as a sole explanation by the observation that pontine section of the brainstem, which excludes the action of the mylohyoideus muscle, does not modify the course of swallowing. There is also considerable evidence against control by proprioceptive feedback: a wide variety of treatments likely to upset any such feedback, including strychnine, moderate asphyxia, cocainization of the pharynx, excision or procainization of participating muscles, fixation of the hyoid mass at extreme rostral or caudal positions, and varying degrees of traction upon the tongue, all fail to produce any change in the temporal organization of swallowing except for minor changes in amplitude and duration of the contraction in individual muscles. Thus, since deglutition can be elicited by a wide range of inputs (see also Doty, 1951), and in view of the evidence against feedback control, the coordination of the movement must be entirely central. The nervous mechanism responsible has been located and its mode of functioning analyzed by R. W. Doty et al. (1967) and Doty (1968).

Although the number of cases in which patterned nervous discharge or movement is known to be entirely independent of patterned input is few, this should not be taken to mean that central patterning is the exception. Rather, as Bullock (1961) stresses, "the output of single neurones and groups of neurones is normally probably always patterned". Usually, however, the central output is modified not only by tonic influences but also by phasic feedback from the periphery. We shall turn now, therefore, to cases in which the patterning is initiated centrally but may be modified by peripheral factors.

Peripheral Influences

In considering the interaction between central and peripheral factors controlling movement patterns, we shall consider a variety of examples drawn from invertebrate and vertebrate animals. They have been chosen to illustrate both the diversity of the peripheral influences and the variations in the subtlety with which central and peripheral control is interwoven.

1. CRAYFISH SWIMMERETS

In the first case, the swimmeret movements of the crayfish (*Procambarus clarkii*), feedback is not necessary for the production of a rhythmic discharge. The rhythmical movements of the swimmerets can be initiated from the abdominal ganglia. When the abdominal nerve cord is completely isolated, rhythmic bursts of electrical activity in the motor roots leading to the swimmerets can sometimes still be recorded (Ikeda and Wiersma, 1964). The activity spreads from posterior to anterior, and experiments involving transection of the isolated cord at various levels indicate that several of the posterior ganglia contain pacemakers which normally drive those in front. Rhythmic discharges can, however, be more readily obtained from the abdominal cord if one of the first segmental roots is left intact. Furthermore, impulses coming down the ventral cord can produce oscillatory movements of the swimmerets at a frequency which varies (within limits) with the frequency of stimulation (Wiersma and Ikeda, 1964). Thus in this case the rhythm can be independent of rhythmical reafferent stimulation from the effectors, though it is influenced by input from head and thorax (Hughes and Wiersma, 1960).

2. *INSECT FLIGHT*

In the well-studied case of the flight mechanisms of the locust, peripheral feedback normally influences the frequency of a centrally determined rhythm, but has little or no effect on its patterning. The alternation of contractions in the musculature used in the upward and downward beats of the wings was earlier ascribed to proprioceptive feedback loops, but it is now clear that such an explanation is inadequate. First, although there is a sensory discharge in nerves from the wings, coming largely from the stretch receptors of the hinge and synchronized with the wing-beat cycle, its removal results only in a slowing of the rhythmical output from the thoracic ganglia: the patterning of the output still resembles that occurring during flight. Second, in the intact locust, flight movements can be reduced by wind on the head. This stimulus is of course non-phasic and can be used to study the output of the system controlling flight in a preparation from which *all* sources of phasic feedback have been removed—for instance, one consisting in essence only of the head and ventral cord. Even in this preparation the discharge in the thoracic motor nerves is phasic and similar to, though slower than, that occurring in the nearly intact animal in normal flight (Fig. 3.6; Wilson, 1961). Third, even a random input to the thoracic ganglia can elicit a fully coordinated output pattern like that occurring in normal flight, in the absence of feedback from moving parts (Wilson and Wyman, 1965). Detailed analyses have shown that wing movement depends on a considerable number of central pacemakers, potentially independent, which are fairly tightly coupled into a dominant frequency which depends on the quantity of the input.

The advantages of such a mechanism over a chain-reflex system can be seen as follows. When the wing speed is such as to produce maximum efficiency, the downstroke muscles begin to shorten just as the wings reach the top position, and vice versa. When the central output has a higher frequency, the upbeat is not over when the downstroke command arrives at the wing muscles. The downstroke muscles thus help to slow the end of the upbeat. A muscular brake such as this at the ends of the strokes permits a higher wing-beat frequency. If the rhythm was controlled by chain reflexes, the contractions of the muscle groups would be fixed relative to each other, and this method of adjusting wing-beat speed would

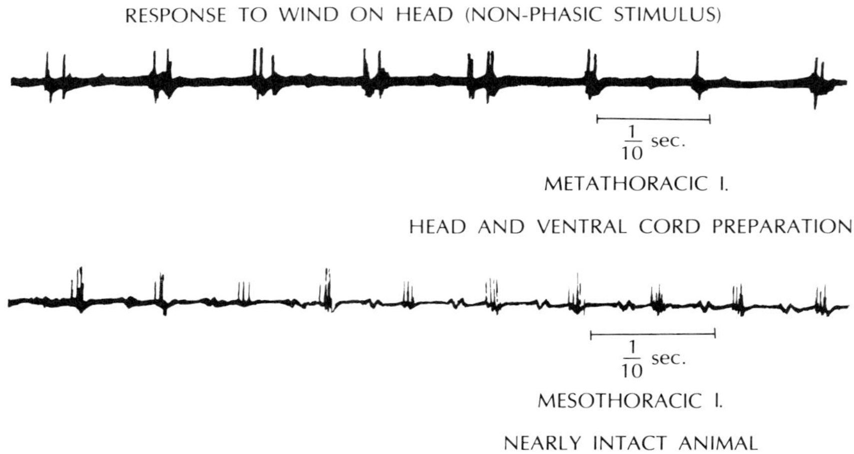

Fig. 3.6. Discharges in motor nerves to locust wings during flight. (From Wilson, 1961.)

Upper strip, discharge in the central stump of metathoracic nerve 1 of a head and ventral cord preparation during response to wind on head; *lower strip,* discharge in the central stump of mesothoracic nerve 1 during flight of a nearly whole animal.

not be possible (Wilson, 1961, 1964; Wilson and Gettrup, 1963; Wilson and Weis-Fogh, 1963; Weis-Fogh, 1964).

The precise mechanisms responsible for the alternation of bursts in the elevator and in the depressor motor neurone groups are still not clear, but Wilson (1966a) has presented a hypothesis which fits all the known facts. Although the individual motor neurones show no inherent tendency to fire in the flying rhythm, excitatory couplings between synergistic motor neurones, whose existence is indicated by evidence from intracellular recordings, could cause them to do so. When activity started in such a group it would spread and reach a high frequency. After a while the frequency would decrease, either through fatigue, accumulating refractoriness, or through a slowly acting collateral inhibition within the group. This would continue until all units were inactive, and they would remain so until one of them recovered sufficiently to fire. This would then excite others, and a new burst of activity start. On this view, then, rhythmicity is due solely to couplings between neurones within functional groups. The couplings would have to be labile, since the relations between muscles may change: the same pair of muscles may be synergists in flight and antagonists in walking. Another view, not yet ruled out, is that interneurones play a part in the rhythmicity: present knowledge of the properties of the motor neurones suggests that motor-neurone interactions cannot by themselves account for the rhythmicity (Kendig, 1968).

The alternation between the elevator and depressor neurone groups presumably depends on inhibitory interactions between the antagonistic groups of motor neurones or interneurones. However, such inhibitory relations are not necessary for the rhythmicity, because rhythmic bursts continue in the depressor motor neurones when the elevator motor neurones are inactive (Wilson, 1966a; Waldron, 1967a).

Although the rhythmicity of the wing beats depends on a central pacemaker system, the mechanism exhibits adaptability adequate to maintain stability in flight. For instance, in the absence of visual feedback many animals roll consistently in one direction, and removal of one wing produces a rolling torque. These are corrected by compensatory changes in the motor pattern if visual feedback is available. Normal flight thus depends on an interaction between a centrally determined "motor score" and a mechanism correcting deviations from a "sensory template" (Wilson, 1968).

Other non-phasic sensory inputs can modify the burst pattern, and Waldron (1967b) has produced evidence that the observed effects can be understood on the basis of one assumption—that input causes slow changes in the "average excitation" of individual motor neurones. Waldron (1968) has also studied how the wing-beat frequency can be influenced by a rhythmically flashing light. Coupling occurs if the frequency of the light is not too different from the initial wing-beat frequency. This coupling depends on a slight acceleration of the cycle by an amount which depends on the timing of the flash relative to the cycle. The phase of the flight cycle then changes relative to the light cycle until the light flashes come at such a point that the acceleration they cause is just enough to produce a frequency of the flight system equal to that of the stimulus. This is then stable, since slight fluctuations in the timing of the flight cycle are self-correcting.

The flight mechanisms of other large insects have not yet been studied in such detail, but a generally similar mechanism seems to operate in large moths and butterflies (Kammer, 1967, 1968). In flies and many other small insects the position is rather different. Here the power comes from a series of myogenic muscles which influence the shape of the thoracic box in which the wings are mounted, and can thus indirectly cause the wings to move up and down. The individual contractions in these muscles are initiated by the stretching of the muscles themselves and not by particular nervous impulses. The oscillatory rhythm is thus determined by the mechanics of the system and by the physiology of the muscles. The behaviour of the individual motor units depends on the average frequency of the nerve impulses reaching them but not on details of sequential pattern or phase. These indirect myogenic muscles can be fully activated neurally even when the wings are still. The aerodynamic power produced depends on another series of muscles, attached directly to the wings, which control a clutching system. The frequency of the system is controlled by muscles which regulate the stiffness of the thoracic box, as well as by tonic changes in the excitation of the indirect muscle system. A major difference from the locust therefore lies in the separation of power and steering mechanisms. In the locust some of the muscles which supply power on the downstroke also regulate the angle of attack of the wing and thus the lift/thrust relationship; while here the direct muscles exert differential effects on the linkages between the indirect muscles, which supply the power, and the wings (Wilson and Wyman, 1963; Pringle, 1965; Nachtigall and Wilson, 1967).

3. *ARTHROPOD WALKING*

An apparently more complex but probably basically similar case is provided by the walking of insects. At first sight insects appear to show a great variety of gaits. Furthermore, if one or two legs are removed, adjustments take place in the position and locomotory rhythms of those remaining, making effective locomotion still possible. The number of different adjustments possible is considerable (Fig. 3.7), so that it is improbable that they arise from previously latent central nervous mechanisms adapted to meet such contingencies. Hughes (1957, 1958) therefore considered the new rhythms to be due to the operation of the same reflex mechanisms which produce coordinated movements in the normal insect—intersegmental and intrasegmental reflexes, coordination among the thoracic ganglia, overall effects from the supra- and suboesophageal ganglia, and so on. In particular, Hughes suggested that there is a mechanism for comparing the forces on the tarsi of the two sides and adjusting the posture until they are equalized.

However, although the gaits used by insects are apparently diverse, certain generalizations are possible. Hughes pointed out that they always involve a wave of protractions running from posterior to anterior in such a manner that no leg is protracted until the one behind is placed in a support-

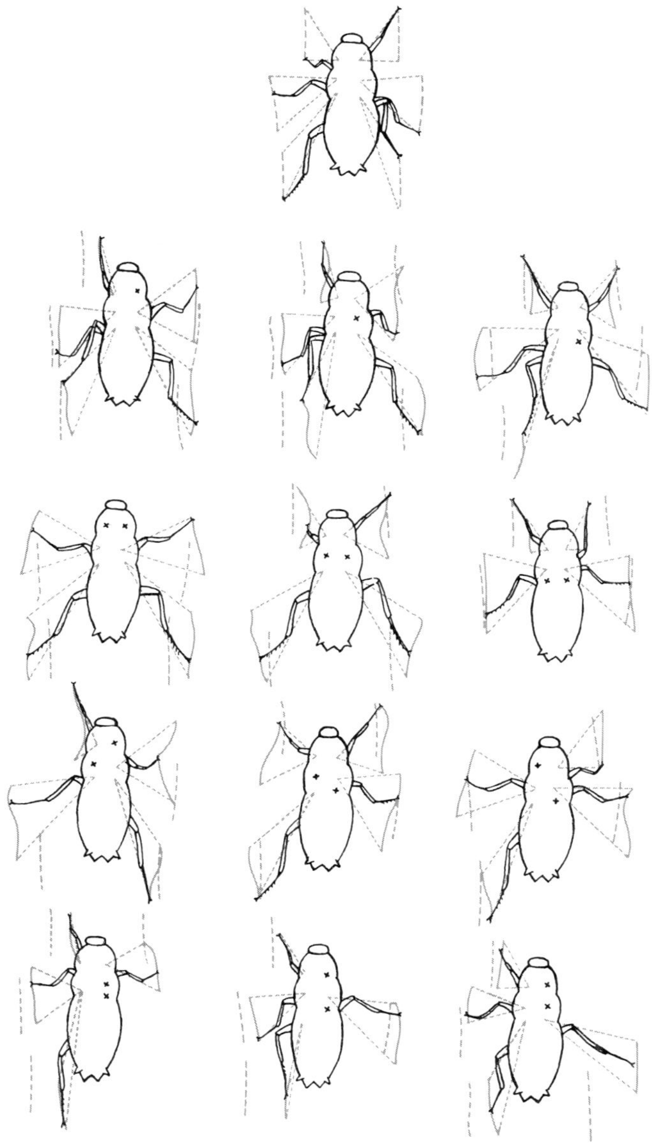

Fig. 3.7. Diagrams show the changes in stride position which occur when different legs are amputated from a cockroach. (Atfer Hughes, 1957.)

Amputation of a leg is indicated by a cross on the thorax corresponding to its position. The positions occupied by the strides of the remaining legs in a normal insect are shown by interrupted lines, and their altered position following these operations is shown by continuous lines. Lines with short dashes indicate the angular movement of these legs, which are drawn for one phase of the cycle.

ing position, and that contralateral legs of the same segment alternate in phase. Starting from this point Wilson (1966b) has shown that nearly all locomotor patterns of intact and operated insects can be described in terms of a simple model which entails the additional assumptions (generally, but not universally, valid) that (*a*) the protraction time is constant and (*b*) the frequency varies through changes in the interval between the steps of foreleg and hindleg, while the intervals between hind and middle leg, and middle and foreleg remain constant. At very low frequencies of leg movement the two sides alternate thus:

$$\begin{array}{l} R_3,R_2,R_1 \ldots\ldots\ldots\ldots R_3,R_2,R_1 \ldots\ldots\ldots\ldots \\ \qquad\qquad\qquad L_3,L_2,L_1 \ldots\ldots\ldots\ldots L_3,L_3,L_1 \ldots \end{array}$$

As speed increases, the variable interval decreases and the sequences of the two sides overlap

$$\begin{array}{l} R_3,R_2,R_1 \ldots\ldots R_3,R_2,R_1 \ldots\ldots R_3 \\ \qquad\quad L_3,\ L_2,\ L_1 \ldots\ldots L_3,\ L_2,\ L_1 \end{array}$$

With a slight further decrease in the interval two patterns are possible:

$$\begin{array}{l} R_3,R_2,R_1,R_3,R_2,R_1,R_3, \\ L_1,L_3,L_2,L_1,L_3,L_2,L_1, \end{array}$$

and

$$\begin{array}{l} R_1,R_3,R_2,R_1,R_3,R_2,R_1, \\ L_3,L_2,L_1,L_3,L_2,L_1,L_3, \end{array}$$

With still greater shortening of the interval, legs 1 and 3 of each side move together

$$\begin{array}{ccccc} R_1, & & R_1, & & R_1 \\ & R_2, & & R_2, & \\ R_3, & & R_3, & & R_3 \end{array}$$

$$\begin{array}{ccccc} & L_1, & & L_1, & \\ L_2, & & L_2, & & L_2 \\ & L_3, & & L_3, & \end{array}$$

This is the most frequently found locomotory pattern and the fastest ever shown by cockroaches. Some species use still further variants—for instance the variable interval between legs 1 and 3 may become negative. Only a few very special cases are known in which the model does not hold—some insects for instance, which use less than six legs. The gaits shown by amputees can all be described in terms of Wilson's model—they are simply speeded-up, leg-deficient gaits of more slowly moving animals.

There is thus no need to postulate the degree of central nervous plasticity which their diversity seems at first sight to indicate.

Furthermore, the fact that the diverse locomotory patterns of insects can be reduced to this simple basic one suggests that they may depend on central pacemakers similar to those found in the crayfish swimmerets (see p. 30). Wilson suggests that the leg movements of each segment depend on coupled oscillators with different inherent frequencies, the metathoracic being the fastest. If the coupling is sufficiently strong, the three will operate at the same frequency, but the slower ones lag in phase, producing the sequential timing of the legs. The phase lag would increase as the frequency of the whole system increases and thus give rise to the various frequency-dependent patterns. However, if locomotion does depend on segmental oscillators in this way, considerable reflex control is superimposed. Proprioceptive reflexes have both tonic and phasic components: although the intersegmental reflexes are always exactly in or out of phase with the moved leg and therefore cannot account for the continuous variation shown in the phase relations between the legs, reflexes are important in intrasegmental control. The leg reflexes are sufficiently fast to effect a modulation of the leg movement cycle by cycle (Wilson, 1965).

An examination of the stepping patterns in tarantula spiders reveals a number of differences from those of insects (Wilson, 1967). In spiders the leg oscillations are more loosely coupled with each other, and antagonism between contralateral legs of the same segment is less apparent. However, Wilson has suggested a model mechanism which could fit both cases. In this the movement of each leg is controlled by an oscillator negatively coupled with both its contralateral partner and its nearest neighbours—the coupling being either central or reflexive.

In summary, Wilson's recognition that the diverse locomotory gaits shown by insects can be understood as variants of a rather simple basic pattern strongly suggests that they may depend on a mechanism which does not possess quite the degree of plasticity which the modifiability of the gaits suggests. This mechanism consists in essence of a series of oscillators, more or less strongly coupled, but influenced also by proprioceptive and other input. Although direct physiological confirmation will probably be difficult, the principle seems likely to be applicable to a wide range of arthropod movements such as the locomotory patterns of crabs (Bethe, 1931) and perhaps the web-building movements of spiders (Szlep, 1952), in both of which amputation does not abolish function.

4. *MAMMALIAN BREATHING*

Another case in which central patterning plays a major part but is subject to both tonic and phasic peripheral influences is mammalian breathing behaviour. If the portions of the paired respiratory "centres" (see Chap. 12) which control inspiration are isolated, they are active continuously and cause sustained inspiration. The expiratory portions are excited by the inspiratory ones and inhibit them: the inspiratory muscles then relax. When the expiratory centre ceases to discharge, the inspiratory one is released from inhibition. These two centres, with an additional expiratory one in the pons, thus provide the basic rhythm. As with the locust, the rhythm may depend on positive coupling between synergistic neurones, an accumulating negative effect on firing in active units, and/or inhibitory coupling between antagonistic groups of units. As in most of the other cases just cited, the activity in some units accelerates during the early part of each burst. Peripheral influences, both tonic and phasic, play an important part in respiration, however, for the rhythm is modified by the action of carbon dioxide on the inspiratory centres and carotid chemoreceptors. Furthermore

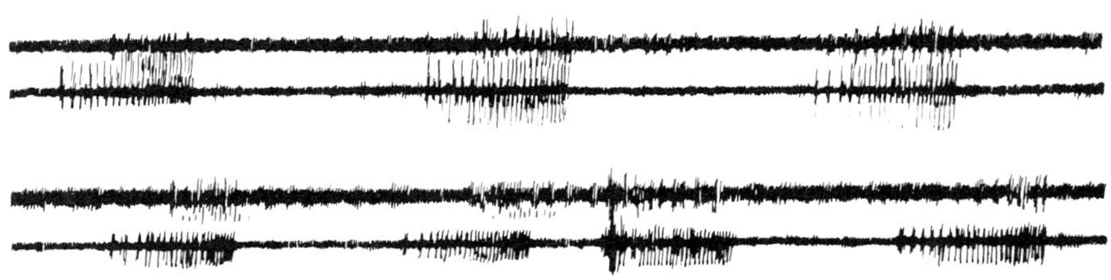

Fig. 3.8. Discharges, in motor roots of abdominal ganglia of aeschnid dragonfly nymph, which control respiratory movements. (Mill and Hughes, 1966.)

Upper strip, normal rhythm; *lower strip,* an extra burst is elicited by stimulation.

stretch receptors in the lungs inhibit the inspiratory centre when strongly stimulated and facilitate it when mildly stimulated (Oberholzer and Tofani, 1960; von Baumgarten and Nakayama, 1964).

5. *DRAGONFLY RESPIRATION*

In other cases, an external stimulus may reset a rhythm which is otherwise endogenous. An example of this, taken from the control of respiration in an aeschnid dragonfly nymph, is shown in Fig. 3.8. In the species concerned the nymph respires by moving water across its gills, which are situated in the modified rectum. The expiratory phase is aided by the contraction of the abdominal respiratory dorsoventral muscles, which receive bursts of impulses, one per respiratory cycle, from the second lateral nerve roots. The pattern of such bursts is shown in the traces in the figure, where the upper trace of each pair shows the expiratory bursts in the second root of the fifth abdominal ganglion, and the lower trace shows the bursts in the corresponding root of the seventh ganglion. Stimulation of the first root of, say, the seventh ganglion between two such bursts produces an additional burst (lower pair, third burst), which is followed by further rhythmical bursts at the normal interval after the extra burst: the phase of the rhythm is thus displaced with respect to the original one (Mill and Hughes, 1966; see also P. L. Miller, 1964).

6. *FISH SWIMMING*

In the rhythmic locomotory movements of vertebrates, peripheral control is normally essential, but the integration of central patterning and peripheral control is usually subtle. In teleost fishes the swimming rhythm may persist in spite of extensive deafferentation of the spinal cord region (von Holst, 1935a), and so it is clear that *all* the normal sensory consequences of each phase of the movement are not necessary for its continuation. As a result of experiments on partially narcotized teleosts, von Holst proposed that the rhythmic passage of the pattern of contraction down the trunk was a consequence of a rhythm generated centrally by cells in the anterior region of the spinal cord and medulla: under partial urethane anaesthesia the respiratory, fin, and tail movements appear to have a common rhythm, but when sensory regulation is possible, separate rhythms are differentiated. In his view, therefore, the rhythm was in the first place independent of afferent inflow, though it could be influenced by it.

Careful studies of dogfish (*Scyllium canicula, Acanthias vulgaris*) by Lissmann (1946a, b) and Roberts (1969) have revealed a complex interplay between an endogenous rhythm and peripheral feedback. Swimming can be analysed into the adoption of an undulatory posture and the propagation of this posture down the body. The static undulatory posture is produced as a reflex in which the spinal cord acts as an entity—in other words the *static* patterning of contraction and relaxation of the trunk musculature is centrally determined. This finding is in agreement with the earlier work of von Holst (1935a, b) on the eel, and of Gray and Sand (1936a, b) on the dogfish. The latter authors cut all nerves to the musculature of about twelve segments of a dogfish and clamped the denervated region rigidly: the swimming movements of the anterior and posterior regions were nevertheless perfectly coordinated. The coordination of the locomotory posture across the denervated region is thus central.

The passage of the static posture down the trunk is, however, another matter. In the first place, weak stimulation of the body during swimming results in modification to the swimming stroke in progress, the precise nature of the modification depending on the site of the stimulus and the phase of the swimming stroke at which it is given. Since the swimming rhythm which emerges after such a stimulus is out of phase with that which preceded it, the change cannot be interpreted as a temporary modification superimposed on a basic pattern but must involve a fundamental change in the swimming rhythm (cf. Fig. 3.8). Second, persistence of the locomotory rhythm depends on afferent excitation: it is abolished by severance of the dorsal roots. It is, however, not necessary for all the sensory roots to be intact—deafferentation of half of the body (anterior or posterior, left or right) still leaves a preparation which shows persistent swimming. With extensive deafferentation rhythmical swimming movements may occur but do not persist for long. With nearly complete deafferentation, exteroceptive stimulation induces a static posture but no rhythmic movements. Lissmann thus suggested that the propagation of the reflex static posture along the body depends on proprioceptive reflexes (Fig. 3.9).

This work has recently been extended by Roberts (1969), who recorded from the ventral motor roots of spinal dogfish immobilised with curare. Under these circumstances the ventral motor output must be independent of phasic sensory input: nevertheless rhythmical discharges still occur (Fig. 3.10). The rhythm does differ from the normal swimming rhythm in a number of ways, however. First, it is slower. Second, the nerves become silent after about an hour, whereas a non-curarized preparation lasts about 10 hr. After the rhythm has disappeared it can be reinstated by unilateral non-phasic stimulation of the type that would normally accelerate swimming. Third, whereas the sigmoid postures of normal swimming involve a time lag between the firing of successive units on the same side, in the curarized preparation this is absent and all units on each side fire synchronously. Roberts thus suggests that the phasic input of normal swimming overrides this spontaneous output, entraining the motor neurones to fire at the right frequency: if a curarized fish is oscillated manually, the motor discharge follows the applied frequency and not the preferred frequency. The phasic input presumably also organizes the longitudinal propagation of contractions down the body. Precisely why deafferentation abolishes the rhythm, as in Lissmann's experiments, while curarization does not, is still not clear.

In the dogfish, the undulatory movements of the body form the basis of locomotion, and the fins show little independent rhythmic movement. In most teleosts, by contrast, the fins have some degree of independence. Medullary preparations of the teleost *Labrus* show rhythmical fin movements which may persist for days. Under the influence of certain drugs the individual fin rays of each fin can beat at different frequencies, though normally they behave as a unit. Similarly each fin can beat in a rhythm of its own but, in addition, rhythms of the different fins influence each other ("relative coordination"). Two main principles are involved. In the first

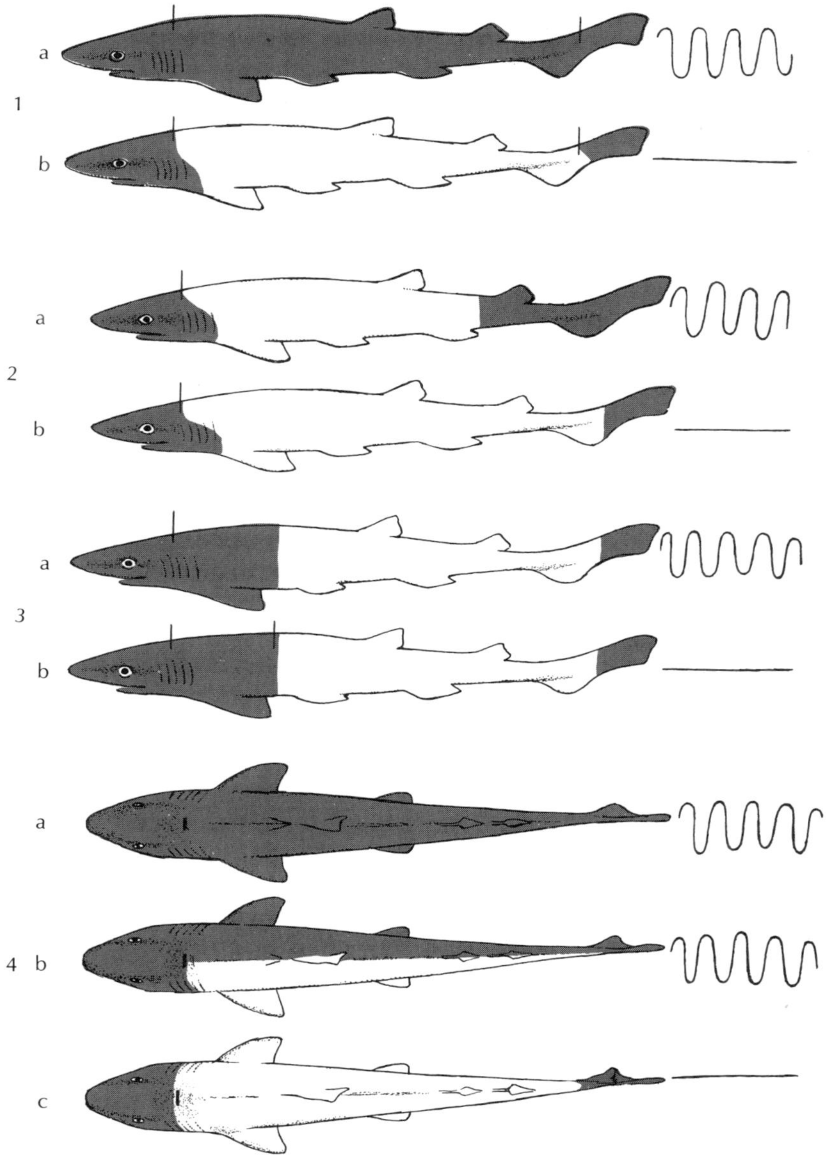

Fig. 3.9. The effect of deafferentation upon the locomotory rhythm of spinal dogfish. (After Lissmann, 1964*a*.)

The white regions represent the desensitized areas of the body. The level of spinal transection is indicated by a thick line. The presence or absence of a locomotor rhythm is indicated on the right.

place, two fins or sets of fins, initially beating with independent rhythms, may gradually fall into step, the periodicity of one being dominant to that of the other. This is known as the "magnet effect", the frequency of one rhythm attracting that of the other. Second, the amplitude of one of two fins beating out of phase with each other may be increased or decreased according to its phase relationship with a beat of the other fin. This "superposition" effect is found also in deafferentated spinal fish and so appears to be independent of peripheral stimuli (see von Holst, 1935a, b, c; 1936, 1939, 1948).

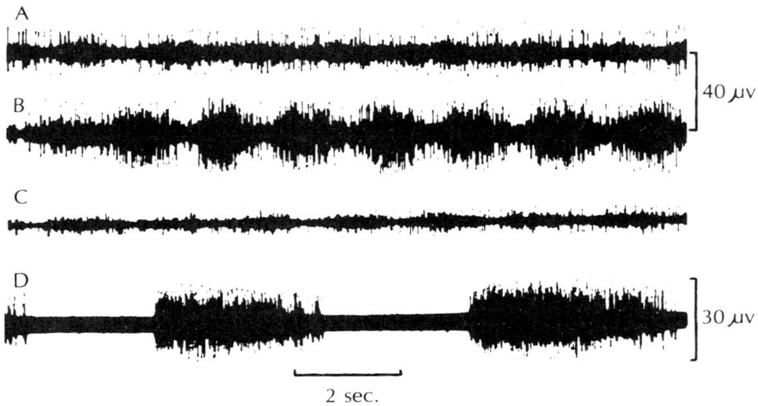

Fig. 3.10. Dogfish swimming. Discharge from motor neurones to the slow muscles of a spinal dogfish when A, lying on the bottom; B, swimming vigorously; C, swimming in an inverted position; and D, paralyzed with curare. (From Roberts, 1967.)

7. CHICKEN LOCOMOTION

A similar type of relationship is found in the coordination of head and leg movements in walking chickens. As the chicken advances, its head moves backwards and forwards relative to the body. The head movement is in either relative or absolute coordination with the leg rhythm, the latter being dominant. In addition, the head movement is elicited by the shifting of the retinal image which accompanies forward locomotion: it is probable that the backward movement of the head is adjusted to the forward speed of the body so that the shifting of the retinal image is minimized over short periods of time. However, if chicks are fitted with ground-glass spectacles, so that shifting of the retinal image is prevented, the head still nods in a rhythmic manner: its movement is thus not wholly dependent on that of the retinal image (Bangert, 1960).

8. *LOCOMOTION OF LOWER TETRAPODS*

In the lower tetrapods central and peripheral factors each play important roles in the integration of the total locomotor pattern. In most species the limbs move in an orderly diagonal sequence, in absolute coordination with each other—left front, right hind, right front, left hind. Each phase of the walking rhythm must involve a characteristic proprioceptive discharge: the question therefore arises, to what extent is this necessary for the rhythmical sequence of limb movement? The walking of the toad has been subjected to particularly detailed analysis. The early experiments suggested that the proprioceptors played only a minor part in coordination, for the normal ambulatory rhythm continues when some of the limbs are deafferentated. However, Gray and Lissmann (1946a, b) showed that the presence of one, but only one, intact spinal nerve is essential for the continuation of rhythmical locomotory movements. So long as one spinal nerve remains intact, even if it supplies only the muscles of the back or the pelvis, the diagonal pattern of limb movement continues, but this pattern is never found unless there is one segment which contains intact both sensory and motor nerves. It is thus clear that afferent input from an intact segment is necessary for continued locomotion. Gray and Lissmann argue that this could be acting in two ways. On the one hand, the rhythmical movements of the limbs could be the result of a rhythmicity originating centrally but requiring a certain threshold level of not necessarily rhythmical impulses from the periphery. Alternatively, the locomotory rhythm could depend on the sensory input also being rhythmical. The first of these possibilities is rendered improbable by the behaviour of a preparation in which the motor nerves to three of the limbs and the dorsal musculature were severed. Under these circumstances stimulation of the deafferentated limbs produced rhythmical movements in the intact limb. Thus, absence of the normal proprioceptive consequences of walking from three of the limbs had practically no influence on the ease with which rhythmic stepping could be induced in the intact limb. But if the fourth limb was then deafferentated, all rhythmical response disappeared: stimulation of the other limbs induced only a monophasic response, and rhythmicity was absent however intense the stimulation (Fig. 3.11). Further experiments showed that proprioceptive impulses from each limb influence the posture in all remaining limbs, producing postures characteristic of specific phases of the ambulatory pattern.

Swimming movements, unlike those of terrestrial locomotion, persist in spinally deafferentated toads, but only so long as the labyrinths are intact. There is suggestive evidence that the acceleration produced by the swimming

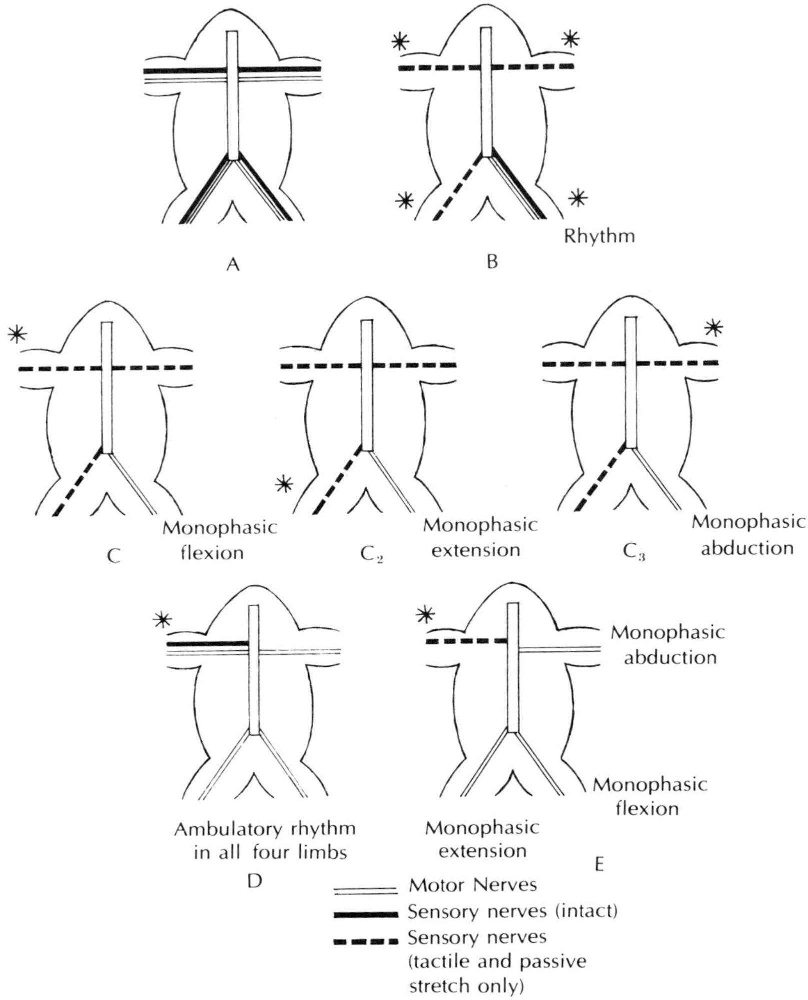

Fig. 3.11. Diagram illustrating the dependence of an ambulatory rhythm in a frog on an intact nerve supply to one or more spinal segments. (After Gray, 1950.)

The site for an external stimulus is shown by an asterisk. In all cases, the back has been completely denervated.

A, the nerve supply to all four limbs is intact and the animal walks normally; *B*, only the right hindlimb has an intact supply, the motor nerves to the other three limbs have been cut. A stimulus applied to any limb elicits an ambulatory rhythm in the right hindlimb; *C*, as in *B*, but the right hindlimb has been deafferentated. The response of the right hindlimb to a stimulus is then monophasic, and its nature depends on the site of the stimulus; *D*, the left forelimb is intact, the remaining three limbs being deafferentated. A stimulus applied to the left forelimb elicits a coordinated ambulatory rhythm in all four limbs; *E*, as in *D*, but the motor nerves of the left forelimb have been cut. A stimulus applied to the left forelimb elicits a characteristic but monophasic response in each of the others.

stroke excites the labyrinths and thereby produces further strokes (Gray and Lissmann, 1947).

In view of these results Gray and Lissmann regarded the earlier view that amphibian locomotion is centrally coordinated as unproved, and emphasized the role of patterned peripheral impulses from all parts of the body in contributing to a general pattern of excitation which, in turn, elicits a pattern of response from the musculature as a whole (Gray, 1950). The peripheral control is thus not merely a matter of local reflexes—sensory input from each limb has some influence on the posture of all the limbs (see also ten Cate, 1965). Not all workers agree with this view. Thus Weiss (1950) believed in the essential importance of central factors in coordination on the basis of experiments in which amphibian limbs were transplanted into abnormal positions. For instance, both forelimbs were reversed, so that their morphologically anterior sides were directed towards the tail. In the normal position, the abductor of the forelimb lies in front and is stretched when the body is moving forwards: it could be this which initiates the swinging forward of the limb for the next step. If the front limb is reversed, the abductor is not stretched when the body moves forwards. Nevertheless, it contracts at the same point in the locomotory sequence as it would have done if the limb had been in its normal position. This results, of course, in the reversed limb moving backwards when the normal limb would have moved forwards. Weiss regarded this as proving that the coordination patterns are determined centrally. His argument, however, seems to neglect the role of the inter-member reflexes demonstrated in Gray and Lissmann's experiments, which could be overriding the intra-member ones in this case.

Further evidence indicating central patterning for movements of each limb has been brought forward by Székely (1968), who implanted a segment of brachial spinal cord into a tunnel in the dorsal fin of salamander larvae and sealed the opening by a limb graft. The limb subsequently showed spontaneous coordinated movements when apparently receiving only non-rhythmic input. Thus a high degree of central patterning is possible, at any rate, in the embryo and as far as single limbs are concerned: Székely suggested a neural model from which the pattern could originate. It remains the case, however, that Gray and Lissmann were concerned with the patterning of all four limbs and not only with that of the muscles within any one of them.

9. *LOCOMOTION AND OTHER RELATIVELY SIMPLE MOVEMENTS OF MAMMALS*

In mammals, sensory input plays an even more important role in rhythmical locomotory movements. Deafferentation of even a single limb usually results in the failure of that limb to show locomotory activity, though the other limbs may move normally (Lassek and Mayer, 1953). Furthermore, cortical stimulation elicits only monophasic responses from deafferentated limbs, even when the intact limbs show rhythmical movements. Most rhythmical movements thus require an intact sensory input. Since severance of the sensory nerves to the distal portions of a cat's limbs does not prevent walking, it would seem that the essential feedback comes from the proprioceptors in the proximal sections (Sherrington, 1910).

Although proprioceptive feedback may be essential for rhythmic limb movement, central integrative processes also play a role. Thus an alternating rhythm of contraction and relaxation can be produced in the deafferentated tibialis anticus and gastrocnemius of a decerebrated cat (Graham-Brown, 1911). Sherrington (1913) produced rhythmical and alternating contractions of the deafferentated extensor muscles of the two knees by simultaneous stimulation of right and left peroneal nerves. Some elements of central patterning are thus certainly possible. Such cases do not necessarily require the postulation of a very complex central mechanism—alternate flexion and extension could, for instance, result from mutual inhibition between the mechanisms underlying extension and flexion, or from activity in each mechanism leading to fatigue, which is subject to recovery with time (e.g., Gray, 1950). Furthermore, such rhythmical movements as do persist in deafferentated limbs could depend in large part on sensory input elsewhere, as discussed above for amphibians. Indeed, in some cases a proper balance between postural flexor and extensor influences may be necessary for the rhythmic movements to appear; in the labyrinthectomized and decerebrate cat, ventroflexion of the head elicits flexion of the forelimbs, dorsiflexion elicits extension, but an intermediate position, approximating to the normal posture, gives rhythmic movements (Pollack and Davis, 1931).

When we come to consider the non-locomotory movements of mammals, a deafferentated limb is normally used little by the animal. In certain circumstances, however, the limb may take part in relatively simple movements. For instance, Mott and Sherrington (1895) described the behaviour of a monkey with a completely deafferentated forelimb. The limb was never used for "voluntary" or skilled movements but was used under duress: for struggling with its whole body. Here the evidence thus suggests that only instance, when recovering from anaesthesia or when held so that it was the more elementary and crude movements do not require proprioceptors in the same limb. Sherrington (1931), however, described one quite complex movement which survived deafferentation: the scratch reflex of the dog, although involving a rhythmic pattern of contraction in about nineteen muscles, can be elicited by simple stimulation from the deafferentated hindlimb, the rhythm of the movement being independent of the frequency of stimulation.

Simple learnt responses can also occur in deafferentated limbs. For example, Górska and Jankowska (1961) and Górska et al. (1961) showed that, in carnivores, simple instrumental acts, established by conditioning procedures, may persist after deafferentation of the limb. Furthermore Taub et al. (1965, 1966a) have shown that even the initial learning of a response may not require the mediation of the peripheral nervous system. Sensation was abolished from one forelimb of rhesus monkeys, and they were then required to flex the deafferentated limb to a stimulus to avoid shock. The conditional stimulus, a click, was so brief that it could not be terminated by the response, and the animal was not permitted vision of its arm. Nevertheless all animals learnt the problem, though they took longer than intact individuals. The authors suggest the operation of a purely central mechanism which returns information to the central nervous system before the movements themselves have occurred—a suggestion similar to that made on the basis of quite different evidence by von Holst (1954), Teuber (1961), Hein and Held (1962), Merton (1964), and others (see pages 96–103). Taub and his collaborators also found a greater resistance to extinction in deafferentated animals than in normal control animals: this difference disappeared when the former group were given indirect information about the effectiveness of the response (Taub et al., 1966b). Further evidence for a role of central processes is provided by experiments showing that subjects can learn to tap a Morse key when kinaesthetic sensation is eliminated by nerve compression block (Laszlo, 1967).

The precise role of central patterning in such crude movements of mammals thus remains problematical, but it is clear that sensory feedback is normally of major importance. We shall see this to be even more the case when skilled movements are considered. It is unfortunate that none of the complicated fixed action patterns used by lower vertebrates in display have yet been analysed from this point of view, since many of them show remarkable constancy of form and at first sight might seem not to require feedback control. It must be remembered, however, that the precise patterning of muscular contraction required in each case will vary with the initial posture. Great tits, for instance, may use a recognizable "head-up threat posture" (see Fig. 16.3*a*) whether perched on a twig, a vertical trunk, or the horizontal ground. Another example is shown in Fig. 3.12: the postures illustrated are all recognizably presentation postures, but they vary both with the position of the animal to which they are directed and with the motivational state of the displaying animal. In such cases the patterning of muscular contraction is not rigidly fixed and could be controlled by a system which produces closer and closer approximation to a required pattern of sensory feedback. With vocal signals the question of orientation to the environment either does not arise or is quite separate from the patterning of the movement itself: as discussed in Chap. 19, deafening sometimes has little influence on the patterning of birds' vocalizations once

Fig. 3.12. Variations in presentation of the Hamadryas baboon: the partner presented to is not represented, but would be to the left in each case. (From Morris, 1967, after Kummer.)

these are fully developed, indicating that auditory feedback plays at most a minor role in the monitoring of the sound. We shall see later that the control of human speech is quite a different matter, feedback playing a major role in monitoring the output.

10. *GENERAL*

In the more elementary cases we have considered, the feedback control could act largely in the manner of a chain reflex, proprioceptive stimuli from the earlier phases of the movement eliciting the later ones. That it is not usually as simple as this is revealed for instance by Gray and Lissmann's studies of dogfish and amphibian locomotion—the sensory control is exerted not through a series of independent local reflexes but through a central mechanism such that altered input from one part may produce changes in all or nearly all the musculature relevant to the response. In fact, the mechanisms of sensory control can be far more subtle even than this. In man, degeneration of the sensory nerves leads to a marked disturbance of all movements. The gait, for instance, is clumsy, the feet being raised too high and stamped down too hard. If sensory information is absent, accurate movement is impossible, especially if the movement is made slowly. Thus, the proprioceptive information is being used not to trigger the next phase of the movement but to grade the movement itself—to control muscle contractions in progress. The mechanism by which this is achieved is considered in the next section.

3.4 THE CONTROL OF GRADED MUSCULAR CONTRACTIONS

Among the proprioceptors serving mammalian striped muscle are muscle spindles which lie parallel to the muscle fibres and share their attachments. In human subjects these do not convey information about the position of the limb, a function which is discharged by sense organs in the joints and skin (Merton, 1964). They do, however, respond to changes in the length of the muscle: extension gives rise to afferent impulses which excite the muscle's own motor neurones. They can thus act to maintain the length of the muscle constant: in the "stretch reflex", extension of the spindle causes an increased excitation of the motor neurone and thus contraction of the muscle. There is, in fact, a self-correcting system here: extension of the muscle sets in course a sequence of events which leads to a reduction in its length. In the language of the control engineer, the spindle provides negative feedback in a closed-loop system (Fig. 3.13).

Graded voluntary movement depends on the fact that the equilibrium length of the muscle spindles is adjustable. The ends of the spindles are contractile, receiving a motor supply (the γ efferents) from the spinal cord.

When the ends of a spindle contract, the sensory part will discharge at the same rate as it would otherwise have done at a longer muscle length. Thus, if the γ efferents cause the ends of the spindle to contract, the sensory portion will be stretched, its rate of discharge will increase, and the muscle stretch reflex will be set off. The muscle will shorten until the increased rate of discharge from the spindle is offset. The sensory portion of the spindle is not, in fact, responding to the absolute length of the muscle but to the difference between the length of the spindle and the

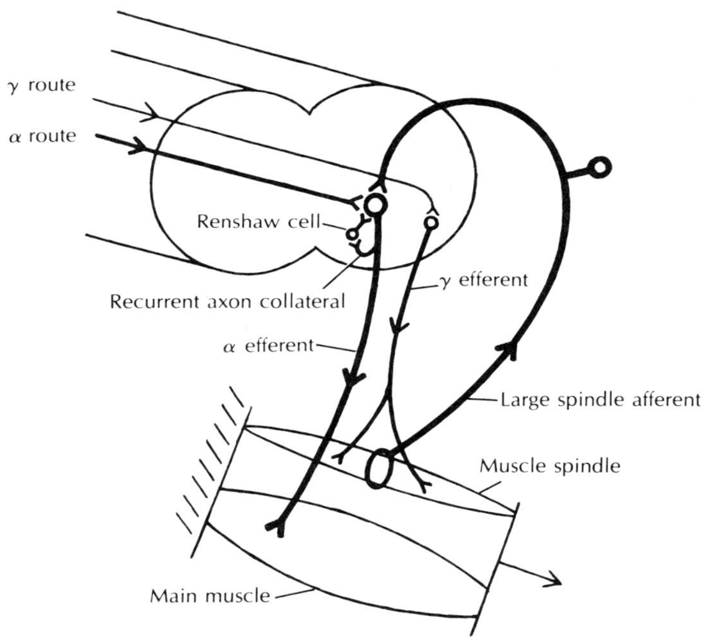

Fig. 3.13. The mechanisms underlying the stretch reflex and involved in the control of graded muscular contractions. (After Hammond et al., 1956.)

length of the muscle. Whether the muscle contracts or relaxes depends in part on the state of contraction of the spindle.

In subsequent chapters we shall have frequent occasion to refer to feedback control of this sort, and it will be convenient to have a general term for the required or optimal stimulation received when the effector organ is in its equilibrium position. A precise English word is hard to find, though terms such as "goal", "target value", or "equilibrium position" are all useful in some contexts. The German term "Sollwert" (literally the "should be value") has a wider applicability. In the present context the Sollwert is achieved when the lengths of muscle and muscle spindle bear a certain relation to each other, but the term might equally be applied to the setting on the thermostat controlling a heating system or to that on the governor controlling a steam engine.

The muscle can thus be caused to contract in two ways: either through the γ efferents and servomechanism or directly. In human subjects either method may be used, depending for instance on previous instructions and the speed of movement required (Hammond et al., 1956).

Another complication is that even the direct route of muscular excitation is subject to servocontrol. Collaterals from the motor neurones run back into the ventral horn of the spinal cord, to form synapses with the small Renshaw cells. When these are excited, they inhibit the motor neurones in the area. The Renshaw cells thus form part of a negative

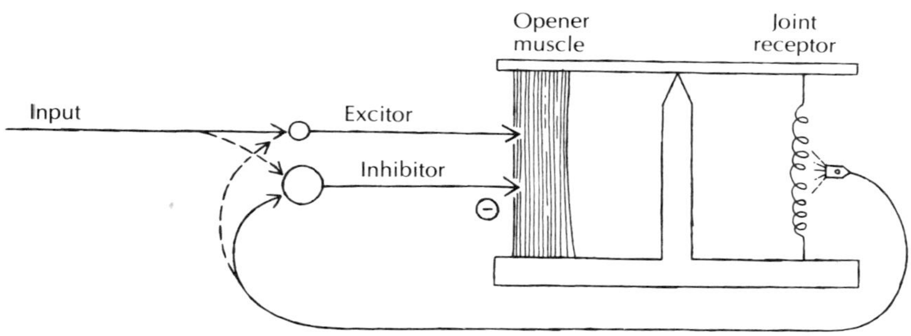

Fig. 3.14. Model summarizing possible organization of the reflex system controlling the opener muscle of the crayfish claw. (From Wilson and Davis, 1967.)

feedback circuit which probably stabilizes the discharge of the motor neurones (Holmgren and Merton, 1954).

Furthermore, striped muscles are also influenced by the tension receptors (the Golgi tension organs), impulses from which tend to inhibit the motor neurones (Granit, 1955a). The precise role these play in the control of muscular contraction is still not clear (review by Merton, 1964; P. B. C. Matthews, 1964).

In passing, it is interesting to compare this mechanism with that controlling the opening of the crayfish claw. Here the degree of contraction of the muscle is controlled in a quite different way—namely by the balance between the activities in separate excitor and inhibitor fibres. This is illustrated schematically in Fig. 3.14. The outputs of the excitor and inhibitor neurones are correlated in frequency for any one kind of input, and they seem to form part of a single motor-neurone pool. The tension developed in the muscle is thus not controlled by a central antagonism: it depends instead on a peripheral balance between impulses in the excitor and inhibitor axons. A stimulus which causes opening, such as touch on the outside of the claw, acts by exciting the inhibitor axon less than the motor one. Conversely a touch on the inside of the claw, which prevents opening, excites the inhibitor more than the excitor. Inputs to excitor and inhibitor are thus weighted differentially according to source: for general body input the excitor has the lower threshold, but if the input is primarily proprioceptive, the inhibitor neurone responds more readily. Feedback from the proprioceptor acts not by inhibiting the excitor neurone but by exciting the inhibitor: in this way it supplements the simultaneous excitation of the excitor neurones of the antagonistic closer muscle. The stability provided by this loop is not clear but is probably greater when the claw is nearly fully open or nearly fully closed (Bush, 1962; Wilson and Davis, 1965).

The generality of this mechanism remains to be elucidated: the control of the tonic muscles of the crayfish abdomen is known to involve central rather than peripheral inhibition (Kennedy and Takeda, 1965).

3.5 SKILLED MOVEMENTS

The movements discussed in Sec. 3.3 were either more or less stereotyped and species-characteristic, or crude and individually specific. There is no sharp dividing line between these and the individually acquired skilled movements of man and other higher mammalian species. However, studies of skilled movements demonstrate a complexity of feedback control through a variety of sensory modalities which exceeds, but is nevertheless almost a logical sequence to, the simpler cases discussed earlier in this chapter. Although they have been studied mostly in man, the mechanisms which have been revealed are undoubtedly variations on mechanisms existing but not yet studied in other animals.

That proprioceptive feedback from the earlier phases of a skilled movement cannot be responsible for the elicitation of the later phases in any simple chain-reflex manner is often apparent from the time relations involved. Thus Lashley (1917, 1951) pointed out that some skilled movements, such as those of a pianist's fingers, are made so quickly that there could be no time for peripheral adjustment. However, the case here depends on the size of the units of movement considered: it remains possible that the playing of a short sequence of notes can be independent of feedback, while the chaining together of successive sequences depends on it (Welford, 1959). In fact the importance of the feedback in the control of skilled movements becomes apparent only when it is cut off or upset. This can most easily be done with exteroceptive feedback. Deutsch (1960) studied the attempts made by singers to sustain a steady note. In practice, the note achieved was rarely steady but subject to small oscillations. By a device which amplified the volume of the voice and delayed its reception by the singer, Deutsch verified the hypothesis that the relative constancy of the note depends on small corrections for error. The singer starts to correct only when he hears his voice straying from the Sollwert and corrects again when he finds that the initial correction has led him into the opposite error (see also Miller and Dollard, 1941).

As we talk, our words are constantly monitored by our ears, and interference with this feedback produces gross distortions of speech. If the speaker is prevented from hearing his own words as he says them, but allowed to do so after a brief delay, his speech rapidly becomes incoherent, with widely fluctuating frequencies and amplitude, and jumbled words. He is, as it were, trying to monitor by a feedback no longer appropriate.

It is evident that such monitoring mechanisms must be of considerable complexity. The disturbance to speech at any instant is a consequence of a discrepancy between feedback assumed to come from what has just been said and the feedback which should have been produced. K. U. Smith (1962) emphasizes that the disturbance caused by delayed feedback is not a simple function of length of delay but varies between individuals and with the nature of the task (see also Lenneberg, 1967).

Interference with the feedback through other sensory modalities produces comparable disturbances. If a subject is asked to write, but instead of seeing his hand directly sees it projected on a television screen, there is no disturbance in his performance so long as the angle of the camera corresponds with that of his own eyes. But if the position of the camera is altered so that the picture that he sees shows him his hand from an angle quite different from that at which he normally sees it when writing, his performance is confused. Delayed feedback causes similar confusion in human operators required to track a target by turning a control. This occurs even when the delay is shorter than their own reaction time. Such results suggest to Smith (1962) that monitoring is a continuous process and cannot be accounted for in terms of unit responses linked into chains —though such a view seems to suggest an approach less likely to be fertile than analytical attempts to disentangle the feedback loops involved (e.g., Chase et al., 1961).

The complexity which must exist in the mechanisms controlling skilled movements is immediately apparent when one considers how the detailed patterning of contractions must depend on the initial posture. One way of conceptualizing the processes involved is to imagine the movements as being made with reference to a "body schema", an organized consequence of past and current sensations stored in some way in the brain (Head, 1920). This concept has been critically discussed by Oldfield and Zangwill (1942a, b): it finds an echo in some more recent work on the control of skilled movement which will be discussed in Chap. 20.

Because skilled movements are made with respect to an initial posture, the feedback systems at the spinal cord/muscle level, discussed in the previous section, can form only one part of the total control system. Proprioceptive reflexes from muscles and labyrinth would be too slow for many purposes, and central activation of the proprioceptive reflexes in anticipation of the strains imposed by the directive components must occur. Postural support must thus be under the continuous control of higher mechanisms. Presumably interlocking control systems with positive and negative feedback are operating at different levels in the nervous system (Jung and Hassler, 1960; Paillard, 1960), complex interreactions among the sensorimotor cortex, cerebellum, and brainstem being involved. The cerebellum, for instance, appears to play a part in comparing an actual movement, as indicated by proprioceptors, with a Sollwert received from the cerebral cortex. We have thus not only a Sollwert, related to reafferent stimulation, at the level of the individual muscle, but also similar systems of control at higher levels, the movement being adjusted at each one (e.g., Lashley, 1952). Some of the possible mechanisms concerned are discussed by Ruch (1951).

As yet, practically nothing is known of how these higher-level control systems might work, but some aspects of the functioning of the mammalian cerebral cortex are at least compatible with multiple control of this sort. For instance, the classical picture of discrete sensory and motor areas of the cortex is now clearly inaccurate: in primates somatic afferents reach the precentral region (generally known as the "motor area") as well as the postcentral "sensory area", and the pyramidal tract seems to originate from postcentral areas as well as the entire precentral area. This close relationship between sensory and motor pathways is clearly compatible with feedback control. Similarly, eye movements can be elicited by stimulation in the "visual sensory" cortex, ear movements from the "auditory" cortex, and so on. Removal of the precentral cortex leads to a disturbance of organized pathways of movement, not in terms of the final solution of the problem presented but in terms of increased awkwardness and clumsiness (Pribram, 1958; Phillips, 1964). To go to the other extreme, it seems that *Octopus* can never learn to make skilled movements with its arms, because it cannot use proprioceptive information about the position of its arms. Presumably, with such flexible arms, the mechanism necessary for it to be able to do so would have to be impossibly complicated (Wells, 1962b).

3.6 CONCLUSION

This chapter has been concerned with the means by which the patterning of movements is controlled. Our principal question has been this: are the sequences of muscular contractions involved in particular movements predetermined from the start in the central nervous system, or are the later phases of the movement adjusted by peripheral stimuli received during the earlier stages? Central patterning certainly occurs in some cases. In others, input from the periphery plays a part in controlling the rhythmicity, phasing, strength, or sequence of the muscular contractions. This input may act in a variety of ways: for instance, it may control intra-member or inter-member reflexes, or it may feed into a control system which makes finely graded contraction possible. In mammals, feedback control plays an essential role in nearly all movements: furthermore, the control may be extremely complex, consisting of a number of interlocking systems.

Thus, as we shall find repeatedly in later chapters, generalizations of universal applicability are hard to find. In the present instance, unifying principles are most likely to come from evolutionary and functional considerations, for the nature and extent of peripheral control are likely to vary among species and with the function of the movement. We might, for instance, expect peripheral control to be more important in locomotion over a rough terrestrial environment than in swimming, in movements involving manipulation of objects than in stereotyped signal movements, and in animals whose motor patterns are modifiable than in those in whom they are more stereotyped. As yet, however, too few species have been studied to provide a sound basis for such generalizations.

3.7 SUMMARY

1. The patterning of complex movements could be determined from the start in the central nervous system or could be adjusted by peripheral (proprioceptive or exteroceptive) stimuli received as a result of the earlier phases of the movement. This chapter is largely concerned with experiments aimed at assessing the relative effects of central and peripheral factors.
2. Each individual and species has a repertoire of "fixed action patterns"—movements whose patterning does not depend on external stimulation, though it may depend on proprioceptive sources of input.
3. Some, but by no means all, fixed action patterns are oriented with respect to the environment. In some such cases the stimuli responsible for the orientation can be distinguished from those responsible for elicitation of the movement in the first instance.
4. The beating of the lobster heart and sound production by a cicada are examples of movements whose patterning is independent of peripheral control. Swallowing, in vertebrates, may be another.

5. Many rhythmic movements, such as the flying of locusts, walking of insects, swimming of fish, and walking of lower tetrapods, depend on central rhythmic processes which can (to various extents) be independent of phasic input but are normally influenced in a variety of ways by peripheral factors.
6. In mammals deafferentation of a limb results in the failure of that limb to take part in any but the crudest movements. However, some learning of new movements is possible even when the limb is deafferentated.
7. The feedback provided by the muscle spindles provides a mechanism for the control of graded muscular movements in vertebrates. Quite different mechanisms are found in invertebrates: that controlling the opening of the crayfish claw is discussed.
8. Skilled movements in man (and presumably also many other vertebrates) depend not only on feedback stimulation from exteroceptors and proprioceptors but also on control systems operating at a number of higher levels.

4

THE EFFECTIVE STIMULI

In his "Stroll through the worlds of animals and men", von Uexküll (1934) described how the mated female tick climbs to the tip of a bush, clinging at such a height that she can drop off onto mammals that run beneath her. She clings there, perhaps for months, without response to the sounds, smells, and other changes about her, until one particular stimulus acts as a signal for her to release her grip. This stimulus is the smell of butyric acid, a product of mammalian skin glands and thus an indication of the proximity of a potential blood meal. Out of the hundreds of stimuli in its environment to which it might respond, the tick selects only this one, "like a gourmet who picks the raisins out of the cake", as von Uexküll puts it.

Stimulus selection of this sort plays a part in all behaviour. At every moment an animal's sense organs are being bombarded by physical energy in many forms. To this chiaroscuro it responds selectively: some configurations of energy change influence its behaviour, others do not. Selection from among the energy changes to which the animal is subjected evidently occurs. Furthermore, since some stimuli influence one type of behaviour and others another, there must be an optimal stimulus situation for each type of behaviour—*Tubifex* worms stimulate a male stickleback to feed, green algae stimulate it to build, another male stimulates it to fight, and a female stimulates it to court. The importance of this selection has been emphasized by many workers (e.g., Lorenz, 1935; Russell, 1943; Tinbergen, 1948, 1951; Broadbent, 1958): in this chapter we shall consider experiments which demonstrate its occurrence, tell us which environmental changes influence the animal's behaviour, and indicate the optimum stimulus situation for eliciting each response.

First, however, we must consider the meaning of the word "stimulus", for it has been used in a number of different ways in the study of behaviour, and this has sometimes caused confusion (Verplanck, 1954). In particular, the word may refer to a part, or to a change in a part of the environment; or it may be confined to something which elicits an observable response. Although it would be an abuse of language to label as a stimulus a physical change to which the animal is actually known to be insensitive, it will become apparent during later chapters that the distinction between producing a response and not producing one is not absolute. An environmental change may, for instance, affect the nervous system but have no immediate effect on behaviour, or its effect may vary with the state of the organism or with the coincidence of other environmental influences. In this book, therefore, the use of the word "stimulus" need not necessarily imply an identifiable response on the part of the organism. On the other hand, the word will be anchored to the physical world and will not be used to refer to hypothetical events postulated to account for changes in behaviour, as is often the case when the term "internal stimuli" is used.

4.1 SELECTIVE RESPONSIVENESS TO STIMULATION

The technique for assessing the effectiveness of a physical character in influencing a given pattern of behaviour consists basically of comparing the responses of the animal to two situations which differ by that character: if the responses differ, then the character is significant. The word "responses" is, of course, used loosely here with intention. It is always necessary to select one or more characteristics for measurement—for instance, frequency, latency, or intensity. These may not be closely correlated with each other, and it is thus advisable to measure several in order to assess the effect of the stimulus (pages 195–197).

In practice, we often start by comparing the naturally occurring situations which do and do not affect the response: this enables us to form an hypothesis about the nature of the effective stimuli. Thus, the observation that a foraging cabbage white butterfly (*Pieris* sp.) visits a series of blue flowers, and occasionally interrupts these visits by fluttering around the blue cover of a notebook, suggests that it responds to the blue colour. In practice, also, we usually narrow the possible sensory modalities before investigating the characters (or stimulus dimensions) important within any one. This procedure may be made unnecessary by field observations of the type just cited or by simple experiments. If we find that great tits will attack their image in a mirror, we can conclude that visual stimuli alone are sufficient for eliciting this response. In other cases, surgical interference may be necessary to delimit the sensory modalities—we can for instance sever the olfactory nerve of a bird and see whether this affects its response to food (Portmann, 1961 and refs. cited), or assess the response of a female rat to her litter when deprived of various senses (Beach and Jaynes, 1956a).

Within each sensory modality we can identify the characters of the total stimulus situation relevant to a particular response by systematically varying the situation and assessing the effect on the response. For instance, male three-spined sticklebacks establish territories in spring and behave aggressively to other males who intrude on them. Intruding females, however, are less often attacked and may be courted. At this time of year the male stickleback has a red belly, and the female has not: by using models, ter Pelkwijk and Tinbergen (1937) showed that this red belly plays a very important role in eliciting attack from the territory owner. Some of the models they used, though only crude imitations of a stickleback, were painted red underneath, while others were very similar to a stickleback but lacked the red belly (Fig. 4.1). The former group of models was attacked much more vigorously than the latter, indicating that the "red belly underneath" was of primary importance in eliciting attack, while the other morphological characters were of minor influence. Behavioural characters also play a part, however, since a dummy is more likely to elicit attack if presented in the head-down posture adopted by a threatening male than if presented horizontally. Similarly, ter Pelkwijk and Tinbergen isolated two features of special importance in eliciting the first phase of the courtship sequence normally given by territory owners to gravid females—the swollen belly of the female and the rather upright position in which she swims in the water (Fig. 4.2).

Lack (1939), investigating the characters of an adult (European) robin (*Erithacus rubecula*) which elicit threat from a territory owner, placed three types of stimulus object in the territories of wild robins—a stuffed adult robin, a stuffed juvenile robin (which has no red breast), and a bunch of red breast feathers lacking the rest of the body. He found that territory-owning males would threaten the bunch of red breast feathers almost as readily as they would a stuffed adult robin, and much more readily than the stuffed juvenile robin. The red breast is thus of primary importance in eliciting threatening behaviour.

The detailed studies of Tinbergen and Perdeck (1950) on the begging of gull chicks provide further examples. After hatching, gull chicks peck at the tips of the parent's bill. The parent regurgitates food onto the ground, picks some of it up in its bill tip and holds it near the young. In the course of its pecking, the young gets hold of some of the food and swallows it. The herring gull, on which the original experiments were performed, has a yellow bill with a red patch near the end of the lower mandible. The importance of the various stimulus characters of the adult's bill in eliciting pecking was studied by using series of cardboard models—within each series the models varied in only one stimulus character. Figure 4.3 shows such a series, the bars indicating the number of responses elicited by each model in a balanced series of tests. The experiments showed that the greatest response was obtained from models which were fairly low, near the chick, moving, long and thin, and pointing downwards, and which

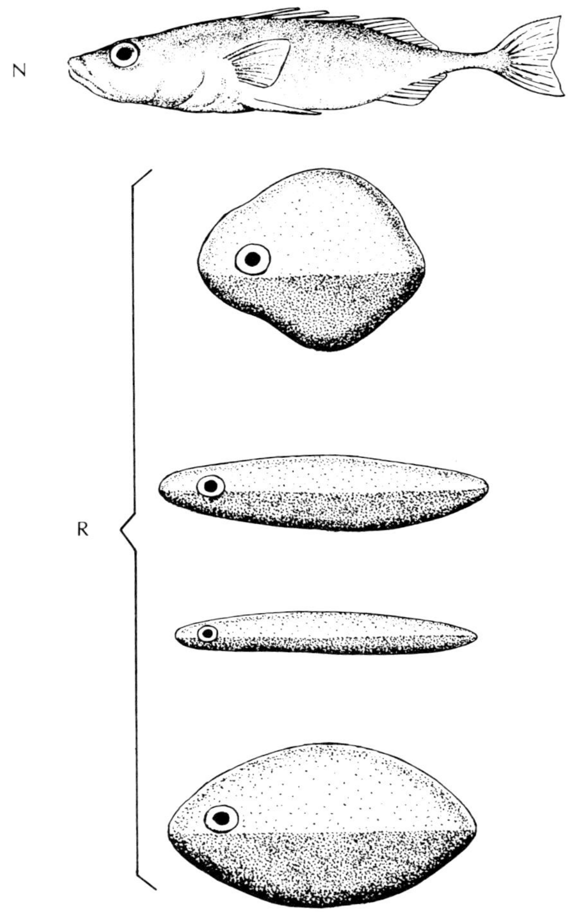

Fig. 4.1. Models used for identifying the stimulus characters important in eliciting the aggressive behaviour of the male three-spined stickleback. The four lower models have a red belly and elicit aggressive behaviour, while the upper model, which lacks a red belly but resembles a stickleback more closely in other respects, is relatively ineffective. (From Tinbergen, 1951.)

had a red patch contrasting with the bill near the tip. In this case, therefore, a number of stimulus characteristics were shown to be important. Others, however, were not: neither the colour of the bill nor the colour of the head was found to influence the response.

In a subsequent study of the laughing gull (*Larus atricilla*) Hailman (1967) found that the most effective stimulus for eliciting pecking is (1) an oblong shape or rod, (2) of about 9-mm width, (3) held vertically, (4) moved (5) horizontally (6) about 80 times per minute. It must (7) contrast

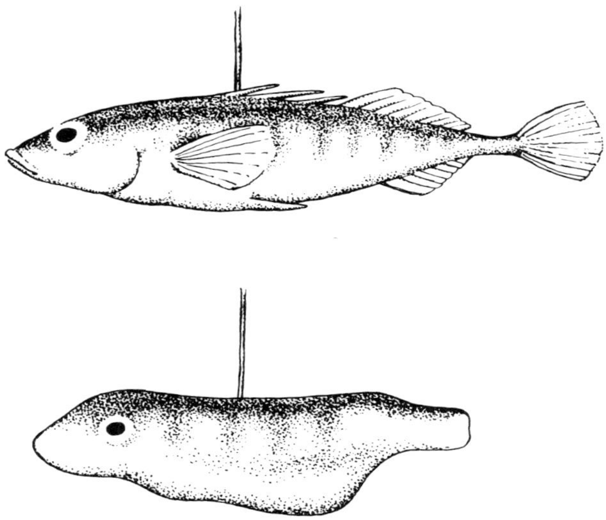

Fig. 4.2. Two models used for identifying the stimulus characters important in eliciting the zig-zag courtship dance of the male three-spined stickleback. The lower and cruder model has a swollen abdomen which is absent in the upper one, and is more effective. (From Tinbergen, 1951.)

with its background and (8) is more effective if darker than the background if not coloured. If coloured, (9) red and blue colours are more effective than intermediate wavelengths. The object is most effective (10) at eye level and (11) when seen by both eyes.

The experiments cited so far have been concerned with visual stimuli, but similar selection occurs in all sensory modalities. Thus individuals of many species respond selectively to the sounds made by their own species (e.g., cyprinid fish, Stout, 1963; cichlid fish, Myrberg et al., 1965; the frog *Pseudacris streckeri*, Littlejohn and Michaud, 1959), and in a few cases it has been possible to isolate the significant characters. An especially clear example is provided by Falls' (1963, 1969) study of the white-crowned sparrow (*Zonotrichia albicollis*). The song of this species consists of a series of clear notes, and it was possible to make artificial songs differing in defined ways from normal ones. These were played to wild individuals,

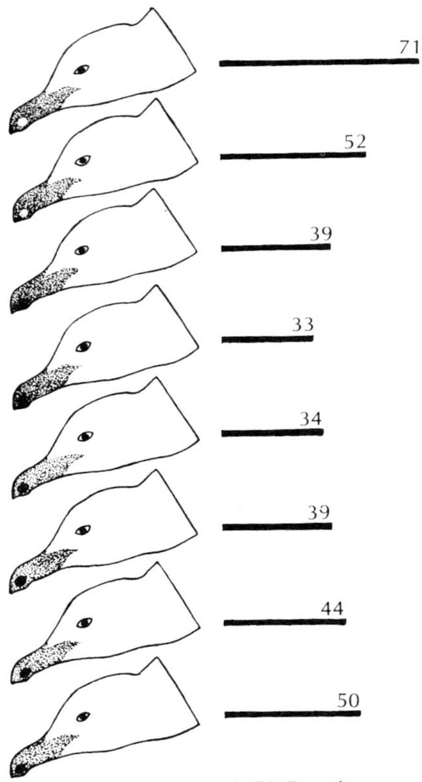

Fig. 4.3. A series of cardboard models of herring gulls' heads with grey bills and spots of varying shade. The lengths of the bars indicate the number of responses elicited by each model in a standardized series of tests. (From Tinbergen, 1951.)

and their effectiveness in eliciting approach, attack, or song assessed. The results showed that the song had to contain unvarying pure tones without harmonics or other overtones. The notes had also to be within certain limits of length, and the intervals between them below a certain maximum. The pitch changes between successive notes were of less importance. Other characters were apparently without influence on the effectiveness of the song. For example the broken nature of some notes, and the slurs which occur in the natural song, were of no apparent significance. The frequency of the notes also could be varied over a considerable range, and within limits variations in loudness and length of the song made no difference.

With chemical stimuli, approach to the appropriate food plant by many phytophagous insects depends on very specific chemical responsiveness: for instance the weevil *Sitona cylindricollis* responds to the substance coumarin, which is given off by the clover on which it feeds (Thorsteinson, 1960; Hans and Thorsteinson, 1961). Similarly Wilson (1962) found that a number of different responses of fire-ant workers (*Solenopsis saevissima*) could be elicited by chemical substances removed from the bodies of other individuals: his experiments demonstrated considerable specificity in the stimulus for each response. These chemical signals are sometimes referred to as "pheromones" (Karlson and Butenandt, 1959).

Many responses can be evoked by cues acting through more than one modality. The begging of the young of both gulls and passerine birds can, at a certain stage, be elicited by both visual and auditory stimuli (Tinbergen and Kuenen, 1939; Tinbergen and Perdeck, 1950); and Beach and Jaynes (1956a, b) showed that the maternal behaviour of experienced rats could be elicited through several sensory modalities. Of course, the finding that a response is influenced through more than one sensory modality is no evidence for supposing that selection is not occurring within each (Lashley, 1938a).

Some further examples of selective responsiveness to naturally occurring stimulus situations are summarized in Table 4.1. In each case some features of the natural object are shown to be of major importance in eliciting the response in question, while others have at most a minor effect. Even in the absence of any precise definition of "a stimulus character", we can say that some responses are elicited by very few stimulus characters, others by many, but selection apparently always occurs. When a response depends much more on some characters than others, these characters are often referred to as "sign stimuli" (Russell, 1943). This term has usually been used to refer to characters whose effectiveness is more or less specific to one response, although the experimental methods described so far in this section provide no evidence on this point (Schneirla, 1952, 1956; Lehrman, 1953): this matter is discussed later (Chap. 5).

Since relational properties are usually important, few sign stimuli can be specified along a single physical scale. Thus, in Tinbergen's experiments on the stimuli releasing aggressive responses from male sticklebacks, the red patch was found to be more effective when on the lower side of the model than when on the upper (see also Baerends, 1957a). An even more interesting case concerns the release of fleeing responses in turkeys by a

TABLE 4.1. Some studies indicating differential responsiveness to aspects of the natural stimulus object for species-characteristic responses.

Species	*Response*	*Characters of Major Importance*	*Characters of Minor Importance*	*Reference*
Silkworm	Male's approach to female	Specific smell of female	Other smells Visual characters	Butenandt, 1955
Water-beetle (*Dytiscus*)	Prey-catching	Chemical and tactile stimuli	Visual stimuli	Tinbergen, 1951
Grayling butterfly (*Eumenis semele*)	Male sexual pursuit	Dark shade Type of movement Distance	Shape	Tinbergen et al., 1943
Fritillary butterfly (*Argynnis paphia*)	Male sexual pursuit	Colour Speed of flicker Size	Black patterning Shape	Magnus, 1958
Mantis	Prey-catching	Distance Type of movement	Size, shape, direction, colour, smell	Rilling et al., 1959
Honey-bee worker	Food-begging	Head carrying specific scent Tactile stimuli from antennae	Body Colour and shape of head	Free, 1956
Honey-bee worker	Stinging	Dark colour Sting venom odour Human sweat odour Movement	General bee odour	Free, 1961
Char (*Salmo alpinus*)	Attack Nest-site selection	Red colour Visual stimuli from gravel	Size, shape Tactile stimuli from gravel	Fabricius and Gustafson, 1954

Darter (*Etheostoma blennioides*)	Egg-laying	Texture	Smell Colour	Winn, 1957
Cichlid fish (*Apistogramma* sp.)	Approach to parent by young	Movement Colour	Form Size	Kuenzer and Kuenzer, 1962 (see also Kühme, 1962)
Turkey	Fleeing from a winged predator	Relation between speed and size	Shape	Schleidt, 1961 Müller, 1961
Domestic hen	Maternal protection of chicks	Distress call	Visual characters	Brückner, 1933
Domestic hen	Mounting	Crouching posture	Male and female characters	Fisher and Hale, 1956/1957 (see also Carbaugh et al., 1962)
Brewer blackbird (*Euphagus cyanocephalus*)	Male mating behaviour	Posture Head and/or tail above horizontal Colour of foreparts	Eye colour Colour of posterior region Wings	Howell and Bartholomew, 1952
Bullfinch (*Pyrrhula pyrrhula*)	Anxiety responses	Visual texture Convex surface Coloured surface	Specific outline of predator	Kramer and von St. Paul, 1951
Herring gull	Egg-rolling	Speckling, size, colour	Shape	Baerends, 1957b, 1959b
White mouse	Retrieving of young by inexperienced female	Cries of young	Visual and olfactory characteristics of drowned young	Noirot, 1964c, in press

flying predator. Lorenz and Tinbergen found that this response could be elicited by a relatively simple moving model, as shown in Fig. 4.4. In birds which had lived in the open and become accustomed to the other birds in the area, the model elicited a fleeing response if moved with the short end in front ("hawk model") but not if pulled the other way ("goose model") (Tinbergen, 1948). Since this case has been rather controversial, it is as well to stress that at this point we are concerned only with the occurrence of differential responsiveness to the two models: its development is discussed later (page 510).

Tinbergen and Kuenen's (1939) study of the stimuli eliciting gaping from nestling thrushes (*Turdus merula*) provides another example of the

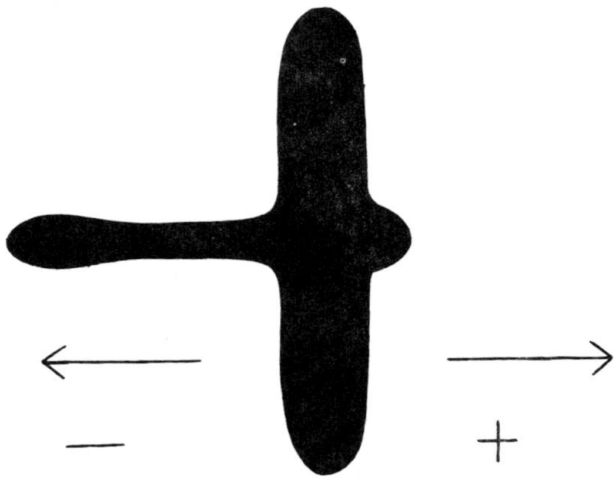

Fig. 4.4. Cardboard model used to elicit escape reactions from young turkeys. It was effective when moved to the right but not when moved to the left. (After Tinbergen, 1951.)

importance of relational characters. The response of birds about 8 days after hatching is both released and directed by visual stimuli, the gape being directed towards the head of the parent. In Fig. 4.5, gaping was directed towards the smaller of the two "heads" of the smaller model but to the larger "head" of the larger model. It thus seems that the stimulus for gaping is characterized not by its absolute size but by its size in relation to the whole.

Relational properties are not the only ones which may make the important features of the stimulus situation difficult to define. We shall see later (pages 84–86) that the organism may respond to both first- and higher-order variables, gradients of physical variables, and rates of change of gradients. Wherever relational or higher-order properties are involved, an attempt to isolate simple physical variables in the stimulus situation can be only a first step. Such difficulties in specifying the important features of the stimulus situation make it important to remember that the "characters" of the stimulus situation studied may be only artificial abstractions of the experimenter: the animal may abstract differently.

Three other points about the technique of such experiments must be emphasized here. First, the strength of the response to any stimulus situation varies with the internal state of the responding animals: a male stickleback behaves aggressively only during a certain season of the year, and then only when on its territory. Even there its tendency to show aggressive behaviour varies with the stage of the breeding cycle. Not only does the

Fig. 4.5. Cardboard models used to elicit gaping responses from nestling thrushes. The nestling's response is directed towards the smaller projection on the left-hand model and the larger projection on the right-hand model, indicating that its orientation is determined not by the absolute size of the projection, but by its size relative to that of the rest of the model. (From Tinbergen, 1951.)

strength of the response to a particular stimulus situation change, but so also does the difference between the response strengths to two models. Thus the models shown in Fig. 4.1 would be equally ineffective in eliciting attack from a male in winter condition, and all might elicit an immediate and maximal response from a male at the peak of his aggressiveness. The skill of the experimenter lies in selecting animals in an intermediate state in which the differences between the models will show up to best advantage. The experimenter should also design his experiments to equalize the effects of the preceding tests between groups. In man, such effects can be important (Welford et al., 1950) and complex (Helson, 1948), and comparable effects occur in animals (Chap. 13).

Second, there is need for extreme care in specifying the difference between two stimulus situations, one of which has been found to be more effective than the other. The difference most obvious to the experimenter may not be what is relevant to the animal.

The third point concerns the distinction between the stimuli which elicit a response and those which orient it. If the test stimuli are presented successively, they reveal primarily the eliciting effects of the different stimulus characters. If, however, pairs of models are presented in choice tests, they may reveal the relative orienting effects of the two stimuli. Where elicitation and orientation depend on different stimulus characters, the nature of the test used therefore requires consideration (Tinbergen et al., 1943; Franck, 1966).

If, as a result of experiments of this type, we are able to specify the characters of a natural object important for a particular response, it may be possible to construct a model in which some of these sign stimuli are exaggerated relative to the natural object. We then have a "supernormal" stimulus object. Thus the egg-rolling response of the oyster catcher (*Haematopus ostralegus*), whereby eggs lying outside the nest are retrieved, is elicited more readily by an egg several times the normal size than by a normal egg (Tinbergen, 1948). Similarly, the pecking response of a newly hatched herring gull can be elicited even more effectively by a red knitting needle with three white bands near the tip than by a three-dimensional plaster head. The needle combines, in an exaggerated form, just those characters, such as the contrasting red/white border and others specified on pages 59–60, which are important in eliciting the response (Tinbergen, 1953).

In all the examples cited so far, the sign stimuli important for a particular response have been more or less the same for all members of a species—all territory-owning male sticklebacks respond aggressively to models with red bellies, all newly hatched herring-gull chicks peck at models which have the characters indicated on pages 59–60. Indeed, the experimental method used in nearly all these studies consisted of presenting models successively to large numbers of individuals and determining statistically which models produced the strongest responses. This method can give information only about stimulus characters effective in eliciting the response from the population, and clear-cut results are to be expected only if the individuals comprising the population behave similarly. In most of the experiments we have been considering so far, similarity between individuals was ensured because responsiveness to the particular features of the stimulus situation was little influenced by such differences in experience as occurred within the group of animals tested.

The use of this experimental method has led to the view that selectivity in responsiveness is characteristic only of responses common to a whole species (Tinbergen, 1951), a view certainly incorrect. Where individuals each respond to different features of the whole, selection from amongst the possible stimulus characters cannot be demonstrated from their grouped data but nevertheless occurs. For instance, Konorski (1962) has shown that, in solving discrimination problems, dogs do not use all the information supplied by the stimuli presented but select certain ones and neglect others. Some further examples of stimulus selectivity specific to an individual are given in Chaps. 6 and 21: we shall see that which cues are selected depends on a number of factors, including the nature of the problem and the previous experience of the individual.

Where a response is influenced by stimulus characters acting through more than one sensory modality, or by more than one stimulus character in any one modality, their effects may supplement each other. In the case of the grayling butterfly (see page 82), the male's sexual pursuit is influenced by the type of movement shown by the model, by the darkness of its colour, and by its distance. Deficiencies in any one of the stimulus characters can be compensated for by increasing one of the others: a pale model close-by is as effective as a dark model at a distance. This fact has been referred to as the Law of Stimulus Summation (Lorenz, 1939; Seitz, 1940/1941, 1943) or Heterogeneous Summation (Tinbergen, 1951), and sometimes applies qualitatively to individually conditioned responses as well as to species-characteristic ones (Baerends et al., 1965). Stated in its strongest form, it would imply that the total number of responses released by the parts of the stimulus situation presented successively is the same as the number released by the whole. This is clearly contrary to the view, basic to Gestalt psychology, that the whole is more than the sum of its parts (e.g., Koffka, 1935). Klopfer (1962) has suggested that the difference may be one of phyletic level—heterogeneous summation applying at lower levels of organization and the Gestalt principle in more highly developed organisms.

There may also be an ontogenetic change, the importance of the Law of Heterogeneous Summation decreasing with age. An experiment supporting this view was reported by Bower (1966). Human infants were trained to turn their heads when the stimulus shown at the upper left of Fig. 4.6 was presented, the reinforcement being a 15-sec "peek-a-boo" from an experimenter. When the response rate had reached a criterion, generalization was assessed by presenting the conditioned stimulus (Fig. 4.6, upper left) and its parts (rest of Fig. 4.6) for 15-sec periods in counterbalanced order until extinction was complete. As shown in Fig. 4.7, with 8- and 12-week-old infants the law of Heterogeneous Summation received remarkably precise confirmation, but at 16 and 20 weeks the whole was found to produce more responses than the sum of those produced by its parts. In a second experiment infants were trained on the parts and subsequently tested on the whole and the parts as before. Once again there was little difference with younger infants (10 weeks), but at 20 weeks the mean number of responses given in the presence of the whole was *smaller* than the mean total number given to the parts. Bower suggests an explanation for these results in terms of an increase with age in the rate at which information can be picked up, plus an increase in the number of items to which the infant can attend simultaneously.

Conclusion

These results have been obtained by one basic method—the animal is presented with stimulus situations differing in known physical characters,

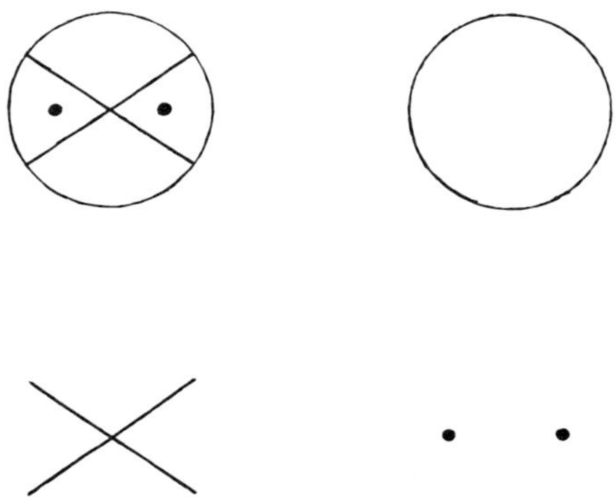

Fig. 4.6. Stimuli for assessing the applicability of the Law of Heterogeneous Summation for human infants. The whole (upper left) comprises the three parts. (After Bower, 1966.)

and a difference in response indicates that these characters are significant for the response in question. The method shows that for each type of behaviour some stimulus characters are important, while others are, at most, of minor significance: stimulus selection or filtering of the potential sensory input always occurs. Of course, some of the stimulus characters found important for one response may also be so for others: their specificity can be assessed only by studying different responses (see Chaps. 5 and 6).

Such experiments show that some aspects of the total stimulus situation are of more importance than others, but they cannot prove that any aspects are of no importance. In fact, territory-owning robins do sometimes threaten juvenile robins, and male sticklebacks sometimes attack models without a red belly (Tinbergen, 1951). A negative experimental result is evidence but not proof that the character is without influence on the response being tested, and further experiments may or may not support such a hypothesis. If the character does appear to be ineffective for one response, it may yet have an influence on others. If it is ineffective for all responses studied, one possibility is that the animal's sense organs are not responsive to it. Before we proceed further, therefore, some methods used for assessing the potentialities of sense organs must be discussed.

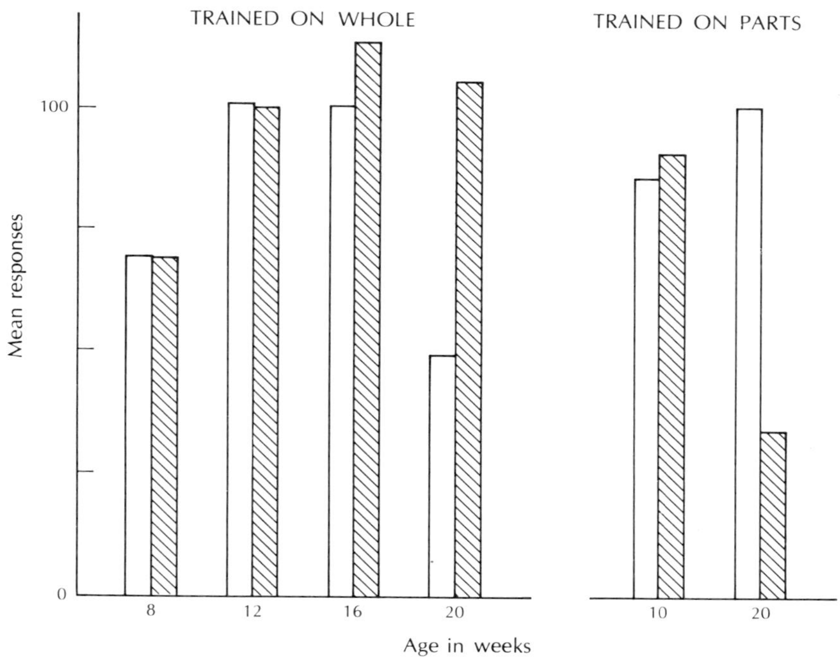

Fig. 4.7. The effectiveness of the whole and the parts of a stimulus configuration after training on either the whole or the parts. Unshaded columns indicate total mean number of responses to the parts, shaded columns the mean number of responses to the whole. (From data in Bower, 1966.)

4.2 LIMITATIONS OF SENSORY EQUIPMENT

No organism is equipped with sense organs suitable for detecting all possible physical changes in the environment. This is obvious enough, but sensory capacities diverge so widely that the characteristics of an object to which an animal responds may be quite different from those by which we recognize that object ourselves.

Physiological Evidence

The most direct techniques for studying the ability of a sense organ to detect changes in the environment involve recording from the sense receptors or sensory pathways themselves. In this way, for example, Sand (1937) established that the lateral-line canals of rays were sensitive to movements in water; Autrum and Stumpf (1950) showed that some ommatidia in a bee's eye could distinguish polarized light from nonpolarized light of the same intensity; and Schneider (1957, 1962) showed that certain olfactory receptors in the antennae of the male silkworm moth were sensitive only to a substance given off by the female and to a few other substances. Similarly Beidler et al. (1955) recorded responses in the chorda tympani nerve to stimulation of the tongue of mammals and showed species differences in sensitivity. For instance rodents responded better to sodium chloride than to potassium chloride, while for certain carnivores the opposite was the case; and while hamsters and guinea pigs were easily stimulated with sucrose, the cat was not. Ralls (1967) used recordings at the level of the inferior colliculus to obtain detailed data on the changes in auditory sensitivity of *Peromyscus* and *Mus* as the frequency of the incident sound was varied. Many further studies are cited in Rosenblith (1961).

Recording from single fibres indicates that many sensory cells are normally sensitive to more than one form of energy: single units in the cat's lingual nerve can be activated by mechanical or thermal stimuli to the tongue (Hensel and Zotterman, 1951), and single units in the lobster statocyst are sensitive to both position and acceleration. At the other extreme, some receptor cells have a highly specific sensitivity. For instance, flies (*Phormia regina, Calliphora erythrocephala*) may drink when a single one of their labellar hairs is touched by water or a sugar solution, but electrolytes inhibit drinking. At the tip of the labellar hair, the dendrites of three to five sense cells are exposed to the environment. Each of these cells has a specific chemical sensitivity, and it seems probable that a cell sensitive to electrolytes is concerned with avoidance, and the others (sensitive to sugar, water, etc.) with ingestion (Hodgson, 1957, 1965; Dethier, 1963; Dethier et al., 1965; den Otter and van der Starre, 1967). The same principle presumably holds true for caterpillars (*Bombyx mori, Galleria mellonella, Philosamia cynthia, Pieris brassicae, Protoparce sexta*), which bear on their maxillae two sensilla styloconia that are organs of taste. Each of these contains four taste cells. Up till now cells have been found specifically responsive to common chemicals such as water, salts, sugars, and acids.

Moreover, depending on the specific insect/food plant relationship, the caterpillars are also equipped with cells responsive to phagostimulants like mustard oil glucosides (*Pieris brassicae*) or repellents like alkaloids and flavonoids (*Bombyx mori*) (Ishikawa, 1966; Schoonhoven and Dethier, 1966; Schoonhoven, 1967). In such cases, therefore, specificity in responsiveness of the organism may depend on specificity in the sense receptors; other examples are given by Liesenfeld (1956) and in a review by Marler (1961). Such limitation of sensitivity is to be regarded as specialized (in an evolutionary sense), a broader range of sensitivity being the primitive condition.

An interesting contrast is provided by a case of selective lack in responsiveness, though it concerns a sensory interneurone ascending from the auditory organ to the brain in a tettigonid grasshopper. McKay (1968) found that over the range investigated (2 to 40 kHz) the evoked response increased with the amplitude of the incident sound, except for an unexpected trough in responsiveness at 15 kHz. This is the dominant frequency of the species song. The physiological characteristics of this neurone suggested that it mediates a withdrawal response.

Behavioural Evidence

Partly because sense cells may be sensitive to more than one form of energy, the use of physiological methods for investigating the limitations of sense organs can be misleading from a behavioural point of view. A change in the pattern of nerve impulses passing along the optic nerve is produced by a blow on the eyeball but must not be taken as indicating that the eye is normally a mechanoreceptor. Further, even when physiological techniques indicate that the sense organ itself is responsive to a certain dimension of the stimulus, we cannot be certain that variation in that feature can influence behaviour: although electrical recording from the cat's retina clearly indicates the presence of colour vision, behavioural confirmation has been obtained only with difficulty (Mello and Peterson, 1964; Sechzer and Brown, 1964; see also review by Cowey, 1968).

For the study of behaviour, therefore, a behavioural method of assessing sensitivity is often a necessary complement to an apparently more precise physiological one. Often, a direct behavioural response to change in one dimension of a stimulus can be used to indicate the sensitivity of the organism. Examples of this are to be found in studies showing that various arthropods respond to the plane of polarization of light (Fig. 4.8): if the plane is rotated, the animal changes its direction of movement (e.g., Bainbridge and Waterman, 1957, and refs. cited). Similarly, if a male moth responds to a place where the female has been sitting, he must be sensitive to some chemical indicator of her recent presence (Butenandt, 1955).

Such a method can, however, be used only where the stimulus change in question already regularly elicits an identifiable response. A method of wider applicability involves a conditioning technique—if an animal can learn to respond differently to two stimulus situations, it must be able to distinguish them. A recent example of this method is the demonstration that certain fishes (*Gymnarchus, Gymnotus,* etc.) are sensitive to changes in the electric field surrounding them (Lissmann, 1958). Preliminary experi-

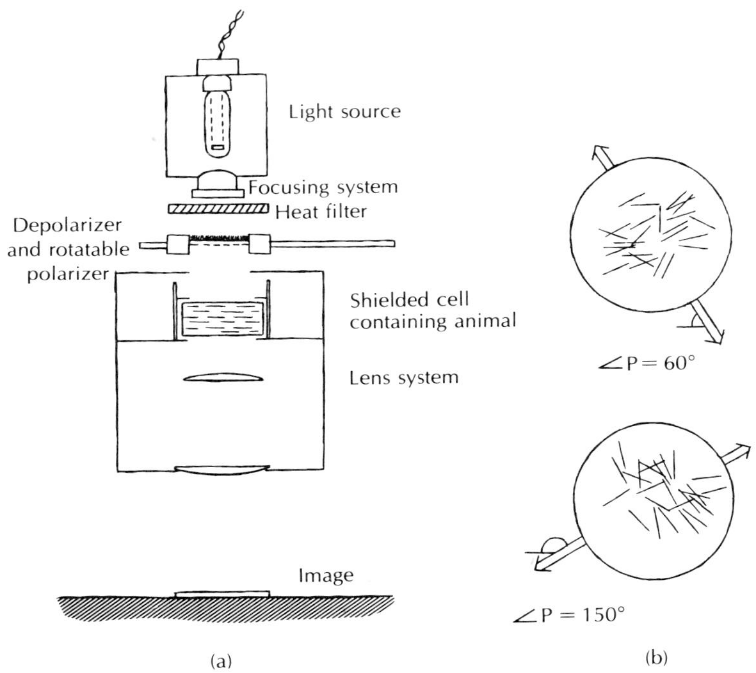

Fig. 4.8. *(a) Apparatus used to expose the small crustacean Mysidium to a vertical beam of polarized light and record the resultant swimming orientation; (b) data obtained on two sample runs.* (After Bainbridge and Waterman, 1957.)

ments showed that they responded by jerking or avoidance responses whenever they approached loops of wire placed in the water, and accordingly conditioning trials were undertaken. In the definitive experiments fish were trained to respond to the presence of a magnet placed behind a screen outside the aquarium. The apparatus is shown in Fig. 4.9. One fish was trained to accept food when the magnet was present and to refuse it for 30 sec when it was not, mild punishment being administered in the latter case for an incorrect response. The other fish was trained in the opposite way. The final tests were carried out in complete darkness to avoid any subconscious optical signals from the experimenter (see also Lissmann and Machin, 1958).

It is often important to know not simply whether a stimulus can be detected or not but also what minimum value or minimum difference along some dimension of the stimulus can be detected. A precise method of doing this involves a "tracking procedure", in which the subject itself continuously adjusts the value of the stimulus to keep it at a chosen level. For instance, the sensitivity of the pigeon to lights of different wavelengths has been

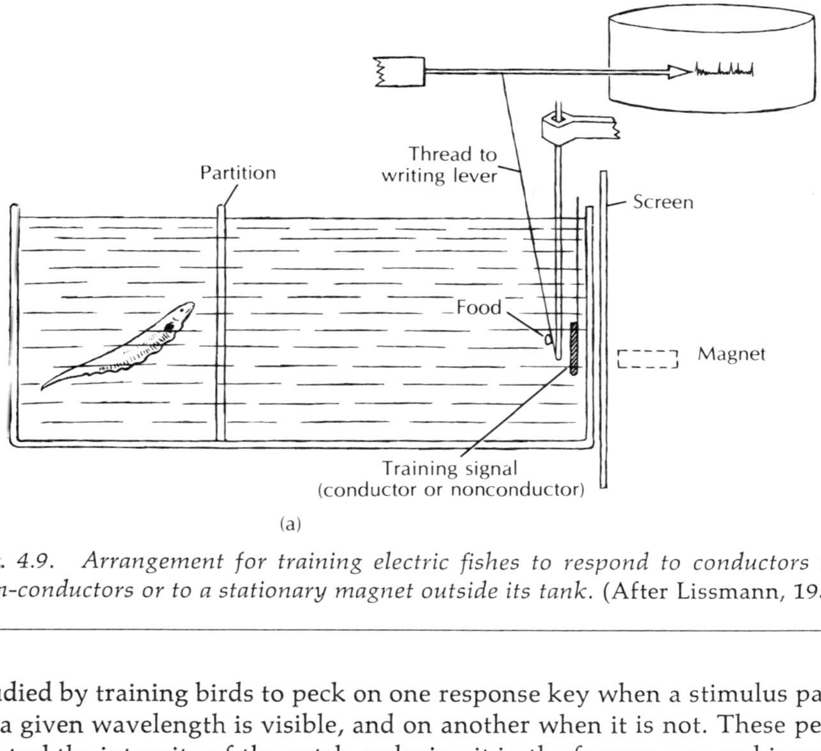

Fig. 4.9. Arrangement for training electric fishes to respond to conductors and non-conductors or to a stationary magnet outside its tank. (After Lissmann, 1958.)

studied by training birds to peck on one response key when a stimulus patch of a given wavelength is visible, and on another when it is not. These pecks control the intensity of the patch, reducing it in the former case and increasing it in the latter. The intensity of the patch thus oscillates up and down across the pigeon's threshold. The apparatus is shown in Fig. 4.10. Pecks at key *A* drive down the optical wedge and reduce light intensity, while pecks at key *B* increase it. Responding is maintained by reinforcing pecks at key *B* with food, but such reinforcement must of course be provided only when the light is not visible. Since the experimenter does not know whether the

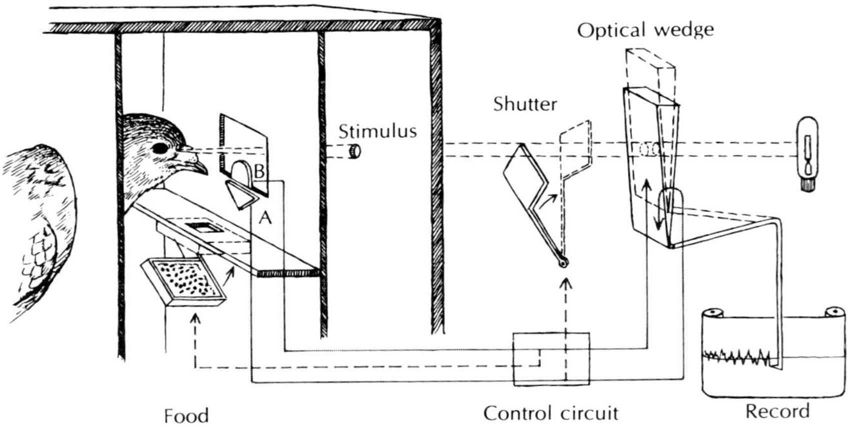

Fig. 4.10. Schematic picture of apparatus for tracking the threshold of a pigeon for light of a given wavelength. For further explanation, see text. (From Blough, 1961.)

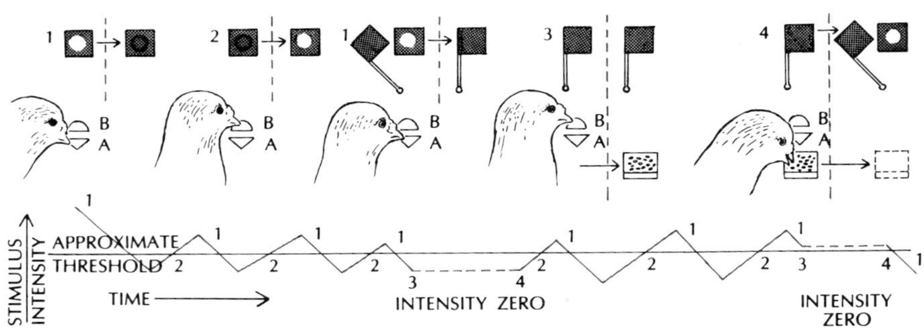

Fig. 4.11. Series of events during tracking runs with apparatus illustrated in Fig. 4.10. (From Blough, 1961.)

The bird pecks key *A* when it can see the light, key *B* when it cannot. Most of the time pecks on key A drive the intensity down (*1*) and pecks on key *B* raise it (2). Occasionally, at random intervals, a reinforcement sequence is introduced: as the bird pecks *A* (1) a shutter closes (3, intensity zero), and pecks on *B* bring access to food (4). Then shutter reopens and tracking resumes.

light is visible to the pigeon or not (and indeed this is what he wants to know), pecks on key *B* produce food only during interpolated intervals during which a peck on key *A* has turned off the light completely by moving a shutter into the path of the light beam. Since pecks on key *A* always diminish the intensity, and subthreshold intensity is the only situation in which pecks on key *B* are sometimes reinforced, responding to key *A* is maintained as well. Although "the bird pecks key *A* 'to turn out the light' and key *B* 'to get food' " (Blough, 1966, page 359), the shutter and reinforcement are inserted into the tracking sequence only sufficiently often to maintain responding. Tracking without reinforcement is shown in the first two diagrams of Fig. 4.11; then pecks on key *A* close the shutter, and finally pecks on key *B* produce food. (For further description and application of this method as well as a review of the study of sensory processes by operant methods, see Blough, 1966.)

It is one thing to show that an animal is sensitive to a particular stimulus variable, but quite another to prove lack of sensitivity. Perhaps the strongest source of evidence that an animal is not sensitive to a stimulus comes from attempts to condition it to avoid a painful shock paired with the stimulus. For example, blinded octopuses can readily be trained to discriminate between pairs of objects differing in texture: the technique involves rewarding the animal with a piece of fish for passing one of the objects to its mouth and punishing it with an electric shock for taking the other. Even after prolonged training, however, octopuses fail to discriminate between similar objects differing only in weight (Fig. 4.12). Since the conditions of the experiments were identical in the two cases, this "can hardly mean otherwise than that *Octopus* is incapable of recognizing objects by their weight" (Wells, 1961). Wells has emphasized that such results do not show that *Octopus* does not respond in any way to differences in weight —it clearly does, because it adjusts the muscle tension in its arms to allow for the weights of the things it handles. Rather it seems that in *Octopus*, as in vertebrates (Merton, 1964), there are two mechanosensory systems, one concerned solely with the local adjustment of muscle tension and the other sending information to higher parts of the nervous system which can be used in learning: the latter does not discriminate weight differences (Wells, 1964b).

Some of these examples show that an organism may be responsive to some types of change in a given form of physical energy but not to others. The human eye is sensitive only to a certain range of light wavelengths, and to these only above certain intensities, but within that range of wavelengths it has considerable powers of discrimination lacking in many other vertebrates. All vertebrates, however, seem to lack the ability to respond differentially to the plane of polarization of the light. As another example, many insects have auditory organs, but these are, on the whole, very insensitive to changes in the frequency of the incident sound. Recordings from the auditory nerve show marked changes related to amplitude variations in the sound but little change when the frequency is varied. For this and other reasons it has been argued that the information contained in frequency changes is lost (e.g., Pumphrey, 1940; Haskell, 1956), though Michelsen (1966) disputes this view on the basis of recordings from some cells in the tympanal organ of a locust (see also Michelsen, 1968).

Even when we know the general nature of the stimuli to which an organism responds, we may not be able to specify the dimensions along

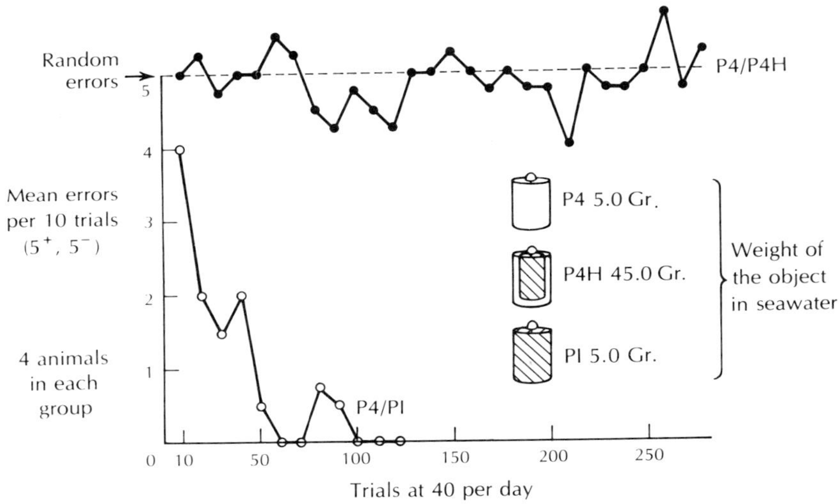

Fig. 4.12. Results of experiments on weight discrimination in octopus. Two comparisons are shown: the first (●) between two objects (P_4 and P_{4H}) of similar texture but differing in weight, and the second (○) between two objects (P_4 and P_1) of similar weight but differing in texture. (From Wells, 1962*b*.)

which these stimuli should be assessed. Thus, *Octopus* makes tactile discriminations between objects held by the suckers on its arms merely by the proportion of sense organs excited in the area of the arm in contact with the object (Figs. 4.12 and 4.13). They thus fail to distinguish between cylinders having longitudinal grooves and ones with transverse grooves, provided the same amount of surface is cut out in each case—a result rather surprising to the human observer, who can readily detect the differences by feel (Wells, 1962b).

The results of studies of the extent to which *Octopus* can distinguish between shapes held in its arms is also rather unexpected. Proprioceptive information about the bending of the arms is apparently not available: instead the animal uses only information about the degree of distortion imposed on the suckers by contact with the surface. This was shown by assessing the response to composite cylinders built up from narrower rods: these caused bending of the arm similar to that produced by a cylinder of the size of the whole, but sucker distortion similar to that produced by cylinders the size of the components. They were treated as being the diameter of the components. Further, since octopuses, trained to reject the smaller and take the larger of two cylinders, without further training also reject the rougher and accept the smoother of two objects of similar size, it

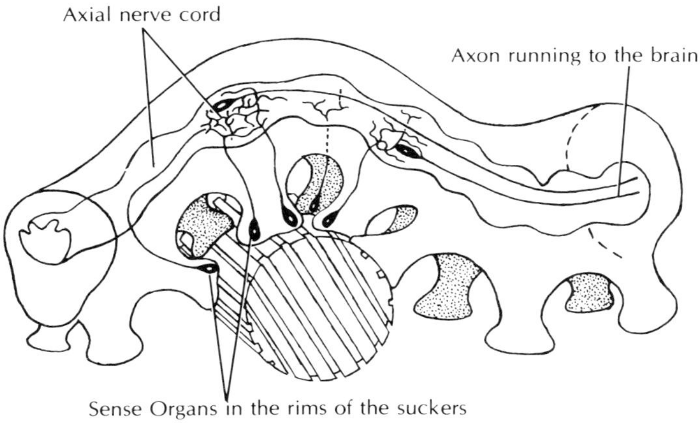

Fig. 4.13. Diagram of the arm of an octopus touching a cylinder of the type labelled P_1 in Fig. 4.12. The proportion of the sense organs in contact will depend on the proportion of the surface of the object that is cut away to form grooves into which the suckers cannot penetrate. (After Wells, 1962*b*.)

seems that differences of texture and surface curvature are measured in the same way: in principle it should be possible to discover a texture equivalent to any shape as far as *Octopus* is concerned (Wells, 1964a, b).

The response of noctuid moths to ultrasonic frequencies similar to those emitted by free-flying bats provides another example of the way in which, even within a modality in which an organism seems to be sensorily well equipped, its behavioural responsiveness may be determined by the limitations of its sensory equipment. Electrical recording suggested that the acoustic information received by the moths could enable them to make a rough determination of the position of a bat 20 to 100 ft away. These studies also predicted that, at closer distances, the intensity of an average bat cry would saturate the sense organs, there would be no differential response from the two sides, and the directional information would be lost. Observations under field conditions confirmed this view: low-intensity ultrasonic sounds caused the moths to fly away from the source, but high-intensity sounds caused a wide variety of manoeuvres which were non-directional except that they carried the insects towards the ground (Roeder, 1962a, 1964).

That an organism's sensory equipment plays a role in determining which of the stimuli to which it is exposed will influence behaviour is, of course, a statement of the obvious, though it is not hard to find cases where this sensory filtering has been wrongly ascribed to more central processes. In the next chapter we shall consider other mechanisms by which the stimuli impinging on an organism are filtered.

4.3 SUMMARY

1. Animals respond selectively to the changes in physical energy to which they are exposed.
2. For any type of behaviour, such selection can be demonstrated by comparing the responses of an animal to stimulus situations differing in known characters. In every case some stimulus characters are found to be significant, others of at most minor importance.
3. The effective stimulus characters or "sign stimuli" cannot always be specified along a single physical scale or even restricted to one sensory modality. Relational properties may be important.
4. In assessing the effectiveness of a stimulus character, the animal's motivational state must be considered.
5. Once the stimulus characters effective for a given response have been isolated, they can be combined in an exaggerated form in a "supernormal" stimulus object.
6. Stimulus selection occurs in all responses, those characteristic of particular individuals as well as those characteristic of a species.
7. The effects of different stimulus characters relevant to the same response supplement each other.
8. The changes in physical energy to which an organism can respond are controlled in part by the limitations of its sensory equipment. These can be studied by both physiological and behavioural techniques.

5

SELECTIVE RESPONSE TO STIMULATION—SENSORY/PERCEPTUAL MECHANISMS

We have seen that not all the energy changes to which an organism is exposed influence its behaviour, and that selection from the potential sensory input is due in part to limitations in the capabilities of the sense organs. However, conditioning experiments show that most animals are capable of responding to, or discriminating between, stimuli which had no influence before training started: thus, further stimulus selection or filtering of the sensory input must occur.

We can distinguish two types of selection from among the stimuli to which the sense organs are responsive as follows. As we have seen (Chap. 4), if we elicit the same response by stimuli differing in known ways, a comparison between the response strengths indicates whether or not the difference between the stimuli is significant for the response in question. By repeating this procedure we can draw up a list of stimulus characters significant for that response. If we do this for more than one type of behaviour, we find that some characters are important in one context but not in others, while some are effective for all, or nearly all, responses we test. In the first case, the effectiveness of the character has some degree of response-specificity, while in the second it does not. In sticklebacks, for instance, the character "red underneath" increases the effectiveness only of models eliciting territorial aggression, but moving models are more effective than stationary ones in producing both sexual and aggressive responses. Similarly, in our own species, some stimulus configurations catch our attention more readily than others: in a visual field, for example, circular forms are more conspicuous than those with irregular outlines, and regularly arranged elements more conspicuous than irregular ones. Stimuli that are louder, brighter, or more intense tend to be more effective for a wide range of responses. In such cases, where the stimulus character affects not just one or a few responses but many or all of the responses in the animal's repertoire, it must owe its effectiveness to filtering mechanisms which are not response-specific but general in their effectiveness. We shall refer to these as sensory/perceptual mechanisms.

That the effectiveness of a character may be valid for behaviour other than the response tested has sometimes been overlooked (Schneirla, 1952, 1956; Lehrman, 1953). Thus many of the stimulus characters which Tinbergen and Perdeck (1950) found to be effective in eliciting the pecking response of herring-gull chicks (pages 59–60) owe their effectiveness to properties of the sensory/perceptual mechanisms common to all visual responses. Movement and contrast are obvious examples, and the greater effectiveness of a vertically oriented beak may be related to a greater conspicuousness of radial, as compared with tangential, movement of an image over the retina (Tinbergen, 1955; Weidmann and Weidmann, 1958; Marler, 1961; Hailman, 1962, 1966; Quine and Cullen, 1964). Again Hailman (1967), investigating the role of movement of the model in eliciting pecking from chicks of two gull species, found that *Larus furcatus* chicks, which have a poorly developed pecten (a structure in the eye thought to aid movement perception) pecked equally often at moving and stationary models, whereas with *L. atricilla,* which have a well-developed pecten, the moving models were preferred. Furthermore he ascribed the most effective position of the stimulus to the restricted view from the temporal foveal area. The wavelength preferences, shown by the greater effectiveness of red or blue models in eliciting pecking, he held to depend on the transmission by the red and yellow oil droplets in the eye, the response rate being determined by the ratio of transmission by the two types of droplet.

In some cases the effectiveness of a character can be ascribed to properties of the sensory/perceptual mechanisms not by assessing its influence on other responses but by a direct study of those mechanisms themselves. In studying the stimuli eliciting the courtship approach of male fritillary butterflies (*Argynnis paphia*), Magnus (1958) used the revolving model shown in Fig. 5.1. As the small drums rotated, their dark and white stripes produced a flickering effect resembling the fluttering flight of a real female, whose brown and black upper surfaces contrast with the partially silver and green underwings. The frame carrying these drums itself rotated round the base, and the apparent movement of fluttering females attracted males in sexual pursuit. Three stimulus characters influenced the responsiveness of the males—colour, size (within limits), and the rapidity of the revolution of the drum. The special effectiveness of a particular orange colour which corresponds to that of the female is presumably specific to this particular response. The increase in effectiveness with size, however, is presumably related to the number of ommatidia stimulated. That the importance of the rapid alternation of stimuli is also related to sensory/perceptual mechanisms is suggested by Magnus's finding that the model becomes less effective if the rate is increased above 125 presentations per sec. This corresponds reasonably well to the flicker fusion frequency of 150 as determined electrophysiologically. Similarly Etienne (1968) found that the predatory strike of larvae of the dragonfly *Aeschna cyanea* could be elicited by a flickering spot of light. Response probability increased as frequency was increased towards 40 Hz but then dropped sharply. The flicker fusion frequency as determined from the electroretinogram for this species

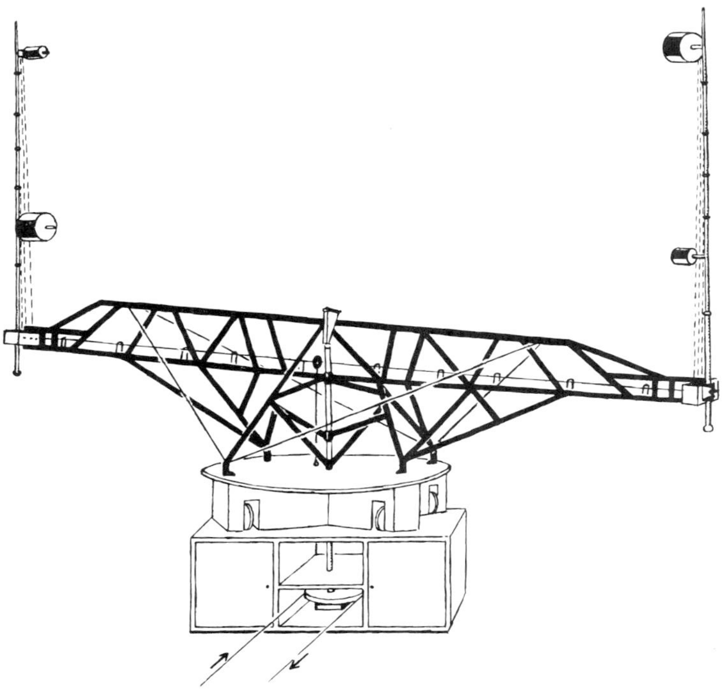

Fig. 5.1. Apparatus used to investigate the stimulus characters eliciting courtship pursuit flight from males of the fritillary butterfly Argynnis paphia. Both horizontal arm and the small cylinders rotate. (After Magnus, 1955.)

is 40 Hz (Autrum and Gallwitz, 1951): presumably this is what sets the upper frequency limit.

That the special effectiveness of certain stimulus characters, as determined by model experiments, can be ascribed to properties of the sensory/perceptual mechanisms, rather than to mechanisms specific to one group of responses, is thus shown by at least two types of evidence—the generality of their effectiveness, and information concerning the functioning of the sensory/perceptual mechanisms themselves. In addition, the very simplicity of some effective stimulus characters suggests that they owe their effectiveness to properties of the sensory/perceptual mechanisms: loudness and brightness are obvious examples (see also Kuenzer, 1968).

The problem of why some patterns of energy distribution are more effective than others for a wide range of behaviour has been studied more in the context of human than animal perception. With human subjects the use of verbal instructions can eliminate the elaborate pretraining necessary for animal experiments and thus avoids many of their inherent difficulties (e.g., Koffka, 1935; Gibson, 1950; Vernon, 1952; Teuber, 1960). Although the perceptual world of animals is often very different from our own (von Uexküll, 1934), the principles which emerge from the study of human perception are often highly relevant to the study of perception in animals. Here discussion is limited to a few problems, all concerned with the relation between the physical properties of the stimulus and its effectiveness in controlling behaviour, which are of special relevance to work on animal behaviour. The first is a caveat: quantitative variations in a stimulus along a simple physical dimension may produce complex changes in its effect on the organism. The second concerns the abilities of animals to discriminate shapes and emphasizes the danger of premature generalizations about the patterns of stimulation an animal can or cannot distinguish. The third, the perception of movement, indicates that the mechanisms which determine the effectiveness of a given stimulus may be considerably more complicated than they appear at first sight. Finally, some physiological correlates of perceptual phenomena will be considered.

5.1 PHYSICAL SCALES AND THE PERCEIVED STIMULUS

One of the lessons to be learnt from the study of human perception is that physical measurements and subjective measurements along a given stimulus dimension do not coincide. For many stimuli the subjective magnitude ψ increases as a function of physical magnitude ϕ above the threshold ϕ_0 raised to a power n. That is,

$$\psi = k(\phi - \phi_0)^n$$

and log ψ plotted against log $(\phi - \phi_0)$ gives a straight line with gradient n. However, the value of n depends on the stimulus dimension. The gradient is only 0.33 for brightness but increases for loudness, smell, vibration, taste, temperature, and heaviness up to a value of 3.5 for electric shock. In other words a large increase in physical intensity is necessary to double the subjective intensity of a light, but only a small increase in physical intensity is necessary to double the subjective intensity of an electric shock (Stevens, 1958, 1961). Although human subjects have been used for most of this work, comparable results have been obtained with animals. For instance, Herrnstein and van Sommers (1962) found that pigeons increased their rate of pecking as a power function of the luminance of the response key.

These "prothetic" continua are concerned with *how much* (intensity); and Stevens distinguishes these from "metathetic" continua, which are concerned with *what* or *where* (quality). In general, while a power function relates psychological magnitude to prothetic continua, a logarithmic function ($\psi = \kappa \log \phi$) relates psychological magnitude to metathetic continua. "At the physiological level, many prothetic continua (loudness, brightness, and so forth) involve the addition of excitation to excitation, whereas some of the metathetic continua (pitch, apparent azimuth, inclination) involve the substitution of excitation for excitation" (Stevens, 1961, page 31). But in any case no particular relationship, good for all stimuli and all modalities, between the strength of a stimulus and its consequences in the central nervous system or on behaviour can be assumed.

Indeed we cannot even assign a subjective sensory quality, like hue, brightness, or loudness, to a particular physical dimension. Hue may depend on the intensity as well as the wavelength of the incident light: with certain exceptions, the perceived hue of all wavebands changes as the intensity of the light is increased. Although this reservation applies primarily to film colours (i.e., colours not obviously belonging to a surface), it could have important consequences for the conduct of experiments aimed at identifying the stimuli relevant for particular responses: the hues of a given series of models could appear to be slightly different on bright and dull days. Similarly, subjective brightness and saturation (or vividness) also depend on both wavelength and intensity of stimulation; while the pitch, loudness, volume, and density of a sound all depend on both its intensity and its frequency. The perceptual dimensions do not correspond completely to physical dimensions, so that it is not possible to produce variation along a single perceptual dimension by variation along a single physical one while other factors are kept constant. Indeed, the response to stimulation can have more dimensions than the stimulus itself, the additional ones arising in part from the properties of the sense organ (for further discussion, see Boring, 1935).

A further complication is that the characteristics of the stimuli on which human perception depends may be more complex than those which have usually been studied by sensory physiologists. The stimulation from the environment usually acts on a sense organ, not just a single receptor cell, so that gradients, ratios, and changes in the pattern of stimulation can function as essential components for discrimination of the situation. For instance, an important visual characteristic of a receding surface is the density of its stimulus elements. In Fig. 5.2 the receding surface *AB* yields a retinal image *ab* in which there is a gradient of texture from coarse to fine, whereas the frontal surface *BC* yields no such gradient. There is considerable evidence that the gradient of texture in *AB* provides one of the means whereby the observer can detect that it is receding. Furthermore, along *AB* the texture changes at a constant rate: if *AB* were curved or discontinuous, however, the rate of change of texture would not be constant. Such higher-order variables are probably among the means by which we detect curvature, discontinuities in surfaces, and so on (Gibson, 1950).

We see, then, that, both with simple stimulus dimensions and with more complex patterned stimulus configurations, physical measurements of

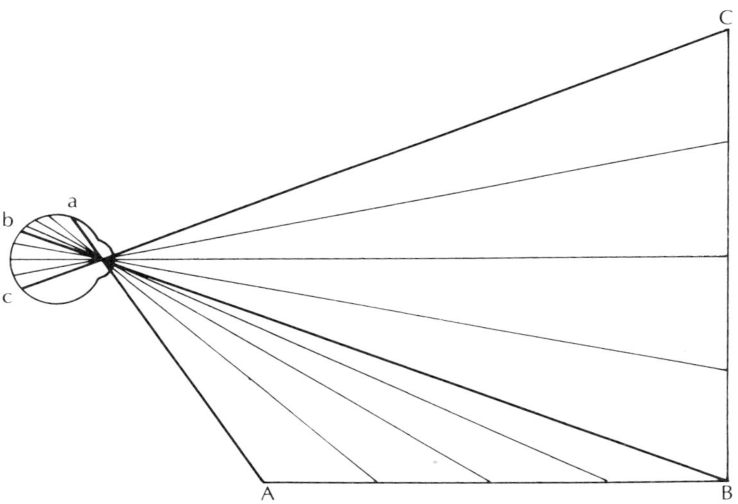

Fig. 5.2. The optical projection of a longitudinal and frontal surface. (After Gibson, 1950.)

stimuli may give a quite misleading impression of their potential effectiveness for influencing behaviour.

5.2 FORM DISCRIMINATION

The response elicited from an organism depends not merely on the total amount of stimulation but also on its patterning—patterning in either space or time or both. We must, therefore, ask not only what patterns of stimulation are likely to elicit what responses, but also what patterns the organism is capable of discriminating and by what criteria they are discriminated. Equally, we may ask the obverse of this—by what properties are patterns of stimulation classified by the organism as similar?

Nearly all attempts to answer these questions for animals have depended on two methods. In order to assess the ability of an organism to differentiate between two patterns of stimulation, the conditioning method already discussed on pages 74–76 is used. The animal is trained to respond differentially to two patterns, and some characteristics of the learning score (e.g., number of errors, number of trials to reach a given criterion of correct choices, number of errors in a specified series of trials) are used as criteria of the difficulty of the task. Secondly, once an animal has been taught to discriminate between two patterns, the extent to which new stimuli are "equivalent" to the original ones can be assessed. If substitution of the new stimulus patterns for the original ones results in no change in, for example, error scores, then the new situation can be regarded as equivalent to the old one: if the precision of the discrimination is reduced, a measure of the extent to which the patterns are not equivalent is thereby provided. A series of such tests then provides information about the dimensions in terms of which the original patterns were discriminated. Both these methods provide information on the abilities of organisms to classify patterns and on the criteria by which they do so: they can also be used to analyse the steps involved.

Such methods could be applied to stimulation received through any sensory modality and patterned in either space or time. In practice, the bulk of the work has been confined to the visual discrimination of shape by animals. The apparatus used by Lashley (1930) for this purpose is shown in Fig. 5.3. The screen had two holes 5½ in. square (*L* and *R*), which could be closed with heavy cards, on which the patterns to be discriminated were drawn. The rats were trained to jump from the stand *S* towards one of the cards. The positive card would fall backwards when the rat jumped against it and allow the animal to reach the feeding platform *FP*. The negative card was fixed, so that if the animal jumped that way, it fell into the net *N*. Some patterns used by Lashley are shown in Fig. 5.4.

Another apparatus widely used at present for primates is the Wisconsin general test apparatus (e.g., Harlow and Harlow, 1962). The monkey is confined in a suitable cage, and the stimulus objects are presented on a horizontal tray to one side of the cage (Fig. 5.5). The objects are placed over small depressions in the tray, where suitable rewards can be hidden. The monkey is trained to reach out and displace one of the objects: if it selects an object which covers a depression containing food, it can eat the food. The apparatus is provided with a device by which the tray can be withdrawn from the animal's reach after it has made a choice. There is usually, also, an opaque screen which can be lowered to prevent the monkey's seeing the tray while it is being baited, and a one-way vision screen to conceal the experimenter.

With other species, rather different testing techniques have been used.

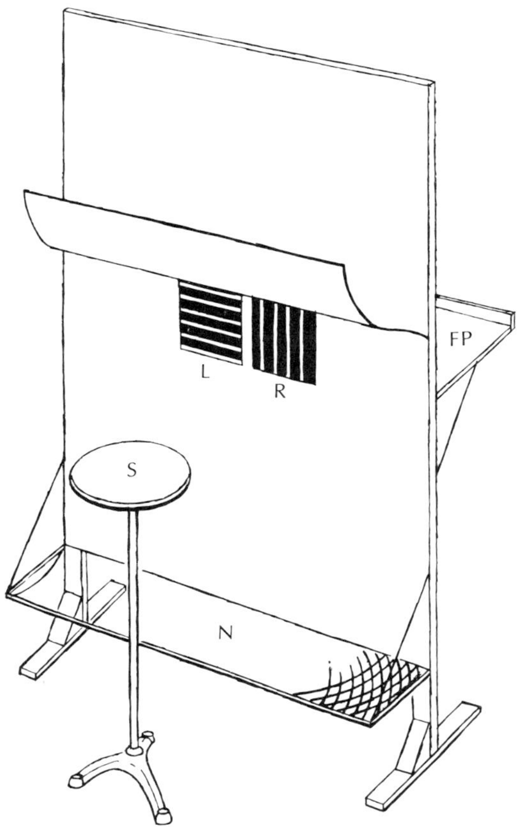

Fig. 5.3. The jumping stand as used by Lashley for testing pattern discrimination by rats. (After Lashley, 1930.)

For instance, *Octopus* can be trained by introducing the shapes one at a time into its tank: the animal is rewarded with food for attacking one of the shapes and punished with electric shock for attacking the other (Young, 1961). Recently, operant methods comparable to that described on pages 75–

Fig. 5.4. Some patterns used by Lashley in testing pattern discrimination in rats. (After Lashley, 1938*b*.)

77 have proved invaluable for some species (Blough, 1966). Other methods are described by Sutherland (1961).

The literature on shape discrimination in vertebrates and cephalopods has recently been reviewed by Sutherland (*loc cit.*, 1963a).* Some of his earlier conclusions can be summarized as follows:

1. *Discrimination of Shapes Differing in Orientation*

(*a*) *Rectangles and Lines.* For a number of species—chimpanzee, dog, rat, various fish and *Octopus*—discrimination between horizontal and vertical lines, and between horizontal and vertical rectangles, is easy; but for some of these species at least, discrimination between opposite oblique striations

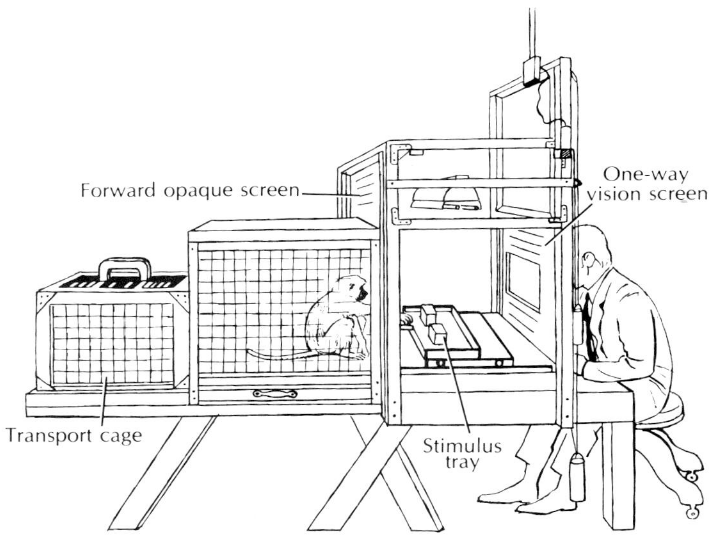

Fig. 5.5. The Wisconsin general test apparatus for studying discrimination in primates. (After Harlow and Harlow, 1962.)

is difficult or impossible. Cats and pigeons (Zeigler and Schmerler, 1965) perform well in oblique discriminations.

(*b*) *Triangles.* Dogs, rats and gudgeon learn to discriminate between upright and inverted triangles, but transfer tests suggest that, while dogs re-

* See also Meng (1957) for the toad *Bufo bufo*; Giebel (1958) for Equidae; Saxena (1960) for trout; Michels et al. (1962) for three squirrels (*Sciurus* spp.); Stichmann (1962) for chickens; Zimmermann (1962) for infant rhesus monkeys; Parriss and Young (1962) and Mackintosh et al. (1963) and refs. cited for *Octopus*: Mackintosh and Sutherland (1963) for goldfish; W. A. Matthews (1964) for various tropical fish; Fidura and Gray (1966) for quail *Coturnix coturnix*.

spond to the whole figure, rats respond primarily to the lower part and gudgeon to the apex (but see Matthews, 1964).

(*c*) *Mirror Images and 180-deg Rotations.* Macaques, rats, and *Octopus* discriminate more easily between figures which are an up-down inversion of each other (e.g., ⊤ and ⊥), than between left-right inversions (e.g., ⊢ and ⊣). There is no evidence that figures which differ only by inversions are more difficult to discriminate than figures with genuine form differences but of comparable complexity.

Although many pairs of mirror images can thus be readily discriminated by a number of species, some such discriminations are difficult, at any rate for fish and *Octopus*. Sutherland has shown that the vertical and horizontal projections (calculated by summing the total vertical and horizontal projections at each point along the two axes) of such shapes are similar (Fig. 5.6, *d*, *e*, *f*) and suggested that differences in the distribution of horizontal and vertical projections may be necessary for discrimination (see pp. 93–94).

2. *Discrimination of Shapes Differing in Form*

(*a*) *Triangles and Circles.* Macaques, cats, rats, chickens, crows and *Octopus* discriminate readily between a circle and a triangle. Monkeys and cats, but not rats, show considerable transfer of training when the triangle is rotated.

(*b*) *Triangles and Squares.* Chimpanzees, macaques, rats, pigeons, various fish, and octopuses make this discrimination easily.

(*c*) *Squares and Circles.* Sheep, cats, crows, chickens, frogs, various fish, and octopuses make this discrimination. For rats, it is easier if the square has its sides oblique (i.e., a diamond shape) than if they are horizontal and vertical. This may be because rats discriminate primarily on the base of the figure.

(*d*) *Crosses and Other Shapes.* Rats, hens, and various fishes can discriminate between crosses and squares, diamonds, rectangles, triangles, etc. (Tree squirrels (*Sciurus*) discriminate random shapes more readily as the number of sides is increased from 4 to 7 [Michels et al., 1962].)

3. *The Properties by Which Shapes Are Discriminated*

Transfer of training experiments has provided considerable data about the properties of the shapes discriminated:

(*a*) *Size.* A wide range of species show good generalization when the sizes of the figures they have learnt to discriminate are altered. This indicates that the shapes are analysed in terms of relative properties which remain constant even when size changes.

(*b*) *Outlines.* Many species show good generalization from solid to outline figures, or vice versa. This shows that the original discrimination was not based solely on differences in area, brightness, or distribution of brightness. Transfer from solid shapes to dotted outlines is moderately good in monkeys and hens but poor in minnows and sticklebacks, though the results obtained depend on the size and frequency of the dots. If the dots are placed only at the corners of the shape, transfer is nearly always poor, though it is better if angles rather than dots are placed at the corners. Comparable results are obtained if the sides are present but the angles omitted.

(*c*) *Background.* Macaques generalize readily when the background of the

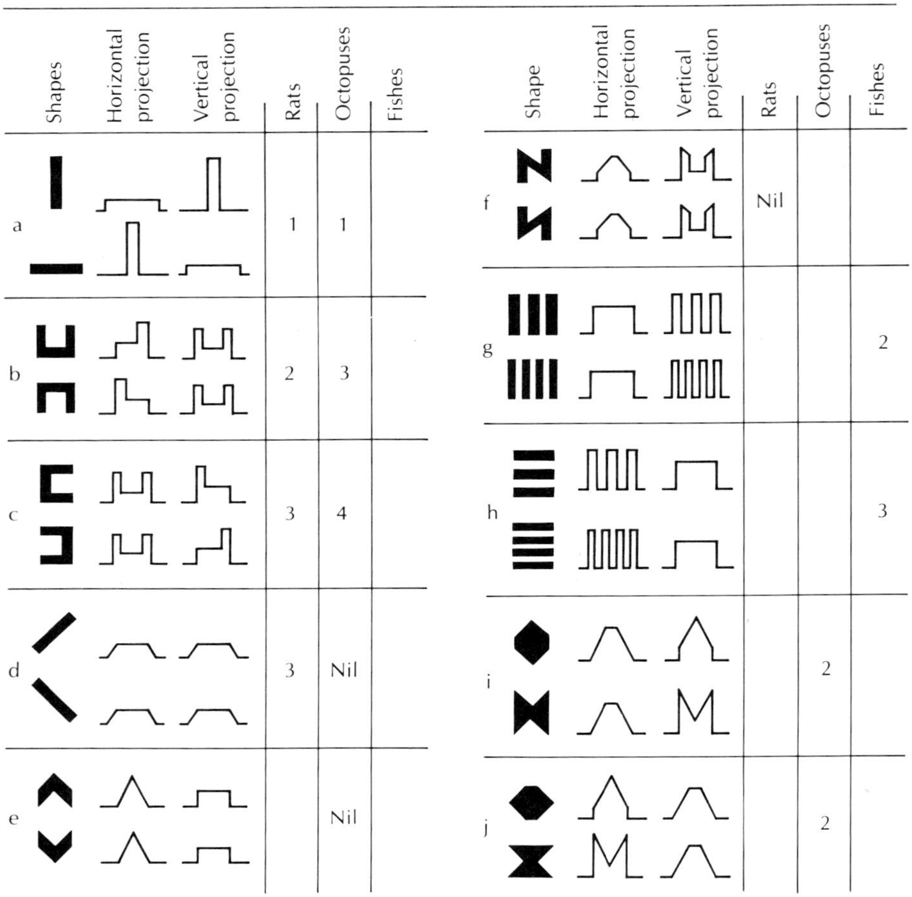

Fig. 5.6. The relation between the abilities of animals to learn visual discriminations and the horizontal and vertical projections of the figures. (After Sutherland, 1961.)

1, 2, 3, 4 and nil indicate decreasing degrees of discriminability.

figures is changed in size and/or shape, but data for other animals are scarce.

(*d*) *Relative Brightness for Figure and Background.* Reversal of brightness relations between figure and background has produced equivocal results, some experiments reporting good transfer, and others little. Species differences play a role here, for some animals (e.g., macaques) tend to identify stimuli by colour or brightness in preference to shape when the two are in conflict. For fishes, the dominance of colour seems to be less marked.

(*e*) *Parts of Figure*. If a discrimination has been learnt in terms of part of a figure only, generalization should occur when that part only is exposed or when the rest of the figure is distorted. This has been shown to occur in some species. Thus, on a Lashley jumping-stand, rats tend to discriminate in terms of the lower half of the figure; while in other types of experiments chickens and fish have been shown to rely on the features which most differentiate the two figures, wherever these occur on them.

(*f*) *Rotation*. Primates, some carnivores, and octopuses have been shown to generalize when the triangle on which they have been trained was rotated (cf. 1 (*b*) above). (In addition, in an operant situation, a pigeon's rate of pecking decreased systematically as an isosceles triangle was rotated away from the training position (Reynolds, 1961b).) For other species there is little evidence of transfer of training when the original figure is rotated. The only certain case in the rat is transference from v + to v ×. In fishes and rats it seems that generalization occurs only if the original shapes differ in the degree to which they are open or closed, i.e., in jaggedness. This suggests that they can discriminate in terms of this property.

In spite of the many experiments conducted, statements of general validity about shape discrimination in animals are hard to make. It is clear that most simple discriminations are readily learnt by the standard laboratory mammals, though there are some surprising failures. For the most part, the results obtained from vertebrates and cephalopods are similar. For instance, 3- to 4-year-old children, just like octopuses, confuse oblique lines oriented in opposite directions although, like octopuses, they can readily discriminate vertical from horizontal lines. Similar results are obtained with U shapes in various orientations (Rudel and Teuber, 1963). Similarities between species in perceptual abilities might be expected on other grounds, for the devices evolved to camouflage shape depend on similar principles over a wide range of species (Cott, 1940).

In spite of these overall similarities there are also clearly some quite marked differences in ability for shape discrimination, even among the vertebrates. Differences are even more marked in the extent to which generalization occurs when the figures are rotated or distorted in various ways and in the extent to which the whole or part of the figure is used.

As mentioned on page 91, Sutherland (1957a) suggested that many of the results on shape discrimination in *Octopus* can be understood on the assumption that the animals analyse shape by counting the horizontal extent at each point on the vertical axis and their vertical extent at each point on the horizontal axis (Fig. 5.6). This hypothesis accounts for many of the experimental findings, though Sutherland (1962) subsequently modified it with some additional suggestions; for instance, that open and closed forms are differentiated by the height and width of the peaks on the horizontal and vertical projections, the horizontal being more important than the vertical.

In a recent paper Sutherland (1969) has extended the data summarized above by comparing the ability of rats, octopuses, and goldfish to discriminate a square from a parallelogram, and to transfer from these to a battery of other shapes. The prediction of his earlier theory that octopuses discriminate more readily between a vertical parallelogram and a square than between a horizontal parallelogram and a square was not confirmed. His results indicated that rats classified the parallelogram as a shape containing oblique contours, and octopuses as one containing thin segments: goldfish were midway between the two. On the basis of this and other data, Sutherland has revised his earlier theory. He suggests that shape recognition involves two stages. In the first local features, such as segments and edges, are extracted, their relative positions being preserved. The next stage involves the "writing down" of a rule describing the shape in terms of its local features. Of course any shape can be described in more than one way: for instance an H is two vertical lines joined by a horizontal line, two ⊤'s on their sides, or a ⊔ above an inverted ⊔. Presumably the simplest rule is chosen—and it is for this reason that shapes which can be described by simple rules are more easily discriminated. Since the description is independent of retinal position, this theory, in Sutherland's view, accounts for intraretinal transfer and size invariance.

The inter-species differences Sutherland accounts for by differences in the relative abundance of different types of receptive units in the visual system. The units postulated are similar to those found in the cat visual cortex (pages 108–114), having an elongated receptive area with an antagonistic area either surrounding it (bar detector) or bordering it (edge detector). He assumes that in the octopus bar detectors are more common than edge detectors while in the rat the reverse is the case. Thus rats distinguish a parallelogram from a square because it stimulates oblique edge detectors, while octopuses depend on the difference in the number of vertical and horizontal bar units stimulated. This would account for rats confusing the parallelogram with other shapes with similar oblique contours, and the octopus confusing it with other shapes with similarly oriented thin segments. It is also necessary to assume that rats concentrate on the bottoms of shapes, and that octopuses store information about shapes in a manner which does not differentiate between left and right—and thus confuse mirror images. Other views on the mechanisms underlying shape discrimination are given by Dodwell (1961, 1962) and by Deutsch (1962).

Relatively little work has been done on form perception in invertebrates other than cephalopods. Some evidence suggests that even the higher arthropods do not use form vision in some contexts in which they might be expected to do so. For instance, in training experiments in which food is used as a reward, bees learn only with great difficulty to distinguish the simple geometric shapes, such as circle, square, and triangle, which are easily distinguished by most vertebrates (though Schnetter, 1968, has shown that their relative preference for stars and circles varies with the figure/ground contrast). In fact, a number of lines of evidence indicate that, in the feeding situation, bees show a marked preference for figures which have a long outline with respect to their surface area (e.g., Wolf and Zerrahn-Wolf, 1937; von Frisch, 1953). Thus bees visit flowers and artificial figures more, the richer the contour they possess, and can readily discriminate figures broken up into black and white areas from geometrical shapes. They show a preference for divided figures and cannot be trained to visit a less subdivided figure for food in preference to a more divided one (Hertz, 1929, 1930, 1931, 1935; Zerrahn, 1933). Presumably, a figure with an uneven outline provides numerous changes of stimulation in the individual ommatidia as the bee passes over them, and it is to this "flicker" effect that they owe their effectiveness. Flicker is undoubtedly also important in many other contexts—for instance in the courtship responses of many insects (see Table 4.1, pages 64–65).

However, in some cases bees have been taught discriminations which appear to be based on patterning (Hertz, 1933, 1935), and bees and other arthropods certainly use form vision in other contexts, for instance in locating their nests. Tinbergen and Kruyt (1938) placed movable landmarks (twigs, pine cones, etc.) near the nests of the hunting wasp *Philanthus triangulum*. When the wasp had had an opportunity to learn the landmarks, some of them were moved and the extent to which the wasp was disoriented assessed. It was apparent that the wasp was responding not to particular landmarks but to the pattern of them all (see Chap. 7 and Wehner, 1967). An ability to discriminate forms is similarly demonstrated by Drees's (1952) analysis of the stimuli eliciting courtship or hunting in spiders and by the ability of dragonfly adults (*Aeschna*) to discriminate almost instantaneously between insects and scraps of paper thrown into the air (Tirala, 1923). Indeed, the highly specific forms evolved by some flowers indicate a considerable ability for form discrimination in the insects which fertilize them; for instance, some orchids resemble the receptive females of certain ichneumonids and are visited only by the males which attempt to copulate with them (refs. in Baerends, 1950; see also Manning, 1956). Even insects which have no compound eyes may be capable of rudimentary form vision through simple stemmata by the use of head movements (Dethier, 1943).

In conclusion, two points from this brief discussion of form discrimination require emphasis. First, an animal's ability to discriminate form should not be assessed from the study of one type of behaviour only: training experiments using the foraging behaviour of bees probably do not reveal their full abilities, because of the overriding influence of flicker in this context. Second, many animals, even higher vertebrates, seem to be incapable of discriminating patterns which appear quite distinct to us. Species may differ markedly in their abilities, and further carefully controlled comparative experiments are urgently needed.

5.3 CONSTANCY AND PERCEIVED MOVEMENT

The question of generalization between patterns is related to that of stimulus constancies. Man sees as unchanging the sensory properties of an object even when the retinal image changes considerably. A cardboard circle of given size and colour retains these apparent characteristics in spite of marked changes in angle of regard, distance, and wavelength of the reflected light. Comparable phenomena have been shown to occur in various animals (refs. in Vernon, 1952; Thorpe, 1963a). It is likely that this ability to perceive constancies depends on the use of higher-order stimulus variables (see pages 85–86) and the relations between the different parts of the sensory field, a view supported by recent findings on the functioning of the retina (pages 103–116). Take, for instance, size constancy: why do objects not look smaller when observed from a greater distance, since their retinal image becomes smaller? One issue here is that the stimulus for perceived size is not given merely by the dimensions of the projection on the retina, but by a comparison between that and the size of the elements of texture in the surrounding space. The latter gives an indication of distance (see pages 85–86), and thus the distance of the object is taken into account in the assessment of size (Gibson, 1959). Higher-order variables are probably important also for animals, since many visual illusions whose effectiveness depends on the surrounding parts of the field are effective for a wide variety of vertebrates besides man (refs. in Thorpe, 1963a).

Constancies involving movement of the sense organ itself present an especially interesting case. If we turn our heads, the image of the world shifts over our retina, but the world appears to stay still. Yet a similar movement of the retinal image produced by movement of the external world gives us the appropriate impression that the external world is moving; our eyes may show nystagmus movements which can be interpreted as an attempt to fixate the moving world. How do we discriminate between movement of an image over our retina produced by change in the external world, and a similar movement produced by turning our heads? A similar problem arises when we consider the orientation of animals to their environment. If a fly (*Eristalis*) is placed on the axis of a vertical cylinder painted with vertical stripes, and the cylinder is rotated, the animal will turn in the same direction; it behaves as though attempting to keep in constant relationship to the world it sees. This is known as the optomotor response (Fig. 7.11). The same movement of the image of the stripes across the eye occurs in ordinary locomotion and does not induce these forced optomotor turning movements. What is the difference between the movement of the image over the retina caused by movement of the outside world and that caused by movement of the organism itself?

It is helpful here to bear in mind a distinction, made by von Holst and Mittelstaedt (1950; von Holst, 1954), between exafferent stimuli, produced by movement in the external world, and reafferent stimulation, produced by the subject's own movement. If an animal moves its eyes or head, the stimulation produced by the movement of the image on the retina constitutes reafference, because it is the result of its own movement. If an object moves in front of the animal the stimulation is exafferent, because it is not a con-

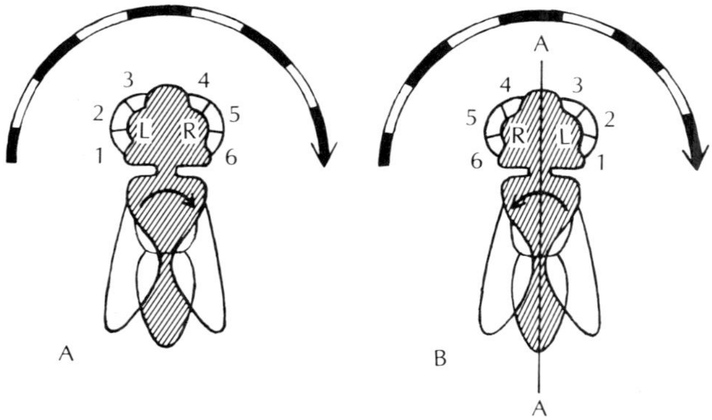

Fig. 5.7. The fly Eristalis in a rotating cylinder. (After von Holst, 1954.)

L, R, left and right eyes; *A,* head in normal position; *B,* head rotated through 180 deg.

sequence of its own movement. Now in some way the insect distinguishes movement of the stripes relative to its body, when this is a consequence of its own movement, from the relative movement which results when the cylinder rotates. This is not due to a mere blocking of the optomotor reflex during self-initiated movements, for if the head is rotated through 180 deg about the axis of the neck, so that the positions of the eyes are interchanged, both types of response are upset. If the cylinder is rotated, the fly now moves in the opposite direction, as might be expected from the reversed order of the retinal elements (Fig. 5.7). If the insect itself starts to move while the cylinder is stationary, it spins rapidly in small circles until exhausted. This would not occur if the optomotor reflex were merely blocked during self-initiated movement: apparently, the reversal of the stimuli received by the fly as a consequence of its movement causes that movement to be increased. Other experiments (von Holst, 1950a) show that if the stimuli received as a result of self-initiated movement are unchanged in direction but increased in strength the extent of the movement is decreased.

On the basis of such results von Holst and Mittelstaedt suggested that these orientational movements are the result of a comparison between the actual afferent stimulation and what should have occurred (i.e., the Sollwert) according to the momentary motivational situation. This is illustrated in Fig. 5.8, where efferent processes are shown in black, and the reafference produced by the movement thus induced in white. Von Holst and Mittelstaedt suggest that self-initiated movements are the results of a "command" (C) from a "higher centre" (*HC*, in *a*) which sets up an "efference copy" (*EC*) in a lower one (*LC*, in *b*). This may then (in *c*) send motor im-

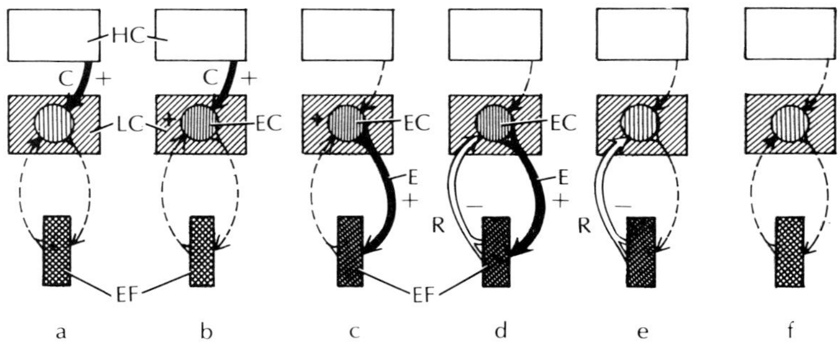

Fig. 5.8. Illustration of the reafference principle. For explanation, see text. (After von Holst, 1954.)

pulses or "efference" (*E*) to the muscles (*EF*). The "reafference" (*R*) produced by the movement is compared with the efference copy in the lower centre (in *d*). If they exactly compensate each other, the movement ceases (in *e*) and a stable condition is achieved (in *f*). If the reafference were too great or too small, the movement would have decreased or increased appropriately. Rotating the head of the fly causes the reafference to be reversed: when the command induces turning to the left, the reafference is that normally appropriate to turning to the right, with the result that the self-initiated movement to the left is accelerated. In the normal fly, rotation of the cylinder produces an unbalanced exafference, and the fly turns to compensate: when the eyes are reversed, the fly turns in the opposite direction (see also Dzimirski, 1959).

The reafference model thus enables us to understand both the behaviour of the normal fly and that of the fly with its head reversed. It involves a comparison between the sensory input actually received and a Sollwert, comparable to that which we have already encountered in the control of skilled movements. In the present case the basic assumption is that the Sollwert for the sensory input is determined by the efference copy (better called the "corollary discharge"; Teuber, 1961) which accompanies the initial command. A comparable model has been elaborated independently by MacKay (1962).

It sometimes happens that an insect does not compensate for its own movement, and the apparent movement of an object in the environment produced by its own locomotion may then elicit an optomotor response, causing the insect to turn towards the object. This results in a primitive type of responsiveness to forms and may constitute the basis of the movements of aphids and locust nymphs towards plants and towards each other (Wallace, 1958, 1959; Ibbotson, 1966; Rowell, 1967; see also Chap. 7).

Similar principles can be applied to perceived movement. Helmholtz (1867) suggested that non-visual knowledge of the movement of the eyes depends on judgement of the "effort of will" required to move the eyes Evidence supporting this view was obtained by Brindley and Merton (1960): when the conjunctival sacs were anaesthetized and visual clues excluded, human subjects were unable to assess the amount or direction of their eye movements unless those eye movements had been produced by the unhindered action of their own eye muscles. Now any particular efferent discharge to the eye muscles will normally result in a particular movement of the image of an external object over the retina. Discrepancy between the displacement of an image as expected on the basis of the efferent discharge (the Sollwert) and that actually received will be interpreted as movement in the external world. Von Holst's model for this is shown in Fig. 5.9 (see von Holst and Mittelstaedt, 1950; Held, 1961; Teuber, 1961; MacKay, 1962): its value can be seen by considering three ways in which the image can move across the retina:

(1) Its movement could be the result of a voluntary movement of the body, head, or eyeball, the external world remaining stationary. In such a case the observer registers his own movement, and the external world is perceived as stable. In terms of the model, we can say that the efference copy or corollary discharge is just matched by reafference from the retina, or that the Sollwert involves just that movement of the image across the retina which actually occurs.

(2) If the eyeball remains still, a similar movement of the image of the external world across the retina caused by an actual movement of the

external world is perceived as such. In this case there is no corollary discharge, the Sollwert is one of no movement of the retinal image, and the perception is produced by the unmatched afferent impulses.

(3) A third possibility is to move the eyeball mechanically by exerting slight pressure with the finger to the side of the eyelid. Here again, the afference is unmatched by a corollary discharge, and the external world appears to move.

An even more convincing demonstration can be obtained if the eye is mechanically fixed. If the subject then intends to move his eyeball, he perceives the environment to move even though there has, in fact, been no movement of the image over the retina. The perceived movement can then

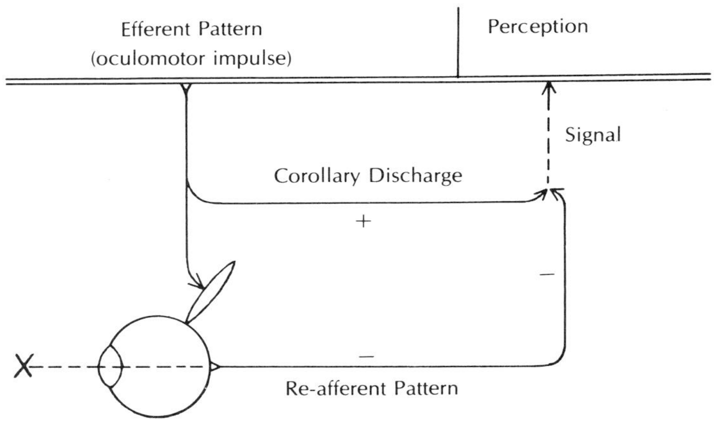

Fig. 5.9. Relation between eye movements and perception. (After Teuber, 1961.)

be understood as a consequence of the corollary discharge unmatched by reafference. In all cases the direction of the perceived movement is that predicted by the hypothesis (see also Gregory, 1958). The complexity of such comparator mechanisms is shown by experiments with human subjects which indicate an interplay between sensory and postural factors in perception; for instance tilting the body influences the relative tactile sensitivity of the two palms (McFarland et al., 1962).

In general, then, the consequences of the movement of a pattern of stimulation across a receptor surface are not immediately predictable: they depend on an interaction with other aspects of the internal state. It is also evident that the perception of movement of the environment and the control of the body's orientation with respect to that environment are in effect the same problem. Orientation of the body with respect to the environment has been studied in some detail in animals and will be discussed in Chap. 7. A consideration of the role of experience in mechanisms of spatial perception and orientation, which involves some further elaboration of this reafference model, is postponed until Chap. 20.

5.4 PHYSIOLOGICAL CORRELATES OF SENSORY/PERCEPTUAL FILTERING

Although the limits of physical change to which an organism is sensitive are determined by the characteristics of its sensory cells, we have seen that, within the sensitive range, some patterns of stimulation are more effective than others. How far can this selectivity within the sensory field be understood in physiological terms? Although our knowledge of the neurophysiological correlates of perceptual phenomena is still scanty, it is growing rapidly (Horn, 1962).

Stimulus Intensity and Quality

In any sensory modality, stimuli of a high intensity are more likely to be responded to than those of lower intensities. This finding is related to two types of physiological observation: first, that increase in stimulus intensity results in the excitation of a higher proportion of the fibres leaving the sense organ; and second, that there is some systematic relation between the stimulus intensity and the frequency of nerve impulses in such fibres (Adrian and Zotterman, 1926; Matthews, 1931; Hartline and Graham, 1932; Fuortes, 1958; Rosenblith, 1961). This latter relationship, though usually logarithmic, may be complicated in a number of ways. In the first place, the repetitive application of a stimulus to a sense organ does not necessarily produce a fixed response, nor indeed one that changes in any simple manner. Instead, the receptor potential may fluctuate, so that changes in stimulus intensity can be detected only on a statistical basis (Ishiko and Loewenstein, 1959; Duncan and Sheppard, 1963). Second, the rate at which sensory endings adapt to a stimulus varies greatly (Adrian, 1928): for instance proprioceptors adapt slowly, touch receptors rapidly. Third, in some receptors there is a considerable resting frequency of discharge: this may increase or decrease according to the direction of stimulation, as in the lateral-line organs of fishes and the vestibular organ of various vertebrates (Hoagland, 1933a, b; Sand, 1937; Löwenstein and Sand, 1940; Gernandt, 1949). Furthermore, some sensory units respond to a decrease rather than an increase in physical energy. In the optic nerve of the frog, for instance, some fibres respond only at the cessation of stimulation and some respond both to an increase and to a decrease of illumination (Hartline, 1938; Barlow, 1953; see pages 106–107). In the auditory pathway of the cat, some (second-order) neurones show a spontaneous discharge which is increased by tones of some frequencies and decreased by others (Galambos and Davis, 1944). For these reasons, and because in most sense organs the inputs from many sensory endings converge onto one nerve fibre, and the impulses in any one sensory fibre pass to a number of post-synaptic cells (e.g., Weddell et al., 1955), an increase in impulse frequency along one fibre cannot be unequivocally interpreted as the result of increased stimulation of a localized receptor; such information can be extracted only from data coming from an area of the receptive surface, and with a knowledge of the response characteristics of the fibres in question. However, in spite of these complexities, the greater the total change in stimulation the greater the change in impulse frequency is likely to be, and it will thus be the more easily detected against the background of "spontaneous" impulse activity.

As discussed on pages 84–85, within any one sensory modality qualities of sensation do not correspond directly to simple physical scales, and the neural mechanisms by which they are assessed are still far from clear. Occasionally one finds a direct relationship between neural activity and the subjective assessment of a stimulus. Thus Borg et al. (1967) compared recordings from the human chorda tympani with subjective assessments of the strengths of standard taste stimuli. A good correlation between frequency of firing and subjective assessment was obtained for citric acid and sucrose solutions for two patients, and for citric acid and sodium chloride for a third. However, the sensory code on which discriminations between different tastes are based is reflected in the relations between the frequencies of discharge in a number of simultaneously active taste fibres, each with a more or less unique pattern of responsiveness to different substances or groups of substances (e.g., Cohen et al., 1955; Erickson, 1963; Pfaffmann, 1963, 1964). It has also been suggested that the discrimination of the various somaesthetic qualities (touch, deep pressure, cold, warmth, pain) may depend on spatial and temporal patterns of discharge (e.g., Weddell, 1955).

Movement Perception: Lateral Inhibition

Objects tend to be more conspicuous when they are moving than when they are stationary. In the first instance, this is a consequence of sensory adaptation: the discharge from most receptors declines with time under conditions of constant stimulation (Adrian and Zotterman, 1926; Hartline, 1938). Indeed, our own visual perception depends upon the saccadic movements of the eyes, which occur about every quarter of a second even when we fixate an object as steadily as possible: in the absence of relative movement between object and retina, some or all of the image disappears permanently (Riggs and Tulanay, 1959; F. W. Campbell and Robson, 1961; H. B. Barlow, 1963). Although retinal ganglion cells which show little or no sensory adaptation have been found in the cat (Kuffler et al., 1957), it is probable that the information they provide is not used in form discrimination.

The greater conspicuousness of moving objects is also helped by another mechanism which has the even more interesting effect of emphasizing edges and increasing contrast between stimulus and background. Hartline (1949) illuminated a single ommatidium of the lateral eye of the king crab (*Limulus*), recording the resultant impulse train from the corresponding fibre of the optic nerve. A few seconds later a nearby ommatidium also was illuminated, and the discharge from the first was found to decrease. When the order in which the ommatidia were illuminated was reversed, the inhibitory effect was found to be reciprocal. Mutual inhibitory relations of this type were found between many pairs of nearby ommatidia in *Limulus:* the effect depended on the integrity of a plexus of nerve fibres lying in the region behind the ommatidia. This effect is known as "lateral inhibition" (Hartline and Ratliff, 1956; Hartline et al., 1956).

Comparable phenomena have since been found in vertebrates by recording from the ganglion cells behind the retina (see Fig. 5.11). These cells are separated from the receptor cells by bipolar cells, and convergence and divergence occur at each stage. Thus each ganglion cell can be influenced from a considerable area of the retina: this area is known as the receptive field of the ganglion cell. Receptive fields of cells near the fovea tend to be smaller than those in more peripheral regions, but in both cases neighbouring receptive fields overlap extensively, so that even a small spot of light affects many axons of the optic tract. In the frog, some of the ganglion cells in the retina give a short discharge of impulses if the illumination either increases or decreases but none under conditions of constant illumination. If the whole retinal field by which such a cell is influenced is illuminated, the resultant discharge is smaller than with illumination of the centre alone—stimulation of the peripheral part of the field inhibits the discharge from the ganglion cell evoked by a stimulus delivered to the centre of its receptive field (Barlow, 1953). Comparable results have been obtained in cats, where some retinal ganglion cells respond with increased firing to illumination of the centre of the receptive field and with decreased firing on illumination of the periphery (Kuffler, 1953; Wiesel, 1960): the effect varies with dark adaptation (Barlow et al., 1957). Similar effects occur in other parts of the central nervous system (Mountcastle, 1957; see below).

This lateral inhibition between peripheral and central areas of the receptive field of a ganglion cell helps to accentuate boundaries between differently illuminated parts of the field. This can be understood as follows.

Consider four cells whose receptive fields lie near a boundary at which a change in illumination occurs (Fig. 5.10). The receptive fields of cells *A* and *B* will be more strongly stimulated than those of *C* and *D*. If there were no lateral inhibition, and response strength depended merely on the area of the receptive field stimulated, the discharge of the four cells would be in the order $A > B > C > D$ and the edges would be blurred. By lateral inhibition, however, a cell whose receptive area is wholly within the area of greater illumination (*A* in Fig. 5.10) will have its discharge reduced by

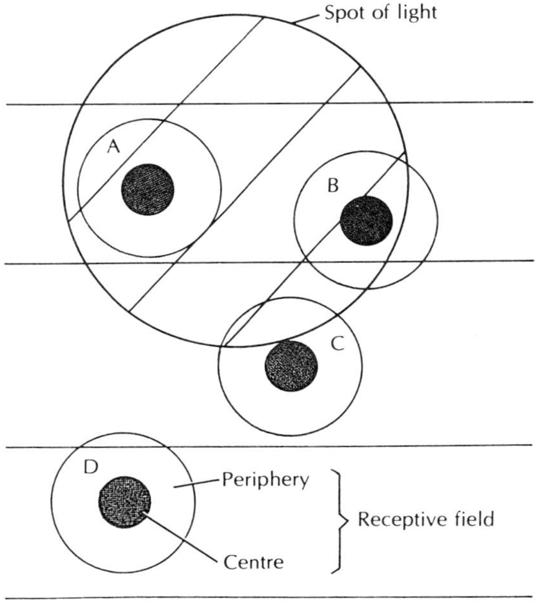

Fig. 5.10. Diagrammatic representation of the sensitive fields of four retinal cells, three of which are influenced by a spot of light. (After Horn, 1962.)

comparison with a cell *B* whose peripheral and inhibitory field is partially outside. On the other hand, the stronger illumination of the peripheral and inhibitory region of cell *C*, whose centre lies in the area of weaker illumination, will reduce its discharge by comparison with a cell *D* lying wholly within the more weakly illuminated area. The discharge from the four cells will thus be $A < B \gg C < D$, and the change in illumination at the boundary will be accentuated. In addition such lateral inhibition will increase the conspicuousness of moving objects as compared with stationary ones (Horn, 1962; see below). Similar principles apply in other modalities: lateral inhibition in the auditory pathways preserves the identity of discrete frequencies sounded simultaneously (Whitfield, 1967), and in the somatosensory system it increases spatial contrast (Gordon and Manson, 1967).

It is not too difficult to see how this concentric-field type of organization of ganglion cells could be produced. On the basis of electron microscope studies of primate retinae, Dowling and Boycott (1966) have suggested a possible "wiring diagram" which could explain many aspects of the concentric-field type of organization in ganglion cells (Fig. 5.11).

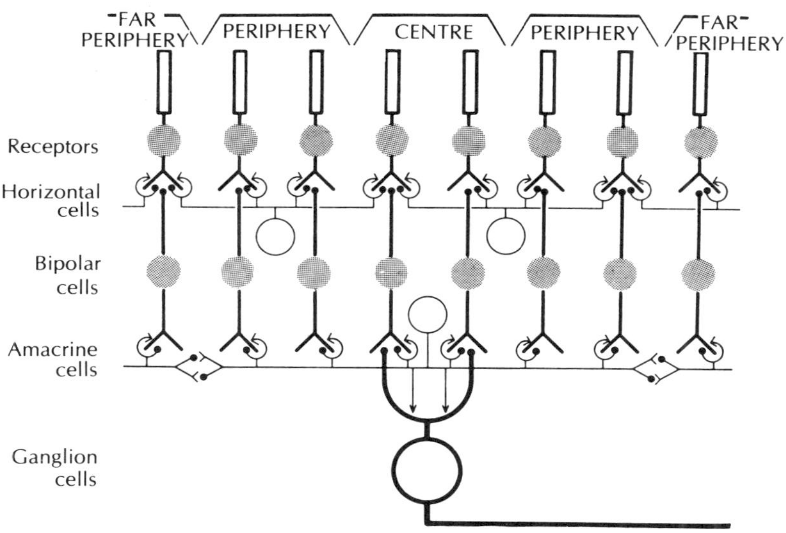

Fig. 5.11. A "wiring diagram" for a ganglion receptive field based on studies of primate retinae. ⅄, *excitatory synapse;* ↓, *inhibitory synapse.* (From Dowling and Boycott, 1966.)

The ganglion cells connect directly with the bipolar cells only in the centre of the field: they receive input from the more peripheral bipolar cells via the amacrine cells. Thus for an on-centre unit, stimulation of the centre of the field, but not of the periphery, causes direct excitation of the ganglion cells. Stimulation of the periphery of the field causes inhibition of the ganglion cell via the inhibitory amacrine–ganglion cell synapses. Simultaneous stimulation of centre and periphery would result in summation, with a weaker ganglion cell response. In animals with colour vision, the antagonistic zones may be colour coded, the centre responding maximally to light of one wavelength and the periphery to another (Hubel and Wiesel, 1964).

Stimulus Selection in the Frog's Visual System

Much of the work discussed so far was based on stimulation of the retina by point sources of light and recording from single fibres in the optic nerve. Like Gibson (e.g., 1959), Lettvin et al. (1959) suggested that this method of stimulation can tell us little about how the eye functions normally. They used, instead, a matt grey hemisphere, 14 in. in diameter, and concentric to the eye, on the inside of which objects could be moved about by means of magnets placed on the outside. In many of their experiments the objects used were a black disc 1 deg in diameter and a rectangle 30 by 12 deg. These investigators identified five types of ganglion cell in the frog's eye:

1. *Sustained-contrast detectors.* These give a sustained discharge if the edge of an object either lighter or darker than its background moves into the field and stops there. There is no response to a change in the general illumination.
2. *Net-convexity detectors.* These respond to a small object passed through the field or to the convex edge of a dark body moved into or across the field. The response is greater, the greater the convexity of the edge until the object becomes less than half the width of the field. It is also greater if the edge is moved jerkily than if it is moved smoothly. If a dotted pattern, with the dots no farther apart than half the width of the visual field, is moved across, there is practically no response; but if any dot moves differentially, the cell responds as if that spot were moving alone. There is no response to changes in the general illumination, and that to a convex edge depends on the sharpness of the edge but not on how much darker the object is than its background. (It has since been claimed that the response characteristics of these units are due to their responsiveness to a small object in the centre of the field, the periphery being inhibitory [Gaze and Jacobson, 1963a].)
3. *Moving-edge detectors.* These respond to a moving edge, whether dark on light or light on dark. The response increases with velocity but is little dependent on general illumination. These are the same as the ON-OFF fibres mentioned above (pages 101–102).
4. *Net-dimming detectors.* These respond to a reduction in illumination. The effect of a moving object depends on its size and relative darkness, but darkening of the periphery of the field has less effect than darkening of the centre.

5. *Absolute-darkness detectors.* This is a small category of fibres with large receptive fields which discharge more rapidly the darker it is.

The receptive fields of these detectors increase from about 2 deg to about 15 deg in the order given above, and the first two types are about thirty times as frequent as the third and fourth.

The first four types each project in an orderly way onto the optic tectum, though each into a different layer. Responses of contrast detectors can be recorded in the superficial layers, the convexity, moving-edge, and dimming detectors at successively deeper layers. The responses of the absolute-darkness detectors are found at the same level as those of the moving-edge detectors. The tectum thus contains four sheets of endings, each containing a map of the retina in terms of one type of detector. Through these sheets project dendritic branches of the deeper-lying tectal cells, each of which is thus influenced by the several properties of one particular area of the retinal field. The image is thus seen in terms of the distribution of combinations of the qualities registered by the detectors (see also Grüsser-Cornehls et al., 1963).

Two points about this description of the functioning of the frog's visual system demand emphasis. First, the hitherto convenient division into sensory and perceptual phenomena, with its implications that the latter are due to something which the higher levels "do" to the messages received from the sense organs, ceases to be useful (see Teuber, 1960, 1961; Grüsser and Grüsser-Cornehls, 1968). It is clear that the message sent to the brain is not simply the coding of a series of spots of light of varying intensity: the ganglion cells collectively transmit a message which is already highly organized and interpreted. Further organization may occur at each successive level, as discussed below.

Second, however large the gap still to be bridged, the fact that the frog's retina can code relatively complex qualities in this way confirms the view that such qualities may form the basis of perceptual phenomena (see pages 84–86). That they form only the basis is indicated, for instance, by the fact that amphibians can learn to make quite precise visual discriminations as a result of experience. Thus toads (*Bufo bufo*) which have been stung by honey bees learn not to attack drone flies which mimic bees (Brower et al., 1960; Brower and Brower, 1965).

At the higher level of the optic tectum the units have even more complex properties which are very difficult to specify. "Newness" units have large receptive fields and respond to objects moving in a jerky fashion: repetition of a movement over the same path 5 to 10 sec later does not cause a second response. Sometimes this habituation can be erased by brief total darkness. "Sameness" units also have very large fields. They respond to a small moving object, but after some latency. They continue to respond so long as the object moves with a slight jerkiness. The cell gives bursts of responses with every discontinuity in the movement. If the target becomes still, the response disappears completely after about 2 min. If it starts moving again, there may again be a short delay before the cell starts to respond. The receptive field of each cell has a small blind spot in which the object elicits no firing. It is almost as though the cell had to have its "attention caught" by the movements of the object (Lettvin et al., 1961).

Stimulus Selection in the Cat's Visual System

In the cat the responses obtained from the retinal ganglion cells are more similar from cell to cell than they are in frogs: characteristically they have two concentric circular zones, the centres responding to either an increase or a decrease in illumination, and the periphery opposing the centre (Kuffler, 1953; Wiesel, 1960). Not all the retinal ganglion cells behave in this way, however, especially in the area centralis, where some cells do not have the centre-surround type of organization and a few of them have properties of directional sensitivity or line detection similar to those found farther along the optic pathways (see below; Spinelli, 1966; Stone and Fabian, 1966; Rodieck, 1967). Furthermore, it has been claimed that stimulation through other modalities may produce changes in retinal receptive fields (Weingarten and Spinelli, 1966).

As a consequence of the opposing centre and periphery, many retinal ganglion cells give a characteristic response to moving figures. Using black or white cardboard against a contrasting background, Rodieck and Stone (1965) found two types of response. "Centre activation" was given by a white figure moved across an on-centre or by a black figure moved across an off-centre. For thin bars the response was unimodal, but for broad bars it was bimodal, with the first peak greater than the second. A suppression of activity occurred as the figure moved into or out of the field (Fig. 5.12). "Centre suppression" was found under the converse conditions.

Although the cat appears to have mostly cells with opposing centre and periphery at the retinal level, greater complexity is found farther along the visual pathways. This is already evident in the lateral geniculate nucleus: as mapped with small spots of light, the cells there appear generally similar to those in the retina, but there is an enhanced capacity of the periphery of the receptive field to cancel the effects of the centre. This has the consequence that the cells in this ganglion are even more responsive to spatial differences in retinal illumination as compared with the illumination itself. Most lateral geniculate cells are driven by only one eye (Hubel and Wiesel, 1961). With moving objects, on-centre cells fire transiently to a white figure entering or a black figure leaving their receptive fields, and off-units the reverse. The response patterns vary with the size, shape, and speed of the stimulus. Yet other cells respond when either a black or a white object enters or leaves their fields: they presumably have inputs from both on- and off-centre units (Kozak et al., 1965).

In the visual cortex the cells again mostly have mutually antagonistic excitatory and inhibitory regions, but a great variety of types of receptive

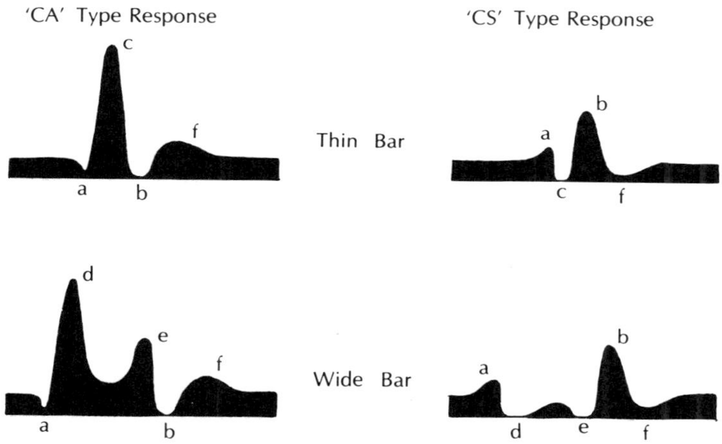

Fig. 5.12. Schematic curves of responses of cat's retinal ganglion cells to movement of bar across field. (After Rodieck and Stone, 1965.)

Left, centre activation. *a*, small suppression as bar of any width entered the receptive field: this component was not always present; *b*, suppression as bar left the field: this was invariably present; *c*, unimodal activation produced by narrow bar; *d, e*, bimodal activation produced by wide bar; *f*, activation sometimes seen here.

Right, centre suppression. *a*, small activation as bar entered field: usually present; *b*, activation as bar left field: invariably present; *c*, unimodal suppression produced by thin bar; *d, e*, bimodal suppression produced by wide bar; *f*, suppression sometimes seen here.

field is to be found. In many of them the response to shapes or moving stimuli seems to bear little relationship to the receptive fields as mapped with small spots of light. Most cells are influenced by both eyes. The majority of the cells can be classified into two main groups.

1. *"Simple" Cells.* These respond to line stimuli (light lines on dark backgrounds or vice versa, and edges) at a particular location and oriented in

a particular direction. As with the retinal cells, the receptive fields, as mapped with small spots of light, are divided into on and off areas: in the former illumination produces an increase in rate of firing, while in the latter it produces a decrease, or an after-response, or both. Here, however, these areas are not circular. Characteristically the central area is long and narrow and is adjoined on each side by larger opposing areas (Fig. 5.13). A slit which is oriented along and just covers the central area will thus produce a stronger response than a slit of similar size oriented so that part of it is covering the antagonistic peripheral region of the

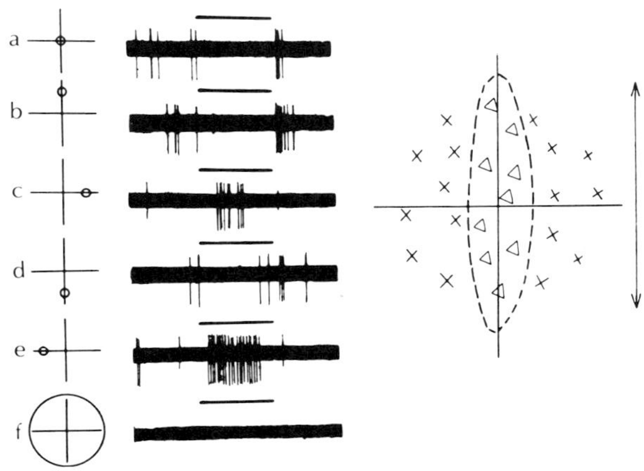

Fig. 5.13. Responses of a cell in the cat's striate cortex to a spot of light. (From Hubel and Wiesel, 1959.)

The complete map of the receptive field is shown on the right: ×, areas giving excitation; △, areas giving inhibition. Axes of this diagram are reproduced on the left of each record, where the circle indicates the area of stimulation. In *a, b, d* the stimulus produced a decrease in activity, with a burst of impulses when the stimulus was turned off. In *c, e* illumination caused an increase in firing.

receptive field. A slit at right angles to the direction of the central area usually produces no response at all (Fig. 5.14). The orientation of the sensory field with respect to the vertical may be temporarily affected by tilt of the body, and then readjusts (Horn and Hill, 1969).

Within this general category, units exist in remarkable variety. Not only may the fields be oriented in any direction, but also the central area may be displaced towards one side of the antagonistic zone to varying degrees. Probably each such cortical unit is fed by a large number of lateral geniculate cells with their centres arranged along a straight line.

2. *"Complex" cells.* These also respond to line stimuli (slits, edges, and bars) in a particular direction. The receptive fields, however, cannot be divided into antagonistic on and off zones. Characteristically an appropriately oriented edge anywhere within the field will cause the unit to respond with an increase or decrease in rate of firing (Figs. 5.15 and 5.16). Unlike simple cells, movement of the stimulus produces a sustained change in rate of firing. It is suggested that complex cells are fed from a large number of

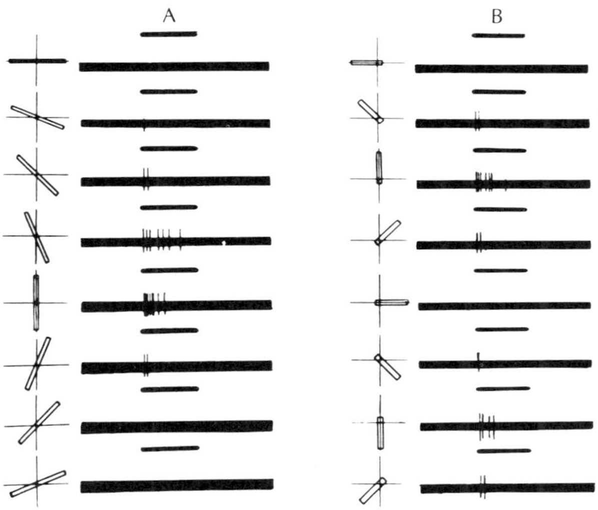

Fig. 5.14. Responses of cat striate cortex cell ("simple" type) to a rectangular bar of light. (From Hubel and Wiesel, 1959.)

In *A* the slit was 1° x 3°, with its centre superimposed on the centre of the receptive field: successive stimuli were rotated clockwise. In *B* the slit was 1° x 5°, with one end always covering the centre of the field. When the centre of the field was stimulated with a small 1° spot, a response was obtained. With these slits a maximum response was obtained when they were in the orientation shown here as vertical. When horizontal no response was obtained.

simple cells with their receptive fields arranged throughout that of the complex cell. This possibility is made more probable by the finding that the cells are arranged in columns in the cortex, all cells in the same column having the same orientation: it is known that most of the intercellular connections run vertically (Hubel and Wiesel, 1959, 1962, 1965a; see also Hubel and Wiesel, 1960, 1968, for data on monkeys).

In addition to these cell types, the visual cortex also contains units with a concentric-field organization, similar to those found more peripherally. This implies that the changeover to the cell types more characteristic of the visual cortex takes place in the cortex itself, rather than between the lateral geniculate body and the cortex (Baumgartner et al., 1965; Horn, personal communication; but see also Spinelli, 1966, cited above). It is,

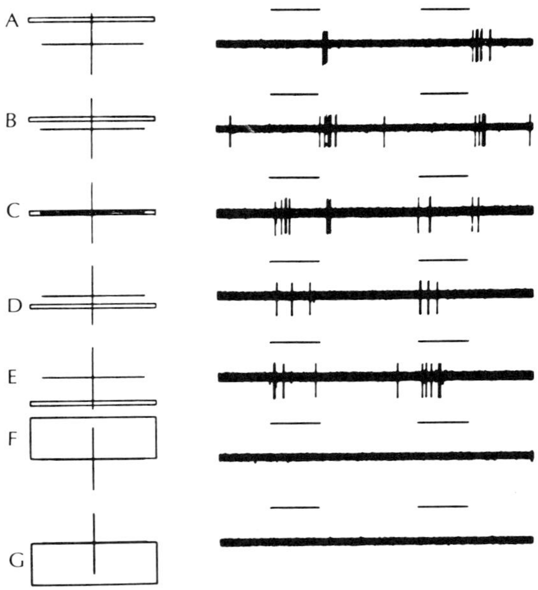

Fig. 5.15. Responses of a cat striate cortex cell ("complex" type) to rectangular bars of light. (From Hubel and Wiesel, 1962.)

Diagrams to left indicate position of stimulus (*A–E*, 1/8 x 3°; *F–G*, 1½ x 3°) with respect to the receptive field, indicated by a cross. In each record, upper line indicates when stimulus is on. Cell gave an 'off' response when stimulus was above the centre of the field, and an 'on' response when below it. Within limits, the precise position of the slit was not critical. The wider stimulus did not give a stronger response, indicating an absence of summation in the vertical direction.

however, difficult to be certain of the precise origin of the responses recorded.

Histological evidence indicates that visual information is transmitted from the visual cortex itself to neighbouring areas for further processing. Here again most cells respond to slits, edges, or bars, and most are "complex". For some ("lower-order hypercomplex") the length of the stimulus line must be limited in one or both directions. The adequate stimulus is thus a line falling within a given region and oriented in a particular way, provided a similarly oriented line does not fall on an adjacent region. For such a cell limited at one end only, a powerful stimulus is two lines at right angles, with one edge optimally oriented across the activating region and the other on the boundary between activating and antagonistic regions. Such cells behave as though they were getting input from two complex cells.

Other cells ("higher-order hypercomplex cells") respond optimally to line stimuli of limited length in either of two orientations at right angles.

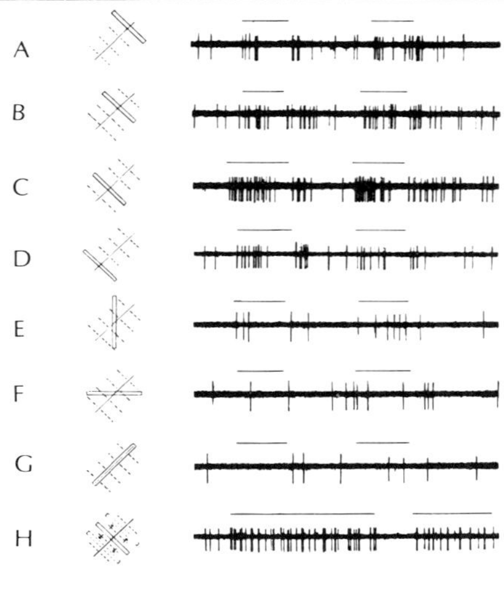

Fig. 5.16. Responses of a cat striate cortex cell ("complex" type) to rectangular bars oriented in various directions. (From Hubel and Wiesel, 1962.)

Diagrams at left show position of stimulus (1/8 x 2½°). Total receptive field was about 2° x 3°. *A–D*, slit oriented parallel to receptive field axis; *E–G*, slit at an angle to the axis; *H*, slit oriented as in *A–D*, is moved to and fro. Horizontal lines above each record indicate stimulus on.

These probably obtain their input from a large number of lower-order hypercomplex cells. A type of columnar arrangement of functionally similar cells like that found in the striate cortex obtains also here (Hubel and Wiesel, 1965a).

Thus the ability to distinguish patterned visual stimuli depends on interactions between neurones at successive stages on the visual pathways. Many even more complex types of interaction remain to be described; but presumably many perceptual phenomena, including stimulus generalization and stimulus constancy, will eventually be explicable in these terms. For instance the simple cortical cells show an increase in selectivity over the geniculate cells in that, while a spot of light is only a weak stimulus, an appropriately oriented line is a strong one. At the next stage, however, a complex cell is fed by simple cells which respond to similarly oriented lines in such a way that any one of the simple cells can fire it: thus the selectivity for orientation is generalized over many positions. Each successive level of integration presumably provides for similar abstraction and generalization.

Some Inter-species Differences in Cell Specificity in the Visual System of Higher Vertebrates

We have seen that the cat, unlike the frog, has comparatively little differentiation of units at the retinal level but considerable differentiation at the cortical. The difference is not, however, one of phyletic level, for the pigeon shows even more retinal differentiation than the frog (Maturana, 1964), and the rabbit much more than the cat. In the periphery of the rabbit retina at least five types of unit occur, each responsive to stimuli with particular characteristics. Some are of the concentric type, with either on or off centres. Others show directional sensitivity. Furthermore while some respond differentially to fast movement, others are excited by very slow but not by fast movement. Direction and speed of movement, as well as localized dimming and brightening, can therefore be abstracted by the ganglion cells in the retina. Thus considerable analysis of the sensory information occurs in the two synaptic layers between the receptor cells and the ganglion cells (Barlow et al., 1964; see also Michael, 1968, for data on the ground squirrel *Citellus mexicanus*).

Even greater diversity is found in the visual streak of the rabbit's retina (Levick, 1967). The five types of unit found in the periphery of the retina are present, but the receptive fields tend to be smaller and oval-shaped with a long axis horizontal. In addition there are three additional types of cell—orientation-selective cells responsive to either horizontally or vertically extended targets, local-edge detectors responsive to the appearance or movement of a contrasting border, and "uniformity detectors" which give a high level of activity in the absence of stimulation.

The preferred directions of the directionally selective units in the rabbit's retina cannot be predicted from a map of on and off responses to stationary spots and are unchanged by reversal of the contrast of the moving spot or black-white border. The directionality can thus not be explained by the asymmetrical arrangement of on and off zones in the receptive field, as suggested by Hubel and Wiesel for neurones in the cat's striate cortex. Barlow and Levick (1965) have suggested that the input for such units comes from a subset of bipolar cells that respond to the corresponding sequence of excitation of two neighbouring retinal regions with which they connect: sequence discrimination could be due to a laterally connecting inhibitory element from one of these regions.

Although Lettvin et al. (1959) point out that the mechanism they describe for the frog is peculiarly suited to its prey-catching and predator-escaping existence, the differences in retinal organization among frog, pigeon, cat, rabbit, and monkey are not easy to understand. The cat has less specificity at the retinal level than the frog but probably has more at the cortical. If this is related to differences in the modifiability of their behaviour, why does the rabbit have marked differentiation at the retinal level? Is it possible that a preponderance of binocular vision demands little retinal differentiation?

Stimulus Selection in the Visual System of Other Animals

The principle that the visual pathways contain units responsive to qualitatively different parameters of the stimulation is of widespread applicability also amongst lower vertebrates and invertebrates. In the goldfish, several types of unit have been detected in the optic chiasma and tectum (Jacobson and Gaze, 1964). In the optic nerve of the crab *Podophthalmus vigil* information is carried by a variety of types of fibre. Among other characteristics these differ according to:

1. The area of the visual field to which the fibre responds: this ranges from the whole to quite small portions of the field.
2. The type of stimulation eliciting the maximal response, ranging from fast-moving large objects to changes in the intensity of stationary illumination.
3. The amount of contrast necessary for maximal response.

In this species there are also numerous fibres running to the eye and responding to visual stimulation of the other eye and/or tactual stimulation of various parts of the body. Visual information is thus integrated with that from other sources in the eye, as well as elsewhere in the brain (Wiersma et al., 1961; Bush et al., 1964; Waterman et al., 1964; Wiersma et al., 1964; see also Wiersma and Yamaguchi, 1967, for the crayfish *Procambarus clarkii*).

Units responsive to complex visual stimulation are known also in insects (e.g., Jander, 1964; Blest and Collett, 1965a, b; Collett and Blest, 1966; Voss, 1967; Bishop et al., 1968). The "novel movement" detectors described by Horn and Rowell (1968) from the locust tritocerebrum show a directional selectiveness in that they are more responsive to stimuli moving in one direction than the opposite. Not only does their response wane with successive presentations of the stimulus (see page 101) but also the directional selectivity shows an effect of primacy in that it is modified in favour of the direction of movement of the first stimulus of a series: such effects last several minutes.

Bat Detection by Moths

A particularly elegant analysis of the manner in which sensory input is encoded has been provided by Roeder's (e.g., 1966a, b) studies of the

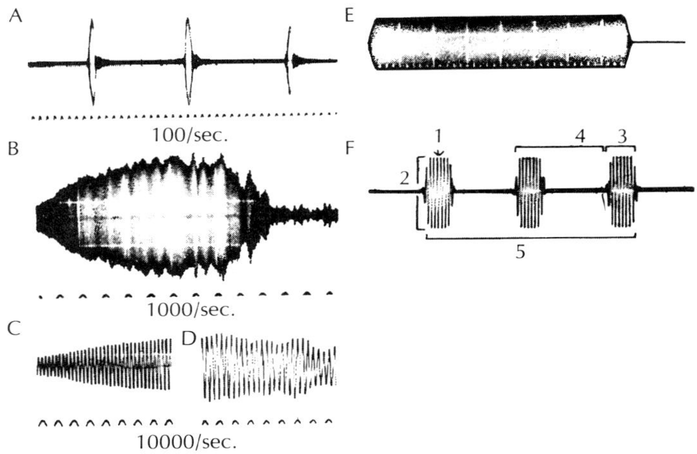

Fig. 5.17 A, record of three ultrasonic chirps from a series made by a cruising bat; B, expansion of one of the chirps in A; C, D, expansion of the initial and final portions of B; E, artificially generated ultrasonic pulse; F, diagram showing the parameters of the stimulus: 1, ultrasonic frequency in each pulse; 2, pulse intensity; 3, pulse duration; 4, pulse repetition rate; 5, pulse train duration. (From Roeder, 1966*a*.)

tympanic organs of noctuid moths. These moths respond to the ultrasonic pulses emitted by bats in two ways: high-intensity sounds produce a variety of manoeuvres carrying the insects eventually towards the ground, while lower-intensity sounds cause the moths to turn and fly away from the source (Roeder, 1962a, b). Each tympanic organ contains two receptor cells, one of which is about ten times more sensitive to ultrasounds than the other. Roeder mimicked the ultrasonic chirps of a cruising bat, whose characteristics are shown in Fig. 5.17 *A-D*, with an artificially generated pulse (Fig. 5.17 *E*). Trains of such artificial pulses have in effect five independently variable parameters (Fig. 5.17 *F*): (1) the frequency of each pulse, (2) the amplitude of each pulse, (3) the duration of each pulse, (4) the inter-pulse interval, and (5) the duration of the pulse train. The extent to which information concerning variations in each of these parameters is transmitted to the central nervous system was assessed by recording the output of the more sensitive of the two receptor calls. Whereas variations in ultrasonic frequency within the sensitive range were not found in the auditory nerve signal, the other four parameters were. Pulse intensity and duration were represented (with some ambiguity) by the length of the train of nerve spikes, the inter-spike time interval, and the latency of the first spike. The interval between pulses and the train duration were accurately correlated with sequences of spikes in the auditory nerve.

The further processing of the information in the auditory nerve was investigated by recording from the central nervous system with micro-electrodes. Among the neurone types found were:

1. Pulse markers, which responded with a single spike to three or four incoming impulses separated by sufficiently short intervals. The response was independent (within wide limits) of sound intensity and pulse durations—it gave only one spike per pulse. The output of such cells thus varied only with two of the stimulus parameters—pulse repetition rate and duration.
2. Train markers, which gave a regular train of spikes at an intrinsic frequency to a series of utlrasonic pulses. This type of neurone, therefore, conveys information only of pulse-train duration, variations in the other four parameters being without influence.

These are only two of the neurone types found by Roeder (1966b), but they indicate the manner in which the signal is transformed in its passage through the central nervous system. For instance, a cell system comparing temporal precedence of pulse-marker spikes from the two sides could derive a measure of sound intensity differences at the two ears, though such a system has not yet been found. Roeder likens the stimulus parameters to keys, each of which is necessary to gain admittance to a door but subsequently becomes superfluous. Thus sounds must be within a certain frequency range to affect the receptor, but frequency is then of no further importance. Similarly a range of pulse intensities and durations is necessary to gain admittance to the pulse-marker neurones, but these parameters are subsequently of no influence. Further analysis of this system is clearly likely to produce results of the greatest interest (see also Payne et al., 1966).

Perception and Eye Movements

Such work on the functioning of the sensory pathways heralds a new advance in our understanding of perceptual phenomena. Although knowledge is still limited, the way now seems open for increasing understanding in neural terms of the manner in which complex spatial and temporal patterns are recognized and responded to. But, as always, progress requires effort at more than one analytical level, and there is still need for further study of the perceptual phenomena themselves. A promising approach involves the use of stabilized retinal images (Riggs and Tulunay, 1959). If the image of a figure is stabilized on the retina it appears and disappears at intervals. If the figure is complex, its parts tend to disappear and reappear intermittently, their visibility fluctuating independently (Pritchard et al., 1960). At one level, this suggests that the perception of the parts is all-or-none, but the perception of the whole is not a unitary event (Hebb, 1963). At a finer level of analysis it suggests that central neurones responsive to different aspects of the stimulus (see pages 108–114) habituate at different rates. The frequency with which the image disappears can be reduced by simultaneous stimulation through other sensory modalities or through the other eye: the visibility of a line stabilized on one retina is enhanced by an unstabilized image on the other more when the latter is of similar line than when it is not, suggesting a field effect in the visual cortex (Cohen, 1961). It seems at least possible that there is a relationship between the stimuli which give the most persistent stabilized images and those which are subjectively most conspicuous. (Although it has been suggested that slippage of the contact lenses played a part in some of these findings [H. B. Barlow, 1963], appropriate techniques can make slippage negligible [Riggs and Schick, 1968].)

Here again, however, we must beware of generalizing between vertebrate groups. The importance of eye movements varies between phyla—in birds, eye movements are usually slight and all distant objects are focused sharply simultaneously (Pumphrey, 1948). In mammals, on the other hand, sharp vision occurs only at the fovea, and scanning movements, together with a short-term retention, give sharp vision over a wide area. Although form vision probably precedes such eye movements in phylogeny, where they are developed to give sharp localized vision, the information from them is likely to be used in the recognition of form (Gregory, in Thorpe 1963a). The use of eye movements to mediate form vision is demonstrated at a different phyletic level by Dethier's (1943) work on the form vision of lepidopterous larvae, which have only a small group of simple "stemmata" on either side of the head.

5.5 SUMMARY

1. Even among the energy changes to which an animal's receptor organs are responsive, some spatial or temporal patterns of stimulation are more effective in influencing its behaviour than others are. The effectiveness of some of these is not linked to particular types of behaviour and depends on properties of the sensory/perceptual mechanisms.
2. In studies of human perception, the relation between the physical magnitude and the subjective magnitude of a stimulus varies with the nature of the stimulus. In addition, a subjective sensory quality (hue, brightness, etc.) may not correspond to a particular physical dimension but to combinations of physical characteristics. In ordinary circumstances most stimuli do not act on single receptor cells but on sense organs: gradients, ratios, and changes in pattern of stimulation are essential components of the situation. For all these reasons physical measurements of stimuli can give a misleading impression of their potential effectiveness for influencing behaviour.
3. Species differ markedly in their abilities to discriminate between simple geometrical forms and in the stimulus characters by which they do so.
4. Organisms must discriminate movement of a retinal image produced by their own movement from that produced by a change in the external world. The "reafference" model of the manner in which this is accomplished involves comparison between the stimulus changes which a movement (or lack of movement) should produce and that which actually occurs. Comparable mechanisms must occur in other types of perception.
5. Physiological evidence concerning the mechanisms of some perceptual phenomena is discussed, with particular reference to selective responsiveness in the visual system.

6

SELECTIVE RESPONSE TO STIMULATION—SELECTION SPECIFIC TO A RESPONSE OR PHYSIOLOGICAL STATE

In Chap. 5 we discussed some examples of the influence of the sensory/perceptual apparatus on the effectiveness of a stimulus in controlling behaviour. Some stimulus characters, however, are effective only when they contribute to the elicitation of particular types of behaviour: in other contexts they are of little significance. Thus we have already seen that a model with a red belly is especially effective in eliciting attack from a male three-spined stickleback, while one with a swollen belly is especially effective in eliciting courtship. The effectiveness of these characters must depend on mechanisms which are more selective than the general sensory/perceptual ones, and have some degree of response-specificity. As another example, Fig. 6.1 shows the effectiveness of models of different colour in eliciting feeding responses and sexual pursuit from males of the grayling butterfly (*Eumenis semele*). The former responses are markedly influenced by the colour of the model, the latter are not: the effectiveness of particular colours is thus response-specific (Tinbergen et al., 1943; see also Crane, 1955).

A note on a terminological point is necessary here. The action of a stimulus whose effectiveness is response-specific usually depends on the animal's being in a particular physiological state. When in that state, it is especially likely to respond to the stimulus in question. When the effectiveness of a stimulus is response-specific, the responsiveness of the organism is thus stimulus-specific.

Even where two different responses of the same species have not been studied, the nature of the stimuli found effective often makes it virtually certain that something more than the general sensory/perceptual mechanisms must be involved. Thus Marler (1961) argues that the results of Drees's (1952) careful study of the stimuli eliciting male courtship in a jumping spider are difficult to account for except in terms of a response-specific mechanism for stimulus selection. "The strongest responses were given to visual patterns with a certain size and orientation, divided into cephalothorax and abdomen, with legs having a particular size and orientation with respect to the body and with the species pattern of black and white shapes on the abdomen. Outside the breeding season the response disappeared and all models evoked either prey-catching or avoidance, according to size" (Marler, *loc. cit.*).

In practice a quite superficial acquaintance with the behaviour of a species reveals the presence of response-specific stimulus selection mecha-

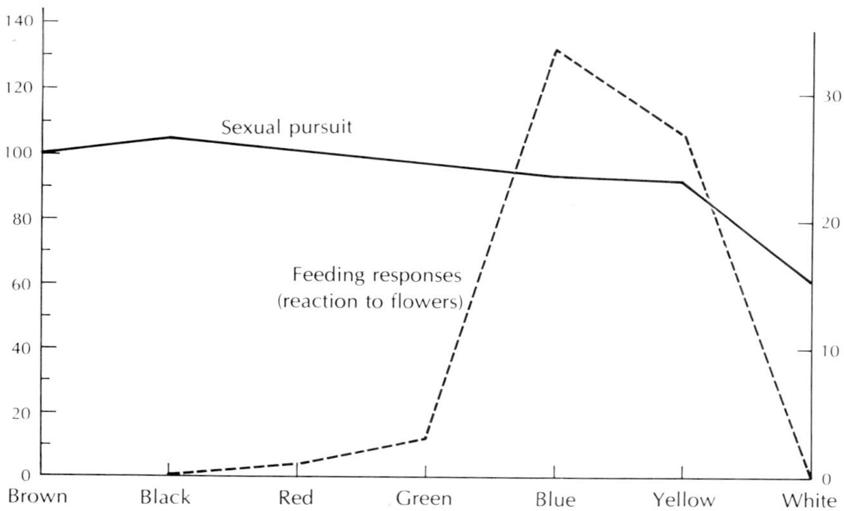

Fig. 6.1. Effectiveness of different colours in eliciting sexual pursuit and feeding responses from grayling butterflies. (After Tinbergen, 1951.)

nisms. Thus, members of nearly every species respond appropriately not only to many objects in their environment but also to many signals from conspecific individuals. Since the responses to these objects and signals are appropriate, each must have stimulus characters specific to the response it elicits (see Chap. 4).

6.1 THE CONCEPT OF THE "INNATE RELEASING MECHANISM"

If stimulus characters are responded to selectively, the organism must have some means by which the selection is achieved. This led to the postulation of a mechanism, termed by von Uexküll (e.g., 1934) and Lorenz (1935) "das angeborene auslösende Schema", and by Tinbergen (e.g., 1951) the "innate releasing mechanism" (IRM). This concept, used to symbolize the specific relations between a particular stimulus character and a particular response, has stimulated research. It can be criticized, however, on two main grounds.

In the first place, the term "IRM" is often used with the implication that the mechanism is specific to a particular response, when the evidence for its existence is based solely on the study of that response. As we have seen, when the stimulus characters relevant to a given response have been identified by comparison of the effectiveness of situations possessing and lacking those characters, they could owe their efficacy to sensory/perceptual mechanisms not specific to the response in question. Two alternatives are thus open. One is to use the concept of "innate releasing mechanism" to cover all mechanisms of stimulus selection: in this case the IRMs for different responses overlap, and each IRM is certainly not a unitary mechanism. Alternatively, as Tinbergen (1955) advocated, the concept can be confined to cases in which particular stimulus characters have been shown to be effective for one response and not for others.

Marler (1961) recognized three types of stimulus selection: that due to the characteristics of the sensory equipment; that due to the perceptual mechanism's greater responsiveness to some patterns of stimulation than others; and finally the filtering which seems to demand the postulation of a selective process specific to one particular response. In the light of recent work, however, the distinction between the first two cannot be upheld. We have already seen that a good deal of "perceptual" filtering may go on in the sense organ itself (pages 103–108), so that a division between "peripheral" filtering in the sense organs and "central stimulus filtering" by the mechanisms of normal "perception" is often misleading. Furthermore, cases are known where a sense organ has become specialized in the service of one particular response, so that it responds only to the stimuli appropriate to that response (e.g., pages 72–73): here, as Marler points out, response-specific filtering is sensory in origin. Thus while the view, that the filtering of incoming information occurs at a number of stages between sense organ and response mechanism, represents an important advance over the earlier view of a unitary releasing mechanism, the classification of those stages given by Marler is an oversimplification.

A second criticism of the "innate releasing mechanism" concept is that comparable selectivity from the total stimulation which impinges on the organism may depend on, or be influenced by, learning: the label "innate" is thus misleading. In addition, as we shall see, stimuli not only elicit responses but also control behaviour in other ways; for instance by orienting the animal or bringing a sequence of behaviour to an end. Selectivity in responsiveness is not confined to stimuli which "release". Thus, while selective responsiveness is a fact, it is not helpful to postulate a unitary "innate releasing mechanism", especially when it is implied that such mechanisms operate only for "innate" behaviour (see Chap. 18), and "releasing" stimuli.

6.2 THE "SEARCH IMAGE" AND "REWARD EXPECTANCY"

Many of the experiments discussed so far have been concerned with stimulus characters effective for all the individuals of a species. It is now necessary to consider some cases where the stimulus characters may be effective only for one animal. In such cases their effectiveness is a consequence of previous experience with similar stimuli.

It is not uncommon for us to miss an object in front of our noses when we are looking for an object with a similar function but different visual characteristics. Von Uexküll, for example, described how he failed to notice a glass carafe on the dining table when he was expecting to see a clay pitcher, and referred to his experience in terms of a "search image" of the pitcher which did not fuse with the carafe.

He also recorded how a hungry toad given one earthworm will subsequently attack earthworm-like objects, but if given a spider it will snap at bits of moss or ants. In a study of the response of jays and chaffinches to the well-camouflaged stick caterpillars of geometrid moths, de Ruiter (1952) found that, once a bird found a caterpillar, it would go on pecking at similar-looking sticks for some time afterwards (see also Beukema, 1968). These cases could be described also as stimulus generalization.

In a posthumous paper, L. Tinbergen (1960) used the concept of a "searching image" in a more precise way as a result of his studies of the hunting behaviour of titmice (*Parus* spp.). He argued that the proportion of a bird's diet made up by a given prey species might be expected to depend on the probability of the bird's encountering that prey while foraging, and thus on the prey density. Field observations of predation by tits on the Hymenopteron *Acantholyda nemoralis* showed that at low (and high) densities the percentage taken fell below expectation. L. Tinbergen ascribed the small amount taken at low prey densities to the insufficient experience of the birds with the prey: with increasing density they learn to find it more efficiently, and the proportion taken suddenly approaches expectation. He thus suggested that the adoption of a "specific searching image" occurred only at a critical prey density and that a certain frequency of encounters was necessary for its maintenance (see also Mook et al., 1960; Holling, 1959, 1961, 1965; Tinbergen et al., 1967).

Gibb (1962) studied the predation of tits on insect larvae which spend the winter concealed beneath the cuticle of ripening pine cones. The birds locate their prey by tapping the cone and then excavate the larvae. Gibb found that predation was very light at low prey densities, increasing abruptly as the density increased: this suggested the operation of something like a "specific search image". In addition, the percentage of larvae taken in areas where the density was very high was below that which might be expected from the extent of predation where the density was lower. Tits look for food while moving continuously from branch to branch or tree to tree. Gibb interprets the unexpectedly low level of predation at very high prey densities as indicating that the birds learnt what prey to expect in a given general area and slackened their search when the expected number of larvae had been taken from the cones, as indicated by the traces of previous attacks. More direct evidence to confirm this possibility would be of great interest.

In these field observations the concept of a specific searching image is used with the implication that the organism is responsive to a particular pattern of stimulation and patterns closely resembling it. However, a little more than straightforward stimulus generalization is implied, because there is also the supposition that the animal does not respond to other stimulus patterns (such as those provided by other food sources) which at other times might elicit the same response. In addition, in our own species at least, there is also evidence that selection can be exercised in favour of members of a class of stimuli which share little in the way of physical characteristics. Had von Uexküll been looking for "something to hold water" he might have found the carafe more easily.

The concept of "reward expectancy" has been used by Tolman and others in a basically similar context. If a monkey is allowed to see some food placed under one of two containers, to which it is subsequently allowed access, it shows some ability to choose the correct one. If a banana is hidden under one of the containers while the monkey is looking, and the experimenter then substitutes for the banana a less preferred food, such as a lettuce leaf, the monkey may reject it and behave as though searching for the banana (Tinkelpaugh, 1928). Similar results have been obtained for rats and chimpanzees (Cowles and Nissen, 1937).

These concepts are, of course, open to some of the objections already raised to the innate releasing mechanism. Indeed searching image comes close to the original concept of an (angeborene auslösende) Schema from which that of the innate releasing mechanism arose (Schleidt, 1962). Furthermore there is another difficulty with these concepts which has not yet been mentioned: they imply the existence of mechanisms in the absence of the stimuli which activate those mechanisms, when the evidence depends almost entirely on what happens when the stimuli are present. Those who take the view that it is legitimate or even desirable to use human experience to interpret animal behaviour will have no difficulty with the view that an animal can have a search image for stimuli not present at the time. Others, however, will feel happier if they do not extrapolate beyond the observable facts, and they are important ones, that animals may show selective responsiveness to particular stimulus patterns and that the effective patterns may be determined by previous experience with them or with related patterns.

6.3 SELECTIVE ATTENTION AND THE SUGGESTED CENTRIFUGAL CONTROL OF AFFERENT PATHWAYS

"Selective attention" usually refers to a limitation of responsiveness by the organism such that some stimulus features control its behaviour over a considerable period, and do so to the exclusion of the other physical stimuli impinging on it at the time. Like the innate releasing mechanism and the search image, therefore, selective attention is used with reference to specificity in responsiveness supposed to be controlled by the internal state: although "search image" is ordinarily used only with reference to stimuli not yet present, there is wide overlap between the concepts.

As Terrace (1966) points out, concepts related to that of "attention", or rather that of "not paying attention", are usually used by psychologists to explain cases in which a stimulus character does not control a response when it might otherwise be expected to do so. For instance Hull (1943) ascribed some cases of lack of correlation between stimulus and response to "behavioural oscillation", a process supposed to be responsible for many of the phenomena usually studied under the heading of attention. Spence (1956) used the term "receptor-exposure acts" in a similar context. In such cases the term may be used as a more or less pure intervening variable—the concept helps to link the measured stimulus to the measured response without any necessary implication about the nature of intervening mechanisms. Spence's term, however, like the "observing responses" of other writers (e.g., Wyckoff, 1952; Zeigler and Wyckoff, 1961), clearly refers to behavioural intermediaries.

In fact peripheral adjustment of the sense organs, so that they are especially effective for one part or category of the stimulus field, plays a large part in the phenomena studied under the rubric of selective attention. For instance, when we attend to a sound, we adjust the position of our heads, and the tensor tympani muscle at the same time sets the tension on the eardrum. The peripheral adjustments in visual attention are more complicated, but include (Horn, 1965):

1. Convergence of the eyes in such a manner that the image falls not only on the most sensitive part of the eye but also on corresponding parts of the two retinae. In the cat, it has been shown that this results in

the activity generated in the two eyes summating at a number of neurones in the striate cortex (Hubel and Wiesel, 1962).

2. The lens of the eye is focused to bring a sharp image on to the retina.
3. The amount of light falling on the retina is adjusted by controlling the diameter of the pupil.
4. Eye movements occur, bringing different parts of the retina into play and thus maintaining activity in the optic pathways.

The precise nature of these peripheral adjustments varies with the species and the sense organs, but it is clear that sense organs are not mere passive transducers of the stimulation which falls on them but are subject to variations in responsiveness. Most vertebrate sense organs receive a sympathetic supply, and in one case at least, the touch receptors in the frog, stimulation of the sympathetic system influences the sensitivity of the sense organs (Loewenstein, 1956; Tucker and Beidler, 1956).

As we have seen, however, selective attention to stimuli with some degree of response-specificity cannot be due solely to general sensory/perceptual mechanisms—processes related to the current motivational state must be controlling the selectivity. There is also considerable evidence that the processes involved in human selective attention are not solely peripheral. This comes in part from experiments in which human subjects are required to select from two or more sets of stimuli arriving simultaneously (Broadbent, 1958). For instance:

1. If a subject is asked questions simultaneously in two different voices and receives visual instructions beforehand about which voice he must listen to, he can subsequently answer questions much better than if the instructions come after he has heard the voices. Since the material discarded depends on experience, this can hardly be a peripheral masking mechanism (Broadbent, 1952).
2. The probability of two messages being received simultaneously depends on the amount of information (in a technical sense) they contain, rather than on their physical characteristics; there is thus more interference between the messages the larger the number of possibilities from which they are chosen. Two messages which contain little information are more likely to be dealt with than two messages which contain much information (Poulton, 1956; other references in Broadbent, 1958).
3. If two speech passages are recorded on the same tape by the same voice, a listener can use his experience of the probabilities with which words or phrases normally follow each other in speech to separate them: this is clearly not sensory (Cherry, 1953).
4. Words of significance to the subject, such as his own name, are more

readily selected from a noisy background than are words without such significance (Howarth and Ellis, 1961).

In selective listening experiments, the selection is not absolute. Cherry (*loc. cit.*) asked subjects to repeat continuously words fed into one ear and ignore words arriving at the other ear. If the voice in the second ear changed during the experiment (e.g., from a man's voice to a woman's and back again), some knowledge of this was retained (see also Broadbent and Gregory, 1963a, b). This again implies that the selective process must be central.

We must, therefore, consider the mechanism of these more central selective processes. It has been suggested by a number of workers (e.g., Horn, 1952; Adrian, 1954) that they may depend on centrifugal influences controlling activity in the sensory pathways. Although centrifugal fibres undoubtedly exist in at least some sensory pathways (e.g., Granit, 1955b; Galambos, 1956), the evidence which earlier appeared to favour their having a role in selective attention has since been challenged. In view of the widespread interest the problem has aroused, it is worthwhile to consider briefly the evidence and the principal areas of contention.

On the one hand, a number of lines of evidence show that centrifugal fibres can control the sensory input. Thus, we have already seen that mammalian muscle spindles contain an intrafusal muscle fibre which has its own motor control through the small ventral root γ efferents. The discharge of the spindle is thus determined both by the state of the skeletal muscle and by the rate of discharge of the γ efferents (pages 49–51 and Fig. 3.13). One consequence of this is that, for any given muscle length, the spindle afferent can show a range of discharge rates depending on the activity of the γ efferents. The activity of the γ efferents is, in turn, regulated centrally, for Granit and Kaada (1952) showed that γ-efferent activity is facilitated by stimulation of the reticular system.

The input from peripheral exteroceptors, like that from muscle spindles, can be modified centrally. For instance, Hagbarth and Kerr (1954) showed that central stimulation changes the volleys of impulses in the spinal cord induced by dorsal root stimulation (see also Hagbarth and Fex, 1959), and Galambos (1956) that centrifugal impulses travelling in the olivo-cochlear bundles can suppress activity in the auditory nerve of cats (see e.g., Desmedt and Monaco, 1961; Desmedt, 1962). Similarly, Jouvet and Desmedt (1956) found that stimulation at some points in the reticular formation would weaken evoked potentials in the cochlear nucleus even in animals whose muscles were paralyzed with a neuromuscular blocking agent (see also Desmedt and Mechelse, 1958). Such studies show a centrifugal control of sensory input, but, of course, they do not necessarily prove that this control operates in selective attention in the intact organism (Livingston, 1959).

In the auditory system, the electrical response evoked by a sound in a number of brain structures, including the cochlear nucleus, has been thought to change when the animal's attention shifts from the auditory to the visual modality. Thus changes in electrical potential in the cochlear nucleus of an unanaesthetized cat, produced by click sounds, were observed to be reduced when the cat was shown a mouse (Hernández-Peón et al., 1956; see also e.g., Hernández-Peón et al., 1957a; Galambos, 1959). Similarly, it was reported that photically evoked responses at the visual cortex, lateral geniculate body, and optic tract were reduced in amplitude when the animal's attention was attracted by non-visual stimuli. Such experiments were interpreted to mean that, when an animal pays attention to stimuli arriving by one modality, non-relevant sensory input in other modalities is reduced near the periphery.

Other workers have not always been able to confirm that selective control of input occurs near the periphery. For instance, Jouvet and Lapras (1959), working with human subjects, claimed that the potentials evoked by flashes of light in the optic radiation were augmented during periods in which the subject was attending to the flashes. But using quantitative techniques to assess the size of the evoked potential, van Hof et al. (1962) found an increase in only one of fifteen subjects.

Even where changes in the size of an evoked potential have been demonstrated, it has been suggested that interpretation of their significance is premature (Horn, 1965). Many of them can be accounted for by changes in orientation of the sense organ or other peripheral changes. Galambos (1960), for instance, has shown that reliable changes in auditory nerve activity are never found in cats attentive to visual, auditory, and other stimuli after their auditory muscles have been cut. In the visual system Naquet et al. (1960) showed that changes in the diameter of the pupil have a profound effect on the evoked potential recorded from the optic tract and lateral geniculate body, so that changes in photically evoked potentials may be due to pupillary changes accompanying fear or convergence. The amplitude of the evoked potential at the visual cortex is less dependent on the diameter of the pupil, but is influenced by whether or not the animal is visually scanning its environment. When it is, the potentials evoked by flashes of light are reduced in size compared with those evoked when the animal is sitting still and not scanning. Thus, if a tone evokes visual searching behaviour, the evoked potential to a flash is reduced. If the tone fails to elicit this behaviour, the evoked potential is not reduced below the control level (Horn, 1960). In addition, there is no simple relationship between the magnitude of the evoked potential and the effective sensory inflow. The former cannot be used unequivocally as an index of the central consequences of a peripheral stimulus (see pages 101–102).

All these factors must be taken into account in interpreting the work of Hernández-Peón and his collaborators. Not all changes in size of an evoked potential in the sensory pathways have to do with selective attention, nor is there any unequivocal evidence to support the view that a reduction in size of an evoked potential in these pathways indicates a selective blockage of unattended signals.

Outside the sensory pathways (e.g., elsewhere on the cortex) the size of the evoked potential does seem to be related to whether or not the subject attends to the stimulus eliciting the response (Chapman and Bragdon, 1964). The significance of this change in terms of neural function is, however, still obscure, since the origin of the potential is not known.

Studies of single cells in the afferent pathways, however, have shown that a small proportion are weakly affected by activity in a heteromodal pathway. For instance, Horn (1963) took recordings from single cells in the cat's visual cortex when the animal was given visual stimuli (flashes) and somaesthetic stimuli (weak electric shocks to the skin). Many of the cells studied responded to both stimuli. In some cases the two types of stimuli had a similar effect, augmenting or reducing the rate of discharge, while in others their effects were opposite. When the stimuli were presented together, the response of the cell was always different from that given to one stimulus alone. Horn cites a number of studies demonstrating a similar effect (see e.g., Akimoto et al., 1961), and gives evidence that the reticular system is involved. If any stimulus, as well as producing a specific response in the sensory pathway appropriate to it, exerts also (via the reticular system) a diffuse influence on the cortex, this would amount to an increase in noise and tend to obscure the pattern of discharge in all pathways. However, the effect, while present, is not strong. Although it could account for the observation that the presence of one stimulus influences the probability of another (see also Worden, 1966) being detected, it seems insufficient to account for the exclusion of information which occurs during selective attention.

On the other hand, the possibility (Horn and Blundell, 1959; Gordon and Jukes, 1962) that centrifugal influence on the sensory input may enhance sensory contrast and hence improve signal-to-noise ratios has recently received experimental support. Dewson (1967) found that the amplitude reduction of the eighth-nerve action potential produced by a click stimulus when the olivo-cochlear bundle was stimulated was not additive with that produced by broad-band noise when both were presented simultaneously: he therefore suggested that the function of the bundle is to reduce signal-to-noise ratios at the cochlea. In support of this view he found that the ability of monkeys to discriminate between human speech sounds in noise was impaired by section of the olivo-cochlear bundle: no impairment was found in the absence of noise (Dewson, 1968).

However, all the phenomena of selective attention cannot be understood in terms of adjustment of sense organs and noise in the sensory pathways. Not only is there no physiological evidence for a blockade of unattended signals in the sensory pathways but also the behavioural evidence indicates that selection can be guided by quite subtle criteria (see pages 123–127). When information is to be discarded by human subjects, it is not discarded at random: the filter can be set to exclude signals defined not only in terms of physical parameters—intensity, frequency, direction—but also in terms of other characteristics (e.g., letters versus numbers). The physiological processes involved are as yet quite unknown.

The magnitude of the problem is apparent if one considers the number of stimulus situations to which an animal or a human can pay attention or respond differentially: it is apparent that the nervous system discriminates a vast number of possible stimulus configurations. The same point arises even more forcibly when one considers the number of situations an animal, and even more so a man, *could* discriminate: it is difficult to imagine that there can be a mechanism for every object that one can recognize (Uttley, 1954), let alone for every one that one *could* (Barlow, 1961).

As a partial solution of this problem, Barlow has pointed out that most sensory messages contain a great deal of redundancy and could be compressed before recognition is attempted. The compression would involve reversible coding in addition to the irreversible filtering we have been discussing so far. If the redundancy in the message was reduced in stages, the complexity of the process would be considerably reduced.

The compression of sensory messages by removal of redundancy involves just those factors by which we recognize it and thus constitutes a step towards recognition. Thus, if small dots are placed on a screen, different random distributions of dots will be difficult to recognize, but an orderly pattern easy. We see here a possible line of attack on the problem of why some patterns are more conspicuous than others (page 81).

As examples of redundancy-reducing mechanisms at the sensory perceptual level, Barlow refers to sensory adaptation (see page 101), lateral inhibition (pages 102–105) and movement-detecting cells (pages 102–108). He thus regards the detection of changes in stimulation, of edges, and of movement which these mechanisms facilitate as being almost incidental consequences of mechanisms whose primary function is the removal of redundancy. In the case of lateral inhibition there is suggestive evidence that the existence of the mechanism is related to the existence in the input of the very sort of redundancy it removes. Thus, lateral inhibition does not occur in the dark-adapted retina but does under conditions of uniform background illumination: it therefore seems to be a consequence of the correlated activity of many receptor cells (Barlow et al., 1957). Comparable mechanisms could exist at higher levels in the nervous system, developing to deal with the specific form of redundancy in the animal's environment, and thus making possible new discriminations. Thus, Barlow suggests that constant features in the stimulus situation cause particular types of redundancy in the sensory messages, and the nervous system then actively modifies its code in order to eliminate them. The rate at which the code is adjusted must, of course, vary at different levels, for perceptual illusions may last seconds, while the effects of wearing inverting spectacles lasts for days. The relationship between Barlow's view of the nervous system actively adjusting its activity to remove redundancy from the input, the von Holst/Teuber/Held models of the comparator unit (pages 96–100 and Chap. 20), and Sokolov's picture of the habituation of the orienting reflex (pages 132–134), will be apparent.

6.4 THE "ORIENTATION RESPONSE" AND RESPONSE TO NOVELTY

When an animal is subjected to a new or unusual form of stimulation, it is likely to show an "orientation response" (or "orientation reflex") in which its sensitivity to stimulation, and especially to the stimulus in question, is enhanced and its readiness for action increased. The orientation response thus includes many of the changes just discussed in the context of selective attention. Originally studied by Pavlov (1955), and recently in detail by Sokolov (1960), it involves first somatic changes (eye movements, body movements, etc.) which result in the sense organs being brought to bear on the source of the stimulus. These are accompanied by vegetative changes (in respiration, heart rate, vascularity, galvanic skin responses, etc.) which are correlates of an increased general readiness to respond. There are also changes in the electroencephalogram, which may have a similar function. Thus Lindsley has produced evidence that the rhythm of electrical changes in the cortex is associated with a cycle of responsiveness to stimulation. In the orientation response slow cortical waves tend to be replaced by irregular activity. In this the different cortical units are presumably no longer synchronized, so that some are always excitable and thus ready to respond to incoming signals (Lindsley, 1961).

In order to understand the further elaboration of the orientation response with continued stimulation, we must consider what happens if a novel stimulus, originally eliciting an orientation response, is repeated at intervals. If the stimulus is without consequence, or has no correlates to which the animal responds, there is first a reduction in those physiological components of the response associated with generalized sensory alerting, so that it becomes limited to components related to the sensory modality of the original stimulus. The "generalized orienting response" thus changes to a "localized orienting response". This localized response then either wanes or becomes replaced by an "adaptive response" which tends to reduce specifically the features of the stimulating situation which elicited the initial response. As a familiar example of the latter, an increase in illumination results in pupillary contraction and thus a diminution in the light entering the eye.

If, on the other hand, the stimulus has consequences or significant correlates, the elaboration of the response depends on their nature. If it becomes a conditioned signal for some other response, then the generalized response becomes changed to a localized one much as before, though the latter shows less subsequent diminution. If the stimulus is of high intensity or painful, the orientation response changes into a defensive response. This may consist in withdrawal or running away, or in a variety of responses which tend to protect the animal from stimulation—immobility, blinking, and so on.

The distinction Sokolov uses among the orientation response (or reflex), the defensive response, and the adaptive response, all of which may be elicited by previously novel stimuli, is a useful one, though of course the categories intergrade. One of the most significant characteristics of the orientation response, and one which distinguishes it from the other two, is the manner in which it wanes with the repetition of the stimulus, provided the latter is without significant correlates. At the behavioural level, this habituation is not merely a consequence of a reduction in sensitivity. Thus, if elicited repeatedly with a light stimulus a little above threshold, the response ceases after a while. If the intensity of the stimulus is then reduced a little, the response reappears. This finding suggests that the waning is specific to the particular stimulus which elicited the response. Another example illustrating this is shown in Fig. 6.2. When a light stimulus subtending 1 deg and focused on a point 40 deg to the nasal side of the centre of the retina was presented at 40-sec intervals, the EEG response faded. If the focus of stimulation was changed by 10 deg, the response reappeared immediately: the waning was thus specific to the point on the retina stimulated.

The stimulus-specificity of the waning of the cortical response has been demonstrated by Sharpless and Jasper (1956). Studying the cortical arousal (see Chap. 9) produced in sleeping cats by sounds, they distinguished two patterns of arousal—a long-lasting one subject to long-term habituation and a shorter one less susceptible. The latter probably depends on the thalamic reticular system. The former is probably mediated by the ascending reticular system and plays a preparatory or protective role: the waning which occurs with repeated elicitation of the response is highly specific to the stimulus (see also Rowland, 1957).

The evidence also suggests that the waning of the orientation reflex is an active process. After the response has disappeared, it can often be elicited by presenting the stimulus when the subject is drowsy, suggesting that the waning is due to an inhibitory process which does not function under these conditions.

On the basis of such findings, Sokolov earlier suggested that the orien-

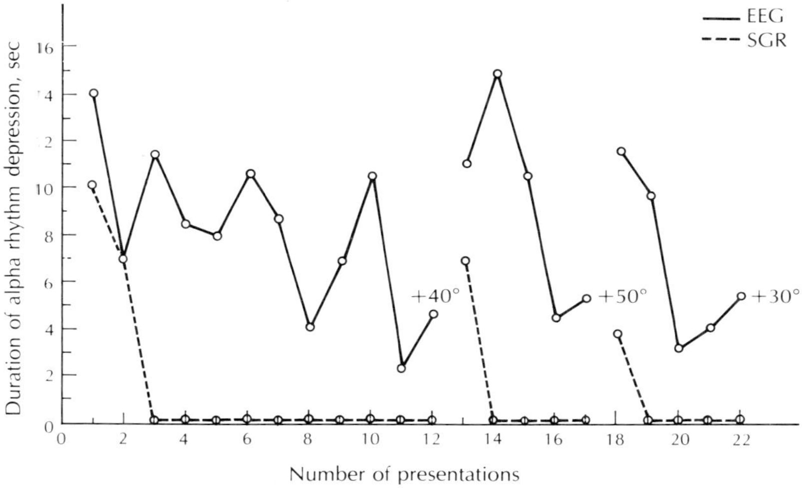

Fig. 6.2. Habituation of α-rhythm and skin galvanic response to stimulation by a light spot focused on a point on the horizontal meridian of the projection perimeter on the nasal part of the retina. (After Sokolov, 1960.)

After the 2nd presentation, the generalized reaction is completely inhibited. The localized orienting reaction (α-depression) is diminished only. A change of position from 40° to 50° (from the central fixation point) evokes generalized reaction (skin galvanic response and increased duration of α-rhythm depression). Repetition of same stimulation evokes decrease in local orienting reflex and complete disappearance of generalized orienting reaction. Change to 30° produces, again, a generalized response and increase of the α-rhythm depression. Habituation is specific to the point of the retina stimulated.

tation reflex appears whenever the sensory input does not coincide with a "neuronal model" previously established in the brain. The force of this suggestion is illustrated by an experiment in which the orientation reflex to a standard sound of a certain duration is allowed to diminish as a consequence of repeated stimulation. If a similar sound of shorter duration is then given, the orientation response appears at the end of the sound, that is, when the sound becomes different from the standard, which had lasted longer. If, on the other hand, a longer sound is given, the response appears at the point at which the stimulus ought to have stopped. Sokolov thus suggested that the neuronal model is set up as a consequence of repeated stimulation, and if at a later time a stimulus is received which does not coincide in all parameters with the model, an orientation response occurs,

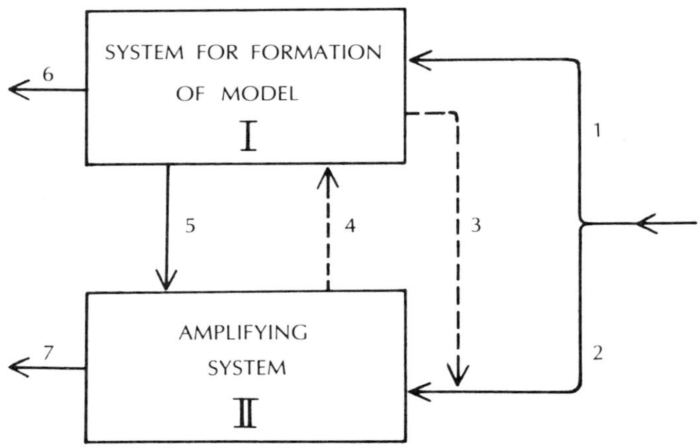

Fig. 6.3. Schema for the orienting reflex. (After Sokolov, 1960.)

I, Modelling system. *II*, Amplifying system. 1, specific pathway from sense organs to modelling system; 2, collateral to amplifying device; 3, negative feedback from modelling system to synaptic connection between collaterals from specific pathway and amplifying system; 4, ascending activating influences from the amplifier upon modelling system; 5, pathway from modelling system to amplifying system (this is the pathway through which the impulses signifying concordance are transmitted from the modelling system to the amplifying system); 6, to the specific responses caused by coincidence between the external stimulation and the neuronal model; and 7, to the vegetative and somatic components arising from the stimulation of the amplifying system.

and the organism's power for obtaining information about the unusual properties of the stimulus is thereby increased.

Sokolov's schema is shown in Fig. 6.3. He supposed that the input reaches both the modelling system (via 1) and the amplifying system (via 2). The latter route can be blocked by the modelling system (3). The amplifying system can increase the discriminatory power of the modelling system (4). If there is coincidence between the input and the model, various specific responses, such as a conditioned reflex, may be elicited (6). If there is no coincidence, impulses pass to the amplifying system (5), and thence to the various vegetative and somatic components of the orienting response.

Such an hypothesis is in keeping with other data on response decrements to alerting stimuli. For instance Bartoshuk (1962a, b), studying heart rate acceleration in human neonates, found that after the response to an 8-sec tone, which changed from low to high frequency, had waned, it could be elicited again by a tone whose frequency changed in the opposite direction. Similarly Engen and Lipsitt (1965) found that after the breathing response of human neonates to a mixture of odours had waned, it could still be elicited by one of the components. Again Broom (1968) found that chicks habituated to a 10-sec light flash repeated with 20-sec intervals showed a behavioural response when flashes of either 5 or 15 sec were interpolated in place of the 10-sec flash (see also Koepke and Pribram, 1966).

The diagram shown in Fig. 6.3 was based on behavioural and autonomic measures and need carry no implications about the particular central nervous structures involved in producing the effects. In fact, electrical correlates of the waning of the orientation reflex, in the shape of a diminution in the potential evoked by a stimulus, have been claimed to occur both in the afferent pathways and more centrally. Hernández-Peón et al. (1957b) claimed that the potential evoked in the cat's cochlear nucleus by an auditory stimulus gradually decreased as the stimulus was repeated. The attenuation of the electrical response, however, usually occurred only after several thousand stimuli had been administered, whereas the behavioural response waned rapidly. Similar results were obtained by Galambos et al. (1956; see also Simons et al., 1966). Later workers, however, have not confirmed this finding. Using very rigorous techniques for the analysis of their data, Worden and Marsh (1963) found no consistent trend in the size of the evoked potentials when stimulation was continued for several hours.

Hernández-Peón et al. (1956) also claimed that the potential evoked in the lateral geniculate body and visual cortex by a flash of light gradually waned. In this case the finding was confirmed by subsequent workers, but it seems likely that it is to be explained by a gradual narrowing of the pupil (Fernández-Guardiola et al., 1961). It thus seems that the evidence for a gradual waning of the evoked potentials in the direct afferent pathways is not substantial or is still disputed (Horn, 1965).

As we have seen, however, there is clear evidence for changes in the electrical activity of the cerebral cortex, as recorded on an electroencephalogram, which parallel the behavioural habituation. When a subject is at rest, with his eyes closed, electroencephalography reveals a high-amplitude rhythm of about 10 Hz. This is replaced by a low-amplitude high-frequency activity when the subject attends to or tries to perceive a visual stimulus.

Novel stimuli are especially effective in blocking the rhythm, but this effect wanes with repetition. Gastaut et al. (1957) and Sokolov (1960) have shown that a novel stimulus initially elicits high-frequency activity from points widely distributed over the cortex, but as the stimulation is repeated, the activity becomes limited to the area of the sensory modality concerned, and then disappears. In this case the course of the waning of the electrical response parallels that of the behavioural response. A further parallel lies in the fact that the waning of the electroencephalographic response is specific to the particular stimulus used. When Sharpless and Jasper (1956) used tones to awaken lightly sleeping cats, they studied the change in the electroencephalogram from the slow waves occurring during sleep to the low-voltage fast activity when the cat awoke. As soon as the sleep pattern had returned, the stimulus was repeated. Gradually the change in electrical activity ceased to occur, and the stimulus had no effect on the EEG pattern or on behavioural arousal. If, after a stimulus of a certain tone had ceased to be effective, another tone was given, it would be effective in arousing the animal. Of particular interest was the finding that, although the EEG changes disappeared with repetition of the stimulus, there was no decrease in the amplitude of the evoked potentials at the sensory cortex, but rather a slight increase.

Sokolov did in fact equate the system for the formulation of the model (Fig. 6.3) with the sensory cortex, and the amplifying system with the reticular formation. On this view, if the input to the cortex matches the neuronal model previously established there, the impulses from the cortex prevent the same input from affecting the reticular formation, to which it otherwise passes along afferent collaterals. He thus supposed that the decrease in the orientation response occurring as a stimulus is repeated results from an active gating process controlled by the output of a matching system.

This aspect of Sokolov's views has been criticized by Horn (1967) on the grounds that response decrement can occur in the absence of the cerebral cortex, and that the phenomena can be accounted for without postulating such elaborate neuronal circuitry. In particular, Horn suggests there is no need to postulate a matching system producing active gating: the phenomena can be accounted for in terms of a "self-generated depression" within the responding system. This view depends on the hypothesis that input of the different stimuli or parameters of the stimulus arrive on different neuronal pathways. Thus in Fig. 6.4*a*, Horn suggests that if the input from cell *A* to *R* attenuates through self-generated depression within it or the pathways which feed it, *R* could still respond to another input, such as that provided by cell *B*. If the input through cell *A* ceased for a while, recovery could occur, so that cell *R* could once again be fired through that pathway.

The two cells *A* and *B* could carry input derived, for instance, from two sensory modalities, from two loci on the retina, or from two cells receiving input from the same locus but differing in directional sensitivity (see pages 109–111). Certain cases require special consideration:

1. *The Elicitation of an Orientation Response by a Weak Stimulus after it Has Become Habituated to a Similar but Stronger Stimulus.*

This appears to pose a difficulty for Horn's scheme, unless different intensities of the same quality of stimulation are carried by different neurones.

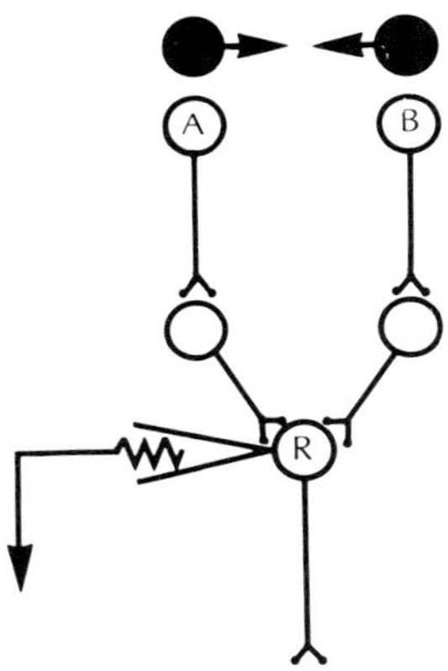

Fig. 6.4. Horn's hypothetical nervous circuits to account for the phenomena of habituation. Each symbol represents either a single cell or a cell system. (Horn, 1968.)

(*a*) Two different inputs *A* and *B* channelled on separate pathways to produce a response through cell *R*. Self-generated depression occurs between *A* or *B* and *R*.

There is in fact some evidence that this is the case: neurones in the cochlear nucleus of the cat and birds show differing thresholds and differing non-linear changes in firing rate with increase in stimulus intensity, such that different cells become active as the intensity changes (Goldberg and Greenwood, 1966; Konishi, personal communication).

2. *Dishabituation*

After the orientation response has waned, it can sometimes be reelicited if an additional stimulus is presented with the test stimulus. Horn suggests that this effect is mediated by the reticular system (see page 223). If the neurones of this system were to hyperpolarize the terminals of cell *A* or cell *F* (Fig. 6.4*b*) the effect might be to increase the amount of available transmitter substance. If the response of cell *F* to cell *A* had failed through synaptic depression, hyperpolarization might permit cell *F* to fire. (It is worth noting that in few cases of behavioural dishabituation is it shown that the response is directed towards the training stimulus rather than the dishabituating one. An even more parsimonious explanation would be to regard the dishabituating stimulus as equivalent to cell *B* in Fig. 6.4*a*.)

3. *The Generalization of the Waning of the Response from One Stimulus to Related Stimuli*

Horn ascribes this to overlap between the neural pathways carrying the two stimuli. Thus in Fig. 6.4*c*, two stimuli of differing frequency affect cells

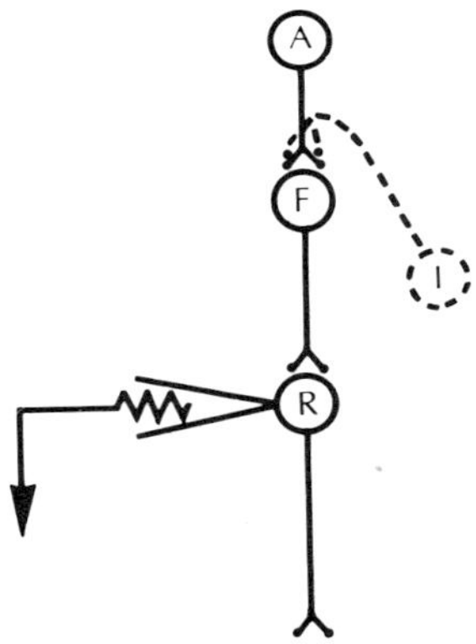

Fig. 6.4. (*b*) Dishabituation. If waning occurs as a result of synaptic depression between *A* and *F*, the response may be elicited again through a hyperpolarizing input on the terminals of cell *A* on *F*.

P and *Q* through different but overlapping groups of cells. If *P* and *Q* both require simultaneous input through three of the *R* cells to fire, *Q* will not fire to the 950-Hz tone when the response to the 1,000-Hz tone has failed through self-generated depression in cells R_1 to R_4.

4. *Time-dependent Effects*

Sokolov's finding that, once the response to a stimulus has waned, it can be elicited once more by a stimulus differing in temporal characteristics from the training one demands, on Horn's hypothesis, that stimuli with different temporal characteristics affect different pathways. Cells with properties which would provide suitable channel-dependent coding for time are not yet known but may well exist.

5. Omitted Stimuli

A more difficult case concerns response to an omitted stimulus from a regular sequence of stimuli to which the organism has become habituated. Cells with the properties necessary for such effects have been found in the

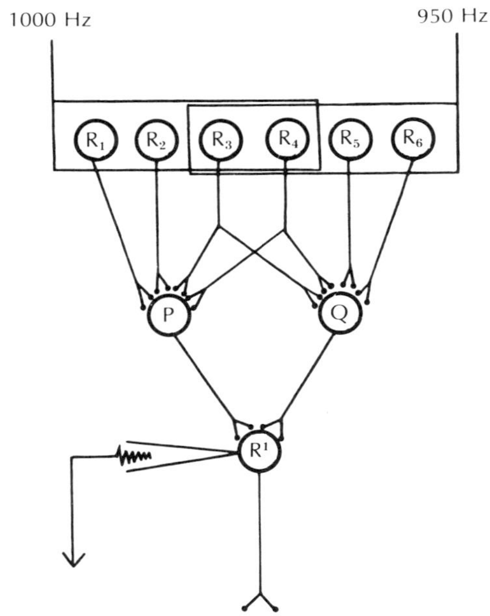

Fig. 6.4. (*c*) Stimulus generalization. The two tones affect cells *P* and *Q*, respectively, by channels which overlap by sharing cells R_3 and R_4.

central nervous system. For instance Vinogradova (1966; see also Horn, 1962) found cells which, responding to a temporal series of stimuli, gave a response when a stimulus was omitted from the series. However, the firing of such cells seems to demand a difference-detecting system. Horn points out that this could be relatively simple: for instance the system shown in Fig. 6.4*d* would detect differences between the activity of cells (or cell systems) *A* and *K*, because *H* and *J* would fire only if the excitatory input reaching them from *A* and *K* exceeded the inhibitory input reaching them from *K* and *A*, respectively: this could only occur if the activities of *A* and *K* differed. Such a system could account for the effects of changing the length of a stimulus if cell *K* had the property of continuing to fire in the manner in which it was originally caused to fire by the training stimulus. Suppose the orientation response fails through depression localized in the terminals of *A* on *F* (Fig. 6.4*e*). If the input into *A* changes, but *K* discharges as before, a difference will be detected by *G*, and *R* caused to fire.

This question has been discussed in some detail because it illustrates both merits and dangers of block diagrams of the type illustrated in Fig. 6.3. It will be noted:

1. The block diagram appears to describe the behavioural facts adequately and parsimoniously.

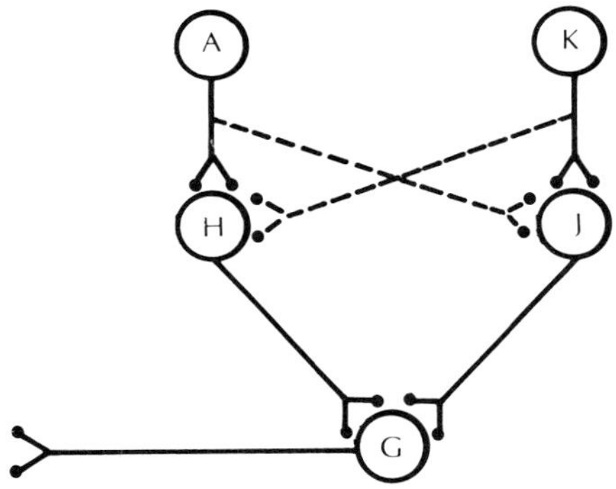

Fig. 6.4. (*d*) A difference-detecting system. Cell *A* has an excitating effect on *H* and an inhibiting effect on *J*, and vice versa for cell *K*. Thus *G* fires only when there is a difference in activity between *A* and *K*.

2. The block diagram is independent of any particular neural interpretation.
3. However, many of the results which the block diagram was designed to explain do not in fact require a comparator mechanism. Although Horn's alternative explanation demands an increase in input channels, it provides great economy in the matching system. This may in fact accord with reality.

With regard to Horn's hypothetical neural mechanisms it must be noted not only that habituation occurs in single cells (e.g., Bruner and Tauc, 1966, a, b) but also that cells acting as novelty detectors (cf. cell *G* in Figs.

6.4*d* and 6.4*e*) have been found in a number of preparations. For instance, Horn and Hill (1966) have found a small region of the midbrain of rabbits where the change in activity produced by a stimulus attenuates rapidly with repetition of the stimulus (Fig. 6.5). The cells in this tectotegmental region were tested with a variety of stimuli, some cells being sensitive to input through only one modality, some to several. Sixty-three of seventy-two cells tested showed an attenuation of response as the stimulus was repeated. As with the electroencephalographic response discussed above, the speed of

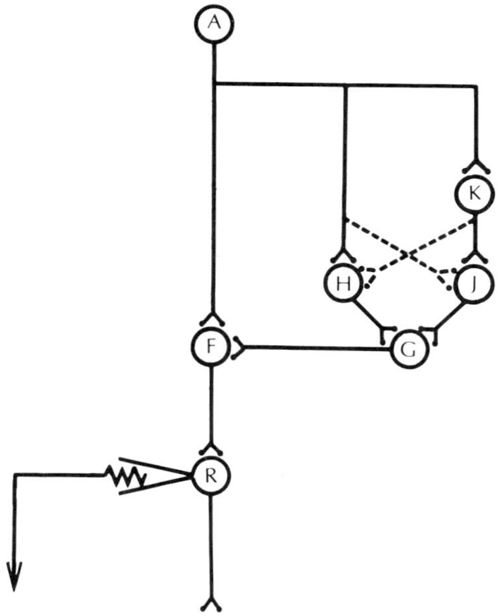

Fig. 6.4. (*e*) System to account for time-dependent habituation effects. The system illustrated in (*d*) is incorporated, and *K* has the property of continuing to produce the output initially elicited from it by the training stimulus.

the waning was comparable with that of a behavioural response, and the waning was specific to the particular stimulus used: when the response of a cell to one stimulus had waned, it would still respond to another novel stimulus. The waning occurred even if the neocortex had been removed. In other species it has been claimed that damage to this tectotegmental region impairs the ability to attend to stimuli, and stimulation improves tachistoscope performance (Horn, 1965). Comparable novelty units have been described in the dorsolateral axial cord of *Octopus*: they habituate to temporal

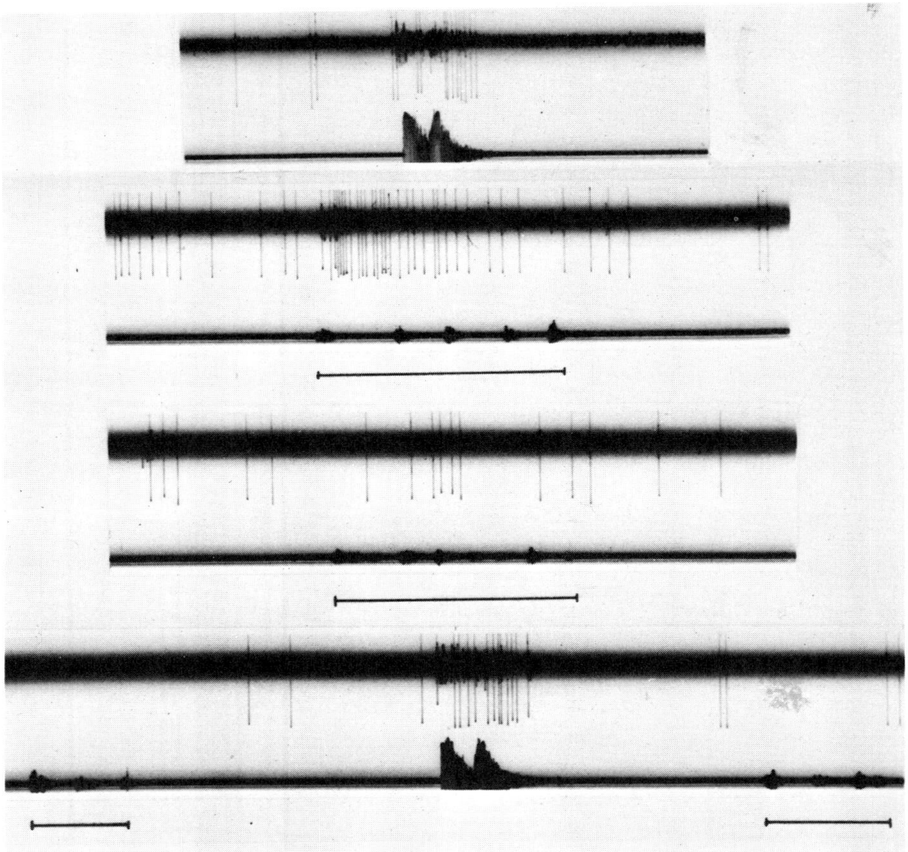

Fig. 6.5. Selective attenuation of response. Rabbit tectal unit. (Horn and Hill, 1966.)

Tactile stimulus (air blown on cutaneous receptive field located on right shoulder) present throughout period indicated by bars. Records 2 and 3 show activity during fifth and seventh presentation, respectively, to tactile stimulus. Between the forty-sixth (record 4 above left-hand bar) and forty-seventh (above right-hand bar) presentations a horn was sounded (middle artifact on lower trace of record 4). The unit responded as vigorously to this stimulus as it had done when it was first delivered (record 1), before any attempt had been made repeatedly to present the puff of air. Note that the brisk response to the horn (record 4) was without effect on the subsequent response to the tactile stimulus. Scale, lower right-hand bar, 0.5 sec.

patterns as well as to maintained stimulation (C. H. F. Rowell, 1966). Other studies are cited by Horn and Hill (*loc. cit.*).

Of special interest in this context are the "novel movement" detectors studied by Horn and Rowell (1968) by recording from the locust tritocerebrum. These were high-order interneurones only weakly responsive to stationary stimuli but strongly responsive to movement in the contralateral field. Most units showed directional selectivity: although movement in either direction along a given axis would elicit a response, one direction was more effective than the other. When a stimulus object was moved repeatedly along a given axis a number of times in succession, the response of the cell waned. Waning often occurred even if several minutes elapsed between successive stimulations, and was specific to the particular axis of movement: if, after waning had occurred, the stimulus was displaced so that it moved along a parallel axis, the response reappeared at a higher level. Full recovery took from 15 min to several hours. Since the background firing rate remained unaltered during habituation, and since the habituation was stimulus-specific, the waning cannot have been due to an inhibitory influence within the cell itself. Horn and Rowell believe the most parsimonious explanation to be a progressive decline in transmitter substance from a store in the synaptic terminals (see also Bruner and Tauc, 1966a, b). The dishabituation which could be produced by stimulation of the contralateral connective (Rowell and Horn, 1968) could be due either to acceleration of transmitter mobilization or facilitation of the post-synaptic membranes.

The conclusion to be drawn from the evidence at present available is, therefore, that only some examples of behavioural habituation may require the postulation of a matching system. On a physiological level, habituation is accompanied by, at most, minor changes in the direct afferent pathways (with the exception of the sensory cortex), but is related to changes in unit activity in the brainstem and to changes in the EEG pattern. Apparently it is with the study of habituation in single cells that future research on the mechanisms underlying the tendency for novel stimuli to capture the attention should be concerned (Horn, 1965). Whether or not such techniques will also provide data relevant to the changes in selective attention to subtle stimulus characters, as discussed on pages 123–131, remains to be seen.

It will already be apparent that the orientation response intergrades with some kinds of exploration in which the animal moves about its environment, either alert to all changes within it, or investigating one object after another, or a combination of these two. The factors of the stimulus situation most conducive to behaviour of these sorts have been reviewed by Berlyne (1960): they include its novelty (the time since it was last encountered or its degree of resemblance to situations encountered previously), complexity, its intensity and contrast, the poverty of the preceding environment, its effective value, and so on. Such behaviour is discussed in more detail later (Chaps. 15 and 24).

6.5 STIMULUS SELECTION—CONCLUSION

As discussed earlier, selection of the stimuli impinging on an organism must occur continuously. In many cases, and especially in simple organisms and the simpler responses of higher organisms, the model of particular stimulus-response connections which are activated whenever the right stimulus happens to come along works well enough. If we are considering the knee jerk or the scratch reflex, the question of stimulus selection raises no problem.

When it comes to the question of responsiveness to particular stimulus characters, we are familiar with man-made devices like locks and telephone dials, which respond selectively to particular stimulus configurations, and some indications of the physiological processes which actually occur in the nervous system are beginning to appear.

However, it cannot be too strongly emphasized that, for behaviour of any degree of complexity, the conception of the organism as passively filtering the stimuli impinging on it becomes inadequate, and mechanical analogies of this type can give a false impression of simplicity.

In the first place, the physical variables are encoded in the sense organs as nerve impulses. The coding may involve the abstraction of higher-order quantities, such as rate of change of illumination (pages 102–115), or rate of change of texture (pages 85–86). At this or a later stage the coding also involves the elimination of redundancy, and this itself may contribute to specificity of responsiveness (pages 130–131).

Second, the effectiveness of the encoded message in inducing a response may in some cases depend on comparison with a preexisting message. The necessity for postulating a process of comparison arises from certain studies of perceptual constancy (Chap. 5), from the response-specificity of selection and the phenomena of the specific search image (page 123), and perhaps from studies of habituation (pages 131–144). Whether comparison at the behavioral level demands postulation of a process of comparison at the neural level is a matter to be determined separately in each case.

6.6 SUMMARY

1. The effectiveness of some stimulus characters is specific to one response or group of responses.
2. The concept of "innate releasing mechanism", which has been used in the past to account for cases of selective responsiveness, can be misleading.
3. Many examples of searching behaviour can be described by postulating a "search image": the animal's further behaviour depends on coincidence be-

tween the stimuli it encounters and the "search image" it has already acquired.

4. There is considerable evidence that the sensory input into the central nervous system is controlled by centrifugal fibres, but the suggestion that these fibres operate in selective attention is controversial. The mechanisms involved in selective responsiveness to particular stimulus characters are thus still not clear.
5. When a novel stimulus is presented to an animal it shows an "orientation response": its responsiveness to the stimulus is enhanced and its readiness for action increased. The further elaboration of this response if the stimulus is repeated depends upon its consequences. If the stimulus is without significant correlates, the response wanes. The properties of this waning can be accounted for on the supposition that the animal forms a "neuronal model" of familiar sources of stimulation: non-coincidence between the stimulation received and a neuronal model leads to an orientation response. The extent to which the postulation of active gating governed by a matching system is necessary to explain the phenomena of habituation at a neuronal level is discussed.
6. Behavioural habituation, as seen in the waning of the orientation response, is accompanied by, at most, minor changes in the direct afferent pathways but is accompanied by changes in the EEG pattern and by changes in the responsiveness of certain cells in the brainstem and other regions of the CNS.

7

ORIENTATION

Some activities of animals are performed without orientation to the environment. At the reflex level, the knee jerk or the ejaculatory reflex, once elicited by the appropriate stimulation, has no special spatial relationship to objects in the environment. Similarly with some examples of more complex behaviour —a particular environmental situation may play a part in the elicitation of a great tit's song, but song is then delivered with no specific orientation.

For the most part, however, animals regulate their activities in spatial relationship with the environment—locomotion is guided towards the proper goals, prey-catching movements are directed towards the prey, courtship towards the female. In this chapter we shall consider how this is achieved.

7.1 THE ORIENTING STIMULI

The first stage in analysing an orientation mechanism is the identification of the stimuli in relation to which the animal is oriented. Often, as in most prey-catching and courtship responses, these are the same as those which elicit the response. As we have seen in Chap. 3, however, in other cases they are not: the flight of a chaffinch to cover is guided by stimuli quite different from the alarm call which elicited the flight in the first place. Another example is provided by the gaping response of nestling thrushes. When the eyes are fully open this is both released and guided by visual stimuli, but the particular visual characteristics differ for the two functions. For an object to release the response three characteristics are important: it must be within certain limits of size, be above the horizontal plane of the nestling, and show movement. The response is oriented towards a protrusion of the object's outline, which has a diameter of about one-third of that of the body: it must also be above the body and closer than it is to the nestlings (see Fig. 4.5). This response illustrates another point—the control of the orientation changes with age. When the nestlings first hatch they are blind, and their gaping is directed vertically: it is released by mechanical or auditory stimuli but oriented by gravity. Not only are the stimuli which release and direct the response different at the two stages, but also the changeover to visual control takes place at different times: there is a period during which the response is released by visual stimuli but still oriented by gravity (Tinbergen and Kuenen, 1939).

Often, as with a moth flying to a candle flame or a male song bird courting a female, the general nature of the stimuli governing orientation is clear, though for a precise physical description experiments with models (as described in Chap. 4) may be necessary. Difficulties can arise when the response is to physical factors outside the range of our own sensitivities. Marine crustaceans (e.g., *Palaemon* sp.) are able to steer a straight course for considerable distances when they are unable to see the sea bed or any other landmarks. Since the course is usually roughly at right angles to the sun, it seems likely that the sun is being used as a compass. Experiments with the species *Mysidium gracilis* have raised the possibility that the sun is not being used directly, but through the polarization pattern. In the laboratory, *Mysidium* tends to orient at right angles to the plane of polarization (Bainbridge and Waterman, 1957) (see Fig. 4.8). Polarized light is known to be used for orientation by a number of other arthropods (e.g., Jander and Waterman, 1960): for instance bees, which normally orient by the sun, can make directional flights when the sun is obscured provided a small portion of blue sky remains, since the light coming from it is polarized in a characteristic fashion.

Another example is the fish *Gymnarchus niloticus*. In avoiding obstacles as it swims backwards, and in identifying its prey from a distance, it shows powers of orientation beyond those which could be mediated by its lateral-line system or its poorly developed eyes. In fact, these fish produce a weak but regular electric pulse: by means of receptors sensitive to changes in an electric field they can detect objects whose conductivity differs from that of water (Lissmann, 1958; Lissmann and Machin, 1958) (see page 74 and Fig. 4.9).

7.2 CLASSIFICATION OF ORIENTING MOVEMENTS

The ways in which animals become oriented with respect to environmental stimuli are so diverse that it is not possible to classify them in any wholly satisfactory way. However, certain categories have proved useful over the years, especially for the orientation responses of invertebrates. These have been set out in detail by Kühn (1919) and Fraenkel and Gunn (1940), whose system incorporates many of the ideas of earlier writers. The following discussion is based largely on their work and that of Kennedy (1945), though it diverges in some details.

Kinesis

The first distinction, between kinesis and taxis, depends on whether the animal's body is oriented with respect to the source of stimulation. In kinesis, which we shall consider first, it is not. The stimulus produces merely a change either in speed of movement or in rate of turning, and this alters the position of the animal in relation to the source of that stimulus.

Woodlice (*Porcellio scaber*) tend to aggregate in moist places. This is due to an influence of the relative humidity on the locomotory activity of the animals: as the humidity increases, the proportion of animals which are motionless increases. The animals therefore become aggregated in the damper regions. A similar aggregating effect occurs when the change in conditions produces merely a retardation of movement, in the same way that cars, travelling on through journeys, become concentrated in towns where their average speed is low. Such an influence on linear velocity is known as "orthokinesis". Similar effects are known in vertebrates: for instance the swimming activity of the ammocoete larva of the lamprey varies with the light intensity (Jones, 1955).

The term "klinokinesis" is used to refer to cases in which the rate of turning is affected by external stimulation. The planarian *Dendrocoelum* tends to aggregate in the darker parts of its habitat: although locomotion continues, the animals tend not to leave dark regions once they reach them (Ullyott, 1936). This is a consequence of two factors. First, the animals change direction frequently as they swim along, and the rate of change in direction is increased by an increase in the light intensity (Fig. 7.1). Second, this response to an increase in light intensity is subject to fairly rapid adaptation: a given light intensity therefore induces a higher probability of turning in an animal not yet adapted to that intensity and moving up the gradient than in an animal moving down the gradient, which has already started to become adapted to even higher intensities. Paths towards the darker region thus tend to be longer than paths away from it, and the animals aggregate there (Ullyott, 1936).

Although this experiment of Ullyott's has been widely quoted, it has apparently not been replicated. However, other examples are known. The human body louse finds its host, in part, by klinokinetic responses to temperature, humidity, and smell. When adapted to an unfavourable environment the louse goes approximately straight (Fig. 7.2). If conditions become more favourable, it continues on this way. But if, having experienced these more favourable conditions, it enters a less favourable zone, it starts to turn. Thus, by keeping straight when the stimulus remains constant or becomes more favourable but turning when it becomes less so, it gradually reaches the most favourable conditions (Wigglesworth, 1941). This case differs from the previous one in that the rate of turning decreases with *increase* in intensity of stimulation, whereas in *Dendrocoelum* the rate of turning decreases with *decrease* in light intensity. In both cases the responses bring the animal into favourable conditions.

Ullyott, *loc. cit.*, and Fraenkel and Gunn, *loc. cit.*, regard adaptation as essential for klinokinetic displacement along a stimulus gradient. Without it, they suppose the animals will turn more at one end of the gradient than at the other, but there will be no resultant movement up or down. Their

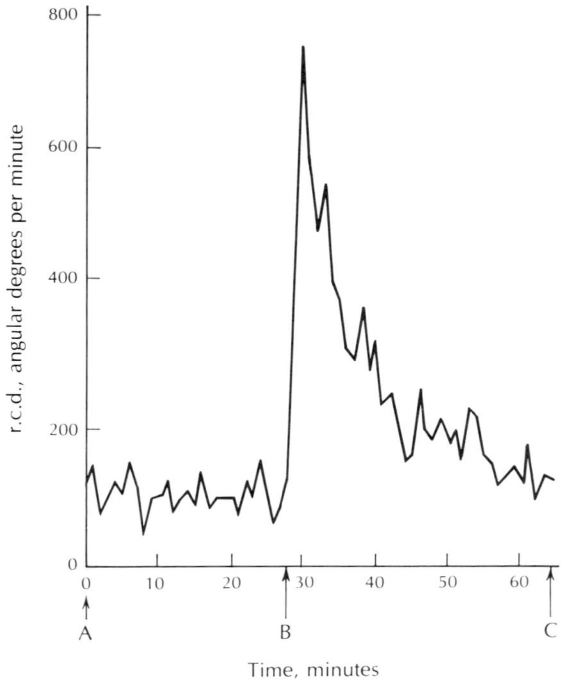

Fig. 7.1. Relation between light intensity and rate of change of direction (r.c.d.) of Dendrocoelum. (After Ullyott, 1936.)

A–B in darkness; *B–C* with light on.

argument here is not convincing and is valid only under certain conditions. Consider an animal which shows no adaptation, so that the rate of turning is a function of stimulus intensity. Suppose that it turns more as the stimulus intensity increases. The probability of turning of an animal which moves up the gradient will be increasing, while that of an animal which moves down will be decreasing. On average, paths down the gradient, or with a component down the gradient, will tend to be longer than paths starting from the same point but with a component up the gradient. The animal will thus tend to shift down the gradient, even though it shows no adaptation.

There are conditions, however, under which this shifting will not occur; namely if the probability of turning is so high that the animal turns so soon that the difference in probability of turning produced by moving up or down is negligible. This is more likely to happen if:

1. The intensity of stimulation changes only slowly with distance.
2. A given change in stimulation produces only a small change in rate of turning.
3. The movement over the ground is slow.
4. The frequency of turning is high.

In such cases there will be less tendency to shift down the gradient,

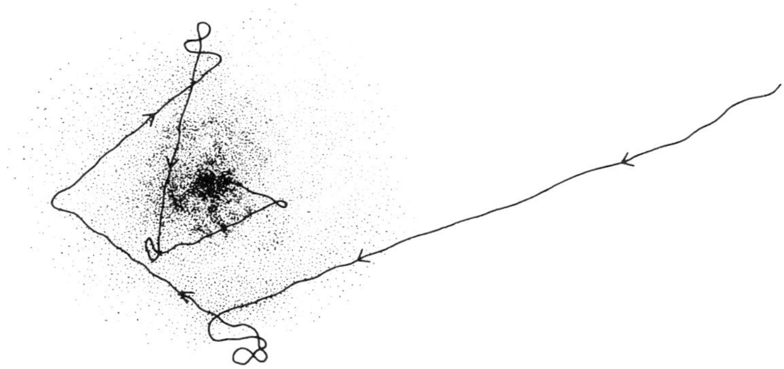

Fig. 7.2. Hypothetical track of a louse approaching the centre of some favourable diffuse stimulus. (After Wigglesworth, 1941.)

and the animal will be likely to remain pursuing a convoluted path in one particular region. Indeed, there may even be a tendency to shift up the gradient, for the following reason. Since the probability of extensive progression in any given direction will be smaller the greater the rate of turning, the animals will tend to stay longer in regions of high stimulus intensity. In a population of animals, individuals which wander into such a region will tend to stay there, and the population may tend to shift up the gradient.

Thus in the absence of adaptation there could be a klinokinetic shift up or down the gradient, depending on the relations among the parameters enumerated above.

The occurrence of adaptation complicates the issue still further. If adaptation is immediate, then the stimulus intensity will be effectively constant everywhere and the rate of turning will not alter along the gradient. No klinokinetic shifting can then occur. However, the tendency to shift down the gradient which, as indicated above, can occur with no adaptation, will be enhanced by some degree of adaptation, as Ullyott suggested (see page 149). Furthermore, the trapping effect that occurs when the rate of turning is high relative to the rate of movement is reduced by adaptation: as adaptation occurs, the probability of turning decreases and the animal makes the longer paths necessary for it to escape. Davenport et al. (1960) were thus correct in supposing that slow adaptation could assist a klinokinetic aggregation near the region where the rate of turning was greatest, though they omitted to specify the other relationships necessary for this to occur. A mathematical treatment is given by Patlak (1953).

In practice changes in environmental conditions often produce both a change in speed of movement (orthokinesis) and in rate of turning (klinokinesis). Surtees (1964a, b) points out that some separation of these effects can be obtained and many of the difficulties inherent in measurements of "rate of turning" obviated if the mean distance travelled in unit time and the resultant linear displacement are assessed. The ratio between these two parameters (distance travelled/linear displacement) is a measure of irregularity of pathway, having a value of unity for a straight path. If the distance travelled per unit time falls, but the ratio is unchanged, the mechanism is orthokinetic. If the ratio changes, however, klinokinesis must be involved.

Jennings's (1906) description of the manner in which *Paramecia* become aggregated in a drop of acid water is well known: the animals reach the drop by chance when swimming in random directions but tend to turn when they reach the boundary on the way out. This "phobotaxis" clearly bears at least a superficial relation to klinokinesis, the change in stimulation as less favourable conditions are approached producing a sudden rather than a gradual increase in probability of turning.

Taxes and Transverse Orientations

In kineses there is no orientation of the body with respect to the source of stimulation: the change in spatial relationship to that source is achieved through quantitative variations in the activity of the animal. In taxes, by contrast, the animal's body takes up a particular direction: the taxis may be combined with locomotion so that the animal moves towards, away from, or at a fixed angle to the source. The term "taxis" is sometimes (but see page 159) restricted to cases in which the stimuli are simple and, in the case of visual stimuli, form vision is not involved.

Klinotaxis

In klinotaxis the orientation of the body does not necessarily require a receptor itself capable of discriminating the direction of the source of stimulation. Provided the receptor is not equally accessible to stimulation from all directions, the animal can compare the intensities of stimulation on the two sides of its body by turning the receptor or receptors first one way and then the other. The comparison is, of course, successive and enables the organism to turn until the two sides are stimulated equally. An example is the response of maggots of various common flies (*Musca, Calliphora, Lucilia*) to light during the period preceding pupation. The photosensitive structures are at the anterior end of the body and are shaded from light coming directly from behind. As the animal crawls, it swings its head first to one side and then to the other. A comparison between the intensities of stimulation falling on the two sides of the body determines the extent of the lateral movements and also their relation to the contractions which bring up the rest of the animal, and thus sets the animal's course (Fig. 7.3).

By observing the behaviour of maggots kept under conditions of dim diffuse light, Gunn (cited by Tinbergen, 1951) showed that the turning depends on successive comparisons between the stimulation on the two sides of the body. As the maggot turned its head to the left side, a second light, hanging centrally above the animal, was switched on momentarily. This was repeated every time the head moved to the left, but not when it moved to the right. As a result, the animal circled towards the right. If the second light was switched on momentarily only when the head was to the right, the circling was to the left. It thus seems that orientation in a horizontal beam is not due to a direct response to the direction of the beam but to successive responses whose strength varies with intensity of the incident light.

In practice, klinotaxes are relatively rare as responses to light, for most animals have eyes capable of discriminating the direction of light directly. They are, however, commonly used as a means of orienting to chemical stimuli; for instance, by the flatworm *Planaria* in orienting to bait (O. Köhler, cited by Fraenkel and Gunn, *loc. cit.*). If a small piece of bait is placed in a dish containing a planarian, the animal soon starts to move about. It then crawls along a convoluted path, tending gradually to approach the bait. This may be a form of klinokinetic reaction. When it gets nearer to the bait it crawls straight towards it, swinging the forepart of its body from side to side: this is apparently a phase of klinotaxis. This more efficient method of orientation may not be possible when the animal is farther from the bait, because there the gradient of stimulation is not great enough.

Tropotaxis

In klinotaxis the animal makes, in effect, successive comparisons of the stimulation impinging on its receptors as they are turned from side to side.

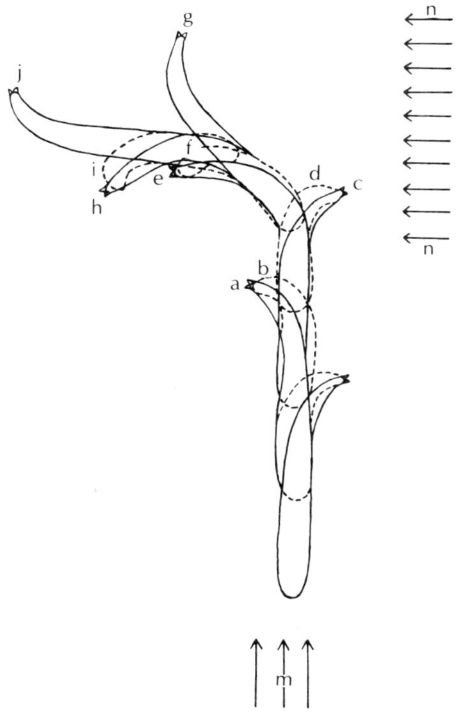

Fig. 7.3. Klinotaxis in the maggot. (After Mast, 1911.)

At *d* the light *m* was switched off and *n* switched on.

Tropotaxis differs in that the comparison is simultaneous: it thus depends on bilaterally symmetrical receptors arranged in such a way that they are stimulated unequally when the animal is not oriented towards or away from the source of stimulation. This simultaneous comparison makes possible orientation to or from a localized source without the lateral swings or wavy movements necessary in klinotaxes: turning occurs if at any instant the intensity of stimulation on the two sides is not the same.

In theory, this type of orientation can be recognized in two ways. First, unilateral elimination of the receptors in a diffuse field of stimulation will cause the animal to turn continuously towards one side: however much it turns, the intensity of stimulation on the two sides cannot be equalized, and so the circus movements continue. Second, if the animal is influenced by two sources of stimulation whose fields cross, it will pursue a course towards or away from a point between the two sources: it seems to move in such a manner that the tendencies to turn to right and to left are just in balance.

An example is given by the crustacean *Armadillidium*, which often behaves in a phototropotactic manner. If blinded unilaterally it makes circus movements (Fig. 7.4), and intact animals usually (though not invariably) go between two lights (Fig. 7.5). Cases of pure tropotaxis, however, are not easy to find, for it is often combined with telotactic responses: these are considered below.

The "dorsal light reaction", by which many aquatic (and aerial, e.g., Goodman, 1965) animals keep the dorsal (or ventral) side uppermost, was distinguished from tropotaxis by Fraenkel and Gunn because the animal moves, if at all, in a plane at right angles to the source of stimulation, rather than towards or away from it. Tinbergen (1951), however, has pointed out that this distinction is based on a failure to separate the orientation component from the locomotion which may accompany it. As discussed earlier (page 146–147), a movement may be oriented by stimuli quite different from those which elicit it, and the two can be analysed independently. In some cases the orientation is independent of any form of locomotion, but this need not imply that it differs from otherwise similar cases which are accompanied by locomotion.

The dorsal light reaction does, in fact, show the characteristics of tropotaxis. The fish-louse *Argulus* sp. normally swims with the dorsal surface to the light. If the light comes from below, it swims dorsal surface down. If the source of light is moved, it either rolls or turns a front somersault to maintain its orientation. If one eye is destroyed it makes continuous rolling and spiralling movements, showing that the dorsal light reaction normally depends on a balance of stimulation through the two eyes (Herter, 1927).

In the fish *Crenilabrus* the orientation is controlled by both a dorsal light reaction and by the labyrinth. If the labyrinths are removed, the fish orients with its dorsal surface towards the light. If the labyrinths are present, however, the position is a compromise controlled partly by the labyrinths and partly by the dorsal light reaction: the nature of the compromise

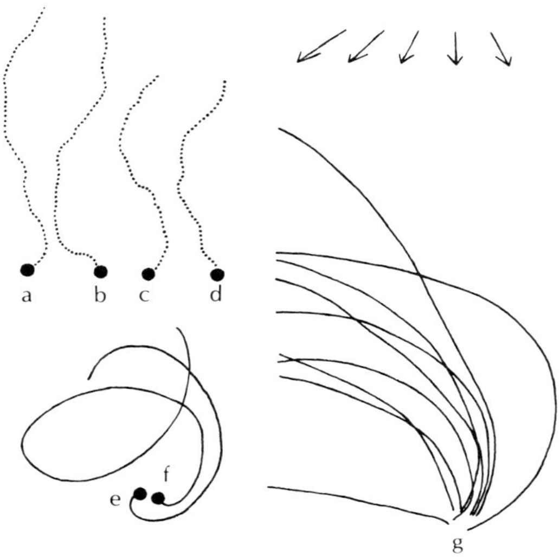

Fig. 7.4. Tropotaxis tracks of photopositive Armadillidium blinded on the right side. (After Henke, 1930.)

a–d, in darkness; *e–f*, with light overhead; *g*, in a beam whose direction is indicated by the arrows.

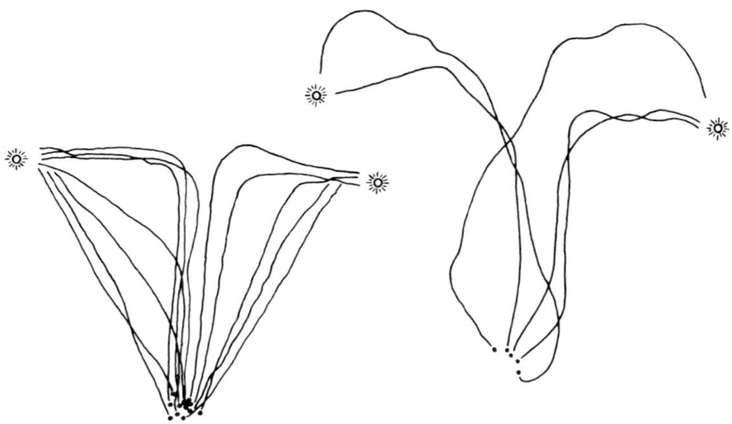

Fig. 7.5. Tropotaxis tracks of photopositive Armadillidium with two equal lights. (After Müller, 1925.)

depends on the intensity and direction of the light (Fig. 7.6). If one eye is removed, the fish at first rolls towards the intact side. Comparable phenomena in amphibians have been studied by Jahn (1960).

There seems, then, to be no justification for Fraenkel and Gunn's distinction between tropotaxis and the dorsal light reaction. However, a convention to restrict the term "taxis" to cases involving locomotion is current in the literature: we do not, for instance, usually describe standing

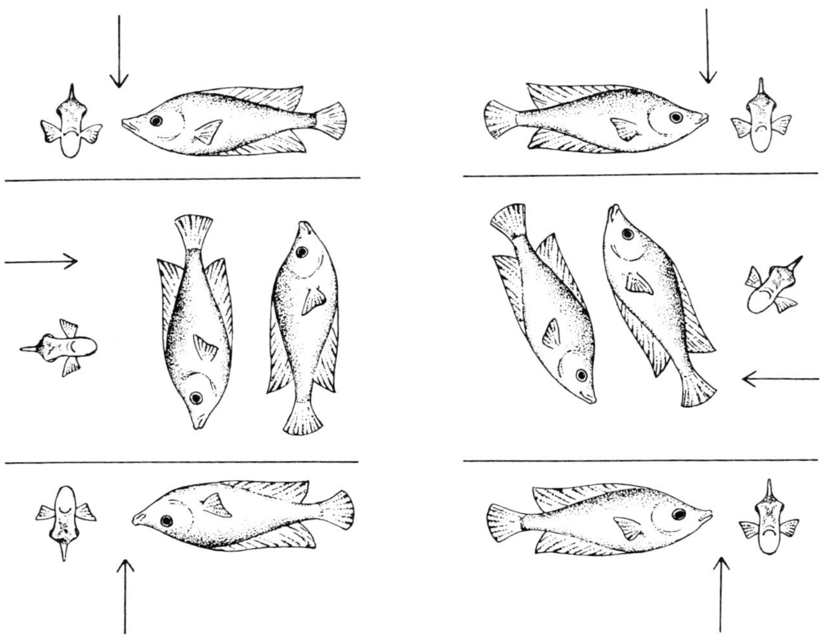

Fig. 7.6. Dorsal light reaction to beams (indicated by arrows) coming from different directions. (After von Holst, 1935*d*.)

Left, labyrinthectomized *Crenilabrus; right,* normal *Crenilabrus.*

upright as "geotaxis". The means by which animals achieve particular patterns of spatial distribution with respect to their environment is of great ecological significance, and it is here that the study of "taxes" has been most actively pursued. Since the argument is solely terminological, there seems no point in discussing it further.

Until recently, it was thought that most tropotactic responses of insects could be understood in terms of variations in the "sensitivity" of the ommatidia. Consider, for example, the orientation of *Eristalis* when exposed to two lights of different intensities. The fly at first orients towards a point between the two lights, and nearer the stronger light than the weaker. On the view that the orientation is due to a balance between the turning effects produced by the light shining on the two eyes, the stronger light shining on, say, the left eye is balanced by the weaker light shining on the right eye. Since the course is nearer to the strong light than the weak, the ommatidia stimulated by the latter are more posterior than those stimulated by the former. Thus the turning tendency produced by the strong light falling on the more anterior ommatidia is balanced by that produced by the weaker light on the more posterior ones. The posterior ommatidia must, therefore, be more sensitive than the anterior ones. By varying the conditions it is possible to map out the sensitivity of the ommatidia: Dolley and Wierda (1929) found that the sensitivity increased fifty-five times over a sector of only 23 deg of the eye. Sensitivity is, of course, an ad hoc hypothesis invoked to explain the experimental facts and is not to be thought of as a property of the ommatidium itself but of the whole turning reaction mediated by it. It has, however, proved useful in accounting for a considerable wealth of detail derived from experiments in which orientation to two sources of stimulation, or orientation of unilaterally blinded animals, was studied. But even in laboratory experiments detailed observations have shown it to be inadequate. Studying the insect *Notonecta,* Lüdtke (Fraenkel and Gunn, *loc. cit.*) found that the variation and plastic adaptation of the orientation required the postulation of variables additional to those involved in a sensitivity map: this shows the explanatory system to be insecure. An alternative method of approach to these problems is discussed later in this chapter.

Telotaxis

Both klinotaxis and tropotaxis depend on balance: the animal orients itself by equalizing the intensity of stimulation on the two sides of the body. Telotactic orientation does not depend on simple balance: if there are two sources of stimulation operating through the same modality the animal orients towards one or the other and not in an intermediate direction. This suggests that the influence of one of the stimuli must be inhibited. An example is shown in Fig. 7.7: the hermit crab orients either to one source or the other. Bees behave similarly: furthermore, if blinded unilaterally, they first show circus movements but then orient towards the light source. The circus movements can be explained as being due to light entering the very sensitive posterior ommatidia: these soon become adapted, and the animal then fixates the source with its anterior ommatidia and moves towards it. However, adaptation in the ommatidia cannot account for the inhibition of the response to objects in the visual field other than those to which the insect orients. This question has been considered by Wallace (1962), who believes that the inhibition here is not on the sensory side but involves mutual exclusion of responses at the motor level.

Kühn (1919) used the term "telotaxis" to describe responses such as the orientation of a dragonfly towards its prey. Fraenkel and Gunn prefer to restrict it to cases in which the stimulus does not involve form vision, on the grounds that it could otherwise be used to include responses which involve complex mechanisms such as "motives, learning and so on which are generally thought to be absent in the elementary reactions which we are considering". The argument here depends on the status accorded to the classificatory system: if it is regarded as concerned with mechanisms, then Fraenkel and Gunn are certainly correct. But since this system is based

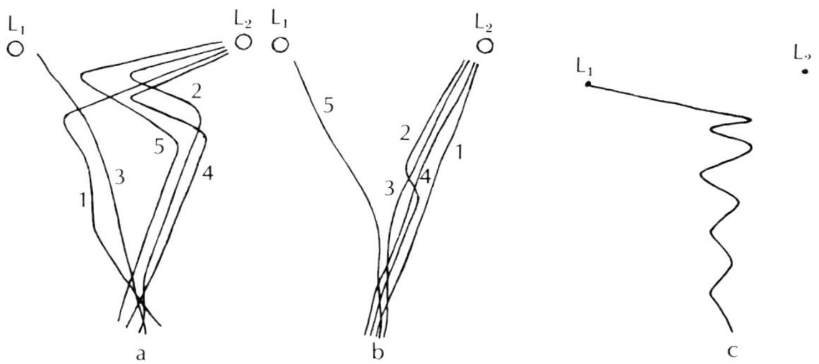

Fig. 7.7. Telotaxis. A two-light experiment. (a), (b), Tracks of hermit crabs; (c) tracks of an isopod Aega. (From Fraenkel and Gunn, 1940.)

for the most part on descriptive observations of the behaviour of the whole animal, it is preferable not to make assumptions about similarities in physiological mechanism between examples of any of the classes of kinesis or taxis. Tinbergen (1951), like Kühn, includes responses to configurational stimuli in this category; for instance the orientation of the gaping response of nestling thrushes towards the parent's head (pages 146–167).

Menotaxis

In the classic cases of telotaxis, the animal orients directly towards or away from the stimulus. Menotaxes, or light compass responses, involve orientation at a constant angle to the direction of a source of stimulation. The path of a homing ant is guided, in part, by the direction of the sun; if the apparent direction of the sun is slowly changed by the use of a mirror, the ant changes course accordingly (Schneirla, 1933).

That menotactic orientation does not depend on a totally independent mechanism but is in some way superimposed on a tropotactic or telotactic response is suggested by Jander's (1957) observations on the ant *Formica rufa*. The menotactic orientation this species uses while foraging develops through a gradual change from a tropotactic response: the continuing in-

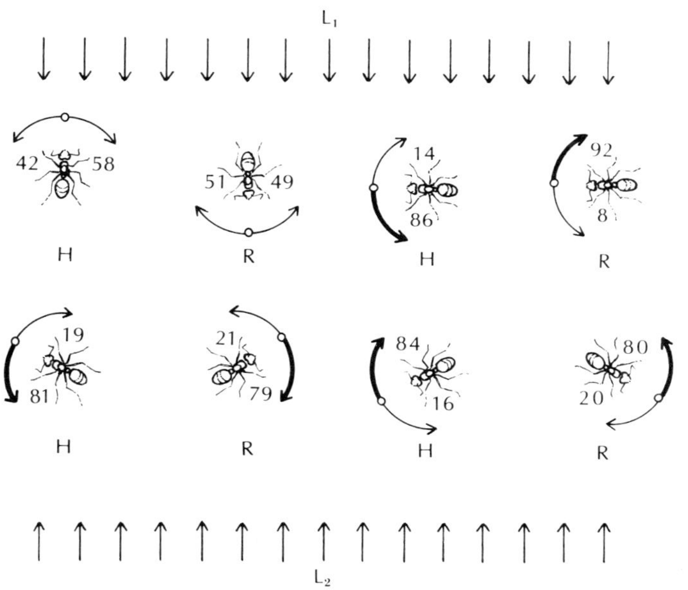

Fig. 7.8. Phototaxis and menotaxis in the ant Formica rufa. (From Jander, 1957.)

The figures indicate the number of times the ants turned to left and to right when the direction of the light changed from L_1 to L_2. *H* indicates outward journey; *R*, homeward journey.

fluence of the latter accounts for the mistakes the ant makes during the early stages of acquiring the menotaxis. Furthermore, once acquired, the menotaxis is in some degree independent from the original tropotaxis: if the sign of the tropotaxis is changed from positive to negative (for instance when the ant starts to return to the nest), the nature of the menotaxis is not altered.

Yet another line of evidence showing that the tropotactic response has a continuous independent existence even while the ant is oriented menotactically can be obtained by switching off the light source by which the animal was orienting and turning on a second source at 180 deg to it. If the ant was initially oriented towards or away from the first light, it is equally likely to turn clockwise or anticlockwise to the second (Fig. 7.8). If the ant had been menotactically oriented to a light in one of the front quadrants, then it usually turns towards the second light source in order to gain its new orientation (bottom left), but if the light was initially in one of the posterior quadrants, it turns away (bottom right). If the animal was originally oriented laterally to the light, it turns towards the light if on an outward expedition and away if returning to the nest. Thus, in each case the direction of the turn made to regain the menotactic orientation seems to be influenced by the sign of the phototaxis (Jander, 1957).

Such experiments suggest that the acquired tendency to orient with respect to a stimulus to one side of the axis is superimposed on a preexisting tropotactic response, in such a manner that the angle between the stimulus source and the body axis is assessed and the turning tendency depends not, as previously, on the difference between that angle and 0 deg (or 180 deg) but between it and the required angle between the body axis and the stimulus (the Sollwert).

Mnemotaxis

As we have seen, Fraenkel and Gunn restricted their classificatory system to cases which did not involve configurational stimuli. Even in invertebrates, however, some orientation responses depend on complex stimulus situations. For instance, the hunting wasp *Philanthus triangulum* uses a number of landmarks simultaneously when returning to its nest (van Beusekom, 1948). Tinbergen and Kruyt (1938) carried out an extensive series of experiments in which the wasp's use of landmarks round the nest entrance was studied. When the wasp had had opportunity to learn the initial layout, the landmarks were rearranged in various ways. The results indicated that the insect's return was guided by the relation of the nest entrance to the whole configuration of landmarks and did not depend on stimulation of special parts of the retina by particular landmarks (see also van Iersel and van den Assem, 1965).

Kühn (1919) distinguished cases such as this as "mnemotaxis", in part because of the configurational nature of the stimuli involved and in part because these stimuli must be learnt. Of course, learning may also be involved in telotaxis and menotaxis, and Tinbergen (1951) emphasizes that these distinctions are of degree only. The apparent simplicity of telotaxis may be only illusory, since it also involves a configurational visual field even though one point in it is singled out. The relations between these categories will be mentioned again later in this chapter.

Utility of the Kinesis-taxis Classification

Many additional examples given in Fraenkel and Gunn's book show how this classificatory system brings order into diverse examples of orientation. This is valuable not only for purposes of description and classification, but also for relating the nature of the orientational responses with the complexity of the sensory equipment and for considerations of their phylogeny (Jander, 1965). Here it is necessary to make two reservations. First, the categories have been worked out largely during laboratory studies on relatively simple animals in simplified situations. In the field, with higher animals, and even in laboratory studies of simple animals in more complex situations, the categories become less useful: the responses are not readily classifiable into one or the other category, and different types of orientation seem to occur together or in close succession, so that the distinctions between the categories fade. This, of course, does not invalidate the system but merely places limits on its usefulness. Second, it is necessary to exercise care before making assumptions about the extent to which the responses of any one category share similar mechanisms. Assignment of a particular case to one or other category depends on an assessment of the role of individual receptor organs, but not on an analysis of all that happens between stimulus and response: it thus tends to oversimplify. The distinctions made in the Fraenkel and Gunn classification are useful insofar as they refer to different types of spatial manoeuvre but can be dangerous if taken to imply types of physiological mechanism.

7.3 ANALYSIS OF ORIENTATION IN TERMS OF CONTROL SYSTEMS

In recent years a considerable advance in the understanding of orientation mechanisms has come not only from attempts to extend the analysis to a neuronal level but also from applications of the methods used for the study of control systems in engineering. We have already seen some preliminary examples of this approach in Chap. 5. To achieve orientation, the animal must have a mechanism which receives information from the outside world, a means of changing its orientation, and a mechanism for relating the two. These can be linked in a variety of ways: it is possible to define the alternatives in terms of information flow diagrams and then to decide between them experimentally. In a similar manner the functioning of each of the subsystems can be analysed in terms of its subsystems and so on. The approach thus aims at specifying the functions of the physiological mechanisms underlying the orientation: a knowledge of the sorts of jobs these mechanisms do should aid in their identification (e.g., von Holst and Mittelstaedt, 1950; von Holst, 1954; Hassenstein, 1959; Mittelstaedt, 1960, 1961, 1962, 1964; Machin, 1964; Kalmus, 1966; Wilkins, 1966a, b).

Open Systems

If the movement produced by the effectors has no consequences which influence the further effect of the external stimulus, the receptor-effector system is said to be an open one. Any orientation mechanism in which the stimulus is over before the movement is completed must be of this type. Male fireflies (*Photinus*) orient fairly accurately to brief flashes of light. Since the fireflies may turn correctly even though they do not start until after the flash is over (Mast, 1912), the output of the system (the change in the orientation) cannot influence the input (the angle between the flash and the body's axis): the system is thus of the open type.

For a basically similar reason, the rapid strike of the forelegs by which mantids capture their prey must also be based on an open system. Information about the position of the prey is received by the compound eyes: the output of the system is the deviation of the strike from the axis of the body. In theory, information about the direction of the strike relative to the prey (i.e., the error) could be fed back into the system by the eyes while the strike is in progress but, in practice, the delay before this could produce any effect would be greater than the time required to extend the forelegs. The extension of the forelegs thus cannot influence the input, and the system is an open one (Mittelstaedt, 1962; see also Maldonado, 1964; Messenger, 1968).

An example of a slightly different type of open system, involving a "mesh", is shown in Fig. 7.9. Here the external change can both affect the receptor organs and produce a physical displacement. The effector, however, produces a compensatory movement which precisely counteracts the deviation which would have been produced by the input. The man standing upright in the small boat sees a wave approaching from abeam; the wave provides both a stimulus, which is received through the eyes, and a force, which tips the boat. The man, however, can bend his leg in such a manner that the effect of the wave is precisely counteracted, and his body remains upright.

It will be apparent that the correct functioning of open systems depends on the precision of their calibration: the movement produced by the effectors must be just appropriate for the measured deviation. In the mantid case this calibration does not depend on previous learning, for young nymphs can capture the first *Drosophila* they encounter. On the other hand, when a man throws a dart or uses the mechanism described above to remain upright in a small boat, the calibration depends on practice. The mesh system has the peculiar characteristic, not shared by other control mechanisms, that it can produce continuous absolute orientation—the effect of disturbing influences can be anticipated and counteracted as they occur. This, however, depends on previous precise calibration (see also Mittelstaedt, 1964).

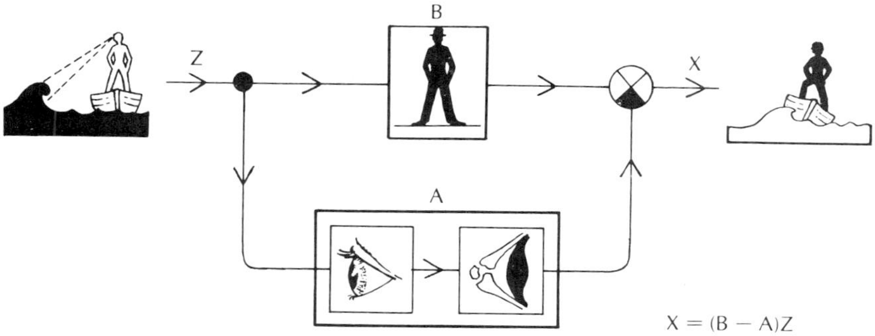

Fig. 7.9. A mesh: the sailor sees the wave coming and takes appropriate action as it arrives. (From Mittelstaedt, 1961.)

Closed Systems—Feedback Control

In many cases the animal can assess its deviation continuously and respond accordingly. The consequences of the initial activity of the effectors can thus be fed back into the system and influence its subsequent course. An example of this is shown in Fig. 7.10. In this case the man maintains his position by observing a stationary object not influenced by the wave. As the wave tilts the boat, he perceives the change in his position and takes the action appropriate for rectifying it.

The basic difference between a mesh and a loop can be illustrated by an experiment in which human subjects are asked to point to a target

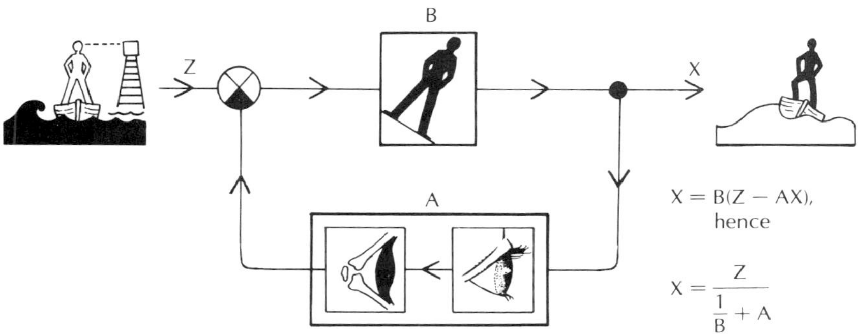

Fig. 7.10. A loop: the sailor perceives the change in his orientation by sighting a distant object and takes action after some tilting has occurred. (From Mittelstaedt, 1961.)

while wearing wedge prisms, which displace the visual field a few degrees: performance is compared between subjects who can and cannot see their own hands. The latter miss the target by an amount equal to the angle of the prism, indicating that the basic pattern is a mesh. The subjects who can see their hands, however, eventually point to the target, indicating that the basic pattern is a loop: the retinal representation of the angle between hand and target (i.e., the error) is not affected by the prism, and

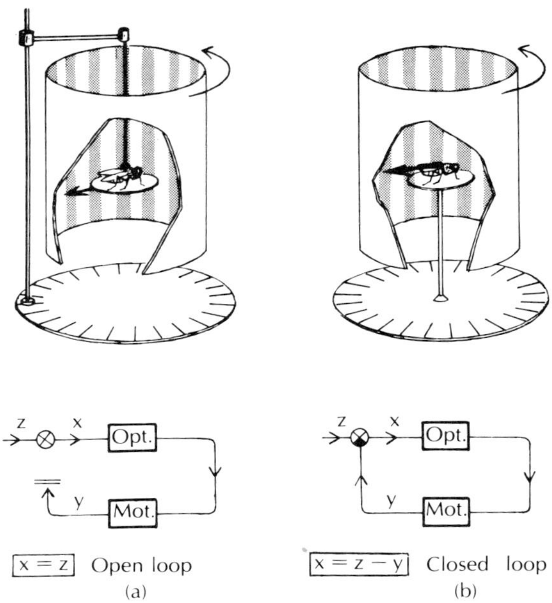

Fig. 7.11. Method for measuring the optomotor response of a fly under (a), open- and (b) closed-loop conditions. (From Mittelstaedt, 1964.)

In (*a*), the fly is fastened by its thorax: its turning movements cause a rotation, which is measured, of the disc below its feet. In (*b*), the fly moves freely on the platform.

the loop can therefore still operate to bring coincidence between the two (Mittelstaedt, 1964).

Another method of distinguishing control patterns is to interrupt the causal flow at some point. In the optomotor response of a normal fly, which we have already seen to depend on a loop (see pages 96–101), the fly's movement causes a reduction in the apparent angular velocity of a cylinder rotated round it. If the fly's body is held still, and its walking movements allowed to rotate a turntable underneath it (Fig. 7.11), there will be no reduction in the relative angular velocity of the cylinder: the loop has been opened. In the open-loop condition the fly's angular velocity greatly exceeds that of the drum (Fig. 7.12), but in the closed-loop condition it approximates to it, at least at low cylinder speeds (Mittelstaedt, 1964).

That the optomotor response may play a part in the directional control of normal locomotion has been shown in the bug *Oncopeltus* by opening the feedback loop under conditions in which any turning bias in normal locomotion can be assessed. Two methods were used to open the loop—keeping the insect in infrared light, and keeping it on a Y-maze globe (Fig. 7.24) in ordinary light: in either case self-produced movement of

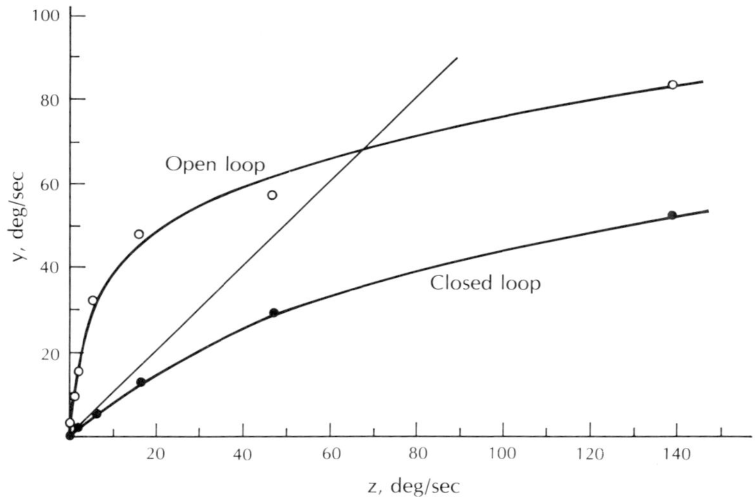

Fig. 7.12. Result of the two experiments illustrated in Fig. 7.11. The angular velocity of the fly (y) plotted against that of the cylinder (z) for open- and closed-loop conditions. (From Mittelstaedt, 1964.)

distant objects across the visual field would not be perceived. Under such conditions turning biases were usually apparent (Wilson and Hoy, 1968).

Another example is provided by the optomotor response of the crab *Carcinus* (Horridge and Sandeman, 1964; Horridge, 1966a). When placed in an optomotor apparatus, the crab tends to turn in the direction of the rotation. If the crab is held firmly by the carapace, the jointed eyestalks move, and their movement can readily be measured. The response occurs over a wide range of speeds of rotation (0.001 deg/sec to 10.0 deg/sec), but the eyes always lag slightly behind the stimulus. The effective stimulus (the "slip" speed *S*) is clearly the difference between the speed of the drum (*D*) and the speed of the eye (E), and the system can be represented as a feedback loop (Fig. 7.13). Since the slip speed is small in comparison with the eye speed, the system must contain an amplifier.

The feedback in this loop results from the movement of the eye, so that the loop can be opened only by preventing eye movement: if that is done, the response of that eye cannot be measured. Fortunately, however, if one eye is blinded or shielded from stimulation, it is driven by the other. Thus the loop can be opened by fixing one eye, blinding the other, and using the movement of the latter to measure the output of the system (Fig. 7.14). Under these conditions the slip speed and stimulus speed will be

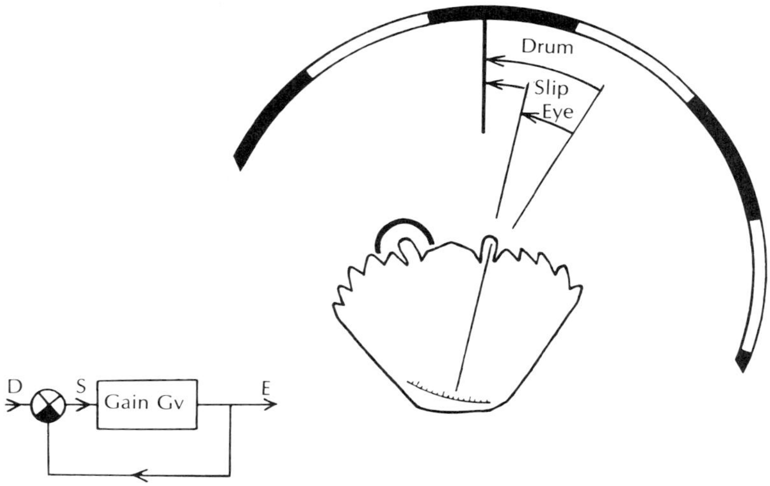

Fig. 7.13. Optomotor response of a crab. The carapace is held firmly, and the left eye shaded or blinded. The eye movement follows that of the drum, but never reduces the velocity error to zero. (From Horridge, 1966*a*.)

the same. The relation between drum speed and eye speed for open- and closed-loop conditions is shown in Fig. 7.15: it will be seen that with the loop open the eye speed may considerably exceed the drum speed, but under closed-loop conditions the eye follows the drum closely over a wide range of drum speeds.

The gain in the amplifier (Gv) will be given by E/S, or $E/(D-E)$. When the loop is opened, $S = D$ and $Gv = E/D$. Thus the gain can be calculated from the observed values of E and D. Some data are shown in Fig. 7.15, and it will be seen that the velocity gain Gv varies with the drum speed. A number of other factors (e.g., the crab's physiological state, other external stimuli) which influence the gain being ignored, it will be apparent that its large variability at some drum speeds (illustrated by the shaded area in Fig. 7.15*b*) does not detract from accurate following of the stimulus in the closed-loop condition. This is a consequence of the relationship given above

$$Gv = \frac{E}{D - E} \qquad \text{or} \qquad E = D\,\frac{Gv}{1 + Gv}$$

Thus as the gain varies between 5 and 25, the eye speed varies only between $\frac{5}{6}$ and $\frac{25}{26}$ of the drum speed. It will be apparent that this analysis in terms of a feedback loop gives an understanding of the manner in which fairly accurate following could occur even though the nervous elements

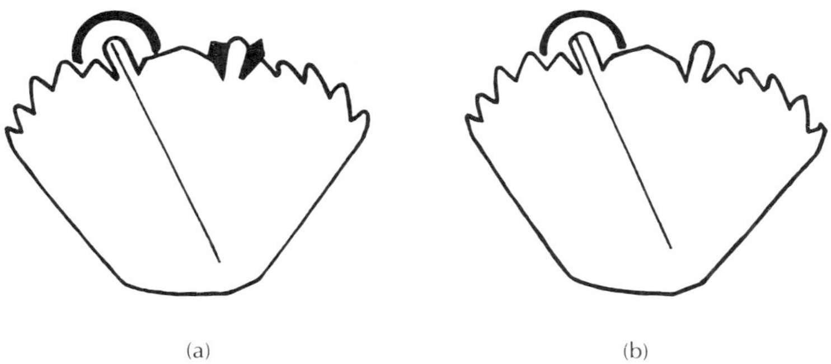

Fig. 7.14. (*a*) *Open;* (*b*) *closed loop situations for optomotor response of a crab. In both, the seeing right eye drives the shaded left eye. The two arrangements differ in that the seeing eye is clamped to the carapace in* (*a*) *but free to move in* (*b*). (From Horridge, 1966*a*.)

controlling the gain were inaccurate and ambiguous to a considerable degree (Horridge, 1966a).

An extension of the system shown in Figs. 7.13 to 7.15 arises from a further finding. A crab is placed in the centre of a stationary striped drum, as before. After a while the lights are turned off and the drum is rotated. The light is restored with the drum stationary in its new position. The crab moves its eye towards (but not to) the position of best correlation between the old and new positions of the drum: they turn less than the extent of drum movement, and turn less and more slowly the longer the dark period. Angles of about 0.1 deg can be remembered for several seconds, and angles of a few degrees for 10 to 30 min. The movement could be regarded as the result of a comparison between old and new positions of the drum or as a

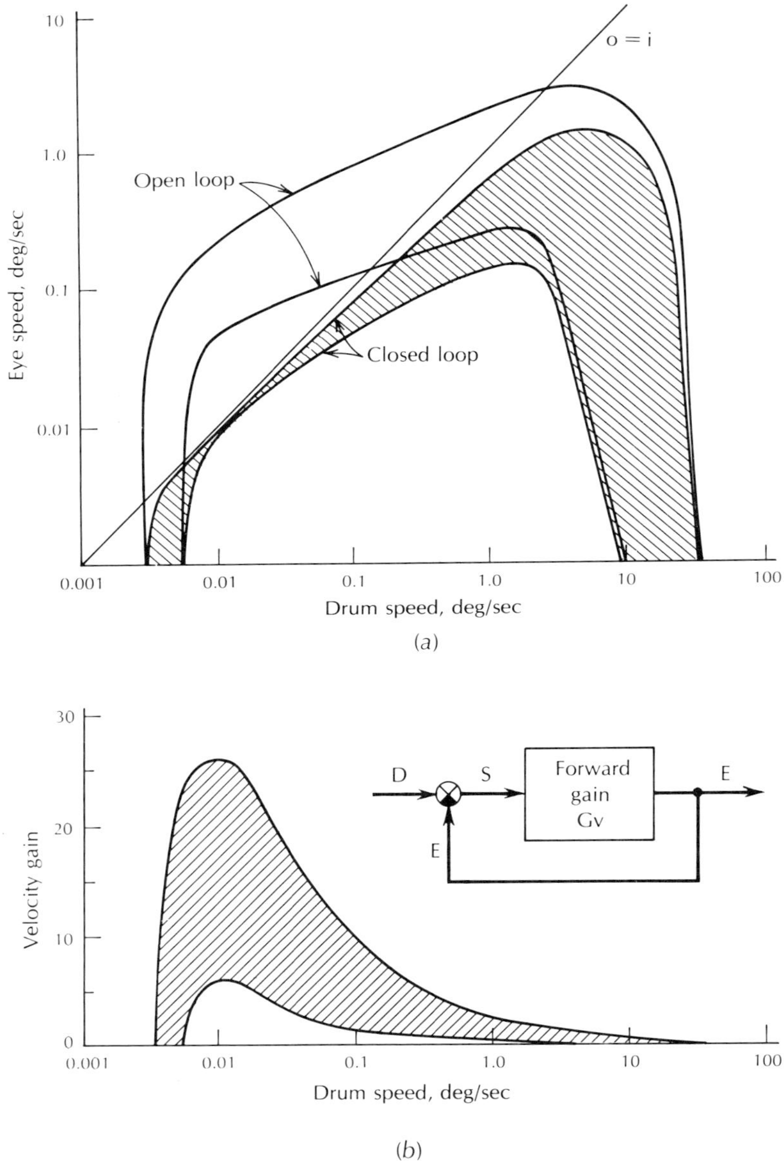

Fig. 7.15. Optomotor response of a crab. (From Horridge, 1966*a*.)

(*a*) The relation between eye speed and drum speed for open (Fig. 7.14*a*) and closed (Fig. 7.14*b*) loop conditions. In each case the upper and lower curves show the approximate range of variation. The straight line shows when eye speed and drum speed are equal. (*b*) The spread of values of velocity gain over the same range of drum speeds. Note how the range of variation of eye speed in (*a*) is small at most drum speeds at which the variation in velocity gain is high.

response to its apparent movement. The latter seems to fit the facts better, so that the mechanism is similar to that of the optomotor response: the mismatch between old and new positions of the stripe is converted to an apparent velocity, which operates as before. Since the moving eye sweeps across the now stationary stripes, a visual feedback operates: if the loop is opened as before by clamping the seeing eye and painting over the moving one, the eye movement can be greater than the drum movement. Comparison of closed- and open-loop conditions show that this negative-feedback loop involves neither marked amplification nor attenuation (i.e., gain is about unity), as must be the case for simple movement perception. The system is illustrated in Fig. 7.16 (Horridge, 1966a).

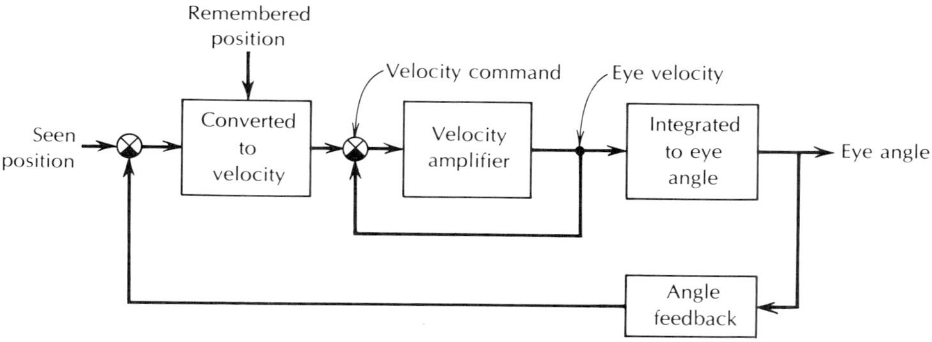

Fig. 7.16. Control system for response of eye in "optokinetic memory" experiment. (From Horridge, 1966*a*.)

The case of the optomotor response thus illustrates the use of a control systems approach in analysing orientation responses. Most cases of tropotactic orientation can be pictured as involving negative feedback in a similar way—the animal turns until the deviation of the relevant body axis from the direction of stimulation is 0 deg (or 180 deg): just so long as any deviation remains, the turning continues.

Many oriented movements depend on interaction between more than one sensory system, and here again an analysis in terms of control loops aids understanding. Von Holst (1950a, b) studied the interaction of gravity orientation and the dorsal light reaction in the maintenance of the normal swimming position in a number of fish species. It was possible to manipulate the external stimuli for the two mechanisms independently by altering the light direction away from the vertical and by altering the gravitational forces in a centrifuge. The resultant orientation of the fish with respect to gravity could be assessed in each case. von Holst suggested that the turning reaction r could be represented by

$$r = L \sin \beta + F \sin \alpha$$

where L = the light intensity
β = the angle between the light and the vertical
α = the angle between the dorsoventral axis and the vertical
F = a constant.

In this analysis it is assumed that the fish tends always to remain upright with respect to light and gravity. The ability of the control systems approach is even more evident in cases where the Sollwert is variable. While explanations of tropotactic orientation in terms of balanced stimulation and variations in the "sensitivity" of the different parts of the sense organ worked reasonably well so long as animals were studied in stereotyped and simple situations, in practice the sign of a tropotactic response can change. The tsetse fly *Glossina,* for instance, is positively phototactic at low temperatures but negatively so at high ones; and some Sphingid moth caterpillars keep the dorsal side uppermost when they first hatch but later reverse this in correlation with a change in the direction of their cryptic countershading (de Ruiter, 1956). Are we to postulate totally different mechanisms, with a reversal of sensitivity gradient (see pages 156–157) for the two cases? The difficulties become even more marked in the case of menotaxis, for instead of having one (or two) preferred directions, the animal can orient at almost any angle to the direction of stimulation. The foraging ant can pursue many different paths, each in a different direction from the sun. Indeed, many arthropods can allow for the daily change in azimuth of the sun: they maintain a constant compass direction while the sun, with respect to which they are oriented, moves across the sky (e.g., von Frisch, 1950). In such cases one could suppose that a large number of mechanisms is available, one for each particular orientation. On this view each possible menotactic orientation to a visual stimulus would depend on the stimulus being located in a particular region of the eye. Each possible region of the eye would thus be the centre of its own system of reflexes which could correct any deviation of the stimulus from that region, only one such system being in action at any particular time. But measurement of the deviation of the stimulus from so many different possible reference points would require an immensely complicated mechanism, and no satisfactory suggestion about how it could be achieved has yet been made.

It would be more economical if the deviation of the source from an anatomically fixed coordinate, such as the median plane of the head, were measured. This could then be fed into a system in which it is compared with a target value (i.e., Sollwert) determined centrally. The control system would then be the type shown in Fig. 7.17. This system is basically similar to that discussed in Chap. 5 in the context of responses to relative movement (pages 96–101). In both cases movement depends on a comparison between input from receptors and a Sollwert, whose value can be varied independently.

For instance, where related species maintain markedly different postures, the differences may depend on the setting of the feedback loop. A few teleosts swim with their heads higher than their tails, the longitudinal axis being more or less permanently inclined to the horizontal. Examination of the utriculus and observation of the behaviour of fish in a centrifuge show that the mechanism whereby this peculiar posture is maintained differs among species. In *Thayeria obliqua* the otoliths are oriented horizontally when the fish is in its normal oblique position. The inclined posture can thus be regarded as due to the peculiar orientation of the otolith organs. Centrifuging can have no effect on the input from the gravity receptors,

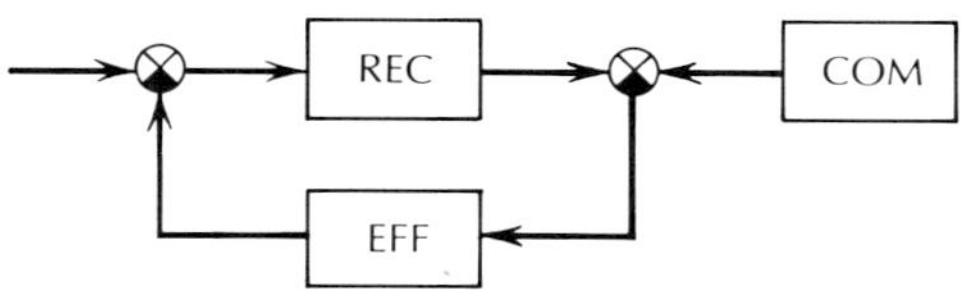

Fig. 7.17. General characters of a control system with variable "Sollwert".

COM is a sub-system which feeds in signals corresponding to the required "Sollwert." The input to EFF thus depends on the difference between the output of REC and that of COM.

and the fish does indeed maintain its posture when gravity is increased to 2.2 g. In *Poecilobrycon eques* the inclination to the horizontal changes with age. Its otoliths are oriented parallel to the longitudinal axis of the body: they are thus inclined to the horizontal. If this species is centrifuged, its posture becomes more horizontal. The change in the weight of the otolith causes a change in the strength of the signal from the sense organ and thus a change in posture. In this species, therefore, the inclined posture is a consequence of a change in the setting of the gravity feedback system, similar to the modification of the ant's orientation which results in the light compass response (Braemer and Braemer, 1958).

Similarly, when the direction of a menotactic orientation differs among individuals or groups of individuals within a species, the differences may be more economically considered as due to changes in the setting of a control system rather than to changes in the nature of the mechanism itself.

For example, individuals from different populations of the amphipod *Talitrus* each tend to move in a direction at right angles to that of their own home beach. Since the preferred direction does not depend on individual experience in the environment, the differences among populations are presumably genetically determined (Pardi, 1960). In the crab *Goniopsis*, by contrast, they seem to be individually acquired (Schöne, 1965).

In practice, the utility of the distinctions among tropotaxis, telotaxis, and menotaxis tends to disappear in an analysis in terms of control systems —as might indeed be expected from Jander's analysis of menotaxis in *Formica rufa* (see page 159). Orientations towards or away from the source of stimulation are merely special cases of orientation at an angle to it. And when the angle of orientation changes, as when an animal which orients by the sun allows for the sun's azimuthal movement and thus keeps a constant compass course (e.g., von Frisch, 1950; Birukow, 1954; Jander, 1957), we can ascribe the change to a synchronized alteration in setting rather than to a succession of new menotactic mechanisms. For orientation movements of this kind to be effective, the effect of the external stimulus on turning must be relatively independent of its intensity: this stands in contrast, for instance, to the dorsal light response (see pages 155–156; Jander, 1963).

Of special interest are cases in which transposition of orientation from one sensory modality to another is possible. For example, if ants (*Myrmica* sp.) or beetles (*Geotrupes* sp.) are forced to change from orientation to a light on a horizontal plane to orientation with respect to gravity on a vertical plane, the orientation to the new stimulus is related to that which had been held to the earlier one. This suggests that there is a mechanism, common to both types of stimulus, for "setting" the orientation with respect to whichever stimulus is controlling it (Birukow, 1954, 1964; Vowles, 1954a, b).

The bug *Velia currens* orients due south by the sun. It allows for the sun's movements, so that when orienting by an artificial light source in the laboratory the angle between the body axis and the direction of the stimulus changes with the time of day. Thus the insect heads to the left of the light in the morning, towards it at midday, and to the right in the afternoon. When orienting with respect to gravity on a vertical surface, there is a similar diurnal change—the insect climbs up and to the right in the morning, up and to the left in the afternoon (Fig. 7.18). The similarity between the rhythmic changes in orientation to light and to gravity, which also parallel a diurnal change in activity, suggest a basic rhythm responsible for all three, which is itself controlled by the natural day–night sequence (Birukow and Oberdorfer, 1959). In such cases the postulation of a central control system involves great economy of hypothesis over the view that a separate mechanism is involved for each menotactic response (see also Birukow, 1954; Fischer, 1961). Similarly the dung-beetle *Geotrupes silvati-*

cus has daily rhythms both of activity and of orientation to the sun: the menotactic mechanism is presumably linked to the mechanism which controls the periodicity of activity (Geisler, 1961).

Such abilities as these enable hive bees to communicate messages about the location of a source of food. If a worker bee discovers a source of food within 150 yards of a hive, she performs a "round dance" on the comb when she returns: she goes round and round in rapid circles, turning first one way and then the other. Other workers try to follow her in her dance and are then stimulated to set out on foraging flights in the vicinity of the hive. They are assisted in this by receiving some of the scent of the food source from the dancing bee.

If the food source is more than 150 yards from the hive, a returning

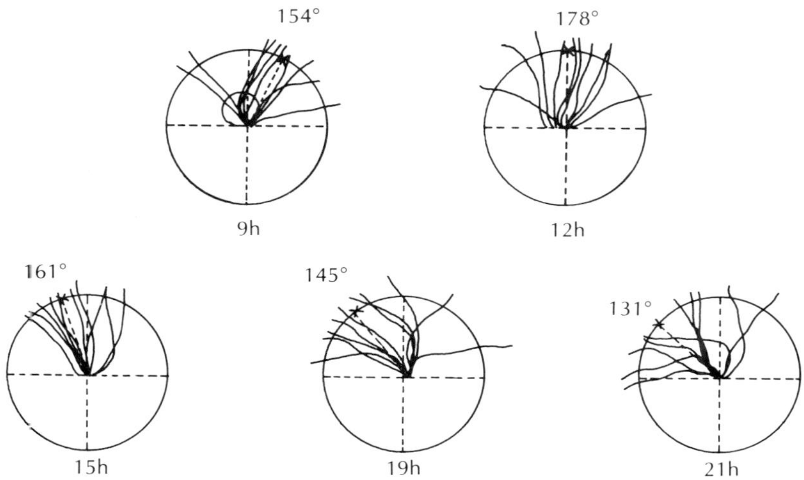

Fig. 7.18. Orientation of Velia currens on a vertical surface at different times of day. (From Birukow and Oberdorfer, 1959.)

× indicates the mean direction.

successful forager performs a "waggle" dance: she moves in a narrow semicircle, turns along its diameter, then describes a semicircle in the opposite direction, and so on. As she runs along the diameter, she wags her abdomen from side to side (Fig. 7.19). The work of von Frisch (e.g., 1953) has shown that this waggle dance conveys information about both the direction and the distance of the food source. The distance is indicated by the tempo of the dance—the further off the source, as indicated by energy expenditure on the outward journey, the slower the dance—or

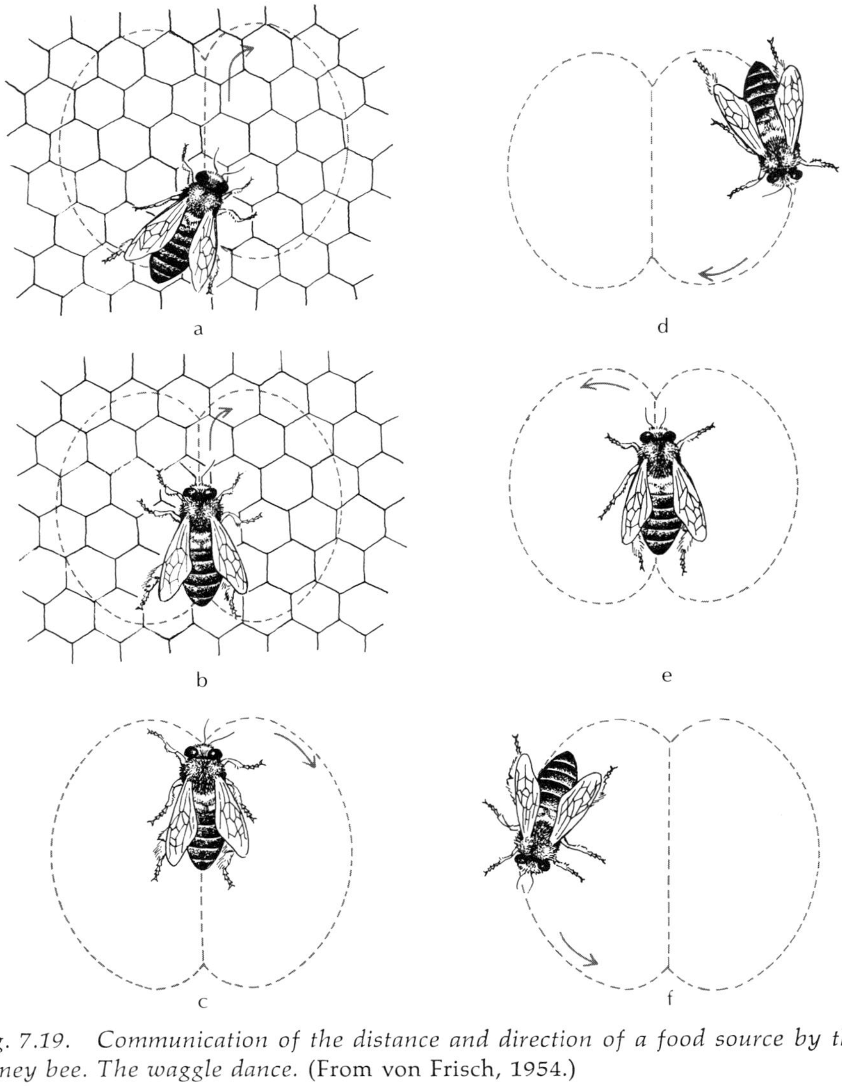

Fig. 7.19. Communication of the distance and direction of a food source by the honey bee. The waggle dance. (From von Frisch, 1954.)

possibly by the accompanying sound, whose characteristics also vary with the distance of the source (Wenner, 1962). The direction is indicated in one of two ways. If the dance occurs on a horizontal surface, the direction of the straight part of the dance is the same as that of the food source. If, as is more usually the case, the dance takes place on a vertical surface, the angle between the straight part of the run and the vertical is the same as that between the horizontal direction of the sun and the feeding place

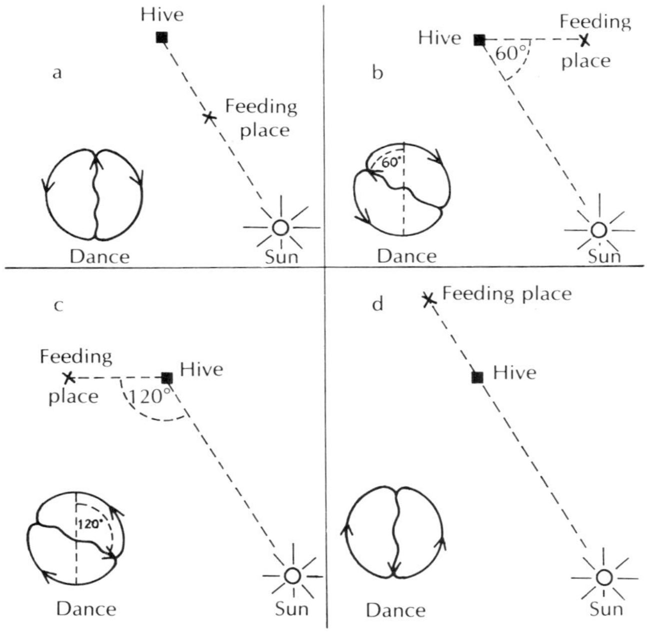

Fig. 7.20. Communication of the distance and direction of a food source by the honey bee. Indication of sun's bearing on a vertical comb surface. (From von Frisch, 1954.)

The small diagrams on the left show the dance as it appears on the vertical comb.

(Fig. 7.20). The accuracy of the information conveyed by the dance can be shown by placing similar food sources at varying distances or directions from the hive: the number of bees visiting each source in two such experiments are shown in Fig. 7.21.[1]

For the waggle dance to be successful, it is not essential that the sun

[1] Some doubt has recently been cast on the accuracy of the directional and distance information conveyed by the dance: see Johnson (1967) and Wenner (1967).

be visible: bees are very sensitive to ultraviolet, and in addition they can orient to the plane of polarization of light coming from a blue patch of sky. Furthermore, forced to take an indirect route between hive and food source by an obstacle, the bearing shown by the dance is the direct "bee line" between the two. More recent work on this problem is reviewed by Thorpe (1963a) and Lindauer (1961): the latter author showed that a

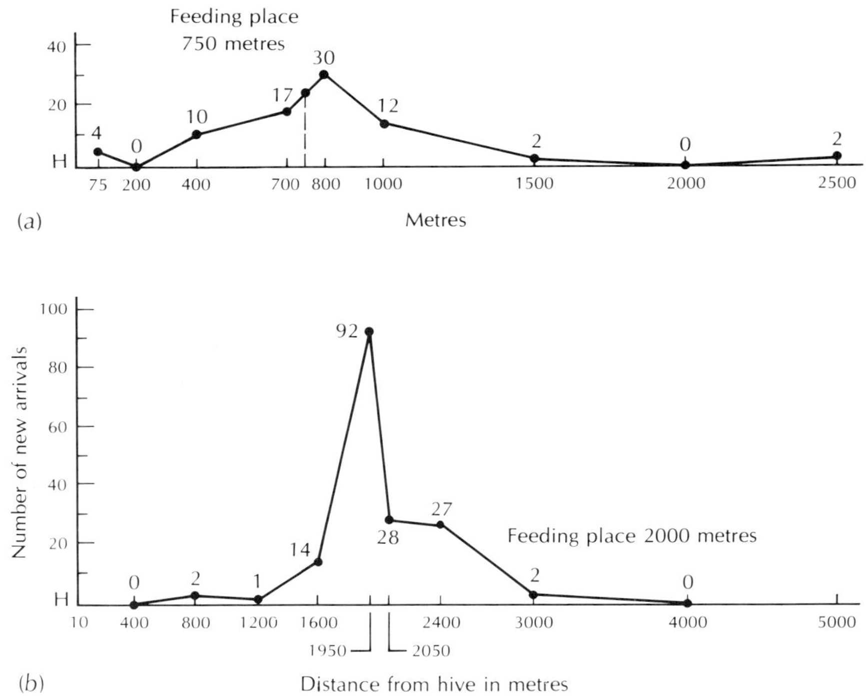

Fig. 7.21. The results of two experiments on communication in the honey bee, with the feeding places 750 and 2,000 metres from the hive. (From von Frisch, 1954.)

similar but more persistent dance is used by scout bees during the selection of new hive sites.

Representation of the orientation mechanism as a control system producing approximation to a Sollwert also helps in the clarification of a number of other findings. Thus, as we have seen, when orientation is controlled by bilaterally symmetrical organs, removal of one of them may upset the equilibrium: if one labyrinth is removed from fishes or frogs, the animal tends to turn towards the operated side (Schoen, 1950; Kolb, 1955). Blinding in one eye produces a tendency to turn towards the intact side in a number of invertebrates—for instance *Acilius* and *Dytiscus* larvae (Schöne, 1951): this can lead to a rotation about the longitudinal axis. Similarly, unilaterally blinded fish swim with the intact side lower (Schoen, 1951; von Holst and Schoen, 1954). However, in many cases such an asymmetrical posture disappears with time—there seems to be a "central compensation" for the sensory asymmetry (Schoen, 1950; Kolb, 1955). Thus the frog *Rana temporaria* recovers normal posture in about 10 days after being blinded in one eye and in about 3 months after removal of one labyrinth (Jahn, 1960). This recovery depends on input through the intact organ: the recovery of unilaterally blinded animals is delayed if the animal is given less than half an hour's light per day. Similarly, the recovery of fish after removal of one labyrinth is accelerated if the gravitational field is increased. The continuing input through the remaining sense organ thus produces gradual adjustment in the central input–output relations (see also Butz-Kuenzer, 1957).

Similar compensation occurs in invertebrates. After unilateral statolith ablation the eyestalks of the crustacean *Palaemonetes* are held asymmetrically. After a time compensation occurs, and the eyestalks are once more held symmetrically. If the other otolith is now removed, the eyestalks once again become asymmetrical, this time in the opposite direction. This also disappears after a time. Thus central compensation occurs both when the input becomes asymmetrical and when the consequent adjustment becomes inappropriate when the symmetry of the input is restored, albeit at a different level (Schöne, 1951). Such a mechanism is in fact essential, since the properties of the statolith change with every moult. The central nervous system adjusts to the discharge from the statolith organ in the light of input from other receptors (Cohen, 1955, 1960).

When orienting movements are controlled by more than one type of sensory input, their relative influences may be determined by non-specific input received through sense organs not concerned in the orientation (Birukow, 1951) or by various aspects of the internal state (von Holst, 1950a, b; Braemer, 1957). Thus, if the geotaxis and dorsal light reflex determining the position of a fish are made to compete, the latter tends to dominate when the fish is feeding, but the geotaxis when the fish is at rest.

The influence of feeding on the geotaxis in this case was studied by von Holst and Mittelstaedt (1950). If a passive fish is tilted, the statolith organ is affected and the fish shows motor movements which bring it back to its normal position. During normal feeding, however, spontaneous tilting often occurs, and it may be asked why the postural mechanism permits such movements and does not immediately tend to pull the fish back to a normal position. One possibility would be that the postural reflex is blocked during self-initiated movements. That this is not the case has been shown experimentally. If the animal is placed in a centrifuge, the force on the otolith is increased. If one then measures the self-initiated tilting movements, they are found to have decreased in magnitude. Thus the righting response is not switched off during "voluntary" movements but regulates them continuously through the afferent feedback. This is, of course, comparable to the mechanisms discussed on pages 96–101.

Another instance of motivational changes in orientation was studied by Schöne (1962) in water-beetle larvae (*Dytiscus* sp.). In this species the orientation of the head to a light source changes with the type of behaviour shown (e.g., swimming upward, swimming downward). The "reference positions" (Sollwert) when the light came from above were determined for a number of behaviour patterns by measuring the angle between the incident light and the dorsoventral axis of the head. Then all but one or two stemmata were covered so that a particular angle of light incidence was forced on the larva. The efforts to regain the Sollwert led to a continuous turning tendency, which could be measured. It was found to vary as a sine-like function of the light angle imposed. The curves for the different reference positions all showed maxima and minima at 90 and 270 deg but differed in their heights above the abscissa: the changes in reference position are thus apparently brought about by a shift of the turning tendency along the ordinate.

More complex problems arise when the sense organs are themselves oriented at an angle to the body. A case in point is the prey-catching response of the mantis. Although, as we have seen, the overall response is of the open type (page 162), the head is freely mobile with respect to the thorax, and thus information concerning the deviation of the prey from the optic axis is insufficient to control the direction of the strike with respect to the axis of the thorax, which carries the striking legs. Mittelstaedt (1960, 1962) has pointed out that three types of mechanism are possible:

1. The head position could be measured proprioceptively and added to the optic information so that the forelegs strike in the appropriate direction (Fig. 7.22,*a*). Although proprioceptors, in the forms of hair plates

---

Fig. 7.22. Three basic solutions for the problem of prey localization. (After Mittelstaedt, 1962.)

OPT, subsystem which yields information about deviation x of the prey from the eye axis; PROP, subsystem which yields information about deviation y of head from body axis; EFF, subsystem which controls deviation y of head according to its input variable w, the "order"; COPY, subsystem which transmits a "copy" of the order w to the optic output; FORELEGS, subsystem which controls the deviation k of strike from the body axis; z, deviation of prey from body axis. (*a*) proprioceptive mesh; (*b*) copy-mesh; (*c*) loop with branch-off.

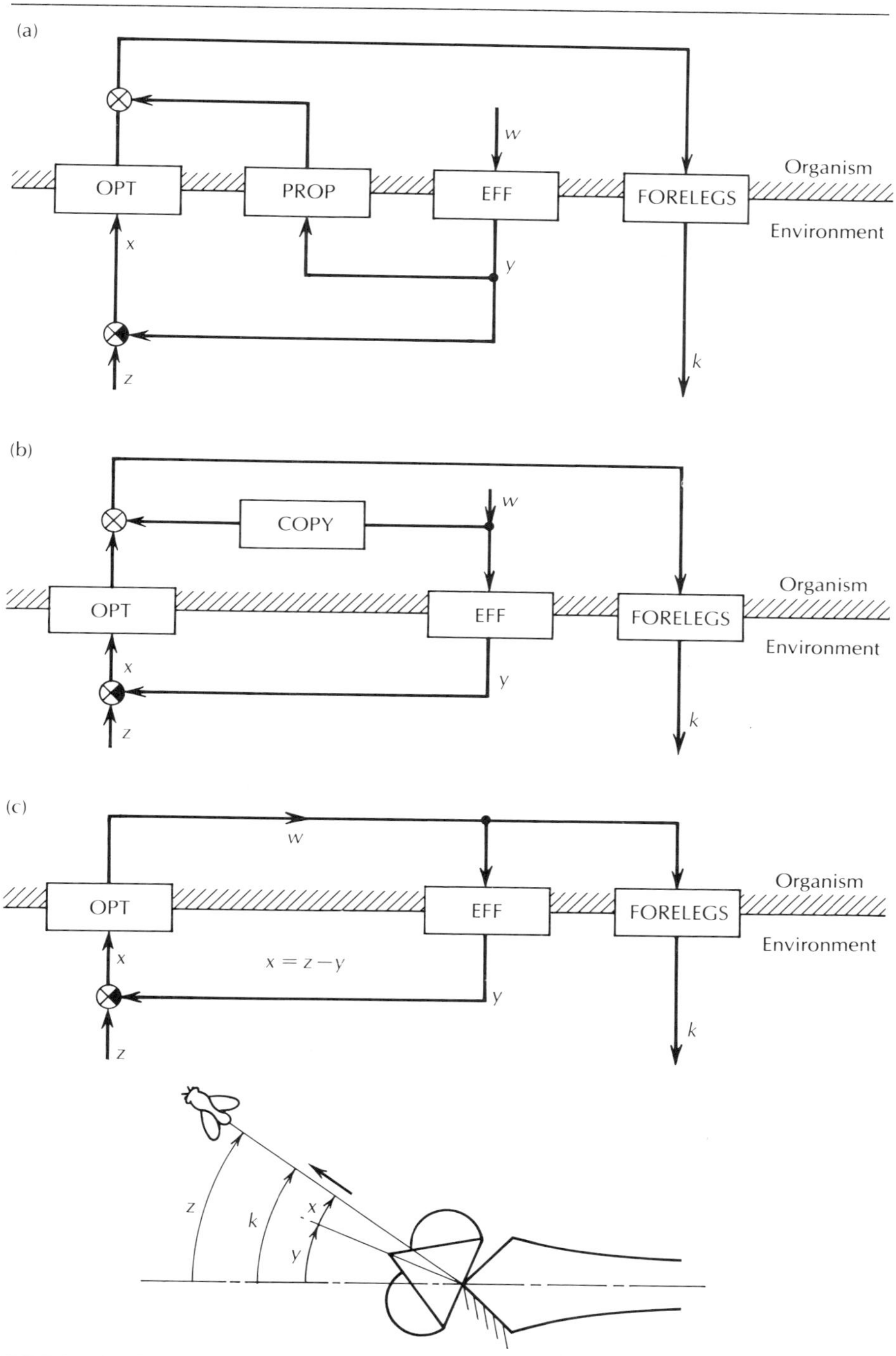
(a)
w
OPT
PROP
EFF
FORELEGS
Organism
Environment
x
y
z
k
(b)
COPY
w
OPT
EFF
FORELEGS
Organism
Environment
x
y
z
k
(c)
w
OPT
EFF
FORELEGS
Organism
Environment
x
x = z − y
y
z
k
z
k
x
y

between head and thorax, are present, this possibility is unlikely: if the head is turned through 10 to 20 deg to one side and fixed to the prothorax, the animal misses its prey to the opposite side. It thus behaves as if it did not notice its head deviation.

2. A copy of the input of the system controlling head orientation could be added to the optic output (Fig. 7.22,*b*).
3. The optic output could be fed back negatively into the system controlling head orientation so that the head turns towards the prey. The final steady state of this optic loop would then determine the direction of the strike (Fig. 7.22,*c*).

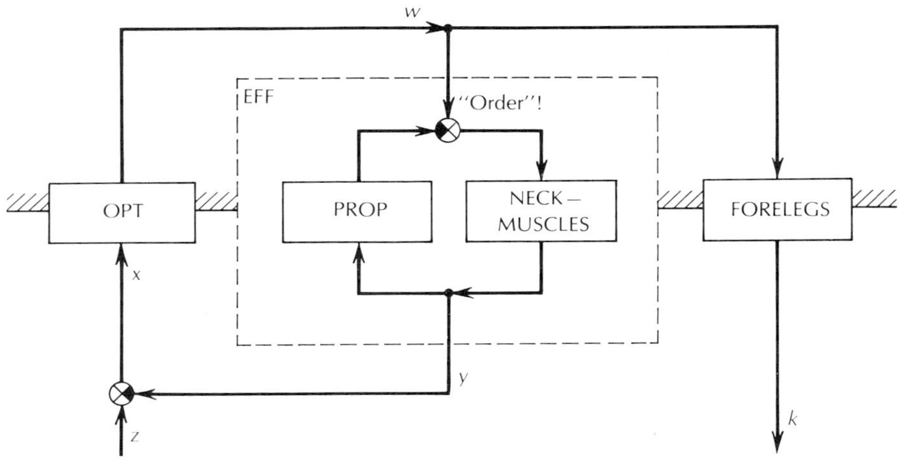

Fig. 7.23. Control pattern of prey localization in mantids. (After Mittelstaedt, 1962.)

The control pattern of EFF (see (*c*), Fig. 7.22) is a loop which turns the head into the position determined by the order *w*.

Since the mantid does, in fact, fixate its prey before striking, the last hypothesis appears the more likely. However, this cannot be the whole story, for unilateral deafferentation of the neck proprioceptors results in an error in striking to the opposite side: the proprioceptors must thus play some role. Such animals also show a permanent head deviation, and it is presumably this which leads to the contralateral error. It therefore seems likely that the proprioceptors and neck muscles form a loop within the optic loop (Fig. 7.23).

On this model the strike involves the following stages:

1. The prey is recognized by the mantis.

2. The optic loop turns the head towards the prey, the contraction of the neck muscles being adjusted according to the information provided by the optic loop. Movement ceases when this "order" is balanced by information from proprioceptors indicating the head's position.
3. The same optic output is then used to determine the deviation of the forelegs from the body's axis.

Basically similar mechanisms are probably widespread in insects: for instance Volkonsky (1939) has described the way in which locusts orient the side of their bodies towards a source of radiation. If an artificial source of radiation is provided, the locust first rotates its head in small jerks and thereby presumably stimulates neck proprioceptors: the orientation of the body is then changed to conform to that of the head. The relationships of head, neck, and body effectors may thus well be similar to that in the mantid.

The largely qualitative control systems approach to orientation movements presented in the preceding pages involves, as might be surmised, considerable oversimplification. The most detailed quantitative treatment is that of Mittelstaedt (e.g., 1963, 1964). He proposes that both the course order (i.e., the required angle between the direction of the stimulus and the heading of the animal, ω) and the deviation between the stimulus direction and the present heading (β) are divided into sinusoidal and cosinusoidal components. The turning reaction r then depends on cross-multiplication of these components:

$$r = K_1 \sin \omega \cos \beta + K_2 \cos \omega \sin \beta$$

where K_1 and K_2 are constants.

One further topic relating to such orientation mechanisms requires discussion. In considering phototactic responses, we have presumed the stimulus to be simple and neglected the manner in which the visual field is analysed. This problem has been studied in the case of optomotor responses of the type discussed earlier. In the usual experimental situation the animal is at the centre of a rotating cylinder painted with vertical stripes: it is subjected to a pattern of retinal stimulation varying in space and time, and we may ask how this is analysed to obtain parameters which can be fed into the control system. To attack this problem, Hassenstein (1959; Hassenstein and Reichardt, 1959) used the beetle *Chlorophanus*. In this case the turning tendency of the beetle was measured by the sensitive "Y-maze-globe" method. The Y-maze globe consists of six pieces of straw joined at four points to form Y junctions (Fig. 7.24). The beetle is suspended and carries the maze by its feet. As it walks, the beetle must remain stationary, but the globe rotates beneath it. As each Y junction reaches the beetle, it must turn one way or the other. The ratio of left to right turns is a sensitive measure of the optomotor turning tendency. Hassenstein used this device in combination with a special method of eliciting the response. The insect was surrounded by a stationary cylinder containing a pattern of vertical slits. Round this was another cylinder, also stationary, with a pattern of stripes on it. Between the two was a third rotating cylinder which contained vertical gaps. The rotation caused a pattern of changing illumination of the vertical slits whose sequence, width, and intensity varied with the precise arrangement of the slits and stripes.

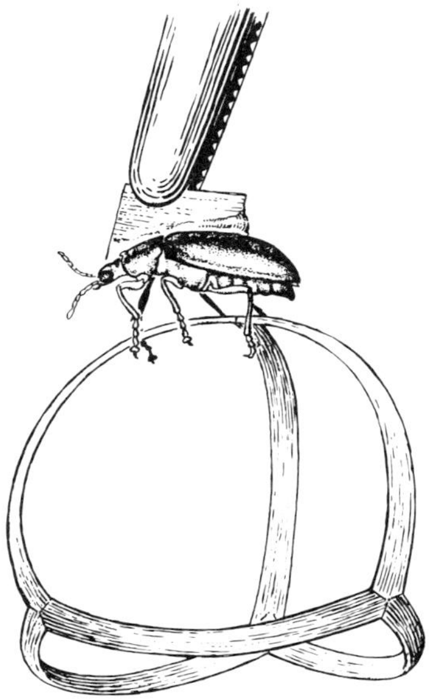

Fig. 7.24. The Y-maze globe, used for studying the orientation of Chlorophanus. (After Hassenstein, 1959.)

This produced an optomotor response in the insect. A typical arrangement is shown in Fig. 7.25.

Hassenstein was able to show that the physiological unit in the eye comprises two ommatidia. The basic stimulus situation for an optomotor response consists of two successive stimuli in adjacent ommatidia: responses to more complex situations are built up from these basic units. Each ommatidium can cooperate only with its immediate neighbour or with the next but one, and there is no interaction between ommatidia separated by more than one unstimulated ommatidium. The optomotor response is strongest when successive stimuli follow each other at ¼-sec intervals, though some response appears even when the interval is 10 sec. If a train of more than two stimuli is used, all possible combinations are assessed by the animal, and the response is increased; thus with three stimuli the response depends on all three possible pairs.

By varying the succession of stimuli, Hassenstein showed that the turning tendency resulting from the movement of the stimuli depends on five parameters: the interval between the successive stimuli, the spatial interval between them, the nature of the series, and the magnitude and sign of the single stimuli being related. The underlying mechanisms are represented in terms of a control diagram in Fig. 7.26, which indicates the simplest possible interrelations which will account for the results of the experiments.

Suppose the stimulus contour passes receptor elements *A* and *B* with

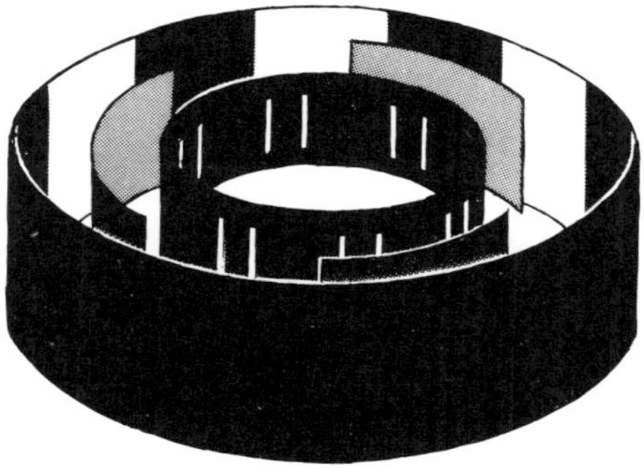

Fig. 7.25. Arena used for studying the orientation of Chlorophanus. See text. (After Hassenstein and Reichardt, 1959.)

a time interval Δt: in a more natural situation the length of this interval would depend on the relation between the movement of the external stimulus and that of the animal. Short signals are induced in the differentiation units *A* and *B*. These can pass to the multiplication units M_A and M_B via two pairs of filters, H_A, H_B and F_A, F_B which have respectively short and long resonance times. If the contour moves slowly, so that Δt is more than the duration of the response in the *F* units (say, more than 10 sec), one of the inputs of the multiplication unit will be zero, and its output will therefore be zero. If, however, Δt is shorter than the time taken by the signals in the slow-reacting elements to fade, M_A is affected at least momentarily by simultaneous signals from both sides. This, of course, occurs only because of the longer resonance duration of the *F* levels, the strength of the signal arriving from F_A being only that which corresponds to the after-effect there. When (and only when) M_A receives simultaneous signals from both sides, its output influences S_A and causes a movement in the direction $A - B$.

Since the H units have a short resonance time, M_B does not receive simultaneous messages through both its inputs and thus has no output. Conversely, if the movement is from B to A, M_B and not M_A produces an output. It will be seen that the faster the contour moves (i.e., the shorter

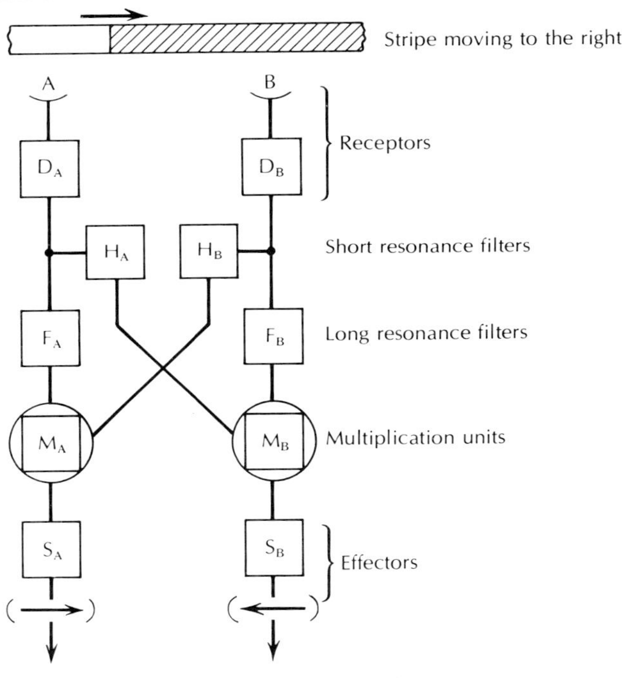

Fig. 7.26. Diagram to illustrate the control of the optomotor response in Chlorophanus. See text. (From Hassenstein and Reichardt, 1959.)

Δt), the less damped is the signal in F_A when the signal from B arrives and thus the greater the output of the multiplication unit.

Although the block diagram is discussed here only in relation to the passage of a simple contour across the eyes, the scheme accounts quantitatively for the insect's response to groups of contours and other complex patterns (Hassenstein, 1959, 1961; Reichardt, 1961).

However, Thorson (1966) has pointed out that such a multiplicative model is not the only one which could account for the data. Using a very precise method for assessing the optomotor response—namely, measuring the neck torque in a locust—he found the Hassenstein/Reichardt model adequate only for the gross gain and phase relationships between stimulus and response. Taking more cognizance of current electrophysiological data, Thorson found that most of the predictions of the multiplicative model could be due also to inhibitory interactions between the receptor channels, and that finer details of the data are better accounted for in terms of intensity-dependent adaptation than in terms of the multiplicative model.

The optomotor response, or rather the optokinetic memory experiment scribed on pages 167–168, has also been used by Horridge (1966b, c) to study the relative importance of the areas of black and white, and the edges of those areas, as stimuli. If the animal is responding to the stripes, a shift of up to half a stripe's width in one direction would be responded to as movement in that direction and vice versa. Locusts respond in this way. If the animal responds to edges, irrespective of which side of the edge is black or white, a shift of three-eighths of a stripe's period in one direction would be interpreted as movement of one-eighth of a stripe's period in the other direction. Crabs respond in a manner which indicates partial dependence on such a mechanism. The dependence on edges can be enhanced by a slight oscillation of the stripes or, presumably, by natural eye or body movements.

7.4 BIRD MIGRATION AND HOMING

Perhaps the most studied and puzzling cases of animal orientation are provided by the long-distance flights undertaken by birds during migration, long-range feeding expeditions, or homing. Bird orientation raises somewhat different issues from the work discussed earlier in this chapter, in part because the nature of the stimuli are not yet known, and in part because it involves more than merely maintaining a given direction. However, space permits only a brief survey here: recent reviews have been given by Kramer (1957, 1961), Adler (1963), Lindauer (1964), Schmidt-Koenig (1965), Matthews (1968), and others.

Any species in which the young birds migrate to winter quarters before the adults do must possess an ability to fly in a constant direction. This ability was demonstrated, for instance, by Schüz (1949): white storks (*Ciconia ciconia*) were taken from East Prussia to be reared and released after the local migration was complete in Western Germany. The recoveries indicated a strong tendency to fly SSE, which would be appropriate for the population from which they were drawn but not for the storks living in the area of the release.

The sensory clues which permit orientation of this type can be studied under aviary conditions. During the migration season aviary birds show intense activity ("Zugunruhe"), during which they tend to head in the same direction as they would if free to migrate. Under these conditions starlings remain oriented only so long as they can see the sky, and the sun is not obscured. If the apparent direction of the sun is changed by mirrors, the orientation of the birds changes accordingly (Fig. 7.27).

The direction of the sun, of course, changes with the time of day, but the migration direction does not. The bird must therefore be able to correct for the movement of the sun—it must have an internal "chronometer". This has been shown by experiment (Hoffmann, 1953; Schmidt-Koenig,

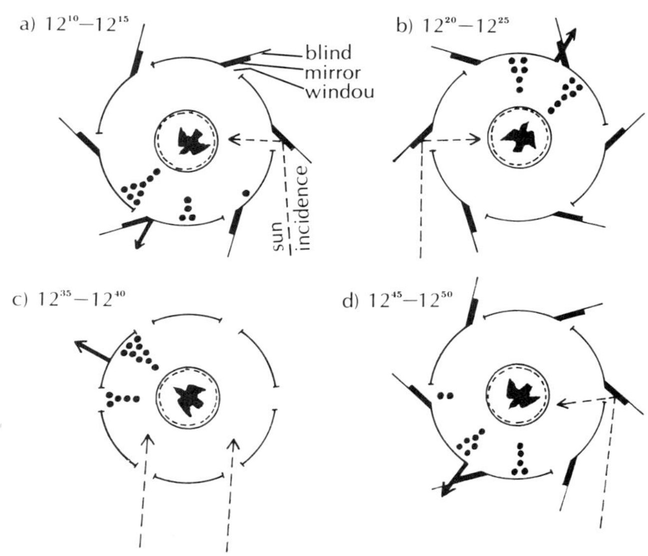

Fig. 7.27. Orientation of starlings when the apparent direction of the sun is changed by mirrors. (After Kramer, 1952.)

Records of an experiment consisting of four different parts. Broken lines show incidence of sunlight; dots average the bird's position in successive periods of 10 sec each. (*a*) Mirrors 135°, left of windows. (*b*) Mirrors 135°, right of windows. (*c*) No mirrors. (*d*) Mirrors as in (*a*).

1965). The chronometer depends on the light–dark cycle and can be upset by providing an artificial cycle out of phase with the natural one. An alternative method is to subject the birds to a change in local time by shifting them in longitude. This can most easily be done near the poles, where the lines of longitude converge: adelie penguins (*Pygoscelis adeliae*) shifted through 120 deg of longitude deviated to the expected side of their preferred northwards orientation when released (Penney and Emlen, 1967).

The exact method by which the sun is used, however, is still uncertain. If the bird merely uses the azimuthal direction, it would have to make allowance for the varying rate at which this changes through the day. Alternatively it has been suggested, as part of a more detailed theory of navigation, that the bird extrapolates the sun's observed path to find the highest point: this is always due south in the Northern Hemisphere, and could serve as a fixed reference point (Matthews, e.g., 1968). The experimental data, however, indicate that starlings can orient using the azimuthal direction alone (Hoffmann, 1953).

Of course, nocturnal migrants cannot obtain their direction by the sun. Using the directional nature of the migratory restlessness (see above), Sauer (1957) showed that various species of warblers, tested during the autumn migration season, oriented in the same way as they would if on migration when they could see only the central part of the starry sky. The birds became disoriented when the stars were hidden by cloud. It is probable, however, that star orientation does not depend on a time-compensated mechanism. Emlen (1967), studying the orientation of the migratory restlessness of indigo buntings (*Passerina cyanea*) caged in a planetarium, found that they maintained their normal migration direction when the skies were advanced 3, 6, or 12 hr from local time. Control experiments showed that the birds were in fact orienting by clues from the planetarium, but they clearly cannot have depended on an internal clock. Emlen suggests they were responding to Gestalt stimuli provided by star patterns—probably those in the neighbourhood of the North Star.

Even these may not be the only mechanisms for maintaining orientation, since it has been claimed that robins in captivity can retain a preferred direction corresponding to the migratory direction with no optical reference points at all (Fromme, 1961): Perdeck (1963) was unable to repeat this result, and the experimental technique has been severely criticized by Matthews (1968). Recently, however, an influence of the magnetic field on the orientation of this species has been claimed (Wiltschko, 1968).

Birds are, however, capable of much more than flying a fixed course. Some of the evidence comes from the study of migrant birds displaced from their normal starting point before release. We have already seen that young storks treated in this way migrate in the direction characteristic of the population from which they were drawn, and a similar result has been obtained with young starlings captured in Holland and released in Switzerland. However, adult starlings similarly displaced flew not in the direction they would have taken if undisturbed but to the place where they had wintered previously: this involved a difference in direction of about 60 deg (Perdeck, 1956). Thus birds can show real goal orientation, and this seems to depend on previous experience of the goal.

The same point is demonstrated by experiments on homing. It is now clear that homing pigeons can direct their flights towards the home loft when released in strange country (Griffin and Hock, 1949). Although it has been suggested that some of these homing experiments can be explained in terms of random search, there are many cases in which this can be ruled out. In some, the speed of return is so rapid that random search could not have occurred, while in others the birds show a significant orientation towards home within a very short period after release (Fig. 7.28).

However, the factors determining orientation on release are complex. Although birds released at a considerable distance from home usually show a strong tendency to orient correctly, birds released within 50 miles may not do so. At such short distances there is usually a directional bias ("nonsense orientation") which obscures the homeward orientation: this is probably usually discarded when the birds have flown a few miles. Recently it has proved possible to plot the tracks of pigeons by attaching small radio transmitters to them. When released in an unfamiliar place, such birds first tend to fly in the same compass direction which got them home on previous occasions, then turn to a direction which takes them to within a few miles of the loft, and then fly in using landmarks (Michener and Walcott, 1967).

Orientation towards home soon after the bird is released from an unknown point implies both that the bird is goal oriented and that it can perform the equivalent of fixing its present position on a grid of at least two coordinates, calculating the course to steer to regain the coordinates characteristic of home, and steering it. This does not mean, of course, that the bird goes through the sequences of calculations necessary for a human navigator who uses, say, a wireless position line, a sextant observation on the sun, nautical almanac, chart, ruler, protractor, and so on. In fact our knowledge of how this is achieved is still so incomplete that we do not know what measurements the bird takes to fix its position on release. It is established that the bird does not respond to forces resulting from the earth's magnetic field (though see page 187), and it is unlikely that it responds to forces due to the earth's rotation. Observations of the sky seem to be important, since pigeons are disoriented when the sky is overcast, and they home less well if confined where they cannot see the horizon from the home loft than if the whole sky is visible (Kramer, 1959).

Matthews (e.g., 1968) suggested that diurnal birds could obtain the necessary information from observations of the sun's arc, provided they could "remember" the characteristics of the sun's arc at home. If, on the basis of observations over a limited period, the arc is extrapolated to the highest point, comparison of the altitude with that at the home will give a measure of the difference in latitude. The angle subtended by the arc joining the observed sun to the highest point would give a measure of longitude if compared with the angle occurring at the same time at home

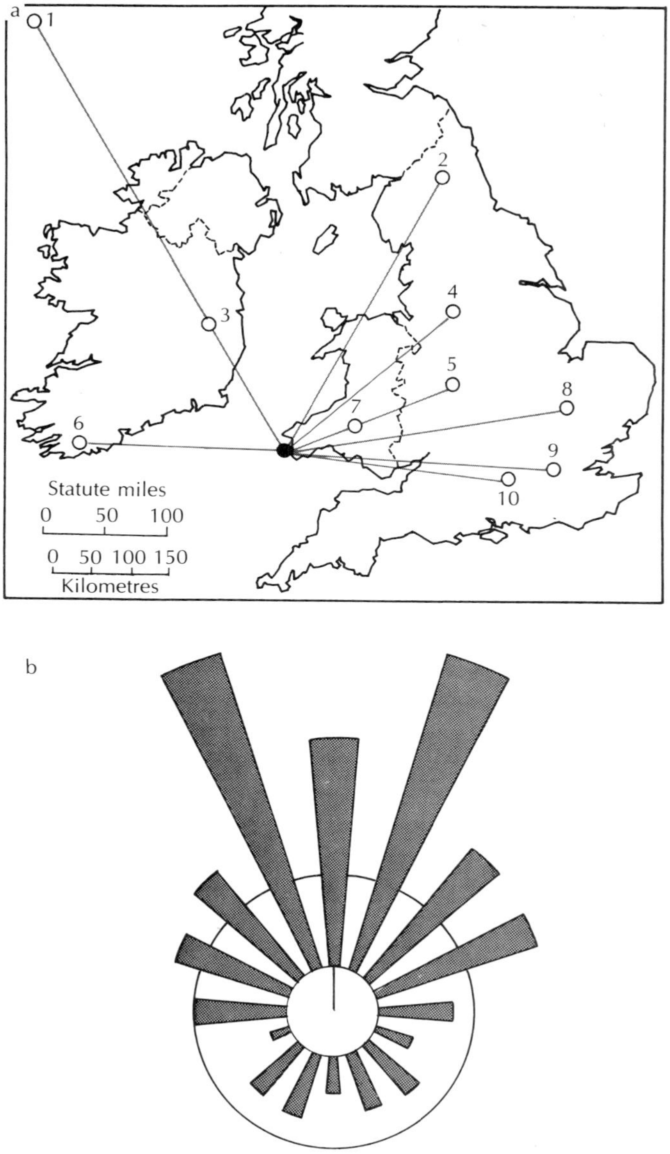

Fig. 7.28. (a) Points at which Manx shearwaters taken from Skokholm Island were released. (b) Orientation of Manx shearwaters on release in sunny conditions. (After Matthews, 1955.)

Results from points 1, 2, 3, 5, 8, 10 in (*a*) superimposed, with the direction of home indicated vertically. The length of each bar is proportional to the number of birds leaving in the direction of the bar.

(Fig. 7.29). Matthews's hypothesis has been criticized by Kramer (e.g., 1957) and Wallraff (1960), chiefly on the grounds that it is extremely improbable that the bird could make measurements of the accuracy which it demands, and that the initial orientation sometimes occurs before the sun can have moved any appreciable distance along its arc. Kramer also argued that Matthews's hypothesis does not explain some of the results of homing experiments. For instance, for any one position of the loft, pigeons

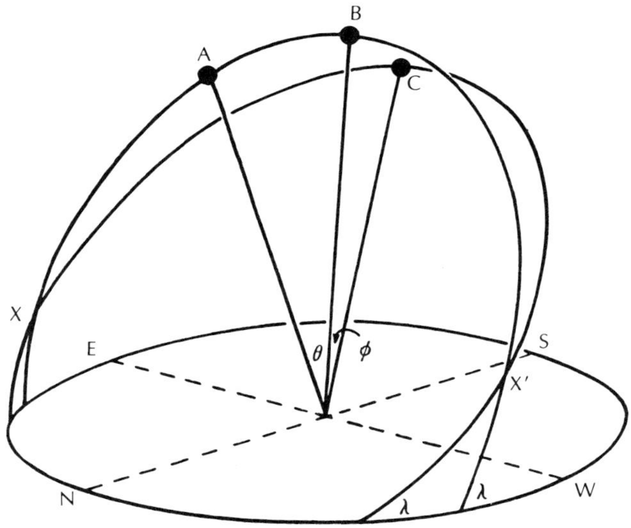

Fig. 7.29. Perspective diagram of the changes in sun arc consequent on a move to the south and west, at noon (home time). (From Matthews, 1955, 1968.)

The altitude of the highest point of the arc *B* is greater (by ϕ) than at home, *C*. The inclination of the arc λ is also greater. The observed sun *A* has not moved so far round its arc (by θ) as it would have done at home. The bird is required to construct the "foreign" arc of extrapolation of the observed movement of the sun at *A*, and to memorize the home arc.

home with greater success when released from some directions than from others. Indeed, if orientation from many release points round one loft is assessed, the deviations from the homeward direction seem highly irregular. Kramer thus considered any hypothesis which entailed the use of the sun for pinpointing extremely improbable. He supposed, instead, that the two steps of homing orientation are independent—the bird first determines its position relative to home by some unknown means and then uses the sun as a compass to orient towards home. Evidence in partial support of this view has been provided by experiments in which the internal clock of pigeons was shifted by subjecting them to a dark–light rhythm advanced or retarded on the natural one. The change in orientation this produced was similar to what would be predicted if the pigeons used the sun only as a compass for orienting their flight and determined their position by some other method (Schmidt-Koenig, 1961).

Matthews (e.g., 1961, 1968) has, however, shown that many of the difficulties raised by Kramer arise from the tendency of birds to orient in a particular direction, usually specific to the stock or population of birds in question, for a short while after release (see above). In addition, in some cases birds may be initially deflected from the home direction towards that taken on previous releases (Graue, 1965). Furthermore it has been shown that pigeons can in fact detect movement as slow as that of the sun (M. E. Meyer, 1964, 1966)—as indeed can crabs (Horridge, 1966d).

Matthews (e.g., 1968) has in fact effectively answered most of Kramer's objections, and there can be little doubt that sun navigation occurs. But the precise parameters of the sun's movement used are still a matter of dispute. Pennycuick (1960) has amended Matthews's scheme in such a way as to overcome some of its difficulties. He suggests that the bird measures the sun's altitude and rate of change of altitude. This, he claims, overcomes the major difficulties of the sun-arc hypothesis, since the accuracy of observation demanded is much less. So far, however, crucial evidence discriminating the two hypotheses is lacking.

It has also been claimed that night migrants show similar navigational abilities when exposed only to the stars. Here, however, the evidence is far less sound. Although it is clear that birds can orient in a given direction by the night sky (see page 186), claims that they can navigate come mostly from planetarium experiments (Sauer, 1957). The number of birds involved in these experiments was however small, and their interpretation has been heavily criticized by Wallraff (1960) and Schmidt-Koenig (1965).

Although most studies of homing ability have used birds as subjects, comparable phenomena are found in other species. Studies of homing in mice, which include references to earlier mammalian work, are given by Lindenlaub (1955, 1960) and Bovet (1960, 1962). Bats have been studied by Mueller (1966). Among other species studied are lizards (Fischer, 1961), toads (Heusser, 1958; Ferguson and Landreth, 1966), newts and frogs (Ferguson et al., 1967; Landreth and Ferguson, 1967, 1968), and fish (Braemer, 1960; Braemer and Schwassmann, 1960; Schwassmann, 1960; Winn et al., 1964). Responsiveness to the chemical composition of the home stream plays a role in salmonid fishes (Hasler, 1960).

7.5 SUMMARY

1. The first step in the analysis of an orientation response is the identification of the stimulus parameters with respect to which the orientation occurs.
2. The means by which animals adjust their position with respect to sources of stimulation can be divided into kineses and taxes. In the former the body is not oriented, the stimulus producing merely a change in speed of locomotion or frequency of turning. Taxes involve orientation of the body: various types have been distinguished by Kühn, and Fraenkel and Gunn.
3. The functioning of the mechanisms underlying oriented movements can profitably be described in terms of control system diagrams.
4. Bird homing and migration present an especially complex series of problems since the orienting stimuli have not yet been identified, and they often involve navigation as opposed to mere orientation.

8

CHANGES IN RESPONSIVENESS TO A CONSTANT STIMULUS—LIMITATIONS OF DRIVE CONCEPTS

A given stimulus does not always evoke the same response. If the external situation is constant, altered responsiveness of an animal must be ascribed to changes in its internal state. This and the following chapters are concerned with the nature of these changes.

In practice, changes in responsiveness are revealed in a number of ways. First, there are changes in the response given when a particular stimulus is presented on separate occasions. During the spring and early summer a caged male chaffinch responds to a stuffed female mounted in the soliciting posture (Fig. 3.1, page 20) with vigorous courtship behaviour, and often also with mounting and ejaculation. The same dummy elicits, at most, a weak response from a male in winter or from a male who has just ejaculated. The internal states of the males in the last two cases must differ from that of a male in spring who has not recently been exposed to such a model. Second, there are changes in the frequency or intensity with which a response is given under constant external conditions: a great tit does not produce its song at regular intervals but in a complicated pattern of bursts of singing which cannot be related to changes in the environment and are, therefore, presumably a consequence of internal changes. Third, the strength of the stimulus necessary to elicit a response of standard intensity varies with time: the longer the period of food deprivation, the more unappetizing the food an animal will accept. Fourth, and related to the others, at different times an animal selects different categories of stimuli to respond to: sometimes those indicative of food, at others a nest site, or a mate, and so on. Finally, the strength of an aversive stimulus which an animal will withstand, or of an obstacle which it will surmount, varies with time: the longer a rat has been without food (within limits), the more highly charged the electric grid which it will cross in order to obtain food.

The nature of the internal changes on which such changes in responsiveness depend is usually considered as part of the problem of "motivation"; other aspects of this problem, such as the integration of separate responses into functional sequences, and the directiveness of behaviour, will be considered later in Chap. 25. Here we shall be concerned only with questions such as "Why does the animal respond more strongly to this stimulus at some times than at others?"; "Why does it respond sometimes to this stimulus, sometimes to that?"; "What determines the beginning and end of each type of activity?"; "Why is the animal active at all?"

Conventionally, changes in behaviour labelled as "motivational" are temporary and reversible, and thus distinguishable from the more permanent changes ascribed to "learning". In practice both temporary and long-term effects may be a consequence of the same behavioural events: if a male chaffinch mates successfully with a female dummy, his responsiveness is temporarily lowered, but there will also be a long-lasting increased responsiveness to such dummies. There is, in fact, no universal agreement about what changes in behaviour should properly be described under the rubric of motivation, and no definition will be attempted here. In general, we shall include changes in response strength or responsiveness which are reversible but last longer than a second or two. It is convenient, however, to exclude changes which can readily be related to peripheral changes in the sense organs (accommodation, adaptation, etc.) and some changes in perception (Chap. 5) or effectors (e.g., muscular fatigue). In addition, discussion of near-permanent changes, whether they are the result of "maturation", "learning", or anything else, is postponed (see Sec. 3). Even with these exclusions, there is no implication that the remaining changes in behaviour, thus grouped as due to "motivational factors", are to be ascribed to a single type of mechanism. Our practice, rather, will be to start with a group of phenomena characterized empirically as involving changes in responsiveness to a constant stimulus (with the restrictions specified above), label them loosely "motivational", and then attempt to analyse the processes responsible. We are thus using the term "motivational" to label a group of phenomena but not to refer to hypothetical entities, factors, mechanisms, or processes underlying the behaviour.

8.1 DRIVE CONCEPTS

Changes in responsiveness to a constant external stimulus indicate the occurrence of temporary conditions internal to the animal which determine the responsiveness. Common speech uses words such as hunger, thirst, or sexual urge to refer to these internal conditions. Provided certain precautions are taken, such "drive" concepts can sometimes be taken over as analytical tools in many branches of experimental psychology, psychiatry, and psychopathology, and they will be used in a number of contexts in this book. This does not mean, however, that they are universally valuable, and it is necessary to define the limits of their usefulness.

In the first place, drive concepts can be useful only if defined independently of the variations in behaviour they are supposed to explain. If they are postulated in numbers equal to those of the types of behaviour in question, they become ad hoc assumptions and lose any explanatory value. A child who is told that a train goes because of the "locomotive force" is left somewhat more puzzled than before, for the argument is circular (Lorenz, 1950). But if we observe that a rat, presented with water, drinks more if it has previously been deprived of water than if it has not, there are circumstances in which it is useful to postulate a "thirst drive". This drive will then be related, on the one hand, to the period of water deprivation and, on the other, to the amount of water the animal drinks when presented with water. In such cases the period of deprivation is often termed an "independent variable", being not affected by the behaviour observed; that is, in this case, by the amount of water drunk. The latter is termed a "dependent variable", and the postulated thirst drive an "intervening variable" (Tolman, 1932).

At this point, the use of a drive concept may not seem a very economical procedure, for it requires the establishment of two relationships (one between hours of deprivation and drive, and another between drive and amount of water drunk) when we may be interested only in one (that between deprivation and amount drunk). Furthermore, in such a case there is no possibility of proving that one intervening variable cannot account completely for the data. Usually, however, we are interested in more than one independent variable and more than one dependent variable. To quote an example used by Miller (1959), an animal can be made "thirsty" by depriving it of water, by giving it dry food, or by injecting it with hypertonic saline. We can assess the consequences of these treatments by measuring the volume of water it drinks in a given time, by recording the rate at which it will perform a learnt response which is rewarded by water, or by assessing the concentration of quinine which it will tolerate in its drinking water. If we were to examine the relationship of each type of treatment to each method of measurement, there would be nine relationships to be established (Fig. 8.1). Alternatively, if we could suppose that each type of treatment influences an intervening variable thirst, which in turn influences each of the response measures, it would be necessary to establish only six relationships (Fig. 8.2). The postulation of the intervening thirst variable has thus led to considerable economy (Miller, 1959).

However, such a procedure can be useful only to the extent that there is a good correlation among the various measures involved: that is, if one type of treatment is more effective than another as assessed by one type of

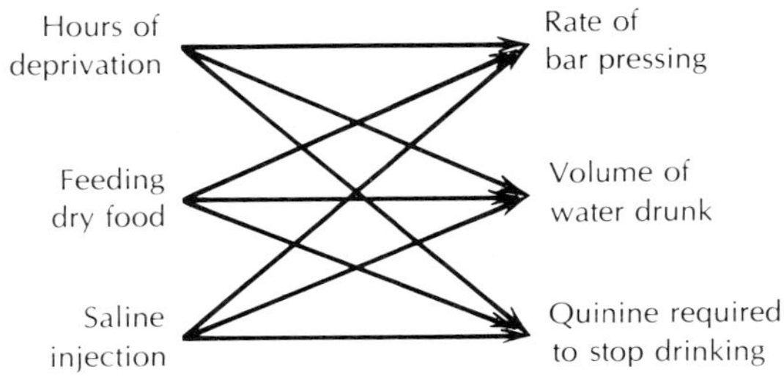

Fig. 8.1. Relationships between three independent and three dependent variables. (After Miller, 1959.)

measurement, it must also be so when assessed by another. Thus, where the fact that the various possible measures are correlated is of primary interest, an intervening drive variable is a valuable tool for research: where the extent to which the correlations are not perfect becomes important, then such a concept is misleading and can be a positive hindrance.

In practice, the extent to which the different measures are correlated varies with the measures chosen and the fineness of the analysis. Some data relating to the present instance are shown in Fig. 8.3 (Choy, cited by Miller, 1956). Rats were given 5 ml of 2 *M* saline by stomach tube, and the resultant "thirst" assessed in terms of the amount of water drunk in 15 min, the number of bar presses reinforced with water on a variable interval schedule given in 9 min, and the concentration of quinine in water

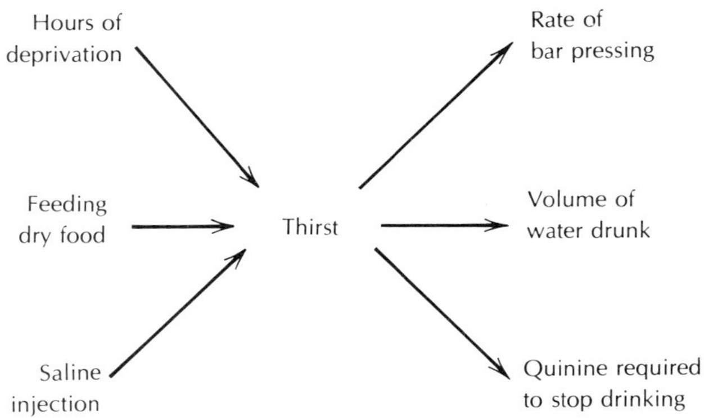

Fig. 8.2. Relationships between three independent variables, one intervening variable, and three dependent variables. (After Miller, 1959.)

that they would tolerate. As the figure shows, the three measures give rather different pictures: bar-pressing, unlike the other measures, showed little change 15 min after treatment; while only the amount of water drunk reached an asymptote after 3 hr. In view of this lack of agreement among measures, we must conclude that the postulation of a single intervening variable is too simple a hypothesis to account for all the changes in behaviour observed. As another example, Slotnick (1967) found only low correlations among various indices of mother–infant interaction in the rat and concluded that maternal behaviour does not represent a unitary behavioural process.

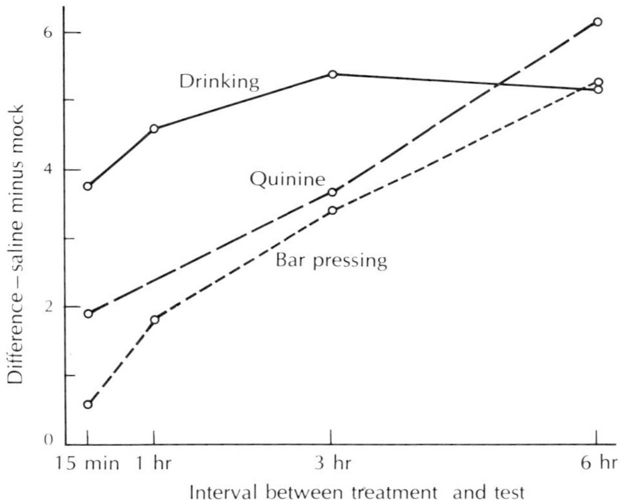

Fig. 8.3. Three methods of assessing the "thirst" of rats given 5 ml of 2-molar saline by stomach tube. The ordinate indicates the amount by which the experimental test exceeded the control test, the nature of the units differing beween the procedures used. (After Miller, 1956.)

The lack of correlation which such detailed studies demonstrate appears also when the central mechanisms are interfered with experimentally. Thus Teitelbaum (1957) found that rats with certain hypothalamic lesions, although hyperphagic, showed less lever-pressing activity reinforced with food than non-operated animals did.

However, this is not all. There are *in theory* an infinite number of ways in which even the amount of water drunk could be measured. We could assess the amount drunk in 1 sec, 1 min, 1 hr, or any other period. According to which we choose we may get very different results. In fact, while a rat is actively engaged in drinking, the rate at which it drinks under standard conditions is very constant. Licking has the characteristics of an elicited reflex, the number of licks in each burst depending on aspects of the stimulus, such as the size of the drop presented (Hulse, 1966). When thirsty rats are given water, they drink more or less steadily for a few minutes and then in short bursts over the ensuing period. For periods of deprivation of between 48 and 168 hr the initial period of virtually continuous drinking is fairly constant (about 8 min), though subsequently the animals which have been deprived longer drink more frequently (Stellar and Hill, 1952). Thus, if we were to assess the amount of water drunk in a period shorter than 8 min, we might conclude that periods of deprivation between 48 and 168 hr were all the same to the rat, though a longer period of drinking would reveal that they were not. Furthermore, our thirst variable would help us not at all in understanding the detailed course of drinking behaviour—for instance, the alternation of bouts of drinking with bouts of sniffing—unless we postulate a series of additional intervening variables: at this stage use of a drive concept tempts us to oversimplify (see also Hilgard and Marquis, 1961).

A similar example is provided by Tugendhat's (1960a) study of the feeding of the three-spined stickleback. After progressively longer periods of food deprivation, the frequency of completed feeding responses increases, but the time spent feeding does not. This is because the individual responses are more rapid and more frequently completed. Any attempt to account for the detailed course of feeding behaviour in terms of a "feeding drive" requires at least a number of ad hoc additional assumptions.

In the cases we have been considering, a concept of a thirst or feeding drive is useful insofar as it can be related to antecedent conditions and subsequent behaviour and insofar as the alternative behavioural measures are correlated with each other. However, a relation with antecedent conditions is not an essential condition for a drive concept to be useful. Sometimes it is convenient to use one to relate the behaviour under study to measurements of other aspects of the animal's organic state, such as hormone concentration or skin resistance, or to other types of behaviour it shows. Hormone concentration, like water deprivation, might be regarded as antecedent to the drive, but skin resistance or behaviour shown in other situations may be dependent variables and may vary with the internal state in a manner similar to the behaviour in which we are interested. Even the hormone concentration might be assessed by means of another dependent variable, such as the size of a cock's comb. If we are concerned with the behaviour of the whole animal the direction of the causal relations is thus often immaterial: drive can be anchored by non-directional mathematical relationships to two sets of data, both of which refer to consequent events. This situation is especially apt to arise in the early stages of the study of a new species. If the animal shows a bizarre movement whose significance is unknown, the first clues may come from its temporal association with other movements. A temporal correlation between movements indicates that they are somehow causally linked, and one possible course is to postulate a common drive—a procedure which is likely to be useful as a first step, but which may require elaboration or modification in the light of further evidence, and whose value will in any case decrease as the behaviour is recorded in finer detail. The more sophisticated use of this approach is discussed in Chap. 16.

Even within the reservations discussed so far, however, the use of drive concepts carries with it further dangers which can be avoided only through constant vigilance (Hinde, 1956a, 1959d, 1960a), and it is necessary to discuss them briefly.

For one thing, the word itself can make difficulties because it has been used by so many different authors in so many different ways. "Drives" may refer to mathematical intervening variables, as indicated above, or to hypothesized entities believed to exist but not yet identified: these may be stimuli or responses, physiological or psychological states, extraneural or intraneural states, or combinations of these. They may be "biogenic", in which changes in behaviour are related directly to changes in the internal state of the animal, or psychogenic, in which they are not.[1] Such a diversity of meanings is of course quite justifiable provided the author keeps within the limits of his own concept, but it does impose a strain on the reader.

The importance of maintaining a clear distinction between intervening variables, which carry no implications about structures or processes underlying the behaviour, and "hypothetical constructs", which do, cannot be underestimated (see, e.g., MacCorquodale and Meehl, 1954). That there must be some physiological basis for correlations between behavioural measures goes without saying, and full understanding of the behavioural correlations demands also understanding of the mechanisms that produce them. But to assume a one-to-one relationship between intervening variables, established on the basis of behavioural relationships, and structures or processes in the nervous system, is naïve in the extreme.

Even within one usage, however, there is a tendency to use drive as a blanket variable—drive concepts are used to provide unitary explanations of a variety of characteristics of behaviour which may depend, in fact, on diverse mechanisms. For instance:

1. The temporal persistence of the effects of stimuli and of activities. Stimuli may be ineffective when first presented and influence be-

[1] Further discussion of these usages will be found in Hilgard (1948); the Nebraska Symposia on Motivation (1953 et seq.); Estes et al. (1954); Peters (1958).

haviour only after repeated presentation. It thus appears that their effects are stored. Further, once evoked, a given pattern of behaviour may persist for some time after the stimulus is removed. Such observations are sometimes ascribed to "persistence of the drive".

2. Temporal grouping of activities. The various patterns of behaviour in an animal's repertoire do not appear at random but in functional groups. This could be simply a chain sequence, where each activity brings the animal into the stimulus situation which elicits the next (see pages 609–618). Another possibility is that the grouping is due to a sharing of internal causal factors which persist over time. In the latter case they may be spoken of as having a common "drive".
3. Directiveness. The statement that behaviour is "directive" may imply either that it continues until a certain goal stimulus situation is achieved, or that variable means (i.e., more than one type of behaviour, characterized objectively) are used to a constant end. Some authors describe such behaviour as being associated with "motives" and not merely with drives (e.g., Peters, 1958): in this discussion the terms "drive" and "motivation" are used interchangeably.

A unitary concept of drive can be taken to imply that these diverse characteristics of behaviour depend on the same features of the underlying mechanism. There is no a priori reason why this should be so, and some reasons for thinking to the contrary will be given later.

In addition, there are theorists whose concern it is to assess the usefulness of the hypothesis that "drive reduction" is essential for learning to occur (e.g., Brown, 1961; Miller, 1961a, b). This approach can lead to the postulation of drives to explain learning; for instance, a "manipulation drive" when it is found that monkeys learn to manipulate mechanical puzzles (Harlow et al., 1950). But at this point there is a danger of circular argument. Drive can, of course, be defined as that which is reduced when learning occurs, but if the drive concept is to be used to explain changes in responsiveness to a constant stimulus, then the question of whether or not learning is necessarily accompanied by drive reduction is an empirical one (see pages 594–603).

Another difficulty that arises when drive concepts are used in the interpretation of the temporal patterning of behaviour concerns the units of behaviour to which they should be applied. For instance, the nest-building activities of canaries can be analysed into gathering material, carrying it to the nest, and sitting (building) in the nest. These activities are all limited to one season of the year, all show related short-term fluctuations, and are all inhibited by stimuli from the nest. We could thus usefully argue that all are governed by a nest-building drive. But the correlation between them is not absolute, and to account for the pattern of changeover from one activity to the next it is necessary to postulate that each is accompanied by a self-suppressing effect (Hinde, 1958a; see Chaps. 25 and 26). This latter could perhaps be taken care of by postulating additional variables which modify the relationship between the nest-building drive and each of these activities. But each of them is itself complex, consisting of component activities which fluctuate more or less together. Are we then to postulate separate gathering, carrying, and sitting drives? If so, where is the process to stop? At this point the concept of a nest-building drive ceases to be useful (cf. Wiesner and Sheard, 1933).

Another aspect of the drive problem that must be mentioned concerns again the temptation to oversimplify when drive concepts are involved. In many theoretical systems it is implied that the independent variables—hormone dosage, hours of food deprivation, external stimuli, or whatever—affect behaviour either in only one way, by influencing drive, or in two ways, by their consequences on drive and their role as stimuli in influencing the nature of the behaviour the animal shows. In practice, such variables affect the organism in diverse ways, and their influence on behaviour is far from simple. Some examples of this will be discussed later (Chap. 10).

Finally, since drive concepts are used to explain activity, drives are often spoken of as though they activated behaviour in the same manner as physical energy can activate a physical system. Indeed in some theoretical systems, such as those of Freud (1915), McDougall (1923), Lorenz (1950), and Tinbergen (1951), some of the properties of physical energy have been imputed to drive concepts, so that drives are referred to as being "stored" or "expended", as "flowing away" or "sparking over". An element of this sort of confusion has also been present in the work of some learning theorists who invoke drive concepts to explain the elicitation of vigorous responses to weak stimuli, assuming that the force with which a rat pushes a bar is correlated with the strength of the drive. Such theorists could get into difficulties if they were to work with types of behaviour like incubation, in which the animal is relatively inactive. This confusion between drives as postulated to explain changes in responsiveness and drives as forms of physical energy has in fact led to serious confusion (see discussion by Carthy, 1951; Kennedy, 1954; Hinde, 1956a, 1960a).

More serious still, however, is the temptation to look for the activating or energizing drive inside the organism. However willing the drive theorist may be to rally physiological data to his aid, his looking inside the nervous system for drives, defined in terms of behaviour, is a logical mistake; and discussion about where drives originate is based on a misconception. Although the drive theorist will not find any drives inside the organism, he will find physiological correlates. He may find, for instance, that changes in the frequency of copulation per day are correlated with androgen levels, and hope to find that minute-to-minute fluctuations in sexual behaviour are correlated with other physiological changes. But as the analysis proceeds he will inevitably find that he is not talking about drives any more: the drive, so useful at an earlier stage, has just ceased to be relevant.

The position taken here, then, is not that drive concepts are always useful or always useless, but that they have a limited range of usefulness and that they can be misleading and dangerous if misused. This, of course, is true of all explanatory concepts. We shall use drive concepts in some of the following chapters in contexts which do not involve physiological data, and in such contexts we shall find that they can have a valuable role in permitting the formulation of unifying principles which cover a variety of observations and experiments. In other contexts, however, it will be more profitable to consider more directly how independent variables affect behaviour, and drive concepts will be superfluous.

8.2 SUMMARY

1. There are no clear criteria for distinguishing changes in behaviour which are to be ascribed to motivational factors from those which are not. There must be no implication that those so labelled are due to a single type of mechanism.
2. Drive concepts are most likely to be useful when they lead to an economy in the number of relationships between independent and dependent variables which must be specified. They can be valuable when a positive correlation between the possible dependent variables is of primary interest, but are misleading when the extent to which such correlations are not perfect becomes important.
3. Drive concepts can be useful for relating dependent variables, with no reference to antecedent conditions.
4. Drive concepts can be dangerous if they are used to explain diverse characteristics of behaviour which could result from quite different aspects of the underlying mechanisms; if they lead to oversimple implications about the manner in which independent variables affect behaviour; or if they are linked with energy models.
5. As physiological analysis proceeds, drive constructs cease to be relevant.

9

GENERAL AND SPECIFIC ASPECTS OF MOTIVATIONAL VARIABLES

We are accustomed to think that each activity of an animal is governed by a number of factors (independent variables) which are more or less specific to that activity. For instance, in most vertebrates male sexual behaviour varies with the concentration of testicular hormones, the presence of certain external stimuli, and so on. Feeding behaviour varies with the period since the animal last ate, with certain consequent internal changes, and again with the external stimuli present, but varies only to a relatively limited extent with sex hormone concentration. It is thus possible to identify for each type of behaviour a number of factors with which its occurrence is more or less closely correlated.

This cannot be the whole story, however, for prolonged food deprivation has some effect on sexual behaviour, albeit a slight one (Sachs, 1965). A variable normally considered relevant to one type of behaviour thus has an effect on the occurrence of another. In this case the effect is negative, and it can be described easily enough in terms of priorities and inhibiting effects: in the food-deprived animal food-searching behaviour is prepotent and sexual behaviour suppressed (see Chap. 17). It is also possible, however, that an activity can be influenced positively by a factor not normally considered relevant to it: thus certain carnivores (Beach, 1947), gulls (Armstrong, 1950), and monkeys (personal observation) are especially likely to show sexual behaviour after a mild disturbance; and painful stimulation, such as that arising from an electric shock to the foot, can elicit fighting (Ulrich et al., 1964) or copulation (Barfield and Sachs, 1968) in a number of mammalian species. Indeed there are factors which influence all the types of behaviour in an animal's repertoire: to cite an extreme example, cockroaches become torpid if the environmental temperature falls below a certain limit. The variables which influence behaviour may thus have either specific effects on a limited group of behaviour patterns or more general effects on more or less the whole repertoire of the organism, or both.

The same problem appears in the ways in which drive concepts have been used. One way we have already discussed by reference to the example of thirst—"drive" is used either as an intervening variable or as a postulated internal state which is linked, on the one hand, to particular preceding circumstances or treatments and, on the other, to a limited number of types of behaviour. On this view it is necessary, in order to account for all the behaviour of an organism, to specify a number of drives: one related to, say, drinking behaviour, another to feeding, another to fear, another to sex, and so on. The relative strengths of the various drives are thought of as playing a primary (though not, of course, exclusive) role in determining both what the organism does and the intensity with which it does it.

Many writers, on the other hand, use a concept of "general drive" which is related only to the intensity [1] of the behaviour shown. On this view the sort of behaviour which appears depends on particular aspects of the stimulation impinging on the animal and on its previous experience in the situation. Thus, whereas only part of the sensory stimulation has an influence in determining what sort of behaviour the animal shows, all "motivational variables" [2] and stimuli operating are supposed to contribute to the general drive and influence its intensity. In Hull's (1943, 1952) formulation, which was the most nearly comprehensive theory of this kind, the intensity with which any behaviour is shown depends on a multiplicative relationship between the "habit strength" appropriate to it and the general drive, the latter depending, in turn, on all stimuli and motivational variables impinging on the organism at the time (see also Spence, 1956).

Brown (1961), one of the more vigorous recent proponents of the general drive view, takes his stand in the first place on a theoretical issue. He

[1] The term "intensity" is intentionally used loosely in this book. Usually it refers to the amplitude or frequency of a movement, but it is also applied to patterns of behaviour which do not involve movement; for instance, incubation can be said to be more or less intense, one criterion being the strength of the extraneous stimulus required to interrupt it. In general it can be said that nearly all statements about intensity can refer to a probability measure of the response in question. They thus usually carry some implication about the causal factors for the behaviour in question and about the relation of these factors to those for other patterns acting at the time, as assessed from observed behaviour.

[2] The term "motivational variables" is used for independent variables believed to be associated with changes in behaviour of the type referred to on page 194, such as hours of food deprivation, sex hormone concentration, and stimuli previously associated with shock.

rejects drive concepts which function in determining both the nature and intensity of activity, primarily because the former function is "traditionally reserved for cognitions or associative tendencies", and argues that, "If both drive and habit are to be included in our theories, then the two should affect behaviour in different ways; otherwise only one construct seems to be required". He then argues that if "drive" is to function only as an activator, it cannot be directed towards any one specific goal or selectively activate one type of associative tendency, since this would involve also a directive function: drive must, therefore, be general.

As an example of the opposite view, Tinbergen (1952) appeals to common sense. He points out that each activity has particular factors associated with it and doubts that "excitement" could ever be "really general".

At first sight the problem should be simple enough to settle empirically. Does change in a motivational variable influence the intensity of one type of behaviour or all? Unfortunately, the issue is not so clear cut as the use of drive concepts would imply. To prove the extreme form of the general drive hypothesis it would be necessary to show that each motivational variable influenced all responses (Fentress, 1968). But an animal can do only a limited number of things at a time, and so a supposedly non-specific effect could only be proved by observations of all specific activities each in a situation appropriate to it. Although this may seem an academic point, it is rare that proponents of a general drive theory have observed even two different activities in the same experiment (see below). In any case changes in motivational factors do not produce a simple effect, either specific or general, but have multiple and ramifying consequences in the systems controlling behaviour. A distinction between specific and non-specific effects of a stimulus change is therefore often difficult to draw; and the question is more profitably phrased in terms of how specific or how general the effects of each factor may be. It is, however, worthwhile to consider some examples of the various types of evidence used to support the view that motivational factors may have a non-specific effect.

9.1 INTERRELATIONS AMONG HUNGER, THIRST, FEAR, AND SEX

Some anecdotal observations on the influence of mild disturbance on sexual behaviour have already been mentioned. Similar phenomena are well known to naturalists and have been quoted especially in the literature on displacement activities: this will be discussed further in Chap. 17. Here we are concerned with laboratory studies which appear to show that one external factor, "need state", or "drive" affects behaviour normally dependent on another. In each case in which an effect has been demonstrated, we must ask whether the evidence supports the view that the effect of the independent variable is in fact likely to be general or limited to only a few activities.

1. *HUNGER ON THIRST OR VICE VERSA*

Miller (1948a) trained rats to run down an alley for water and then tested them when they were not thirsty. They ran faster and drank more if they were hungry when tested than if they were not. On a simple drive view, therefore, it seems as though the irrelevant drive of hunger facilitated running to water. Again, although a number of experiments have shown that either strong hunger or strong thirst impairs the acquisition and reduces the resistance to extinction of a task based on the other, there is some evidence that low levels of the "irrelevant" thirst can enhance performance of a task performed under hunger. The effect was asymmetrical, however, since irrelevant hunger did not produce a similar effect on a task performed under thirst (Bolles and Morlock, 1960).

Bacon and Bindra (1967) obtained evidence for a relationship between hunger and thirst in a different way. A conditioned stimulus (CS) was first paired with an unconditioned stimulus (water) for thirsty animals. They then assessed the effect of the CS on the acquisition of an instrumental response (running) trained with either hunger and food reinforcement or shock and shock avoidance. In both cases the acquisition of the instrumental response was facilitated. Since the "incentive-motivational" properties of the conditional stimulus (see pages 259–260) acquired under one set of motivational conditions affected responses under a different set, they argue that it must create a central state with widespread influences.

The interpretation of such data is, however, not straightforward. Hunger and thirst are known to be closely related, and an effect of one on behaviour relevant to the other need not by itself imply any wider generality. Both hunger and thirst are influenced by peripheral stimuli from mouth, throat, and stomach, and some of these may be common to both states. Indeed the influence of the one on the other involves both short-term effects consequent upon peripheral changes (e.g., Verplanck and Hayes, 1953; see also Warden, 1931; Kutscher, 1967) and longer-term effects, in which feeding increases thirst and thirst, in turn, inhibits feeding (McFarland, 1965a, b; McFarland and L'Angellier, 1966; see also pages 248–254).

The relations between hunger and thirst have also been studied in another sort of experiment aimed at evaluating the concept of a general drive. If a given set of conditions influences a general drive state, then differences between animals under these conditions will be due, in part, to individual differences in the relationships between deprivation conditions and drive. If the concept of a general drive state is to be useful, these individual differences should be valid for all types of deprivation. In fact, when rats are tested under various conditions of food and water deprivation, the individual differences have little generality across deprivation conditions (Anderson, 1937, 1938; Bolles, 1959).

2. *FEAR OR ANXIETY*[3] *ON HUNGER AND THIRST*

Although mild disturbance or shock usually disrupts the current behaviour, a number of studies have shown that electric shock will increase the response of rats to food or water, provided it is administered in a situation different from that in which they are tested (Siegel and Siegel, 1949; Amsel and Maltzman, 1950; Siegel and Brantley, 1951; Muenzinger et al., 1952; Miller, 1959). In some of these studies the dependent variable was the amount of food or water ingested, in others the speed of running in a situation in which the animal had previously been reinforced. In either case the direction of the effect may depend closely on the conditions: if the shock is administered in the test situation it may lead to a reduction in the amount ingested (Amsel, 1950).

A careful study by Tugendhat (1960a, b) of the influence of electric shock on the feeding behaviour of three-spined sticklebacks is of special interest here. This author did not confine her attention to one measure of behaviour, and she recorded the behaviour of the fish in both the "home area" and the "food area" of the tank. These were partially separated by a partition, and the fish were shocked only in the latter. After electric shock the behaviour while feeding was that normally characteristic of more severe deprivation—the number of completed feeding responses per time spent feeding was increased. But shock also decreased the amount of time spent feeding. Over the whole test, therefore, shock might produce either an increase or a decrease in the total number of prey animals eaten, the latter being more likely with high shock intensity.

Although these experiments are compatible with some sort of general drive hypothesis, other explanations are possible. For instance, fear-producing stimuli are likely to produce peripheral effects, mediated by the autonomic system. Some of these, like dryness of the throat in land animals, could affect ingestion. The hypothesis that pain or fear affects feeding and drinking through these peripheral changes, though perhaps not likely (Siegel and Brantley, 1951), cannot be ruled out.

Another possibility is that the consequences of shocks affect the incentive value of the food rather than the drive state of the animal. Strongman (1965) found that brief (3 sec) shocks (2.6 milliamp) increased the food intake of rats during the next hour, but longer durations of shock (30 and 300 sec) decreased it. If the food was adulterated with quinine, however,

[3] See pages 348–349 for note on the way in which these terms are used.

food intake was decreased at all shock durations. Strongman points out that if the shock affected some general state, such as "anxiety", which in turn affected other types of behaviour, such as eating, it would be expected to enhance intake whether or not the food was adulterated. He therefore suggests that there is an incentive/anxiety interaction which could be considered as a threshold mechanism. With low anxiety, thresholds are lowered so that there is an increase in responding with positive incentives and a decrease with negative. At higher anxiety levels, when feeding is reduced even with non-adulterated food, Strongman regards the hunger drive as temporarily blocked.

In other cases, reinforcement may play a role. For instance, Sterritt (1962), whose study is in some respects comparable to that of Tugendhat, studied how electric shock influenced the feeding of pigeons. The shock was administered for 5 sec in each minute and was not under the bird's control. The birds ate more when the shock was on than was eaten by unshocked control subjects during a comparable period. However, when the shock was off the experimental birds ate less than the control birds did and consumed less in toto. The increased eating during the period of shock was interpreted in terms of "superstitious" learning (Skinner, 1948): eating was reinforced because it was followed by shock reduction (see also Fowler, 1963; Sterritt, 1965).

Finally, it must be noted that electric shock is an extremely artificial form of stimulation and may have diverse effects other than painful ones, especially at low intensities (Beach, personal communication).

3. *HUNGER AND THIRST ON STARTLE OR AVOIDANCE RESPONSES*

The data here are somewhat conflicting. Earlier reports indicated that rats run faster to escape from pain or fear when moderately hungry or thirsty than when satiated. Even the amplitude of the unconditioned startle responses given to a loud noise is greater in rats motivated by both hunger and conditioned fear than if motivated by either separately. Furthermore, rats which are hungry show greater startle responses than recently satiated animals (Fig. 9.1, Meryman, quoted by Brown, 1961). However, Griffiths (1962) found that hunger decreased reactivity to shock, and Blanchard and Blanchard (1966) found that food-deprived animals vocalized less but showed no difference in flinch and jump responses to shock as compared with satiated animals. No doubt the conflicting data are due in part to the differences in the deprivation and testing conditions used.

Positive results from such experiments are more difficult to interpret in terms of specific peripheral effects than are the experiments showing an effect of fear- or anxiety-inducing stimuli on feeding or drinking, though it remains possible that peripheral conditions (e.g., in the throat) common to hunger or thirst and anxiety could act as conditioned stimuli for the startle response.

4. *FEAR, ANXIETY, AND ACTIVITY ON SEXUAL BEHAVIOUR*

In addition to anecdotal observations, a number of experimental studies have demonstrated a positive effect of fear- or anxiety-producing stimuli

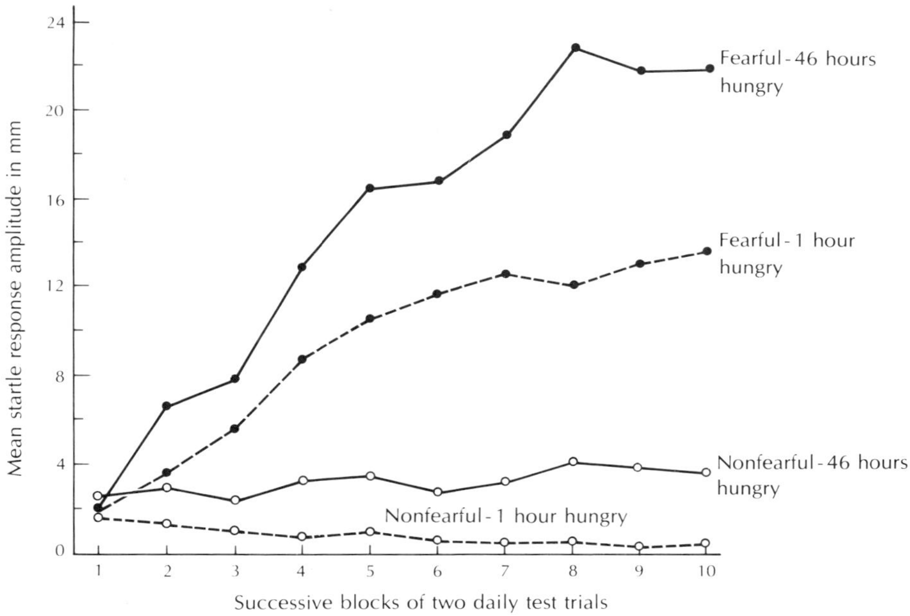

Fig. 9.1. Startle-response amplitude as a function of fear, no fear, intense hunger, weak hunger, and their combinations. (After Brown, 1961.)

Startle responses to a pistol shot were tested in four groups, two of the groups being rendered fearful by being given an electric shock in the test situation once a day.

on sexual behaviour. Beach et al. (1955) found that the administration of electroconvulsive shock to male rats decreased the rapidity with which sexual arousal occurred; for instance there were longer delays before mating started and a slower recovery from ejaculation. At the same time, however, there was an increase in reactivity to the stimuli derived from coitus—a greater rate of intromission, a decrease in the number of intromissions before each ejaculation, and a higher number of ejaculations per test. After considering various possible explanations, they considered an effect of the shock treatment on the functioning of the autonomic system, with consequent repercussions on sexual behaviour, as the most likely. In a later paper Beach and Fowler (1959b) found that anxiety engendered by previous experience of being shocked in the situation reduces the amount or duration of sexual stimulation necessary to produce ejaculation in male rats. A similar effect is produced if mild (but not strong) shock is given each time the male mounts the female. Barfield and Sachs (1968) also found a positive effect of electric shock on the copulatory behaviour of male rats: they interpreted their results in terms of "general arousal".

Larsson (1963), also studying copulatory behaviour in rats, found that two 1-min periods of handling by the experimenter during a copulatory sequence resulted in an increase in sexual activity and a reduction in the number of intromissions before the first ejaculation in older males but not in younger ones. He suggests that the absence of an effect in the younger animals was due to their sexual behaviour being near the maximum already. Larsson also refers to earlier studies showing that the sexual activities of old males is greater in the active part of the light/dark cycle and also while they are copulating in a group (see also Larsson, 1956).

5. *HUNGER ON SEXUAL BEHAVIOUR*

Jarmon and Gerall (1961) did not find that food deprivation led to an elevation of sexual behaviour in guinea pigs with the scoring system they used. Sachs (1965) found that it had a mild inhibitory effect with rats, though even after 8 to 9 days' starvation, 11 of 16 male rats ejaculated without marked changes in most aspects of their sexual behaviour.

6. *SEXUAL FACTORS ON FEEDING AND DRINKING BEHAVIOUR*

It has been claimed that rats trained to run a maze when hungry or thirsty subsequently ran faster when tested during oestrous induced by exogenous hormones than at other times (Felsinger, quoted by Miller, 1959), but the study has apparently not been repeated. In any case such a finding must be considered with reference to many other studies showing variations in running wheel activity and other types of behaviour with the oestrous cycle. These indicate that the oestrous cycle is accompanied by profound physiological and behavioural changes which go beyond the alterations in sexual receptivity. Although a correlation between running speed to food and sexual receptivity could be interpreted as due to an influence of sex hormones on a "general drive", it would still be necessary to specify the generality of the effect.

7. *CONDITIONED FEAR ON STARTLE RESPONSES*

As indicated above (Fig. 9.1), fear has been found to augment the amplitude of startle responses. However, Kurtz and Siegel (1966) found that, whereas the magnitude of startle responses of rats to a loud noise was augmented by a conditioned stimulus previously paired with shock to the feet, it was not so augmented by one previously paired with shock to the back. They thus suggest that the former result is due to a postural adjustment minimizing the severity of foot shock.

8. *CONCLUSION*

In Hull's conception, all need states and stimuli were thought of as contributing towards a general drive, which then influenced all behaviour. In principle, no one of the studies just quoted could give conclusive support for this hypothesis, because the effects on at most two types of behaviour were assessed in each case. In fact the evidence shows that need states or stimuli normally relevant to one type of behaviour can augment others but do not always do so. Further progress demands more sophisticated experiments to determine the range of inputs which will augment specific types of behaviour, the range of behaviours augmented by each input, and the conditions under which these effects are to be found. Furthermore it is necessary to consider inhibitory as well as facilitatory effects: we shall later consider further evidence that factors for one activity may facilitate a second when they are of moderate strength but lead to its being inhibited when stronger (Fentress, 1968; see page 416).

When a facilitatory effect on an apparently irrelevant activity is found, it still has to be established that it is best considered in terms of a concept of general drive. Bindra (1968) prefers to think in terms of a limited number of central states, which he links to terms such as "fear", "anxiety", "frustration", and "excitement", each capable of affecting a variety of responses. Some alternative possibilities—peripheral mediators, major changes in endocrine balance, reinforcement, and so on—have been mentioned already. The possible role of the reticular system will be considered later. No one of these is likely as a general explanation (e.g., Rescorla and Solomon, 1967): all have explanatory value in some instances.

Miller (e.g., 1959) has explored yet another possible theoretical explanation. Whereas Hull supposed that the various motivational variables had completely generalized effects, Miller suggested that drives have some of the same properties as strong stimuli, drive generalization occurring as a special case of stimulus generalization. Thus finding that rats, trained in a runway when hungry but tested when satiated, would run faster if given an electric shock just beforehand, Miller suggested that the habit learnt when hunger stimuli were present generalized to pain. Since the internal states are so different in the two cases, Miller suggested that the generalization might be based on similar states of muscular tension occurring in the two conditions. As Miller himself points out, no definitive experiments which could decide between his hypothesis and Hull's have yet been done.

In summary, then, studies of the interrelations among different activities provide some evidence that motivational variables normally considered relevant to one group of activities can sometimes affect others also. Some additional instances will be mentioned later (pages 214–223 and 406–418). Although such cases have been interpreted in terms of a general drive influencing all types of behaviour, it seems doubtful whether such a concept is useful beyond the most superficial level of analysis.

9.2 ASPECTS OF THE GENERAL SENSORY INPUT

A second type of evidence concerning general effects of independent variables comes from studies in which responsiveness to a variety of stimuli, or performance in a skilled task, has been shown to vary with the general level of sensory input falling on the receptors. Such effects have been suspected for a long time in insects, where the ocelli (simple eyes) often seem not to mediate specific responses but to accentuate responses to light stimuli received through the compound eyes. The effects, however, vary with the light intensity, and it seems likely that their role is more one of modulating responsiveness to visual stimuli according to the overall light intensity (e.g., Dethier, 1963; Kerfoot, 1967; Jander and Barry, 1968).

However, such effects certainly occur in vertebrates. One type of evidence comes from studies showing that input otherwise apparently irrelevant to the response in question can postpone the waning of response strength which usually occurs with repeated elicitation. One of the earlier detailed studies is due to Birukow (1951). If a frog is placed on a table tipped rhythmically once every 5 sec, the animal makes compensatory movements which tend to maintain its orientation in space. These compensatory movements continue for long periods with no decrement. The response is mediated primarily through the semicircular canals: in an optically homogeneous field, or in darkness, the responses are weaker but still show no decrement. However, if the optic nerve is severed, the response does slowly wane. Apparently the input from the retina, even under conditions of darkness, enables the responsiveness to increase again to its previous level between successive tilts of the platform, whereas in the absence of this input it cannot. A similar effect is produced by deafferentation of other parts of the body; for instance, severance of dorsal roots 8 to 10. Proprioceptors and other enteroceptors work in the same way as the eyes in maintaining the "tone" of the neuromotor apparatus. Similarly Ewert (1967b), studying the waning of the orientation of the common toad (*Bufo bufo*) to prey objects, found that optical and tactile stimulation administered between sequences of prey stimuli facilitated recovery from habituation.

Comparable effects can be found at a physiological level. For instance Rowell and Horn (1968), recording from neurones in the insect tritocerebrum, found that the response to a visual stimulus gradually waned. It could, however, be restored by stimulation of the contralateral connective (see also Chap. 13).

Such studies deal with the influence of diverse sensory inputs in maintaining a response. A greater bulk of work has been concerned with the effects of stimulating one sensory modality on the sensitivity of another. Chernetski (1954) found that a tactually elicited leg flexion in the frog *Rana clamitans* was facilitated by the short sound of a bell immediately before the tactile stimulus. The sound alone produced no response. Since this intersensory facilitation was reduced by sympathectomy, Chernetski suggested that it depended on sympathetically elicited somatic sensory inflow. Intersensory effects are, however, complex (see, e.g., London, 1954, for a review of Russian work). For instance, although a number of investigators have found that auditory stimulation improves visual sensitivity in human subjects, some workers find a facilitatory effect of visual stimulation on auditory sensitivity, and others an inhibitory one: the direction of the effect may vary with the frequency of the sound (Sheridan et al., 1966). Thus no generalization claiming that stimulation through one modality either increases or decreases the sensitivity of others can have general validity. Furthermore, it has been shown that sensitivity thresholds vary not only with stimulation through other sensory modalities but also with various other aspects of the state of the body. Thus, skin sensitivity varies with the menstrual cycle in women (Herren, 1933) and can be increased by exogenous hormones in canaries (Hinde and Steel, 1964; see also page 231).

Neurophysiological evidence for these effects has also been obtained: stimulation of one sensory system may lead to a change in the electrical response evoked by stimulation normally considered appropriate to another part of the central nervous system (see page 129). For instance, Gerard et al. (1936), working with cats, found that the potentials evoked in the brainstem by the tick of a watch were larger when the eyes were simultaneously exposed to light than when the animal was in darkness. By recording from single cells Horn (1963) has shown that the activity of the cells in the visual cortex of the cat can be affected by stimulation of the skin as well as by retinal illumination. In this case some of the cortical cells responded in the same direction to the two types of stimulation while others responded in opposite directions: the effect is thus neither simple nor general.

9.3 LEVEL OF ACTIVATION AND OPTIMUM INPUT

These cases have been concerned primarily with the effects of the level of total stimulation on an organism's snsitivity to, or responsiveness to, particular classes of input. We must also consider the effect of non-specific stimulation or motivational factors on a number of other physiological variables—muscular tension, skin resistance, the pattern of electrical activity in the brain, the heart rate, and so on. These measures show some correlation with activity of almost any sort and change with some degree of consistency in one direction with increased stimulation of the organism and in the opposite with decreased stimulation. In human subjects they are related not only to stimulation in the ordinary sense of the term but also to changes in motivating conditions, to activity or readiness for activity, problem-solving, and so on (Duffy, 1962).

Such findings have led to the suggestion that the non-specific effects of external stimulation, motivational factors, or self-engendered activity are such as to produce a state which can be scaled along a continuum from drowsiness to extremes of alertness and responsiveness. Concepts such as "vigilance" (Head, 1920), "the specific vigilance reaction", "psychological arousal", and "activation" (e.g., Duffy, 1951, 1962; Bindra, 1959; Malmo, 1959; Lindsley, 1960), as well as general drive, have been coined to describe the changes involved.

The relations among these concepts have recently been discussed by Duffy (1962). On her view, the various physiological measures mentioned above can be used as indices of the "level of activation" of the organism, which is produced either by stimuli or other extraneural effects, or by mental activity. This "level of activation" is supposed to be related to the rate at which energy is released by metabolic activity in the tissues, the inadequacy of measures of total metabolism as indicators of the level of activation being ascribed to the influence of the local energy reserves within the body. "Activation", as assessed by the various physiological indices mentioned, has in fact been shown to be related to reaction time, speed of movement, intensity or extent of response, and various other measures of performance: a well-known example is the way in which the extent of the knee jerk is increased when the fist is clenched.

The concept of a "level of activation" or "arousal" has a strong subjective appeal, but, as with so many other concepts used in the study of behaviour, the limits of its usefulness must be defined. It could be valuable over a wide range of situations only if the various indices by which the level of activation is assessed were highly correlated. This is not in fact the case. Lacey (1966) has cited much evidence that complete dissociation between arousal or activation, as measured by behavioural, autonomic, and electrocortical signs, can be produced in a number of ways, including pharmacological techniques and localized lesions in the central nervous system. For instance, atropinized dogs show a pattern of electrical activity in the cortex characteristic of sleep but complete behavioural arousal (Wikler, 1952). Again, Feldman and Waller (1962) showed that cats with bilateral lesions of the posterior hypothalamus cannot be behaviourally aroused, although desynchronization of the cortex—an index of physiological arousal—can readily be produced. The converse dissociation was produced in some degree by bilateral lesions in the midbrain reticular formation. These and other studies show that the various indices of arousal—autonomic, electrocortical, and behavioural—are separable experimentally.

Furthermore, evidence for such dissociation is also available for normal human subjects not subjected to drugs or other experimental insult. In fact, as Duffy herself points out, even the correlations between somatic indices of activation are not always high, in part because the response to any particular stimulus situation is nearly always patterned to meet the demands of that situation. Thus Lacey et al. (1963) showed that patterns of heart rate and skin conductance responses depend on the nature of the situations to which human subjects are exposed: they offer evidence that heart rate increases with tasks requiring rejection of environmental information (e.g., mental arithmetic) and decreases with tasks requiring its intake (e.g., exposure to flashes, with requirement merely to report on their colour or pattern), while skin conductance increases with both. Lacey (1966) cites a number of other studies also leading to this view (see, for example, Malmo et al., 1951; Ax, 1953; Malmo, 1959; also Wilson and Bartels, 1968) and offers an explanation of the lack of correlation between cardiac and conductance measures, supported by physiological evidence, in terms of a feedback mechanism from the cardiovascular system on cortical activity (see also Flanagan, 1967).

Using a somewhat different approach, Brady et al. (1969) recorded various autonomic indices during "emotional conditioning" produced by shock administered to the feet of rhesus monkeys. The course of conditioning and extinction was monitored by the suppression of ongoing lever-pressing for food. Initially conditioning was accompanied by decreases in both heart rate and blood pressure, but in the later stages of conditioning the reverse effects were found.

Furthermore, the direction of response to a stimulus may depend on the state of the animal before stimulation. Bridger (1960) studied the heart rate changes induced in human newborns by an air puff. In any one individual the puff could produce an increase if the initial heart rate was low and a decrease if it was high.

It is thus clear that activation processes are multidimensional, the correlations between the indices emphasized by Duffy being in part a consequence of the limited number of situations studied. Thus attempts to measure activation or arousal along a linear scale overlook qualitative diversity: the concept is useful only so long as the dependent variables are correlated, and thus only within very limited groups of situations.

In a recent discussion of this problem, Berlyne (1967) has conceded that the various indices of arousal are imperfectly correlated with each other, but believes that the lack of perfect correlation does not matter so long as arousal is regarded only as a dimension and not as a process in the central nervous system. He regards attempts to be overprecise about the arousal dimension as premature, since we are not yet ready for more than an ordinal scale of measurement. He therefore takes a view about arousal similar to that taken here about drive—the concept has a certain usefulness at a preliminary stage of analysis, but its range of usefulness is limited. Certainly, many students of behaviour who do seem to find a concept of arousal useful use it only at a qualitative or barely quantitative level of analysis. Mason (1967), for instance, uses this concept to interrelate his observations on social behaviour and play in chimpanzees—but it remains an open issue how far the concept will continue to be useful as analysis proceeds to more refined levels.

A major difficulty with the concept "level of activation" is thus the extent to which the several indices of activation fail to correlate with each other—a difficulty which becomes more important as analysis proceeds. A second difficulty comes from experiments in which the total sensory input is varied beyond the normal experimental limits. Most of this work has been concerned with reduction in the input and has used human subjects. Characteristically they are confined in soundproof and anechoic chambers, wearing gloves to reduce tactile stimulation and restrict movement (sensory isolation); or they are kept under conditions (e.g., wearing translucent goggles) in which the stimulation falling on their receptors is deficient in patterning but relatively little changed in amount (perceptual isolation). Periods ranging from a few hours to several days under such conditions produce deficiencies in a wide variety of intellectual, perceptual, orientational, and physiological tests: these deficiencies are often accompanied by hallucinations and dream-like states in which coherent thinking is difficult. The subjects sometimes respond to the decrease in stimulation by an increase in "activation", as indicated by the physiological indices discussed above (Bexton et al., 1954; Solomon et al. [eds.], 1961): there can thus be no simple relationship between sensory input and "level of activation".

Such studies demonstrate that normal behaviour disintegrates when the general sensory input falls either in amount or patterning below a certain level. It would, however, be premature to interpret the dysfunction as a consequence merely of the quantitative level of sensory input. It is not even clear that fewer impulses pass up the sensory nerves during sensory isolation, for while some cells in the cat's retina fire more frequently in the light than in the dark, for others the opposite is the case, and in some no change can be detected (see page 108). Furthermore, as discussed in Chaps. 5 and 20, there are reasons for thinking that regular relationships between different types of sensory input may be necessary for integrated behaviour: it may be the absence of these relations, rather than the low quantitative level, that is responsible for most of the behavioural abnormalities accompanying sensory or perceptual isolation (e.g., Schultz, 1965).

The concept "level of activation", therefore, has limits to its usefulness. It also carries with it a danger which derives from the readiness with which it, and similar concepts, are held to be congruent with the concept "general drive" (e.g., by Brown, 1961; see also discussion by Berlyne, 1967). In the first place, such a view is not fully in harmony with the data. For instance, food and water deprivation, which would be held to *increase* general drive, is associated with a *decrease* in heart rate, and vice versa (O'Kelly et al., 1965). But even more important is the fact that any such argument involves a missing link. The evidence on which Duffy's concept "level of activation" was based was concerned primarily with showing that there are a number of consequences common to many and diverse motivational variables. This is not the same as saying that they reflect a general state causal to all possible activities.

That it may in fact be useful to think in terms of general factors or states which influence a wide range of activities is indicated by two types of evidence advanced by Lát (e.g., Lát and Gollová-Hémon, 1969; see also Lát, 1963). The first involves the study of individual differences. Rats which explore vigorously also respond intensively in a number of other situations (feeding, defence, sexual activity, grooming, etc.). A common factor related to all these activities can thus be presumed. This factor is subject to genetic influences and can also be modified temporarily or permanently by dietary and endocrinological factors, and temporarily by exteroceptive influences. The second type of evidence involves correlations between different types of behaviour and a physiological index. Rats are placed in an activity recorder and their various maintenance activities (e.g., rearing, locomotion, sniffing, cleaning, sitting, and sleeping) recorded (Fig. 9.2). By weighting differentially the various types of behaviour a "behavioural index" can then be computed which shows a very good correlation with hippocampal theta activity (Fig. 9.3). There is thus a relationship between the type of maintenance activity shown and the theta-wave activity (though which way the cause–effect relationship goes is of course not proved). On the basis of these two sources of evidence, Lát postulates a "non-specific excitability level" (NEL) which influences the intensity of all types of behaviour. Once again, it must be stated that this is not the same as saying that all independent variables which influence the intensity of any activity influence that of all activities.

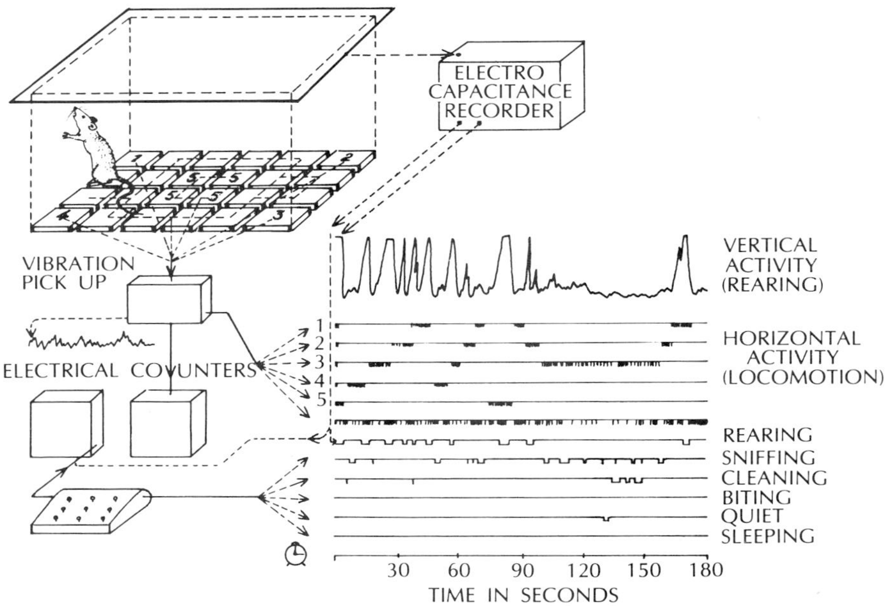

Fig. 9.2. Apparatus for recording activity of rat. (From Lát and Gollová-Hémon, 1969.)

Locomotion horizontally is recorded by the "vibration squares" which give information about the place, intensity and duration of locomotion, rhythmic scratching, etc. Vertical activity (e.g., rearing) is registered by an electrocapacitance recorder. All other types of behaviour are recorded manually on an event recorder.

Fig. 9.3. Correlations between the "non-specific excitability level" (NEL, sensu Lát), defined behaviourally, and hippocampal theta (slow wave) activity. (From Lát and Gollová-Hémon, 1969.)

Left and right indicate different trials on the same rat. Durations of specific responses in successive 20-sec periods are indicated on the ordinates. These are weighted according to the ordinate scales, and the NEL represents the sum of the weighted scores. The dotted line represents the sum of rearing reactions and number of squares crossed during successive 20-sec intervals: this provides a simple measure.

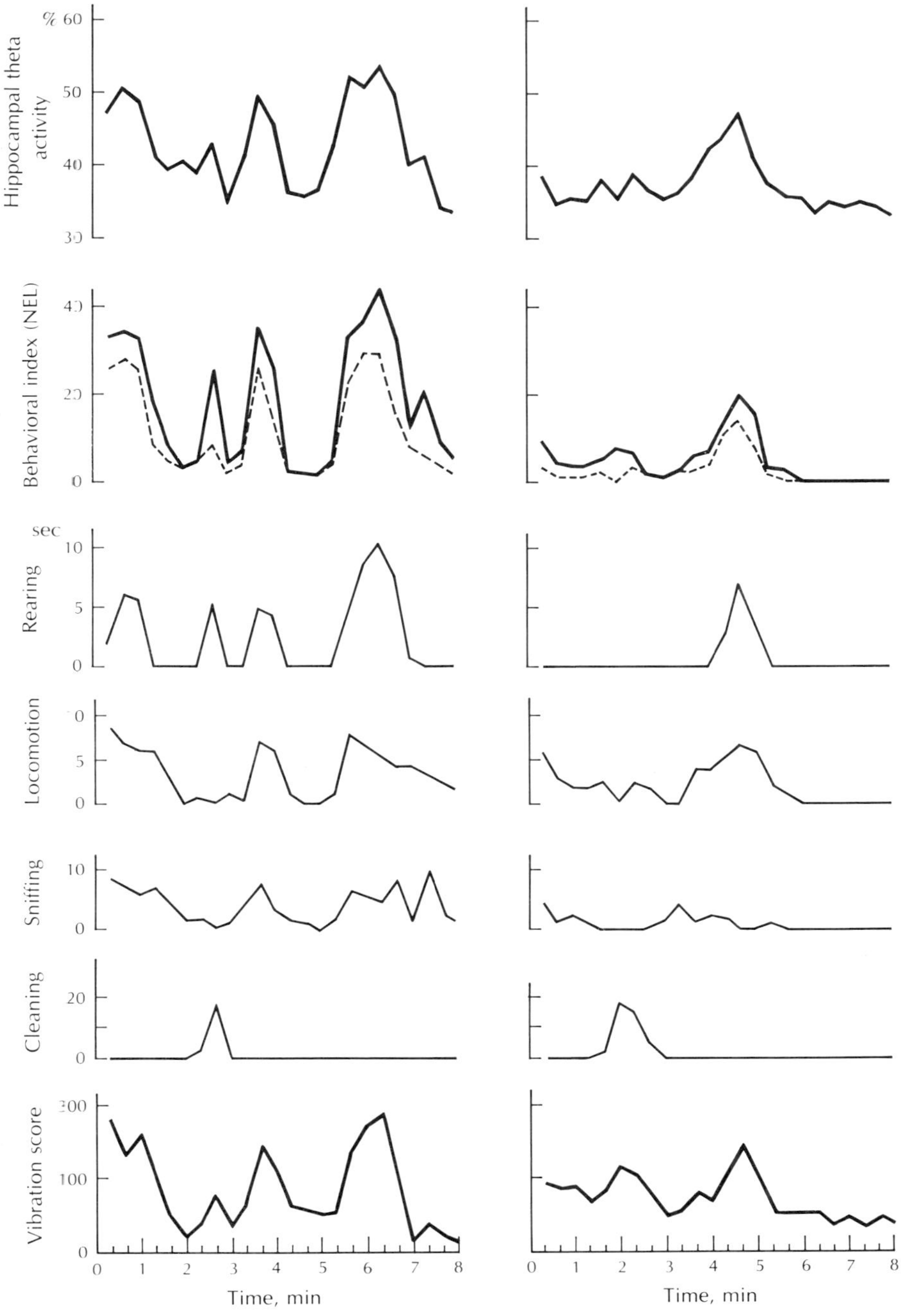
Hippocampal theta activity
%
Behavioral index (NEL)
Rearing
sec
Locomotion
Sniffing
Cleaning
Vibration score
Time, min
Time, min

In spite of their drawbacks, such concepts as "level of activation" have thrown into relief the important possibility that there are optimum limits of input for responsiveness, efficiency, or performance. On this view the relation between quality of performance and activation has the form of an inverted U (Schlosberg, 1954; Duffy, 1962; Hebb, 1955; Leuba, 1955). So far we have been concerned with evidence for the rising arm of the inverted U, with increase in the dependent variables or improvement in performance as the input increases. What is the evidence for a decrease in performance as the input increases still further?

Some of it comes from common experience. Many human activities depend on a tendency to seek moderately high levels of stimulation—much of the entertainment industry depends on this fact (Hebb, 1949). The level of stimulation sought, however, is either only moderately high or else soon relieved. Very intense stimulation is experienced as unpleasant, and performance under such conditions deteriorates, as in combat (Marshall, 1947) or disaster (Tyhurst, 1951). There is also experimental evidence that sensory overload, resulting from intense multisensory input, may lead to deterioration in performance and grossly abnormal motivational states (Lindsley, in Solomon et al., 1961).

Data supporting an inverted U-shaped relationship between motivational factors and performance are also available from studies of animals. Various measures of performance, such as pressing the bar in a Skinner box for food reinforcement, are related in a curvilinear manner to the period of deprivation. In rats, the rate increases up to about 24 hr of food deprivation, but then falls off. Bélanger and Feldman (1962) argue against the usual view that the falling off is due to inanition, and suggest that it is due to a decrease in efficiency when "arousal" passes a certain level.

The effect of intense motivational factors varies with the difficulty of the task. Broadhurst (1959) tested rats on a brightness discrimination which had to be made under water. The rats were first kept submerged for periods of 0 to 8 sec, before being released to swim through a T maze. If they turned correctly they could come to the surface to breathe. For easy discriminations, performance was little affected by the period of submergence before release, but for a difficult discrimination it was better with a period of 2 sec than 0, 4, or 8 sec (see also pages 224–225).

Such studies also provide some support for the "Yerkes-Dodson Law" that the optimum motivation for a learning task decreases with increased complexity of the task. Although useful as an empirical generalization, this has too often been taken to imply that different types of motivation when increased beyond an optimum affect performance deleteriously by similar mechanisms. In fact the effects are probably produced in diverse ways—for instance by overvigorous responding in the case of hunger and by competing emotional responses in the case of shock or air deprivation (see Bolles, 1967).

Another type of evidence which seems to indicate that there is an optimum level of input comes from studies in which animals are able to administer intracranial electrical stimulation to themselves by performing a suitable operant task. Olds (e.g., Olds and Milner, 1954; Olds, 1961) has shown that rats will press a lever to give themselves electrical stimulation through electrodes permanently implanted in certain areas of the brain ("reward areas"). Elsewhere ("punishment areas") electrical stimulation may be aversive. In some areas both effects are found—for instance subjects will alternate between a response which turns the stimulation on and one which turns it off, or will both work to turn stimulation on and also to escape from continuously delivered trains of stimuli (Bower and Miller, 1958; Roberts, 1958; Hodos, 1965; Poschel, 1966). Reynolds (1958) found that the rate of bar-pressing shown by a rat was related in an inverted U-shaped fashion to voltage magnitude: with increase in the intensity of stimulation the rate of bar-pressing increased to a maximum and then decreased. Such experiments must not be interpreted too lightly, for large differences in input voltage or train length may involve qualitative changes in the physiological nature of the stimulus. Over a considerable range of train length for a constant current stimulus, Prescott (1964) found that rats adjusted their response rates so that they received a more or less constant amount of electricity. Further, pulse trains longer than those self-selected may still not be aversive (Keesey, 1964).

The rate of responding for intracranial stimulation may be increased by food deprivation (Brady et al., 1957; Goodman and Brown, 1966). It is also *positively* correlated with the variations in activity which accompany the day/night rhythm and the oestrous cycle (Prescott, 1966, 1967). This does not favour the hypothesis that the rat adjusts its bar-pressing to receive a *constant* optimal amount of general input. Prescott showed that the changes in rate of responding for intracranial stimulation are not artifacts of the oestrous or diurnal changes in activity but reflect correlated changes in the reinforcing properties of electrical stimulation. One possible hypothesis here is that the change in motivational state involves a change in the optimum level of input, that is, in the level of input (specific or general) for which the animal works. A similar hypothesis could be applied to some of the results of experiments on light-reinforced bar-pressing (see Chap. 24).

The view that animals seek a level of input which varies with their state receives some support from a different type of experiment. One way in which an animal can control the input into its central nervous system is to alter its general activity. It is well known that the opportunity to run in a running wheel is reinforcing for rats, the animals tending to run so that a given amount of activity occurs in each 24 hr (Kagan and Berkun, 1954). The amount of activity varies with the motivational state: for instance it changes with the oestrous cycle (Richter, reviewed Munn, 1950). Kavanau (1963), working with deermice (*Peromyscus crinitus*), has shown that the activity level, as measured by a running wheel, varies according to the motivational state at the time. These animals will learn to press levers to operate an activity wheel, whether the lever unlocks a previously locked wheel, or starts or stops one driven by a motor. The patterns characteristic of spontaneous free-wheel activity tend to be repeated when the operation of the wheel requires an instrumental act, whether that act operates a motor-driven wheel or unlocks a free-running one. However, the deermice do not accept motor-driven running which is initiated by the experimenter. If they have been trained to operate the motor with on and off levers, they will stop the motor if it is started by an outside agent, and vice versa. Thus they resist any change in activity imposed on them from outside.

If there is an optimum level of input, then we should expect there to be mechanisms for controlling both the level of input and its central consequences. Some of the experiments quoted in preceding paragraphs show that excessive input can be aversive, and the possible existence of an arousal-inhibiting system within the central nervous system has been suggested, though the evidence is yet slender (Berlyne, 1960; Delius, 1967).

In summary, there is clear evidence that behaviour becomes inefficient outside certain limits of sensory input. The mechanisms involved, however, are far from clear. The dysfunction which accompanies a low level of input is certainly not necessarily due to inadequate "activation", for the indices of activation may be high. Indeed, in many cases the operative factor may be not the low total level of input but the absence of ordered relations between movement and consequent feedback (Chap. 20). Likewise the effects of excessive stimulation may be due to interfering responses, distraction of attention (page 125), and so on. Moreover, although normal functioning requires the level of sensory input to be within certain limits, there is little hard evidence to suggest that, within those limits, an increase in any type of input will necessarily increase the intensity of behaviour: the extent to which any particular pattern of behaviour is shown depends primarily on factors specific to that pattern. Finally, whereas animals have a "preferred level" for many kinds of exteroceptive and proprioceptive input, additional motivational factors such as hunger, which should increase the level of activation or general drive, may in fact increase the amount of stimulation sought. On the hypothesis that the animal seeks an optimum level of general input, they would be expected to produce a compensatory decrease. In practice, the preferred level of input (specific or "general") seems to vary with the motivational state.

9.4 THE RETICULAR SYSTEM

In recent years interest in the role of general sensory input in determining responsiveness and the efficiency of performance has been brought to a focus through studies of the ascending reticular activating system and the diffuse thalamocortical projection system in mammals. These include diffuse columns of nerve cells extending through the lower brain as far as the thalamus and consisting of a network of cells with short fibres: their functions include the maintenance of cortical and behavioural (see pages 214–222) arousal by relaying general sensory input to the cortex (e.g., Moruzzi and Magoun, 1949; Lindsley, 1956, 1961; Jasper et al., 1958; Worden and Livingston, 1961).

As the principal sensory pathways travel up through the brainstem they give off collaterals to these systems, which, in turn, have a diffuse projection onto the cortex (Fig. 9.4). Several sources of evidence indicate that activity in any sensory system, besides causing activity in its specific cortical projection area, also has a diffuse effect on the cortex. Thus electrical stimulation of the reticular formation of an animal in light sleep or light anaesthesia can produce arousal, as indicated both by a change in the EEG record from the synchronized slow waves characteristic of such states to low-amplitude fast waves, and by the behavioural response: comparable arousal can be produced by sensory stimulation. [As might expected, the response of the cortex to electrical stimulation of the reticular system differs in some ways from that which occurs with stimulation of specific sensory nuclei; for instance, there is a difference in the manner in which the cortical potentials change with depth into the cortex (Jasper, 1961).] Lesions in or barbiturate anaesthesia of the ascending reticular system lead to a lack of behavioural responsiveness to external stimulation: the primary sensory pathways can conduct to the cortex, but perception or discrimination does not result. Stimulation of any peripheral nerve gives evoked potentials throughout the reticular formation. Most of its units are activated by a variety of inputs, though there is also evidence for a degree of specificity (see below).

Thus behavioural arousal, cortical EEG arousal, and reticular activity usually go together, and much of the evidence indicates that the cortical arousal, mediated by activity in the reticular formation, is a necessary accompaniment to a stimulus for a behavioural response to occur. Further evidence indicates that reticular activity can increase the efficiency of responsiveness. For instance, when Lindsley (e.g., 1961) presented pairs of light flashes separated by a 50-millisec interval to a cat, the cortex gave only a single evoked response to each pair. When the reticular system was electrically stimulated, however, the cortex showed a separate response to each flash for about 10 sec after the cessation of reticular stimulation.

However, as we have seen, dissociation between the indices of behavioural and physiological arousal sometimes occurs. Furthermore, the relationship between the level of input into the reticular system and corti-

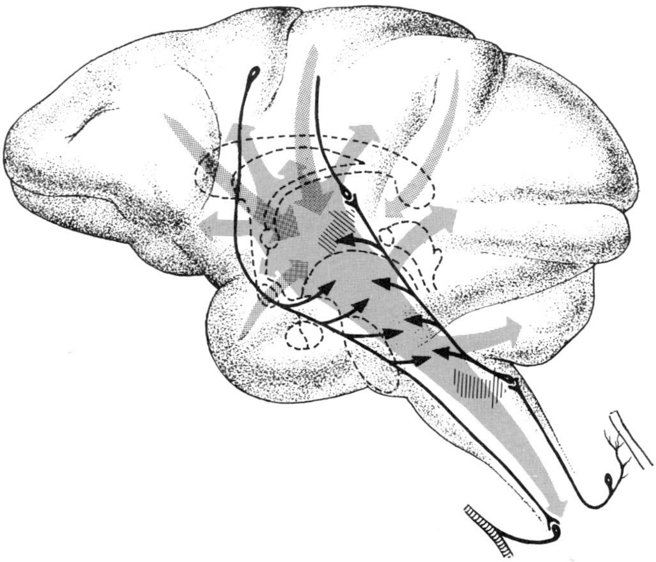

Fig. 9.4. Diagrammatic representation of the relationships of the reticular formation. (After Worden and Livingston, 1961.)

The thin-line arrows indicate collaterals from sensory and motor pathways to reticular formation. The thick arrows entering the formation indicate corticofugal influences. The arrows leaving the reticular formation indicate its ascending influences to cortex and cerebellum, and its descending influences to motor mechanisms and (hatched arrows) central sensory relays.

cal functioning is not a simple one. Strong electrical stimulation of the reticular system produces not only arousal but also apparent discomfort, fear responses, and fleeing. Although moderate activity in the reticular formation aids cortical functioning, a very high level of activity in the brainstem decreases sensitivity (e.g., Lindsley, 1956). Thus, Fuster (1958) found that, while tachistoscopic perception in monkeys was enhanced by mild stimulation in the mesencephalic region, the contrary effect could be produced if the strength of the stimulus was increased (see also Wilson and Radloff, 1967). Stimulation of the reticular system itself may have both facilitatory and inhibitory effects on the responses of cortical cells (e.g., Akimoto et al., 1961).

The reticular system receives its input not only from the main sensory pathways but also from the cortex, the limbic system, the basal ganglia, and the cerebellum. Thus electrical stimulation of the monkey cortex can elicit evoked responses from the reticular system—responsiveness is mostly diffuse, but there is some evidence for specificity. Through its effect on the reticular system, localized cortical stimulation can activate the entire cortex.

The reticular system can also be influenced by the endocrine state of the animal. Sawyer (quoted by Dell, 1958) assessed the effects of injections of oestrogen and progesterone on the threshold for electrical stimulation of the reticular formation in the female rabbit, using cortical arousal as the criterion. He found that this threshold diminished to about 10 percent of its initial value at the time when oestrous behaviour appeared. It thus seems that certain hormones, acting via the brainstem reticular system, may affect the responsiveness of the cortex.

The output of the system passes not only to the cerebral cortex but also to the cerebellar mantle, brainstem, and spinal cord. It has facilitatory and inhibitory effects on the spinomotor outflow: the more rostral portions augment spinal reflexes and voluntary movements, while the more caudal parts have an inhibitory effect. For instance, a cat's reaction time in removing its foot from a metal plate when given an electric shock is shortened by increasing the sensory input or by direct stimulation of the reticular system (Isaac, 1960).

Although it was thought formerly that the system was quite nonspecific, evidence of differentiation within it is now accumulating. For instance, particular cells in the reticular formation may be responsive to one sort of stimulation but not to other sensory modalities. Furthermore, the anterior part, lying in the thalamus, is structurally more complex than the rest, and the nuclei within it are connected to specific parts of the cortex (e.g., Jasper, 1949). The relationships between cortical and subcortical structures are clearly intricate, and the cortical arousal effect is probably only a crude description of one of the functions of this system (cf. page 129). The amorphous structure of the reticular system, with extensive and diffuse connections between neurones, would make the identification of functional entities within it a task of extreme difficulty: the use of chemicals to which it is selectively responsive provides a hopeful line for current studies (e.g., Grossman et al., 1965; Grossman and Grossman, 1966). Livingston et al. (1954) describe the role of the reticular system as that of a "transactional link" for all parts of the nervous system.

Although the realization of the importance of the reticular system has led to a greatly increased understanding of brain functioning, as yet it provides no easy answer for questions of motivation. The apparently non-specific action of the reticular system has frequently been quoted in support of a concept of general drive (e.g., Hebb, 1955), but the evidence at present available does not warrant it. There is clear evidence that the reticular system influences sensory/perceptual mechanisms and reaction times, and that reticular activity must be between certain limits if functionally integrated behaviour is to occur. But there is as yet no satisfactory evidence that increase in reticular activity within those limits results in an increase in the intensity or frequency with which a particular pattern of behaviour is shown. It can be argued at most that any non-specific aspects of reticular functioning would be compatible with a general drive theory of behaviour, not that they require one. Another suggestion, that the reticular system acts as a homeostat regulating input–output relations (Lindsley, 1960), is also not yet substantiated: the role of the reticular system in controlling input is still not clear (see Chap. 6).

9.5 CONCLUSION ON GENERAL VERSUS SPECIFIC EFFECTS OF SENSORY INPUT AND MOTIVATIONAL FACTORS

In summary, there is clear evidence that the proper integration of behaviour requires that the level of ordered sensory input be between certain limits. On the other hand, there is no clear evidence that the intensity of behaviour is determined equally by all the motivational factors and stimulation acting at the moment, as a general drive hypothesis demands. In the present state of the evidence, the reasonable assumption is that, within limits, the intensity of any one type of behaviour depends only on motivational factors more or less specific to it, and that others are of relatively minor importance. Just how general or how specific the effects of any particular stimulus or motivational state may be is a matter for empirical investigation.

9.6 SUMMARY

1. Most factors that influence an animal's behaviour have a limited sphere of influence; they affect some types of behaviour and not others. There is, however, also evidence for more general effects. Whereas some theorists use drive concepts specific to particular types of behaviour, others postulate a "general drive", which is supposed to energize all the patterns of behaviour in an animal's repertoire.
2. Studies of the interrelations among hunger, thirst, fear, and sex provide some evidence which could be interpreted in terms of a general drive, but alternative explanations are also possible in most cases.

3. The various sensory inputs to which an organism is exposed interact with each other.
4. Changes in various physiological indices (e.g., skin resistance, heart rate) are correlated with activity of many sorts and have been used as indices of the "level of activation" of the animal. The utility of this concept is limited: the correlations between the different indices are often not high.
5. The concept of "level of activation" is not necessarily congruent with that of "general drive".
6. A number of lines of evidence indicate that responsiveness, efficiency, or performance deteriorates if the sensory input falls outside certain limits. Animals show behaviour which results in the stimulation remaining within these limits.
7. Motivational factors may influence the level of stimulation for which an animal will work.
8. The brainstem reticular system has a general effect on behaviour, and the functioning of other parts of the brain require reticular activity to be within certain limits. Evidence drawn from studies of this system, however, does not demand a general drive theory of behaviour.

10

THE SPECIFIC EFFECTS OF HORMONES AND OF STIMULI

In this chapter we shall sidestep the drive issue (see Chaps. 8 and 9) and consider two categories of the independent variables which might be described as "motivational factors" (see page 194)—hormones and stimuli. For the moment, discussion will be limited to the more commonly studied activities—feeding, drinking, and various aspects of reproductive behaviour: the extent to which it is possible to extract principles of motivation common to these and other activities will be considered later (Chap. 15).

10.1 HORMONES

Most of the work on the influence of hormones on behaviour has been concerned with reproductive behaviour, and it is here that the problems are most clearly defined. Evidence for a hormonal influence on the reproductive activities of vertebrates is of three main types (Beach, 1948; Young, 1961a):

1. The occurrence of reproductive activities is correlated with enlargement or secretory activity of the endocrine glands. In many seasonally breeding animals the gonads are enlarged, and certain gonadal cells show secretory activity, during just that period in which the reproductive activities are most common. For instance, Bullough (1942) showed that the appearance of various aspects of reproductive behaviour in the starling (*Sturnus vulgaris*) is related to the growth of the gonads in the early part of the year. This species normally breeds in the spring. There are two races which breed in Great Britain. One is sedentary: the gonads of these individuals develop precociously during the autumn, and this is associated with an increase in sexual activity. In the other race, which leaves the country in the autumn to winter farther south, there is no autumn recrudescence of the gonads and no autumn sexual behaviour.

2. The frequency or intensity of sexual behaviour is reduced after gonadectomy. The rapidity of this decrease, however, varies considerably between species (see page 550).
3. Injection or implantation of certain hormone preparations increases the frequency of reproductive responses. For instance, Noble and Zitrin (1942) showed that injections of testosterone propionate into male chicks led to the appearance of the sexual behaviour patterns of the adult cock—crowing, copulating, and so on—while female chicks treated with oestrogen squatted for treading males. Similar results have been obtained with mammals: prepuberally castrated male rats fail to show sexual behaviour, but after administration of androgens they do so. Similarly, the sexual receptivity of spayed female rats is restored by treatment with combinations of oestrogen and progesterone. Many similar results are given in reviews by Beach (1948), Eisner (1960), Guhl (1961), Lehrman (1961), Young (1961a), and others.

In general, male sexual behaviour of higher vertebrates is induced by androgens, while female sexual behaviour is induced by oestrogens and progestagens, alone or in combination. However, perhaps because these are all steroids of similar chemical constitution, it is not always easy to define clearcut differences between their actions. Thus under certain circumstances oestrogen may activate components of male behaviour in males or females as well as normal female behaviour in females (e.g., Beach, 1948; Yaschine et al., 1967), while androgens can play a role in inducing female behaviour in females (see pages 237–238). Generalizations applicable to all vertebrates are hard to make, for in some species sexual and aggressive behaviour, as well as parental behaviour, are influenced directly by pituitary hormones (e.g., Aronson, 1965). Indeed, it is not even possible to generalize about which patterns of behaviour are under hormonal control and which are not: many of the responses made by female mice to their young, for instance, are elicited equally easily from primiparous females, females pregnant for the first time, virgin adult females, adult males who have seen their mate taking care of a litter, and inexperienced adult males. It is thus improbable that differences in hormonal state play a significant role in controlling these responses (Beniest-Noirot, 1958; Noirot, 1964a; see also Leblond and Nelson, 1937). More recently Rosenblatt (1967) has shown that non-pregnant, ovariectomized, and hypophysectomized female rats, and intact and castrated male rats, will show maternal behaviour to pups. In most rodents maternal behaviour is elicited more readily from recently parturient females than from virgins or males (e.g., T. E. Rowell, 1961; Richards, 1966a).

Hormones affect behaviour in diverse ways. Here we shall not consider the general effects of hormones on growth or metabolic functions but shall concentrate on six possible mechanisms by which hormones may influence behaviour. The first four of these were listed by Lashley (1938a).

Hormonal Influences on the Development of the Nervous System

One mechanism of hormone action suggested by Lashley involved an effect on the growth of nervous connections. As we shall see later (Chap. 22), sex hormones do indeed exert a profound influence on the development of the nervous system at an early stage, especially during perinatal or prehatching life (Young, 1965, and refs. cited on pages 543–545). There is, however, little indication that a hormonally influenced growth of new nervous connections is the factor determining the change in behaviour which occurs at the onset of sexual maturity. Thus male hamsters show some elements of the mating pattern even when castrated and adrenalectomized before puberty, indicating at least that the sensorimotor mechanisms involved can develop independently of the increased androgen production at this time (Warren and Aronson, 1957). Similarly, the castration of immature male pigeons does not completely prevent the development of copulatory behaviour (Carpenter, cited Beach, 1951b; see also Schwartz and Beach, 1954).

Peripheral Effects

In some cases hormones influence behaviour by causing changes in a peripheral organ which mediates sensory input to the central nervous system. For instance, in the copulation of rats, ejaculation depends on stimulation received through the penis during a series of intromissions. Beach and Levinson (1950) have shown that one of the effects of androgens is to increase the density of cornified papillae between the epithelial folds of the glans penis, and that there is a close correlation between the appearance of the papillae and sexual behaviour. Thus one of the ways in which androgen influences the sexual behaviour of rats is by altering the sensitivity of the penis. Other changes in the penis, for instance in the penile musculature, which may result from an alteration of the androgen level, are reviewed by Rosenblatt and Aronson (1958a, b; see also Aronson and Cooper, 1966).

The influence of the pituitary hormone prolactin on the parental feeding of doves is due, at least in part, to a peripheral effect. Doves, unlike other birds, feed their young on a "milk" formed by a sloughing of the epithelium of the crop. Prolactin causes the enlargement of the crop, which normally takes place before the young hatch. Lehrman (1955) has shown that sensory stimulation from the enlarged crop plays an essential role in the induction of parental feeding. If this afferent stimulation is cut off by the application of a local anaesthetic to the crop region, the proportion of adults which will feed squabs is reduced, whereas local anaesthesia elsewhere does not prevent parental feeding. Prolactin is thus instrumental in inducing parental feeding by its effect on the crop. (These results have been criticized by Klinghammer and Hess (1964), who found that some feeding by adult ring doves could occur in the absence of prolactin-induced crop engorgement. The difference between the two experiments may lie in the fact that Klinghammer and Hess's birds were brought into contact with young during incubation, whereas in Lehrman's experiments the squabs were placed in the nest of a non-incubating bird. This suggests that the peripheral effect of prolactin is on the tendency of the adult to orient towards the young from a distance.)

Another example of hormones influencing behaviour through their peripheral effects comes from studies of the nest-building behaviour of domestic canaries. In this species the stimulation the female receives from the cup of the nest, which she herself has built, has an important influence on many subsequent aspects of reproductive behaviour: it affects, for instance, the amount of nest-building behaviour shown, the selection of nest material, the date of egg-laying, and so on. The effectiveness of stimuli from the nest cup is enhanced by the development of a brood patch —the female loses the feathers from her breast, which becomes vascular and also more sensitive to tactile stimulation (Fig. 10.1). These changes in the brood patch are under hormonal control. The hormones thus influence behaviour in part through their effect on the brood patch (Hinde et al., 1963; Hinde and Steel, 1964; Steel and Hinde, 1964).

Another example which has been cited in this context is the effect of oestrogen on the dominance of chimpanzees. Although oestrogens lower the dominance status of male castrates, they raise that of female castrates. It has been suggested that the latter effect is due to the oestrogen injections producing engorgement of the sex skin: Birch and Clark (1950), utilizing the fact that the female's swelling can be temporarily suppressed with progesterone, found that the dominance status of the female varies with the degree of swelling when the oestrogen level is kept constant and high. However, other explanations in terms of central interactions between oestrogen and progesterone are possible in this case.

A slightly different case is that in which a hormone produces a change in behaviour not by affecting a particular organ but by influencing general bodily functions. Nest-building by rats, for instance, is increased by thyroidectomy, gonadectomy, adrenalectomy, and hypophysectomy (Richter, 1937). Since nest-building is known to be closely influenced by temperature, it seems possible that the effect is partly due to the decrease in body temperature which follows removal of these structures (Kinder, 1927).

Specific Central Effects

Perhaps the most important type of hormonal influence on behaviour involves effects on specific mechanisms within the central nervous system. One of the first demonstrations of a central effect involved the induction of oestrous behaviour in hamsters. Kent and Liberman (1949) found that, if hamsters were first primed with oestrogen, oestrous behaviour could

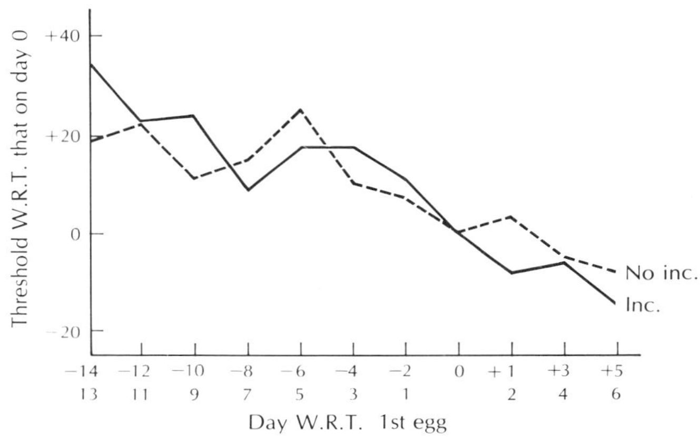

Fig. 10.1. Changes in the threshold of tactile sensitivity of the canary brood patch around the time of egg-laying. (After Hinde, Bell, and Steel, 1963.)

Ordinate, tactile threshold; *abscissa,* date with respect to laying of first egg. Data for birds which subsequently did and did not incubate their eggs are plotted separately, but do not differ significantly.

be induced by the introduction of progesterone into the brain ventricles in dosages which were insufficient when administered subcutaneously.

Although there is some evidence that sex hormones can affect the spinal mechanisms mediating copulatory behaviour (Hart, 1967a, 1968), in mammals such central effects are exerted primarily on the hypothalamus. Thus female cats with the central nervous system sectioned in or posterior to the lower midbrain fail to show integrated sexual behaviour, but oestrogen facilitates the mating response to sacral and vaginal stimulation if the transection is rostral to the hypothalamus (Dempsey and Rioch, 1939; Bromley and Bard, 1940). Comparable studies have been made of the clawed toad (*Xenopus laevis*). The male uses three clasping patterns to maintain his position on the female during mating: a "tonic clasp", which in normal mating may be maintained for several hours on end; "backward kicking", given in response to movements of the female and serving to keep the male in place; and "forward kicking", given if the male is stimulated by another object (such as the experimenter's fingers) and serving to push that object away. These patterns are elicited by tactile stimuli to the pads and throat skin, by stretch stimuli to the forelimbs, and so on. They can still be elicited if the brain is transected in the anterior medulla but not if it is transected either more posteriorly or in the midbrain. Evidently transections in the posterior medulla eliminate the coordinating mechanisms, whereas transections anterior to the medulla leave functional mechanisms which are inhibitory (Hutchison and Poynton, 1963). Intact *Xenopus* are normally sexually inactive in the laboratory, but they can be made responsive by an injection of chorionic gonadotrophin (Russell, 1954). Variations in hormone level, however, have little effect on the behaviour of the males transected in the anterior medulla. Thus the hormones presumably act on the regulatory mechanisms in more anterior parts of the brain and not on the control mechanisms in spinal cord and medulla (Hutchison and Poynton, 1963; Fig. 10.2).

This direct action of hormones on the central nervous system has been demonstrated elegantly by Harris et al. (1958), working on the induction of sexual receptivity in the female cat by oestrogens. In this case it had already been shown that stimuli from the genital area were not necessary for the manifestation of female sexual behaviour, for total deafferentation of the pelvic erogenous zones failed to inhibit oestrous behaviour. Harris and his collaborators were able to localize the site of action of the hormones more precisely by implanting very small quantities of crystalline stilboestrol di-*n*-butyrate into the brain. When such implants were made in the posterior hypothalamus of spayed cats, there was full development of sexual behaviour even though the genital tract remained anoestrous. Furthermore, similar implants made subcutaneously or at other sites in the central nervous system did not induce sexual behaviour. It could be concluded, therefore, that the hormones influence behaviour by affecting hypothalamic structures.

Comparable data are available for rats. Implantation of oestradiol in the medial-basal preoptic and anterior hypothalamic regions of spayed rats produced behavioural sexual receptivity, though implants placed elsewhere in the hypothalamus, or subcutaneously, were ineffective (Lisk, 1962a, b). Similarly, ovariectomized female rabbits implanted with oestradiol benzoate in the ventromedial–premammillary hypothalamus showed oestrous behaviour in 1 to 2 days. Implantations of blank tubing, cholesterol, or progesterone within this area were without effect. Although pure oestrogen produced a similar effect if placed anywhere within a larger hypothalamic area, this was probably due to spread of the hormone (Palka and Sawyer, 1966a; Palka et al., 1966): estimates of the distances to which steroids diffuse from an implant in this general area vary from 1 mm (Michael, 1961a) to 2–2.5 mm (Lisk, 1962a; Palka and Sawyer, 1966a).

Although the relationship between sex hormones and sexual behaviour is less close in males than in females, at any rate in mammals,

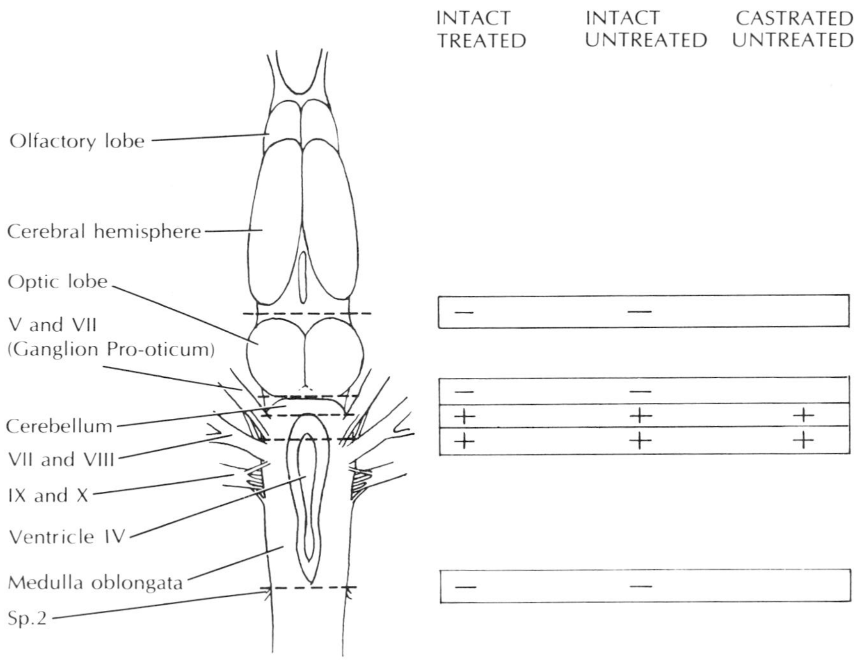

Fig. 10.2. Effects of brain transections on the sexual clasping patterns of intact Xenopus laevis males treated with human chorionic gonadotrophin (HCG), intact untreated males, and untreated castrates. HCG, through its effect on testis development, produces full sexual activity and development of nuptial pads. +, clasping patterns displayed fully; —, no clasping. (From data and drawing by J. B. Hutchison.)

there is also evidence that androgens may influence male sexual behaviour by acting on specific loci in the central nervous system. Thus localized androgenic implants have been shown to induce male copulatory behaviour in castrated rats (Davidson, 1966b; Lisk, 1967a) and capons (Barfield, 1964), and courtship in castrated male doves (*Streptopelia risoria* Hutchison, 1967b, 1968). In all these vertebrates the sensitive area lies in the preoptic and anterior hypothalamic regions. In doves progesterone induces incubation behaviour if implanted in the preoptic nuclei and lateral forebrain systems, and suppresses courtship both in these regions and in the anterior hypothalamus (Komisaruk, 1967).

Evidence suggesting that hormones may affect central nervous structures directly has been obtained in a quite different manner by Olds (e.g., 1961). Electrodes were implanted into rats' brains and connected in such a way that the rat could stimulate its own brain by stepping on a pedal. When the electrodes were in certain positions the animals would stimulate themselves regularly for long periods of time (see pages 220–221). Olds claims that this positive area is itself differentiated into subareas, in some of which performance is improved by, for instance, hunger, and in others by higher androgen levels. In the latter cases the effectiveness of a standard electrical stimulus to particular areas of the central nervous system is enhanced by the hormone treatment.

Although such data indicate that steroids affect relatively limited regions of the central nervous system, the effects are not always absolutely specific to one particular hormone. Thus Palka and Sawyer (1966b) found that hypothalamic testosterone implants in rabbits induced oestrous behaviour differing only slightly from that induced by oestrogen; and Hutchison (1968) reported that hypothalamic implants of oestrogen, as well as of testosterone, induced the nest-soliciting courtship pattern from male doves. Hutchison argues that such results are to be understood in terms of the similarity in molecular structure between these two steroids (see also Fisher, 1956).

In considering the means by which steroids produce these central effects, a number of points must be borne in mind. In the first place, a positive effect on behaviour does not necessarily imply a positive effect on neural activity—the hormone may act by removing an inhibitory influence. Thus the lordosis response can be elicited from newborn guinea pigs but disappears a few hours after birth and cannot be elicited again until puberty. Beach (1966) has summarized evidence that in the neonate this response forms part of the excretory pattern, which at this age is normally elicited by stimulation from the mother. As micturition and defaecation come under the control of internal stimuli, the spinal mechanisms come under inhibitory control from the brain. In adult animals this inhibition is removed by the influence of ovarian hormones.

A second point is that hormones may act not only by increasing the probability of particular patterns but also by producing specific inhibitory effects. For instance, oestrogens inhibit aggressiveness in female golden hamsters (Kislak and Beach, 1955); and, to take an invertebrate example, a hormonally mediated consequence of the filling of the receptaculum seminis is responsible for the postcopulatory decline in receptivity of female grasshoppers (Haskell, 1960). In the ring dove both the bowing-coo and the nest call are androgen-dependent. If castrated birds are injected with both androgen and progesterone, the effect of androgen on the bowing-coo is abolished, but its effect on the nest-coo is unaffected. Thus progesterone selectively inhibits the reactions of some neural mechanisms to androgen, perhaps by insulating specific neural sites against it (Erickson et al., 1967). A similar suggestion has been made by Zucker (1966) in considering the interactions of oestrogen and progesterone in controlling sexual receptivity in guinea pigs, though this is probably not the only way in which progesterone reduces receptivity (Zucker, 1968).

Another complication is that the central effects of a hormone may be multiple. Thus Michael (1961b) has suggested that the mechanism underlying full receptivity in female cats is separate from, and requires a higher hormone level than, that underlying sexual reflexes (see also Hart, 1967a, b). More detailed data illustrating a similar point for rats will be considered in Chap. 15.

The precise manner in which steroids affect nerve cells is still a matter of dispute. Autoradiographic techniques indicate that labelled oestrogens are taken up by cells scattered over a wide region of the hypothalamus, as well as by the median eminence and pituitary (e.g., Michael, 1961a). The behavioural effects of oestrogen, however, are usually not manifest for several days after hypothalamic implantation (but see Komisaruk, 1967), suggesting that long-term changes in the state of neurones may be involved. Progesterone injections, however, can produce rapid results. Lisk (1966d) claims that rats primed with oestrogen mate within 1½ min if injected intravenously with progesterone, suggesting very rapid action of progesterone on "oestrogenised" neurones. It has also been shown that, a few minutes after progesterone has been injected into rats, there are changes in the responsiveness of neurones in the central nervous system to stimulation of the vaginal cervix. This has been attributed to a selective depression on the effects of stimuli from the genital tract (Barraclough and Cross, 1963; Cross and Silver, 1966). However, Komisaruk et al. (1967) recorded EEG and unit activity simultaneously and found that progesterone influenced the activity only of neurones whose activity was linked to the general (EEG) arousal level, and not that of independent neurones. They thus reinterpret the earlier findings in terms of a non-specific effect of progesterone on arousal as indicated by the EEG, and suggest that the apparent specificity for vaginal stimulation was due merely to its greater dependence on arousal level as compared with the other stimuli used. Oestrogen, however, can enhance the response of preoptic units to cervical stimulation and depress their response to pain and cold (Lincoln and Cross, 1967): this does seem to imply a specific effect.

Non-specific Central Effects

That hormones may affect behaviour by producing a non-specific effect revealed for instance through a change in EEG activity was mentioned above. The effect may be mediated through the reticular activating system (see pages 223–226).

Indirect Central Effects

The hormone-sensitive systems considered so far have been concerned in the immediate control of various aspects of reproductive behaviour. A second type of hormone-sensitive system, demonstrated in a number of mammals, has longer-term effects on behaviour. It serves to monitor the circulating levels of gonadal steroids and to control them by influencing the release of gonadotrophins from the pituitary. Such systems are found in the region surrounding the median eminence. For instance, implantation of oestradiol in the arcuate nucleus of rats results in a decrease in the diameter of the nucleoli of the neurones in that area and in gonadal atrophy, suggesting a functional hypophysectomy (Lisk and Newlon, 1963; review by Lisk, 1966c). Comparable effects have been obtained in lizards (Lisk, 1967b).

Effects Mediated by Changes in Effectors or Social Stimuli

Another route by which hormones influence behaviour involves changes in effectors, such as the hormone-induced hypertrophy of the brachial musculature used by frogs in amplexus (Aronson and Noble, 1945), or the development of the penis in mammals. Of particular interest are cases in which the hormones influence structures which are likely to be used in social communication, such as the cock's comb or the colour of the beak in many birds. In mammals, changes in olfactory stimuli may be of considerable importance. For example, in rhesus monkeys it has been established that sex hormones affect female receptivity and female attractiveness differentially. Oestradiol administered subcutaneously has a marked effect on the tendency of the female to initiate copulation, but oestrogen administered intravaginally leads to an increase in the tendency of the male to mount the female. In the latter case treatment of the female may cause the male to attempt to mount when the female is not receptive: the effect is probably mediated through a change in olfactory stimuli produced by the female (Michael and Saayman, 1968). Males will press a lever to gain access to a female treated intravaginally with oestrogen, but will not do so if they are rendered anosmic (Michael and Keverne, 1968). Injection of the female with low doses of testosterone increases the frequency with which she attempts to initiate sexual behaviour but does not increase her attractiveness for the male, and the frequency with which he mounts her remains low (Herbert and Trimble, 1967; Trimble and Herbert, 1968). Progesterone treatment of ovariectomized females pretreated with oestrogen causes loss of both their receptivity and their attractiveness (Michael et al., 1968). Similar effects are known in invertebrates: a hormone from the corpus allatum causes the female cockroach to produce a pheromone which attracts the male (Barth, 1962).

We see, then, that hormones can influence behaviour in a number of different ways. It is indeed safer to make the initial assumption that the effects of hormone treatment are multiple than to presume that the hormone has only a simple effect. It must also be stressed that a hormonal influence on one type of behaviour may have consequences ramifying through much of the rest of the animal's repertoire. Michael and Herbert (1963) found a relation between the stage of the menstrual cycle and both mounting and grooming behaviour in rhesus macaques. The amount the female grooms the male reaches a minimum, and the amount the male grooms the female a maximum, in midcycle. This does not necessarily indicate a direct influence of hormones on the mechanisms for grooming, for the fluctuations in grooming may depend on changes in social status (see T. E. Rowell, 1963).

The finding that a response is influenced by hormones does not mean that its occurrence is solely a function of hormone concentration. For one thing, in many species behaviour which is later hormonally influenced occurs during a short period of early development when there are no apparent endocrine changes. Infantile sexuality is not peculiar to man—many birds show incomplete courtship and nest-building behaviour when a few weeks old, long before their gonads attain breeding condition (e.g., Curio, 1960). Such observations must, however, be interpreted with caution. For one thing, the presence of some hormonal basis for such precocious behaviour has not been ruled out. For another, as we have seen, the lordosis pattern in infant guinea pigs is sexual neither in function nor in causation but forms part of the elimination pattern (see page 235).

Again, experience often has an important influence on hormonally mediated behaviour. For example, progesterone is much less effective in inducing incubation behaviour, and prolactin in inducing parental feeding behaviour, from inexperienced ring doves than from experienced ones (Lehrman, 1955, 1962; Lehrman and Wortis, 1960). Further examples are given in Chap. 22.

Furthermore, the occurrence of hormone-influenced behaviour is also a function of many other aspects of the physiology of the animal. Grunt and Young (1953) have shown that the level of sexual activity attained by individual male guinea pigs before castration was correlated with that shown after the operation when the animals were receiving standard androgen treatment: thus the level of responsiveness to a given hormone level is a characteristic of the individual. Beach and Fowler (1959a) obtained a similar result for rats, and Whalen et al. (1961) concluded that individual differences in copulatory behaviour were due to differences in ability to make use of circulating androgens rather than to the androgen level itself. Similarly, Goy and Young (1956/1957) have demonstrated strain differences in the responses of female guinea pigs to sex hormones, and Stewart and Palfai (1967) showed that, after castration, individual dominance status in rats was restored by standard androgen therapy.

Even when it seems certain that the hormone is producing its effect on a particular locus in the brain, it must be remembered that other parts of the central nervous system may influence the strength of the response. Removal of parts of the forebrain from male rats leads to a reduction in sexual responsiveness correlated with the amount of tissue removed. In experiments by Beach (1940, 1944), ablation of 60 percent or more of the neocortex led to a complete cessation of sexual behaviour, whereas ablation of less than 20 percent had no effect. However, the effects of cortical ablation can be offset by the administration of androgens—the more cortex that has been removed, the more androgen is needed to restore sexual behaviour to a standard level, though if the total lesion exceeds approximately 60 percent, no amount of hormone will compensate for the injury. A similar result was obtained with pigeons (Beach, 1951b).

Beach (1958b) has suggested that there are phyletic trends towards independence of sexual behaviour from hormonal control and towards increasing importance of the cerebral cortex. Thus the sexual behaviour of cats is more affected by cortical ablations than that of rats. However, the issue is not a simple one: more recent studies show that the deficit is not solely a function of the amount of tissue removed, for frontal lesions produce more severe effects than do more posterior ones in cats (Zitrin et al., 1956) and rats (Larsson, 1964). Using spreading cortical depression induced by application of potassium chloride, Larsson (1962) also found that functional decortication prevented the copulatory activity of male rats. Unlike the results obtained by Beach for female rats, Larsson found a differential effect on some components of the female's copulatory behaviour.

In the foregoing discussion we have considered only the routes by which the internal hormonal state influences responsiveness to stimuli, and not the influence of environmental stimuli on the hormonal state. It is, of course, well known that the gonadal condition of cyclically breeding animals may be influenced by external factors, and the particular example of domesticated canaries is discussed in Chap. 25. Another type of case which has been studied in some detail recently is the influence of "stress-inducing" social and other environmental stimuli on the functioning of the pituitary–adrenal cortex system. Discussion of this lies outside the scope of this book, but some aspects are mentioned in Chap. 22. (See, e.g., Clarke, 1953; Harris, 1966; Chitty, 1957; Barnett, 1958, 1963.)

10.2 STIMULI

During the spring, male three-spined sticklebacks leave the sea and migrate upriver to their breeding grounds. This migration is due to a hormonal change involving primarily the pituitary and thyroid glands (Baggerman, 1958). During this period, they do not behave aggressively if they encounter other males and they do not show nest-building behaviour. When they reach a suitable site for breeding, their behaviour changes. They start to patrol a limited area, to attack other males, and to dig a pit in the sand. Evidently these new forms of behaviour depend on stimuli from a suitable territory as well as on an appropriate hormonal state.

Observations of this kind have been used (e.g., Lorenz, 1950; Tinbergen, 1951) as a basis for distinguishing between stimuli which elicit particular responses when the animal encounters them ("releasing" or "eliciting" stimuli), and those which act continuously to determine its specific state of responsiveness ("motivating" or "sensitizing" stimuli). Stimuli from an intruding male would come into the former category, stimuli from the territory into the latter. Often, however, there is little evidence that releasing and motivating effects depend on different types of mechanism, and the distinction seems to depend on the nature of the stimuli rather than on the way they act: motivating stimuli are often continuously present, whereas releasing stimuli appear more suddenly.

A similar distinction is, however, operationally useful in the study of learning, where it is usually couched in terms of "eliciting" and "discriminative" stimuli. In "classical" conditioning, a response (e.g., salivation) comes to be elicited by a new stimulus (e.g., a tone) as a consequence of its elicitation by a reinforcing stimulus (e.g., food) immediately following the presentation of the new stimulus: there comes to be a more or less one-to-one relationship between the new stimulus and the response, and the term "elicit" is clearly appropriate. In "operant" conditioning a response (e.g., pecking a key) comes to be given more often in the presence of a stimulus if it is followed by a reward (e.g., food). The stimulus permits the animal to "discriminate" between conditions in which the response is and is not likely to be followed by reinforcement, but there is no one-to-one relationship between the stimulus and the response, the frequency with which the response occurs depending in part on the schedule of reinforcement. This is discussed further in Chaps. 24 and 25.

There is a further apparent similarity between such discriminative stimuli and "incentive-motivational" stimuli. If a stimulus is paired with a reinforcer (e.g., food), it will subsequently facilitate the acquisition of an instrumental response (e.g., running) reinforced by food (e.g., Bacon and Bindra, 1967; see pages 206 and 259–260): it is thus said to have "incentive-motivational" properties. However, for the establishment of a discriminative stimulus, it is necessary that the animal have differential experience of reinforcement in the presence of the stimulus situation and no reinforcement in a different stimulus situation (with certain exceptions: see Terrace, 1966). By contrast it is usually assumed that for the establishment of incentive-motivational properties, like that of conditioned reinforcing properties (see Stein, 1958), mere pairing is sufficient. Whether the three cases (conditioned eliciting, discriminative, and incentive-motivational effects) differ in mechanism is, however, an open issue.

Of course, where external stimuli act by influencing the endocrine state there is a clear case for distinguishing this mode of action from direct elicitation of a response. However, it must be remembered that the same stimulus may both elicit an immediate response and (directly or indirectly) produce a change in internal state. For instance, stimuli from hatched chicks not only elicit parental behaviour from black-headed gulls (*Larus ridibundus*) but also cause a progressive reduction in readiness to incubate (Beer, 1966a).

We shall discuss here the role of stimuli which have an immediate influence on responsiveness to other stimuli, without implication about the mechanism by which this occurs. They may produce their effects by continuous direct action or by inducing durable central nervous or hormonal states. Furthermore, they may result in an increase or a decrease in responsiveness. Discussion of longer-term "reinforcing" effects is postponed to Chap. 24.

Augmenting Effects

One example of an augmenting effect has already been mentioned (pages 230–231)—stimuli from the enlarged crop increase the readiness of experienced adult doves to feed squabs: the application of a local anaesthetic to the crop region reduces responsiveness (Lehrman, 1955).

The precise extent to which internal sensory stimuli contribute to the responsiveness of mammals in feeding, drinking, and sexual contexts is still controversial. Although it seems clear that stomach contractions are correlated with subjective feelings of hunger in man, they are not essential for the occurrence of feeding behaviour in adult mammals. Tsang (1938) removed the contractile main bulk of the stomach from seven rats, and then tested for activity and maze performance. For the first 5 days the maze performance of the operated animals was little different from that of controls, but in the next 2 days they appeared to be very poorly motivated, and their activity was reduced. Bash (1939a, b) approached the problem by severing the vagus nerve and destroying the sympathetic nerve trunks. Many of the rats, although making a good surgical recovery, at first refused food and, in some cases, starved to death. Those which recovered maintained good health and increased in weight but showed abnormalities in the diurnal rhythm of feeding. Bash thus suggested that gastric contractions play an important role in feeding behaviour but are not essential: other contributory factors can take their place. Morgan and Morgan (1940a, b) found that resection of the vagus nerve failed to eliminate the augmentation of hunger by insulin, but rather enhanced the effect. They interpreted this as indicating that the nutritive conditions of the organism are more important in the control of hunger than are stomach contractions, a conclusion in harmony with a number of studies indicating that weight loss is often closely related to subsequent food intake (Bolles, 1967). From these studies it is clear that, while stimuli from the stomach play a role in normal eating behaviour, they are not essential for it to occur. Similarly, peripheral stimuli from the mouth/throat region do not play an essential role in the initiation of drinking behaviour, central osmoreceptors being largely responsible for indicating water deficit (e.g., Verney, 1947; see pages 253–254).

Similar data are available for sexual behaviour. Normal copulatory responses are shown by female rats after removal or denervation of the uterus and vagina (Ball, 1934) and by female cats after denervation of the genitalia (Bard, 1935). Male rats deprived of testes, prostate gland, and seminal vesicles copulate normally after androgen treatment (Beach, 1947; Beach and Wilson, 1963). Again, the behaviour of mammectomized rats shows that mammary engorgement does not play a critical role in governing maternal behaviour in that species (Moltz et al., 1967).

Such experiments indicate that input from some of the organs normally concerned with a response is not essential for that response; but they do not show that such input has no role. Other experiments demonstrate that some as yet undiagnosed characteristics of the states of hunger and thirst can act as cues for a differential response, but do not tell us whether the cues are central or peripheral. For instance, Hull (1933) showed that rats could learn to take one of two possible paths when hungry and the other when thirsty, though the process of acquisition was slow and performance not perfect (see also Wickens et al., 1949; Bolles, 1958). Similarly, Amsel (1949) showed that some rats could learn to escape a painful stimulus by running one way when hungry and the other way when thirsty (see also Jenkins and Hanratty, 1949; Levine, 1953).

These cases have been concerned primarily with stimuli operating on the surface of or inside the body: stimuli with more distant origins often have motivating effects. The influence of stimuli from the territory on the stickleback's aggressive behaviour in spring has already been mentioned briefly (page 240). Another case emphasizes an additional point, namely that, just as a given hormone may influence a particular pattern of behaviour by affecting more than one part of the internal mechanism, so also may stimuli have multiple effects. The parental fanning of the male three-spined stickleback, which serves to aerate the eggs in the nest, normally increases from day to day until the eggs are 6 days old: it then falls fairly steeply (Fig. 10.3*a*). Since the initial increase is steeper the more eggs that are present in the nest, and since fanning increases in water poor in oxygen and rich in carbon dioxide, it is probable that the initial increase in fanning activity is due to the increasing metabolism of the eggs. However, such a direct effect is not the only way in which the eggs influence fanning activity. If the eggs are replaced, when a few days old, by a clutch of fresh eggs, the curve of fanning has two peaks, one just before the day on which the original clutch would have hatched and the other the day before the new clutch hatches. Thus stimuli from the original clutch must produce an effect which continues for some days after their removal (Fig. 10.3*b*). Stimuli from the eggs therefore not only have an immediate influence on fanning activity but also produce a long-term effect: the latter is presumably mediated by a change in some internal factor. Since van Iersel also showed that the later the time at which the young eggs are substituted for old ones the smaller the influence of the new clutch, this internal factor presumably becomes progressively stronger with time relative to the effect of the new eggs (van Iersel, 1953).

In most of the cases cited so far, the motivating effect is on one response only or on a small group of responses. Territorial stimuli, however, influence a number of different activities in the male stickleback, and we shall see later that the sharing of common motivational factors by a number of different patterns plays an important part in the integration of behaviour (Chap. 25).

Consummatory Stimuli

Some stimuli have inhibitory effects on behaviour. For instance, deafened turkeys kill their own young immediately after hatching, just as they would a predator approaching the nest: normally this response is inhibited by the vocalizations of the young (Schleidt et al., 1960). Similarly many rats consistently kill mice, but such individuals usually do not kill rat pups: the differentiation seems to be due to an inhibition of killing produced by an odour from the rat pups (Myer, 1964). To cite an invertebrate example, the parasitic wasp *Trichogramma evanescens* will not lay its eggs on the eggs of host species which have already been parasitized: such eggs carry the smell of the previous female and are avoided (Salt, 1937).

Such inhibitory effects may operate to bring sequences of behaviour to an end, serving as part of a self-regulating system—the pattern of behaviour is terminated by stimuli which the animal encounters as a result of that behaviour. Such stimuli are often referred to as "consummatory stimuli". An example is again provided by the stimuli from the territory of the stickleback, for these stimuli are not only "motivating"

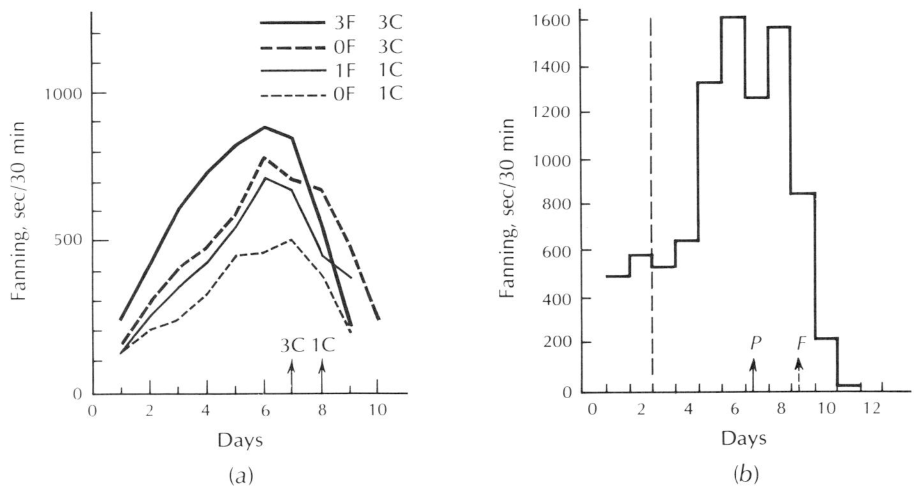

Fig. 10.3. (a) Changes in the amount of parental fanning shown by male three-spined sticklebacks on successive days after the eggs were laid. The figures indicate the number of clutches present, C; and fertilized, F. The arrows indicate the average hatching days of the young.

(b) Frequency of fanning of a male three-spined stickleback which had its original clutch replaced by a fresh one on the second day of the parental cycle. Hatching day of original clutch and foster clutch shown by arrows labelled P and F, respectively. (After van Iersel, 1953.)

for aggressive and nest-building behaviour, but also "consummatory" for migration. A suitable habitat brings migration to an end even though the hormonal state which gave rise to it is still (presumably) present. Similarly, the presence of a female decreases the probability of appearance of many of the activities of male song birds which function in pair formation, including song. The presence of another member of the species brings to an end the appetitive behaviour of a bird separated from its flock (Hinde,

1952) or a fish from its school (Keenleyside, 1955). Stimuli from the mother, or from siblings, similarly inhibit the "distress" calls of young domestic chickens (Collias, 1952; Kaufman and Hinde, 1961), rhesus monkeys (Harlow and Zimmermann, 1959), and chimpanzees (Mason and Berkson, 1962; Mason, 1967). Such stimuli are effective even when the distress calls are induced by electric shock or other fear-producing situations. Other examples are discussed by Prechtl (1952), Moynihan (1955), Baerends (1959b), Igel and Calvin (1960), Beer (1961), and Wiepkema (1961). Clearly the detailed mechanisms are diverse.

One of the best analysed cases of a reduction in responsiveness brought about by a specific external stimulus is given by Sevenster-Bol (1962) for the courtship of the three-spined stickleback. The male of this species, having built a nest, courts egg-laden females who enter the territory with a "zigzag dance" (see page 607). The courtship culminates with the female's swimming through the nest and depositing her eggs there: the male follows, fertilizes the clutch, and then usually attacks the female. In an experimental situation, the frequency of zigzagging by the male to a standard stimulus is reduced immediately after fertilization and his aggressiveness is increased. Evidently some aspect of the culmination of courtship in fertilization produces a reversal of the relative probabilities of the zigzag dance and aggressive behaviour. In order to isolate the factors responsible, Sevenster-Bol varied (1) whether or not courtship occurred, (2) whether or not fertilization occurred, and (3) whether or not the clutch was retained. The reversal of the zigzag dance/aggression relationship was found to be primarily a consequence of the presence of a fresh clutch of eggs in the nest and to a lesser extent of stimuli from the spawning female, but not of the occurrence of fertilization (Fig. 10.4). Thus the act of fertilization and the discharge of sperm play no significant part in the subsequent decreased tendency to show the zigzag dance to females. In addition, it was found that the recovery of the zigzag dance/aggresssion relationship occurred about twice as fast in the absence of a clutch as in its presence: the preceding occurrence of courtship and fertilization did not affect this recovery. Stimuli from the clutch thus have a suppressing effect on courtship behaviour.

In this case, recovery occurred in an hour or so: a second type of reduction in sexual responsiveness in the three-spined stickleback, studied by van Iersel (1953), occurs after several fertilizations and has a recovery period of several days. This indicates that the processes underlying the decrease in sexual responsiveness are not unitary and differ in their recovery periods (see Chaps. 13 and 15).

In *Drosophila*, also, female receptivity is inhibited by two mechanisms. There is first an effect of copulation itself, probably mechanical, which wears off after 48 hr. In addition, there is an effect from the presence of live

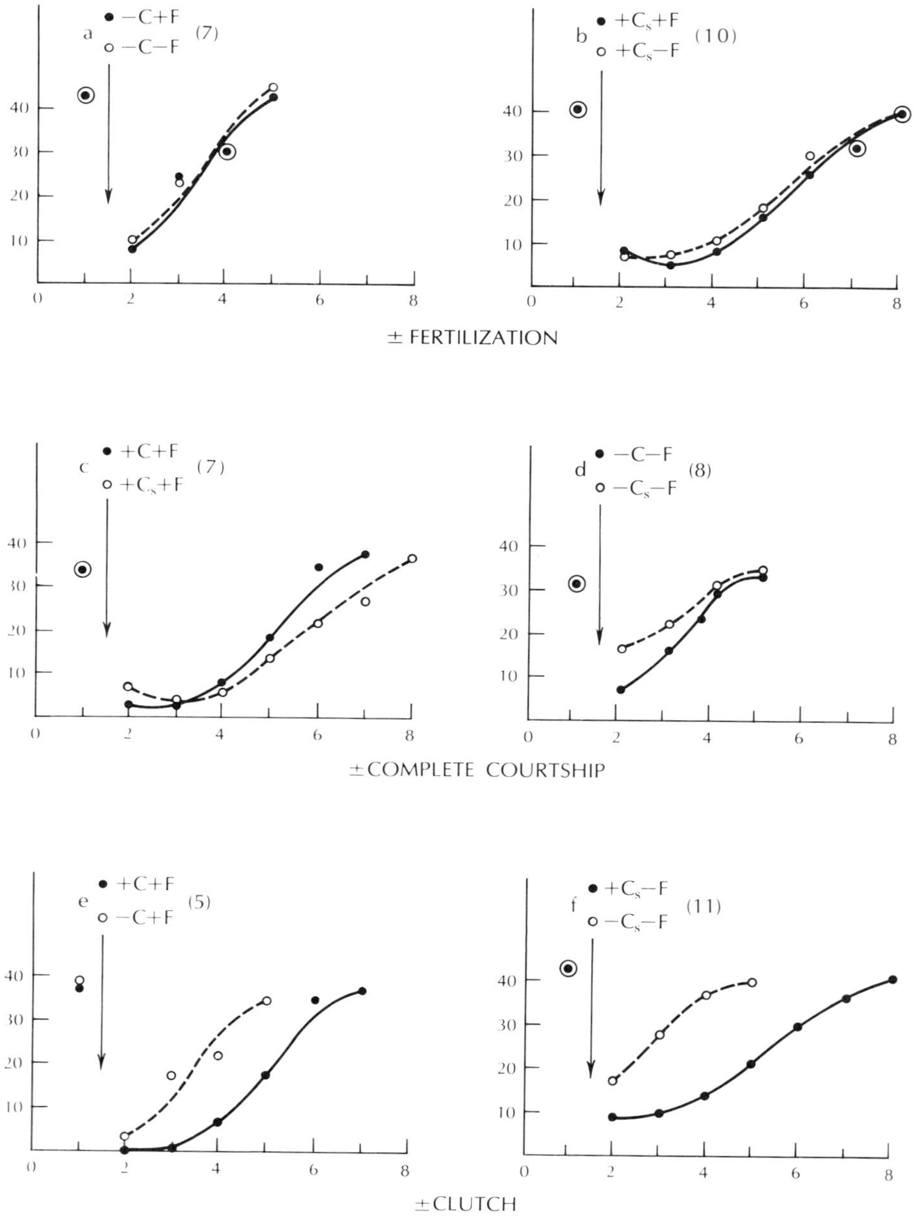
a
• −C+F
o −C−F
(7)
b
• +Cs+F
o +Cs−F
(10)
± FERTILIZATION
c
• +C+F
o +Cs+F
(7)
d
• −C−F
o −Cs−F
(8)
± COMPLETE COURTSHIP
e
• +C+F
o −C+F
(5)
f
• +Cs−F
o −Cs−F
(11)
± CLUTCH

sperm which lasts until these are exhausted by egg-laying, usually about 8 to 10 days (Manning, 1967b). In the grasshopper *Gomphocerus rufus* the female's change from sexual willingness to defensive behaviour is due to the presence of the spermatophore in the spermathecal duct: the stimulus is probably mechanical (Loher and Huber, 1966).

Consummatory stimuli share many properties with those which augment responsiveness. Just as positively motivating stimuli may act continuously to augment specific responsiveness, so may consummatory stimuli act continuously to decrease the probability of certain responses. Just as one set of positively motivating stimuli may affect several types of response (e.g., the stickleback's territory), so also may consummatory stimuli (e.g., the influence of stimuli from the stickleback's eggs on courtship behaviour). Just as any one response may be influenced by several motivating stimuli (e.g., the stickleback's aggression is influenced by stimuli from the territory, by temperature, light, etc.), so any one response may be reduced by several consummatory stimuli (e.g., the distress calls of the chick by contact, warmth, visual stimuli from the mother, etc.).

In both cases there is a procedural difficulty in the way of establishing that the effect is a direct one. A stimulus may be consummatory for one activity but elicit another. Its inhibitory effect on the former could thus occur only because it increases the probability of the latter, or vice versa. Usually, however, these possibilities can be differentiated by assessing the predictability of the sequence of the two patterns. If the stimulus switches off the first activity, but the second does not regularly follow, its effect on the first is more likely to be direct. For instance, the threatening behaviour of one herring gull to another may be brought to an end by the second bird's fleeing, in which case the threatening bird is very likely to follow, or by the second bird's attacking, in which case the threatening bird flees. In these cases the new situation acts by eliciting a second activity (chasing or fleeing), and it is this which interrupts the threatening. But if, when threatened, the second bird adopts an appeasement or submissive posture (see Chap. 16), the threatening bird may cease to be aggressive and change to any one of a number of activities—feeding, preen-

---

Fig. 10.4. The number of zigzags shown by male sticklebacks (ordinate) in "sex tests" in successive trials (abscissa) given at 10-min intervals. (After Sevenster-Bol, 1962.)

C, a clutch is laid in the nest by a female; *F*, the male fertilizes a clutch in the nest; C_s, the male is given a clutch of eggs obtained by stripping a female, no female being introduced into his tank. The dots indicate tests with the supposedly greater amount of "sex-reducing factors," and the open circles those with the supposedly smaller amount. Figures in brackets indicate number of cases.

ing, and so on. Here, the new situation seems to have a directly inhibitory effect on the threatening behaviour, and what follows depends on other factors in the situation (Tinbergen, personal communication; see also Sparks, 1965).

A related point is discussed in detail by Wiepkema (1961). Sevenster-Bol, *loc. cit.*, considered that the frequency of zigzag dances shown by a male stickleback was a measure of its sexual tendency. She therefore concluded that the decrease in the frequency of zigzags produced by stimuli from a new-laid clutch indicated a reduction in the sexual tendency as such. Wiepkema points out that, in practice, stimuli from the eggs result in an increase in some other sexual activities; for instance, creeping through the nest and ejecting sperm. Sevenster-Bol's interpretation of the mechanism underlying sexual behaviour was thus oversimple because sexual behaviour was assessed in terms of only one response. Wiepkema further suggests that the shift in behaviour following perception of the eggs is a consequence of a change in the relative attacking and fleeing tendencies of the male (see Chap. 16), i.e., his "A/F ratio". On this view the influence of the egg stimuli on the frequency of the zigzag dances is not direct, but a consequence of an increase in aggressiveness. Wiepkema presents comparable data for the bitterling (*Rhodeus amarus*), where egg-laying increases both the sexual and the aggressive tendencies of the male: since the aggressive tendency affects the sexual movements differentially (see page 388 and Fig. 16.13), the consequences depend on relations between the two tendencies.

10.3 THE CONTROL OF FEEDING AND DRINKING

As we have seen, the role of internal stimuli in augmenting eating and drinking behaviour in mammals is not clear, but they do act as consummatory stimuli. An animal usually stops feeding long before the concentration of nutrients in the bloodstream has reached peak level and before tissue needs are satisfied peripherally. The cessation of feeding is, in fact, brought about by stimuli received during eating or soon afterwards. These inhibitory influences are multiple and due to a variety of consequences of ingestion producing effects with different time courses (Sharma, 1967).

The more immediate effects have been studied in mammals by severing an animal's oesophagus and bringing the two cut ends to the surface. If the animal eats, the food drops out of its throat and can have no direct effect on the stomach ("sham feeding"). Alternatively, food can be introduced into the stomach directly. It is thus possible to separate the effects of stimulation in the mouth/throat region from the effects of changes in stomach contents. Using this technique, James (1963a) found that dogs sham-ate an amount which was larger than, but varied with, their deficit. Similarly, Lorber et al. (1950) found that sham feeding produced an almost immediate cessation of gastric motor activity in dogs. Berkun et al. (1952) showed that milk administered by mouth reduces the hunger of rats (as measured either by a consummatory or an instrumental response) more than milk introduced directly into the stomach (see also Kohn, 1951). Oral factors thus play a role in bringing feeding to an end.

This oral effect is probably mediated by the taste receptors: saccharin, which is sweet tasting but non-nutritive, is effective in reducing the hunger of rats when administered by mouth but not when injected directly into the stomach (Miller, 1957). Proprioceptive stimuli consequent upon chewing and swallowing may also be important.

The particular oral stimuli produced by the food play a part also in the control of specific hungers (Young, 1949; Lát, 1967a). Using salt-deprived adrenalectomized rats, Nachman and Valentino (1966) allowed some animals to drink 10 ml of 0.4 *M* sodium chloride solution and gave others a similar amount by stomach tube. Two hours later the stomach-loaded animals drank much more of the solution than the animals which had been allowed to taste the salt, even though the stomach-loading procedure had raised the plasma sodium concentration to normal levels. Oral stimuli are probably important in a similar way in the control of normal feeding. If more than one type of food is available an animal usually does not satiate itself on only one of the possibilities, but distributes its choices among the various food items. In both birds (Vince and Warren, 1963) and rhesus monkeys (Weiskrantz and Cowey, 1963) the extent of this distribution is a function of experience: in the latter case the primary effect of experience is to decrease the rate at which the animal becomes satiated with a new kind of food.

Although oral and pharyngeal factors may play a role in controlling eating, they are not essential—at any rate in rats. Epstein and Teitelbaum (1962) fitted rats with chronic gastric tubes which bypassed the receptors in mouth, nose, and pharynx (Fig. 10.5). The rats were able to feed themselves by pressing a bar: this resulted in the delivery of a small volume of liquid diet into the stomach. In this way they could regulate their own food intake by direct intragastric ingestion (Fig. 10.6). The animals regulated their food intake and body weight with precision for periods of 17 to 44 days, responding appropriately to dilution of the liquid food or to changes in the volume of the individual stomach loads. That stimulation of the oesophageal receptors had in fact been successfully eliminated was checked by the finding that the adulteration of the food with quinine did not affect the intake. Thus postingestion factors must be sufficient to control intake.

There is, in fact, considerable evidence that feedback from the stomach plays a role. Janowitz and Grossman (1949) found that the amount of sham feeding which occurred before satiation greatly exceeded the food deficit, indicating that although oral factors alone can produce cessation of feeding, they are less effective than oral and gastric factors together. Similarly, Berkun et al. (1952) found that milk injected directly into the stomach is more effective than saline (see also Kohn, 1951; Stellar et al., 1954; Smith and Duffy, 1955, 1957). However, it is usually found that the stomach must be preloaded with an amount of food greater than the food deficit for eating to be completely inhibited. The precise nature of the gastric effects is not clear: both chemical effects and mechanical stretching may be involved (Miller, 1957). Although bulk in the stomach depresses eating

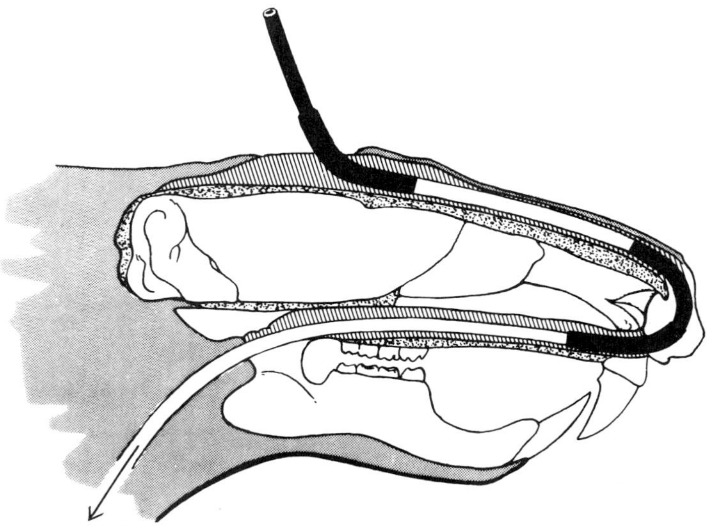

Fig. 10.5. Nasopharyngeal gastric tube used for delivering liquids to rat's stomach without stimulation of the oropharyngeal receptors. (After Epstein and Teitelbaum, 1962.)

behaviour (Smith and Duffy, 1957; Smith et al., 1962), its effectiveness varies with the nutritive value of the food (Adolph, 1947), and may be effective only when the physical limits of the system are approached. Miller (1957) has evidence that the action of a balloon, inflated in the stomach, in reducing feeding behaviour is negatively reinforcing, and suggests that it is related to the induction of nausea rather than to satiation. A further factor is the osmotic changes which follow stomach-loading: Mook (1963) found that the amount of sham feeding varied inversely with the osmotic pressure of solutions simultaneously entering the stomach.

The interaction between oral and postingestion factors is complex. In rats with both oesophageal fistulae and gastric cannulae, Mook (1963) showed that the influence of a particular oral stimulus varied with the nature of the substance reaching the stomach. For instance, intake increases with increasing concentration of glucose tasted if water enters the stomach, but decreases if hypertonic saline enters the stomach. The manner in which the immediate stimuli from the food, the amount ingested, and the

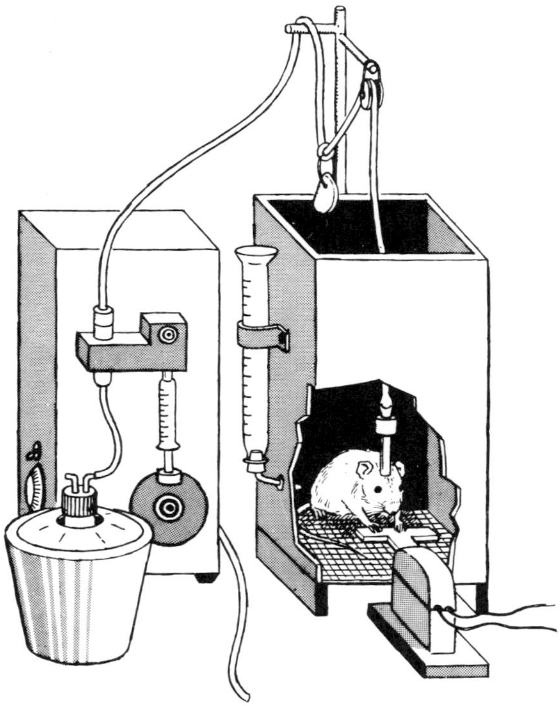

Fig. 10.6. Schematic drawing of the apparatus used for studying intragastric self-feeding by rats. (After Epstein and Teitelbaum, 1962.)

previous nutritive state of the animal interact has also been studied by Collier and Willis (1961).

Although signals from the oral, pharyngeal, and gastric regions clearly play a large part in the control of feeding, they have little effect on overall food intake. Gastric factors thus limit meal size, but overall intake is controlled by events subsequent to the arrival of food in the stomach.

Once absorption has occurred, changes in blood composition may play a role. It has been suggested, for instance, that glucoreceptors are sensitive to the rate of sugar utilization and control overall food intake. The evidence, however, is conflicting. Janowitz et al. (1949) found that injections of glucose failed to produce a depression of food intake, and other workers have found that blood from hungry or satiated animals, injected into others, has no differential effect on their rate of eating (Siegel and Dorman, 1954). Smith and Duffy (1957) could find no evidence that blood sugar plays any direct role in the control of hunger, and Epstein (1960) found no change in food intake after hypothalamic injections of glucose. On the other hand, Bash (1939a, b) found that blood from a satiated animal inhibits gastric mobility in a hungry one, and Davis et al. (1967), that the food intake of hungry rats was reduced after their blood was mixed with that from satiated rats (see also page 253).

McCleary (1953) believed that his finding, that both the nature and amount of substances with which the stomach was preloaded influence the rate of ingestion immediately afterwards, also suggested a role of blood composition in controlling eating. Coppock and Chambers (1954) investigated this further by delivering food directly into the bloodstream. If a rat is given a direct intravenous injection of glucose whenever its head is in a certain position, it keeps its head there for longer than if it is given just saline. The results of a similar experiment with dogs and rabbits, however, produced no such clear-cut effect: the rabbits did spend more time in places where they had been injected, but this could have been a consequence of secondary reinforcement (Chambers, 1956). The evidence is thus not consistent but lends some support to the view that blood nutrients may play a part in controlling eating, at least in some species (see also Steffens, 1967).

The central receptors involved in these responses may be located in the hypothalamus, which may be influenced directly by changes in glucose concentration and osmotic pressure (e.g., Verney, 1947; Brobeck, 1960; Anand, 1961; Cross, 1964). Glucoreceptors may also be located in the liver (Russek, 1963).

It has also been suggested that feeding is inhibited by the extra heat produced in assimilating food and preparing it for oxidation by the tissues. After feeding there is a rise in temperature in the centre of the body and an even greater rise in skin temperature, which is correlated with the decrease in appetite. Furthermore, the effectiveness of various foodstuffs in producing satiety is related to the heat produced in preparing them for oxidation. It may be that the influence of changes in glucose concentration on feeding behaviour is related to the correlated changes in heat production (Adolph, 1947; Brobeck, 1955; Hamilton, 1963). More recently Collier et al. (1967) have shown that rats fed isonitrogenous diets varying only in carbohydrate content adjusted their intake in terms of the ratio of calories to body weight. These authors therefore regard the tendency to recover body weight up to the genetic potential to be the governing factor. As a further and related possibility, Kennedy (1953) suggested that the long-term regulation of feeding is a function of the animal's fat reserves. These various views on the control of feeding are not, of course, mutually exclusive. Furthermore, the problem of the long-term control of feeding brings us back to the problem of the factors which immediately motivate it (see pages 241–242). The data cited suggest that the most useful course is to think of long-term factors (e.g., loss of body weight, state of fat reserves, blood composition) producing eating whenever eating is not temporarily inhibited by the shorter-term consummatory factors. In this context it is significant that Davis et al. (1967) found that the intake of food by hungry rats was reduced after their blood had been mixed with that of satiated rats, but they obtained no evidence for the existence of humoral factors that increase food intake (Davis, 1965).

A comparable complexity exists in the control of drinking behaviour. Oral factors have some inhibitory effect, for sham drinking (i.e., drinking when the oesophagus has been severed and the two ends brought to the surface of the neck) does, after a time, lead to a cessation of drinking behaviour: in dogs this does not occur until the animal has drunk much more than its water deficit, and the satiation is very temporary. In rats, a role of oral factors in controlling drinking behaviour has been shown in a different way. Thirsty rats will lick at an air jet, and this "air-drinking" results in an abnormal loss in weight through evaporation of saliva: it nevertheless produces a reduction in water consumption in a test given immediately afterwards. The air blast cools the mouth, just as drinking water does, and the response is presumably due to a failure to discriminate between the immediate sensory consequences of drinking water and those of drinking air. Air-drinking acts as a reward for thirsty rats but not for satiated ones (Hendry and Rasche, 1961). There are, however, differences between the consequences of air- and water-drinking, for air produces the immediate sensory effects of water but not the long-term ones (Treichler and Hamilton, 1967).

Stomach distension, whether produced by a balloon or by liquid, and changes in blood composition also play a role in controlling drinking, though their relative importance varies among different mammalian species (Bellows, 1939; Towbin, 1949, 1964; O'Kelly, 1954; Montgomery and Holmes, 1955; Moyer and Bunnell, 1962). Clark et al. (1961) have shown that a thirsty monkey will learn to press a bar to get isotonic saline injected into his bloodstream but will not press it when satiated with water. Again, intravenous water injections reduce the subsequent drinking behaviour of rats more than intravenous saline injections do (Novin et al.,

1966). Although specialized osmoreceptors may be involved (Verney, 1947), there is also evidence that the osmotic pressure of the body fluids, the chloride concentration in the blood, and the volume of the cells in the body may all influence drinking behaviour (Adolph et al., 1954; Corbit, 1965a, b). Drinking is also affected by injection of minute amounts of saline into the brain itself, indicating that there is a direct effect on the central nervous system (see pages 275–277).

As mentioned earlier, there are complex interactions between drinking and eating, and drinking is markedly influenced by previous experience in the situation (see review by Bolles, 1967; Grossman, 1967).

In these cases, then, it is clear that the cessation of eating or drinking is controlled by a number of different consequences of the behaviour. The manner in which these interact with the positive factors to determine whether or not the animal eats is demonstrated by brain stimulation experiments. Miller (1960) found that, although thoroughly satiated rats can be induced to eat more by brain stimulation, as ingestion proceeds, progressively stronger electrical stimulation is necessary to elicit eating (see Chap. 12).

That the various inhibitory effects are not merely additive but differ in kind is indicated by differences in their time relations. Towbin (1949) found that the satiation of thirst which follows sham drinking (i.e., when no water enters the stomach) is only temporary, and the animal soon starts drinking again. The study by Stellar and Hill (1952), mentioned previously on pages 197–198, illustrates the same point. When thirsty rats are given water, they drink steadily at a constant rate for a few minutes and then drink in short bursts over the ensuing period. For periods of deprivation of between 48 and 168 hr, the initial period of continuous drinking is virtually constant (about 8 min), though subsequently the animals which have been deprived longer drink more frequently. It thus seems that the initial period of rapid drinking is brought to an end by the resultant of all consummatory factors then acting, which consist predominantly of the immediate consequences of drinking, and not by the satisfaction of tissue needs. The former produce only a temporary inhibitory effect, so that drinking is quickly resumed if the water deficit is not satisfied. It is significant that various measures of the changes in readiness to commence eating or drinking as deprivation increases are better correlated with each other than with measures of consumption (e.g., Fig. 8.3): the latter depend also on postingestional factors which do not influence the former (Bolles, 1967).

This discussion of the control of eating and drinking has been confined to mammals, but comparable data are beginning to be available for other groups. In the three-spined stickleback, for instance, at least two factors govern the rate of feeding—one gastric and the other systemic (de Ruiter and Beukema, 1963). However, doubt has been cast on the existence of a gastric feedback system in birds, for pigeons trained to peck a key showed extinction of the response when it produced food by stomach fistula (Angermeier et al., 1964): other explanations are of course possible.

In the blowfly (*Phormia regina*) also, feeding is regulated by feedback consequences of imbibing of food, though the mechanisms are quite different. In this species feeding is elicited by excitatory input through the tarsal chemoreceptors, which initiates proboscis extension, and through the labellar chemoreceptors, which initiates spreading of the oral lobes and sucking. Initially, feeding continues so long as this input is maintained: if the receptors become adapted, feeding ceases temporarily until recovery has occurred. Since the critical level of adaptation is reached more rapidly when the sugar concentration is low, the fly feeds on dilute solutions for a shorter time than on more concentrated ones.

However, prolonged cessation of feeding cannot be due to sensory adaptation. Although the receptors disadapt within a few minutes, the threshold of stimulation necessary to elicit the response rises during feeding and remains high after feeding has ceased for many hours. This threshold elevation, which varies with the type of food eaten, is regulated by the effects of ingestion. These come from several sources. The first involves the stretch receptors in the foregut. The food is passed to the foregut from the crop at a rate governed by a negative feedback relationship from the blood osmotic pressure such that increases in the latter above a certain level slow food transport. A stretch receptor in the wall of the foregut sends impulses via the recurrent nerve to the brain, where it inhibits input from the peripheral chemoreceptors: section of this nerve produces hyperphagia. This mechanism is augmented by information from stretch receptors in the body wall which travel via the ventral nerve cord, and monitors distension caused by the fullness of the crop. In addition a locomotor centre in the thoracic ganglion inhibits feeding behaviour. The long-term regulation of sugar intake is apparently regulated by the stimulating power (not the nutritive value) and osmotic pressure of the sugar (Dethier and Bodenstein, 1958; Gelperin, 1966; Dethier et al., 1965, 1968; Dethier, 1966; Dethier and Gelperin, 1967; Gelperin and Dethier, 1967).

An even simpler mechanism operates in the blood-sucking bug (*Rhodnius*). This animal imbibes blood by means of an abdominal pump. The pump ceases to operate when the abdominal pressure reaches a critical level, at which point the emptying stroke of the pump is not initiated. Thus the size of the meal depends in part on the ease with which the outside of the animal can be stretched (Bennet-Clark, 1963).

In these pages, the role of consummatory stimuli in bringing behaviour to an end has been discussed. In conclusion it must be said that in some of the cases cited, the stimuli play a dual role, having both incremental and decremental effects on the response in question. For instance, ingestion of saccharin solution by mouth, but not by stomach, reduces a rat's subsequent consumption of sucrose solution (Miller, 1957). The taste of saccharin thus has consummatory properties. Nevertheless, rats drink more saccharin solution than water, so that the taste of saccharin also augments ingestive behaviour. Similarly, rats will work harder to obtain a moderately strong glucose solution than a weaker one (McCleary, 1953; Guttman, 1954), and it is well known that the taste of food may have an "appetizing" as well as a consummatory effect. Conflicting results of experience are discussed further in Chap. 13.

10.4 SUMMARY

1. Evidence for a hormonal influence on reproductive activities is of three main types—correlations between endocrine activity and the behaviour, the effects of removal of endocrine organs, and the effects of hormone treatment. Generalizations applicable to all vertebrates are hard to make.
2. Hormones may affect behaviour by influencing the (early) development of the nervous system; by causing changes in a peripheral organ which mediates sensory input to the central nervous system; through specific central effects; through non-specific central effects; by central effects influencing the levels of other hormones; and by changes in effectors or social stimuli.
3. Any one hormone may produce effects of several kinds and may even influence one response in a number of different ways.
4. The occurrence of a hormone-influenced response is likely to depend also on many other factors internal and external to the animal.
5. The distinction between the releasing and motivating effects of stimuli, which has been made sometimes in the past, does not necessarily imply different types of mechanism or points of action.
6. Various internal sensory stimuli play a role, but not an essential one, in the motivation of eating, drinking, and sexual behaviour in mammals. External stimuli often have positive motivational effects.
7. When an activity is terminated because the animal encounters a particular stimulus situation as a result of that activity, that situation is referred to as a "consummatory stimulus".
8. Eating and drinking in mammals are controlled, in part, by consummatory stimuli arising from the several consequences (oral, pharyngeal, gastrointestinal, metabolic, etc.) of ingestion. Their effects decay at different rates.

11

THE MODE OF ACTION OF MOTIVATIONAL FACTORS UPON RESPONSIVENESS

In Chap. 10 we discussed two categories of motivational factors—hormones and stimuli—and considered some evidence as to where and how they produced their effects. We may now continue this topic by considering more precisely some of the behavioural consequences of such factors.

Under natural conditions, most hungry animals show appetitive behaviour which involves visiting places where food is likely to be found. They are also especially likely to respond to stimuli indicative of food. The question thus arises, is the response a consequence of increased activity bringing the animal to places where suitable stimuli are encountered, or is the activity the result of increased responsiveness to stimulation? And if an increase in responsiveness occurs, is it general or limited to specific types of stimulation?

The traditional view that deprivation conditions or other motivational variables have a direct energizing effect on general activity was based on numerous observations of changes in the activity of rats and other species with hunger, the oestrous cycle, and so on (Richter, 1927). That the effect is a direct one and not due to an increased responsiveness to stimuli previously associated with reinforcement was supported by the finding that wheel-running or stabilimeter cage activity increases with deprivation even in the absence of a previous history of reinforcement in the apparatus (e.g., Richter, 1927; Finger, 1965; Wright et al., 1966).

The view that deprivation affects activity is, however, a superficial one. In the first place, "general activity" is a blanket category and lumps together many qualitatively different types of behaviour (Bindra, 1961; Richards, 1966b): careful examination shows that the increase in activity is limited to particular patterns of behaviour. Second, the change in activity may vary with the nature of the deprivation: food-deprived rats show an increase in activity but water-deprived ones may not (Bolles, 1965; Fig. 11.1). Furthermore, the consequences of deprivation on activity vary with age, experience, and the nature of the apparatus (Weasner et al., 1960; Bolles, 1963, 1965; Draper, 1967; Gross, 1968). Not surprisingly, there are also marked species differences (Campbell et al., 1966). Thus the relation between motivational factors and activity is far from simple: different variables have qualitatively differing effects, which may be positive or negative, and vary with the individual's experience.

The influence of the nature of the apparatus on deprivation-induced activity is of special interest. Changes in activity tend to be greater when the activity is measured in an apparatus in which the activity itself produces a large change in stimulation than when the animal gets relatively little

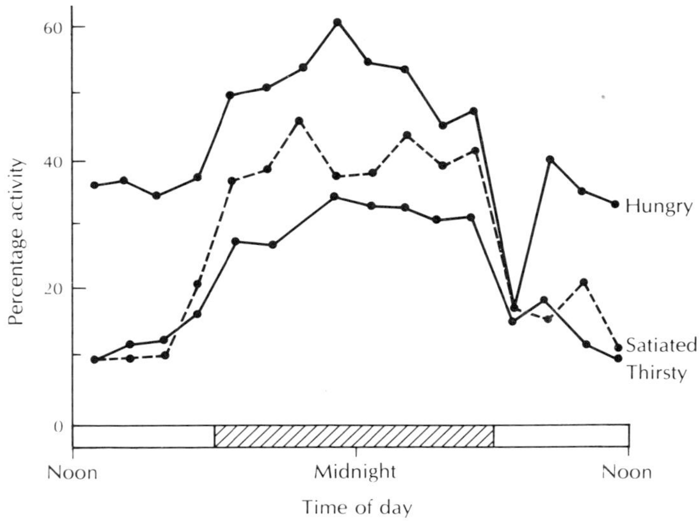

Fig. 11.1. Diurnal activity cycles of hungry and thirsty rats, and satiated controls. Thirsty rats were less active than the controls at night but differed little by day. Hungry rats were more active at all times. (From Bolles, 1965.)

feedback. For example, activity changes in rats with both food deprivation (Weasner et al., 1960) and the oestrous cycle (Eayrs, 1954) are more marked in a running wheel than in a stabilimeter cage. In the former case the rat runs inside a wheel which turns as a result of its own activity, while in the latter the floor merely tilts slightly as the rat moves: there is thus a considerable difference in the stimulation the animal receives from its own movements. As discussed earlier, therefore, the motivational changes seem to be associated with changes in the reinforcing value of the activity rather than to cause changes in activity directly. Kavanau's (1963) finding that wheel-running was positively reinforcing for deermice if voluntarily initiated but aversive if imposed by the experimenter similarly suggests it has reinforcing consequences only up to a certain level of activity, which varies with the state of the animal.

The view that food deprivation has a direct effect on activity was in fact challenged by Campbell and Sheffield (1953). They suggested that the deprivation produced an increase in responsiveness to stimulation rather than in activity per se. In their experiment, food-deprived rats showed only a little more activity than a control group if kept under very constant conditions, but considerably more activity if kept in a more variable environment. It thus seemed that the deprivation increased activity by lowering the threshold of responsiveness to environmental change. Hall (1956) failed to get adequate confirmation of this, but Campbell (1960) obtained a similar result for water-deprived rats, and DeVito and Smith (1959) for food-deprived monkeys. However, even if environmental conditions are carefully controlled, deprivation brings some increase in activity. Teghtsoonian and Campbell (1960) found that food-deprived rats kept in a sound-isolated, stimulus-controlled environment showed an increase in activity up to 10 percent above the predeprivation level, though this was much less than the 400 percent shown by rats kept under ordinary laboratory conditions. It thus seems that food deprivation causes some increase in activity even under constant conditions, though the increase is much greater in a variable environment. The effect may be mediated through adrenaline release and its consequent effect on the reticular activating system.

Given that the deprivation-induced increase in activity is largely dependent on external stimulation, the next question concerns whether it is specific to particular types of stimulation or more general. In other words, does exposure to any sort of stimulus variation lead to an increase in activity in deprived animals, or does it have to be, for instance, a stimulus change previously associated with reinforcement? There is in fact evidence that the increase in activity induced by stimulation in hungry rats occurs especially with stimuli previously associated with food. Sheffield and Campbell (1954) found that stimulus change was associated with an increase in activity only if feeding followed immediately and not if there was a time lag of 1 or 2 hr. These results were confirmed and extended by Amsel and Work (1961). It thus appears that the response to stimulus change is selective, those stimuli previously associated with food being particularly effective (see also Bolles, 1963; Baumeister et al., 1964). However, the presence of stimuli previously associated with reinforcement is not a necessary condition for an increase in activity with increase in motivational factors: for example, Bindra (1968) points out that female rats in oestrous show a marked increase in activity even though it has never been associated with reinforcement.

Where the stimuli have previously been associated with reinforcement, their effectiveness in augmenting activity has been ascribed to the classical conditioning of a central state to the unconditioned stimuli: the efficacy of such conditioned stimuli in augmenting activity depends on the motivational (e.g., hunger, thirst) condition of the animal (Campbell, 1960; Bindra and Palfai, 1967; Bindra, 1968). The rapidity with which a change in incentive affects performance makes it convenient to classify these effects as "motivational" (see page 194). For instance, not only do rats run faster along an alley to large amounts of food than to small ones but also, if the amount of reinforcement is changed, running speed changes almost immediately to that appropriate to the new level (Crespi, 1944): however, although the amount of reinforcement affects running speed, it does not usually influence learning in a discrimination situation. Such effects are thus often described as "incentive motivational".

Whereas some learning theorists use "incentive motivation" as an intervening variable additive to other sources of general drive (e.g., Brown, 1961; see Chap. 9), others consider drive and incentive to be parallel and supplementary sources of motivation (e.g., Spence, 1956). Bindra (1968), for instance, distinguishes between "drive-induced" and "incentive-motivational" effects on activity. The former is a direct effect of the deprivation, the latter of immediate stimulation. Incentive-motivational effects depend on previous association between the stimuli and reinforcement and, in Bindra's view, only these are exploratory or appetitive. This last point, however, is not necessarily valid, for Kamback (1966) found that the increase in activity shown by food-deprived rats involves an increase in object-directed behaviour, even with rats tested in visual and auditory isolation in an environment of which they had had no previous experience. The effect of deprivation on the "exploratory" aspects of activity thus seems not necessarily to depend on a previous history of reinforcement in the experimental setting.

Mendelson (1966) used an ingenious technique to assess the relative importance of hunger in facilitating food-seeking behaviour and of hunger in facilitating the response of eating and thereby providing incentive motivation on subsequent trials. He first showed that lateral hypothalamic stimulation can substitute for food deprivation as a motivational condition for the acquisition of a *T*-maze response for food (see page 283). This provided him with a means of switching hunger on and off at will. If the stimulation was switched off as the rat entered the goal box, the rat subsequently chose randomly. By contrast, if stimulation was not turned on until the rat entered the goal box, it consistently made food-directed choices. It thus seems that for the maintenance of persistent food-seeking behaviour "it is not sufficient for rats to be hungry at the choice point and for their expectation to find food in a given goal box to be confirmed;

they must be hungry when they encounter food in the goal box". Thus hunger selectively facilitates the consummatory responses, not the instrumental ones. The instrumental responses, it seems, are facilitated by "incentive motivation".

The influence of the motivational state on responsiveness to stimulation of a specific sort has been demonstrated in a quite different way by MacDonnell and Flynn (1966b). Electrical stimulation of certain hypothalamic regions in the non-anaesthetized cat can induce attack behaviour (Chap. 12). Such stimulation is associated with the appearance of sensory fields for the head-orienting and jaw-opening movements used in biting (Fig. 11.2). During stimulation, light tactile stimulation along the lip line produces jaw-opening, and similar stimulation of a wider area above the mouth produces head orientation. The extent of the fields is a function of the intensity of stimulation.

In the studies cited so far the animals were kept in a relatively impoverished laboratory environment and the types of activity they could show were limited. The principle that motivational changes are associated with selective changes in responsiveness to stimulation is immediately apparent under more natural conditions. Thus the appetitive behaviour of at any rate experienced animals is rather specific—a song bird shows one type of appetitive behaviour when patrolling its territory, another when searching for food, and another when collecting nest material. Even when the motor patterns are similar in the different cases, they are directed towards different situations and involve responsiveness to different stimuli.

Indeed, selective responsiveness to stimuli is often the most important feature by which different types of appetitive behaviour can be categorized. A great tit hops over the ground in much the same way whether feeding or looking for nest material, but it responds to different stimulus objects (see also Chap. 6). Similarly, the adults of some species of cichlid fish, while in the parental phase, may guard or lead *Daphnia*, normally their favourite food, as if they were their offspring (Baerends, 1957a): in this state they respond to the characteristics which *Daphnia* share with the young rather than to their food characteristics.

The mechanism by which the motivational state affects responsiveness to stimuli can be different from that by which it affects activity. In the blowfly, food deprivation is associated both with changes in the threshold concentration of sugar solution which will elicit feeding and with locomotion. After feeding with mannose or fucose, locomotor activity increases steadily with time, whereas the sugar threshold increases to a peak some hours after feeding and then declines. In this species, therefore, a change in activity is merely one of the effects of deprivation: since activity and threshold-lowering are not well correlated, they are presumably controlled by different mechanisms. Since activity is affected by injection of fluid into the haemocoel, dilution of blood may be the critical factor in its control (Evans and Browne, 1960). However, Green (1964a, b) has evidence that a hormone from the corpus cardiacum is involved, its release being

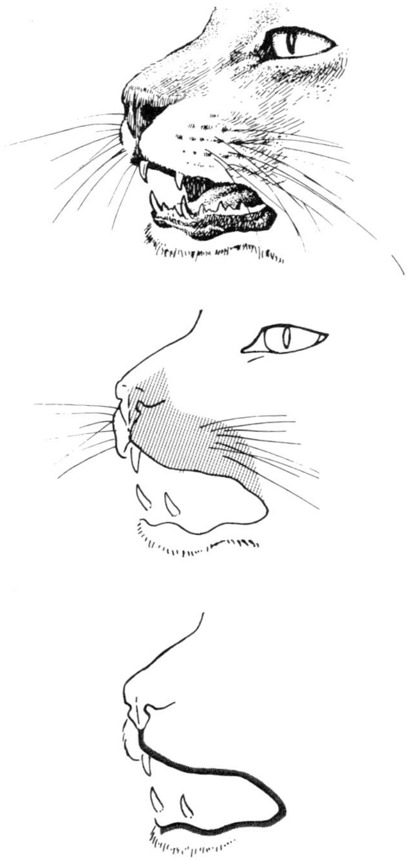

Fig. 11.2. Effect of attack-eliciting hypothalamic electrical stimulation on sensory fields for reflexes related to biting. (After MacDonnell and Flynn, 1966.)

Top, the cat's muzzle; *centre,* maxillary sensory field for head-orienting responses during relatively intensive stimulation (mandibular field not yet accurately known); *bottom,* maximum extent of sensory field for jaw opening during relatively intense stimulation.

affected by receptors in the foregut after stimulation by food (see also Dethier, 1966).

We must next consider the mechanism by which the motivational state affects responsiveness to stimuli. One possibility is that the sensitivity of the sense organs themselves changes with the motivational state. Although the suggestion that selective attention is mediated by centrifugal control of activity in the sensory pathways is not well substantiated (Chap. 6), the longer-term changes in hunger or sexual arousal could still depend on sensory factors. The evidence, however, is against such a view. In the blowfly, the time relations involved in the adaptation of the taste organs are quite different from those of hunger itself (Dethier and Bodenstein, 1958). In mammals, more direct evidence is available. Meyer (1952) found that 34 hr of starvation produced no changes in the salt, sweet, or bitter thresholds of human subjects. Adrenalectomised, and therefore salt-deficient, rats have a low preference threshold for salt solutions, but Pfaffmann and Bare (1950) found no difference in the gustatory nerve thresholds between normal and adrenalectomised rats (see also Carr, 1952). Similarly, sodium-deficient rats select sodium chloride solutions in preference to water or solutions not containing sodium chloride, but there is no difference between deficient and non-deficient rats in the frequency of discharges produced in the chorda tympani by sodium chloride. Nachman and Pfaffmann (1963) thus conclude that taste preferences are governed by a central mechanism which is affected directly by sodium levels, though the possibility that there is a change in the patterning of the peripheral discharge even when the overall frequency is constant remains open. It has also been suggested that the contrast with the salt concentration in the saliva bathing the receptors is a significant variable (Pfaffmann, 1963).

A similar conclusion has been reached from studies of sexual behaviour. Intact male rats choose the arm of an olfactometer containing a receptive female in preference to that containing a female in anoestrus, while prepuberal or castrated males do not (Le Magnen, 1952). However, gonadectomy does not affect the ability of either males or females to distinguish between the odours of sexually active and inactive members of the opposite sex when running through the apparatus for water. Furthermore, the olfactory threshold for oestrous female urine is not affected by gonadal state (Carr and Caul, 1962). It thus seems that the sex hormone affects the stimulus–response relation rather than the ability to discriminate the stimulus. The evidence on this problem is, however, not wholly consistent: steriods injected into both men (Le Magnen, 1953) and women (Beidler, 1961) can affect their sensitivity to particular odours.

In the case of patterned visual stimuli, where patterns relevant to many different types of behaviour are received through the same sensory elements, motivational changes in responsiveness can hardly depend on changes in the sense organs themselves. In general, therefore, the effect of motivational factors on changes in responsiveness to particular stimuli is usually described as a sensitization of particular stimulus–response relations. A food-deprived rat eats food when a satiated one ignores it; a male chaffinch attempts copulation with a female dummy in winter only if treated with androgen. Often, as in this last case, the motivational factor influences not just one response but a group of functionally related ones (see Chap. 25).

Crook's (1964b) observations on the nest-building of weavers show that hormonal action may also influence the effectiveness with which movements are oriented. When building motivation is low, the stitching movements result in structures without the characteristic nest form, consisting of an unorganized webbing over the supporting twigs. Only in more active building are the building movements properly oriented. Thus with increasing nest-building vigour (presumably correlated with changing hormone levels), the building increases not only in amount but also, because the movements become properly adjusted to the structure, in qualitative effectiveness. Since the male may build simultaneously on several nests at different stages, this does not involve merely a change to a different type of movement: the building behaviour is adjusted to each of the nests in hand at the moment. Many comparable examples are given by Howard (1935).

Sensitization of stimulus–response relations implies that there is an inverse relationship between the strength of the motivational factors and the strength of the stimulus required to elicit a response of given strength. An example is given by the work of Baerends et al. (1955) on the courtship of the male guppy (*Lebistes reticulatus*). In this species the coloration of the male provides an index of his tendencies to attack, flee from, and behave sexually towards the female. In Fig. 11.3 the colour patterns of the male are arranged along the abscissa in an order representing increasing tendency for the male to show sexual behaviour. The effectiveness of a female in eliciting courtship increases with her size: this is plotted on the ordinate. The curves represent the relations between internal and external factors required to produce particular patterns of behaviour: "posturing", "incomplete sigmoid", and "sigmoid". The same pattern may thus appear with a wide range of combinations of internal and external factors.

The scaling of the abscissa in Fig. 11.3 was necessarily indirect and depended on the relative frequency with which copulation and two of the courtship patterns ("sigmoid" and "sigmoid intention", see Figs. 3.3 and 25.2) were associated with the different colour patterns. However, over much of the range, the curves are parallel.

Rather more quantitative data on the relationship between the effect of a stimulus and the current motivational state have been obtained for the attack readiness of a cichlid fish (*Pelmatochromis subocellatus*) and the chirping of a cricket (*Acheta domesticus*) by Heiligenberg (1965b, c; 1966). In the latter case standard chirps were played to a captive cricket either every 2.5 sec or every 0.625 sec, and the resultant effect on the cricket's own chirping recorded. Fig. 11.4 shows the relation betwen the cricket's chirp rate in the 15 sec before stimulation started and the average rate in the first two 15-sec intervals afterwards. The increase in the cricket's chirp rate depends on the strength of the stimulus but not (within limits) on the prior level of activity.

The data in Fig. 11.3 show that the same response pattern may appear with a range of combinations of internal state and external stimuli.

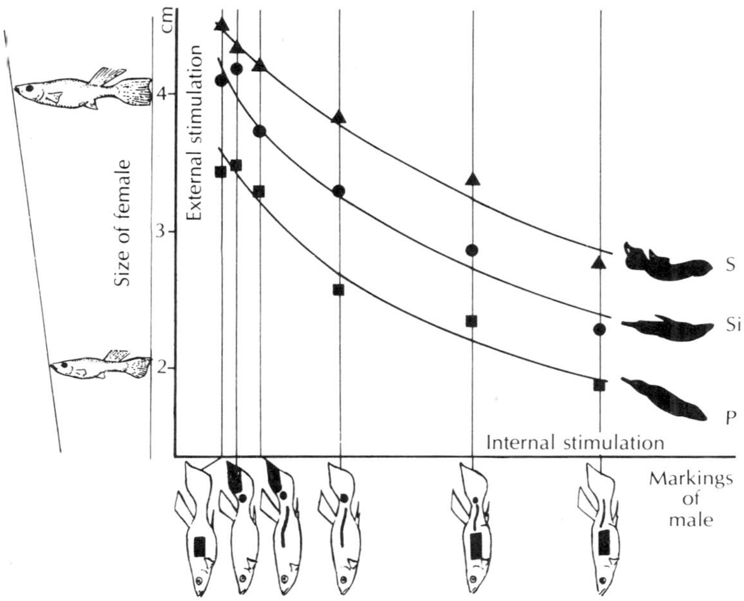

Fig. 11.3. The influence of the strength of the external stimulation (measured by the size of the female) and the internal state (measured by the colour pattern of the male) in determining the courtship behaviour of male guppies. (From Baerends et al., 1955.)

Each curve represents the combination of external stimulus and internal state producing a particular type of behaviour, namely, posturing *P*, the sigmoid intention movement *Si*, and the fully developed sigmoid *S*.

Even when stimulus characters cannot readily be measured along a simple physical scale, a similar principle applies. Thus we have already seen that the different stimulus characters effective for a given response (page 69) can supplement each other—deficiency in one can be compensated by increase in another. It is as though the several stimulus characters fed into a common pool to produce a certain quantity of stimulation. The consequence is that, as the motivational factors increase, objects sharing fewer and fewer characters with the natural or optimal object are adequate to produce a response of given intensity: the range of effective stimuli thus increases. For instance, Prechtl (1953) showed that the range of stimuli which would elicit begging from the nestlings of passerine birds increased with hunger,

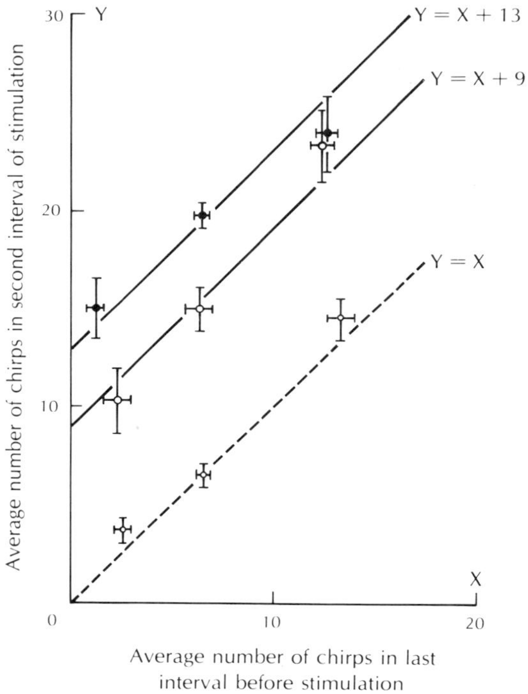

Fig. 11.4. Relation between number of chirps given by cricket in the last 15-sec interval before stimulation and the first two 15-sec intervals after stimulation. (From Heiligenberg, 1966.)

Black circles, data with stimulus rate of one chirp per 0.625 sec; *large open circles,* one chirp per 2.5 sec; *small open circles,* no stimulation.

and Beach (1942b) found that the lower the copulatory threshold of male rats the greater the range of stimulus objects with which they would copulate. Similar effects occur in men, for ambiguous or non-existent stimuli are interpreted as related to food more by hungry subjects than by satiated ones (McClelland and Atkinson, 1948; see also Vernon, 1952).

So far, three ways in which motivational factors affect behaviour have been mentioned—they may increase activity, influence selectivity in responsiveness to stimuli, and they may also be spoken of as sensitizing stimulus–response relations. We shall see later that these do not provide an adequate description of all the ways in which motivational factors affect behaviour. In the senses discussed on pages 624–632, much behaviour is goal-directed; and goal stimuli have reinforcing properties. The motivational factors act by determining the goal (see Chap. 25).

SUMMARY

1. The increase in activity which often follows deprivation may be largely a consequence of an increased responsiveness to external stimulation rather than a direct effect of deprivation on activity.
2. The changes in activity which accompany the oestrous cycle in rats may be related to a change in the reinforcing value of the activity-produced stimulation, rather than a direct effect of the oestrous cycle on activity.
3. An increase in the motivational factors for a given type of behaviour is accompanied by selectivity in responsiveness to stimuli.
4. This selectivity is usually not due to an influence of the motivational factors on sensory mechanisms.
5. Motivational factors can also be said to sensitize stimulus–response relations. There is an inverse relation between the strength of a stimulus and the strength of the motivational factors required for a response of a given intensity.
6. Motivational factors can also be described as determining the goals of behaviour.

12

MOTIVATIONAL PROBLEMS AND THE CENTRAL NERVOUS SYSTEM

When a male chaffinch wakes from its roosting place in early spring, it may sing for a while, patrolling its territory intermittently between bouts of singing. During this period it is likely to attack or threaten any other males it sees. It may then fly down from the trees and feed on the ground for a while: often it feeds in proximity to other males, whom it tolerates at a distance of a few feet. After a period of feeding it is likely to fly up into a bush to preen and then return to singing. It shows, in fact, periodic changes from being concerned primarily with one group of activities to being concerned with another.

Such changes in behaviour presumably depend on changes in physiological state, each state having some degree of persistence in time. Nevertheless, they seem to be independent of changes in many of the continuously acting extraneural factors known to influence the behaviour in question: thus the chaffinch sings only when in a certain hormonal condition and while on its territory, yet neither of these alters when it flies down to feed. This suggests that the activities are not influenced directly by these extraneural factors but depend on the state of central nervous mechanisms which mediate their effects.

Similarly, when a stimulus is presented to an animal, its response is not constant but changes with time. Further, if such a stimulus is removed, a state of changed responsiveness to that stimulus, revealed if it is presented again, persists for some time (e.g., in Chap. 13). If extraneural factors are constant, such changes must be ascribed to the neural mechanisms mediating the response.

For such reasons various authors have postulated a "central excitatory state", or "central motive state". This is regarded as carrying the effects of the extraneural factors, and of influences from other parts of the central nervous system, for some time, and as predisposing the organism to respond either more vigorously in general, or else in particular ways to particular stimuli (Sherrington, 1906; Lashley, 1938a; Beach, 1942b; Morgan, 1943, 1959; Stellar, 1954, 1960; Dethier et al., 1965, 1968; Bindra, 1968).

When it was first introduced by Sherrington, the concept of "central excitatory state" was used mainly to refer to changes in excitability which outlasted the stimulation of an afferent nerve by a few milliseconds. For instance, contraction of the tibialis anterior of the dog can be elicited by stimulation of more than one cutaneous nerve. If just-threshold shocks are administered to different nerves the response is greatest when the stimuli are synchronous, and falls rapidly when the interval between them increases to about 20 millisec. Apparently, the first stimulus creates a central state which lasts for about this time. Although the experiments by Sherrington and his collaborators were concerned mostly with reflex responses within the spinal cord, the concept has been taken over to account for similar effects in more complex responses, where the time courses are quite different.

A central state postulated to account for changes in behaviour need not, of course, be related to the condition of any particular anatomically localized part of the central nervous system. However, it is convenient to start by considering the sorts of evidence used to implicate particular parts of the central nervous system in motivational changes. The studies cited are mostly on mammals, and many (though not all) of them point to hypothalamic involvement. Later we shall see that, although the hypothalamus does indeed have an essential role in many types of change in responsiveness, the state of responsiveness of the animal is not to be ascribed solely to the state of particular mechanisms in the hypothalamus. Rather, complex networks which ramify through much of the central nervous system are involved. The concept of a "central excitatory state", useful in the early stages of analysis, fades into the background as that analysis proceeds.

12.1 EVIDENCE FOR THE INVOLVEMENT OF PARTICULAR CENTRAL NERVOUS REGIONS

1. Localized Ablations

Numerous studies have shown that localized ablations in the hypothalamus produce specific effects on behaviour. For instance, in the rat, destruction of an area in the medial part of the hypothalamus, including the ventromedial nucleus, results in hyperphagia and obesity, whilst lesions in a more lateral area lead to aphagia, the animals starving to death in the midst of plenty. The two areas have thus been said to contain "satiation" and "feeding" areas, respectively (Anand and Brobeck, 1951; Teitelbaum, 1955; Miller, 1957). As we shall see later, eating induced by hypothalamic lesions, though similar to normal eating in most respects, differs in that it involves increased sensitivity to taste stimuli. Localized hypothalamic lesions can affect also many other types of behaviour (e.g., Phoenix, 1961), though in only a few cases (e.g., sex; Lisk, 1966a) is there evidence for opposed excitatory and inhibitory mechanisms.

2. Electrical Stimulation

Electrical stimulation of particular regions of the hypothalamus may produce highly specific effects which are, in general, the reverse of those of ablation (Hess, 1954). For instance, stimulation of the lateral hypothalamus of rats induces feeding behaviour, whereas stimulation of the ventromedial region stops it (Anand and Brobeck, 1951; Larsson, 1954; Brobeck, 1960; Anand, 1961). An example is given in Fig. 12.1 (Smith, 1961), which shows data from a rat with electrodes in both lateral and medial areas. The strength of stimulation was the same for both: it remained constant for the first half-hour of each experiment and was then increased slightly in the case of the lateral electrode and decreased for the medial electrode. The animal was starved for 18 hours before the beginning of each test. In spite of this, consumption was markedly depressed during the first half-hour of stimulation through the medial electrode. With the lateral electrode, consumption was increased most markedly during the second half-hour, since this was the period during which eating ceased in the control condition. The differences between the 6-hr points were not significant. Unlike lesion-induced eating, that induced by electrical stimulation does not involve changes in sensitivity to taste stimuli (Tenen and Miller, 1964).

Other types of behaviour elicited by hypothalamic stimulation through implanted electrodes include drinking, attack, rage, fleeing, cooling, micturition and defaecation, sleep, and motor activity. It is of particular interest that electrical stimulation can elicit the activities of looking for a sleeping place, lying down, and sleeping: there are in fact at least two kinds of sleep in higher mammals, controlled by different regions of the mid- and hindbrain (Nauta, 1946; Hess, 1954).

As an example, some diagrams of the opossum brain, showing the sites at which various types of behaviour were induced, are shown in Fig. 12.2 (Roberts et al., 1967). These data were obtained by implanting a number of electrodes in each animal and stimulating through these in turn. Testing was carried out in an observation box containing a variety of goal objects (e.g., food, water, nesting material) to permit the display of a variety of responses. The particular behaviour shown varied with the site stimulated.

Earlier mammalian studies were reviewed by Akert (1961), and there appears to be close anatomical correspondence between marsupial and placental mammals (Roberts et al., 1967). Comparable results have been obtained with birds (Åkerman et al., 1960; Åkerman, 1966a, b; Brown, 1969).

Although the behaviour produced by electrical stimulation sometimes consists merely of stereotyped movements, the effect usually seems to be a physiological one. For instance, it often depends on the availability of an appropriate object—food, water, a mate, a board to gnaw, according to the point stimulated: when the appropriate objects are absent, only searching behaviour appears (e.g., Miller, 1957; Vaughan and Fisher, 1962; von Holst and von St. Paul, 1963; Levison and Flynn, 1965; Roberts and Carey, 1965;

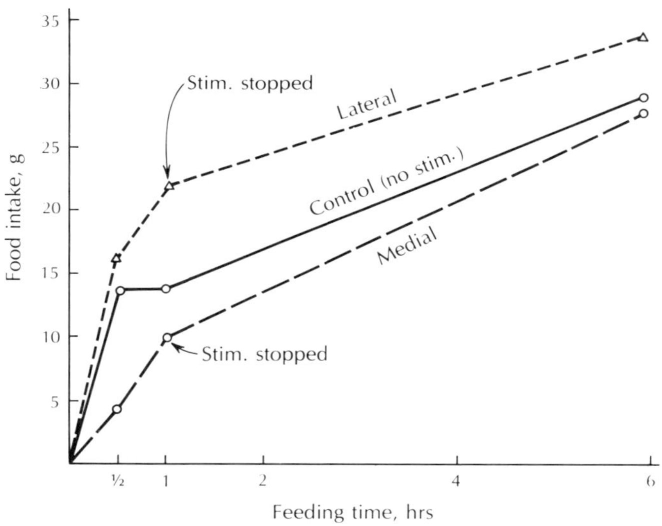

Fig. 12.1. Effects of stimulation of the lateral and medial hypothalamus on food intake. (After Smith, 1961.)

The intensity of stimulation was set at a value which did not produce somatic motor responses. It remained unchanged for the first half-hour. For the second it was increased slightly for the lateral placement and decreased for the medial placement.

Harwood and Vowles, 1966; MacDonnell and Flynn, 1966a, b; see also C. H. F. Rowell, 1963, 1965, for an invertebrate example).

Furthermore, integrated sequences of movements may be involved. Some examples of the consequences of intracranial electrical stimulation on the behaviour of pigeons is shown in Fig. 12.3: the patterns are similar to those seen in normal social contexts and are well integrated, the different parts of the body adjusting their posture and tonus appropriately. Similarly, electrical stimulation of part of the hypothalamus of the cat produces first

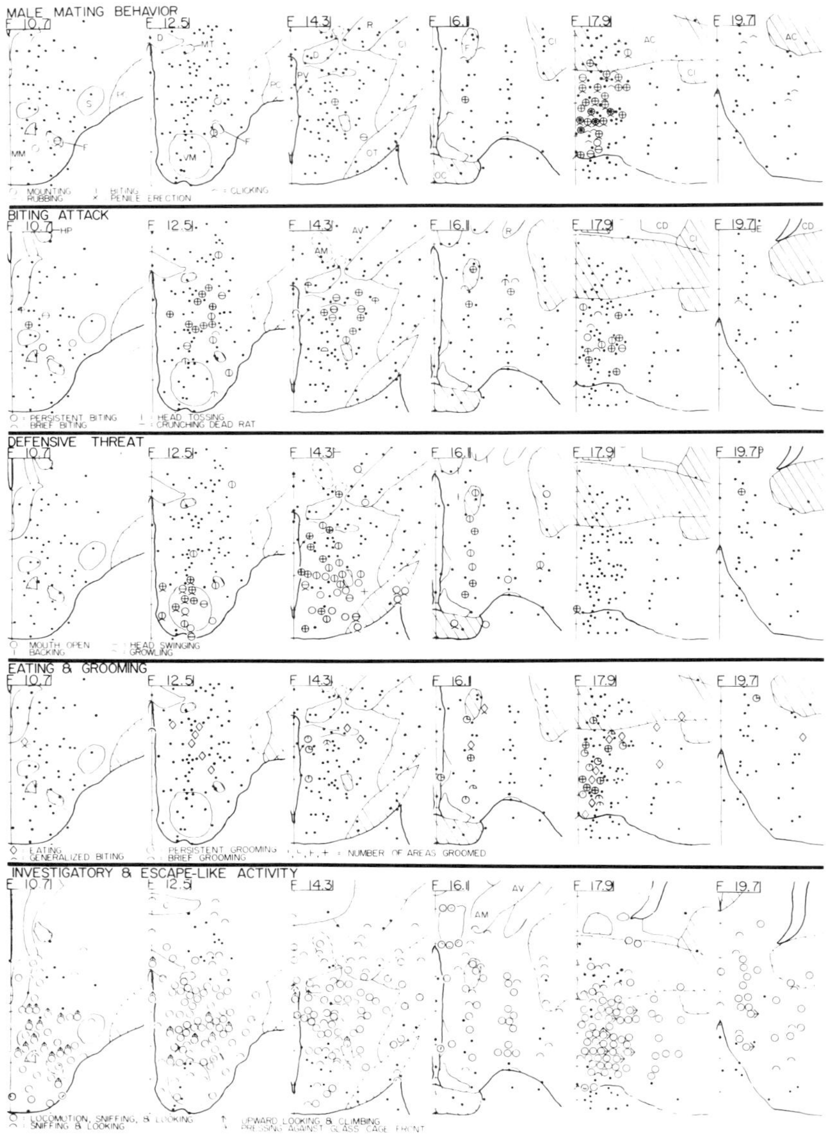
MALE MATING BEHAVIOR
F 10.7
F 12.5
F 14.3
F 16.1
F 17.9
F 19.7
MOUNTING
RUBBING
BITING
PENILE ERECTION
= CLICKING
BITING ATTACK
PERSISTENT BITING
BRIEF BITING
HEAD TOSSING
CRUNCHING DEAD RAT
DEFENSIVE THREAT
MOUTH OPEN
BACKING
HEAD SWINGING
GROWLING
EATING & GROOMING
EATING
GENERALIZED BITING
PERSISTENT GROOMING
BRIEF GROOMING
= NUMBER OF AREAS GROOMED
INVESTIGATORY & ESCAPE-LIKE ACTIVITY
LOCOMOTION, SNIFFING, & LOOKING
SNIFFING & LOOKING
UPWARD LOOKING & CLIMBING
PRESSING AGAINST GLASS CAGE FRONT

behavioural arousal, then a defensive position and certain vocalizations, leading subsequently to hissing, spitting, and attack. This behaviour is accompanied by the autonomic changes which normally accompany such behaviour in the intact animal. The aggressive behaviour ceases as soon as stimulation is discontinued, but a lowered threshold may persist (Hess, 1954; review by Akert, 1961). Often quite complex chains of behaviour are involved. For instance electrical stimulation in the thirst or feeding area will cause a goat to show a previously learnt instrumental response which has been rewarded by water (Andersson and Wyrwicka, 1957; Wyrwicka et al., 1960). The induced behaviour may interact with naturally occurring changes in behaviour, so that summation or mutual inhibition occurs, and the animal may even show behaviour whose goal appears to be the diminution or inhibition of the movement evoked (Delgado, 1962).

By remotely controlled electrical stimulation of a variety of (not necessarily hypothalamic) sites in the brains of monkeys living in a group situation, Delgado (1963, 1964, 1966, 1967) has also shown that the complex sequences of behaviour induced may be adjusted according to the dominance status of other animals in the group, and that induced expressive movements are responded to by other individuals. In one experiment a subordinate monkey learnt to press a lever which caused the caudate nucleus of the boss monkey to be stimulated by radio and inhibited his aggressiveness. Such observations provide compelling evidence that electrical stimulation produces effects closely similar to that associated with naturally occurring activity of the brain regions in question: it will, however, be necessary to return to this question again later.

That the effectiveness of intracranial electrical stimulation may vary with time, indicating the occurrence of internal changes in responsiveness, has already been mentioned (page 221). Von Holst and von St. Paul (1963) describe such changes as involving a change of "mood". They point out

---

Fig. 12.2. Electrical stimulation of hypothalamus and preoptic areas of opossum (Didelphys virginiana). (From Roberts et al., 1967.)

Locations of positive and negative points for motivational responses are plotted on frontal plane diagrams of the hypothalamus and preoptic area located 10.7 to 19.7 mm anterior to the interaural axis. Negative points are indicated by small solid circles. *AC*, commissura anterior; *AM*, nucleus anterior medialis thalami; *AV*, nucleus anterior ventralis thalami; *C*, nucleus centralis; *CD*, nucleus caudatus; *CI*, capsula interna; *D*, nucleus dorsalis hypothalami; *F*, columna fornicis; *H*, hippocampus; *HA*, nucleus habenularis; *HC*, commissura hippocampi; *HP*, tractus habenulopeduncularis; *LH*, nucleus lateralis hypothalami; *MM*, corpus mammillare; *MT*, tractus mammillo-thalamicus; *OC*, chiasma opticum; *OT*, tractus opticus; *PC*, pedunculus cerebri; *PV*, nucleus filiformis pars paraventricularis; *R*, nucleus reticularis; *S*, nucleus subthalamicus; *SE*, area septalis; *SM*, stria medullaris; *TH*, thalamus; *VM*, nucleus ventromedialis hypothalami.

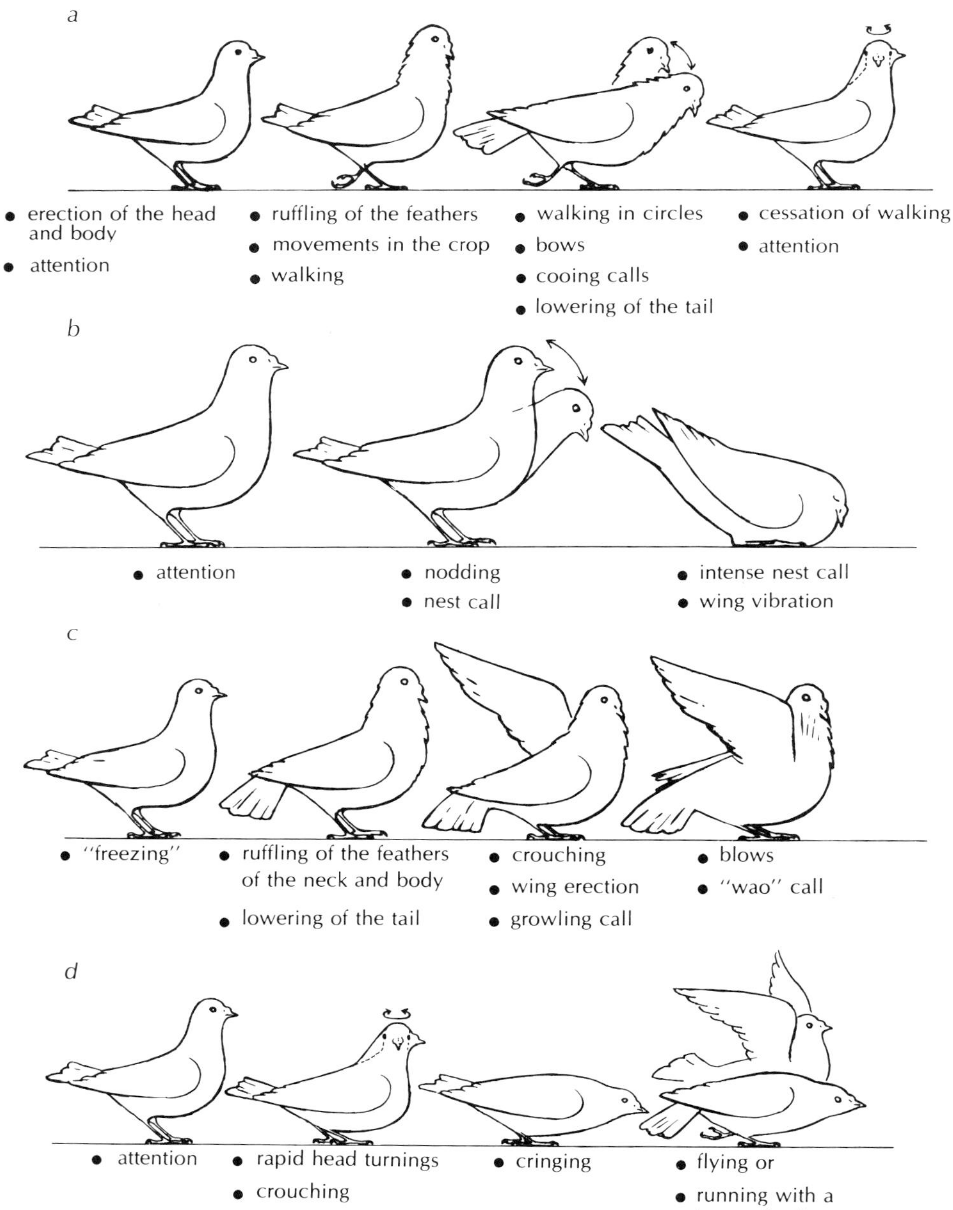

Fig. 12.3. Intracranial electrical stimulation in the pigeon. (From Åkerman, 1966*a*, *b*.)

(*a*) Bowing; (*b*) nest demonstration, produced by 30 sec of stimulation in the preoptic area; (*c*) defensive threat, produced by 30 sec of stimulation in the ventral diencephalis paraventricular gray; (*d*) escape, produced by 30 sec of stimulation in the lateral hypothalamus.

that the tendency for an animal to show one type of behaviour has inertia and is only slowly interrupted by stimuli for other responses. For instance, in Fig. 12.4 the previously present "cackling" mood is interrupted only briefly by an electrical stimulus which induces "watching-out" but more effectively by one which induces "sitting". However, a brief stimulus for sitting is insufficient to change the mood for long, and the cackling mood reappears. Only after a prolonged stimulus does its disappearance persist. Presumably, therefore, the electrical stimulus induces a change in the organism which outlasts the stimulus (see also Harwood and Vowles, 1967). Usually, such effects are not of very great duration, but in some cases

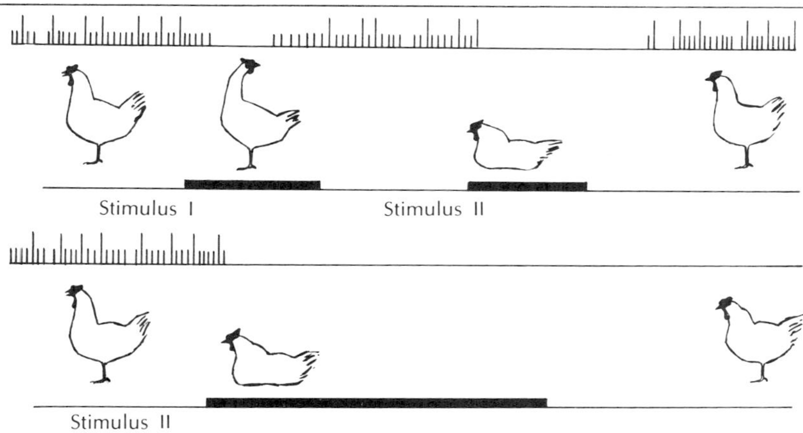

Fig. 12.4. Results of stimulating the brain of a fowl through two separate electrodes. (After von Holst and von St. Paul, 1963.)

Initially, the bird was cackling, as indicated by vertical marks on top line. The first stimulus produces "watchful staring", and a brief interruption of the scolding. The second stimulus produces "sitting down in a sleepy mood" and a more prolonged interruption of scolding. Nevertheless, the scolding reasserts itself. Only after a prolonged stimulus through the second electrode does the scolding disappear completely.

effects lasting even longer have been produced by stimulation known to have been in the hypothalamus: Delgado and Anand (1953) found increases in food intake the day after stimulating the lateral hypothalamus of cats. Smith (1961) obtained a similar result with rats.

3. The Administration of Drugs, Hormones or Metabolites to Localized Areas

This may produce specific effects on behaviour. The "respiratory centre" in the hypothalamus is a well-known case (see page 37–38); the influence of oestrogens on the hypothalamus of cats, discussed previously, is another (see page 233). The injection of minute quantities of hypertonic saline into the hypothalamus of goats produces an appropriate increase in drinking behaviour (Andersson, 1953). Miller has extended this work using cats, showing that they will work to obtain water after injections of hypertonic sodium chloride into the brain; this indicates that the injection produces the same general characteristics as thirst (Miller, 1959, 1961a; and pages 278–284).

Although studies of lesioned animals indicate that the anatomical areas controlling feeding and drinking overlap extensively, pharmacological experiments indicate chemical specificity in the two systems. Grossman (1960) found that adrenergic substances introduced into the feeding area induce eating, and cholinergic substances drinking. The time course of drinking induced by the cholinergic substance carbachol indicates that it ceases as a result of feedback from drinking in a manner similar to drinking induced by deprivation. Furthermore, whereas water introduced into the stomach reduces carbachol-induced drinking, water by mouth is even more effective. Thus carbachol-induced drinking resembles normal drinking (Quartermain and Miller, 1966; see also Khavari and Russell, 1966). However, eating induced by hypothalamic injections of norepinephrine differs from eating induced by food deprivation and resembles that produced by hypothalamic lesions in involving increased sensitivity to taste stimuli (see pages 279–281; Booth and Quartermain, 1965). The effects appear to be complex, however, and there are interactions between the deprivation condition, site of stimulation, and chemical used: Hutchinson and Renfrew (1967) found that norepinephrine in the preoptic area during food deprivation produced an increase in drinking, whereas carbachol in the anterior hypothalamic area during water deprivation produced a decrease in drinking.

As we have already seen (pages 252–253), eating and drinking may be controlled in part by changes in heat production, and it is therefore not surprising to find that eating and drinking have also been obtained by thermostimulation in the anterior hypothalamic preoptic area (Andersson and Larsson, 1961). Heating hypothalamic loci produces also thermoregulatory responses in mammals and birds (Ranson et al., 1937; Åkerman et al., 1960; Hammel et al., 1960; Baldwin and Ingram, 1967), and neurones responsive to temperature changes have been identified (Nakayama et al., 1963; Hellon, 1967; see also Cross, 1964). Feldberg and Myers (1965) produced temperature changes in cats by injections of amines into the anterior hypothalamus, and Myers (1966) has shown that pumping fluid from the third ventricle of one monkey into that of another causes the recipient to show thermoregulatory responses appropriate to the temperature of the donor. Using intraventricular injections, Myers and Yaksh (in press) found that aminergic substances evoke eating and temperature changes, but cholinergic compounds produce only changes in temperature. The complexities arising from experiments on the chemical activation of hypothalamic mechanisms are emphasized in a recent review by Myers and Sharpe (1968), and the mechanisms involved in thermoregulation have been discussed by Bligh (1966).

Recent work has shown that intracranial (though not necessarily hypothalamic) infusion may produce not merely temporary changes in behaviour but also long-lasting effects. Rats with ethanol chronically infused into the cerebral ventricles showed sudden and lasting oral preferences for moderately high ethanol concentrations: histological examination revealed no morphologic alterations in the brain (Myers, 1964).

4. Electrical Recording from Specific Brain Regions

Another technique involves the recording of changes in the electrical activity of specific brain regions during experimental manipulations. For instance, Brobeck et al. (1956) found that appetite depressants produced EEG changes similar to those seen in barbiturate anaesthesia in all parts of the hypothalamus except the medial nuclei, in parts of which activity increased. Stomach distension also causes activity in the ventromedial nucleus (Sharma et al., 1961), but hyperglycaemia causes an increase in ventromedial activity and a slight decrease in the lateral regions (Anand et al., 1962).

5. Intracranial Self-stimulation

Another possible line of evidence for centres in the hypothalamus comes from experiments on the reinforcing effects of self-administered intracranial electrical stimulation. Olds (1958, 1962) has claimed that, within the areas of the brain from which reinforcing effects are produced (see pages 219–221), there are subareas in which the effects are enhanced by androgen or food deprivation (see also Brady, 1961): this seems to be true for birds as well as mammals (Goodman and Brown, 1966). That the medial and lateral hypothalamic areas do in fact control self-stimulation in rats in a manner similar to that in which they control feeding is supported by a study of Hoebel and Teitelbaum (1962). Stimulation of the satiety area inhibited not only feeding but also self-stimulation through an electrode placed in the lateral area. Conversely, anaesthetization of the satiety centre led both to more feeding and to more self-stimulation in the lateral area (see also Fonberg, 1967). Furthermore Spies (1965) found that rats given a choice between food and self-stimulation maintained their preference for the latter with increasing food deprivation if the electrodes were in the lateral hypothalamus but not if they were elsewhere.

However, other evidence indicates that eating induced by externally controlled electrical hypothalamic stimulation and the behaviour involved in producing intracranial self-stimulation may depend on different mechanisms. For one thing, the electrical thresholds for the two effects differ. In addition, amphetamine, which decreases hunger by many behavioural tests, raises the electrical threshold for inducing eating and lowers that for self-stimulation (N. E. Miller, 1960; Prescott, 1964).

A relationship between self-stimulation and sexual behaviour is suggested by Miller's (1961a, b) finding that electrode placements in or near the hypothalamus which produced ejaculation were more likely to prove rewarding in a self-stimulation experiment than those which did not (see also Olds, 1962; Herberg, 1963). Caggiula (1967), using electrode placements in the posterior hypothalamus, found that the self-stimulation rate increased with testosterone treatment. Electrical stimulation to the same site caused the male to copulate with an oestrous female or to press a bar to gain access to a female: the sexual activity continued after ejaculation.

One possibility in such cases is that self-stimulation occurs only if a neural "drive system" and a "reinforcement system" are active simultaneously. Coons and Cruce (1968) found that rats would press a bar for intracranial stimulation with a current only slightly above the threshold necessary to induce eating if food was near the bar, but not without food. At higher current strengths self-stimulation was maintained without food. They suggest that, under these conditions, the current spreads to activate feedback appropriate to the drive stimulated. With more intensely stimulated drive mechanisms, wider ranges of sensory feedback may be reinforcing. On this view the sensory feedback can be fairly non-specific, self-stimulation occurring because reinforcing stimuli of some sort are nearly always present (see also Chap. 24).

12.2 THE NATURE OF THE CONSEQUENCES OF CENTRAL NERVOUS MANIPULATIONS

Taken as a whole, the types of evidence cited above indicate clearly the existence of specific mechanisms of great importance in the control of functional sequences of behaviour, such as those involved in eating, drinking, and sexual behaviour. It is, however, necessary to consider some possible objections and qualifications.

One possible objection to experiments involving electrical stimulation of the brain is that the stimulus does not produce a specific response but results in a general activation. This has been effectively answered by Coons et al. (cited N. E. Miller, 1960). Electrodes were implanted into the brains of rats in such a manner that stimulation induced eating only. If such animals were tested when satiated with food but mildly thirsty, they would leave the water spout to go to the place where they had previously learnt that food was hidden. If the effect of the stimulation was merely to activate an otherwise dominant but subthreshold type of behaviour, the rats should have continued to drink instead of going to food.

A related point concerns the extent to which the hypothalamic circuits activated by such stimulation are rigidly bound to particular types of motivated behaviour. Valenstein et al. (1968) stimulated rats in the presence of stimulus objects for eating, drinking, and gnawing. Each animal usually showed only one type of behaviour. The object for that behaviour was then removed, and the animal given intracranial stimulation with the same parameters but in the presence only of the other objects. A new type of behaviour, directed towards one of the remaining objects, usually appeared: furthermore, it persisted at least to some degree even when all the objects were again available. Thus although the great bulk of experiments involving intracranial electrical stimulation demonstrate that the effect of stimulation in one locus is repeatable, and that stimulation in different loci produces different responses, these experiments suggest some degree of plasticity in the relationship between the hypothalamic circuit and the behaviour which appears.

Although central stimulation or central lesions can influence complex chains of behaviour as a whole, they may not affect all aspects of a given type of behaviour equally. A hypothalamic lesion which causes a rat to overeat, and thus seems to increase feeding motivation, also causes a reduction in bar-pressing rewarded by food, a reduction in tolerance of quinine, and a reduction in speed of running to food: the lesion thus does not have similar effects on all aspects of the behaviour (Miller, 1957; see also Choy, quoted by N. E. Miller, 1960; Ehrlich, 1963).

The consequences of hypothalamic lesions are in fact complex and may involve something other than the mere elimination of one group of behaviour patterns or one type of control. Most evidence is available for feeding behaviour. As we have seen, rats with ventromedial hypothalamic lesions overeat. They cannot, however, be said simply to have a stronger hunger motivation: although they eat more than normal animals, they are less ready to work for food in an operant situation.

A more detailed examination of the behaviour of these hyperphagic animals is revealing. After operation they first go through a "dynamic" phase during which they gain weight and become "obese". Their food intake then falls until it is only a little above the normal level, and weight remains static at its new high level (Fig. 12.5). When fed on a liquid diet, such animals overeat by eating larger meals than normal but not by eating more often. This suggests that a short-term satiety mechanism is upset. However, the stabilization of body weight at the new level suggests that the ventromedial lesions affect also the setting of a mechanism regulating food intake relative to body weight. If obese animals are starved for several days and then given free access to food, they overeat until their body weight regains its previous high level and then again levels off. Contrariwise, if normal animals are artificially caused to become fat, ventromedial lesions have little influence on food intake. Teitelbaum (1961) suggests that the hypothalamic satiety centres are sensitive to some correlate of obesity, perhaps

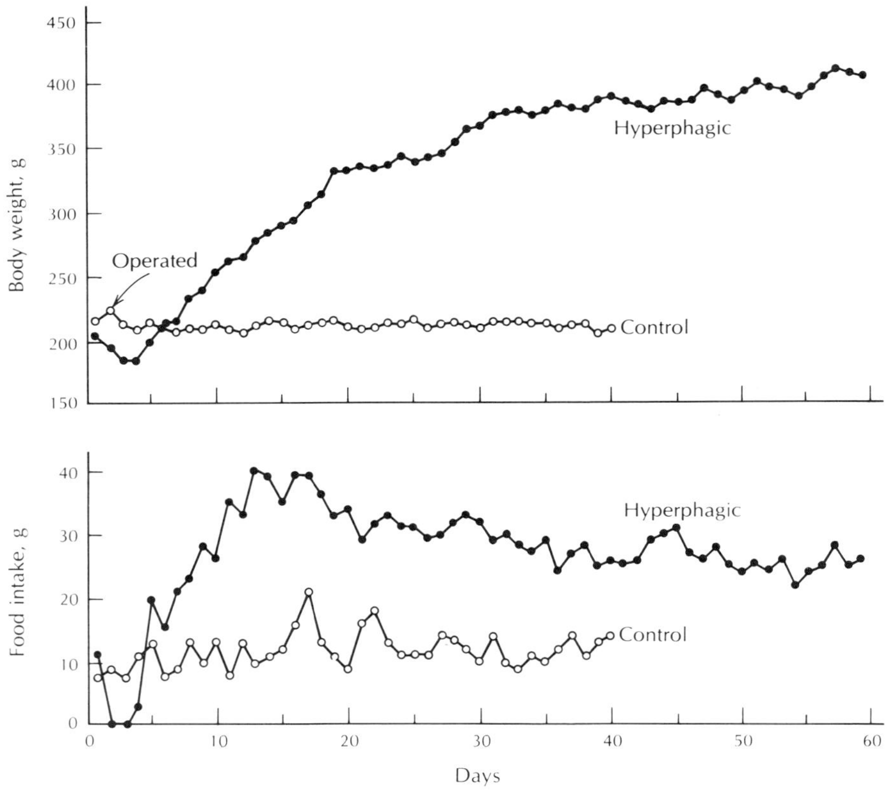

Fig. 12.5. Post-operative body weight and daily food intake of a hypothalamic hyperphagic rat compared with an unoperated control. (From Teitelbaum, 1966, after Teitelbaum, 1955.)

to a circulating metabolite related to the state of fat deposits. If ventromedial hyperphagic rats are given solid pellets, they not only eat larger meals than normal rats but also eat more often: the pellets provide fewer calories per meal than the liquid diet.

One factor which affects the size of the meals of ventromedial hyperphagic animals is an increased sensitivity to the oral stimulus properties of the food. Thus hyperphagic animals decrease their intake when the food is adulterated with quinine at a concentration which does not affect normal animals, but increases in the palatability of the food exaggerate the overeating (Teitelbaum, 1955; Corbit and Stellar, 1964). However, hypothalamic hyperphagia is not solely a consequence of increased responsiveness to taste stimuli. In hypothalamic rats that can feed themselves intragastrically (see Figs. 10.5 and 10.6) some hyperphagia persists in spite of the absence of oropharyngeal sensations, though high levels of obesity are not reached: the addition of small oral incentives produces further overeating and rapid weight gain (McGinty et al., 1965). That hyperphagic animals regulate their caloric intake in a manner which does not depend on oral stimuli from the food has in fact been demonstrated recently by Thomas and Mayer (1968). Although hyperphagic rats eat large meals, they can, like normal rats, maintain the caloric intake during each meal within a stable range of values even when they are receiving a gastric infusion which markedly alters the average caloric density of liquids entering the stomach. The rapidity of the response to gastric infusion suggests a role of gastric chemoreceptors. Changes in meal size and caloric input are accommodated by changes in the frequency of meals in both normal and hypothalamic hyperphagics, though at different levels.

In any case it is not yet certain whether the hyperphagia is *necessarily* correlated with increased sensitivity to the food or whether these are independent effects: Graff and Stellar (1962), using small ventromedial lesions, found evidence that the neural mechanisms for hyperphagia and finickyness are separate, though overlapping in area. The precise nature of the increased finickyness of hypothalamic hyperphagic rats is thus a matter of dispute. Since it involves also increased aversion to quinine water in relieving thirst, it is not specific in hunger (Corbit, 1965b; Krasne, 1966; Nachman, 1967). It does not involve an increase in sensitivity to quinine but rather intolerance for an aversive substance (Mook and Blass, 1968). Although the finickyness is related to a disturbance of the satiety mechanisms, its nature can be understood only in relation to the particular component of satiety which has been reduced and its time course (de Ruiter, 1967a). Furthermore, as Teitelbaum (1966) points out, ventromedial lesions have a number of consequences besides increased food intake—"they increase the animal's emotional irritability and viciousness, interrupt its sexual behaviour, and decrease its general level of activity" (see also Smith et al., 1961). It is thus clear that ventromedial lesions have a number of effects probably potentially independent, including lowered feeding motivation as measured in an operant situation (see page 196), an interference with a satiety mechanism which leads to overeating, and increased sensitivity to food which may be part of more general affective changes.

The consequences of lateral hypothalamic lesions are also complex. They always produce both aphagia and adipsia, but recovery can occur. If left alone, operated animals starve to death in the midst of plenty. They can, however, be kept alive artificially. After a while they then pass into a stage in which they will accept palatable foods, though not in sufficient quantity to maintain body weight. They still refuse to drink. Later they will regulate food and water intake on wet palatable foods but not on dry food and water. Later still they become able to eat dry food and water normally.

Rats which have recovered from lateral hypothalamic aphagia sufficiently to maintain themselves still show a rather specific deficit. Although they respond to cold or to dilution of their food by increasing their intake, they do not so respond to insulin-induced hypoglycaemia. This could mean that some of the factors controlling intake act directly on the lateral hypothalamic mechanism and not via the ventromedial satiety centre. It also implies that the effects of cold or dilute food are not normally mediated by hypoglycaemia (Epstein and Teitelbaum, 1967).

The effects of lesions in or stimulation of the lateral hypothalamus also depend on the precise locus. For instance, stimulation in the far-lateral regions produces both eating and grid-crossing for food by satiated animals, but stimulation of the midlateral region, only the former (Morgane, 1961).

Although the consequences of lateral hypothalamic lesions appear to be motivational, other interpretations have been suggested. Baillie and Morrison (1963) investigated the possibility that rats rendered aphagic by lateral hypothalamic lesions might be suffering from lack of ability to eat. They found that animals conditioned to obtain food by lever-pressing continued to press the lever after operation. Furthermore, rats conditioned to feed themselves through an intragastric injection by lever-pressing continued to do so after operation even though showing aphagia and adipsia when offered food and water by mouth. They therefore concluded that the lateral hypothalamic lesions produce predominantly a motor failure—a true aphagia. However, Rodgers et al. (1965) found that recovering lateral-hypophagic rats, given the choice of oral feeding with highly palatable foods or intragastric injections, all resumed the former before the latter. They suggest that Baillie and Morrison's results were due to the use of an insufficiently palatable diet.

Such studies of the effects of lesions on the feeding behaviour of rats indicate that any idea of a simple "feeding mechanism", or even of conflicting feeding and satiety mechanisms, involves oversimplification. We must similarly beware of oversimple interpretations of stimulation experiments: although the behaviour elicited by intracranial stimulation often seems to resemble normal behaviour (see pages 275–276), we must still ask whether it shares all the properties of normally motivated behaviour.

Detailed studies by Miller and his coworkers (reviewed N. E. Miller, 1960, 1961a, b) have shown that this problem is far from simple. On the one hand, in many experiments the consequences of central stimulation seem very similar to those of normal hunger. For example, central stimulation causes a satiated rat to eat solid foods or drink liquid ones, elicits learned food-seeking habits, and causes a mildly thirsty animal to stop drinking water and go to find concealed food. Furthermore, the threshold of stimulation required is raised by the appetite-reducing drug dextroamphetamine (cf. page 278); and cessation of stimulation acts as a reinforcement for the learning of a *T* maze just as the food which terminates hunger does (see also Harwood and Vowles, 1966). However, the rewarding effect of cessation of stimulation can be obtained from points which do not elicit eating, and there are threshold differences between the effect of stimulation on eating and its rewarding effect; the reward effect and eating may thus be independent effects of the stimulation (Krasne cited by Miller, 1960; Tenen and Miller, 1964).

Again, Miller and his coworkers found that central elicitation of the cat's "alarm" reaction has all the functional properties of externally elicited pain and fear: (1) the stimulation can motivate, and its termination reinforce, trial-and-error learning; (2) it can be used to establish a conditioned response; (3) it can be used to condition an emotional disturbance to a distinctive environment and, after this, during trials without any further stimulation, the cats will learn to escape from that environment; (4) it can serve as a punishment to teach hungry cats to avoid food. Similarly, the central readiness for attack produced in cats by hypothalamic stimulation has both motivational and cue properties. For example, when the readiness to attack is aroused by central stimulation in cats, or the readiness to gnaw in rats, the animals learn to respond appropriately in a *Y* maze to reach an object that would elicit the behaviour in question (Roberts and Kiess, 1964; Roberts and Carey, 1965). On the other hand, centrally evoked rage responses, apparently similar to those produced by peripheral pain, differ from them in that they cannot easily be conditioned; a cat in which the flight response was induced by central stimulation showed rapid escape learning but failed to show avoidance learning. Miller comments, "In investigating the motivating effects of central stimulation of the brain, it is unsafe to take anything for granted".

Finally, it remains to emphasize that the response elicited is often characterized not by the motor movements of ingestion but by the sensory feedback obtained. When a rat with an electrode implanted in an area where eating is induced is stimulated, it will not drink water but will drink sugar-water or milk, even though satiated. Apparently the stimulation engenders a tendency to respond in a way which brings a particular type of sensory feedback (N. E. Miller, 1960). As we have seen, cats that attack objects as a result of hypothalamic stimulation are, similarly, not indiscriminate in the objects they attack (Levison and Flynn, 1965), and objects appropriate to the behaviour elicited can serve as reinforcers (Roberts and Kiess, 1964; Roberts and Carey, 1965).

12.3 THE COMPLEXITY OF THE CENTRAL NERVOUS MECHANISMS

It will be clear that the central nervous mechanisms under discussion do not act as the bases for "central excitatory states" where excitation is stored up ready to be released in specific activities. Rather it would seem that they form parts of mechanisms through which extraneural motivating factors (Chap. 10) can produce a facilitation of stimulus–response relations. More than this is involved, however, for the behaviour may be varied so that the appropriate stimuli are encountered—in other words, the goals are set (Chap. 25).

While the hypothalamic structures in question are often referred to as centres, it is clear that they are not anatomically localized, but diffuse. The points from which any particular pattern of behaviour can be elicited by electrical stimulation are intermingled with points whose stimulation leads to different responses (Fig. 12.2). In view of the diverse functions assigned to the hypothalamus, the multiple influences to which it is subject, and the effectors it influences, an interpretation of hypothalamic function in terms even of functionally distinct centres is certainly an oversimplification. It is necessary to stress in this context that some responses are common to many functional categories of behaviour. Cardiovascular changes, for instance, occur in the defence reaction (fight, flight, and fright), adaptations to external heat or cold, sexual behaviour, and so on. It will also be apparent that the control of different activities (e.g., eating, drinking, and thermoregulation) are interlinked in complex ways (Teitelbaum, 1966).

Furthermore, the fact that a majority of experiments have involved the hypothalamus, and have shown it to be important for the behaviour studied, must not be taken to imply that responsiveness depends solely on the state of a hypothalamic mechanism—clearly other parts of the central nervous system also are involved. We must consider a few examples indicating how extensive are the areas of the brain in which experimental manipulations can produce the sorts of effects we have been discussing.

Hyperphagia in rats can be produced by forebrain lesions as well as lesions in the hypothalamus (Richter and Hawkes, 1939). Rats with septal lesions show increased licking activity for sucrose solution which is independent of deprivation or sucrose concentration, and seems to involve an increased incentive value of the reward (Beatty and Schwartzbaum, 1968). The effects of adrenergic and cholinergic substances on eating and drinking behaviour are found also in the ventral amygdala, and those of cholinergic substances in the septal area (Grossman, 1964a, b).

Again, although the hypothalamus clearly contains mechanisms of basic importance for sexual behaviour, many other parts of the brain are also involved. For instance, Beach (1940, 1944) showed that removal of parts of the cerebral cortex resulted in a behavioural deficit which increased with the extent of the cortical damage. The change in behaviour, however, did not consist in the omission of particular parts of the sexual pattern, but in a general reduction of sexual motivation. This could be compensated for by administration of androgen—the greater the cortical damage, the more androgen was required to restore the sexual behaviour to the normal level. It seemed as though the cortex was exerting an excitatory influence on sexual behaviour synergic with that exerted by the androgens. Similar results were obtained after ablations of the telencephalon of pigeons (see Chap. 10).

To cite another example, the so-called rage response of cats can be abolished by hypothalamic lesions, evoked by stimulation in the hypothalamus, amygdala, and midbrain, and can be shown by cats lacking a telencephalon. The normal response, however, depends on complex interrelations between telencephalon and diencephalon. For example, whereas removal of the neocortex may result in placidity, ablation of parts of the rhinencephalon results in an increase in ferocity. The rage response of hypothalamic cats not only lacks direction, but also has a very low threshold and practically no after-response: unlike normal rage, it ceases almost as soon as the eliciting stimulus is removed (Bard and Mountcastle, 1948). The contribution of the higher levels seems to be concerned with the release and regulation of the response, while the mechanism of expression depends primarily on the hypothalamus (see Akert, 1961; MacDonnell and Flynn, 1968). Even this, however, may represent an oversimplification, for acute lesions in the anterior part of the system have no effect on the threshold for electrical stimulation applied more posteriorly, but coagulation in the midbrain and hypothalamus at first eliminates elicitation of threat behaviour by more anterior stimulation. Two weeks later, however, anterior stimulation can again elicit threat behaviour but at a higher threshold (Brown and Hunsperger, 1963). Recent reviews are given by Brady (1960) and Stellar (1960), and the special importance of the limbic system has been stressed by Pribram (e.g., 1961) and Weiskrantz (1964).

In the ring dove (*Streptopelia risoria*) Harwood and Vowles (1967) found that agonistic behaviour could be elicited by electrical stimulation in the forebrain, hypothalamus, and midbrain, but the nature of the effect differed markedly among these regions. That elicited by forebrain stimulation was not stimulus-bound, was very variable, and had prolonged and variable after-effects. By contrast midbrain stimulation yielded agonistic behaviour which was stimulus-bound, stereotyped, lacked after-effects and affective components, and involved no interaction with external stimulation: presumably it was due merely to excitation of a motor system. Stimulation in the anterior hypothalamus yielded very variable behaviour in which it seemed to the observer that the bird was hallucinating a potentially dangerous object. Clearly electrical stimulation in these different regions excites functionally distinct parts of the networks controlling agonistic behaviour.

Another case is provided by the feeding behaviour of pigeons. This depends on a network of structures at several levels of the brain. It appears to originate in the Vth cranial nerve and its sensory nucleus, and includes both striatal and limbic components. Lesions of any of these components produce disturbances in feeding behaviour, and in some cases it has been possible to dissociate experimentally sensorimotor and motivational effects (Zeigler et al., 1967; Zeigler and Witkowsky, 1969).

Even in lower vertebrates the brainstem mechanisms are not merely command centres determining peripheral motor responses, but are also integrators of sensory input and proprioceptive feedback. In *Xenopus,* both clasping and feeding are coordinated by spinomedullary systems, and occur in response to the appropriate input even in animals in which the central nervous system is transected just anterior to the Vth cranial nerves. However, the elicitation of clasping and feeding, and therefore the functioning of their respective control systems, depends on the integrity of the sensory as well as the motor innervation of the hindbrain (Hutchison, 1964; see also Hutchison, 1967a).

In conclusion, the fluctuations in behaviour which occur independently of changes in extraneural factors, and the persistence of the effects of stimulation, indicate the existence of central states of considerable endurance. However, the postulation of "central excitatory mechanisms" showing long-term changes in responsiveness can lead to great oversimplification if pressed too far. The mechanisms concerned with changes in responsiveness to a constant stimulus involve complex networks ramifying through the central nervous system and involving functionally distinct parts. It is clear that they do not store excitation but may be affected by both excitatory and inhibitory factors, the latter often resulting from feedback. In some cases behavioural regulation is achieved by mutually inhibitory central mechanisms, some excitatory and others inhibitory. This has long been established for breathing (Chap. 3), and we have seen evidence also for the feeding (pages 269–272) and sexual (page 270) behaviour of rats. In the case of feeding, the feedback loops may be of much greater complexity than has been indicated here (de Ruiter, 1963).

12.4 SUMMARY

1. Under natural conditions animals show changes in behaviour which presumably depend on changes in physiological state, each state having some degree of persistence in time. Nevertheless, the changes seem to be independent of continuously acting extraneural factors (e.g., hormones, external stimuli) known to affect the behaviour in question. This suggests that the activities are not influenced by these extraneural factors directly, but depend on the state of neural mechanisms which mediate their effects.
2. Physiological evidence for such mechanisms is provided by studies involving lesions of the central nervous system, stimulation of the CNS by electrical and chemical stimuli, recording from the CNS, and perhaps by experiments involving self-administered intracranial stimulation.
3. In some cases the behaviour produced is stereotyped and abnormal, but in others it shows properties similar to those of normally motivated behaviour. Several lines of evidence indicate that the central mechanisms operate by influencing sensitivity to stimulation rather than by controlling particular movement patterns.
4. In some cases control depends on mutually inhibitory and excitatory mechanisms, but their nature and interactions are far more complex than such labels imply.
5. The central nervous mechanisms concerned are diffuse and not limited to particular parts of the central nervous system.

13

SHORT-TERM CHANGES IN RESPONSIVENESS CONSEQUENT UPON EXPERIENCE

When an animal eats, its tendency to continue eating is gradually reduced. We have seen in Chap. 10 that this can be ascribed at least in part to influences from the food ingested. But, even when there is no obvious new source of stimulation comparable to that provided by ingested food, the performance of a response is often associated with an increase or decrease in the probability that it will appear again: activity is accompanied by positive and negative feedback effects. Thus if the begging response of a young passerine bird is elicited by an appropriate stimulus a number of times in succession, the bird becomes unresponsive: the waning of responsiveness is a consequence of the presentation of the stimulus and must be due either to a direct effect of perceiving the stimulus, or to proprioceptive or humoral effects of responding, or to some direct consequence of activity in the nervous mechanisms underlying the response.

In this chapter we shall consider the properties of the changes in responsiveness which occur in such circumstances, concentrating for the most part on short-term effects: more permanent effects are discussed in Sec. 3. Changes in responsiveness of the type discussed are important both in the design of experiments and to the theoretician, and yet they are often ignored. To emphasize how ubiquitous they are, we shall consider examples from a wide range of species.

13.1 WANING IN THE ABSENCE OF MUSCULAR FATIGUE AND SENSORY ADAPTATION

A waning of responsiveness could, of course, be a consequence of metabolic fatigue: thus in *Drosophila* flying may wane independently of walking, but the waning is due to the depletion of carbohydrate reserves necessary for flying but not for walking (Wigglesworth, 1949).

Muscular fatigue is, however, unlikely to be a major factor if, after waning, the same effectors can be used for a different response. When the skin of its back is stimulated, a frog will make a wiping movement over the spot with its leg. Franzisket (1953) studied the waning of this response by giving series of 100 stimulations, each series taking about 17 min. When such series were given for a number of days in succession, the proportion of stimuli which elicited responses increased from about 10 percent to 60 to 90 percent. If two series were then given one after the other in the same locus, the number of responses to the second series was much reduced. However, such a reduction in the second series was not found if it involved elicitation of a different response, even though both responses used the same groups of muscles; for instance, wiping and pulling up the suspended hindlimb when pinched. As another example, Eikmanns (1955) studied the waning of the orientation movement made by a toad to a fly dummy presented to one side of its body axis. This movement is normally a preliminary to catching the prey, but is nearly identical in form to the optomotor response made when the toad is placed in the centre of a striped cylinder which revolves about it. After a long series of presentations of the prey dummy the orientation response wanes, but the optomotor response can still be elicited at full strength. Further evidence against the view that muscular fatigue plays any significant role in such cases of waning comes from experiments howing that the decrement is specific to one stimulus or sensory modality: some examples are cited below (pages 290–295).

In some cases adaptation of sense cells may play an important part in the waning of a response. For example, the feeding bouts of certain flies last until the chemoreceptors become adapted (see page 255). More usually, however, sensory adaptation can be ruled out. Thus after one type of response to a stimulus has waned, the same stimulus may yet elicit a different response. If a reflexogenous zone round the mouth of a hungry human newborn is stimulated, the head is turned so that the mouth moves towards the point stimulated. The response wanes with successive stimuli, but its place may be taken by another response (turning away) mediated by the same sensory region of the skin and presumably by the same sense organs (Prechtl, 1958). In other cases adaptation of the sense organ cannot be an important issue, because the reduction in responsiveness applies also when the same response is elicited through a different receptor cell. The twitching response of the earthworm, which can be elicited by thermal, mechanical, or galvanic stimulation, wanes with successive stimuli: the decrease in responsiveness to stimuli in a given modality affects also neighbouring segments and thus cannot be due to changes in the sense organs themselves (Kuenzer, 1958; see also Franzisket, *loc. cit.*). Similarly, the waning of the scratch reflex of dogs (Sherrington, 1906) and of the wiping response of toads (Kuczka, 1956) affects all points within several millimetres of the point stimulated.

Sensory adaptation is also ruled out in experiments in which a stimulus, the response to which has waned, again elicits a response when presented in a different context. For instance, Engen and Lipsitt (1965), studying the breathing responses of human neonates to odour, found that when the response to a mixture of two odours had waned, it could be elicited again when only one of them was presented: clearly the initial waning cannot be regarded as sensory fatigue (see also Bartoshuk, 1962a, b).

Another source of evidence that the waning can be independent of changes in both the sense organs and the effectors comes from studies in which the peripheral nerves or central nervous system are stimulated electrically. Thus in a study of the flexion reflex produced by brief trains of shocks in spinal cats, Thompson and Spencer (1966) found that waning was not due to sensory adaptation or muscular fatigue, and that it took place after the afferent input terminals and before the output motor neurones—that is, in the course of transmission through the interneurones (see also Wickelgren, 1967a, b). At a more complex level, von Holst and von St. Paul (1963) elicited a variety of integrated patterns of behaviour by stimulation through electrodes permanently implanted in the brains of unanaesthetized chickens. If more than one electrode was implanted in the same animal, the same response could sometimes be elicited from different loci. Repeated stimulation in one locus led to a reduction in responsiveness which affected that locus only: stimulation through the other electrode was still fully effective. The waning process must have occurred between the electrode and the final common path.

To summarize, although muscular fatigue and sensory adaptation may be involved in many cases of waning in responsiveness, the evidence cited shows that changes occur also between the sensory cells and the effectors and that it is often these changes that play the major role.

13.2 STIMULUS SPECIFICITY OF DECREMENTAL PROCESSES

Even where adaptation of the sensory cells is unlikely to be a factor, it is sometimes clear that if a response can be elicited by more than one stimulus situation the waning is specific to the particular stimulus (or modality) used. We may consider some examples drawn from a variety of species ranging from annelids to vertebrates:

1. The waning of the twitching response of the earthworm (see above) is at first specific to the sensory modality through which it is elicited. The specificity decreases if stimulation is prolonged (Kuenzer, 1958; see also Clark, 1960).

2. *Drosophila* spp. show an orientation response both to a localized light source and to gravity. The response to either wanes after about 100 elicitations, but the response to the other is unaffected (Dürrwächter, 1957).
3. The waning of the prey-catching response of the water bug *Notonecta glauca* is specific at least to the side of the body on which the stimulus is presented. If a stimulus is presented 300 times on the left side the proportion of responses elicited falls off markedly, but responsiveness to stimuli presented on the right side is unimpaired (Wolda, 1961).
4. The equilibratory response of the frog *Rana esculenta* when the substrate is tilted is mediated by both eye and labyrinth. Responsiveness wanes when the animal is stimulated successively through one modality, but the animal still responds to the other (Butz-Kuenzer, 1957).
5. If a small dummy is presented to a toad (*Bufo bufo*), the animal turns towards it. This movement is the first stage of catching prey. If the stimulus is presented successively so that it is in exactly the same position relative to the toad each time, the response wanes. The waning of the response is specific to the characteristics of the dummy (size, colour, etc.) and to the particular spot on the retina stimulated (Eikmanns, 1955).
6. A toad makes wiping movements directed at any spot on its skin touched with the hand. The waning which occurs with repeated stimulation is specific to the particular area stimulated and, in some degree, to the frequency of stimulation. This case is especially interesting since there is evidence that the stimulus-specific waning can occur even if no response is given. Sudden stimulation with a bristle does not elicit wiping but does lead to a reduction in responsiveness to subsequent stimulation of the same spot by hand (Kuczka, 1956).
7. The gaping response of 5- to 7-day-old chaffinches can be elicited both by shaking the nest and by imitating the call of the parents. If a sequence of one of these types of stimulus is given every 2 sec until it ceases to be effective, 10 to 13 responses can be elicited. If, however, they are given in alternate short sequences, the young gape 40 to 46 times (Prechtl, 1953).
8. The waning of the gobbling response of the turkey is specific to the particular sound stimulus used to elicit it, though the response to similar stimuli is also affected (Schleidt, 1954).
9. The "mobbing" response given by chaffinches to non-flying predators can be elicited by a wide range of stimulus objects, including owls, dogs, stoats, etc. It wanes gradually if the stimulus is presented continuously (Fig. 13.1). Some recovery occurs if the stimulus is removed, but if the response is elicited again even 24 hr later, the response on the second occasion is still reduced in strength in comparison with

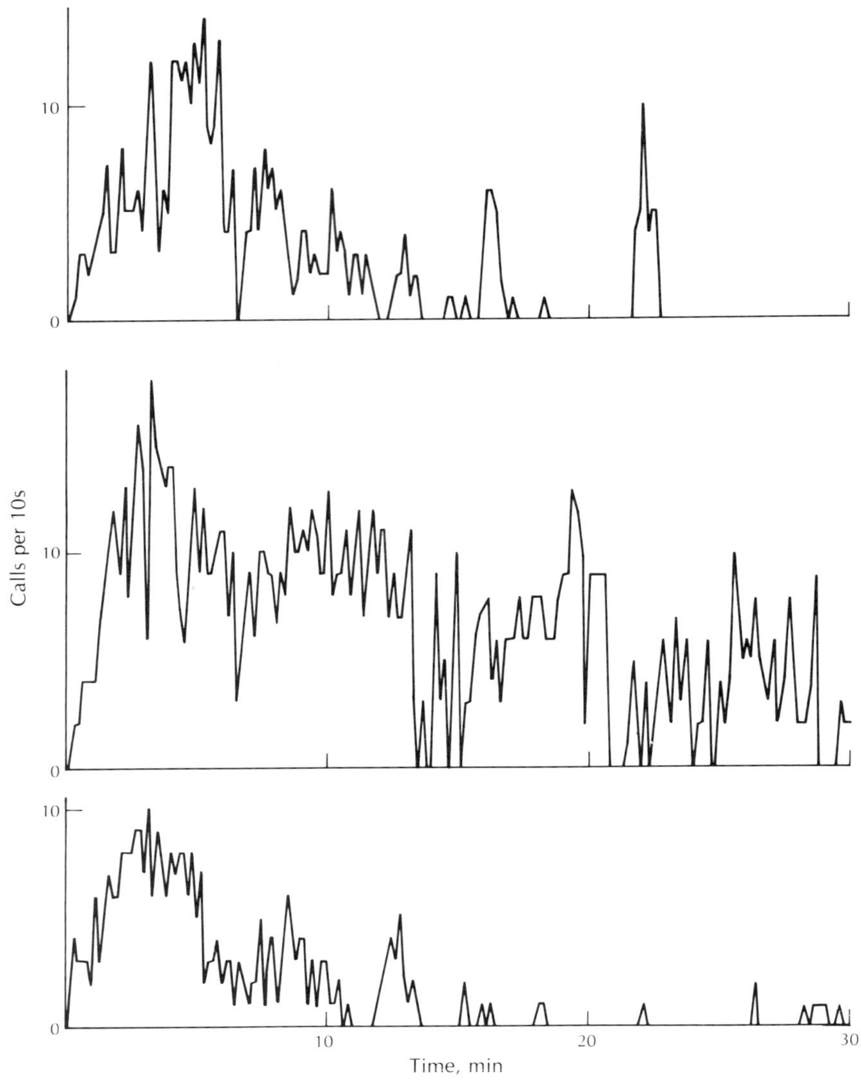

Fig. 13.1. Rate of calling by chaffinches when mobbing a stuffed owl. Data for three individuals are shown. (From Hinde, 1954*a*, by permission of the Royal Society.)

Ordinate, number of calls per 100 sec; *abscissa*, minutes since stimulus presented.

experimentally naïve birds. This reduction in responsiveness is usually greater if the second stimulus object is the same as the first, but some decrement occurs even if it is not. The stimulus-specificity of the decrement increases with the length of the initial presentation (Fig. 13.2; Hinde, 1960b).

10. If female mice are given ten closely successive presentations of a

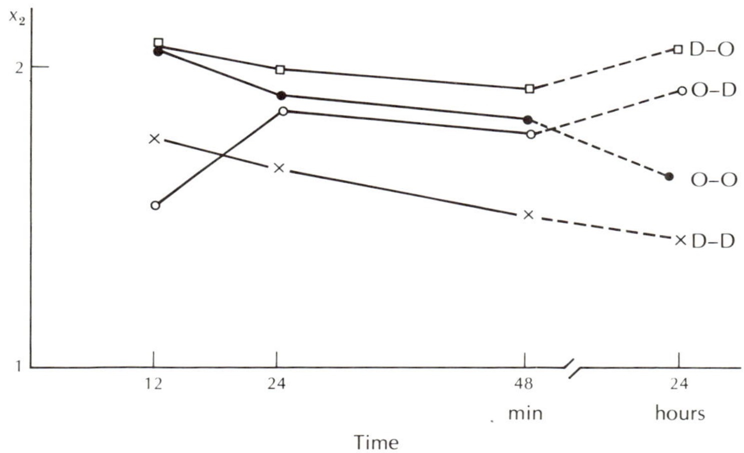

Fig. 13.2. Rate of calling of chaffinches to a stimulus after presentation of a stimulus eliciting the same response for 12 min to 24 hr, 24 hr previously. (From Hinde, 1960*b*, by permission of the Royal Society.)

Two stimuli were used, an owl, *O*, and a dog, *D*, presented in the order indicated. The ordinate indicates the strength of the response to the second stimulus (x_2 = $\log_{10}$ number of calls given in first 6 min of second presentation). The mean strength of the response x_1 to the first stimulus presented was 2.28 for owl and 1.98 for dog. The abscissa indicates the length of the initial presentation. The interval between the two presentations was 24 hr. The response on the second presentation was less if the stimulus used was the same as that used on the first occasion than if it was not (*O–O* versus *D–O* and *D–D v. O–D*), the extent of this stimulus specificity increasing with the length of the initial presentation.

1-day-old baby, the number of subjects showing licking and nest-building responses falls off on the successive presentations. If a 10-day-old baby is given immediately afterwards, the proportion of females showing these responses increases. Since a 10-day-old baby is a weaker stimulus than a 1-day-old baby for naïve subjects, at least most of the waning during the first ten presentations must be more or less specific to the stimulus of a 1-day-old baby (Noirot, 1965).

11. The waning of the directed head-turning response of the human baby is specific to the area stimulated. If the left-hand corner of the mouth is stimulated until the reactions disappear, a new response series of normal length can be elicited by stimulation of the other corner. In this case there is a suggestion that waning does depend on the elicitation of the response [cf. (6) above]: at any rate a long series of stimulations given to a baby which is sleepy or engaged in some other activity does not result in a decrement in the response to stimuli given just afterwards when it is awake and not preoccupied (Prechtl, 1958).

12. A change in responsiveness specific to a particular stimulus is probably also responsible for the phenomenon of "spontaneous alternation". If entries into the two arms of a *T* or *Y* maze are equally rewarded, rats will tend to alternate between them on successive runs. This could be explained either in terms of an alternation of responses (turning left and turning right), or in terms of an alternation of the stimuli to which the animal was responding. Montgomery (1952) and Glanzer (1953a, b; 1958) distinguished between these possibilities by using a cross-shaped maze. The animals were started from the north (say) arm with the south arm blocked off, and from the south arm with the north arm blocked off, on alternate trials. At the choice point, therefore, the animal could enter the arm it entered last time, or the opposite one, or it could turn the same way as last time, or the other way. The rats tended to alternate the arm entered (see review by Dember and Fowler, 1958). Glanzer explained this result on the hypothesis that each time the rat receives a stimulus it becomes less likely to respond to it, some recovery occurring when the stimulus is absent. The distinction between stimulus alternation and response alternation is not quite so clear cut, however, for the way the rat turns is influenced by the stimuli it received after turning on the previous run as well as by those it received before turning: these can include proprioceptive stimuli received in the course of the response of turning (Walker et al., 1955). Alternation can also involve a weak tendency to avoid the odour trail of the previous trial (Douglas, 1966). Sutherland (1957b) showed that in some circumstances the alternation concerns the stimuli or places they are approaching rather than the stimuli actually at the choice point. Because the animal is tending to respond to the more novel stimulus on each occasion, alternation is often discussed in the context of exploration (pages 351–355).

13. Sharpless and Jasper (1956) used the arousal reaction of the EEG in sleeping cats as a dependent variable in assessing the waning of

responsiveness to a tone. They found that the waning was specific to the quality, modality, and pattern of the stimulus (see also Koepke and Pribram, 1966).

14. In their studies of the activity of single cells in the brainstem of rabbits, Horn and Hill (1966) found cells whose responsiveness waned with successive presentations of a stimulus, though they remained responsive to other stimuli. Other examples are cited in Chap. 6.

Many other examples of a decrement specific to a particular stimulus are given, for example, by Sherrington (1906). Thus, in a number of cases in organisms ranging from earthworm to man, waning of responsiveness after exposure to a stimulus has some degree of specificity to the particular stimulus used. However, when an animal is satiated with food, it refuses all kinds of food; and when a mature male rat has ejaculated a large number of times, it refuses all females (but see page 330). In such cases the waning appears to apply to the response as a whole rather than to any particular stimulus. Response-specific waning has been claimed also in a number of other cases. In their studies of the feeding and mating behaviour of jumping spiders, Drees (1952) and Precht and Freytag (1958) found that waning of the fixation response (i.e., orientation towards the stimulus), as a result of repeated presentation of one stimulus in one place, was accompanied also by a decrease in responsiveness to other objects in other places (see also Kolb, 1955).

It is, however, difficult to make an absolute distinction between stimulus-specific waning and waning of the response as a whole. In some of the cases quoted above, the decremental effect generalized to other related stimuli as waning progressed: apparently response-specific waning may thus differ from stimulus-specific waning only in degree. In other cases in which the waning of responsiveness to one stimulus involves also decreased responsiveness to another, there is, in fact, a common element between the two stimulus situations. As we have seen, the mobbing response given by chaffinches to owls was found to wane if a stimulus was presented continuously (Fig. 13.1): the decremental effect influenced also the responsiveness to other stimulus objects, such as dogs, stoats, or snakes, which elicit the same response (Fig. 13.2). There was, however, a common element shared by the successive presentations, namely, the "circumstances of the experiment" (Hinde, 1954b, 1960b).

13.3 DECREMENTAL PROCESSES AND STIMULUS INADEQUACY

Stimulus specificity of a decremental effect suggests that the waning could be related to some inadequacy in the stimulus which elicited the response —perhaps the waning is comparable to the extinction of a conditioned response elicited successively without reinforcement, and would not occur in an adequate situation. Certainly, in a number of cases the waning is more rapid with a weak stimulus than with a strong one. The twitching response of the earthworm, referred to above, wanes more rapidly to a weak stimulus, and longer series of strikes can be elicited from mantids by prey dummies moved jerkily than by dummies which move smoothly, and by dummies with wings and legs than by dummies without. Indeed, mantids cease to strike at some prey dummies behind glass in 2 or 3 min but will continue to strike at a live fly for several hours (Rilling et al., 1959). Similarly, the courtship of the hymenopteron *Mormoniella vitripennis* wanes more rapidly when the male is courting non-receptive females than when they are receptive (Barrass, 1961), and the mobbing response given by chaffinches to predators wanes more slowly with a strong stimulus than with a weak one. However, in this last case at least, the more rapid waning with the weaker stimulus is an aspect of the weaker response elicited: the regression lines relating the initial response strength to the rate of waning for different individuals are the same for both stimuli (Hinde, 1960b).

In any case it is clear that waning occurs even with a natural stimulus object. This mobbing response wanes with time even if it is elicited by a live owl (Fig. 13.3 and Hinde, 1954a, b). Similarly, the responsiveness of a female mouse to her litter is reduced as a consequence of previous exposure to the young: virgin females exposed to babies only once a day give some maternal responses more readily than mothers living continuously with their own young (Noirot, 1964a, b). As a further example, "spontaneous alternation" in a two-choice situation (see above) occurs even when both alternatives are reinforced.

The hypothesis that decreased responsiveness is due to deficiencies in the stimulus situation and would not occur with the natural stimulus object cannot be tested for activities like eating or mating: the natural object would elicit further responses (e.g., intromission) which would alter the situation. Waning in responsiveness to objects eliciting fear responses might well not occur if the object made an attack: this, however, is not evidence for a stimulus-deficiency hypothesis, for it could be argued that the decremental effect still occurs but is masked by an incremental one induced (or augmented) by the attack.

In view of such examples, it can be said that at least many cases of waning responsiveness cannot be ascribed to stimulus deficiency.

13.4 THE RATE OF WANING

In general, waning of responsiveness to a repetitive stimulus is more rapid the weaker the stimulus (but see above) and the greater the frequency of stimulation. It is also said to become increasingly rapid in successive stimulus series (review by Thompson and Spencer, 1966), though this may be in part a consequence of the weaker response elicited (see above).

The rate at which waning occurs varies between responses—while some are persistent, in others the decrement occurs rapidly. The retrieving response of mice, elicited by the high-frequency calls of the young, can be elicited 148 times in succession (Zippelius and Schleidt, 1956; see also Noirot, 1965), while the directed head movements of the human neonate

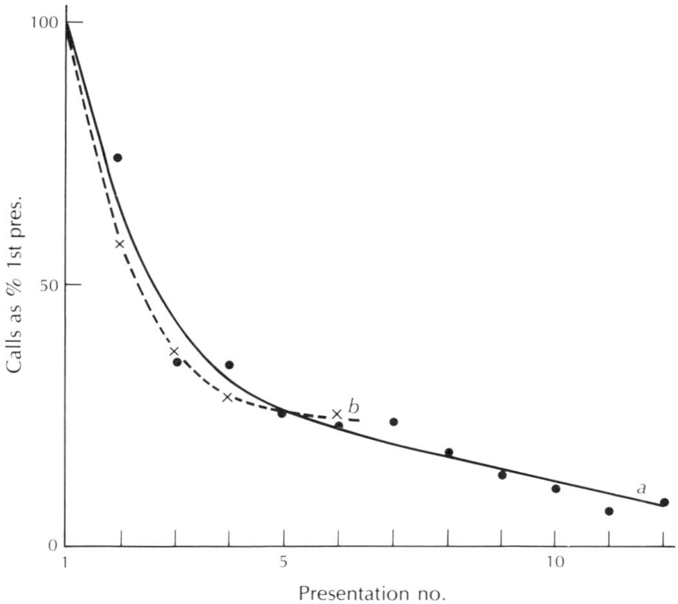

Fig. 13.3. Calling by chaffinches while mobbing live little owl (continuous line) presented for 20 min a day, or stuffed tawny owl (broken line) presented for 3 min a day. (After Hinde, 1954*b*, by permission of the Royal Society.)

The number of calls given by each bird were expressed as percentages of the number given on the first day; the ordinate represents the group mean.

disappear after a few dozen (Prechtl, 1958). Precht and Freytag (1958) have shown that various responses of jumping spiders differ considerably in their persistence with successive stimulation, as well as in their recovery rates. Similarly the different activities involved in the hunting and killing behaviour of cats wane on repetition at different rates and differ also in the speed with which they recover (Leyhausen, 1965).

For any one response the rate at which waning occurs may depend on a variety of factors; thus with the photopositive response of *Drosophila* it varies with the rearing conditions (Dürrwächter, 1957), but with the directed head movements of human babies it varies with the exact point stimulated, the baby's posture, sleepiness, and so on (Prechtl, 1958). In the waning of the orientation response of *Bufo*, as studied by Eikmanns (*loc. cit.*), the response can be elicited in successive series separated by rests of standard length (e.g., 5 min). The number of responses in each series wanes progressively. The ratio between the number of responses in two successive series gives an index of the rate of waning. This waning quotient oscillates, apparently as a result of at least two processes—the taxis-specific waning and recovery. These can be affected differentially—the former by certain brain extirpations, the latter by altering the lengths of the pauses (Ewert and Birukow, 1965; Ewert, 1967a).

Lát and Gollová-Hémon (1969) have studied individual differences in the waning of the "non-specific excitability level" (see pages 217–218) when rats are placed in a strange environment for 10-min sessions once an hour. A few individuals show a high initial level which wanes slowly. The majority show an intermediate initial level and more rapid waning. These authors stress that the waning does not occur smoothly but shows clear intrinsic oscillations whose amplitude is an individual characteristic. Such oscillations are probably common to most examples of waning responsiveness (e.g., Fig. 13.1), though they are often neglected or swamped in group data.

13.5 RECOVERY—THE NATURE AND MULTIPLICITY OF DECREMENTAL PROCESSES

After waning has occurred, the response can often be elicited again if the stimulus is given in the presence of a previously irrelevant stimulus. This is often referred to as "dishabituation" and used as evidence that the waning is due to a process of inhibition which the new stimulus "disinhibits". At least in many cases, it is better regarded as a superimposed "response sensitization" which overrides the processes underlying the response decrement (Thompson and Spencer, 1966; Sharpless and Jasper, 1956). For example, in the case of the waning of the flexion reflex in spinal cats, Spencer et al. (1966) have evidence that the waning is due to alterations in synaptic efficiency within the interneuronal pathways, and not to the sustained discharge of inhibitory neurones (see also Bruner and Tauc, 1966a, b; Krasne and Roberts, 1967).

Even in the absence of an additional stimulus, however, recovery usually occurs with time. Sometimes it is rapid and complete. After waning of the strike response of mantids to prey dummies, recovery occurs mostly in 5 min and is complete after half an hour (Rilling et al., 1959). When the polysynaptic flexion reflex of spinal cats is elicited by brief trains of shocks, waning occurs in about 10 min and recovery in about 26 min (Thompson and Spencer, 1966). In his detailed study of the courtship of *Mormoniella vitripennis*, Barrass (1961) found that even when a male had courted twenty females in succession, recovery was considerable in 1 hr and complete in 24. The movement used by toads in turning towards prey recovers from total waning to about 50 percent of the initial level after 8 hr, and completely after a day (Eikmanns, 1955). The waning of the optomotor response of the turtle *Pseudemys scripta* recovers in 24 hr (Hayes et al., 1968). Similarly, after a moderate degree of stimulation the twitching response of the earthworm recovers completely in 24 hr, though after strong stimulation for several successive days recovery is slow, reaching only 20 percent after 10 days (Kuenzer, 1958).

In other cases, however, complete recovery takes a considerable time, if it occurs at all: usually the initial stages are fairly rapid, but the rate then slows or ceases. To cite some examples, the prey-catching response of *Notonecta* recovers from a long series of stimulations rapidly during the first few minutes, but then more slowly, and is not complete after 24 hr (Wolda, 1961). The escape reflex of the worm *Branchiomma* recovers only to about 80 percent after 24 hr (Krasne, 1965). Drees (1952) and Precht and Freytag (1958) found that several of the hunting and courtship responses of hunting spiders show rapid recovery during the first few hours after waning, followed by a period of only slow recovery: the courtship dance, however, is exceptional in showing almost linear recovery with time. The wiping response of the toad recovers to 56 percent of the initial level after 60 min but thereafter very slowly (Kuczka, 1956). The waning of the response of the guppy *Lebistes reticulatus* to a repeated shadow takes about 48 hr to recover completely (Russell, 1967a). The "chink" calls, given by a chaffinch while mobbing a stuffed owl, cease almost completely after a 30-min exposure. If the stimulus is then removed, recovery reaches about 55 percent after 30 min but the remaining decrement is virtually permanent (Hinde, 1954b). The Siamese fighting fish (*Betta splendens*) will display to its own mirror image. If permitted access to a mirror for a number of days the response, as measured by the frequency of "challenges", falls off. The effect of various recovery periods after a 10-day exposure to a mirror were assessed. Recovery after periods of 15 min to 24 hr was not statistically significant, but after periods of 2 days and 4 days it was (Clayton and Hinde, 1968; see also Baenninger, 1966). When there is a phase of slow recovery like this, or a permanent decrement, the effects of successive bouts of stimulation may be cumulative. An example is shown in Fig. 13.3: if an owl is shown to a chaffinch daily, the response falls off progressively.

A period of rapid recovery followed by a period of only slow recovery, or no recovery at all, suggests that at least two processes are involved in the initial waning. This is reminiscent of the extinction of conditioned responses, where the response decrement has often been ascribed to two factors, one temporary and the other permanent (e.g., Hull, 1943). Even this, however, involves oversimplification, for decremental processes resulting from any one response may have recovery periods ranging from seconds to years. In the case of the chaffinch's owl-mobbing response these processes can be classified into three groups according to whether the effect lasts for seconds (see pages 304–305), or for minutes, or is near-permanent (see above). At this point the distinction between "motivational" changes and "learning" becomes obscure, for the temporary changes are usually studied under the former rubric, while the permament ones come within most definitions of learning. Thorpe (1963a) reserves the term "habituation" for relatively permanent decremental changes of this sort (see page 577), but this usage is not general amongst physiologists.

Evidence that more than one process may be involved in the waning of responsiveness comes also from the fact that different aspects of the response may wane, or recover from waning, at different rates. In studying the cardiac acceleration produced in human neonates by sounds, Bartoshuk (1962a, b) gave series of 40 trials at 60-sec intervals. The prestimulus heart rates for trial 2 were higher than those for trial 1, indicating that at this point the effect of trial 1 lasted for at least 60 sec. By the end of the series, however, the prestimulus heart rate had returned to the initial value, although there was still an immediate response to the stimulus. Thus the tonic response waned more rapidly than the phasic one (cf. Sharpless and Jasper, 1956). On a much longer time scale, the "tail flip" response of goldfish (Rodgers et al., 1963) and the fleeing response of ducklings (Melzack, 1961) wane quite rapidly, but in both cases an orientation response towards the stimulus is very persistent (see also Evans and Browne, 1960; Moyer, 1963; Russell, 1967a).

That more than one decremental process may be involved is also supported by physiological evidence. In the cockroach, the waning of the evasive response produced by air blasts on the anal cerci is largely due to the lability of the giant-fibre/motor-neurone synapse, but some slow waning also occurs in the receptor cells (Roeder, 1962b). In the withdrawal response of the earthworm, clear evidence for a dual process is available. This response, elicited by tactile stimulation at one end and mediated by the giant fibres in the ventral nerve cord, wanes with repeated stimulation. By electrical recording from various points between sense organs and effectors, Roberts (1962) showed that failure of transmission occurred between the sensory neurones and the giant fibres, and between the giant fibres and the motor neurones. Rapid fatigue to repetitive stimulation was not shown in the muscle, in the neuromuscular junctions, or in the junctions in the motor-neurone tracts. A similar result had previously been obtained for the worm *Nereis* by Horridge (1959a). If the waning of such very simple responses is due to changes at at least two points, how much more complex may be the changes underlying waning in responsiveness among vertebrates? It is thus not unexpected that experiments involving electrical stimulation of the mammalian brain should demonstrate waning processes with diverse time constants. In general, the response evoked by stimulation of the motor cortex wanes in a few seconds, while that evoked by stimulation of the motor pathways lasts for up to an hour. Other areas, such as parts of the hypothalamus, show no fatigue after 72 hr of continuous stimulation (Delgado, 1962).

The evidence cited in this section thus indicates clearly that an apparently simple behavioural decrement may be the result of diverse underlying processes.

13.6 INCREMENTAL PROCESSES AND THEIR INTERACTION WITH DECREMENTAL ONES

A further complication comes from the fact that the elicitation of a response may be accompanied by a tendency towards an increase in response strength instead of, or as well as, a tendency to a decrease. The rapid withdrawal response of the earthworm which, as described above, wanes on repeated elicitation, shows a tendency to increase in the early stages of repetitive stimulation. The facilitating effect lasts for several seconds, and occurs at the giant-fibre/motor-nerve junctions (M. B. V. Roberts, 1966). The wiping response of frogs can be elicited from a large reflexogenous zone. If frogs are stimulated with a bristle within this general area 100 times a day for 12 days, the proportion of responses gradually increases. Only when the stimulus is repeatedly applied to exactly the same spot does waning occur (Fig. 13.4; Franzisket, 1963; Kimble and Ray, 1965). Again, if the croak reflex is elicited a large number of times by finger pressure on the back, it may start to occur spontaneously. Furthermore, stimulation with bristles may then elicit croaking instead of the more usual foot-wiping (Franzisket, 1963; see also Kuczka, 1956). With the directed head movements of the human baby, intense reactions occur more quickly the higher the frequency of stimulation used: the early stimuli increase responsiveness to the later ones (Precktl, 1953). By analysis of the behaviour sequences during the courtship of certain glandulocaudine fishes, Nelson (1964a) obtained evidence that the male's courtship was self-facilitatory. After the maternal behaviour of adult mice to 1-day-old babies has waned as a consequence of successive presentations, they will respond more strongly to a 10-day-old baby than will animals with no such immediately preceding experience (Noirot, 1964c; see also page 293).

The positive or excitatory effect which accompanies a response may continue to mount after the response itself is over. In the male rat, ejaculation occurs only after a number of intromissions. If the interval between intromissions is forcibly increased, the number of intromissions required for ejaculation is decreased. The full effect of each intromission does not seem to be realized for 0.5 min, and since intromissions normally follow each other at intervals shorter than this, the maximal effects of each are never attained under normal conditions (see page 328).

As with the processes which lead to a decrease in responsiveness, the various incremental processes accompanying one type of response may

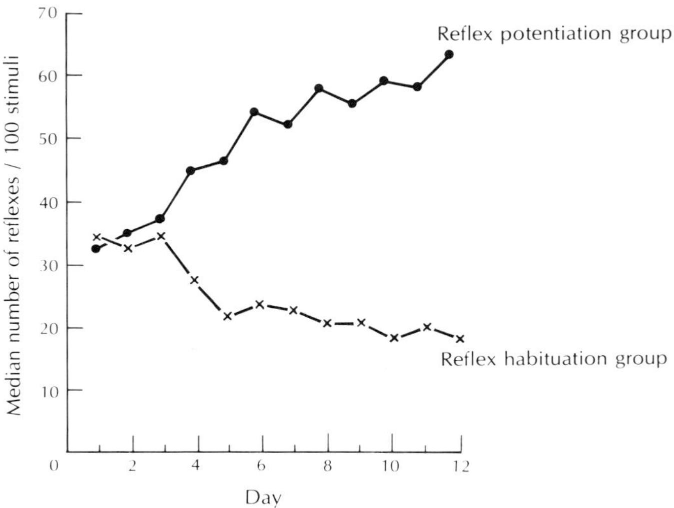

Fig. 13.4. Wiping response of frogs. Median number of wiping reflexes elicited per 100 stimuli. (From Kimble and Ray, 1966.)

In the potentiation group, stimuli were administered within the reflexogenous zone in the middle of the back. In the habituation group, stimuli were delivered to a precise locus within that area.

have decay periods varying from seconds upwards. The calls accompanying the mobbing response of the chaffinch will again serve as an example. In one series of experiments two stimulus objects were used, a stuffed owl and a toy dog, the former eliciting the stronger response. Each experiment consisted of two stimulus presentations in which the two objects were used in all possible combinations (Figs. 13.1, 13.2, and 13.5).

The consequences of an initial stimulus presentation on responsiveness on a second occasion when the interval between presentations was short ($\frac{1}{2}$ to 5 min) or long (24 hr) were considered separately. Although the number of calls given in the second presentation was usually less than the number given in the first, there was considerable evidence that the decremental processes concealed less powerful but nonetheless influential incremental ones. In the case of the long-term effects (24-hr rest interval between presentations) this evidence was of three types:

1. Although the number of calls given in the second presentation was reduced, the latency of the first call was also reduced.
2. With most stimulus combinations, the longer the initial presentation the smaller the response on the second. With the sequence owl–dog, however, the reverse was the case. Apparently in this case an incremental effect resulting from presentation of the stronger stimulus increased with the length of that presentation more rapidly than did the concurrent decremental effect (Fig. 13.2).
3. Another experiment involved tests with two owl models. Although the models looked similar, one had eyes and the other did not: they therefore differed markedly in effectiveness. When the more effective model was shown first, the response to the weaker one, presented 24 hr later, was actually greater than that shown by naïve birds.

For the short-term effects (i.e., interval $\frac{1}{2}$ to 5 min), the evidence for an incremental effect was as follows:

1. In every stimulus presentation the rate of calling increased gradually, reaching a peak only after 2 or 3 min (Fig. 13.1).
2. The response strength on a second presentation did not always increase smoothly with the length of the rest interval but showed marked and significant fluctuations (Fig. 13.5). These could be accounted for in terms of the interaction of incremental and decremental effects having different decay periods.
3. Since the response strength shows an initial increase at the beginning of each presentation, it seems that the incremental effects build up more rapidly than the decremental ones. Any incremental effect is therefore more likely to be evident if the initial presentation is short, when the effects of decremental processes are minimized. If the first presentation lasted only 2 min and the interval between presentations was varied between 3 sec and 24 hr, the response in the first minute of the second presentation was usually greater with rest intervals of $2\frac{1}{2}$ or 5 min than with longer or shorter rest intervals. This, again, is consistent with the suggestion that there are interacting incremental and decremental processes. Presumably, the former are masked by

the latter after very short rest intervals, revealed by the decay of the decremental effects after $2\frac{1}{2}$ to 5 min, and subsequently decay themselves (Hinde, 1954b, 1960b).

For this response there is thus strong evidence for incremental effects which are relatively permanent, and also for others which decay over a period of a few minutes. This is not all, however. The calls are given in phrases of 1 to 6 calls, the intervals between phrases being considerably longer than the intervals between the calls in a phrase. Describing it in

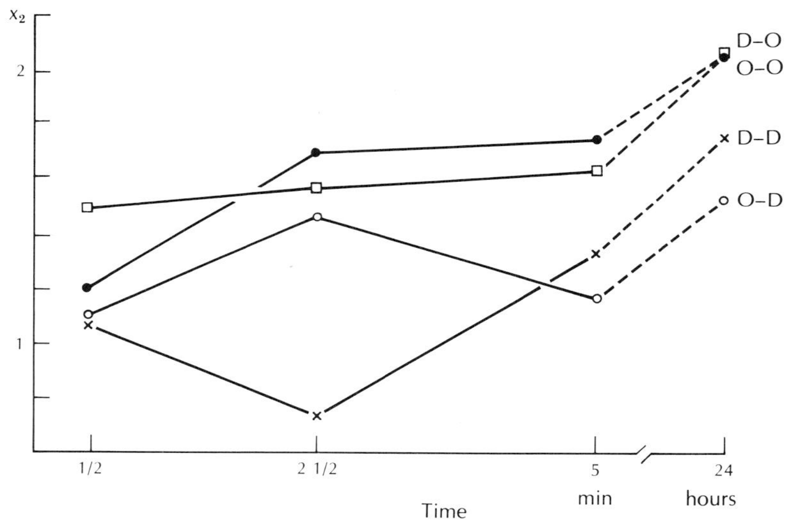

Fig. 13.5. Influence on the rate of calling given by chaffinches in mobbing a predator of a 12-min previous presentation. (After Hinde, 1960*b*, by permission of the Royal Society.)

Conventions as in Fig. 13.2, but here the abscissa indicates the interval between the presentations.

one way, a single call is likely to be followed immediately by another, but a short series is likely to be followed by silence. This raises the possibility that each call is associated with very short-term incremental and decremental effects, which build up and decay at rates which, though different, are measurable in seconds.

A comparable case, involving very short-term incremental and decremental effects, concerns the singing of the chaffinch. Each male has a repertoire of songs and, during an outburst of singing, gives a sequence of songs of one type followed by a sequence of another. The intervals between songs within a sequence (that is, between songs of a similar type) tend to be shorter than those between sequences. Since intervals of less than 3 sec between songs are extremely rare, it would seem that each song is accompanied by an inhibitory effect on singing which is dissipated with time. That this inhibitory effect is a consequence of performance is suggested by the fact that the interval tends to be shorter after the occasional incomplete song. Now if this inhibitory effect were specific to the song type just uttered, alternation between song types would occur more frequently than on a random distribution of song types: in fact the opposite occurs. There is, thus, a tendency to repeat the song just uttered rather than to switch to another type: this we may describe as a facilitative effect. The singing pattern of the chaffinch thus implies, as a minimum requirement for explanation, inhibitory and facilitative effects consequent upon performance (Hinde, 1958b; see also Isaac and Marler, 1963). In the domestic turkey the frequency distribution of intervals between "gobbling" is affected by deafening, indicating a role of auditory feedback in the inhibitory effect (Schleidt, 1964; see also Konishi and Nottebohm, 1969).

That incremental and decremental effects occur simultaneously is not *necessarily* an indication of great complexity in the underlying mechanism. Post-tetanic potentiation and low-frequency depression can occur in a monosynaptic ventral-root response of a spinal animal: which occurs depends on the frequency of stimulation (Thompson, 1967). Again, Hughes and Tauc (personal communication), working with the mollusc *Aplysia*, stimulated the right giant cell, whose axon makes direct synaptic contact with the left giant cell, and recorded from the latter. The recording showed a biphasic synaptic potential with an initial depolarizing phase and a longer-lasting hyperpolarizing phase. With repetitive stimulation the post-synaptic potentials summated to produce a hyperpolarizing effect at low frequencies and a depolarizing effect at high frequencies of stimulation. Thus, depending on the frequency of stimulation, a single synapse displays the properties of both inhibition and excitation. As another example, Horn and Rowell (1968) showed that the response of single cells in the locust tritocerebrum waned with the repeated stimulation provided by the movement of a visual stimulus, but they showed an incremental effect in that the unit responded preferentially to the direction of movement of the first of a group of stimuli.

These studies demonstrate not only that incremental or facilitatory effects with widely varying time courses may accompany the performance of any one type of response, but also that the observed behaviour depends on a complex interaction between the various incremental and decremental effects. As another example, the mobbing behaviour of the blackbird

(*Turdus merula*) involves two types of call which are given in flight or when the bird has a strong tendency to fly. In a detailed study of the moment-to-moment fluctuations in calling, Andrew (1961a, b, c, d) found it necessary to postulate at least two variables. The "action potential" is increased with the tendency to fly and is reduced by calling but recovers rapidly. Its level determines which of the two calls is given. The "general threshold" for calling is increased by calling and thus rises progressively during a bout: it decays between bouts of calling. Andrew has elaborated a detailed model along these lines and has shown that a number of predictions made from it are supported by the data.

Another aspect of the complexity of the processes that occur is illustrated by the sucking behaviour of human newborns, where a complex interaction between the effects of preceding strong and weak stimuli has been demonstrated (Lipsitt and Kaye, 1965). Two stimuli were used, a piece of ¼-inch tubing and a bottle nipple. Infants were given fifty successive trials, each lasting 10 sec with an interval of 30 sec between trials. Two groups of infants received either the one stimulus or the other throughout: the nipple elicited about twice as many sucking responses as the tube. A third group was given five trials with one stimulus and five with the other, alternately. Whereas the infants given one stimulus throughout showed no consistent change in the number of responses given on each trial through the series, the third group tended to give more responses to the nipple and fewer to the tube as the series progressed, though the difference was not clearly significant (Fig. 13.6). Furthermore, within each block of five trials with the nipple the response rates started low and finished high, but with the tube it started high and finished low. That this cannot be due merely to a lasting effect of the previous stimulus is indicated by the finding that the response rate at the end of each nipple series was higher than that shown by the group given the nipple throughout, and that at the end of each tube series the response rate tended to be lower than that of the tube group: some sort of contrast effect seems to be involved.

13.7 STIMULUS SPECIFICITY OF INCREMENTAL PROCESSES

While most decremental processes seem to be specific to the particular stimulus used to elicit the response, the evidence concerning incremental processes is meagre. As described above, the proportion of wiping responses given by spinal frogs to stimulation with a bristle increases on successive days. The response can be elicited by stimulating both medial and lateral areas of the back, but a long series of stimuli applied to one area does not involve increased responsiveness to stimulation of the other (Franzisket, 1963). In this case, therefore, there is some degree of specificity to the locus of the stimulus.

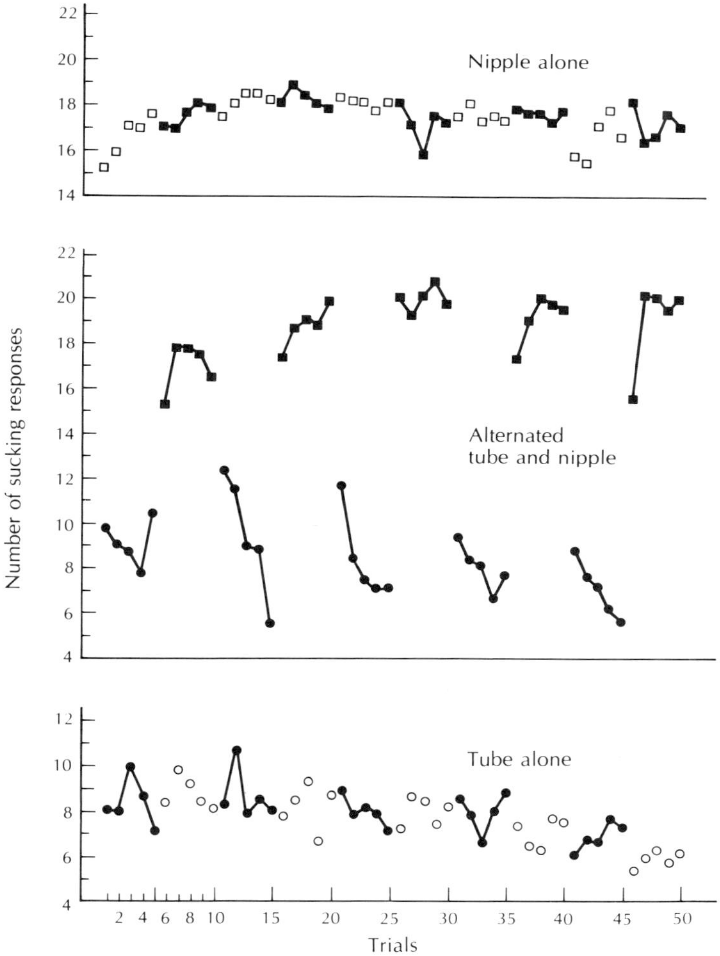

Fig. 13.6. Number of sucking responses given by human newborns in series of trials with a bottle nipple (top); a tube (bottom); and alternating runs of five trials with each (centre). (After Lipsitt and Kaye, 1965.)

Each run of five trials is shown by a different symbol whether or not the stimulus changed.

In other studies, however, the contrary seems to be the case; for instance the exposure of an animal to a frightening stimulus often results in an increased responsiveness to a wide variety of stimuli eliciting similar types of behaviour ("pseudo-conditioning"; Hilgard and Marquis, 1961). Since, in this case, the behaviour concerned involves avoidance, fear, or startle responses, the effect may be mediated by a changed pattern of activity in the autonomic system.

Another case, in which the incremental effect appears not to depend on any common stimulus characters between the initial "priming" experience and the test situation, concerns the maternal behaviour of mice. A much higher proportion of adult mice show certain maternal responses to a live 1-day-old baby than to a drowned one, and there is strong evidence that this is due to the ultrasonic cries emitted by the former (Noirot, 1964b, c; 1965). Responsiveness to a drowned baby is, however, increased by previous exposure to a live one, even if there is an interval of several days in between. What is more, the effect occurs even when the adult has no opportunity to make tactile contact with the live baby. Experiments in which virgin females were exposed to young, confined inside tins with small holes and providing thereby only auditory and olfactory stimuli, indicated that these stimuli increased subsequent maternal responses to a drowned baby. While the dead baby may have provided olfactory cues in common with those provided by the live baby, it was the auditory stimuli provided by the latter which were probably of prime importance. There can thus have been few stimulus characters in common between the two situations.

Turning to neurophysiological studies, there is again some evidence that incremental effects may not be specific to the particular stimulus used. We have seen that when stimulation by two electrodes in different loci of the brainstem of chickens elicits the same response, the waning which follows stimulation of one of them is specific to the locus stimulated. If, however, both loci are stimulated simultaneously with a voltage which would have been subthreshold when applied to only one, a response may occur. The summation must occur at or beyond the point where the effects from the two loci meet (von Holst and von St. Paul, 1963).

It must be emphasized that the few cases which provide evidence about the stimulus-specificity of an incremental effect are extremely diverse and certainly involve quite different mechanisms. Generalizations would thus be premature.

13.8 INTERACTION BETWEEN RESPONSES

The performance of one response may affect the ease with which others can be elicited. On the one hand, the decrement in one response may involve also decrement in others. In jumping spiders, waning of the "jump" is accompanied by a reduced readiness for "approach", and vice versa (Precht and Freytag, 1958). The attack readiness of the cichlid fish *Pelmatochromis subocellatus* is decreased either by attack biting or by digging (i.e., biting the substrate) (Heiligenberg, 1965b). On the other hand, decrement in one response may be related to an increase in the probability of others and even to the appearance of another response to the same stimulus. A number of examples are given by Prechtl (1958). When the directed head-turning movement of the human neonate has waned, the same stimulus may elicit turning away. When the gaping of young chaffinches to a shaking of the nest has waned, crouching occurs. This is called by Prechtl "central reset". Sometimes both incremental and decremental effects occur. In a study of aphid behaviour, Kennedy and Booth (1964) have demonstrated that flying can have both incremental and decremental effects on settling responses: which predominates depends on a variety of factors including the period spent flying and the stimuli for settling (see pages 618–620).

In a number of cases there is evidence suggesting that waning involves a gradually increasing probability that the response in question will be interrupted by other types of behaviour. Thus the gradual waning of drinking which occurs after a previously deprived rat is given water involves an increase in the lengths and frequency of the pauses in drinking, during which the rat is occupied in other activities (Stellar and Hill, 1952). Whalen (1963b) makes a similar point with the mating behaviour of the cat (see also Hurwitz, 1957, and Chap. 17).

13.9 CONCLUSION

In considering these examples of changes in responsiveness to a stimulus, we have drawn evidence from diverse responses of animals at all phyletic levels. The accuracy of the generalizations made must therefore be tailored to the diversity of the data. However, it is clear that in any particular case our initial assumption should not be that there is a single process underlying the change in responsiveness: rather we must expect processes with varied characteristics, presumably occurring at different points in the underlying nervous mechanisms. Furthermore, in the absence of evidence to the contrary, it would be safe to assume that exposure to a stimulus will be accompanied by processes (with varied time characteristics) leading to an increase in subsequent responsiveness and also to processes leading to a decrease. The actual response (or lack of it) observed on a subsequent occasion will be a result of these varied processes. Similar conclusions relating to the performance of a skilled task by human subjects have been formulated by Ammons (1947), though, with more complex tasks, repeated performance affects the integration of the individual responses as much as (or as well as) the actual occurrence of the responses themselves (Bartlett, 1953).

Further understanding will depend on an analysis aimed at separating the various processes involved and assessing their characteristics. The studies cited above suggest that it may be useful to classify them according to their decay periods, according to whether they produce incremental or decremental effects, according to the extent to which they are specific to the particular stimulus used to elicit the response, and according to whether they depend on actual performance of the response in question. In the case of complex responses, however, such a classification must involve a considerable degree of approximation and is only a first step. Further progress demands greater standardization between investigators over such matters as the method of stimulus presentation, and the precise way in which rate of waning is measured, than has hitherto been the case. Given that, this is an important and promising field of research, for the gap between behavioural and neuronal phenomena is not unbridgeable (Horn and Hinde, in press).

13.10 SUMMARY

1. Response to a stimulus is associated with changes in responsiveness when the stimulus is presented again. In many cases these are not attributable to changes in the sense organs or effectors.
2. The decremental consequences of responding are usually more or less specific to the particular stimulus which elicited the response.
3. The decremental effects cannot be ascribed solely to inadequacy in the stimulus situation.
4. The rate at which waning occurs varies considerably between responses.
5. The rate of recovery also varies: often, a period of rapid recovery is followed by a period of slow recovery or of no further recovery. This suggests that in such cases more than one decremental process may be involved. Physiological evidence supports this view.
6. Incremental processes may accompany and interact with the decremental ones.
7. The incremental effects are often not stimulus-specific.
8. A decremental effect on one type of behaviour may be accompanied by either incremental or decremental effects on others.

14

SPONTANEITY AND RHYTHMICITY

In the preceding chapter we saw that successive or continuous presentation of a stimulus may result in a waning of responsiveness. Recovery from the decremental processes involved occurs "spontaneously" with the passage of time. This raises the more general question of the extent to which changes in behaviour can be spontaneous.

Semantic difficulties arise easily here. "Spontaneous" is used in the present context to refer to changes in the output of a system without a known corresponding change in input. Whenever spontaneity is discussed, therefore, it is necessary to specify the system under consideration. Behaviour of the whole animal may be called spontaneous though the change in central nervous activity controlling it depends on changes inside the body. Similarly, the output of the nervous system may be spontaneous when that of its components is not, and the spontaneous firing of a nerve cell depends on complex chains of events within it. In each case, of course, spontaneity of the system will depend on conditions external to the system remaining within certain limits.

Since recognition of spontaneity depends on an absence of correlation between input and output, the distinction between spontaneous and environmentally elicited is one of degree only. It depends, in fact, on the length and variability of the sequence of events within the system which preceded the activity and which are uninfluenced by extrasystemic factors. If the chain of internal events is long enough, the activity appears to be spontaneous.

At the behavioural level, spontaneity thus consists of changes in behaviour not closely related to changes in exteroceptive stimulation and is a commonplace in animals from protozoans (Jennings, 1906) and sea anemones (Batham and Pantin, 1950) to mammals. We have already seen that, as motivational factors increase in strength, less external stimulation is required to elicit a given response (Chap. 11): in the extreme case, no stimulus is necessary, and the response appears spontaneously. A number of dramatic cases have been described by Lorenz (1935) as "vacuum activities". For instance, a well-fed starling, which had no opportunity to catch flies for some time, suddenly went through all the movements of searching for a fly, catching it, and killing it, although no fly was discernible to the observer. Similarly, canaries deprived of nest material will perform the movements of weaving material into a non-existent nest (Hinde, 1958a). In such cases it is the performance of a fixed action pattern out of context which easily catches the observer's attention: similar changes in appetitive behaviour, as when the starling searched for a fly, are more commonplace and may go unremarked.

The lack of a sharp dividing line between the spontaneous and stimulus-elicited activities of the whole animal is also illustrated by Franzisket's (1963) studies of decerebrated and intact frogs. The croak reflex is normally elicited by blunt pressure on an area of the back. Stimuli were given at 5-sec intervals 500 times a day for 3 weeks. After this training, spontaneous croaking occurred about twice a day, though it was never observed over a period of several weeks before the training. As a result of frequent stimulation, therefore, a reflex response became spontaneous. On a different time scale, von Holst and von St. Paul (1963) used electrical stimulation of the brain to induce "geckering" in hens. If the stimuli were repeated at short intervals, the threshold voltage required to elicit the response decreased. If they were sufficiently rapid, the threshold fell to zero, that is, the reaction became spontaneous.

The question of the spontaneity of the nervous system, that is, changes in output independent of changes in either exteroceptive or interoceptive stimulation, merges with that of the patterning of nervous discharges necessary for the coordination of complex movements, as discussed in Chap. 3. If a simple input produces a patterned discharge, the patterning of the output is at least in part determined centrally. If the input is tonic and the output rhythmical, then the rhythmical changes involve spontaneity as well as central patterning, even if the rhythmicity is merely a matter of fatigue and spontaneous recovery. Thus many of the cases discussed in Chap. 3 as evidence for central patterning are relevant here also. Of course, whenever the output of a system is rhythmic, there must be some feedback (in a broad sense) effect whereby the consequences of the activity affect its own repetition, but in spontaneous rhythms this is inside or part of the system (see discussion by Grey Walter, 1959).

The relations between the systems controlling the patterning of particular movements and the rhythmicity in their appearance are summarized in Fig. 14.1 (Bullock, 1961). The timing of the rhythm may depend on either peripheral or central factors.

1. The timing cues come from the periphery. In such cases the patterning of the response itself may be determined centrally, even though elicita-

tion depends on external factors: for instance, an object approaching the eye results in a patterned discharge leading to an eye blink (Fig. 14.1*A*). If there is sensory feedback from exteroceptors or proprioceptors, discharge may be rhythmical (Fig. 14.1*B*): in this case the initial input may just serve a triggering function.

2. Timing cues from central pacemakers. This case differs from the previous one only in that a central pacemaker is present which, as it were, rings an alarm at intervals (Fig. 14.1*C* to *E*). In practice, the distinction between the systems shown in Fig. 14.1*B* and *C* may be hard to make: if feedback starts the next cycle before the central pacemaker, the existence of the latter is concealed. In most cases the external feedback modulates the spontaneous rhythm, which would be altered but not

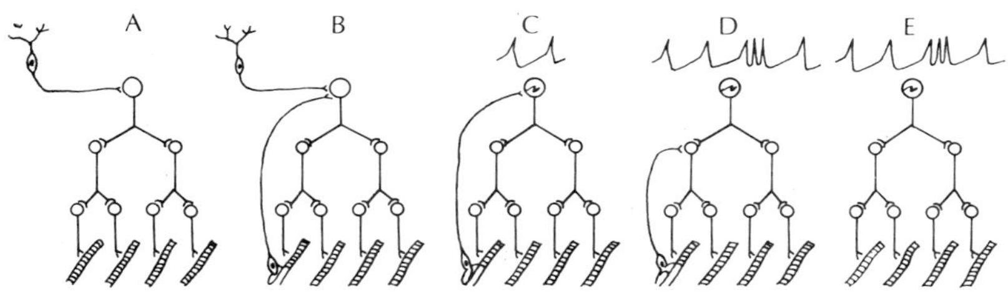

Fig. 14.1. Diagrams of hypothetical types of nervous mechanisms producing patterned movement. (After Bullock, 1961.)

The three levels of neurones represent branching chains in whose junctions integrative properties may alter the actual impulses and deliver them to the effectors. *A* and *B* are shown with receptors; *C*, *D*, and *E* with spontaneous pacemakers. *B* and *C* have proprioceptive feedback acting on the trigger neurone, while in *D* feedback influences only the shaping of the pattern.

broken by its removal (see Chap. 3). Of course, the operation of the central pacemaker may itself depend on activity in internal feedback loops.

In some cases the feedback may act not on the pacemaker but on its followers: it will then modulate the form but not the frequency of the rhythmic output. Bullock quotes the disturbances in walking in a tabetic or blinded man as an example (Fig. 14.1*D*). In any case the feedback is likely not to be simple, as implied in the diagram, but to involve a complex mixture of positive and negative, specific and diffuse, fast and slow adapting influences. Absence of any sort of feedback control external to the pacemaker appears to occur in some rhythms of short periodicity (Fig. 14.1*E*), such as that produced by the electric organ of gymnotid fish (Lissmann, 1958). Others, like the lobster's cardiac ganglion and the sound production mechanism of the cicada, have been mentioned already.

In fact, the occurrence of rhythmic processes in the central nervous system, after it has been experimentally isolated from peripheral influences, has been demonstrated in a considerable number of cases; for instance, in the ventral nerve cord of *Dytiscus* (Adrian, 1931) and in a slug ganglion (Hughes and Kerkut, 1956). Horridge (1959b) has shown that the rhythm of swimming movements in medusae is of nervous origin, and P. L. Miller (1960) that the respiratory movements of the desert locust are controlled by a pacemaker in the metathoracic ganglion whose activity continues in the isolated nerve cord, though its activity may be modified by carbon dioxide. Other examples were given in Chap. 3.

These neurogenic rhythmical discharges have also been studied in the embryos of fishes and birds. In these species there is an interval between the establishment of motor innervation of the muscles and the completion of the reflex circuit on the sensory side. During this interval spontaneous movements occur which are often rhythmic, even though reflex responses cannot yet be elicited (Tracy, 1926; Hamburger, 1963). In chick embryos spontaneous movements start at $3\frac{1}{2}$ days and continue throughout embryonic life. The activity occurs in cycles each consisting of a period of activity followed by a period of quiescence: the average length and composition of the cycles vary considerably with age. If the spinal cord is interrupted by extirpation of several segments, either at the cervical level or between wing and leg, spontaneous activity continues independently in each section. Thus potential pacemakers are present at each level, though it is probable that the anterior ones are normally dominant: spinal embryos show 10 to 20 percent less activity than intact ones at all ages between 8 and 17 days. This is not merely a mass effect of neural tissue: selective removal of brain tissue shows that specific influences from different brain regions appear at different periods during incubation (Hamburger, 1964; Hamburger et al., 1965; Hamburger et al., 1966; Decker and Hamburger, 1967; Hamburger and Oppenheim, 1967).

Of course, such examples do not imply that the spontaneous activity of individual neurones is necessarily rhythmical—it may be random (e.g., Werner and Mountcastle, 1963). In any case the rhythmic output of a cell group depends on certain tonic steady-state conditions. For instance, Kerkut and Taylor (1958) showed that the isolated abdominal ganglia of a crayfish (*Astacus fluviatilis*), the nerve cord of a cockroach (*Periplaneta americana*), and the pedal ganglion of a slug (*Agrolimax recticulatus*) all show a spontaneous discharge. The discharge, however, depends markedly on temperature: if the temperature is changed and the preparation given a few minutes to settle down, a new steady rate of discharge appears. For each preparation there is an optimum temperature at which the rate of discharge is maximal, though this optimum may vary with the temperature at which the animal has been kept. Sudden changes in temperature produce marked fluctuations in the rate of discharge (see also Magni, 1967).

More recently, techniques of recording from single units, and the use of microelectrodes which penetrate the cell body, have taken this study of the electrical activity of the nervous system down to the cellular level (e.g., Bullock, 1961; Roeder et al., 1960). Many individual cells show spontaneous discharges, originating from certain regions of the cell (the pacemaker loci) which may influence part or all of the rest of the cell. Some cells have more than one such locus, with different temporal characteristics. Such changes in the state of a cell may affect its responsiveness to influences from outside and may even affect other neurones electronically, as well as generate spike discharges (see also Chaps. 5 and 6 and Bullock, 1958, 1959a, b; Bullock and Terzuolo, 1957).

Most of the cases of rhythmic output from the nervous system have involved a periodicity measurable in seconds or parts of seconds—quite a different order of magnitude from the changes in behaviour which we see in the whole animal, with a time course of minutes, hours, or days. Nevertheless, rhythms of comparable time periods, which appear to be spontaneous, are well known. The polychaete worm *Arenicola marina* lives in a U-shaped burrow. A burst of feeding movements occurs every 6 to 7 min, and every 20 to 60 min there is a sequence of tailward creeping, forward irrigation, tailward irrigation, and defaecation. Such a cyclic pattern could result from responses to regularly recurring peripheral events—empty gullet, full rectum, and so on—but the evidence strongly suggests this is not the case. The outbursts apparently begin without external stimulus or biological need and subside without satisfaction. It is highly probable that there are two biological "clocks", one for each of the principal rhythms mentioned above. The brain is unnecessary for the continuance of the rhythms, but the ventral-nerve cord is essential, and the spontaneity depends in part on the wall of the oesophagus (Wells, 1950, 1955).

Whether the periodicity of a rhythm is measurable in fractions of a second or seconds, as with nerve cells, in minutes, as with this polychaete worm, or in hours, days, weeks, or months, the problems of analysis are initially similar. Some of the most interesting data have come from studies of the rhythms with a periodicity of approximately 24 hr which form the basis of the day–night changes in activity shown by most animals. Such rhythms may continue for weeks or months undamped, and with only a small change in periodicity, when the animals are kept under constant conditions in the laboratory (Fig. 14.2). For instance, the coefficient of variation of the period of activity cycles of canaries under such conditions is only 0.2 to 4 percent (Wahlström, 1964). Such constancy of the rhythm in the absence of external pacemakers indicates that the rhythm is endogenous. There is, of course, a difficulty in proving the null hypothesis that changes in external factors are having no effect (Brown, 1960). However, it is possible to upset all the external factors temporally related to the earth's cycle of rotation by transporting animals in latitude. For example,

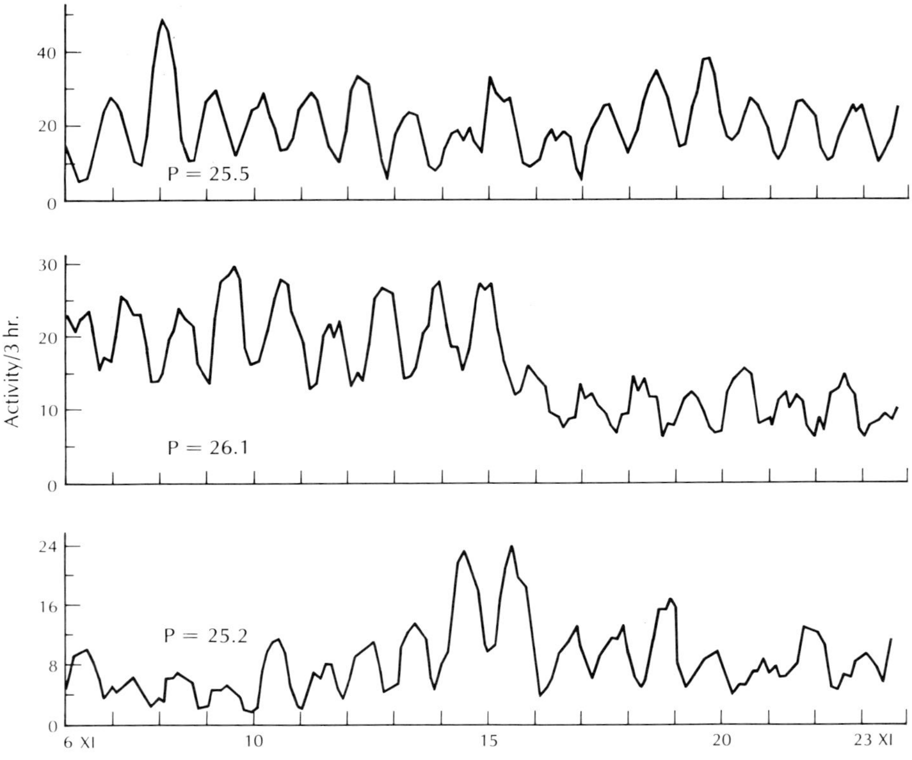

Fig. 14.2. Activity changes in mice kept for two generations under constant conditions. (From Aschoff, 1955.)

Ordinate, activity score per 3 hr; abscissa, date; *P*, the length of the period in hr.

experiments involving the translocation of bees from Paris to New York under constant environmental conditions showed that the diurnal rhythms in the bee's behaviour persisted after the move (Renner, 1957). Another source of evidence for the independence of the rhythm from external factors comes from experiments with groups of animals in which the individuals show different but internally consistent cycles (e.g., Hoffmann, 1959). But the most convincing type of evidence comes from the fact that, under controlled conditions, the periodicity deviates slightly from 24 hr. Usually, as in the parts of the records shown in Fig. 14.3, which were taken under constant light conditions, the deviation is small but consistent from day to day. Clearly such a "circadian" rhythm could not depend on environmental factors which recur with a precise or average periodicity of 24 hr (see Aschoff, 1960, 1963; Pittendrigh, 1960).

This does not mean, of course, that external factors are without influ-

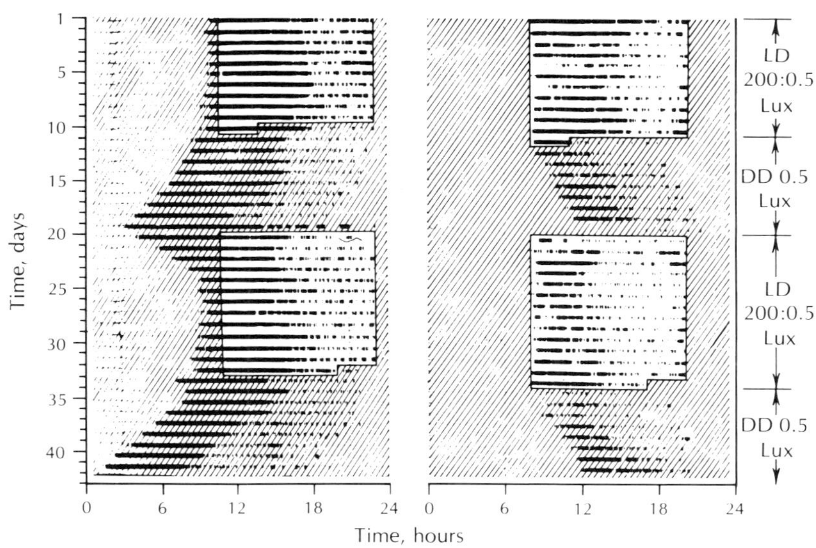

Fig. 14.3. Activity of two chaffinches kept in either a light-dark cycle (LD) with 200 lux in L and 0.5 lux in D, or in constant conditions with a dim illumination of 0.5 lux (DD). Shaded area, 0.5 lux; white area, 200 lux. (From Aschoff, 1967, after Aschoff and Wever, 1966.)

ence. In most diurnal species the frequency of the rhythm, the ratio of activity to rest time, and the absolute amount of activity are all greater in bright than in dim light. With nocturnal species the reverse is usually the case, the amount of activity decreasing with increase in illumination (Aschoff, 1960, 1963). There are, however, some exceptions to this generalization (Hoffmann, 1965), and Kavanau (1968) claims that it is merely a consequence of the unnaturally high light intensities used in many experiments; at low levels of illumination deermice show an increase in activity when the illumination is increased. Such results can be understood on the assumption that the rhythm depends on a circadian oscillator whose "level" with respect to the activity threshold varies with the illumination (or the difference between the illumination and a preferred level) (Fig. 14.4). Since such oscillators show a positive relationship between level and frequency, all the three measures mentioned above would be correlated. Temperature

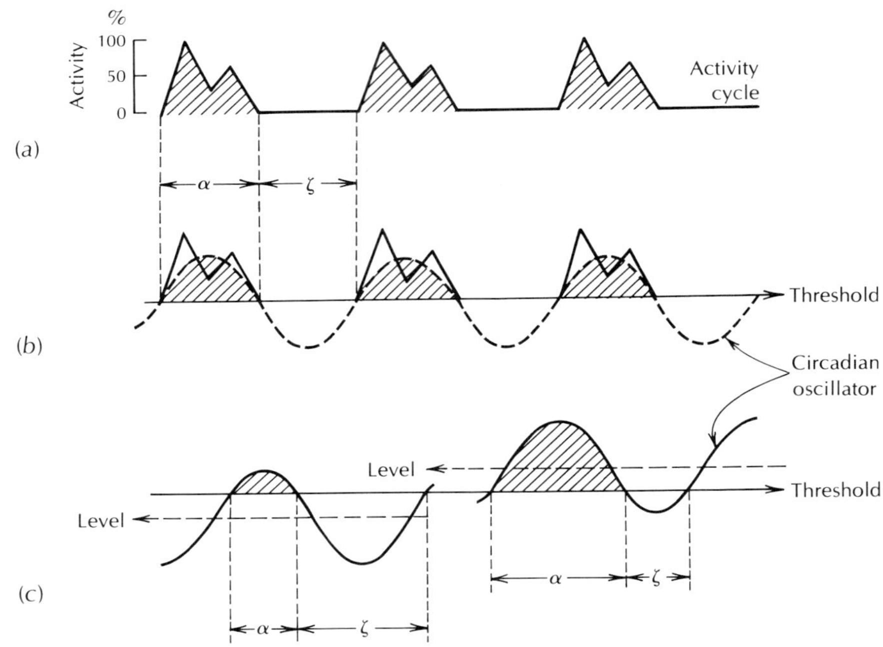

Fig. 14.4. A, schematic representation of an activity rhythm with activity-time α and rest-time ζ; B, its description by an oscillator which passes through a threshold twice per cycle; C, changes in the α:ζ ratio described by changes in the position of the oscillator's level (dashed line) relative to the position of the threshold. Shaded area, activity. (From Aschoff, 1967.)

may also have an effect on circadian rhythms, though it is usually a surprisingly small one (e.g., Kalmus, 1934; Brown and Webb, 1948).

Under natural conditions circadian rhythms are synchronized to exactly 24 hr by means of periodic factors in the environment (e.g., Bruce, 1960; Harker, 1964). These are usually known as Zeitgebern (Aschoff, 1958). Light and temperature changes are of particular importance in this context. If a rhythm has been proceeding under constant conditions, it may take some time to become entrained when a Zeitgeber is provided (Fig. 14.3), and the phase relationship varies with the frequency of the self-sustained oscillation (De Coursey, 1961; Aschoff and Wever, 1962b). There are of course limits to the periodicity of Zeitgebern with which circadian rhythms can become tenrained: for instance, domestic mice can be entrained to day lengths varying from 21 to 27 hr (half-light, half-dark in all cases), but with light–dark cycles shorter or longer than this the animals revert to an approximately 24-hr rhythm (Tribukait, 1956). It is possible that entrainment to rhythms differing even more from 24 hr could be achieved if gradual transitions between light and dark were used (Kavanau, 1962).

Entrainment of the circadian rhythm by a Zeitgeber provides a fixed-phase relationship between the circadian rhythm of the organism and the periodicity of the environment. Light onset appears to play a more important part in resetting the rhythm than light offset in diurnal species, and vice versa (De Coursey, 1961), and the onset of activity often bears a more constant relationship to dawn than the offset bears to sunset (Fig. 14.5). Aschoff (1967) has pointed out that, in considering the phase relationship between Zeitgebern and activity, differences in level make it preferable to compare the midpoints of the two oscillations rather than the beginning or end of activity. Thus the two activity cycles shown in Fig. 14.6b both show a phase-angle difference of 30 deg to the sun's movement when midpoints are compared but inconsistent differences if beginning and end of activity are compared to sunrise or sunset.

Since the frequency of the circadian rhythm usually increases with light intensity (see page 317), and since the phase-angle difference varies with the frequency, diurnal animals would be expected to show a greater positive phase-angle difference in summer than in winter: field data bear out this prediction (Aschoff, 1967).

Thus a variety of observations on diurnal activity rhythms, relating particularly to birds, can be understood in terms of a single endogenous oscillator coupled to an internal Zeitgeber. In other cases, however, such a model is too simple, and it seems necessary to postulate two systems, one strictly endogenous and the other more responsive to external stimuli (e.g., Brown, 1959; Pittendrigh, 1960). For instance when mice are exposed to very long or very short days, the morning peak of activity keeps time with the Zeitgeber while the evening peak maintains a roughly 24-hr periodicity (Tribukait, 1956). Physiological studies of human subjects also suggest that more than one mechanism is involved.

The anatomical and physiological bases of circadian clocks are little known. Harker (e.g., 1964) claimed that the circadian activity of the cockroach *Periplaneta americana* is controlled by an endogenous neurosecretory cycle in the suboesophageal ganglion, partly controlled by a second clock responsive to light signals through the ocelli. However, S. K. de F. Roberts

(1965, 1966) has cast doubt on these conclusions, and Brady (1967a, b), only partially able to reconcile the conflicting results, suggested an "electrical pacemaker in the brain co-ordinating rather ephemeral neuroendocrine rhythms in the nerve-cord ganglia".

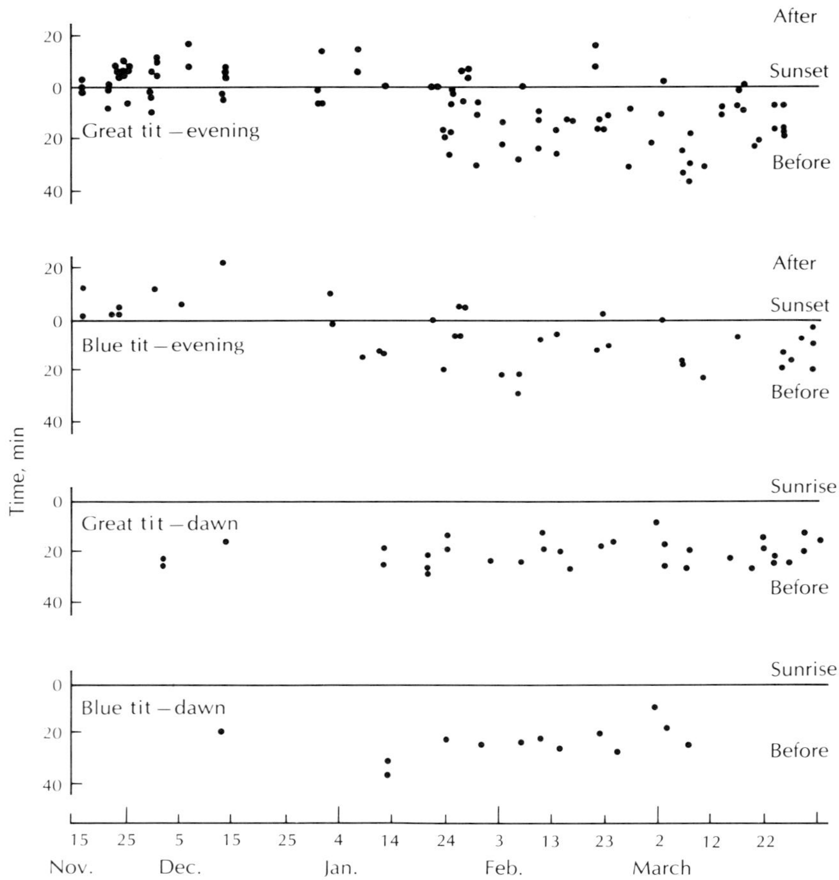

Fig. 14.5. Times of leaving and entering rooting holes by blue and great tits (Parus caeruleus and Parus major) during the winter months. (From Hinde, 1952.)

SUMMARY

1. "Spontaneity" implies change in the output of a system without a known corresponding change in input.
2. Examples of spontaneity at the behavioural level are commonplace. There is no sharp dividing line between spontaneous and stimulus-elicited activities.

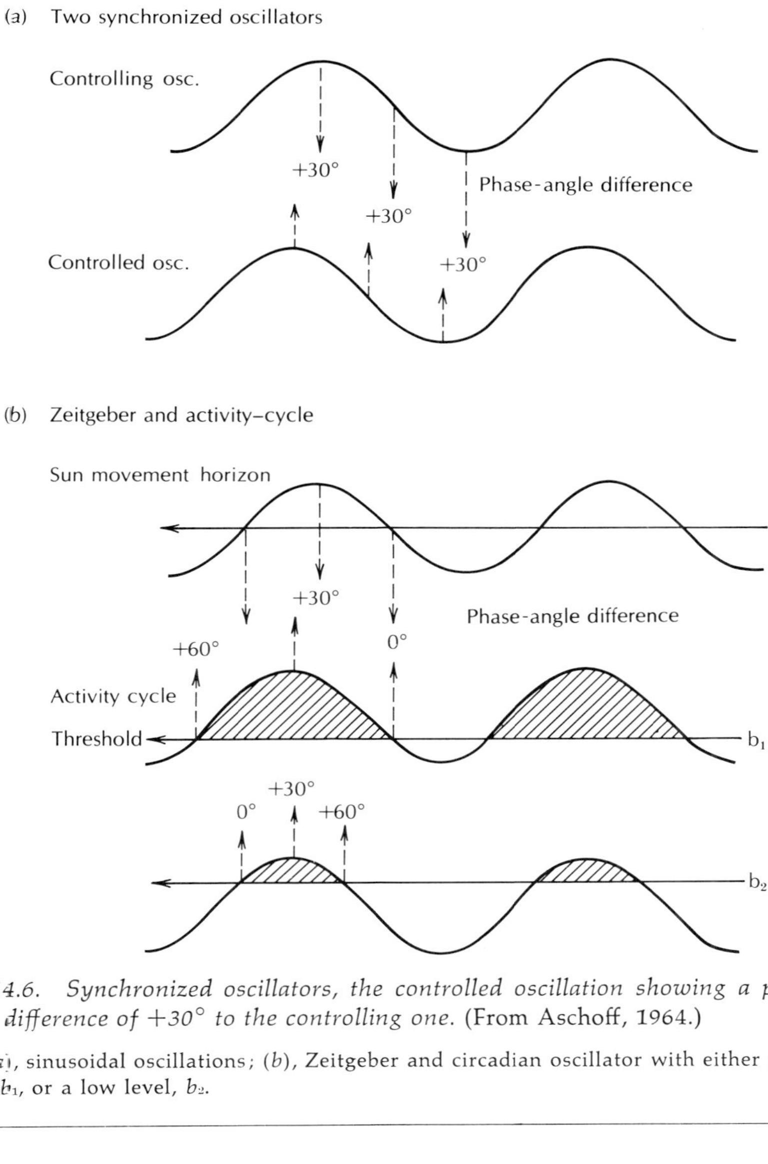

Fig. 14.6. Synchronized oscillators, the controlled oscillation showing a phase-angle difference of +30° to the controlling one. (From Aschoff, 1964.)

(*a*), sinusoidal oscillations; (*b*), Zeitgeber and circadian oscillator with either a high level, b_1, or a low level, b_2.

3. Rhythmical output from a system may depend on external timing cues, often resulting from feedback consequences of its activity, or on an internal pacemaker. The activity of the latter may of course depend on its own internal feedback.
4. Rhythmic activity in the isolated nervous system has been demonstrated in a number of cases. Spontaneous rhythmic activity is also well known in single cells.
5. Spontaneous rhythms in the behaviour of the whole animal having periodicities measurable in minutes, hours, or days have been studied in many animals. The nature of "circadian" rhythms is discussed.

15

DIVERSITY IN MOTIVATIONAL SYSTEMS

In the preceding chapters we have seen that short-term changes in responsiveness to external stimuli and in behaviour must be ascribed to changes internal to the organism. Even at the behavioural level, these changes are complex and, except in the preliminary stages of analysis, cannot be represented by a unitary intervening variable, such as drive (Chaps. 8 and 9). The changes in responsiveness are influenced by extraneural factors (Chaps. 10 and 11) and by the consequences of the activity itself (Chaps. 10 and 13), as well as by the immediately eliciting stimuli (Chaps. 4, 5, and 6). As we analyse towards a more physiological level we see that these factors influence complex networks within the central nervous system, whose parts affect different aspects of responsiveness (Chap. 12). The state of these networks can vary independently of changes in extraneural factors (Chap. 14), and they can influence behaviour by determining the goals to which it leads as well as by increasing the probability of particular stimulus–response sequences (Chaps. 11, 12, and 25).

So far, however, the diversity of the behaviour considered has been limited, and we must ask whether general principles applicable to other types of behaviour can be found. In this chapter, therefore, we shall summarize and extend the preceding ones by considering the changes in responsiveness found in a few examples of functional categories of behaviour. In doing so, our concern is solely with relatively short-term changes in behaviour: problems of integration and development are considered later.

15.1 EATING, WITH SPECIAL REFERENCE TO RODENTS

Various aspects of the control of eating behaviour in rats have already been mentioned: the present discussion is intended merely to summarize these so that eating can be compared with other types of behaviour, and to put them in a wider context.

Eating in rodents is a consequence of interaction between central excitatory and inhibitory mechanisms: these mechanisms are themselves far from unitary, and their parts affect both the motor patterns of feeding and responsiveness to stimuli from food. The states of these mechanisms are influenced by multiple factors, including the inhibitory consequences of ingestion, whose effects decay with diverse time courses (Chaps. 10 to 12). Experiential factors also play a role, including those determining the temporal rhythms in behaviour (Chaps. 14 and 22).

When the excitatory factors are predominant, the animal shows a type of appetitive behaviour which is likely to bring it into the presence of food. This is, in part, a direct consequence of deprivation, but results largely from an increased responsiveness to external stimuli (pages 260–264). While showing this appetitive behaviour, the animal is especially sensitive to stimuli likely to be indicative of food (Chap. 11), the precise stimuli depending on the availability and palatability of different kinds of food (e.g., Holling, 1958, 1959, 1965), specific nutritional requirements (page 249), previous experience, and other factors. The stimuli from the food may be general to a wide variety of food substances or, especially in some invertebrates, characteristic of one type of food only (pages 72–73; Thorsteinson, 1960). These elicit activities such as catching, preparing, and eating, the details varying with the species and the circumstances (review by de Ruiter, 1967b).

When a mammal eats, feedback effects from the mouth/throat region and the stomach, and from changes in blood composition, swing the balance in favour of inhibition. These decremental processes take effect before the basic tissue needs are relieved, and decay with varied time courses (Chap. 13). Furthermore, at any rate in some species, they may affect the consummatory behaviour of eating independently of the appetitive behaviour which usually leads up to it: it is for this reason that hunting behaviour is often shown by satiated animals (Lorenz, 1935; Hinde, 1953b).

The central excitatory and inhibitory mechanisms are not simple, for the various dependent variables by which feeding behaviour can be assessed are only loosely correlated with each other (pages 196–197). The diversity of the factors which control the relative attractiveness of different foods and the amount eaten can hardly be overemphasized. For instance, in rats, novelty of the food may lead either to an increase (Welker and King, 1962) or a decrease (Barnett, 1963) in the amount eaten: in adult monkeys the latter effect predominates but may be overcome by the sight of another animal taking the food (Weiskrantz and Cowey, 1963), though young animals are more likely to pioneer the exploitation of new food resources than are older ones. Social factors have a facilitating influence on feeding in puppies (James, 1953; James and Gilbert, 1956), opossums (Platt et al., 1968), birds (Lorenz, 1935; Katz, 1937; Tolman, 1964, 1967a, b, 1968; Turner, 1964), and other species; and the amount of food eaten may also be influenced by the amount present (Ross et al., 1962).

15.2 MALE SEXUAL BEHAVIOUR IN RATS AND OTHER SPECIES

The male rat's sexual behaviour involves reflex responses, such as penile erection and thrusting movements of the pelvic musculature, which probably depend on neural circuits in the medulla and spinal cord: in the rat erection and flips of the glans penis can be elicited from spinal animals (Hart, 1968b), and in other species these components can be elicited both from adult animals after complete cord transection and from foetal or neonatal animals before the higher brain regions are fully functional. However, the organization of these reflexes into a functionally effective pattern depends on the more anterior parts of the brain. There is some evidence that the role of these more anterior regions is in part inhibitory (Beach, 1967; Hart, 1967a, b).

Male sexual behaviour depends also on the presence of circulating androgens. Apart from their influence during development (pages 543–545), these produce their effects both centrally, by influencing the hypothalamus and perhaps other parts of the central nervous system (pages 232–237), and peripherally, by affecting the genitalia (pages 230–231). They may produce their central effects in part by modifying the inhibitory action of higher neural mechanisms (Beach, 1966, 1967) (page 235). The central effects of the hormone are augmented by influences mediated by the cortex, which, in turn, presumably depend on experience (pages 239, 550). Even when hormonal levels are high, the male is not sexually responsive all the time: periods of sexual responsiveness are succeeded by periods during which the animal is engaged in other matters. Presumably, these changes in responsiveness are determined in part centrally, in part by stimuli received as a result of sexual behaviour itself, and in part by the incidence of factors specific to other types of behaviour (pages 396–400).

From time to time the male shows appropriate appetitive behaviour and is especially responsive to external stimuli of the type normally provided by a receptive female (e.g., Carr et al., 1965, 1966). The appetitive behaviour includes locomotor exploration and sniffing at the female (Dewsbury, 1967a). After a while mounting occurs. This leads to a brief intromission, after which the male dismounts. A series of such intromissions with brief pauses between is concluded by one in which ejaculation occurs. There is then a longer interval (the post-ejaculatory refractory period) before another series of intromissions starts. The preejaculatory intromissions not only contribute to the achievement of an ejaculatory threshold in the male, as discussed below, but also contribute towards the successful fertilization of the female (J. R. Wilson et al., 1965).

Intromission involves two components. The first consists of a number of shallow thrusts around or just into the vaginal opening, and the second of a single thrust deep into the vagina, followed by withdrawal and dismounting. Lidocaine anaesthesia of the penis prevents or delays full intromission, though mounting continues at a high rate (Carlsson and Larsson, 1964; Adler and Bermant, 1966; see also Cooper and Aronson, 1962, for data on cats).

The time relations of these copulatory sequences, though differing among strains, have provided important evidence concerning the consequences of feedback from intromission and ejaculation. If a male is allowed access to a female for an hour, the number of intromissions per ejaculation at first falls and, in some cases, later rises, while the post-ejaculatory refractory period rises throughout (Beach and Jordan, 1956; Larsson, 1956). The consequences of intromission and ejaculation are thus by no means simple. In order to explain such results, Beach and Jordan postulated the existence of two mechanisms: an arousal mechanism which mediates the increase of sexual excitement leading to copulation, and a copulatory–ejaculatory mechanism which mediates the behavioural components of copulation and seminal emission.

The state of the arousal mechanism was held to be indicated by the latency after the female is introduced before the first intromission, and by the refractory periods between one ejaculation and the next intromission (Fig. 15.1*a* and *b*). Since the refractory periods increase after each successive ejaculation, it was supposed that a series of ejaculations has a cumulatively inhibitory effect on the arousal mechanism. If the male is allowed to copulate to sexual exhaustion (criterion 30 min without mounting), the latency, as shown by a subsequent test, remains high for several days: it was thus suggested that the arousal mechanism does not regain its maximum responsiveness for about a week.

The state of the copulatory mechanism was held to be indicated by the number of intromissions necessary to produce ejaculation and by the latency between the first mount of each series and ejaculation (Fig. 15.1*c*). Since the number of intromissions per ejaculation falls in the successive ejaculations of a series, it seems that intromission or ejaculation has a sensitizing effect on the copulatory mechanism (see also Larsson, 1958). The number of intromissions necessary to produce ejaculation returns to the resting level after about 90 min of sexual inactivity.

The view that two separate mechanisms are involved was also supported by endocrinological evidence. Beach (1942b), Beach and Holz-Tucker (1949), and Whalen et al. (1961) found that, with increasing androgen dosage, the range of stimulus objects eliciting mounting from male rats increased and the latency before mounting decreased. The number of intromissions before ejaculation, however, showed no correlation with androgen level.

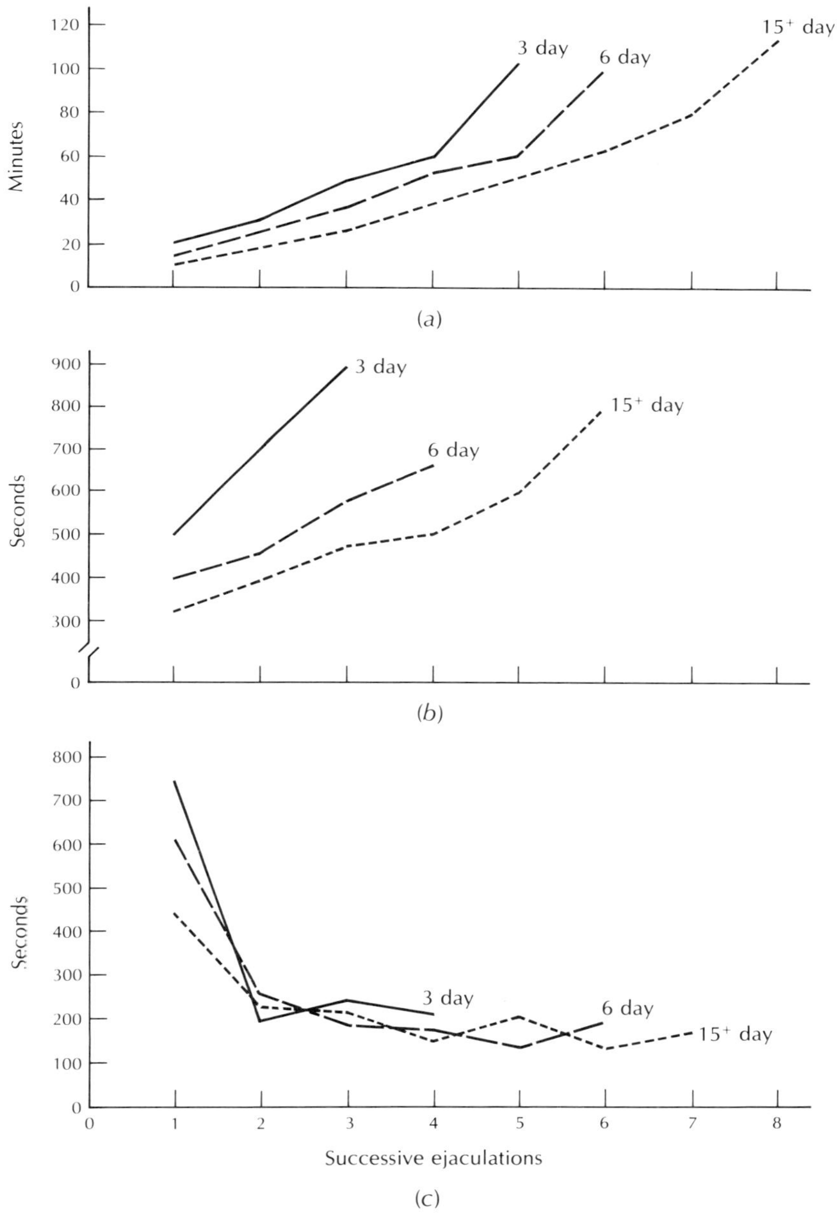

Fig. 15.1. The course of sexual behaviour in the male rat. The animals were given either 3, 6, or 15 days' recovery, since complete sexual exhaustion (i.e., 30 min without mounting a receptive female). Abscissa in each case indicates serial number of ejaculation. (From Beach and Jordan, 1956.)

(*a*) Time from beginning of test at which successive ejaculations occurred. (*b*) Duration of refractory periods following successive ejaculations. (*c*) Successive ejaculation latencies (i.e., time from first mount to occurrence of ejaculation).

They therefore concluded that sexual responsiveness, as measured by mount or intromission latency, is directly related to the level of circulating androgen. By contrast, the ejaculatory process, while requiring a certain minimum androgen level, is not affected by additional amounts of hormone. This view is also supported by evidence on the effects of castration. After castration, intromission latency rises progressively, but the ejaculatory process is temporarily facilitated: the latter could be a non-specific consequence of the surgery (Davidson, 1966a).

A number of other lines of evidence indicate that the mechanisms underlying erection and seminal emission have some degree of independence from those mediating other aspects of copulatory behaviour. The former lie in the spinal cord and can function independently of higher parts of the nervous system. They can be activated by genital stimulation, although the occurrence of "spontaneous" emissions shows that this is not essential. However, their discharge does not produce all the effects of normal ejaculation as produced by coitus: in particular, emission produced by electrical stimulation does not produce the same effects on subsequent mating behaviour as does ejaculation *in copula*. Furthermore, seminal discharge is not essential for the normal behavioural consequences which follow ejaculation (Beach, 1967).

The time relations of the excitatory state which follows intromission have been subjected to more detailed analysis. The number of intromissions required to produce ejaculation is reduced by enforcing a delay after each one, provided the delay exceeds the normal intercopulatory interval of 45 to 60 sec and does not exceed 3 to 5 min. With intervals of more than 10 min ejaculation is prevented, though some males go on copulating for several hours, making up to 40 or 50 intromissions (Larsson, 1956). There would thus seem to be an excitatory process, consequent upon intromission, which increases for some time after intromission and then decays.

Beach and Whalen (1959b) traced the decay of this excitatory effect in more detail by permitting a male rat 1, 4, or 7 intromissions and then enforcing a rest by removing the female for a period of from 7.5 to 120 min. The excitatory effect from the initial series of intromissions which still remained when the female was returned was assessed by comparing the number of intromissions then required for ejaculation with those required in control tests. The results showed that the excitatory effects of successive intromissions were cumulative and continued to rise for some minutes after the last intromission. After 30 min the number of additional intromissions required had returned to about the number which would have been necessary if the series had not been interrupted: some excitatory effect remained even after 120 min (see also Bermant, 1964).

This further analysis showed that the effects of a series of intromissions and ejaculations are more complicated than the original scheme suggested. The effect of one series on the ejaculation latency (and the intercopulatory interval) of a second persists for much longer than does its effect on the number of intromissions required to reach ejaculation. These two measures can thus not be regarded as indices of the state of a unitary mechanism (Beach and Whalen, 1959a, b). In addition, the view that ejaculations have an inhibitory effect on the arousal mechanism is open to question. If intromissions are prevented after the first ejaculation for an enforced rest interval, the mount and intromission latencies in the second series are shorter than those in the first: this suggests a stimulatory rather than a depressive effect on the arousal mechanism. Beach and Whalen (1959a) thus suggest that an initial series of intromissions ending in ejaculation has several more or less independent effects.

1. A sensitization of the copulatory mechanism, resulting in a reduction in the number of intromissions required for a second ejaculation.
2. A sensitization of the arousal mechanism, tending to produce shortened mount and intromission latencies and thus a reduced ejaculation latency.
3. Non-sexual fatigue, which temporarily masks (2) but is cumulative over successive ejaculations and thus leads to progressively longer post-ejaculatory intervals.

It is not clear why this last "fatigue" effect is specified as non-sexual. In view of the data in Chap. 13, it is possible that ejaculation has both positive and negative effects on sexual arousal, which decay with different time constants. The use of a concept of non-specific fatigue in this context has also been questioned by McGill (1965) on the basis of observations on the sexual behaviour of mice. This species has a conspicuous ejaculatory pattern in which the male clutches the female and falls to one side. McGill postulates a third mechanism ("ejaculatory mechanism") which discharges at ejaculation: its consequent depression delays activation of the arousal mechanism. On this view successive ejaculations result in progressively greater depression of the ejaculatory mechanism and thus an increase in the time necessary for its recovery.

Still further complications remain. One is introduced by the observations of Peirce and Nuttall (1961a) that the mean duration of copulatory contacts increases through a series of intromissions: this suggests that, in addition to the sensitization effect, the reflex becomes more sluggish. Here again is a suggestion of simultaneously produced positive and negative effects. In addition, the inter-intromission intervals within a copulatory series at first increase and then progressively decrease as ejaculation is approached (Dewsbury, 1967b, 1968).

Soulairac (reviewed Larsson, 1956) suggested from pharmacologic evidence an even more complex mechanism, namely that the primary regulation of ejaculation, copulatory behaviour, and intromission, and the refractory period, are due to different groups of factors. This is based on the observations that: (1) the frequency of ejaculations, but not the number of intromissions per copulation or the refractory period, is influenced by certain disturbances in the endocrine balance; (2) certain neuroexcitatory drugs influence the frequency of intromission, but not ejaculation, and have only a slight influence on the refractory period; (3) certain other drugs which influence enzyme activity in the central nervous system prolong the refractory periods without directly affecting the other two characters.

In Chap. 13 we saw that a waning of responsiveness is often specific to the particular stimulus rather than to the response, and a similar process may operate in sexual behaviour. In guinea pigs (Grunt and Young, 1952), rats, and many other mammals (J. R. Wilson et al., 1963), substitution of a new female when a male has become sexually satiated in a series of matings produces a renewal of sexual behaviour. The substitution of a new female, especially of one not recently mated, is more effective than the removal and replacement of the one with which the male has just been copulating. Fisher (1962) found that, after sexual exhaustion, flashing lights and intermittent tones coupled with reintroduction of the same female also led to reactivation of the male's sexual behaviour, but the effect was small compared with that of a new female. Changing the environment thus has little effect in rats, though it has in some other species (see also Beach and Ransom, 1967). Changing females had less effect on the capacity scores of old male rats than of young ones—perhaps they had become sophisticated about novelty. Fisher also showed that the number of intromissions required per ejaculation rose with the new female, as though the male were beginning to copulate again. Fisher's data on this point were not substantiated by Hsaio (1965), who found that the sexual scores associated with the additional ejaculations elicited by a new female were characteristic of those late in a mating sequence: this latter result was confirmed by Bermant et al. (1968; see also Krames et al., 1967; Hsaio, 1969).

If a male is allowed to copulate to satiation with one female, his readiness to mate with the new female (as indicated by intromission latency) decreases with time, reaching a minimal value after 24 hr. However, at all time intervals up to 24 hr the number of ejaculations achieved with the second female is smaller than the number achieved with the first, the frequency of total failures to ejaculate being greatest at 6 to 18 hr after satiation. Furthermore the interval from satiation to test is inversely related to the initial test post-ejaculatory interval but directly related to the median test intromission latency, whereas these two measures should co-vary as measures of sexual arousal on Beach's original model. These and other data presented by Bermant et al. (1968) show that the latter is inadequate to describe all aspects of rat copulatory behaviour.

Fowler and Whalen (1961) suggested that the effect of changing the female was due to attempts by the first female to repulse the male. Other studies have shown that, after a mounting, the female rat stays away from the male for a short period. The duration of this period varies with the nature of the previous mounting: after mounts without intromission, mounts with intromission, and mounts with ejaculation, the median intervals are 13, 60, and 170 sec (Peirce and Nuttall, 1961a, b; see also Bermant, 1961; Kuehn and Beach, 1963). However, in a recent study male rats were tested with five oestrous females simultaneously: there was no indication that a female which had been mated was any more negative to the male than an unmated female, and the males copulated indiscriminately throughout the test (Beach, personal communication). In the cat, the behaviour of the female certainly is of importance. Substitution of a fresh female after the male has reached a criterion of sexual exhaustion leads to renewed intromissions. However, if the new female has just been mated with another male, the male under test achieves fewer intromissions before criterion than in the normal series. When a sexually fresh male is allowed access to a sexually exhausted female, little mating results (Whalen, 1963b). Finally, it must be emphasized that, in the few species which have been studied from this point of view, individual preferences between potential mates are known to be important (e.g., dogs, Beach and Le Boeuf, 1967; monkeys, Herbert, 1968).

Data for the course of sexual exhaustion are also available for a number of other subhuman vertebrates. A cumulative effect of successive ejaculations has been found in all species capable of successive ejaculations within a limited mating session (Beach, 1966b). The inadequacy of a unitary concept of sexual arousal has also been stressed by Rosenblatt and Aronson (1958a, b) and by Whalen (1963a, b) as a result of their studies of cats. Hamsters (Beach and Rabedeau, 1959), mice (McGill, 1962b, 1965), and deermice (Clemens, 1967) are qualitatively similar to rats. Hystricomorph rodents (e.g., guinea pig, chinchilla) show some differences from rats and other myomorphs: in particular, ejaculation is usual on the first intromission (Bignami and Beach, 1968). Data are also available for dogs (Beach and Le Boeuf, 1967), sheep (Pepelko and Clegg, 1965), rabbits (Rubin and Azrin, 1967), monkeys (Michael and Saayman, 1967; Herbert, 1968), and man (Masters and Johnson, 1965): the probable role of olfactory stimuli in monkeys has already been mentioned (pages 237–238). A few data for chaffinches are given by Hinde (1959c), and studies on domestic cocks are reviewed by Guhl (1962). Other references are cited by Schein and Hale (1965); see also Hildreth (1962) for *Drosophila.*

Mammalian male sexual behaviour has been discussed in some detail because the studies of Beach and his collaborators exemplify so well the manner in which humoral factors, external stimuli, and feedback factors interact. The hormone has multiple effects, central and peripheral, which determine the general level of sexual activity over long time periods, but the external stimuli and feedback effects which decay with varied time courses determine moment-to-moment responsiveness. It will be apparent that, for analysis of these interactions, the concept of a unitary central excitatory state needs very considerable elaboration.

15.3 NEST–BUILDING OF FEMALE CANARIES

The sequence of nest-building behaviour considered here consists of gathering nest material, carrying it to a nest site, and sitting in the nest cup building the material into the structure. The latter involves a small number of stereotyped building movements. Like rat sexual behaviour, this is influenced by hormone levels—as the hormone levels increase, the sequence of behaviour is repeated more often and is more likely to be completed (Warren and Hinde, 1959).

The course of building behaviour is much influenced by stimuli received as a result of building. The female becomes more sensitive to these stimuli as a consequence of hormonal changes, probably mediated peripherally (pages 231 and 633–640). Thus, as in the rat, sex hormones have both central and peripheral effects.

The stimuli in question come primarily from the nest which the female herself has built. They affect a number of different aspects of nest-building behaviour. These are considered later (Chaps. 25 and 26), but they include the frequency of the nest-building movements which the female makes while sitting in the nest cup, the selection of nest material, and the actual amount of building behaviour (Figs. 25.7, 26.1, and 26.3). In addition stimuli from the nest cup accelerate reproductive development and thus a new hormonal condition not conducive to building (Fig. 26.2).

Since the feedback effects from some sorts of nest structure differ from those produced by others, there must be an optimum nest structure which is maximally effective for each of these functions. In the normal course of events the inhibitory effects on building produced by the finished nest contribute to the cessation of nest-building, which is due also to the hormonal changes occurring at the time (Hinde, 1965, 1967; Hinde and Steel, 1966).

We see that, as in rat sexual behaviour, the hormonal state affects multiple central mechanisms as well as producing peripheral effects, and that there are multiple feedback effects from the behaviour, producing both positive and negative, short- and long-term effects. These are discussed further on pages 633–640.

The nest of the canary is a relatively simple one, but even those of species which build complex structures depend on relatively few stereotyped movements. The complexity depends, in part, on changes in the manner in which these movements are utilized as the structure grows. This principle is illustrated in Figs. 15.2 and 15.3, which show how the village weaver (*Ploceus cucullatus*) builds its retort-shaped structure. At each stage in the building of the outer shell the bird perches in the same place but orients its weaving to the edge of the existing fabric (Crook, 1960c, 1964b; Collias and Collias, 1962, 1964).

15.4 SHELL SELECTION BY HERMIT CRABS

In nest-building, the behaviour is influenced by stimuli from the nest whose effectiveness varies with their difference from those provided by a completed nest. A comparable mechanism has been studied also by Reese (1963) in his investigations of the selection of shells by hermit crabs (*Pagarus samuelis* and *Calcinus paevimanus*). The behaviour of hermit crabs towards the shells in which they live is independent of previous experience of shells. It involves a sequence of responses, each elicited by different properties of the shell, mediated to some extent visually but principally via the mechanoreceptors. In normal circumstances the properties of the shells release successive motor patterns each of which brings the crab into the situation for the release of the next. The behaviour thus proceeds as a chain until the crab has entered and righted the shell. The chain is not a rigid one, and the elements can be elicited independently. Once the shell has been entered the crab may go on showing the appetitive behaviour which normally serves the function of finding new shells. The amount of appetitive behaviour shown (as assessed by the amount of motor activity and responsiveness to shells) depends on the characteristics of the shell it has entered. The total stimulus value of the shell is determined primarily by two properties—weight and internal configuration—but others may also contribute. Only if the stimulus value is adequate does the appetitive behaviour and responsiveness to other shells decline. Thus, like nest-building, the behaviour continues until inhibited by the adequate consummatory situation. Similar principles operate in the land hermit crab, *Coenobita rugosus* (Kinosita and Okajima, 1968).

15.5 AGGRESSION

At least some of the heat engendered by discussions of aggressive behaviour has arisen from the difficulty of defining terms. Aggressive behaviour as used in a psychiatric or social sense often includes not only fighting but also a wide range of behaviour which could also be labelled

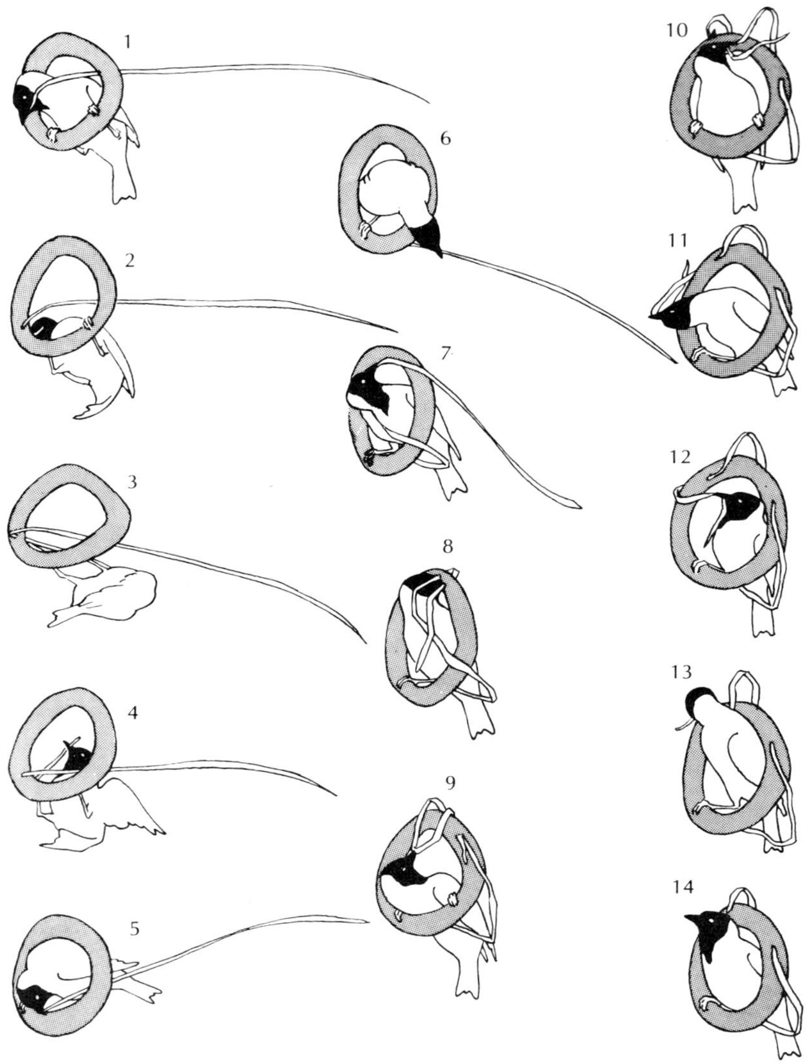

Fig. 15.2 A typical sequence of movements by a male village weaver-bird as he weaves a single strip, torn from a leaf-blade of elephant grass, into his ring. (From Collias and Collias, 1962.)

as self-assertive or competitive. In the study of animal behaviour the term is usually used in a more restricted sense: for reasons which will be mentioned in a moment, this is not easy to define precisely but refers to behaviour directed towards another individual which could lead to physical injury to the latter and often results in settling status, precedence, or

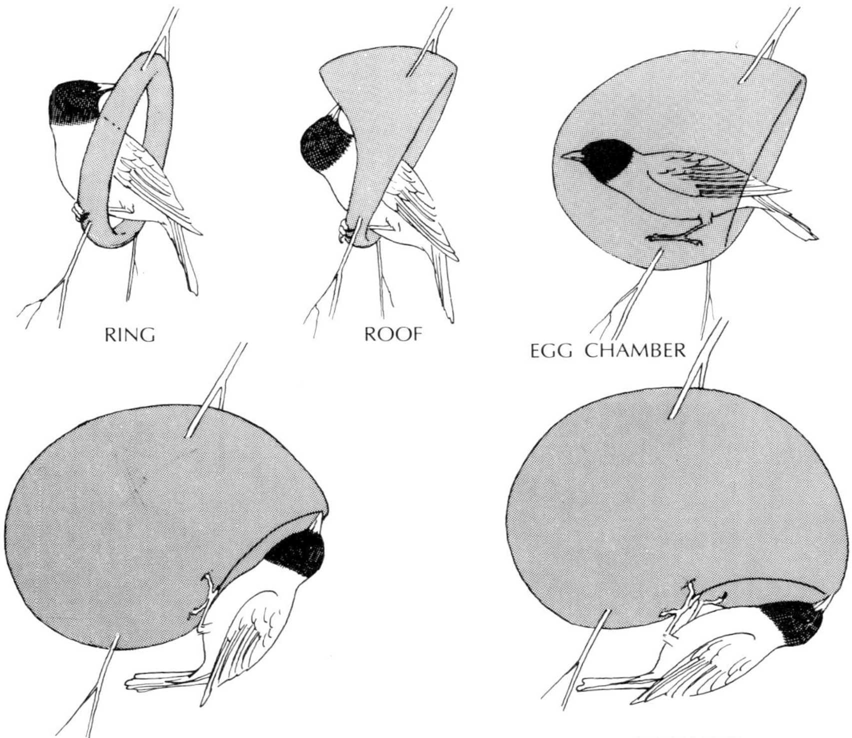

Fig. 15.3 Normal stages in nest-building by a male village weaver-bird. (From Collias and Collias, 1962.)

The diagrams show successively the building of the ring and roof, the completion of the egg chamber, and the building of the antechamber and entrance. Notice how the male perches in the same place while building throughout.

access to some object or space between the two (see also Feshbach, 1964; Kaufmann, 1965). The difficulties of definition arise for the following reasons. First, inflicting injury on another individual is usually associated with a risk of sustaining injury, and so an attack on a rival is often associated with self-protective behaviour. In fact, as discussed in the next chapter, a tendency to attack is often associated with a simultaneous tendency to flee, and many threat postures are associated with both these tendencies. Furthermore threat postures likely to be followed by an attack on the rival intergrade with submissive ones. Thus attacking, threat, submissive, and fleeing behaviour form a complex, often referred to as "agonistic behaviour" (Scott and Fredericson, 1951). Within this complex, the limitations of aggressive behaviour are hard to specify. Second, many aggressive interactions involve only threat and related behaviour; actual physical injury is rare. Third, it is by no means always easy to specify what a fight is about: this is discussed further below. Finally, much behaviour which results in settling status or precedence does not involve inflicting physical injury (e.g., bird song): such behaviour is usually not described as aggressive.

In spite of such difficulties in definition, it is usually preferable to distinguish intra-specific fighting from inter-specific conflict as it appears in the predator–prey context. In any one species they are elicited by different external stimuli, depend on different internal states, usually involve some different movement patterns, and probably depend on different neural mechanisms (Hutchinson and Renfrew, 1966).

By definition then, aggressive behaviour is normally directed towards another individual. In addition it is normally elicited by stimuli from another individual: these may be visual (e.g., Tinbergen, 1951), auditory (e.g., Thorpe, 1961b), or olfactory (e.g., Mackintosh and Grant, 1966; Ropartz, 1968). Much aggression is thus primarily the consequence of the proximity of another individual: some studies analysing the stimuli involved were reviewed in Chap. 4. But there has been much dispute over the extent of the role played by other factors, internal or external to the animal, in the instigation of aggression. Part of the difficulty here has arisen from tendencies to overgeneralize: it seems certain that aggression can arise in more than one way, and generalizations about *the* cause of aggression are inevitably misleading.

Some workers have emphasized the importance of immediate external factors, in addition to the proximity of another individual, in triggering aggression. Thus aggressive behaviour often occurs in situations which can be described as "frustrating". McDougall (1923), for instance, supposed that aggression is instigated by interference with some other form of activity. This view has been worked out in detail by Dollard et al. (1939; see also Miller, 1959) and has been applied especially to the human case. That the same principle is applicable also to animals, at any rate in the laboratory, is illustrated by an experiment in which pigeons were taught to peck a key under a procedure in which periods of food reinforcement alternated with periods of extinction. At the beginning of the extinction periods the experimental pigeon would attack a nearby pigeon or even a stuffed model. This occurred even in pigeons which had lived for 9 months since independence in cages which prohibited physical contact with other pigeons. The duration of attack increased with the number of preceding reinforcements but was reduced by prior satiation: attack duration decreased with time since the last food reinforcement (Fig. 15.4). The experimenters (Azrin et al., 1966) regard "the transition from food reinforcement to extinction as an aversive event that produced aggression". Rats and squirrel monkeys also show an increase in aggression at the onset of extinction, the extinction-induced increase in the rate of the previously reinforced response and the increase in aggression occurring together (Azrin et al., 1966; Thompson and Bloom, 1966; Davis and Donenfeld, 1967; Hutchinson et al., 1968b). Cole and Schafer (1966) report an apparently similar phenomenon in cats. Aggression induced by morphine withdrawal (Boshka et al., 1966) can perhaps also be regarded as a response to frustration. There has been little work on frustration-induced aggression in other species, but several indices of aggression in paired encounters between domestic fowl are greater when the birds are food-deprived and food is present but not available (because covered by perspex) than in any of these four conditions: not deprived/food available, not deprived/no food, deprived/no food, and deprived/food available (Duncan and Wood-Gush, in press).

There have also been few systematic observations on frustration-induced aggression in the field, but van Lawick-Goodall (1968) records that chimpanzees may behave aggressively when an artificially provided supply of bananas becomes exhausted. She also saw, on a number of occasions, one animal threaten its grooming partner when the latter stopped grooming. The defensive fighting shown by cornered animals of many species should also perhaps be regarded as induced by the frustration of withdrawal.

It has also been shown that fighting between rats can be elicited by an aversive stimulus, such as an electric shock. If two rats are confined together over an electric grid, one or both animals are likely to show attack responses towards the other when the shock is turned on (e.g., Azrin et al., 1964a; Ulrich, 1966; Ulrich et al., 1966). Shock-induced aggression is decreased by castration and influenced by previous social experience (Hutchinson et al., 1965): in these respects it appears to resemble naturally occurring aggressive behaviour. Comparable data are available for squirrel monkeys (Fig. 15.5; Azrin et al., 1964b; Azrin et al., 1965a; Hutchinson et al., 1968a). Furthermore squirrel monkeys can be trained to pull a chain which produces an inanimate object which can be attacked. In the absence of electric shock, little chain-pulling occurs, but its frequency increases when shock is administered (Azrin et al., 1965b). Aggression sometimes appears to be induced by fear in chimpanzees. For instance, van Lawick-Goodall (personal communication) records that free-living chimpanzees were at first somewhat afraid of individuals partially paralyzed in a poliomyelitis epidemic, and then attacked them. Another adult male was exceptionally aggressive while its foot was injured.

Most of the aggression seen under natural conditions appears not to be instigated by some immediately preceding frustrating or painful experience, but to be a direct response to the proximity of another individual—proximity either to the animal itself, as in "individual distance" and "dominance order" aggression, or to an individual in relation to an object, such as a nest box, or to an area, such as a territory. In every case the rival is liable to elicit fleeing and avoidance responses as well as aggression (see above and Chap. 16). Before considering such cases further, we may

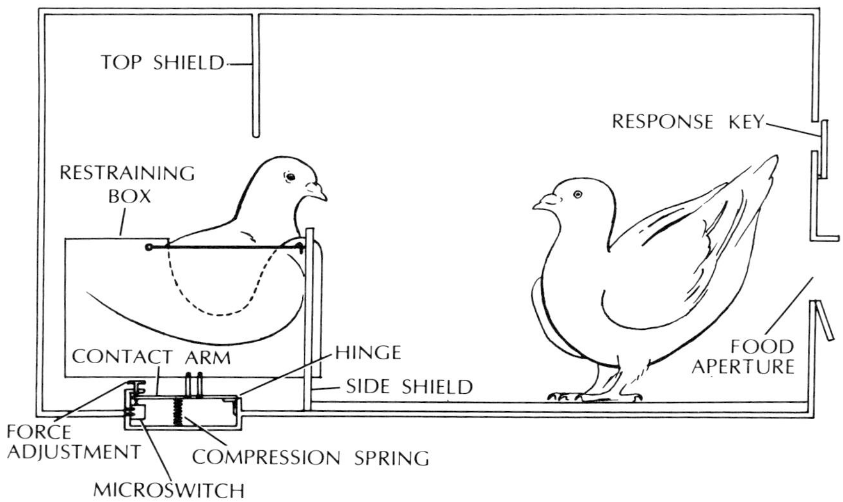

Fig. 15.4 (a) *Schematic diagram of apparatus for measuring extinction-induced aggression in pigeons. The shields prevent the experimental pigeon from getting behind the target pigeon.*

consider their relation to fighting precipitated by an immediately preceding aversive or frustrating event.

Those aversive stimuli which elicit fighting also elicit fleeing and avoidance behaviour (e.g., Ulrich, 1966): the parallel to proximity-induced aggression is clear. The distinction between the latter and frustration-induced aggression is also not so absolute as might appear. The original frustration/aggression hypothesis has been amended by Miller with the suggestion that frustration instigates a number of responses, one of which is aggression. Frustration is thus regarded merely as increasing the likelihood of aggressive behaviour, while the actual appearance of aggression depends on other factors—external stimuli, inhibitory factors, and so on; and as in proximity-induced aggression, the attack is directed towards another individual. Furthermore, the definition of what constitutes "frustrating conditions"

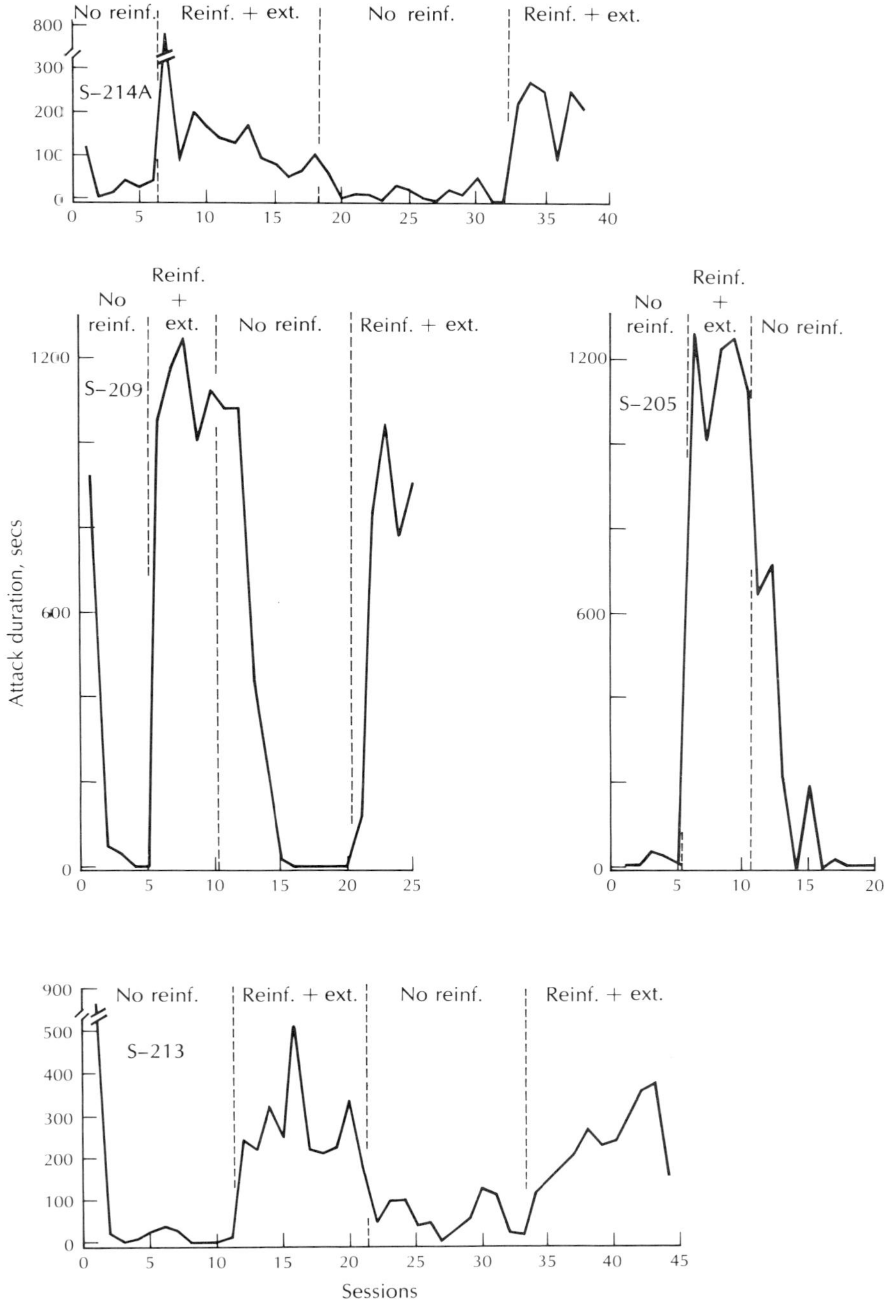

Fig. 15.4 (b) *Sample data for four pigeons. The conditions consist of alternating periods of food reinforcement and extinction within each session, or of no food reinforcement. The attack behaviour decreases with time since the last food reinforcement.* (From Azrin et al., 1966).

can be somewhat flexible (e.g., Berkowitz, 1963), and it would be easy to coin a definition such that a territory-owning great tit was "frustrated" by an intruder. The inclusion of incompatible response tendencies as a source of frustration (Brown and Farber, 1951) widens its meaning still further. Thus the differences between aggression immediately preceded by an aversive or frustrating event and that apparently elicited solely by the proximity of another individual may not be as great as might appear.

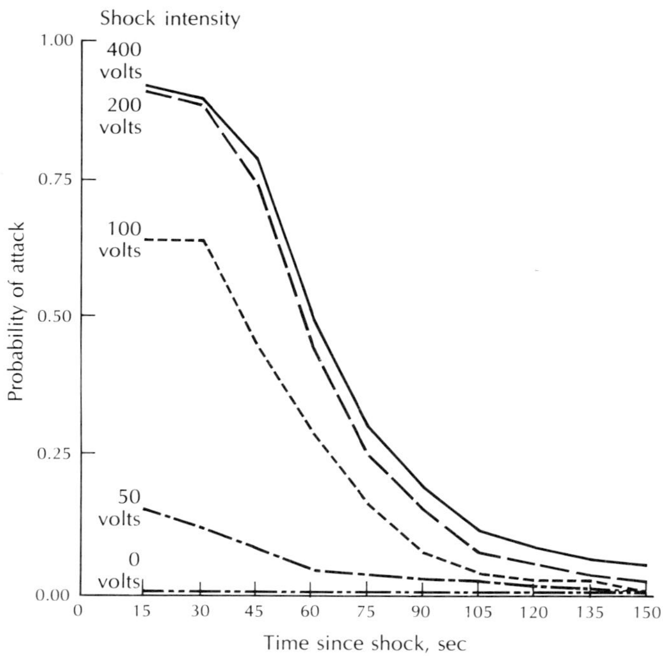

Fig. 15.5 The probability of an attack upon an inanimate object (ball) by squirrel monkeys as a function of the intensity of shock delivered through tail electrodes. (From Azrin et al., 1964*a*.)

Turning now to the latter, we may consider the extent to which aggressive behaviour is influenced by the internal state of the individual concerned. That there is some effect is obvious from the fighting shown by passerine birds (e.g., chaffinch, *Fringilla coelebs;* yellowhammer, *Emberiza citrinella;* great tit, *Parus major*). These show some intra-specific aggression in the winter flock: this often occurs near food and involves attacks on other individuals a few inches or feet away. Such winter fighting differs at least in degree from that seen in the reproductive season, when territories are being established. Here the fighting is not related to food: rather any conspecific male seen within an area perhaps an acre or two in extent elicits aggression. The motor patterns used in the two contexts are similar, except that certain threat postures may occur in territorial fighting more often than in winter fighting.

In most species, fighting over territories occurs only in the spring, when the gonads are active and sex hormones are being released. Its nature and intensity may vary with the stage of the breeding cycle (e.g., Vowles and Harwood, 1966, for data on doves). However, territorial fighting does not depend solely on hormone level. In cold weather, or early in the season, the bird may fluctuate from territorial to flocking behaviour over a period of a few minutes: at one moment the male may be feeding quietly on his territory in the company of a dozen other individuals, tolerating them at a distance of a foot or two, but at the next he will fly up to a song post and attack any bird within a hundred yards. Correlated hormonal changes are unlikely, and so we must presume that there are temporary central states independent of the more enduring hormonal ones. In other species such states are often initiated by, but outlast, a particular incident of aggressive behaviour (e.g., Lagerspetz, 1964, for data on mice). Comparable states can be elicited by central stimulation in both birds (von Holst and von St. Paul, 1963; Brown, 1969) and mammals (e.g., Delgado, 1966, 1967; see also pages 270–275).

Although the difference between winter and territorial fighting in passerine birds is thus clearly correlated with a difference in internal state, the relation between the internal state and the intensity of fighting behaviour is not obvious. Field observations on a number of avian species show that fighting in the flock is more common when the weather is cold and food is scarce, unless conditions are very severe (e.g., Hinde, 1952; Stokes, 1962b). This, however, does not necessarily mean that the factors which cause an increase in feeding behaviour (e.g., a period of deprivation) also influence the threshold for fighting, as has been implied (Hinde, 1952). In a study of captive chaffinches, Marler (1956c) assessed the "50 percent distance", that is, the distance between two individuals at which there was an even chance of their tolerating each other. This distance was 7 cm for females and about 25 cm for males. The incidence of aggression was then observed when the males were feeding simultaneously from perches fixed 20 cm apart and subjected to periods of starvation varying from 45 to 180 min. The period of food deprivation did not affect the proportion of encounters leading to fights. It did, however, increase the readiness of subordinate birds to approach dominant ones at the food, and thus also the frequency with which the tolerance distance was infringed. Andrew (1957b) obtained a similar result in studies of captive yellowhammers, and suggested that the increase in the number of aggressive encounters which occurred during deprivation was due to the increase in activity and, thus, in the frequency of encounters at local sources of food (see also Blurton-Jones, 1968).

In fighting over breeding territories in these species, it is not mere proximity but the presence of another bird within a particular area which elicits aggression. The territory owner patrols his territory, flying from tree to tree and being particularly responsive to stimuli indicative of intruders. He ignores a nearby rival outside the territory but flies off to attack a more distant one who is within its boundaries. A stranger on the boundary may be attacked when a known neighbour is tolerated, but the neighbour may be attacked if he appears on a boundary remote from his own territory (Falls, 1969). Only in the early stages of territory establishment is the aggressiveness not tied to a particular locality. Often the probability or intensity of attacks on rivals decreases with distance from an object of particular significance, such as a nest site (von Haartman, 1957), though in the colonial nesting black-headed gull the proximity of the intruder to the territory owner is more important than his distance from the centre of the territory (Patterson, 1965). Comparable observations have been made on territorial fishes (van Iersel, 1958; van den Assem, 1967).

On the basis of their observations, Marler (e.g., 1957a) and Andrew, *loc. cit.*, emphasized the role of the external stimulus situation in eliciting aggression, regarding the internal state as of minor importance. Insofar as the internal state affected aggressiveness, it did so not by affecting the probability of aggressive behaviour when an encounter occurred, but by affecting the probability or duration of encounters. However, even if the state of hunger makes no difference to the threshold for aggression, the fact remains that, while behaving gregariously, other birds are attacked only if they come within a few inches, but an individual behaving territorially attacks other birds whenever they are seen within a certain area. The internal states which lead to flocking behaviour, on the one hand, and territorial behaviour, on the other, must have some influence. This may lie in giving significance to an external object with reference to which the fighting occurs. In the flocking situation, mere proximity is a cause of aggression, but in the territorial situation it is not the spatial relationships between the combatants which are important but rather those between the rival and some external object—song post, nest site, etc. On this view the internal state can be regarded as determining a situation which the fighting serves to achieve—"no birds within a foot of me" or "no birds near my trees". It influences the relevance of stimuli rather than the intensity of the behaviour (Hinde, 1956b).

The generality of these data, obtained for the most part from passerine birds, remains to be assessed. In other species there is some evidence that food deprivation may have a more direct effect on aggressive behaviour. In pigeons aggressiveness can be influenced by food reinforcement, the duration and temporal patterning of the attacks varying with the schedule (Azrin and Hutchinson, 1967). Presumably the incidence of such aggressive behaviour would vary with food deprivation. In some fishes (e.g., Albrecht, 1966) and in hermit crabs (*Calcinus tibicen,* Hazlett, 1966) starvation augments fighting. In the former case the reverse relationship also holds, stimuli for fighting also promoting feeding behaviour. However, Marrone (1965) could find no evidence that hunger influenced the (territorial) fighting of the Siamese fighting fish, and Lester and Cheses (1968) found that 24-hr food deprivation reduced rats' chances of winning paired encounters. Of course general conclusions about the effect of hunger on fighting, valid across a wide range of species, are not to be expected. Even within one species, careful analysis should precede generalization. For instance, in young junglefowl hunger increases the incidence of aggressive pecking during fighting but does not affect the frequency of fighting itself (Kruijt, 1964).

There is, however, no doubt that social stimuli may affect aggressiveness. The readiness of a cichlid fish to attack other (younger) individuals, measured by the number of attacks per minute, is increased by presenting another male or a dummy behind glass (Heiligenberg, 1965c). The aggressive behaviour of mice may be reduced by the presence of a female (Levine et al., 1965a), but that of male rats appears to be increased (Barnett et al., 1968): no doubt the difference is in part one of testing conditions. Male zebra finches (*Poephila guttata*) show more aggressive behaviour if another individual is present in an adjacent cage, especially if that individual is familiar (Caryl, 1969).

We may now turn briefly to the more detailed course of aggressive sequences. When birds attack a dummy, their aggressive behaviour waxes and wanes over a period of a few minutes. There are thus decremental (and probably incremental) consequences of responding. Since the response to a dummy wanes after a while, these decremental effects are presumably cumulative. With a real rival, however, aggressive encounters may go on for a long time: presumably incremental effects override the decremental ones. Under these circumstances, aggression may be brought to an end when the rival adopts a submissive posture (see page 681) or when he disappears. In the latter case the termination of aggressive behaviour depends not on the achievement of a new consummatory situation, as with feeding, sexual, or nest-building behaviour, but on the removal of the eliciting situation. In this respect it resembles fleeing (see below).

A further issue concerns the extent to which fighting can be "spontaneous". Does its occurrence depend on the chance proximity of another animal when the individual concerned is in the appropriate state, or do animals go out and actively seek for fights? Although some writers (e.g., Marler, 1957a; Scott, 1962) treat aggression as depending solely on external factors, others (e.g., Lorenz, 1966) regard it as "spontaneous" and inevitably finding expression. This is clearly an important issue, but hitherto discussion seems to have involved confusion among several questions not necessarily related to each other:

1. *DO CHANGES IN AGGRESSIVE BEHAVIOUR DEPEND SOLELY ON EXTERNAL FACTORS?*

The answer here is clearly in the negative: we have already mentioned cases in which aggressive behaviour varies with the hormonal condition or stage of the breeding cycles and with other shorter-term fluctuations in internal state.

2. *DO ANIMALS GO OUT AND ACTIVELY SEEK FOR FIGHTS?*

This question could be answered in the affirmative if animals showed appetitive behaviour in which they visited places where encounters with other individuals were likely, and showed a lowered threshold for responding aggressively to them. This seems to be the case with territorial birds, who patrol their territories apparently searching for rivals and visit the sites of previous encounters. Marler (e.g., 1957a; Marler and Hamilton, 1966), however, regards such patrolling as having a more general function than aggression, and the seeking for fights as merely "temporary, stemming from relatively recent aggressive experience". Although he quotes a few other examples of animals in unstable dominance hierarchies apparently seeking out fights, Marler regards the evidence for appetitive behaviour for aggression as inconclusive.

However, it is clear that the opportunity to behave aggressively can act as a reinforcer. Thompson (1963, 1964, 1969a; Goldstein, 1967; Hogan, 1967) has shown that domestic cocks and Siamese fighting fish (*Betta splendens*) will learn an instrumental response which results in the appearance of the stimuli usually provided by a rival male (Fig. 15.6). The greater the effectiveness of the stimuli in eliciting fighting responses, the greater the reinforcing effect: whether it is due to the stimulus situation itself or to the behaviour elicited thereby cannot be determined. Thompson (1969a) also found that not only stimuli for aggression but also a stimulus light, which had previously been paired with the presentation of a model, were reinforcing. Such effects are not limited to lower vertebrates: Beach (personal communication) has shown that male C-57 mice will learn an instrumental response if this brings them into a situation where they can attack another mouse. The observation (see above) that patrolling territory owners visit the sites of recent encounters could clearly be accounted for in such terms.

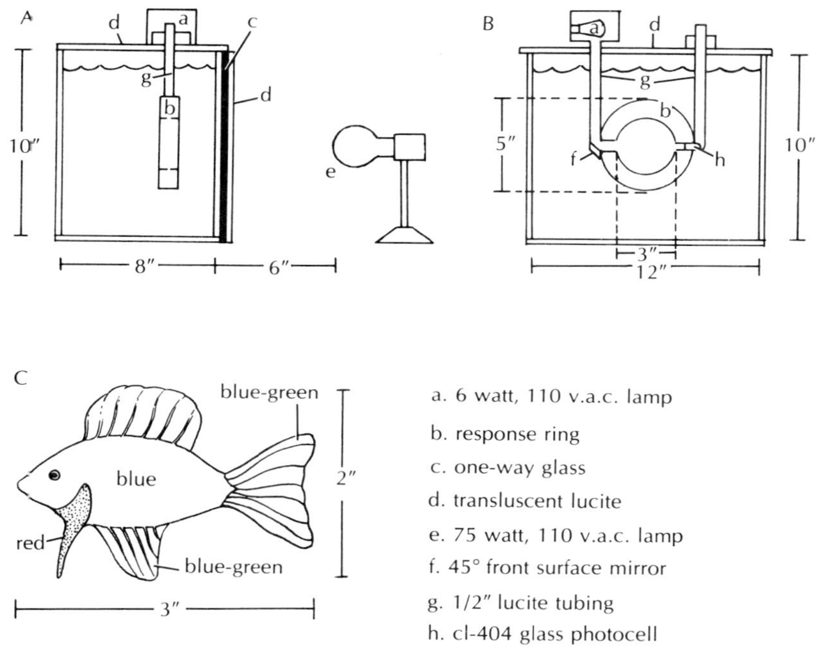

Fig. 15.6 Reinforcing effects of visual stimuli for aggression in Betta splendens. (From Thompson, 1963.)

A, Side view of tank, showing response ring, one-way mirror and lamp. The mirror was presented by switching off the lamp. *B*, front view of tank facing the mirror. Responses were recorded by the beam of light reflected from the mirror *f* onto the photocell *h*. *C*, Model of male.

Thompson and Sturm (1965a; Thompson, 1969a) also found that stimuli previously paired with the elicitation of the threat behaviour of the fighting fish came themselves to elicit the display [1]: the lowered threshold for aggression shown by patrolling territory owners at places where fights had occurred previously could also arise in this way.

[1] It was incidentally of interest that the four display components conditioned at somewhat different rates, and the relative rates were replicable across subjects. This suggests that each component is unitary and individually manipulable.

Such studies show clearly that a wide range of species can show appetitive behaviour for aggression in the laboratory, and strongly support the plentiful and highly suggestive field data indicating that the same occurs in nature.

3. *IS AGGRESSIVE BEHAVIOUR SPONTANEOUS?*

Spontaneity normally refers to a change in output of a system without a corresponding change in input (Chap. 14). Insofar as, for example, two

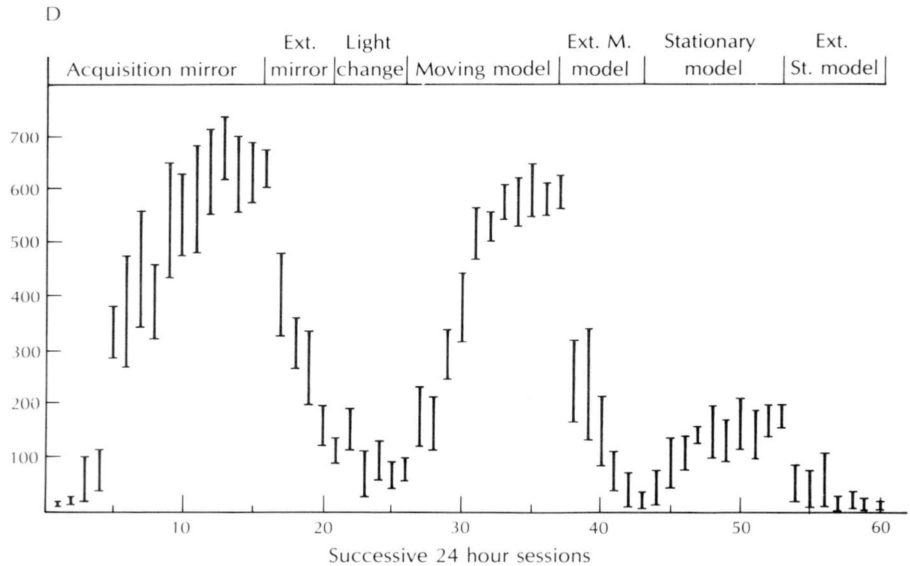

Fig. 15.6

D, Range of responses per 24-hr session for four fish. The response to the mirror was acquired and then extinguished. There was then a period controlling for the light change associated with mirror presentation. Response to a moving model was then acquired and extinguished, and then that to a stationary model.

fish left together may fight intermittently and not continuously, the fighting could be called spontaneous: each is continuously exposed to stimulation from the other, but responds only some of the time. Since similar fluctuations occur in the aggressive behaviour of one individual towards a dummy, they cannot be due to changes in the external situation. Such fluctuations are presumably due to short-term waning in responsiveness consequent upon responding to the stimulus, and to subsequent "spontaneous" recovery. Defined in this way, spontaneity of aggression clearly occurs, but the issue is not a particularly interesting one.

4. DOES THE TENDENCY TO SHOW AGGRESSIVE BEHAVIOUR INCREASE WITH TIME SINCE THE LAST ENCOUNTER?

It is important here to specify the time courses of the processes in which we are interested. On the one hand, there are a number of studies showing that animals reared in isolation may be exceptionally aggressive (see Chap. 22): in such cases, however, the aggressiveness seems to be at least a semipermanent characteristic of the individual, and the problem is a developmental one, scarcely relevant to the present issue. On the other hand, it is a commonplace that temporary isolation, for a period of minutes or hours, may lead to a lowering of the threshold for aggression: for instance, Siamese fighting fish, living with little aggression in a community tank, are likely to threaten each other if removed and placed in isolation for a while before testing (Simpson, personal communication). There seems, however, to have been no detailed study of the time course of such effects, and there seems to be no evidence that the threshold for aggression continues to fall up to the point at which it occurs spontaneously. In another study Siamese fighting fish, which had become habituated to their own mirror image by 10 days' exposure to it, were found to recover slowly and partially over a period of 2 days (Clayton and Hinde, 1968), but the length of the initial exposure period makes generalizations from this experiment of doubtful value in the present context. On general grounds an aggressive encounter is likely to have multiple consequences on subsequent aggressive behaviour, positive and negative, and decaying with varied time courses (Chap. 13). Which predominates will depend on the length, nature, and outcome of the initial encounter: if the immediate consequence is an increase in aggressiveness, this may decay with time; but if it is a decrease, the aggressiveness may recover.

5. DOES AGGRESSIVE BEHAVIOUR INEVITABLY FIND EXPRESSION?

In conformity with his hydraulic model of motivation, Lorenz (1966) has emphasized the role of an internal drive for aggression which inevitably finds expression. The evidence he brings forward, however, bears on the questions posed above rather than on this one. For instance, he cites a number of cases of animals responding aggressively towards a suboptimal stimulus object (e.g., a member of another species) in the absence of the natural object (a member of the same species): this, however, is relevant to questions (1) and (3) above, but a positive answer is not necessarily relevant to this one. Lorenz also gives evidence pertinent to question (4), but his data are in fact all drawn from other types of behaviour: although the thresholds for feeding or sexual behaviour do usually fall with increasing deprivation, aggressive behaviour may be affected quite differently. In any case, it must be emphasized that a positive answer to questions (1) to (4) implies a positive answer to question (5) only if one assumes an aggressive drive which accumulates with time and energizes fighting behaviour. The shortcomings of such an energy model of motivation have been discussed earlier (Chap. 8): if we do not make the assumptions it entails, then there appears to be no evidence that aggressive behaviour inevitably finds expression. Even if it is shown that the probability of aggressive behaviour increases with time, it must be shown that this increase continues until normally quite inadequate stimulus objects are attacked before we can conclude that aggressive behaviour inevitably finds expression. That suboptimal stimulus objects are sometimes attacked in the absence of the normally adequate ones is not in itself sufficient evidence: this could be a consequence of spontaneous *fluctuations* in the tendency for aggression which would not be continued until the threshold was nearly zero.

This issue is important if the control of aggressive behaviour is an aim. If aggression is the product of a drive which inevitably finds outlet, then the only course is to see that that outlet is a harmless one. But if aggression is not like this, it is preferable to focus on ontogeny, searching for environmental conditions which minimize the subsequent tendency to aggression.

It is apparent that generalizations about the bases of aggression which are more than truisms are hard to make. It is also clear, however, that the control of fighting behaviour differs in a number of ways from that of, say, feeding behaviour, and generalizations about control made on the basis of the one can be applied to the other only with great reservations.

Finally, the question of dominance relations must be mentioned. In many social groups, the outcome of aggressive interactions is predictable. Some individuals are consistently dominated by others, while others consistently dominate. The result is a "dominance hierarchy", which may involve a linear sequence from the despot to the most subordinate animal or may include triangular relationships such that A dominates B, B dominates C, but C dominates A. Other variations on this theme are also possible —for instance there may be one despot, with fighting suppressed amongst all other individuals. First studied extensively in chickens (Schjelderup-Ebbe, 1922), similar organisations have since been found in a very wide range of species (e.g., Allee, 1938). It is important to recognize that dominance in one context is not necessarily related to dominance in another and that there is no necessary relationship between dominance and leadership. These are empirical matters, to be established in each particular instance (e.g., Varley and Symmes, 1966; van Lawick-Goodall, 1968). Furthermore, while the name "dominance hierarchy" implies an organisation imposed from above, this is by no means necessarily the case. In her study of captive baboons, Rowell (1966) found the hierarchy to be maintained by the behaviour of the low-ranking rather than of the high-ranking animals.

She stresses that subordination is not mere lack of dominance: "the subordinate animal . . . cautiously observes and maintains a hierarchy, while a dominant one could almost be defined as one which does not 'think before it acts' in social situations". Although dominance and subordinance of course depend on interactions between individuals, Rowell rightly corrects the usual emphasis on the initiative of the dominant animal.

15.6 FEAR RESPONSES

The label "fear responses" is applied here to a diverse group of behaviour patterns, including both those leading to withdrawal from or avoidance of the stimulus (e.g., locomotion away from it and self-protective responses) and immobility responses, such as freezing and crouching. The case for classifying withdrawal and immobility responses together rests largely on the similarities in the stimulus situations which elicit them. Recently, however, Hogan (1965, 1966) has suggested that, at any rate in young animals, they should be regarded as separate and mutually inhibitory systems of behaviour. Pointing out that many exceptions can be found to the generally accepted view that the stimuli eliciting the two groups of responses differ only in degree, he suggests that fear (i.e., immobility) and withdrawal are in fact elicited by different aspects of external stimulation. Withdrawal occurs to stimuli at high intensity, or to particular stimulus situations, the response to which may or may not depend on learning. Fear, on the other hand, is elicited by stimuli that are strange, novel, or surprising, and is associated with fixation of the source. Fear, in Hogan's view, is associated with the orientation response, progressive inhibition of movement, and sleep: in the chicks which he studied it is also associated with shrill calling (see also Glickman and Hartz, 1964; Miller and Murray, 1966, and refs. cited).

Hogan's views are supported by experimental data on the responses of chicks to mealworms, and certainly help to bring order into a previously neglected group of responses. As he points out, however, immobility and withdrawal responses are probably independent only in young animals: he suggests, though without entering into details, that with experience some of the control of withdrawal is transferred to factors primarily controlling fear. In what follows, the term "withdrawal" will be used in Hogan's sense, behaviour involving inhibition of activity (Hogan's "fear") will be referred to as immobility responses, and the term "fear" will be retained in its original loose sense for the whole group: this would seem to involve least conflict with existing usage. It will be apparent that the causal bases of fear responses are related to those of the orientation response (Chap. 6), aggression (see above), and exploration (see below). We shall later discuss its role in a number of ambivalent behaviour patterns, such as the displays used in threat and courtship (Chaps. 16 and 17).

As implied above, the situations which elicit fear responses are varied. For most species some stimuli have special effectiveness in eliciting avoidance behaviour and may be effective the first time they are encountered: for instance, stimuli characteristic of predators prevalent in the area, or the alarm calls of conspecific individuals. In addition, most species respond to generalized stimuli which could precede danger, such as a circular shadow suddenly enlarging (Schiff, 1965; Hayes and Saiff, 1967), or indeed practically any stimulus at an unusually high intensity. In addition, fear responses may be elicited by strange objects or novel situations (Hebb, 1946; Thorpe, 1963a). Hogan's view that these last elicit immobility responses and that the earlier ones elicit withdrawal certainly has many exceptions in the behaviour of adult animals: the "seep" alarm call of passerine birds elicits an immobility response from conspecific individuals, and strange objects may produce withdrawal (or approach; see below).

Thus fear responses can be produced either because the stimuli impinging on the organism differ from the familiar or because they comprise a particular pattern which specifically elicits fear: it can be elicited by either discrepancy or concordance. A similar point is illustrated by Andrew's (1964a) study of chick calls. This author interprets the causal basis of the calls in terms of "stimulus contrast". Stimuli are said to have "contrast" if the animal pays marked attention to them. There seems to be a considerable danger of circularity in such a definition, but the interesting point here is that contrast, according to Andrew, may occur either because the stimulus differs markedly from the familiar (i.e., lack of correspondence with a "neuronal model"), or because it serves as a conditioned stimulus signalling, for instance, the presence of food (i.e., correspondence with a neuronal model). Andrew shows that the same response (the twitter) can be evoked in either of these ways, that is, by either correspondence or lack of correspondence. Another response (the peep) is produced by even more intense stimulus contrast (see also Andrew, 1969).

Within any one species, which of the various responses indicated above is elicited, and with what intensity, depends on the nature of the external stimulus and on various aspects of the internal state, including experiential factors, hormonal state, and so on. Social factors also are important. Fear responses of young monkeys are reduced by the presence of the mother or mother-surrogate (Harlow and Zimmermann, 1959), and those of adult rats by the presence of social companions, especially in group-reared individuals (Morrison and Hill, 1967). When extraneural factors are apparently constant, the animal may pass through phases when it is more or less likely to flee. This implies the existence of central nervous states with some degree of persistence which influence responsiveness, a finding consistent with the fact that the behaviour can readily be elicited by central electrical stimulation at a wide variety of loci (pages 270–275).

Once avoidance behaviour has been elicited, it may persist until the stimuli are no longer present: in this respect it resembles aggressive behaviour. If persistent avoidance fails to achieve this result, however, habituation occurs; the previously fear-inducing object loses these properties, and the behaviour wanes, to be followed by exploration. Several studies show that, when fear responses are elicited by the sudden appearance of a new stimulus object, the fleeing response wanes much more rapidly with successive presentations than does the orientation response which is also evoked (see page 300). Melzack (1961), on the basis of experiments in which mallard ducklings were shown models of flying predators, describes this by saying that there is a progressive organisation of behaviour in which the "emotional disruption" involved in fleeing is replaced by non-emotional orienting responses (see also Russell, 1967a, b). However, a difference in rate of habituation between withdrawal responses and the orientation (immobility) responses is also in harmony with Hogan's view that these involve two different systems (see above).

15.7 EXPLORATION

In all higher animals a sudden change in stimulation will usually elicit a movement of turning towards the source, a response described already in Chap. 6 as the "orientation response". They will also approach and examine strange objects with whatever sensory equipment is available to them. In a strange environment, they may move around and examine all of their surroundings—in other words, explore. Even within one species the types of behaviour which come within the broad category of exploration or investigation are diverse, and it is difficult to give a more precise definition than that they are such as to familiarize the animal with its environment or with a source of stimulation. Although certain forms of play also result in increased familiarity with the object, exploration and play are not to be equated. With a strange object exploration must precede play, and it decreases as familiarity increases: in children, as exploration changes to play the emphasis changes from "what does this *object* do?" to "what can *I* do with this object?" (Hutt, 1966; see below).

The extent to which it is profitable to erect subcategories of exploratory behaviour is an open issue. It seems at least worthwhile to discriminate between the orientation response, which is associated with immobility, and active exploration, which involves movement with respect to the area or object under investigation. The two are of course incompatible and mutually inhibitory. Both monkeys and children, as well as lower forms, when released into a strange area or confronted with a strange object, alternate "passive staring" with "active exploration" (Berlyne, 1960; Hutt and Hutt, 1965). It is also sometimes convenient to distinguish between "inspective" and "inquisitive" exploratory behaviour according to whether the objects or characteristics concerned are yet present (Berlyne, 1960). Beyond this point, the most profitable type of subdivision might be in terms of the sensory modalities involved.

Since the extent to which exploratory behaviour is evoked varies with the animal's previous experience of the situation, it would seem to be a function of the discrepancy between the perceived situation and the familiar world as experienced in the past. One experiment may be cited to illustrate this. Dember (1956) allowed rats to explore the stem of a *T* maze: the arms were painted respectively black and white, but entry to them was prevented by transparent glass screens. The rats were removed and replaced shortly afterwards with the glass removed and both arms the same colour: they tended to enter the one that had changed colour. As might be expected, such characteristics of a stimulus situation as complexity and intensity, in addition to novelty, may increase its effectiveness in eliciting exploratory behaviour (Berlyne, 1960).

It will be apparent that fear behaviour and exploratory behaviour are overlapping categories. Some types of stimulus change can evoke either type of behaviour, and both types include orientation responses: indeed, any particular orientation response is likely to be classed in one category or the other according to the behaviour which accompanies it. There is even some evidence to suggest that preceding electric shock enhances the tendency of a rat to explore stimuli previously associated with that shock rather than to explore elsewhere (Hudson, 1950; Halliday, 1966, 1968). The latter author (see also Lester, 1968) has even gone so far as to suggest that rats explore situations because those situations evoke low levels of "fear" (used here to refer to a motivational state, not to a category of behaviour). His principal evidence comes from experiments showing that exploration of an elevated maze continues far longer than that of an enclosed maze, and is correlated with longer-lasting fear as indicated by defaecation scores. Assuming that the enclosed maze and elevated maze become equally familiar after a given time spent exploring them, Halliday argues that exploration is correlated more closely with fear than with novelty. However, this is questionable on at least two grounds—the relation between fear and defaecation is at least complex (see pages 535–536), and there may well be more to learn about an elevated maze than about an enclosed one. There is also evidence suggesting that fear and exploration are independent: factor analysis of the scores of rats in a battery of tests yielded an "emotional reactivity" factor, with high loadings from defaecation and other behaviour associated with fear, and a "field exploration" factor orthogonal to the first (Whimbey and Denenberg, 1967). Furthermore a number of studies show that fear may inhibit exploratory behaviour: rats which are shocked explore less than control rats which receive no shock (e.g., Montgomery and Monkman, 1955; Baron, 1964).

Some of the difficulties which have arisen in discussions of the relations between "fear" and "exploration" have come from using these words to refer to hypothetical unitary drives, for this carries the implication that they can only be independent, mutually inhibitory or mutually facilitatory. As used here, they refer to categories of behaviour. As we have seen, the categories partially overlap. Some responses (such as orientation) are common to both, while others are mutually compatible (e.g., visual inspection and autonomic indices of fear). Some responses from the two categories are, however, incompatible. The behaviour which appears then depends in part on the magnitude, intensity, or strangeness of the eliciting stimulus. A small change in the environment elicits investigation, a major one fear. Chaffinches and great tits keep their distance from owls, real or stuffed, but approach and investigate objects with only a few owl-like characteristics (Hinde, 1952, 1954a; Marler, 1956b). Movement of a strange object may precipitate withdrawal, though it may enhance the probability that a moderately familiar one will be investigated (Menzel, 1964a). With intermediate degrees of strangeness, the animal may vacillate, alternately approaching and avoiding the object (see Hebb, 1946; Melzack, 1952; Hebb and Thompson, 1954; Hinde, 1954a; Montgomery, 1955; Menzel, 1964b). The relationship between the familiarity of the environment and that of the object is also important here: a strange object encountered in a strange environment is investigated less, partly because other aspects of the environment must also be explored, partly because it is in a sense less strange since there is no background of the familiar, and partly because the animal is more likely to freeze or show fear responses which conflict with exploration (e.g., Russell, 1967b).

The probability that a given situation will elicit exploration rather than fear responses depends also on various aspects of the internal state. For instance, when a young monkey, reared on a cloth-covered artificial mother, is put into a room containing strange objects, it may freeze and cover its eyes, or run about screaming. If the artificial mother is present, however, the infant clings to it and its fear is allayed. In these circumstances it then soon starts to make exploratory forays about the room (Harlow and Zimmermann, 1959). Apparently the presence of the mother influences the internal state in a way which sways the relative prepotence of avoidance behaviour and exploration. Comparable data are available for human children (e.g., Hutt et al., 1965; Ainsworth and Wittig, 1968).

The response of passerine birds to owls, mentioned above, provides another example of the influence of the internal state. At a few weeks of age chaffinches and great tits have a high threshold for fear responses. At this age a stuffed owl evokes approach and investigatory behaviour, with few signs of fear. A few weeks later the birds keep several yards away from such a dummy, the tendencies to approach and avoid being more or less in balance at that distance (Hinde, 1954a). The internal state apparently changes with age, though the manner in which this change depends upon experience is not known. Strain and species differences are also important here: wild rats are less likely to examine and more likely to avoid strange objects placed in their cage than are tame rats, a difference correlated with the larger adrenal glands of the wild forms (Barnett, 1958, 1963).

Even if it starts to examine a novel object, an animal may vacillate between investigation and other activities such as feeding and grooming. The precise effect of hunger on exploratory behaviour varies with the degree of hunger and the conditions of testing. Hunger usually decreases exploratory behaviour (Chapman and Levy, 1957), but hungry rats are more likely to leave a familiar environment and explore an adjacent strange one than are satiated animals (Fehrer, 1956; see also Bolles and de Lorge, 1962; and pages 257–260). It may be suggested, however, that in the latter case the animals would show enhanced responsiveness to stimuli likely to be indicative of food rather than to novel objects: the behaviour is therefore better described as appetitive to feeding rather than exploratory.

Thus, although at first sight the incidence of exploratory behaviour seems to depend primarily on stimulus factors, the internal state of the animal is important in several ways. First, the external situation's effectiveness depends on its relation to the familiar as established by experience. Second, the internal state influences whether a given degree of discrepancy elicits fear or exploration, and the intensity of either of them. Third, other types of motivation may conflict with exploratory behaviour (see also Walden, 1968). This importance of the internal state is consistent with the finding that an exploratory or investigatory mood can be elicited by brain stimulation and may persist for some time after the stimulus ceases (von Holst and von St. Paul, 1963).

Whether or not the tendency to show exploratory behaviour increases with preceding sensory deprivation is not clear. Charlesworth and Thompson (1957) kept rats in environments impoverished to varying degrees before giving them opportunity to explore a novel chamber, but found no large differences in the latency with which they entered the chamber or the time they spent there. On the other hand, Hill (1958) found that rats showed a period of increased motor output after confinement (see also Woods, 1962), Butler (1957b) found that deprivation increased the tendency for rhesus macaques to show visual exploration, and Levison et al. (1968) found that young rhesus monkeys reared in a chair in a complex visual environment worked harder to see slides the longer they had been inside a translucent hood. However, Haude and Ray (1967), using a different technique, found no such effect. Clearly, even when an effect of preceding deprivation on exploratory behaviour is found, a number of alternative explanations may be possible, and it must not be assumed that the tendency to explore increases with time for the same reasons as does, say, the tendency to eat.

Exploratory behaviour may be extremely persistent. A wild rat encountering a new object in familiar territory will spend minutes and even hours investigating it before resuming its normal maintenance activities (Barnett, 1963). Of greater interest is the accumulating evidence that rats will learn to overcome obstacles on pathways to novel situations and show, in a familiar situation, persistent appetitive behaviour which will lead to an opportunity to explore. Thus rats will cross an electrified grid in order to reach a maze containing a variety of objects (Nissen, 1930), and behaviour which leads to an opportunity to explore is likely to be repeated. Rats will run repeatedly down an alley which has an unfamiliar goal box at the end (Mote and Finger, 1942), and run faster when the stimuli in the goal box are changed repeatedly than if they are not, provided that the rats are satiated (Chapman and Levy, 1957; see also Schneider and Gross, 1965). Similarly, rats will learn to take that turn of a Y maze which leads to a large open box containing rectangular blocks or a Dashiell maze (Montgomery, 1954; Montgomery and Segall, 1955). Krechevsky (1937) ran rats in a maze where they had the choice of two paths, both of which led to food. The sequence of turns in one of these paths was varied from trial to trial, while the other remained unchanged. The animals took the former more often than the latter. In such experiments the complexity of the stimulus objects cannot easily be separated from their novelty (Berlyne and Slater, 1957), but the data suggest that it is primarily novelty per se which augments exploration (Schneider and Gross, 1965). Monkeys confined in a dimly lit box, with a window giving a view of the outside world, will spend a high proportion of their time looking through it and will also learn to perform a simple task which opens the window (Butler, 1953; Butler and Harlow, 1954). Not surprisingly, in view of the above discussion, monkeys work better if the result is an opportunity to see figures in motion (Butler and Woolpy, 1963), but will not learn the task if it results in fear-provoking stimuli—the sight of a large dog or fear calls from a monkey colony (Butler, 1957a).

Such experiments imply that exploratory behaviour is not merely elicited by new stimuli, continuing until "satiation" (Glanzer, 1953a). Some have argued that exploratory behaviour is peculiar in that the animal seeks to increase the stimulation impinging on it. However, we have already seen that, although some types of behaviour (e.g., aggression) are usually brought to an end merely by the elimination of the eliciting factors, others (e.g., feeding, sexual behaviour, nest-building) involve seeking for consummatory stimuli, and it is these new stimuli which bring them to an end. If we wished to link exploratory behaviour to these we could point to the fact that it involves the elimination of the discrepancy between an internal representation of the familiar situation and the central consequences of perceiving the novel one, just as nest-building involves elimination of the discrepancy between the stimuli from the incomplete nest and those from a completed one. However, a basic difference lies in the fact that elimination of the discrepancy is a consequence not of a change in the stimuli impinging on the animal but of a change in the "neuronal model" so that it conforms to them. The novel stimuli are assimilated so that they cease to be novel. We have already discussed the usefulness of the concept of a neuronal model in Chap. 6, and shall return to the question of exploratory behaviour in Chap. 24.

15.8 PLAY

Play is yet another term which has given rise to controversy through difficulties of definition. The kitten playing with a piece of wool, or the dog chasing a ball, are common enough domestic examples which find parallels in the majority of mammals and many lower species. But just what it is that such cases have in common to justify the use of a common term is hard to specify. The difficulty probably arises in part from lack of thorough analyses: in very few cases, so far, has the so-called play behaviour of a species been thoroughly described and analysed, and its characteristics compared with those of other behaviour of the same species. When more data are available, the category of play may well turn out to be even more heterogeneous than at present appears. In the meanwhile, however, a wide variety of species show behaviour involving motor patterns clearly adapted for use in other contents (e.g., aggression, locomotion) but nevertheless giving the human observer the impression of a lack of "seriousness". It must be emphasized that such impressions are inevitably subjective: that young puppies do not harm each other when they fight could be partly due to lack of equipment, as well as to lack of seriousness. Furthermore we do not always label it as play when we see an animal performing aggressive movements without inflicting pain: an adult male rhesus monkey often "bites" females which are squabbling without harming them—behaviour which aids social stability within the group. Nevertheless the category of play cannot be avoided, and it is worthwhile considering some of its characteristics (see especially Lorenz, 1956; Marler and Hamilton, 1966; Loizos, 1966, 1967):

1. *INCOMPLETE SEQUENCES*

The motor patterns used in play are usually ones which occur frequently in other functional contexts in the species concerned—for instance hunting and killing prey in kittens (Leyhausen, 1956; Egar, in prep.); fighting, sex, and manipulating activities in rhesus monkeys (Harlow and Harlow, 1965); galloping in young horses (Tyler, 1969). Nevertheless the functional sequences of behaviour are often incomplete—the colt gallops, stops, and gallops back; the rhesus infant mounts without intromission. In polecats (*Putorius putorius*) four of the agonistic patterns are absent from aggressive play—two extreme forms of attack ("sustained neck-biting" and "sideways attack") and two extreme fear patterns ("defensive threat" and "screaming") (Poole, 1966).

2. *ELABORATION OF NEW MOTOR PATTERNS*

Occasionally animals will develop new patterns which are specific to the play situation and seem to be functionless outside it. For instance Eibl-Eibesfeldt (1967) had a tame badger which learnt to turn somersaults.

3. *INTERMINGLING OF FUNCTIONALLY DIVERSE TYPES OF BEHAVIOUR*

While "play" behaviour often involves motor patterns used in other contexts, functionally diverse motor patterns are often mixed in the same sequence. Thus hunting and sexual behaviour are mixed in the play of the mongoose (Rensch and Dücker, 1959), aggressive and sexual behaviour in the social play of rhesus monkeys (Harlow and Harlow, 1965; see also Ludwig, 1965; Eibl-Eibesfeldt, 1967).

4. *THE INTENSITY OF THE MOVEMENTS MAY NOT BE IN KEEPING WITH THAT OF THE SEQUENCE AS A WHOLE*

The individual movement patterns used in play are often incomplete: for instance in rhesus monkeys the aggressive rushes are frequently not carried home, the teeth are not closed in biting. Alternatively some movements may be exaggerated in comparison with their normal functional context: this is especially the case with the bounds and leaps often seen in locomotor play (e.g., in rhesus monkeys and kittens). Often individual movements are repeated a number of times, instead of leading on to the next link in the sequence as would happen in other contexts. Or the sequence may be re-ordered, elements appearing late in the normal sequence preceding those normally occurring earlier.

5. *SOCIAL PLAY IS USUALLY RESPONDED TO AS THOUGH NOT "SERIOUS" BY OTHER INDIVIDUALS*

In some species it is said to be preceded by a special signal indicating its nature—a "play-face" in rhesus monkeys (Altmann, 1962), a half-crouch in dogs and cats (Loizos, 1966).

6. *THE BEHAVIOUR PATTERNS ARE ELICITED BY A WIDE RANGE OF STIMULI*

The movement patterns often involve objects which would be inadequate to elicit those patterns in their normal functional context. Kittens play with balls of wool, young kestrels swoop on sticks and other inanimate objects (L. Tinbergen, cited in Thorpe, 1963a).

7. *AMONGST HIGHER MAMMALS, THE PLAY OF THE YOUNG IS OFTEN ELICITED BY AN OLDER ANIMAL*

The lioness twitches her tail, stimulating the cub to play with it (Schenkel, 1966): female chimpanzees tickle their infants, rolling them over and mock-biting them (van Lawick-Goodall, 1967).

No one of these characteristics of play-behaviour is general to all behaviour which is called "play". Some of them occur in other contexts. Thus incomplete chains of behaviour are well known from the hunting behaviour of satiated animals, such as carnivores (Lorenz, 1935) and birds (Hinde, 1953b): whether or not this should be called play is a matter of opinion. Again, the intermingling of functionally diverse types of behaviour is seen in the responses of adolescent or adult female rhesus monkeys to infants not their own—they will switch rapidly from maternal to grooming, aggressive, or sexual behaviour (Rowell et al., 1964). Nevertheless taken together the characteristics of play indicate that while the motor patterns may be similar to those seen in other functional contexts, the motivational state is not. Play activities occur in the absence of the long-lasting motivational factors normally necessary for them: aggressive or sexual patterns occur when the animal seems not to be aggressively or sexually aroused. They may also be terminated without the consummatory situation being reached—the mounting attempts of young monkeys may not lead to either intromission or ejaculation: presumably they wane through other consequences of performance (see Chap. 13). On the other hand, play activities may recur although the consummatory situation is repeatedly presented (Leyhausen, 1965).

Play activities are readily interrupted by other activities: they may be immediately abolished by hunger or fear. Nevertheless play is not merely the consequence of the absence of high-intensity motivational factors for anything else. Play-fighting in primates and carnivores can occur with great intensity and yet still be accompanied by signs of friendliness, such as tail-wagging (Eibl-Eibesfeldt, 1967). Furthermore it is accompanied by an inhibition specifically of biting. Locomotor and manipulatory play can be extremely persistent, even though it does not lead to a food reward. Locomotor play may also involve movement patterns specific to that context.

But perhaps the clearest demonstration that play is not merely what the animal does in the absence of anything else to do comes from a study in which chimpanzees were allowed to choose food or social stimulation, the latter involving either play or petting. When the animals were tested after feeding, play was preferred to food, and even when deprived (no food for at least 15 hr), play was chosen in about 40 percent of trials (Mason, 1965a).

The factors controlling play behaviour are thus far from clear and certainly complex (see, e.g., Inhelder, 1955; Holzapfel, 1956a, b; Tembrock, 1958). What is certain, however, is that play behaviour cannot be shrugged off. Many mammals spend a high proportion of their time playing, especially, but not exclusively, when young. Whether or not the various lower vertebrates also show play behaviour is a question which must be answered not with a yes or no, but with specification of the ways in which the behaviour in question differs from the behaviour of the same species in other contexts and resembles the play behaviour of other forms. The bases of play behaviour would certainly repay study and would throw light on the nature of behavioural control in many contexts.

15.9 CONCLUSION

This brief survey demonstrates that no easy generalizations about "motivation" are possible. In the first place, the conventional categories of behaviour are heterogeneous; to ascribe an underlying unitary drive to a category defined in functional terms inevitably breeds trouble. Second, statements about matters such as the effects of deprivation on motivation, the extent to which behaviour is spontaneous, or the role of the performance of a stereotyped "consummatory act" in bringing behaviour sequences to an end, cannot be equally valid for the different types of behaviour shown by one animal, let alone for those of different species. The control of each type of behaviour is a separate problem: only when an adequate sample of behaviour types has been analysed can generalizations be attempted and the level of analysis at which they are useful assessed. The immediate aim for students of behaviour must therefore be not to fabricate a general theory of "drive", but to provide precise analyses of a wider range of individual cases.

15.10 SUMMARY

Work on the factors controlling a number of types of behaviour shows that the relative importance of extraneural factors, of spontaneity, of feedback effects consequent upon performance, and so on, varies greatly from case to case. It is thus not profitable to search for general theories of drive applicable to all types of behaviour in all organisms.

16

CONFLICT—COURTSHIP AND THREAT

16.1 CONFLICT

In Chap. 12 we saw how a chaffinch may successively sing and patrol its territory, then come down to feed on the ground, then fly up to a bush to preen, and then perhaps sing again. It engages first in one set of activities and then in another. Comparable observations could be made on any other species and show that the different activities in an animal's repertoire do not appear at random, but in an orderly manner. In implying that the chaffinch was always engaged in only one thing at a time, however, the description involved considerable oversimplification. Let us consider another case.

When a flock of great tits is feeding in winter, the individuals are usually spaced out over the ground. If one bird comes too close to another, threatening and fighting occur. The gregarious tendency which holds the flock together is thus balanced in part by aggressiveness engendered by proximity (pages 333–342). As each bird feeds, it remains on the alert for approaching danger and for rivals, and frequently interrupts its feeding to look around: the efficiency of its feeding is thus limited by its watchfulness. From time to time the flock moves in a rather straggling fashion from one feeding site to another: the later birds to leave hastily finish the seed they were eating and fly rapidly to catch up with the flock: here feeding behaviour conflicts with the gregarious tendency. In such a flock, therefore, the birds do tend to engage first in one type of behaviour and then in another, but they are, in fact, rarely single-minded: for much of the time a tendency to show one type of behaviour is in conflict with tendencies to show others.

Before proceeding it is necessary to be a little more precise about the terms "tendency" and "conflict". In everyday speech, to say an animal has a tendency to behave in a particular way means that we have evidence that it is likely to do so. If it is likely to behave in a particular way, causal factors for the behaviour in question must be (potentially) present. The use of "tendency", therefore, implies that such causal factors are present. Causal factors for a particular type of behaviour may, however, be present even when the behaviour is unlikely to appear: a hungry animal in the presence of food may not partake if a strange object is nearby. In such cases we can say that the animal has a tendency to show the response (e.g., to feed), but the factors are too weak, or inhibited (Hinde, 1955/1956).[1] The use of the term "tendency" is suitable in the present context because it obviates the need for specifying the nature of the causal factors involved, though subsequently we may wish to analyse them in terms of hormone level, strength of stimulus (cf. Chap. 10), or other variables.

If tendencies for two incompatible types of behaviour are simultaneously present, we may speak of them as being in conflict. In the case of the great tits in the flock, stimuli from another individual may produce either approach or withdrawal. The distance apart of individuals in the flock depends largely on a balance between these two: at the point of balance both tendencies are presumably present but are in conflict and mutually inhibit each other. When the flock moves, the later individuals are influenced both by factors which normally elicit feeding behaviour and by factors which induce following. These are incompatible, and the resulting conflict modifies the feeding behaviour. It will be apparent that an animal which appears to be engaged in only one activity may yet be in conflict, the effect of causal factors to do something else having been totally suppressed by the causal factors for the ongoing behaviour. "Conflict", however, is not used to refer to a hypothetical or physiological state, but to incompatible response probabilities.

Just because most activities are self-terminating, either by bringing the animal into a consummatory situation (Chap. 10), or by removing the eliciting factors (e.g., Chap. 15), or in some other way (e.g., Chap. 13), most conflicts are soon resolved. But how quickly this occurs depends in part on the nature of the conflict itself. This can be seen most readily in conflicts which involve simultaneously tendencies to move in different directions.

When the two tendencies in conflict involve approaching two different objects some distance apart, we can label the situation an "approach-approach" conflict, without, of course, implying that the incompatibility is merely a physical one. In such a case the animal may reach a point in between where the tendencies to approach each object are in balance. Its position will then be unstable, for the tendency to approach either goal increases with its proximity. Any slight departure from the point of balance

[1] Blurton-Jones (1968) prefers to reserve the term "tendency" for "observed likelihood of occurrence" and to use "potential tendency" in cases such as this. Berlyne (1960) speaks of the response as being "aroused", but confusion with "arousal" as used in a non-specific sense (Chap. 9) seems likely.

towards one goal will result in an increased tendency to approach that goal and a decreased tendency to approach the other. The conflict is thus easily resolved. The "vicarious trial-and-error" behaviour shown by an animal at the choice point in a maze can also be regarded as involving an approach-approach conflict: it often involves intention movements along each alley in turn. On the other hand, an animal placed between two objects both of which it strives to avoid ("avoidance-avoidance conflict") is in a relatively stable situation. Since the tendency to avoid either object is likely to increase with its proximity, movement towards either is likely to lead to a return to the point of balance. Continued movement is possible only along a line at right angles to that joining the two objects.

A more interesting case occurs when the animal has simultaneous tendencies to approach and avoid a given object or situation. Suppose we train a rat to run up an alley for food, and then give it an electric shock at the goal box. When subsequently placed in the alley it may run a little way up, hesitate, turn back, and oscillate about a point some distance from the goal, or hesitate there grooming its fur or cleaning its paws. Its tendency to approach the food is balanced by a tendency to avoid further shock ("approach/avoidance conflict"). Another example is shown in Fig. 16.1. The mobbing response given by chaffinches to owls involves conflicting tendencies to approach and to fly away (Hinde, 1954a; Marler, 1956b). The relative strengths of these two tendencies vary with the distance between the chaffinch and the owl. If the chaffinch is placed in a small cage, it flies equally often to the front and back of the cage while mobbing when the cage is about 17 ft from the owl. If the cage is farther from the owl, most of the chaffinch's flights are to the front of the cage, and vice versa.

Such situations have been the subject of a rigid analysis by Miller (1959). In the case of the rat in the runway he points out that, for a position of stable balance to occur, the tendency to avoid the end of the runway must decrease with distance from it more rapidly than the tendency to approach. This is illustrated in Fig. 16.2. The gradients are represented as straight lines for simplicity, but any lines with a continuous negative slope steeper for avoidance than for approach at every distance would do equally well. It is apparent that if it was the approach line which had the steeper slope, no stable balance point could be reached: the animal would either go right away or right up to the goal.

Miller verified this model by systematically enumerating certain postulates, drawing deductions from them, and verifying these experimentally—an example of precise systematic theory-building. The crucial issue of the steeper slope of the avoidance line was deduced by Miller from the postulate that fear is a learned drive. On this view, stimulation from the end of

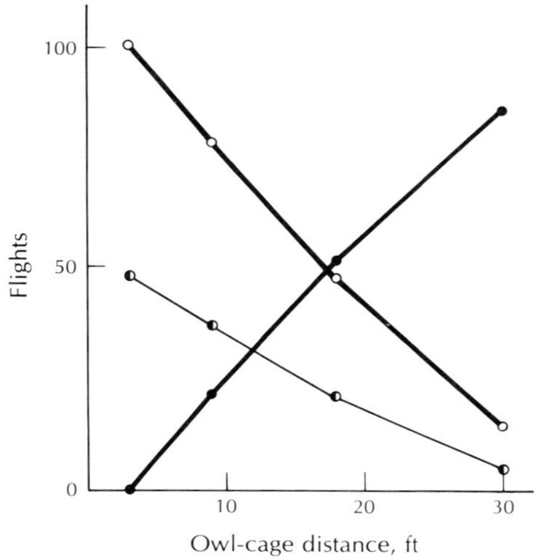

Fig. 16.1 Proportion of flights made to front (—●—) and to back (—○—) of cage; and total number of flights (—◐—) made by female chaffinches, with stuffed tawny owl placed at various distances from cage. (After Hinde, 1954*a*.)

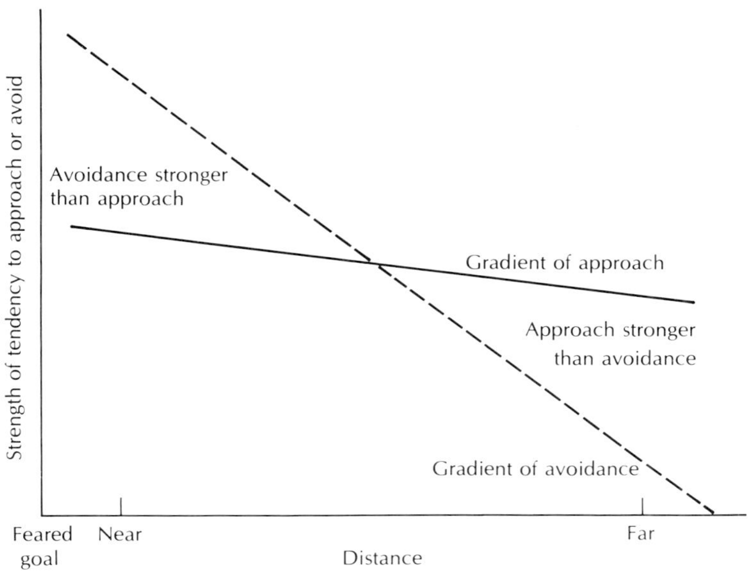

Fig. 16.2 Simple graphic representation of an approach/avoidance conflict. The tendency to approach is stronger far from the feared goal, while the tendency to avoid is stronger near the goal. (From Miller, 1944.)

the alley affects avoidance in two ways: by arousing fear motivation and eliciting the specific responses of avoidance. Thus, increasing distance from the end of the alley weakens the avoidance response in two distinct ways. Hunger, on the other hand, is thought of as depending on internal physiological factors: increasing distance from the stimulus therefore does not affect the "drive" and affects approach only in that the eliciting stimuli become weaker. The gradient of avoidance is therefore steeper than that of approach. (There are, of course, assumptions here which could be questioned.)

The view that avoidance is affected by a change in the stimulus in two ways leads to a further deduction which can be tested. If rats are taught to run down an alley to avoid shock and then tested without shock (1) in the original alley and (2) in a different one, they should respond much less in the second situation because the change in the stimulus situation will weaken both the fear motivation and the learned response of running. On the other hand, if a similarly trained group is tested in both situations with the shock on, the drive will not be thus affected. The avoidance behaviour of these animals should therefore decline less in the second situation than did that of the first group. This was verified experimentally.

As yet there seems to have been no application of this method to cases in which both approach and avoidance depend similarly on the same external stimulus, as would seem to be the case in the attack/flee or curiosity/flee conflicts discussed later in this chapter and in the mobbing response of the chaffinch. Miller's system has, however, been extended for use in studies of the effects of drugs, and also to cover some types of neurotic behaviour (Dollard and Miller, 1950). Its generality has been questioned by Hearst (1962) who, in a different situation, found steeper generalization gradients for a food-rewarded response than for a shock-avoidance one: the importance of internal factors in motivating avoidance may be a crucial issue here.

Another method of studying conflict behaviour has been described by Hearst and Koresko (1964). Rats are presented with two stimulus conditions: in the first, milk rewards and electric shocks occur intermittently independently of the rat's behaviour, while in the other neither occur. The rat could change the stimulus condition from that occurring at the moment to the other by pressing a lever. The subjects oscillate between the two conditions, the nature of the oscillations varying with factors such as food deprivation and the strength of the shock. If the relative frequencies of food and shock in the first condition are varied, the subjects initiate and terminate the stimulus most often at intermediate relative frequencies of food and shock rather than when conditions are either very favourable (e.g., only food possible) or unfavourable (e.g., only shock possible) (Hearst, 1967).

Until such time as a conflict is resolved the animal's behaviour is more or less influenced by both sets of causal factors. Before discussing the types of behaviour that may result, we must consider how to recognize a conflict situation. How can we tell that an animal has tendencies to behave in more than one way at a time? In most of the cases discussed so far, the existence of a conflict situation is recognized by the experimenter because he is familiar with the animal's previous history and knows that two sets of factors (for instance deprivation of food and stimuli previously associated with running to food, on the one hand, and stimuli previously associated with shock, etc., on the other) are present, either of which alone would lead to a completed sequence of behaviour. If a new observer were brought into the situation at this point, would he be able to tell what the rat was up to? By observing the rat's behaviour, would he be able to deduce that it was in a conflict situation, and the nature of the two types of behaviour which were incompatible? Comparable situations frequently present themselves in nature, and before the behaviour can be analysed it is necessary for the observer to detect the presence of conflicting tendencies, and to describe their nature.

As examples, we shall consider the analysis of threat and courtship behaviour in birds and fishes. Such behaviour often seems difficult to understand because of the variety and complexity of the movements which follow each other in rapid and apparently meaningless succession: this is because many of the behaviour sequences are not completed, and their nature is thus not apparent to the observer. It becomes comprehensible only when the occurrence of conflict is recognized. A description of two actual cases will pave the way for a more critical assessment of the sort of evidence that can be used.

16.2 THREAT AND COURTSHIP IN BIRDS AND FISHES

During the spring, great tits settle in pairs on territories which are defended primarily by the male. Along the boundary between territories is a narrow no man's land where prolonged skirmishes take place. Actual combat is rare, but each bird attacks and flees in turn, and in between shows a bewildering sequence of postures and movements. Among these is a "head-up" posture in which the bird stretches its head upwards, craning its neck from side to side and holding its body almost vertical (Fig. 16.3a). This posture may be followed by a flight away from the rival, it may just subside, or it may lead to a rather hesitant flight towards the rival.

A clue to the basis of the behaviour which occurs during these skirmishes is given by the observation that they are practically limited to the boundary region. If one male meets another on his own territory, he immediately attacks and the intruder flees. If, however, this territory owner is surprised as an intruder on his neighbour's territory, the roles are reversed—the bird which is now the territory owner attacks and the one which is now the intruder flees. The boundary area is thus one in which each bird's tendencies to attack and to flee are more or less in balance: sometimes one may predominate and sometimes the other. Since the head-up posture is followed sometimes by approach and sometimes by a flight away, it appears to be associated with a rather finely balanced conflict between the two tendencies. Similar arguments indicate that many of the other varied postures and movements used in the skirmish are also associated with conflicting tendencies to attack and to flee. Once the skirmish is seen

Fig. 16.3 (a) *Head-up threat posture used by great tits in disputes over territories;* (b) *head-forward threat posture used by blue tits in disputes over territories and in fighting over food.* (From drawings by Yvette Spencer-Booth.)

to be associated with two incompatible tendencies, it becomes more comprehensible (Hinde, 1952).

Chaffinch courtship will serve as a second example. Early in the breeding season, male chaffinches begin to threaten other birds in the flock. Females ready to form a pair do not fly off but adopt a fluffed posture: the male then gradually desists from his aggressiveness and begins to show courtship behaviour. Whereas at first he used a threat posture with his body horizontal and pointed directly towards the female (Fig. 16.4a), now his body is oriented laterally towards her (16.4b). This is related to a change in dominance: the male was at first dominant to the female, but now he gradually becomes subordinate to her, and she drives him away from food. In the later stages of his courtship, which may not be reached until some weeks after the initial pair formation, the male approaches the female only with hesitation and uses small pattering steps and a zigzag walk, holding his body upright and somewhat withdrawn from the female (Fig. 16.4c). Many of his attempts at copulation are thwarted by his fear of the female.

It seems, therefore, that the female may elicit attack and fleeing behaviour from the male as well as sexual behaviour: what he does depends both on the stimuli she presents and on his own internal state. Early in the season his behaviour to the female is primarily aggressive. As the tendency to behave sexually increases, the balance between his tendencies to attack and flee from the female changes in favour of the latter, and he behaves as if afraid of her. This is not due simply to a hormonal change or a general lowering of aggressiveness, for his readiness to attack other males is increasing at this time. Later still, when he begins to make copulation attempts, the main conflict is between tendencies to approach the female for copulation and to flee from her. Finally the former predominates and he mounts the female, but as soon as copulation is over he flees and gives the same call as that given in the presence of a flying predator (Hinde, 1953a; Marler, 1956b).

Correlated with these changes, the postures used by the male alter as the season progresses. At first his head-forward threat posture, oriented directly towards the female, differs little from that used in winter fighting. In the early stages of courtship this is modified in such a way that the male's body is lateral to the female: this can be regarded as a compromise between approaching and fleeing. Later still he approaches the female for copulation with his body in a vertical position comparable to that adopted by a bird about to flee from a rival in front of it.

Similar principles apply to the female when the male attempts copulation. The female does not always show sexual behaviour: she may fly off or attack. This is shown in the case of the greenfinch in Fig. 16.5: out of 102 "hovering approaches", only 25 led to successful copulation. Indeed, in 7 cases the female even reverted to aggressive behaviour after she had allowed the male to land on her back. Evidently proximity of the sex partner may elicit aggressive or fleeing as well as sexual behaviour from the female, just as with the male.

These examples of threat and courtship must not be taken to imply that the course of social encounters is simpler than is actually the case. Many additional factors are involved, such as the particular traits of the

Fig. 16.4 (a) *Head-forward threat posture of chaffinch used in fighting over food. The body is oriented towards the rival.* (b) *Courtship posture used by male chaffinch. The body is lateral to the female.* (c) *The posture used by a male chaffinch approaching a female immediately before copulation. The body is fairly upright, and the approach hesitant.* (From drawings by Yvette Spencer-Booth.)

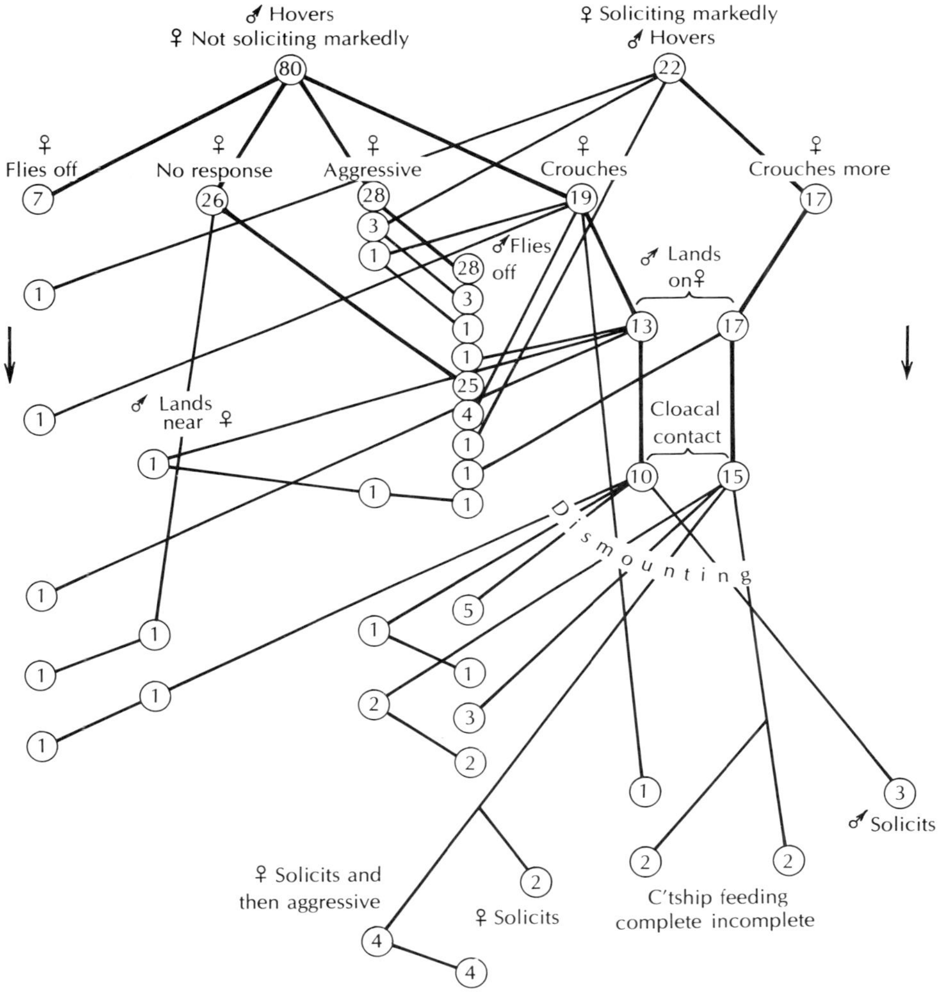

Fig. 16.5 Sequences of behaviour involved in copulation attempts by male greenfinches. (From Hinde, 1954*c*.)

The figures indicate the number of times each situation was observed. The sequences start with the male hovering over the female, who may (right) or may not (left) have been soliciting previously. Read from top downwards.

individuals concerned. Baeumer (1959), discussing fighting in domestic fowl, distinguishes three different types of fighting—between adult cocks, between hens, and between cockerels and hens holding positions of higher rank: the relative status of the individuals has an effect on the encounter which is superimposed on, or influences, the fluctuations in tendencies to attack and to flee. But the evidence that the diversity of behaviour shown in threat and courtship can be understood in terms of ambivalence among a relatively small number of behavioural tendencies is considerable.

In the cases discussed so far three groups of factors have been postulated: one of these if acting alone would lead to attack, one to fleeing, and one to sexual behaviour. It is suggested that the other types of behaviour which occur are linked with particular strengths or relative strengths of the tendencies to behave in these three ways. Thus, each threat posture is associated with a particular range of strengths and relative strengths of tendencies to attack and to flee. Similarly, in a wide range of species pair formation is initiated by aggressive behaviour directed by a member of one sex (usually the male) to other individuals. Changes in the subsequent courtship behaviour can be understood in terms of changes in tendencies to attack, flee from, and behave sexually towards the sex partner.

Although tendencies to attack, flee, and behave sexually are by no means the only ones involved, similar arguments have been applied not only to threat and courtship in a wide variety of species but also to behaviour shown in many other contexts—nest-relief ceremonies, submissive behaviour, distraction and mobbing displays to predators, and so on. In each case recognition that the behaviour involves conflict, and identification of the incompatible patterns, lead to greater understanding (Tinbergen, 1952, 1959).

16.3 NATURE OF THE EVIDENCE USED IN ANALYSING COURTSHIP AND THREAT DISPLAY

Much of the evidence used to support this view is based on field observation, and it sometimes involves a modicum of circular argument. It is therefore worthwhile to examine its nature in a little more detail. It can be classified as follows:

1. *THE SITUATION*

Often, some evidence can be drawn from the situation in which the behaviour occurs. We have already seen one example of this in the case of territorial fighting—threat postures are used primarily on the boundaries, where there is reason to think that tendencies both to attack and to flee from the rival are in balance. Similarly, if a given display occurs only between potential mates and never between rivals, there is good reason for thinking that a sexual tendency is involved.

2. *THE BEHAVIOUR WHICH ACCOMPANIES THE DISPLAY*

While displaying, an animal may edge towards or away from the object of its display. Oscillations or hesitation betray the presence of conflict, and the direction of movement indicates which tendency predominates at the moment.

In some species the external colouration can be used in a similar way, for changes in the motivational state are mirrored by changes in external colouration produced either by vasoconstriction/dilation or by chromatophores. Examples have been described from many phyla (e.g., Cephalopoda, Wells, 1962b; fishes, Lissmann, 1932; Baerends et al., 1955; Forselius, 1957; Keenleyside and Yamamoto, 1962; Neil, 1964; lizards, Klausewitz, 1953; birds, Armstrong, 1947; subhuman primates, Goodhart, 1960; Wickler, 1967b; and man, Darwin, 1872; Masters and Johnson, 1965). If a change in colouration is associated with (for example) a change in the probabilities of attacking and fleeing, and also with a change in the frequencies with which various display movements are used, an additional source of evidence as to the nature of the latter is available. A complex case has been analysed in detail by Baerends et al. (1955).

3. *SEQUENTIAL AND TEMPORAL CORRELATIONS*

If a particular display is sometimes followed by an attack, and sometimes by fleeing, it is presumably associated with conflicting tendencies to attack and to flee. The underlying assumption here is that the tendency changes more slowly than the overt behaviour, so that the two patterns are likely to share common causal factors (internal or external to the animal) if they are closely associated in time (cf. Chap. 25). On this view, the relative strengths of the conflicting tendencies (that is, here, the relative probabilities that each will find overt expression) can be assessed by counting the relative frequencies of, e.g., attacks and fleeing movements which follow the display.

This method was first used by Moynihan (1955) to substantiate his view that the various threat postures of the black-headed gull were associated with different absolute and relative strengths of the tendencies to attack and flee from the rival, and has proved a powerful tool (see also, e.g., Stokes, 1962a, b; Kruijt, 1964; Blurton-Jones, 1968). However, some general difficulties in interpreting "sequential associations" must be pointed out.

First, many of the changes in behaviour shown by a displaying animal are consequences of changes in the stimulus situation presented by its rival or mate. Apparently important differences between postures seen in the natural situation may in fact be due to differences in the position or orientation of the animal to which the display is directed (e.g., Brown et al., 1967). This difficulty can be circumvented by recording sequences of behaviour given to motionless dummies, or by selecting for analysis only those sequences in which the behaviour of the second animal did not change.

Another difficulty with this method is that many displays are seldom followed by pure expressions of one or other of the tendencies with which, according to other sources of evidence, they are associated. In one series of observations it was found that the pivoting display (Fig. 16.6) of the goldfinch was associated with agonistic behaviour in 68 percent of 152 cases observed and with sexual behaviour (courtship feeding or copulation) in only 8 percent. Nevertheless, other evidence indicates that a sexual element is important in the display. Pivoting is associated with a call seldom heard in agonistic contexts, is much more common between mates than between flock birds, increases in frequency as the breeding season progresses, and is associated with a drooping of the wings which also occurs in other types of behaviour in which the sexual tendency is more conspicuous. Such cases show that understanding of any display demands the use of at least several of the lines of evidence discussed here (e.g., Blurton-Jones, 1968).

Yet another difficulty is that movements may be associated together in time not because they share causal factors in any ordinary sense, but because they share a low priority among the organism's activities and can occur only in the absence of response tendencies of greater priority. This may be one reason why different "toilet" or "comfort" movements tend to be associated together. Such movements rarely interrupt other types of active behaviour but are themselves often interrupted (Andrew, 1956a; see also Chap. 25).

A variant of this method involves recording not temporal sequences but correlation within chosen intervals. The latter are usually continuous (e.g., 1 min compared with another) but may be situationally defined (e.g., all territorial combats compared with fighting at a feeding table or displays at a nest box). Temporal correlations have been used with success (e.g., Beer, 1961, 1963a, b; Blurton-Jones, 1968) but suffer from two disadvantages. First, the result obtained varies with the time interval chosen (see pages 604–605). Second, it gives less information than sequential analysis about the organisation of the behaviour. Thus if *A* follows *B* as often as *B* follows *A*, they are likely to share causal factors, whereas if *B* follows *A* but not vice versa, it is probable that *A* helps to cause *B* (i.e., a chain response, see page 609). This information is lost with temporal correlations (Blurton-Jones, 1968).

4. THE NATURE OF THE DISPLAY ITSELF

In many analyses of display, the postures are regarded as the basic units. Each posture, however, can be analysed into components—raising of the wings, opening of the beak, spreading the tail, and so on. Further under-

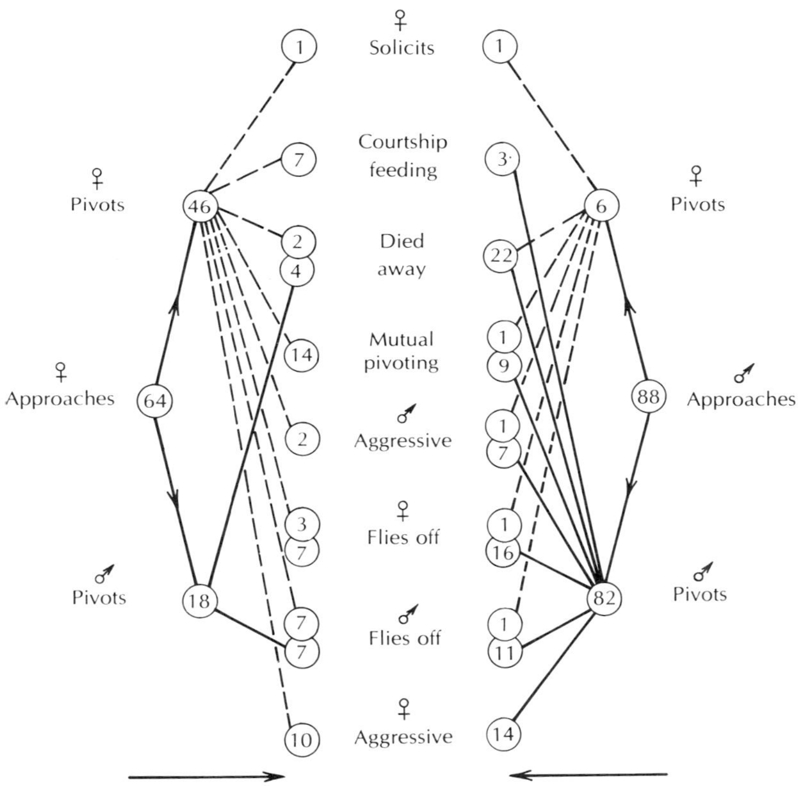

Fig. 16.6 Sequences of behaviour associated with the "pivoting" display of goldfinches. (After Hinde, 1955/1956.)

The figures indicate the number of times each sequence was observed. Read from sides to centre.

standing of displays has been obtained by using these as the units, each posture being regarded as a combination of components.

In some displays there may be changes in the intensity of particular components and in the number of components present, but the components occur only in certain combinations: *A, AB, ABC, ABCD,* and so on. An example is the "soliciting" posture used by the female chaffinch before copulation: successive stages in the posture are shown in Fig. 3.1, page 20.

Another such case, which has been analysed in more detail, is the aggressive display of the Siamese fighting fish, *Betta splendens* (Fig. 16.7).

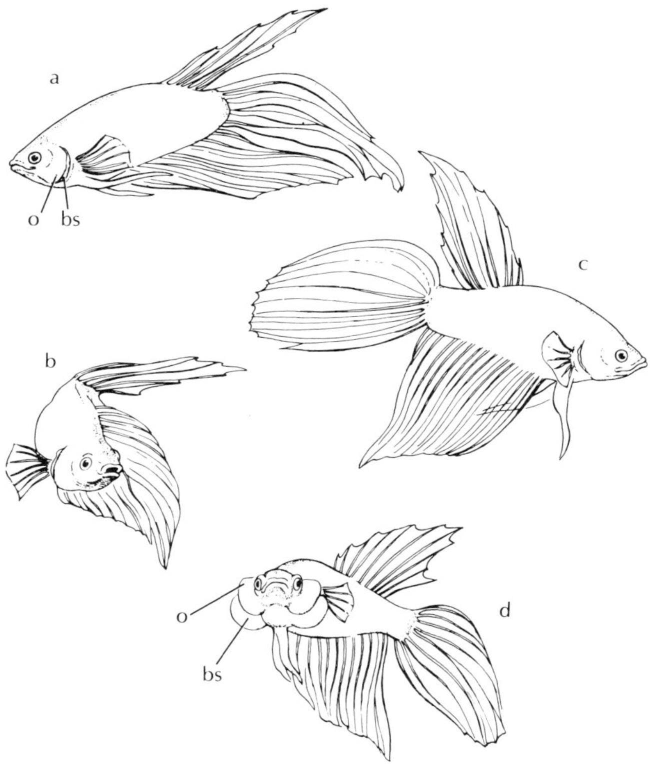

Fig. 16.7 Threat display of Siamese fighting fish. (From Simpson, 1968.)

(*a*), (*b*) Non-displaying fish; (*c*), (*d*) displaying fish; *o*, operculum; *bs*, branchiostegal membrane. In (*c*) the fish is broadside, with pelvic fin nearest rival raised. In (*d*) it is facing the rival, with gill-covers raised.

Here Simpson (1968) has shown that a number of measures of the display to a puppet, including the proportion of a test spent with the gill covers erected, are intercorrelated. This type of relationship among different components or measures of a display suggests that all vary with the same tendency, but appear at different threshold levels.

At the other extreme the components could, in theory, occur in all possible combinations. For that to be the case, there would have to be as many independently variable groups of causal factors as there are components, and there would be no recognizable "postures", i.e., commonly occurring groupings of components.

In practice, most displays lie somewhere between the two. In a study of the agonistic behaviour of the blue tit (*Parus caeruleus*) Stokes (1962a) assessed the correlations among nine different components. Of the thirty-six possible two-component combinations, a strong correlation ($p < 0.01$) between the components was found in thirty-one. The correlation was in some cases positive; for instance a bird with fanned tail usually raised its wings as well. In others the correlation was negative—an erect crest never occurred with wings raised, tail fanned, beak open, or beak down. Stokes made a more detailed analysis of five of the components—position of crest, body, nape, wings, and orientation to the second bird. Out of the forty-seven possible combinations of the measures of these five elements, only twenty were actually observed, and eight of them accounted for 576 of the 629 observations in which all these components were recorded. Such results justify the use of crude postures in the preliminary analysis of displays, but indicate that further refinement can be achieved by studying the components independently.

The relation between particular components and subsequent behaviour is not usually a precise one. Stokes (1962a) studied the relations between the presence of individual components in the agonistic displays of blue tits at a winter feeding station and whether they subsequently attacked, fled, or remained where they were. When a bird raised its crest or fluffed its body feathers it subsequently fled on 90 percent of occasions and never attacked. But for all other components the probability of subsequent attack, escape, or staying was 52 percent or less.

This relatively small correlation between particular components and subsequent behaviour was due in part to interaction between the components. For instance, in an otherwise non-aggressive situation raising of the nape feathers was correlated with an increased likelihood of attack, but when a bird was already in an aggressive posture, the raising of the nape was associated with a reduced incidence of attack. However, as we have seen, these components tend to occur in particular combinations, and these combinations show more reliable relationships with subsequent behaviour. One combination led to escape on 94 percent of the occasions, another to staying on 79 percent, and another to attack on 48 percent. Basically similar conclusions were reached by Dunham (1966) from a study of the rose-breasted grosbeak (*Pheucticus ludovicianus*).

There are also other types of evidence which suggest that particular components are associated with one or other tendency:

(*a*) Some components are in fact elements of one of the main types of behaviour occurring in the conflict situation. In the "upright" threat posture of the herring gull the carpal joints are raised, the neck is stretched up and forwards, and the bill pointed downwards: these are components of attacking by wing-beating and pecking from above (Fig. 16.8). On the other hand, a lateral rather than head-on orientation to the rival, sleeking of the feathers, an upward pointing of the head, and an upward instead of obliquely forward position of the neck are components of escape. The posture is, in fact, a combination of these components of attack and fleeing, and the preponderance of one or the other can be understood in terms of the relative strengths of the

Fig. 16.8 (a) *Aggressive and* (b) *anxiety forms of the upright threat posture of the herring gull.* (After Tinbergen, 1959.)

associated tendencies (Tinbergen, 1959). However, a reservation must be made here. The assumption is that the components have the same causal basis in the display as in the other contexts in which they appear. This is often a useful starting point but is obviously not necessarily true in the case of locomotor components. Similarly the head-down display of great tits (*Parus major*) resembles both hammering-in-attack and the hammering movement used in opening nuts: further evidence is needed to decide to which it is more closely related (Blurton-Jones, 1968).

(*b*) *The postures in which the components occur.* If a component occurs in postures which, on other sources of evidence, are known to be threat postures depending on conflicting tendencies to attack and flee, but never occurs in courtship postures, it presumably does not depend on the sexual tendency.

(c) *The circumstances in which it occurs most often or most markedly.* If a posture varies from one extreme, that is very often followed by fleeing, to another, that is very often followed by copulation, and if certain components are absent in the pre-fleeing form but present in the pre-copulation form, they can be said to be associated with the sexual tendency. Similarly, a component of a threat posture which is more marked when the bird is inside its own territory than when it is outside it, is likely to be associated with attack.

(d) *The components with which it is associated.* Vertical stretching of the neck in gulls provides an example here. This may be an intention movement of striking a rival or an intention movement of mounting a mate. The other components with which the neck-stretching is combined provide evidence as to its significance in any particular case (Tinbergen, 1959; see also Stokes cited on page 375).

(e) *Comparative evidence.* Sometimes a comparison of closely related species provides an additional source of evidence. For instance, the facts that the skuas do not raise the carpals in the aggressive upright threat posture and do not use wing-beating in fights give additional weight to the view that wing-raising is an aggressive movement in gulls, which do both (Tinbergen, 1959).

Using one or more of these methods, the relations between display components and behaviour tendencies have now been assessed in a number of birds and fishes. Although the quantitative relationships may vary with the season (Stokes, 1962a), the evidence that many components are associated with a particular behavioural tendency is strong. Furthermore, the relationships are similar in closely related species (Andrew, 1961e; Stokes, 1962b).

However, it cannot be argued that the absence of any components associated with a particular tendency is evidence for the absence of that tendency. For instance, Kruijt (1964) has argued from other types of evidence that certain displays of the Burmese red junglefowl, though consisting entirely of agonistic components, depend in part on sexual factors. He believes that the latter exert a stabilizing influence on the attack/flee conflict and thus permit expression of the associated ambivalent postures.

5. FACTOR ANALYSIS

The methods discussed so far provide a basis for the understanding of many types of behaviour whose motivation is at first sight obscure. More recently further refinement of method (3) has been achieved by applying the techniques of factor analysis. The method consists of: (*a*) obtaining detailed recordings of sequences of identifiable elements of behaviour; (*b*) assessing the frequency with which each of these elements follows (or precedes) each other element, and hence determining the correlation coefficients between them; (*c*) determining whether these correlations could be deduced from correlations between the elements and a smaller number of hypothetical variables. If they can, then these new variables can be regarded as factors associated with two or more of the behavioural elements and can be said to explain the original correlations.

Wiepkema (1961) has analysed the behaviour of the bitterling in this way. The male fish defends a territory round a freshwater mussel, to which he will admit only ripe females. Such females are led to the mussel and lay their eggs within its gills. The male then ejects sperm while skimming over the siphon of the mussel.

Wiepkema recorded the occurrence of twelve identifiable movements when a territorial male was presented with another male, an unripe female, or a ripe female. The correlation coefficients between these movements were calculated as indicated above and then subjected to factor analysis. It was found that three factors accounted for about 90 percent of the total common variance, further factors having no practical significance. The results could therefore be expressed in terms of a three-dimensional model (Fig. 16.9) in which the dependent variables were represented as vectors. In this model the correlation coefficient between any two activities is represented by the cosine of the angle between them: obtuse angles thus indicate negative correlations. The three common factors are shown at right angles to each other, their position being chosen so that each corresponds as closely as possible to a group of vectors. The length of each vector indicates the extent to which it can be explained in terms of the common factors introduced, and its projection onto each factor indicates the extent to which that factor is represented in it (i.e., its factor loading).

It will be seen that the vectors of four activities (head-butting, chasing, turning beats, and jerking) were closely grouped around the positive side of factor 1. These movements have an obvious aggressive function, and thus factor 1 can be called the aggressive factor. Vectors for four non-reproductive activities, which include fleeing, are closely grouped around factor 2; while those for skimming and three other movements used in courtship lie close to factor 3. Factors 2 and 3 can thus be described as non-reproductive and sexual factors, respectively.

Some activities (e.g., head-butting, chasing, fleeing, skimming) are almost pure measures of one or other of the factors, since their loadings on the others are almost zero. Others, while having high positive loadings on one factor, also have positive loadings on another: they thus depend on two or more factors. For instance, turning beats and jerking have positive

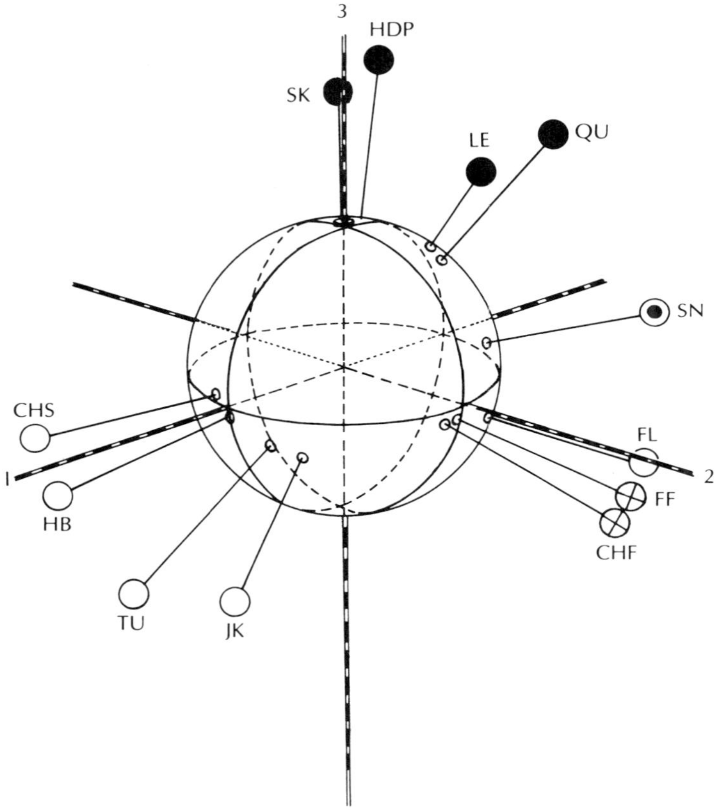

Fig. 16.9 The vector model of the behaviour of the male bitterling as obtained from a factor analysis of the correlation coefficients between successive activities. (After Wiepkema, 1961.)

The patterns of behaviour are abbreviated as follows: *CHS*, chasing; *CHF*, chafing; *FF*, finflickering; *FL*, fleeing; *HB*, head butting; *HDP*, head-down posture; *JK*, jerking; *LE*, leading; *QU*, quivering; *SK*, skimming; *SN*, snapping; *TU*, turning beat. The positive side of the aggressive factor is indicated by 1, the positive side of the sexual factor by 3, the positive side of the non-reproductive factor by 2. The vectors of the 12 variables are determined by the projections (i.e., factor loadings) of these variables on the three main axes.

loadings for both factors 1 and 2 and can thus be regarded as expressing both aggressive and non-reproductive tendencies. The latter is in fact probably multiple, the various non-reproductive activities being grouped together on the diagram because they occurred largely in the absence of aggressive or sexual behaviour. This, incidentally, exemplifies the way in which the results of a factor analysis require information from other sources for their full interpretation (see also Overall, 1964). Furthermore it is important to emphasize that the results of the factor analysis will be influenced by the decisions taken about the situations to be studied and the types of behaviour sampled.

Such an analysis enables the activities whose vectors have positive loadings for two factors to be arranged in a series of increasing loading by one factor and decreasing loading by the other. For instance, the agonistic movements can be arranged in a sequence of fleeing–jerking–turning beat–head-butting–chasing which corresponds to an increase in the ratio of the aggressive factor to the non-reproductive (fleeing) factor (see Fig. 16.13).

Used in this way, the factorial method is a valuable adjunct to the methods used previously for assessing multiple motivation. Another example is provided in the study by Baerends and van der Cingel (1962) of the snap display of the heron (*Ardea cinerea*). It remains true, of course, that the method is liable to the various difficulties discussed under (3) above. Furthermore, by itself it gives no information about the nature of the factors.

In the examples mentioned, the correlations between activities were assessed from a study of sequences of behaviour. Another possibility is to assess the strengths, intensities, or durations of different activities in a large number of individuals. The analysis is then based on the correlations between these measures. For example, Bell (1960) examined thirty-seven measures of behaviour of human neonates: from the factor analysis five orthogonal factors emerged and were labelled Arousal, Depth of Sleep, Sensitivity–Strength, Oral Integration, and Foetal Position. In this case the factors referred to characteristics of the individual rather than of its motivational state.

6. *INDEPENDENT MANIPULATION OF FACTORS CONTROLLING THE DISPLAYS*

The most direct approach to the analysis of displays involves experimental manipulation of the supposed underlying tendencies. For example, Blurton-Jones (1958/1959) identified the stimulus characters that would cause a pair of tame Canada geese (*Branta canadensis*) to attack him and flee from him: when he combined these characters, the geese threatened him. Later he applied this method systematically in a study of the threat displays of the great tit (*Parus major*). Using hand-reared birds, he found stimulus objects which would elicit attack (a pencil held through the wire of the cage), fleeing (a small lamp bulb), and feeding (food held in forceps). These increased three groups of co-varying responses which other lines of evidence, similar to those discussed in the preceding paragraphs, had shown to be controlled by three separate sets of causal factors. He then used them to create artificial conflict situations. For example, addition of the fleeing-evoking stimulus to the attack stimulus reduced the amount of attacking and increased the amount of display. If the level of attacking was not too high, the fleeing stimulus also increased the proportion of one of the threat displays (the head-up; see Fig. 16.3a): observational evidence had indicated that this display was related to a stronger tendency to flee than the other displays.

We see, then, that many of the complicated activities shown by fishes and birds during threat and courtship can in large part be understood on the view that tendencies for two or more incompatible types of behaviour are present.[2] Although none of the methods used is in itself conclusive, most studies have used more than one, and the results have been mutually confirmatory. Blurton-Jones's (1968) study of great tit threat displays probably represents the most critical and nearly comprehensive application of these methods: he found considerable agreement not only between the different observational approaches but also between observational and experimental ones.

As emphasized on pages 360–361, it must be remembered that the "tendencies" are merely intervening variables: the utility of this type of analysis does not imply that they can be regarded as entities (cf. pages 194–202). Furthermore, this type of analysis is not universally profitable in the understanding of displays: for instance in some cases all aspects of the display are correlated, so that there is no need to postulate two groups of factors controlling it (page 374); Simpson, 1968; see also Willis, 1967). Nevertheless analyses in terms of conflicting tendencies have certainly enriched our understanding of much display behaviour.

Although the cases discussed so far have involved principally conflicting tendencies to attack, flee, and behave sexually, these are by no means the only possibilities. For example great tit threat displays may be influenced by a tendency to "stay put", whatever the nature of the latter. A

[2] The following is a representative selection of studies. Fish: Morris (1954b, 1958a); Baerends et al. (1955); Forselius (1957); Oehlert (1958); Wickler (1958); Barlow (1962a, b, 1963); Keenleyside and Yamamoto (1962). Birds: Hinde (1952, 1953a, 1954c, 1955/1956); Tinbergen (1952, 1953, 1959); Morris (1954a, 1957b, 1958b, c); Moynihan and Hall (1954); Moynihan (1955, 1956, 1958a, b); Baggerman et al. (1956); Marler (1956b); Wood-Gush (1956); Andrew (1957a, 1961e); Kunkel (1959, 1962, 1967); Crook (1960a, b, 1963); Lind (1961, 1962); McKinney (1961); Immelmann (1962); Rüppell (1962); Stamm (1962); Stokes (1962a, b); Brockway (1963); Delius (1963); Kruijt (1964); Kahl (1966).

similar point was made earlier by Tinbergen (1959), who showed that one display, "facing away", is performed by a kittiwake (*Rissa tridactyla*) when fleeing is opposed by a tendency either to attack, or to stay on an (owned) nest, or to form a pair-bond with another bird (see also Fischer, 1965). Even in courtship, tendencies to nest-build, sing, behave sexually, beg, look around, and probably others are often involved (e.g., Andrew, 1957a, 1961e; Kunkel, 1962; Stamm, 1962; Crook, 1963). Other displays, such as the distraction and mobbing displays given to predators (Armstrong, 1950, 1954; Hinde, 1954a; Simmons, 1955) and nest-relief ceremonies, have also been shown to depend on ambivalence. Such conclusions, however, are not to be overinterpreted as implying that all displays, let alone all signals used in intra-specific communication, depend on conflict. When a bird sings, or feeds its mate, conflict is of minor significance.

Furthermore, even when conflict is involved, the interpretation of the signal by the recipient may be influenced not only by the signal itself but also by its context. For instance, many gulls use the "head toss" display in three different situations: in food-soliciting by adult females, as a precopulatory display by both sexes, and in hostile encounters (Tinbergen, 1959; Moynihan, 1962). The common factor is the occurrence of a tendency to escape in conflict with some other tendency which involves approach. The appropriateness of the response by the recipient presumably depends on the context.

16.4 THREAT AND COURTSHIP IN OTHER VERTEBRATES

The extent to which such principles are applicable also among the amphibia is not so clear (e.g., Gauss, 1961), but the courtship of one species has been interpreted in terms of an "interaction of drives" by Rabb and Rabb (1963). In some reptiles, also, courtship clearly involves aggressive components (e.g., Eibl-Eibesfeldt, 1955b; Kästle, 1964).

Sufficient work has been done with mammals to show that ambivalence plays a large part in many of their displays. For instance, Fig. 16.10 shows how the threat posture of the domestic cat changes with tendencies to attack and flee from the rival; the precise nature of the posture changes with the absolute and relative strengths of the two tendencies (Leyhausen, 1956). Figure 16.11 shows the results of a similar study of the positions of the elephant's trunk (Kühme, 1961, 1963). Tembrock (1962a, b) regards the threat displays of foxes as due to a superposition of attacking and defensive tendencies under the influence of social inhibition, and Walther (1958, 1960) found dominance changes to play an important role in the pre-mating behaviour of okapi. All the rodent species studied by Eibl-Eibesfeldt (1953), Grant and Mackintosh (1963), and Grant (1963) showed ambivalence among fleeing, aggression, and mating tendencies in social encounters (see

Fig. 16.10 Threat and fear expressions in cats. In each section fear increases from above downwards, and aggressiveness from left to right. (From Leyhausen, 1956.)

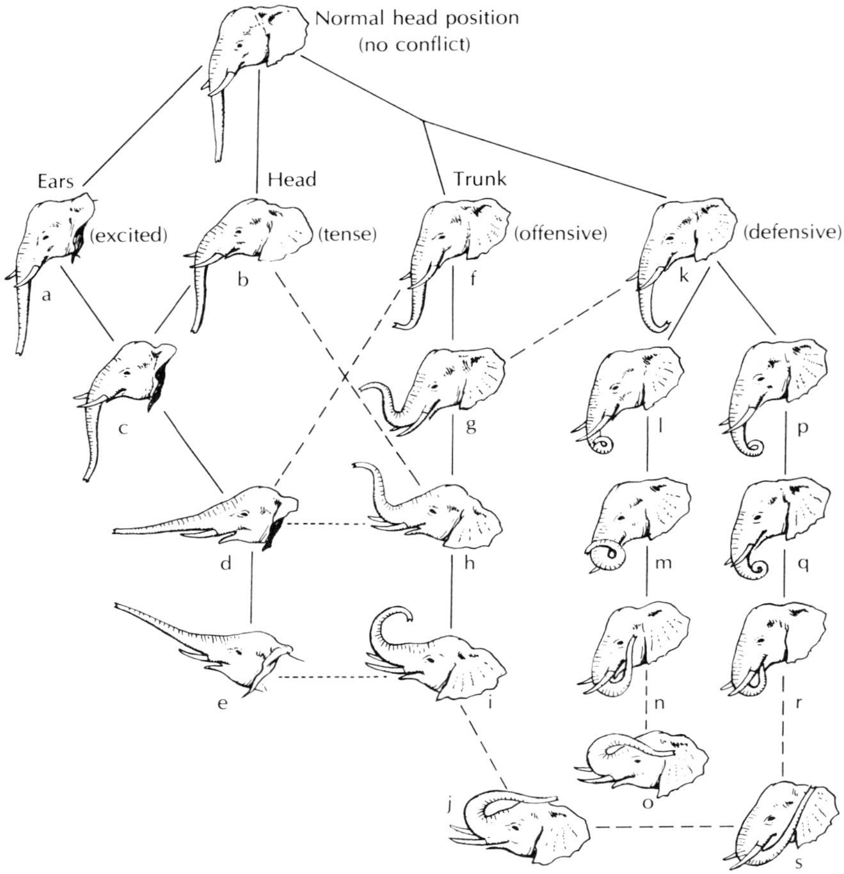

Fig. 16.11 Positions of head, trunk, and ears shown by elephants in aggression, fear, and threat. (After Kühme, 1963.)

Kühme interprets the postures as follows: Aggressiveness increases $a \rightarrow d$. A forwardly directed trunk *f* is associated with activity or rage, a backwardly directed one *k*, inhibition or fear. Position *i*, used in threatening by the bull, indicates some slight inhibition of aggressiveness as compared with *d* or *e*. Similarly, courage preponderates more in *n*, fear more in *p*, *q* or *r*; and the upward bending in *j* and *o* indicates more courage than *s*. Raising of the ears or head in association with these positions indicates increased aggressiveness.

also Peirce and Nuttall, 1961a, b). Other mammalian display movements which have been interpreted in terms of ambivalence include those of voles (Johst, 1967), the moose (Geist, 1963), Grant's gazelle *Gazella granti,* (Walther, 1965), and various primates (Hinde and Rowell, 1962; Rowell and Hinde, 1962; Rowell, 1962; van Hooff, 1962, 1967; Marler, 1965): in the latter, facial expression becomes of special importance (van Hooff). Whereas in fish and birds the different signal movements used in display tend to be discrete (see pages 401–404), in mammals they often intergrade. An example from the vocalizations of rhesus monkeys in agonistic situations is shown in Fig. 16.12: such intergradations presumably permit communication of more subtle items of information (e.g., Marler, 1965).

This is not to say, of course, that all movements used in display and social communication by mammals are associated with conflict. In some cases frustration rather than conflict seems to be involved. In others, such as the scent-marking of many mammals, and grooming, it is unnecessary to invoke either conflict or frustration (Andrew, 1962).

Furthermore, Andrew (1963b) suggests that some movements given in conflict situations convey information about the nature of that conflict, not because their components are associated directly with one or other of the conflicting tendencies, but by a more indirect means. For instance, lemurs raise their tails while threatening, and this could be said to indicate low fear and high aggression. However, when all the situations in which tail-raising occurs are considered, it is seen to be associated with a "high degree of facilitation of postural reflexes": since subordinate animals tend to show loss of tone in such reflexes, it appears in threat only in animals likely to attack.

Andrew (1963b, c) also argues that many primate facial expressions are in fact derived (in evolution) from protective responses which serve to protect the major sense organs and other sensitive areas against possible noxious effects from the source of stimulation: he believes their causation not to have changed in evolution. Such protective responses are, of course, related to fleeing. Andrew also suggests that many of the calls given by primates are associated with alerting (orientation, see page 131) responses which serve to bring the major sense organs to bear on an object. The evidence for the relevance of protective responses in this context has been questioned by van Hooff (1967) and, while it may be that such responses play a more important role in the communicatory gestures of primates than in those of lower species, it remains true that many other primate expressive movements, both gestures and vocalizations, are associated with particular ranges of tendencies to attack, flee, and behave sexually.

The view that threat behaviour has a dual motivation has been criticized by Brown and Hunsperger (1963) on the basis of experiments which involved elicitation of agonistic behaviour by direct electrical stimulation of the central nervous system of cats. Finding that threat could be elicited by stimulation of only one area of the brain at a time, they argued that it therefore cannot have a dual motivation. In fact, this could be interpreted on the view that either "motivational" mechanisms for both attack and fleeing, or coordination mechanisms for threatening, come within the area affected by the electrical stimulus. Most of the rest of Brown and Hun-

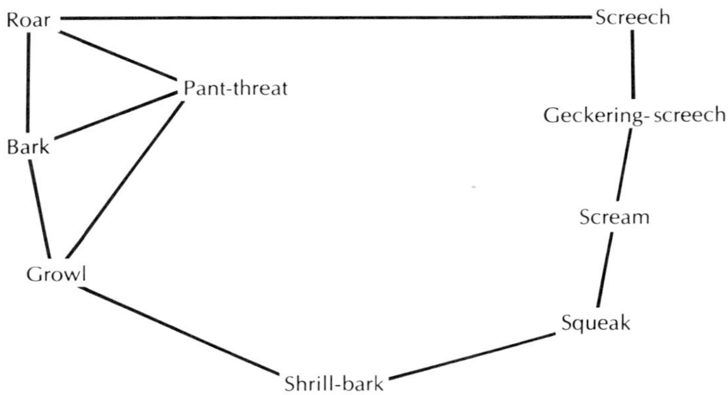

Fig. 16.12 Relationships of certain noises made in agonistic situations by rhesus monkeys. The lines indicate named noises between which intermediates have been recorded. (After Rowell, 1962.)

The noises are made in the following situations. *Roar,* made by a very confident animal, when threatening another of inferior rank; *Pant-threat,* made by a less confident animal wanting the support of the rest of the group in its attack; *Bark,* given by a threatening animal insufficiently aggressive to move towards the other; *Growl,* given by a mildly alarmed animal; *Shrill-bark,* the alarm call of the species, probably given to predators in the wild. *Screech* typically has an abrupt pitch change up and then down, made when threatening a higher-ranking animal and when excited or highly alarmed. *Geckering screech,* made by an animal being threatened by another; *Scream,* made by a losing monkey in a fight while being bitten; *Squeaks,* made by a defeated and exhausted monkey at the end of a fight.

sperger's data are consistent with an interpretation in terms of dual motivation. They found that threat-attack, threat, and escape were elicited from related areas; that the response became less aggressive, and included more fleeing components, the more posterior the point stimulated; and that the extent of the aggressiveness of the threat elicited from any one locus was influenced by the strength of the electrical stimulus and by the external object presented. All these could be accounted for in terms of partially overlapping systems for attack and escape, with other factors influencing the precise balance produced by stimulation in any one locus.

In any case, rather different results for chickens were obtained by von Holst and von St. Paul (1963). They found that, with a particular electrode placement in the brainstem of a domestic cock, stimulation produced motor unrest. If a stuffed polecat was presented with the electrical stimulus it was vigorously attacked. If the electrical stimulation was continued beyond the attack, the cock immediately changed to fleeing. Thus stimulation in one locus, together with the external stimulus, produced both attack and fleeing.

Brown and Hunsperger also used another argument. The view that threat has a dual motivation is associated with an assumption of mutually inhibitory relations between attack and fleeing. They found that stimulation in the escape area augmented or hastened the threat-attacks obtained by simultaneous stimulation in the brain area appropriate to them. They write: "present results, therefore, indicate that *mutual* inhibition between threat-attack and escape zones does not occur". Such negative results must be interpreted with much caution. For one thing, Brown and Hunsperger did not obtain "pure attack" from any of the loci they stimulated, though apparently it has been obtained by other workers (e.g., Wasman and Flynn, 1962; MacDonald and Flynn, 1964; see also Akerman, 1966a, b for data on pigeons). Furthermore, Adams (1968) recorded action potentials related to fighting behaviour in particular cells in the midbrain of cats. For another, considerable variations occur in the results of experiments involving simultaneous stimulation. Egger and Flynn (1963) found that amygdaloid stimulation at current strengths apparently ineffective when administered alone might either facilitate or suppress hypothalamically induced attack behaviour, depending on the locus stimulated.

Brown and Hunsperger's views have also been criticized on logical grounds by C. H. F. Rowell (1964; see also Brown, 1964). It is, of course, desirable that physiological data should be used in formulating hypotheses about behaviour (see Chap. 1), but caution is necessary before an interpretation valid at the behavioural level is revised in terms of physiological data whose own interpretation involves considerable basic assumptions.

16.5 THE NATURE OF THE CONFLICTS IN DISPLAY

So far, we have assumed some degree of incompatibility among tendencies as assessed by groups of responses; for instance among all attacking patterns and all fleeing patterns and all sexual patterns. This assumption is based ultimately on the view that the patterns of behaviour under study can be classified in groups and that a series of situations could be found in which all patterns from one group become less common or less evident, while those from another become more so—as, for instance, attacking elements diminish and fleeing ones increase as a male crosses his territorial boundary into foreign ground. However, we have also seen that elements of behaviour patterns associated with different tendencies can occur together; for instance in display postures. The question thus arises, should the incompatibility be regarded as occurring between the overall tendencies, or between more basic elements of behaviour? The issues here echo some of those discussed in Chap. 8 in the consideration of unitary drive concepts.

Andrew (1956a), criticizing the older view that conflict situations involve incompatibility between tendencies to give groups of responses, argues that it is "more accurate and more useful in predicting behaviour to say that tendencies to give two groups of responses are present, and that some of the responses of each group are incompatible with each other". He bases this view on evidence showing that the incompatibility between two responses does not depend on the "drives" they express. Thus, a passerine bird flicks its tail while hopping through twigs, when it has simultaneous tendencies to move by flight and by sidling along, and also when perched near a food dish at which a feared superior is feeding. In the latter case it has tendencies to approach (to feed) and to fly away (a fear response). The common factor in the two situations is that each involves a conflict between two locomotory movements, one at least of which involves flying. Similarly, the calls given by a blackbird (*Turdus merula*) during the approach/avoidance conflict involved in mobbing an owl depend on a tendency to fly in the presence of a frightening object but not on whether it flies towards or away from it (Andrew, 1961a–d).

Wiepkema (1961) also criticizes the view that incompatibility exists between the major drives or tendencies. He points out that, in many of the studies on which this view was based, the drives were assessed in terms of only one variable; for instance the frequency of zigzags was held to indicate the sex drive of a male stickleback. The results of his own detailed analysis of the reproductive behaviour of the bitterling is shown in Fig. 16.13. Some external and internal factors were found to be common to groups of responses designated as sex, attack, and fleeing. The corresponding internal mechanisms are labeled *S*, *A*, and *F*. The radiating lines indicate the relationships assessed from a study using factor analysis (pages 377–380) and other techniques. The letters down the right side refer to the various behaviour patterns observed. Some of the relationships were positive, some neutral, and others negative; for instance an increase in the aggressive factors increases the frequency of chasing (*CHS*); decreases those of quivering and leading (*QU*, *LE*), both movements which occur in early courtship; and does not affect the purely sexual movements of skimming (*SK*). Thus the relationships cannot be explained by postulating inhibitory effects between major tendencies each of which is associated with a limited number of behaviour patterns.

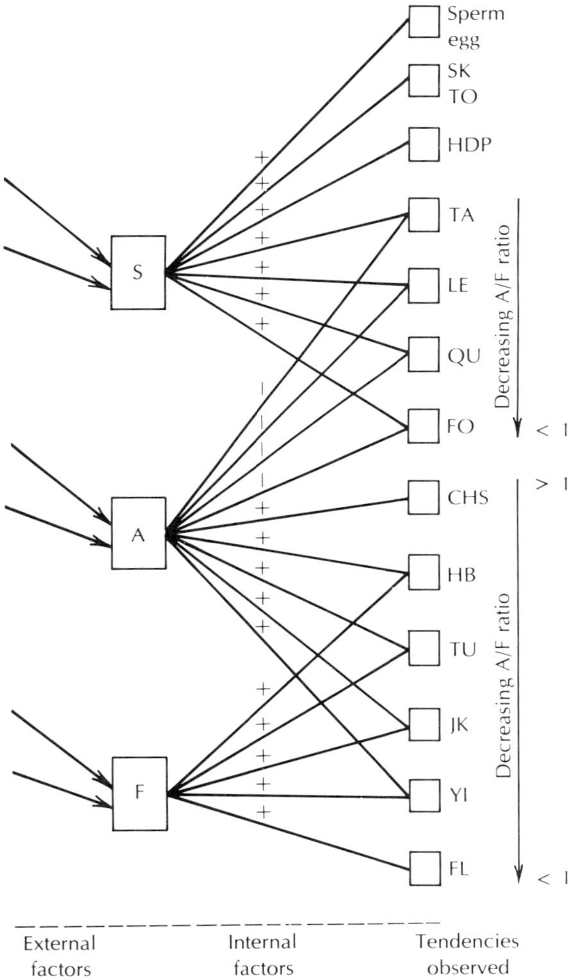

Fig. 16.13 Wiepkema's diagram of the reproductive behaviour of the bitterling. (After Wiepkema, 1961.)

S, A, F symbolize the internal mechanisms that underlie the temporal association and frequency of the different movements shown on the right. (The abbreviations for patterns of behaviour are given in the key to Fig. 16.9, to which should be added the following: *TO*, touching; *TA*, tail bending; *FO*, following; *YI*, yielding.) The arrows pointing to *S, A*, and *F* symbolize, the external factors which stimulate these mechanisms. The agonistic movements are arranged from above downwards in decreasing order of the *A/F* ratio (see page 380) as revealed by factor analysis. In accord with the experimental evidence, increase in aggressive tendency is associated with an inhibitory effect on some but not all sexual activities.

The dangers of generalizing about the nature of the conflicts associated with display are exemplified by Blurton-Jones's (1968) observations on the great tit. All four displays studied (head-down, horizontal, head-up [see Fig. 16.3a], and wings-out) depend on a stimulus for attack rather than on stimuli which evoke any kind of approach. Normally they occur when attack is prevented by simultaneous factors for fleeing: these can inhibit attack even though they do not lead to hopping away. Indeed, in most cases the factors for fleeing appear to act merely by preventing attack: the amount of threat posturing can be increased either by associating the attack-eliciting stimulus with a stimulus for fleeing or by making it inaccessible behind wire netting. The latter, however, does not increase the proportion of head-up displays: this suggests that the head-up differs from the other displays in depending specifically on fleeing rather than merely on the blocking of attack. The stimulus for fleeing does not operate merely by taking the bird out of reach of its opponent: another stimulus which tends to attract it away from the attack stimulus (e.g., food at the other side of the cage) does not have the same effect as the fleeing stimulus in reducing attacks, but does have a similar effect in increasing most threat displays. In examining the range of circumstances in which the display components occurred, Blurton-Jones concluded that some depended on conflicts involving specifically fleeing, attacking, or staying, rather than the locomotor movements these involve: thus crest-raising depends on the stimuli for fleeing and not on locomotion. By contrast "turns" depend on conflict between locomotion in different directions, regardless of the causes of the locomotion (cf. Andrew, cited above). Finally, some components depend on conflict between one group of factors and any other which blocks them: for instance crest-raising is associated with conflicts between fleeing and staying or approaching, and not between fleeing and specifically attacking.

In any case, one must be constantly on guard against overinterpretation in this type of analysis. The tendencies we have been discussing are merely intervening variables, and although experience has shown that the method is useful in bringing order to apparently complex display behaviour, its success depends on the correlations between the units (successive items of behaviour or simultaneous components of displays) studied. Where such correlations are absent, then we must work with smaller groups of units or with components. To cite another example, the triumph ceremony of geese, which occurs after one member of a group has just attacked a real or apparent enemy (Fig. 16.14), involves a conflict between two responses or response tendencies. The first, rolling (Fig. 16.14, top left and 5), is itself associated with a conflict between tendencies to escape and to stay near the other individual(s). Its elicitation depends on the proximity of a friendly partner and on the presence of a stranger. There is often an additional tendency, such as aggression, associated with it. The other, cackling (Fig.

Fig. 16.14 Triumph ceremony of grey-lag goose (Anser anser). F, *the direction of a real or apparent foe.* (From Fischer, 1965.)

16.14, 6), varies with the tendency to stay near the group or mate: the posture resembles that seen in real attack (Fig. 16.14, 2), though in this species it is not associated with a tendency to attack. Thus the triumph ceremony as a whole is associated with tendencies to show two movement patterns, one of which is itself also associated with conflicting tendencies (Fischer, 1965; see also Symons, 1966, for an analysis of the factors determining spine-raising in sticklebacks).

The question of incompatibility between movement patterns will be discussed again in the next chapter: for the moment it must be emphasized that incompatibility even between responses does not necessarily imply physical incompatibility between the actions of the effectors. Even when we describe a conflict with terms such as "approach/avoidance", they are descriptive labels only: the essential inhibition may be a physiological one. When we find incompatibility between responses, the physiological mechanism need not be on the response side: it could involve, for instance, the mechanisms of selective attention (Chaps. 5 and 6).

Further, as we have seen, incompatibility is a relative matter. When the two tendencies are of low or medium strength, the activities associated with each which occur during a given time interval may be independent of the strength of the other, but if one tendency is strong it may exert positive inhibitory effects on the activities associated with the other (Wiepkema, 1961; see also pages 396–400). The relationships found thus depend in part on the precise conditions of study.

16.6 VICTORY IN SKIRMISHES

In natural skirmishes between two territory owners, actual attacks are often rare, and the result of the skirmish is settled without physical injury to either participant. What is it, then, about the behaviour of each individual which determines which one will be the victor? Sometimes the result can be predicted from the start by the human observer on the basis of the relative size or vigour of the combatants, but often it is impossible to do so. One such case, involving aggressive display between Siamese fighting fish, has been analysed in some detail by Simpson (1968).

As noted above, in this species the several aspects of the threat display are well correlated with each other, so that it is unnecessary to postulate conflicting tendencies to understand the display movements themselves. A skirmish involves each fish alternately facing and turning broadside to the opponent (Fig. 16.7). Facing is nearly always accompanied by erection of the gill covers: while broadside the fish may flicker their pelvic fins, and beat and flash their tails. When two males separated by a transparent screen fight, they spend most of their time with one fish facing and the other broadside, some time with both broadside, and only a little time with both facing. Within the general outline of alternating bouts of facing and broadside, there are regularities in the relations between the components. For instance, the frequency of tail-beating increases over the time that the rival's gill covers are held erect, and there is often a temporal relation between the peak in the tail-beating frequency and the moment when the rival lowers its gill covers again. This suggests that each fish responds to the other's tail-beating by lowering its gill covers and itself turning broadside.

During a skirmish, there is a gradual increase in gill cover erection, tail-beating, and biting. This occurs even in fish displaying to their own mirror image. There is a great deal of reciprocity between the behaviour of two rivals such that when one beats its tail, the other does too, and when one bites, the other bites back. At first the proportions of time that each individual spends with its gill covers erect increase in parallel: only towards the end of the fight does one outstrip the other (Fig. 16.15). For the greater part of the fight each fish in some way matches the duration of its gill cover erections with those of its partner. Simpson suggests that this comes about by each fish "responding, by altering its subsequent readiness to lower its gill covers, to the duration of its partner's gill raising bout, and perhaps also to how the partner has responded, by lowering its gill covers, to the fish's broadside bout. For example, if the partner originally lowered its gill covers in response to one tail beat by the fish, but if now the fish has to beat its tail twice before the other partner lowers its gill covers, the fish itself may become correspondingly less responsive to the partner in terms of the time when it lowers its own gill covers in the presence of its partner's display. At the end of the display, one of the fish finally continues to increase the durations of its gill-raising bouts, in spite of further increases in the other's broadside display, and the other 'gives up' when it can no longer control the partner's gill cover erections". This brief account omits many of the complexities revealed by Simpson's detailed study but is perhaps sufficient to show that "winning" involves something more subtle than a mere show of strength, for each fish responds to the timing of its rival's display relative to that of its own.

16.7 SUMMARY

1. To say an animal has a tendency to behave in a particular way implies evidence that causal factors for the behaviour in question are present. In nature, more than one tendency is usually present at a time: such a situation is described as a conflict situation.
2. Most conflicts are soon resolved.
3. Some conflicts can be characterized in terms of incompatible directions of movement. The point of balance in an approach-approach conflict is un-

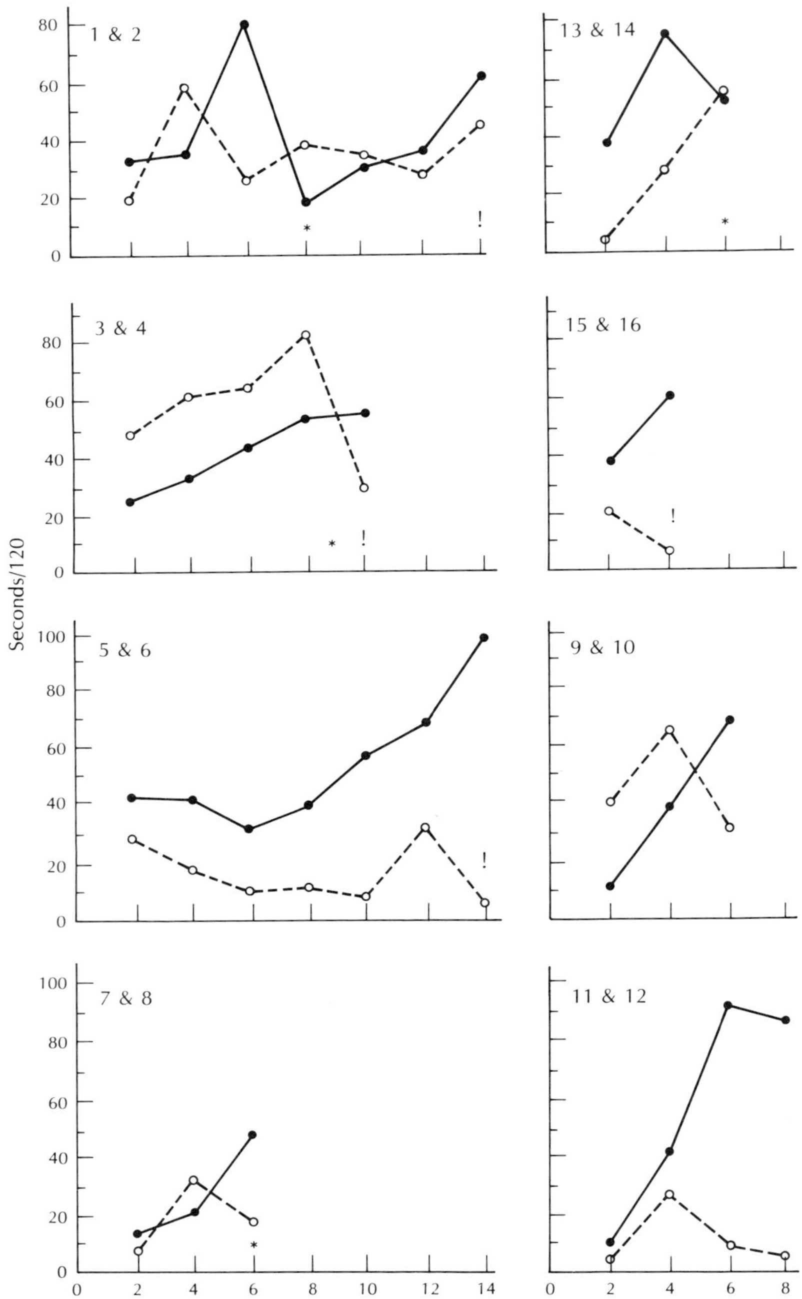
1 & 2
13 & 14
3 & 4
15 & 16
5 & 6
9 & 10
7 & 8
11 & 12
Seconds/120
0
20
40
60
80
100
0
2
4
6
8
10
12
14
0
2
4
6
8

stable, but those in avoidance-avoidance or approach/avoidance conflicts may be stable.

4. The threat and courtship postures of fishes and birds can be understood in terms of conflicting tendencies to behave in incompatible ways. Characteristically, threat involves tendencies to attack and to flee, while courtship involves tendencies to attack, to flee from, and to behave sexually towards the mate. In many species, however, other tendencies also play a part.
5. Evidence that threat and courtship postures occur when there are conflicting tendencies comes from study of the situations in which they occur, the behaviour which accompanies the display, the behaviour which occurs before or after the display, and from the components of the display. The technique of factor analysis can be used to assess evidence drawn from the study of the behaviour preceding or following the display. The two tendencies can also be manipulated experimentally.
6. Similar principles apply to mammals.
7. This approach to the study of display movements is essentially a behavioural one. Caution is necessary in postulating relationships with neurophysiological data.
8. To understand the behaviour shown in conflict situations it is necessary to consider both the extent of incompatibility between tendencies to show groups of activities and also that between particular activities or between the components of particular activities.
9. The manner in which one individual emerges as victor from a skirmish between two Siamese fighting fish is described.

Fig. 16.15 Skirmishes between pairs of individuals of the Siamese fighting fish. (From Simpson, 1968.)

Each figure refers to one skirmish, and shows the seconds per 2-min period spent with gill covers erect by winners (solid points and continuous lines) and losers (open points and broken lines) in successive 2-min periods. *, the first 2-min period with a rapid exchange of bites; !, the 2-min period was incomplete, but the time spent was corrected to a proportion.

17

BEHAVIOUR IN CONFLICT SITUATIONS

When animals are used for behavioural experiments in the laboratory, the physiological condition of the animal is usually controlled: often attempts are made to satisfy all the basic needs except one, which is used by the experimenter to channel the behaviour. In nature, as we have seen, such a situation is unusual. Most of the time causal factors for more than one type of behaviour are present, and often the balance between them is a fine one. In this chapter we shall consider various types of behaviour which occur in conflict situations. The categories used are useful for purposes of discussion, but it will be apparent that they overlap in a number of ways. Discussion will be limited to immediate responses: the longer-term effects of exposure to conflict situations have recently been discussed by Caïn (1959), who reviews also the literature on animal neurosis (see also Weiskrantz, 1968c).

17.1 INHIBITION OF ALL BUT ONE RESPONSE

Undoubtedly the commonest consequence of the simultaneous action of factors for two or more types of behaviour is the suppression of all but one of them. That behavioural inhibition of this type occurs is a matter of common observation. Thus the appearance of a flying predator causes a feeding great tit to dash for cover, abandoning its food. Since great tits do not usually fly off in this way unless alarmed, it is reasonable to presume that the great tit would have gone on feeding if the predator had not appeared. In other words, causal factors for feeding are presumably still present but overridden by the effects of the predator. Behavioural inhibition is thus said to occur when the causal factors otherwise adequate for the elicitation of two (or more) types of behaviour are present, and one of them is reduced in strength because of the presence of the causal factors for the other.

Since it is usual for causal factors for more than one type of behaviour to be present, some degree of behavioural inhibition probably occurs all the time. The amount of time an organism spends in many of its activities is thus restricted by the necessity for doing other things; and one of the most important consequences of a change in one of the motivational variables discussed in Chaps. 10 to 15 is an influence on the order of prepotency among the organism's possible activities. For instance, the proportion of the day blue tits spend feeding rises from about 70 percent in the summer to nearly 90 percent in midwinter when food is scarce, temperatures are low, and the days short: there is a correlated decrease in the time spent resting and preening (Gibb, 1954). Thus a consequence of winter conditions is to increase the priority of feeding behaviour relative to the others. A more detailed example on a different time scale is provided by Cotton's (1953) study of the influence of different periods of deprivation on the speed with which rats run down a runway for food. Over all test trials, running speed was found to decrease with increasing deprivation time. However, if trials in which the animals showed other types of behaviour, such as sniffing in the runway, were excluded, running speed was little affected by deprivation. This result indicates that the most important effect of deprivation was to increase the priority of running to food. Using a slightly different procedure, in which animals were both trained and tested under the different drive conditions, Cicala (1961) showed that deprivation affected both the running speed and the priority of running.

A similar point is made by Bindra (1961). Starting from the observation that every organism has a large repertoire of "acts" (e.g., walking, grooming, head-turning, sniffing), he suggests that the outcome of competition between these acts can be determined from a knowledge of their relative probabilities, those with a higher probability on the average occurring sooner. On this basis, Bindra describes performance in a learning situation in terms of the elimination of irrelevant responses (e.g., sniffing, sitting, and grooming, in a task involving running along an alley), and an increase in the relative probabilities of the relevant ones. Of particular importance in the test situations used in the laboratory are the various responses to novelty—freezing, fleeing, exploring—which compete with those necessary for the solution of the problem. Preexposure of the animal to the situation reduces these novelty responses (Claus and Bindra, 1960). A similar hypothesis to account for the effect of earlier experience on fighting in mice has been advanced by King and Gurney (1954) and supported by Banks (1962) (see page 542.)

Just because behavioural inhibition involves the suppression of an activity, and because that suppression may be partial rather than absolute, detailed analysis may be necessary to reveal it. Some examples are provided by the experiments of van Iersel (1953) and Sevenster (1961) on the interaction of sexual and parental behaviour in the three-spined stickleback. The number of zigzag courtship dances given to a standard female, and the proportion of time spent fanning, were among the dependent variables chosen for study. In fanning, the fish takes up a position in front of the nest and pointing towards its entrance, and moves its tail as though swimming forwards while moving its pectoral fins as though swimming backwards. As a result a current of water is directed towards the nest and the eggs are aerated. Considering some short-term experiments, the relation between fanning during a 15-min observation test and the number of zigzags given in a test immediately following is shown in Fig. 17.1. This indicates considerable independence between the two tendencies, but also a degree of negative correlation; low numbers of zigzags are found only with large amounts of fanning. This suggests that a strong

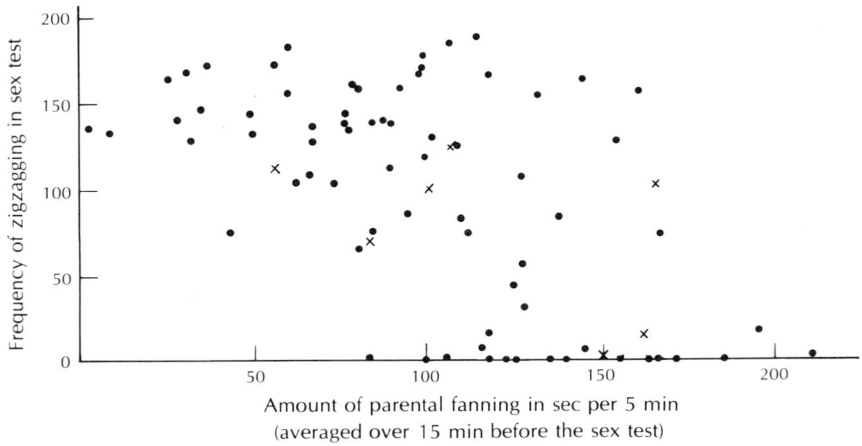

Fig. 17.1 Relation between amount of parental fanning and frequency of zigzag courtship shown in a subsequent sex test by three-spined sticklebacks. (After Sevenster, 1961.)

tendency for fanning is associated with an inhibitory effect on zigzagging. That the reverse effect also occurs was demonstrated as follows. If two zigzag tests are given in succession, the number of zigzags in the second is usually greater than in the first. This indicates that the initial exposure to the standard female (and/or performance of the zigzag response) has a positive effect on the tendency to show zigzags which persists to the second test (cf. pages 301–308). Comparison of the amount of fanning before and after the sex test gives an indication of the effect of this increased courtship tendency on fanning: if the number of zigzags in the sex test is considerable, the subsequent fanning is decreased (Fig. 17.2). Actually the depression of fanning is even greater than is apparent from the figure, for it overrides a tendency towards an increase in fanning due to its interruption by the sex test. Thus an increase in the courtship tendency is accompanied by a decrease of fanning even after the sex stimulus has been removed.

There have been few studies of the detailed course of inhibition between complex responses, but Heiligenberg's (1963, 1965a) analysis of the interaction between aggression and fear in a cichlid fish indicates that analyses of further cases would be profitable. The two activities are mutually inhibitory: the more often the fish attacks conspecifics the less it responds to stimuli for fear responses (see pages 365–382), and vice versa.

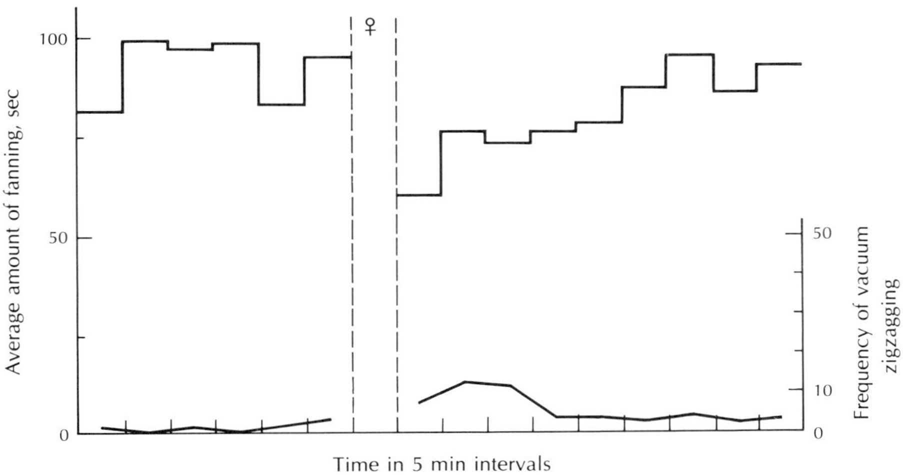

Fig. 17.2 The effect of a sex test (with more than 20 zigzags) on the parental fanning of male three-spined sticklebacks. A female was present for the period between the vertical discontinuous lines. (After Sevenster, 1961.)

The responsiveness to stimuli for fear responses is enhanced by their successive presentation. If it rises sufficiently, there is an immediate cessation of attack behaviour. Analysis shows that the responsiveness to stimuli for fear responses falls gradually to its previous level with a half-life of about 5 min. Attack returns suddenly at its previous level: this happens the sooner, the higher its previous level.

Behavioural inhibition involves suppression of one or more activities while another predominates, but what are the consequences of the conflict for the behaviour which does appear? This depends in part on the criteria by which it is assessed, but in terms of frequency measures it too usually suffers a decrement. Indeed, an acute conflict may result in the inhibition of all possible responses. The paralysis of fear is well enough known in human subjects: a comparable phenomenon has been studied by Rigby (1954). He trained rats to obtain food by performing one response in the presence of a light and to avoid shock by another response in the presence of a buzzer. When both light and buzzer were presented simultaneously, many of the animals sat motionless.

In some cases, however, there is evidence that the behaviour not suppressed may actually be augmented. If sticklebacks are fed in only one part of their aquaria and are then given one or two electric shocks there, they subsequently spend less time in the danger area, but while there they feed more intensively (cf. page 207). These changes are not a consequence specific to electric shock: similar increases in intensity of feeding combined with a reduction in time spent feeding were shown by fish in conflicts between feeding and migratory behaviour, or feeding and fleeing from rivals (Tugendhat, 1960b).

The concept of behavioural inhibition is, of course, a loose one, and diverse types of mechanisms may be involved. One may fail to leave one's room to go to a meeting when the clock strikes because of failure to notice it or refusal to respond appropriately: both would qualify as behavioural inhibition. Competition for the mechanisms of selective attention, as discussed in Chaps. 5 and 6, competition for the effectors, and perhaps competition for more than one type of intervening mechanism, may all be involved. Where selective attention is the issue, all responses associated with the relevant object may be possible and all responses associated with others suppressed. Thus a great tit behaving territorially may show any of the sorts of aggressive behaviour elicited by other great tits but will ignore nest holes. At the other extreme, competition for effectors may result in inhibition only of particular responses or components of responses (see below).

17.2 INTENTION MOVEMENTS

Behavioural inhibition is often not complete, so that the behaviour is merely reduced in intensity or frequency or appears in an incomplete form. This may consist of the initial phases of movements or movement sequences, usually referred to as "intention movements". For instance, the take-off leap of a bird before flying consists of two phases: first it crouches, withdraws its head and raises its tail, and then reverses these movements as it springs off (Fig. 17.3). These incomplete movements may be repeated several times before take-off when the bird is about to fly up from rest. In conflict situations they (or a modified form of them) may occur repeatedly for much of the time that the bird remains in the situation (Daanje, 1950).

17.3 ALTERNATION

We have seen that, just before copulation, a male chaffinch has conflicting tendencies to approach and avoid the female. He does not, however, stay still at the point of balance, but alternates between the two patterns, sometimes edging a little closer and then fleeing away. Another well-known example of such alternation is provided by the zigzag courtship dance of the three-spined stickleback (Fig. 25.1, page 607). When a pregnant female enters a male's territory he swims in a zigzag course. One leg of this course is an intention movement of attacking the female, the other of leading her to the nest (Tinbergen, 1951; van Iersel, 1953). Similar alternation behaviour is to be seen in a rat placed in an approach/avoidance conflict

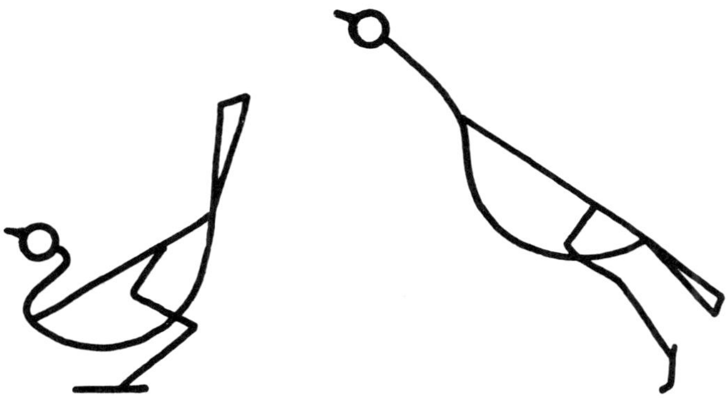

Fig. 17.3 The first and second phases of the take-off jump by a bird. (After Daanje, 1950.)

as described above: the animal vacillates about a point some distance from the goal.

17.4 AMBIVALENT BEHAVIOUR

Sometimes intention movements appropriate to the two tendencies are combined into a single pattern which contains components appropriate to both. Thus, a half-tame moorhen (*Gallinula chloropus*) offered food may make incipient pecks towards it, and even swallowing movements, while simultaneously keeping its distance or even edging away: components of both feeding and fleeing are shown simultaneously. A number of examples of ambivalent postures have already been considered (Chap. 16): they are often a patchwork of components, each of which depends on only one of the conflicting tendencies. Since both tendencies are influencing the effectors, the primary competition between them cannot be for the mechanisms of selective attention but must be nearer the final motor pathways: indeed such postures occur especially when both tendencies are elicited by the same object.

Although they are associated with incompatible tendencies, there is reason to believe that some ambivalent postures may remain relatively unchanged over a considerable range of intensities of the associated tendencies. There is, of course, a methodological problem here. At the behavioural level of analysis tendency is implied from frequency measures—the frequency with which an activity occurs in given circumstances, or of what follows it. It would not be possible to measure the tendency at any particular instant, even if it were meaningful to do so. Nevertheless it seems reasonable to assume that at some level the causal factors for a particular

Fig. 17.4 Oblique and forward threat postures of black-headed gull. By counting ciné-frames, Tinbergen showed that the oblique appeared in 32 frames and the forward in 19 frames to every 5 in which the transition postures appeared. (From Tinbergen, 1959.)

activity change over a continuous range. The threat and courtship postures which an animal shows, however, often do not. If an animal has two or three postures, it is usually possible to say at one instant which one it is showing, and intermediates are rare. For instance, the blackheaded gull (*Larus ridibundus*) uses both "oblique" and "forward" threat postures (Fig. 17.4). Analysis of ciné-frames shows that the postures themselves are much more frequent than intermediates (Tinbergen, 1959): it thus seems reasonable to suppose that each posture is associated with a range of values of the conflicting tendencies. This view is also supported by the observation that in many cases practically the same threat posture may be shown when the animal is both edging slightly towards and slightly away from its rival (but see pages 370–371).

Another line of evidence that the form of such postures may be constant over a range of strengths of the underlying causal factors comes from a simultaneous assessment of form and frequency. Whereas some ambivalent postures change in form as they become more frequent, others are similar over a wide range of frequencies. An example is the courtship posture of the male cut-throat finch (*Amadina fasciata*): whether it is given at low or high frequency its form is almost identical. In other words, the increase in the causal factors which gives rise to the increase in the frequency of the posture makes little difference to its form (Morris, 1957a). A slightly different case is illustrated in Fig. 17.5: the frequencies of the bow-coo displays of the collared dove (*Streptopelia decaocto*) and necklace

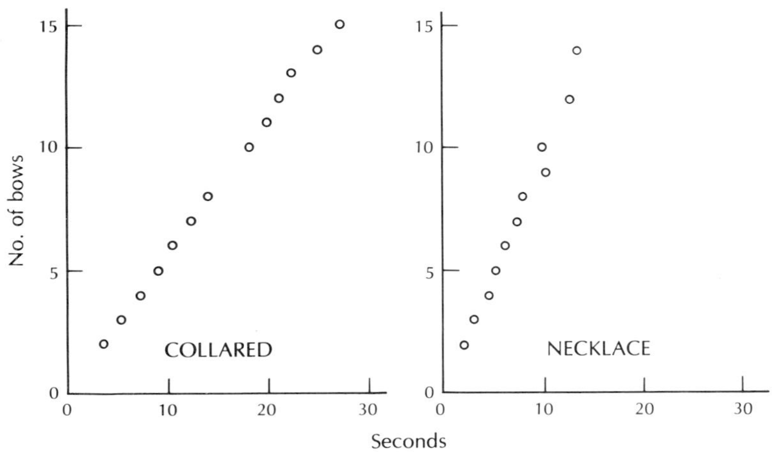

Fig. 17.5 The relation between the mean duration of bouts of bowing display and the number of bowing cycles in the bout for collared and necklace doves. (After Davies, 1965.)

dove (*Spilopelia chinensis*) do not vary with the length of the bout of displaying (Davies, 1965).

In practice, many movements other than those occurring in conflict situations show a constancy of form which is independent of frequency. While "fanning" the eggs in the nest the three-spined stickleback moves its fins with a fairly constant frequency and amplitude which vary little with the length of the bout or the proportion of time spent fanning (van Iersel, 1953). By contrast, the fanning movements of the river bullhead (*Cottus gobio*) vary considerably in frequency and amplitude (Morris, 1954b). Morris (1957a), who first called attention to such relationships, described it as the development of "typical intensity".

Morris (1957a) apparently regarded the ruffled courtship posture of the cut-throat finch as not ambivalent, and used the term "typical compromise" to refer to ambivalent postures which remain constant in form over a wide range of the conflicting tendencies. Comparative evidence would, however, suggest that this posture, like so many other courtship postures, is ambivalent, and in any case there seems no justification for the use of two terms.

Typical intensity in ambivalent postures often seems to be the result of natural selection acting to increase the efficiency of the posture as a social signal. A signal which is constant from one occasion to the next is more easily recognized than one which varies. Other ways in which ambivalent postures have been modified in evolution to this end are considered in Chap. 27.

17.5 COMMON COMPONENTS—COMPROMISE BEHAVIOUR

When a hungry bird wishes to fly towards a food source, which it also has a tendency to avoid, it flicks its tail vigorously. This tail-flicking is an intention movement of take-off which would be appropriate to both approach and avoidance. This has been called "compromise behaviour" (Andrew, 1956a)—the animal shows elements common to both tendencies.

Compromise behaviour, as used here, thus differs from ambivalent behaviour in that the one element is associated with both tendencies, whereas ambivalent behaviour is a patchwork of elements from both. However, the two intergrade. Thus the bunting *Emberiza schoeniclus* has a call "eee" which expresses fear and is given during flights, and a female soliciting call "ti-ti-ti". When attacked by her mate the female may give "eee" calls which change into "ti-ti-ti" calls as she alights. At the moment of changeover she gives a call in which the individual notes resemble the fear call but the repetitive sequence resembles the soliciting call. In Andrew's opinion this could be described either as compromise behaviour or ambivalent behaviour. It will be apparent that compromise behaviour, like ambivalent behaviour, involves competition for effector rather than receptor mechanisms.

It must be added that the term "compromise behaviour" suggested by Andrew, though now used widely, is not altogether a happy one. The cases he cites involve the appearance of components or characteristics common to the two otherwise incompatible patterns: this selection of "common components" hardly comes within the meaning of compromise in its everyday sense.

"Compromise" is more appropriate when the behaviour shown is in some sense intermediate: for instance, when gravity and the dorsal light response conflict for the control of the orientation of a fish, the result is an intermediate position (Fig. 7.6, page 156), the precise nature of the compromise depending on various aspects of the internal state (von Holst, 1950a). The approach/avoidance conflicts involved in courtship or threat often result in a circling movement round the object of the display, another example of compromise. That the result of conflict between orientational mechanisms is not always a compromise is shown by Birukow's (1958) experiments on the beetle *Geotrupes silvaticus:* this species orients to either light or air currents, but does not take up an intermediate position when both are present.

17.6 AUTONOMIC RESPONSES

Perhaps because so many cases of conflict are associated with agonistic tendencies, they are frequently accompanied by autonomic responses. Defaecation and urination in situations involving conflict or frustration are indeed often used as measures of "anxiety" or "emotionality" (see pages 535–536). In other cases, however, the automatic basis of the behaviour observed is less easily recognized. For instance, many species of bunting show "cooling responses", i.e., panting, feather-sleeking, and wing-raising as normally used with high environmental temperatures, after fleeing. This could be due to the muscular exertion of fleeing, but such an explanation cannot hold in all instances. For instance as part of courtship, the male flies rapidly after the female in "sexual chases" at a stage when he has, in fact, a considerable tendency to flee from her. During the chase male and female undergo similar degrees of muscular exertion, but after it is over, the male, but not the female, shows cooling responses. Furthermore, if the male attacks the female despite his strong fear of her, the attack is usually followed by cooling behaviour. At other times buntings show warming responses in fear, as indeed does man. Andrew (1956a) has suggested that the occurrence of such temperature-regulatory responses in fear is to be ascribed to the changes in peripheral circulation consequent upon autonomic activity and the resultant warming or cooling of the temperature receptors (Morris, 1956). Like other types of behaviour given in conflict situations, these autonomic responses may become ritualized in evolution to serve as social signals (see Sec. 4).

The effects of "stressful" stimulation of various types are of course mediated in part by the autonomic system and can lead to relatively long-term effects, such as ulceration of the gastrointestinal tract. There has been some dispute about whether the possibility of avoiding the stressor ameliorates its effects. In a recent experiment Weiss (1968) found that rats which could perform a "coping response" to avoid or escape electric shock developed less severe physiological symptoms of stress than animals which received the same amount of shock but had no possibility of controlling it. This substantiated earlier work with rats but contradicted that with monkeys. Differences in experimental procedure as well as in species may account for this.

17.7 "DISPLACEMENT ACTIVITIES"

Animals in conflict situations sometimes show behaviour which appears to be irrelevant to any of the tendencies in conflict. For instance, in many aggressive or sexual situations passerine birds may wipe their beaks or preen their feathers, show drinking or feeding behaviour, or engage in some other activity which seems unrelated to the context. Such activities were clearly not evolved for use in aggression or courting, and their occurrence requires explanation in causal terms. Comparable examples have been described in many other groups, including arthropods, fishes, and mammals (e.g., Tinbergen, 1940, 1952; Kortlandt, 1940; Crane, 1957). First labelled by Kirkman (1937) as "substitute activities", they are now usually grouped under the heading of "displacement activities" [1] (Zeigler, 1964).

In the first instance, it must be recognized that the "irrelevance" of displacement behaviour is a relative term, open to misinterpretation through anthropomorphism. For instance, in mouth-breeding cichlids and many other teleosts many motor patterns of fighting not only are homologous with those of feeding, but also retain motivational relations with them: stimuli eliciting fighting also promote feeding activities, and vice versa. Thus activities of one group appearing in the context of the other are not so causally irrelevant as they might appear to be on functional grounds (Albrecht, 1966; see also Heiligenberg, 1965). Since we shall see later that the category of displacement behaviour is causally heterogeneous, it seems better to tolerate a certain looseness in definition at this stage.

Motivational relationships of the type just described could account for only a very small proportion of the cases which have been labelled as displacement activities. How then is the occurrence of apparently irrelevant activities to be understood? It was originally suggested that they depend on causal factors quite different from those which produce the same activities in their normal functional contexts. It is of course plain that the total set of circumstances which obtain when a starling preens his wet plumage is different from that when he preens during a fight, but more than that was claimed. The displacement activities were distinguished as depending on motivational factors other than those normally activating them, the suggestion being that the factors denied their normal expression because of

[1] The term "displacement" is thus used here in a sense different from that found in the psychoanalytic literature. Psychoanalytic "displacement" is discussed here as "redirection."

thwarting or conflict "sparked over" to activate another type of behaviour. Such an explanation depends on an energy model of motivation, and its deficiencies have been discussed by a number of authors (e.g., Kennedy, 1954; Hinde, 1956a, 1960a; van Iersel and Bol, 1958; Bindra, 1959).

Even when the "sparking over" explanation was still in vogue, it became apparent that the occurrence of any particular displacement activity was influenced by the presence of those same factors which determine its occurrence in its normal functional context. Thus in an approach/avoidance conflict fighting turkeys may drink if water is available but feed if food is there (Räber, 1948; Armstrong, 1950; Tinbergen, 1952; Andrew, 1956a, b). Even the posture of the animal could influence its subsequent behaviour (Tinbergen, 1952; Lind, 1959). Subsequently, further evidence along the same lines accumulated; for instance van Iersel and Bol (1958) found that rain increases the frequency of both "normal" and "displacement" preening in terns.

Not only external factors but also internal ones may influence the occurrence of displacement activities; for instance most of them are activities which the organism performs frequently and are elicited fairly easily (Bindra, 1959).

If the occurrence of an apparently irrelevant pattern of behaviour is influenced by the same causal factors which influence it in its normal context, are such factors sufficient to explain its occurrence, and what is the significance of the conflict? These questions became relevant as soon as the shortcomings of an energy model of motivation were recognized: if the displacement activity is not motivated by "sparking over", does the thwarting of other patterns play any role in its causation?

Five ways in which the conflict might influence the occurrence of the displacement activity have been suggested:

1. *AUTONOMIC ACTIVITY*

In some cases the displacement behaviour is a consequence of autonomic activity aroused by frightening stimuli or other aspects of the external situation. We have already seen an example of this in the cooling responses shown by buntings in sexual chases. The autonomic activity engendered by the conflict situation may also provide stimuli for somatic activities which appear at first sight to be irrelevant. For example, vasomotor or pilomotor activity may provide skin stimuli which, in turn, elicit scratching or preening, and dryness of the throat may elicit drinking (Andrew, 1956a; Morris, 1956).

2. *THE DISINHIBITION HYPOTHESIS*

In some cases the conflict plays a more permissive role. When mutual incompatibility prevents the appearance of those types of behaviour which would otherwise have the highest priority, patterns which would otherwise have been suppressed are permitted to appear. This suggestion, now known as the "disinhibition hypothesis", was first made by Andrew (1956a): the "displacement" preening which buntings show in conflict situations can be regarded as a consequence of peripheral stimuli which normally induce preening but whose effect is easily suppressed by factors for other activities. In a conflict situation these other activities cannot occur, and preening is possible.

A similar hypothesis was put forward independently by van Iersel and Bol (1958) on the basis of a quantitative study of preening by terns (*Sterna* spp.). They showed that intensive incubation inhibited preening: preening during incubation occurred only when the incubation tendency was relatively weak. Preening which occurred in association with aggressive or escape behaviour they labelled as "displacement" on the grounds that it was associated with conflict and was often frantic or hurried. Van Iersel and Bol assessed by independent means the strengths of the conflicting tendencies and found that displacement preening was most likely to occur when they had a certain relation to each other. For instance, when returning to incubate after an alarm the distance at which the bird lands from the nest can be regarded as a consequence both of its tendency to incubate and its tendency to flee. The latter is also indicated by its posture, four qualitative grades of which were distinguished. When birds were only weakly alarmed, displacement preening was most likely to occur when the bird landed far from the nest, i.e., when the tendency to incubate was also weak (Fig. 17.6). On the other hand, strongly alarmed birds were most likely to show preening when they landed near the nest. With an intermediate degree of alarm displacement preening was most likely to occur at an intermediate landing distance: in other words, the displacement activity was most likely to occur when the conflicting tendencies were both very strong, both weak, or both of intermediate value. On the basis of these and other observations, van Iersel and Bol argued that when the conflicting tendencies are "effectively equal", they are mutually inhibited, and this results in a removal of their inhibitory action on other activities such as preening.

C. H. F. Rowell (1961), studying the preening shown by chaffinches in an approach/avoidance conflict, confirmed the essential elements of this disinhibition hypothesis. The chaffinches flew to and fro between perches, occasionally showing a few preening movements as they stopped there. The tendencies to approach and to fly away, when not in conflict, were correlated with a low frequency of preening: during conflict, preening occurred when there was equilibrium between the two tendencies. As with the terns, Rowell found that "displacement" preening was increased by wetting the plumage. He also found that it increased with the duration of equilibrium between approach and avoidance (as measured by the time spent on particular perches), that is, with the opportunity for the external stimulation to produce its effect.

Whereas van Iersel and Bol ascribed the occurrence of preening primarily to a constantly present though fluctuating "preening drive", Andrew, *loc. cit.*, and Rowell, *loc. cit.*, laid more emphasis on external stimulation. All these authors, however, assume that some degree of external stimulation for preening is constantly present. Of course, the argument here easily becomes circular, for the presence of external stimulation for preening is normally

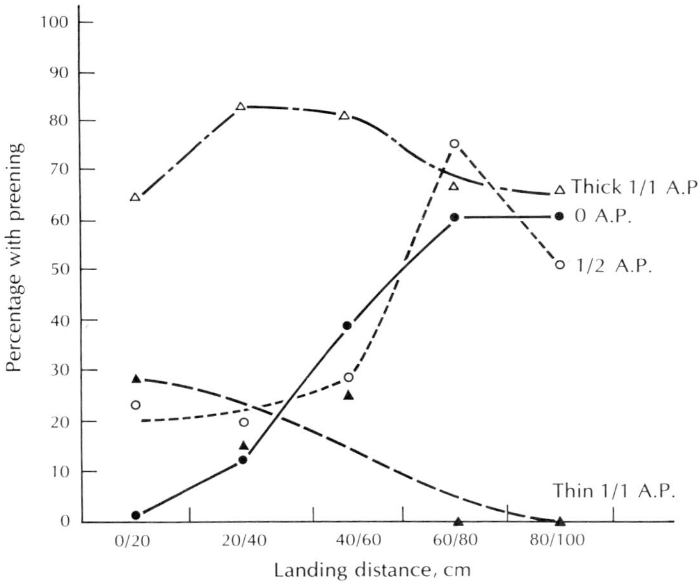

Fig. 17.6 "Displacement preening" by terms. The relationship between the probability of preening by birds approaching their nests and the distance away from the nest at which they land. (After van Iersel and Bol, 1958.)

The four sets of data are for birds showing increasing alarm, as indicated by their posture, in the order 0 A.P., ½ A.P., thick ¼ A.P., thin ¼ A.P.

revealed only by the occurrence of preening: preening is, however, augmented by increasing the external stimuli.

Even more detailed evidence in support of the disinhibition hypothesis has been given by Sevenster (1961) in a study of the three-spined stickleback. One activity involved in the male's care of the eggs is "parental fanning", an activity which, as we have seen, serves to aerate the eggs. Fanning also occurs, however, before the eggs are laid—in particular, during the sexual phase which occurs after the nest is completed but before eggs are laid, and during courtship at any time. This latter "displacement fanning" seems not to serve any useful function (though see below) and was formerly ascribed to the sexual "drive", when "thwarted", finding "outlet" through parental activities.

When Sevenster counted the number of zigzag (courtship) dances and the number of bites given during tests in which the male was presented with a female or another male (see pages 397–399), he was able to show that the male's sexual or aggressive tendency is raised for some minutes after the model is removed. By further observations during this period, Sevenster showed that the sexual tendency has a strong inhibitory effect on parental fanning, and the aggressive tendency a weak one.

The relations between the sexual tendency, the parental tendency, and the amount of "displacement" fanning shown were then investigated as follows. Using males in the parental phase, the amount of fanning in a standard period was used as an index of the parental "drive". The fish were also given a "sex test": a female was presented and the number of zigzag dances and periods spent fanning assessed. Figure 17.7 shows that at low levels of zigzagging the average amount of "courtship fanning" was correlated with the amount of parental fanning. This is consistent with Sevenster's finding that displacement fanning is influenced by the same factors that influence parental fanning (van Iersel, 1953), namely, the number and age of eggs in the nest, the presence of CO_2 in the water, and certain internal factors. Figure 17.7 also shows that the average amount of courtship fanning was negatively correlated with the number of zigzags: this is ascribed to an inhibitory effect of the sexual drive on the parental activities (see pages 397–399). At high levels of zigzagging the average amount of courtship fanning decreased only so far as the minimum value found in the sexual phase, about 15 sec per 5-min courtship. Sevenster concluded that, at this point, the sexual factors were not inhibiting the parental activities: if they had been, the amount of fanning would have been correlated with the strength of the parental factors, and an increase in sex drive would have resulted in a further decrease in fanning. In other words, for those 15 sec the parental fanning is temporarily disinhibited. In fact, both sexual and aggressive factors can inhibit fanning: the greatest amount of disinhibition occurs when both have intermediate strengths (Fig. 17.8).[2] For instance, this displacement fanning fluctuates in a cyclical

[2] This view, it may be noted, differs slightly from that of van Iersel and Bol. The latter showed that effective equality between the conflicting tendencies is more probable the greater the absolute strengths: when it occurs, it results in a negation of the inhibitory effect on preening. On Sevenster's view, if one tendency is strong its inhibitory effect on the "displacement" activity is bound to be large: the total inhibition due to both tendencies will be minimal when both have intermediate strengths. The views could be reconciled by making certain assumptions about the strengths of the tendencies in the two studies; for instance that Sevenster was working with a range which included "stronger" tendencies than van Iersel and Bol.

fashion, the cycle being associated with changes in the balance of sexual and aggressive tendencies: at the stage in the cycle at which the sexual tendency reaches predominance, fanning falls off abruptly.

In other cases there is suggestive evidence that the strengths and relative strengths of the tendencies which are in conflict influence which displacement activity is shown. In his study of junglefowl (*Gallus gallus*) Kruijt (1964) compared the incidence of various displacement activities in the winners of fights with their incidence in the losers. A number of differences were found and are presumably to be ascribed to the different attack/flee ratios in the two cases (Fig. 17.9).

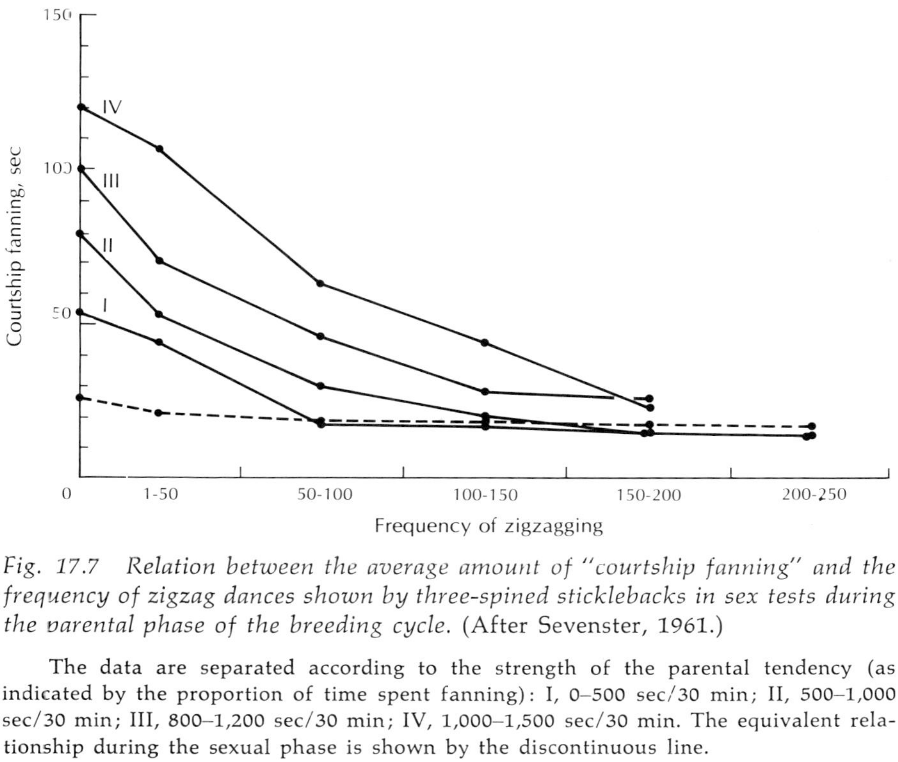

Fig. 17.7 Relation between the average amount of "courtship fanning" and the frequency of zigzag dances shown by three-spined sticklebacks in sex tests during the parental phase of the breeding cycle. (After Sevenster, 1961.)

The data are separated according to the strength of the parental tendency (as indicated by the proportion of time spent fanning): I, 0–500 sec/30 min; II, 500–1,000 sec/30 min; III, 800–1,200 sec/30 min; IV, 1,000–1,500 sec/30 min. The equivalent relationship during the sexual phase is shown by the discontinuous line.

Fentress's (1968) study of grooming in voles provides only partial support for the disinhibition hypothesis. After a frightening stimulus these animals flee and/or freeze, groom their fur, and then walk around or continue with other maintenance activities. The apparently irrelevant grooming, therefore, occurs in the transitional state between freezing or fleeing and the other types of behaviour. Fentress was able to vary the

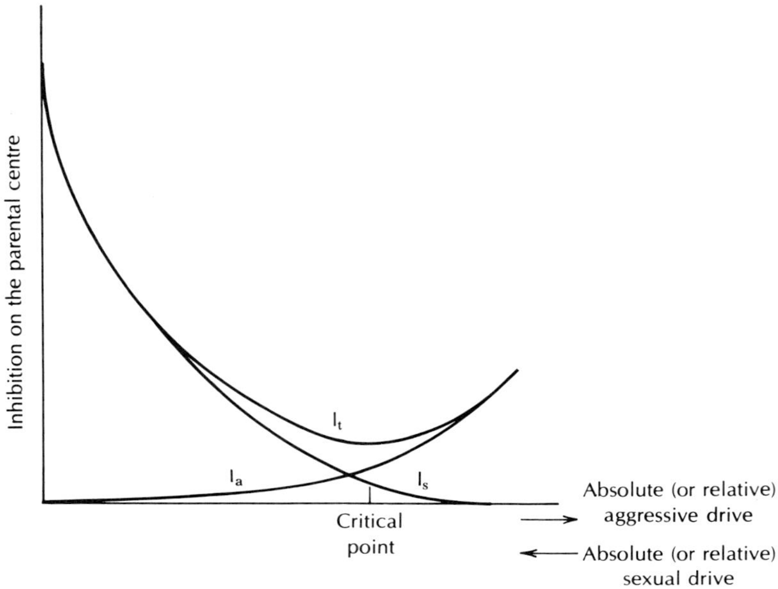

Fig. 17.8 Sevenster's schematic representation of variation in total amount of inhibition on parental fanning as a result of interaction of sex and aggression. (After Sevenster, 1961.)

I_s, the inhibition due to sexual factors; I_a, that to aggressive factors; I_t, the total. The mechanism controlling fanning is termed the "parental centre".

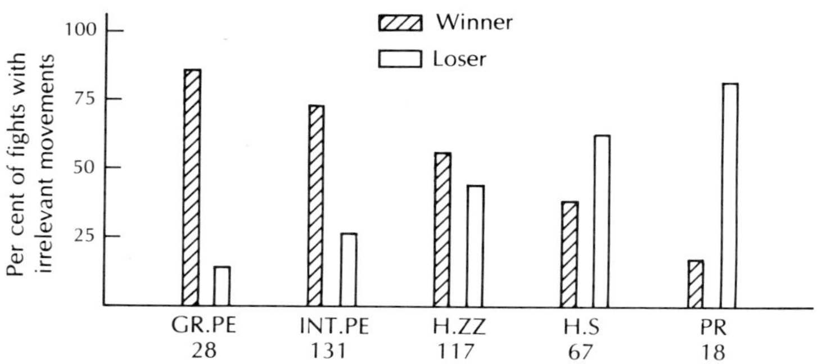

Fig. 17.9 Apparently irrelevant movements made during fighting by Burmese red junglefowl. The distribution of different movements between winners and losers. (After Kruijt, 1964.)

GR.PE, ground pecking; *INT.PE*, intention ground pecking; *H.ZZ*, head zigzagging; *H.S.*, headshaking; *PR*, preening.

amount of freezing shown in a number of ways; for instance by manipulating the strength of the stimulus, or the rearing conditions of the animals. The stronger the tendency to freeze, the greater the delay after the stimulus before the transitional state was reached and grooming appeared. A particularly interesting comparison was that between *Microtus agrestis* and the more timorous *Clethrionomys britannicus:* as shown in Fig. 17.10, grooming was much slower to appear in the latter species. However, the disinhibi-

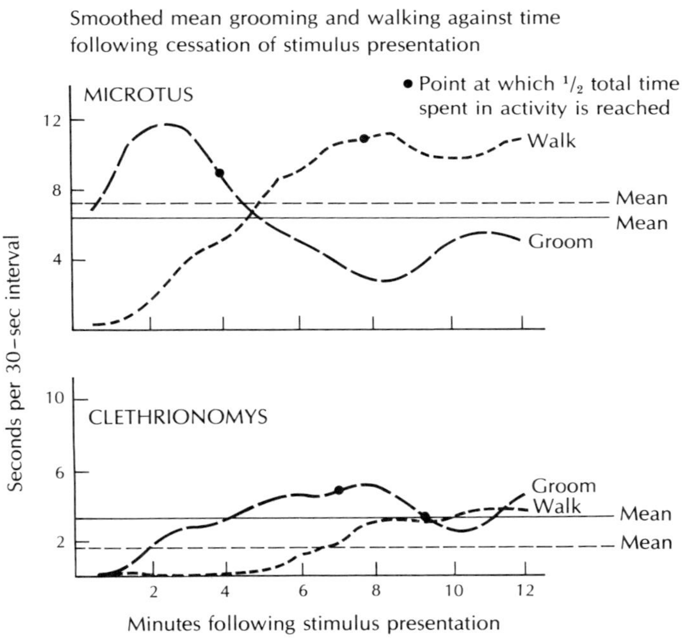

Fig. 17.10 The effect of a fear-eliciting stimulus on grooming and on locomotion in two species of vole. Ordinate, time spent in each activity per 30-sec interval after stimulation. (After Fentress, 1965.)

tion hypothesis proved inadequate in accounting for the detailed course of grooming. For example, the total amount of the apparently irrelevant activity of grooming shown in the standard time of 12 min was greater in the conflict situation than in a comparable control period for *Microtus* but less for *Clethrionomys.* This would not be predicted from the disinhibition hypothesis without additional assumptions. Furthermore, when individual records were analysed, there was a number in which grooming was not followed by walking: these can be accounted for on the disinhibition hypothesis only on the assumption that some activity other than walking was in conflict with freezing and inhibiting grooming. Thus Fentress suggests that the disinhibition hypothesis accounts for the overall occurrence of grooming but may need to be supplemented by further considerations: these are discussed below.

The disinhibition hypothesis has elements in common with the view advanced earlier by Kennedy (1954) that displacement activities were analogous to the phenomenon of positive induction, as described by Sherrington (1906), in which the reciprocal inhibition of two spinal reflexes lowers the threshold for a third. This comparison must not be pressed too far, however, because of the obvious differences in complexity between the mechanisms.

Von Holst and von St. Paul (1963) also advanced a similar hypothesis in describing the behaviour of a cockerel with two electrodes in the brainstem, stimulation of one causing fleeing and of the other, sleep. When hungry, the elicitation of a slight fleeing tendency causes the bird to stop eating. If the sleeping stimulus is then given, it may feed briefly before sleeping. They ascribe this to the liberation of feeding when sleeping inhibits fleeing.

All these authors thus agree that "displacement" activities are not to be sharply distinguished from the same activities appearing in their normal functional context, though Sevenster points out that displacement fanning is characterized by an almost complete dependence on the interaction of sex and aggression, but normal fanning is not. In all these cases the so-called "displacement" activity occurs at a point of balance between two other tendencies which would otherwise dominate the scene. Rowell, *loc. cit.*, emphasizes that preening occurs not only at the points of balance in an approach/avoidance conflict, but is also especially probable at other changeover points; for instance between resting and feeding (see also Fentress, cited above). Data pointing to a similar conclusion are available for grooming in the rat (Bindra and Spinner, 1958; Bolles, 1960), comfort movements in sticklebacks (Tugendhat, 1960b), and various other activities (Räber, 1948; Armstrong, 1950; Tinbergen, 1952; Moynihan, 1955; Andrew, 1956a, b; Schmidt, 1956; van Iersel and Bol, 1958; C. H. F. Rowell, 1961).

3. *EFFECTS OF NON-REWARD*

A variant of the disinhibition hypothesis has been advanced by McFarland (1965a, c; 1966). He pointed out that activities labelled as displacement have been recorded in situations in which it is difficult to postulate conflict between two behavioural systems—for instance with physical thwarting of appetitive behaviour, or with absence of the object of consummatory behaviour—in other words, in situations of frustration. McFarland points out that one factor in the incompatibility between two activities is that the animal cannot pay attention to the eliciting stimuli for both: which activity is dominant depends in part on the stimuli to which it pays attention. He thus suggests that frustration also involves a change in attention. When frustrated, the animal switches attention to stimuli other than those eliciting the ongoing activity: in other words, disinhibition involves a switch of attention.

McFarland's hypothesis was worked out principally as a result of observations on displacement pecking by doves frustrated from drinking. His evidence includes experiments showing that doves do in fact learn more about incidental aspects of a discrimination situation (and thus presumably pay more attention to them) when partially rewarded (and thus partially frustrated) than when consistently rewarded; that such partially rewarded doves respond more readily to previously irrelevant stimuli than do consistently rewarded ones; and that birds which are most resistant to extinction, and which are therefore believed to have paid most attention to incidental cues during training, show most displacement pecking.

McFarland's view is of interest in that it involves the suggestion that the mechanism behind displacement behaviour is related to that which Sutherland (1959) suggested operates in discrimination learning (see pages 586–590). On the other hand, we have already seen that the concept of attention has dangers in that it can too easily explain anything. In any case, the generality of McFarland's hypothesis remains to be assessed.

4. *POSSIBLE EFFECTS OF NON-SPECIFIC AROUSAL*

Another possible factor in the causation of displacement activities is a non-specific influence resulting either from the conflict or from other factors in the situation (Bindra, 1959; Hinde, 1959a). We have already seen (page 209) that a variety of apparently irrelevant activities, such as sexual, grooming, and preening behaviour, may be elicited by mild extraneous stimuli (see also McKinney, 1965). Such effects could be mediated by the brainstem reticular system: in many situations in which displacement activities occur the degree of "arousal" (Chap. 9) is high, and it is possible that this could influence which activities occur and, perhaps, the intensity with which they are shown. Bindra (1959) has suggested that only activities with high habit strengths are likely to occur when the degree of arousal is high, and this is consistent with the frequency with which activities like feeding and preening occur in conflict situations.

From a different theoretical viewpoint, Miller and Stevenson (1936) noted that when rats, previously trained to run down an alley for food, were given a number of trials without food, activities such as sniffing and grooming increased and were often given in an explosive manner. They suggested that absence of reward led to conflict between the learned response of eating and that elicited by the empty food dish, and that conflict produced proprioceptive stimulation which facilitated the apparently irrelevant activities. Essentially similar hypotheses were subsequently put forward by a number of other learning theorists (e.g., Brown and Farber, 1951; Hull, 1952; Amsel, 1958), the implication usually being that a "general drive" (see page 204) was augmented. The literature has been reviewed by Brown (1961).

We have already seen (Chap. 9), however, that the evidence for such non-specific motivational effects is far from conclusive. Furthermore, many apparently irrelevant activities are characteristically incomplete (Tinbergen, 1952): if they were influenced by motivational factors determining strong incompatible tendencies, the opposite would be expected.

In his study of the responses of voles to a moving overhead stimulus (see page 411), Fentress (1968) suggested that the stimulus had effects which were neither specific nor completely general. He found that the stimulus was most likely to elicit fleeing, rather than freezing, if the animals were, or had just been, engaged in locomotor activity. Since the stimulus also caused a momentary increase in other ongoing activities (e.g., grooming), he suggested that it had a non-specific facilitatory influence on ongoing behaviour. In his view, therefore, the effects of the overhead stimulus can be understood only if facilitatory relations on the several activities are postulated in addition to the inhibitory ones demanded by the disinhibition hypothesis. Thus he suggests that diverse motivational inputs may "feed into" displacement activities, but only when they are of moderate intensity. At higher intensities the seemingly irrelevant patterns of behaviour are replaced by more specific reactions to the inputs, so that displacement behaviour is reduced.

5. *OPERATION OF A DE-AROUSAL SYSTEM*

A somewhat different hypothesis is advanced by Delius (1967). Stimulating the forebrain and brainstem of gulls, he found a group of behaviour patterns (e.g., preening, staring down, yawning) which often appeared together when particular loci were stimulated: sometimes sleep appeared as well. Loci which did not produce preening also did not produce these other patterns. He suggests that these patterns, which appear often as displacement activities, reflect the activation of a more or less unitary system leading to "de-arousal". On this view, their appearance in conflict situations is due to a mechanism for arousal homeostasis brought into action by the arousal resulting from the conflict. There is, however, no further evidence to support this suggestion.

The explanations advanced for the occurrence of activities outside their usual functional contexts, in circumstances in which they appear irrelevant, are thus almost as diverse as the activities themselves. The evidence that some of these explanations are valid in at least some cases is substantial, and some of them do not invoke new principles additional to those already in use for behaviour not classed as displacement. What then remains of the category of displacement activities?

In part this depends on the level and type of analysis in which it is to be employed. For the field worker, concerned with the initial descriptive phase of the behaviour of a previously unstudied species, it will undoubtedly continue to be useful. The student of the evolution of behaviour, interested in the evolutionary sources of display movements (Chaps. 27 and 28), will also continue to employ it. But for causal analysis, the term can no longer be used to imply a category of instances of behaviour whose causation involves principles not shared by other types of behaviour. Their common feature is that they occur in situations of conflict or frustration. Conflict, however, is almost ubiquitous; and the difference between activities occurring as "displacement" activities and those occurring as "transitional" activities (see Rowell and Fentress, quoted above) is only one of degree. In detailed causal analysis of a particular case, the utility of the concept fades away.

This view is emphasized strongly by Beer (1963a, b) on the basis of his study of the occurrence of nest-building behaviour during the incubation phase of blackheaded gulls. Previous authors had shown that such nest-building activities were associated with situations in which sitting on the eggs is hindered, or in which the eggs or nest provide the "wrong" feedback for sustained sitting. On the basis of such observations the nest-building had been ascribed to frustration of the incubation behaviour (Kirkman, 1937) or inadequate consummatory stimulation (Bastock et al., 1953). Now the types of nest-building behaviour shown are "sideways building" and "collecting", and the incubation activities "settling" and "shifting" the egg in the nest. By analysing the sequences of behaviour shown, Beer found that the evidence for common causation drawn from analysing sequences of behaviour cuts across this grouping based on function. For instance, "sideways building" was more closely linked in time to "rising and settling" than to "collecting", and "settling" was more closely linked to "sideways building" than to "shifting". It is thus not useful to talk about the interaction of drives labelled in terms of their functional consequences, and judgements made about what is "irrelevant" or "out of context" from a functional point of view may have no causal validity.

So far displacement activities have been discussed as though they were by-products of conflict or frustration, mere epiphenomena extraneous to the mainstream of the organism's activities. In at least one case, however, there is strong evidence that this is not the case. As we have seen, the courtship of the three-spined stickleback involves a chain of activities in which the male first performs a zigzag dance and then leads the female to the nest (Fig. 25.1). The zigzag dance involves an element of aggression towards the female. Wilz (1967) observed that leading was often interrupted by "pricking", an activity in which the male pushes the female towards the surface. This is more likely to occur if the male has just previously been stimulated by another male, less likely if another female has been presented. As a result of the pricking, the female waits a short distance away, and while she does so the male usually goes to the nest and performs various activities at the nest, such as "fanning" and "creeping through", which, because of their irrelevance, are usually labelled as "displacement activities." When the male returns to the female, his tendency to behave sexually is greater, relative to his tendency to attack, than before. By preventing the male from performing the displacement activities at the nest—for instance by preventing "creeping through" by blocking the nest entrance—Wilz was able to show that the motivational change depended on the performance of the displacement activities. This change is essential for the completion of the courtship sequence. Whereas aggressiveness is at a premium in territorial defence, males must change to a more sexual state on the arrival of a female: the "pricking" causes the female to wait while the male's performance of these displacement activities facilitates the motivational change.

17.8 REDIRECTION ACTIVITIES

Here the motor patterns appropriate to one of the conflicting tendencies are shown but are directed onto an object other than that which initially elicited them. This is often seen in winter flocks of birds: one individual is supplanted at a food source by a superior and, instead of retaliating, turns its aggressiveness onto an inferior bird. Similarly male blackheaded gulls, whose tendency to attack their mates is inhibited, may attack other birds (Moynihan, 1955). In herring gulls, pecking during aggressive encounters is often redirected onto objects in the environment (Tinbergen, 1959).

The distinction between redirection activities, displacement activities, and responses to suboptimal stimuli is not always clear-cut. Thus redirected pecking, as described for the herring gull, occurs also in junglefowl: in the early stage of the development of fighting behaviour the ground-pecking during fighting is indistinguishable from that during normal feeding, but later it has the vigorous quality characteristic of aggressive pecking. Thus it seems that irrelevant feeding and aggressive pecking occur in combination with each other. The vigorous ground-pecks in aggressive situations often result in swallowing, and when the bird is fighting a coloured stick it pecks more at food particles of a colour similar to that of the stick than those of a different colour. In such cases the categories of "displacement", "redirection", and "responses to suboptimal stimuli" seem to merge (Kruijt, 1964). Ficken and Dilger (1960), describing cases of redirected copulation in passerine birds, also suggested that redirection was a special case of reaction to subnormal stimuli.

Such findings would perhaps have been expected in terms of the conceptual scheme elaborated by Miller (1959) to account for the way in which the new target of the behaviour is determined. He postulates that the inhibited response generalizes onto other objects according to their similarity with the one which originally aroused it. The inhibiting response generalizes in the same way but with a steeper gradient. The inhibited response is thus most likely to generalize onto an object of intermediate likeness to the original one.

17.9 SEXUAL INVERSION

Certain types of sexual inversion among animals appear characteristically in conflict situations. Among birds, species vary widely in the extent to which the behaviour patterns of courtship and copulation are characteristic of both sexes, and sexual "inversion" is a matter of degree. The adoption of the female copulatory posture by males can, however, reasonably be described in this way. It occurs when the female is unresponsive to the male's courtship in the zebra finch (Morris, 1954a) and after copulation attempts in carduelines (Hinde, 1955/1956, and refs. cited). This can be understood on the assumption that sexual arousal in either sex involves an increased tendency to show both male and female patterns, though male patterns normally have a higher priority in males, and vice versa. If the behaviour of the characteristic type is thwarted, the other may be shown. Similar principles apply to other species (Barraud, 1955; Morris, 1955, and refs. cited).

The question of sexual inversion is, however, a complex one. In mice and many mammals there is a correlation between dominance in an aggressive context and masculine sexual behaviour by females (Mainardi, 1965), and in other species conditions of rearing play an important role.

17.10 REGRESSION

The literature on regression in animals is now fairly extensive, but there has been little attempt to analyse the processes underlying it. Indeed, the phenomena commonly lumped under this heading are probably physiologically heterogeneous: there is no a priori reason for thinking that reversion by an adult animal to a more juvenile form of behaviour has anything in common with trying again a solution found earlier to be successful in a problem situation. Some cases of juvenile behaviour in adults imply that the mechanisms controlling the juvenile pattern remain present but latent in the adult and become active when the adult pattern is thwarted (e.g., Holzapfel, 1949).

17.11 IMMOBILITY RESPONSES

When wild-caught great and blue tits are held in the hand, they sometimes lie on the outstretched palm for a minute or longer, their eyes open but bodies limp. They fly immediately at a sharp noise, or if they are thrown into the air. Similar responses to handling or stroking occur in a wide range of animals and are often accompanied by analgesia: they have been ascribed to frustration or conflict (e.g., Armstrong, 1947; Ratner and Thompson, 1960). The relation of such responses to the "freezing" postures used when escaping from a predator has not been investigated.

17.12 RESPONSES TO FRUSTRATION

When an animal, engaged in a sequence of behaviour, is unable to complete it because of a physical barrier, the absence of an appropriate stimulus link, or for similar reasons, its behaviour may take a number of forms. One possibility is the appearance of investigatory behaviour or trial and error, which may lead the animal into a situation which makes possible the completion of the chain. Another is a response to a normally inadequate stimulus situation, making possible the completion of the sequence albeit possibly in a nonfunctional manner. In other cases the animal may show one of the types of behaviour already discussed as appearing in conflict situations, such as displacement or aggressive behaviour (see page 336).

One feature which seems more frequent in frustration than in conflict situations is an increase in the vigour of the behaviour being shown at the time. This is well exemplified in an experiment by Marzocco (1951). Rats were trained to obtain food by depressing a bar in a Skinner box. After preliminary habituation and training in bar-pressing, the animals were given three rewarded trials followed by a series of unrewarded ones. The mean force exerted on the bar rose from 28 g in the first four trials to 35.9 g in the responses following the first unrewarded one. The force exerted then fell off gradually but remained above the mean level in the initial four trials for about ten trials after the first unrewarded one. The increase in force was found to be related to the length of food deprivation at the time of frustration. However, the interpretation of these results is not clear. Although it is possible to regard the increase in vigour as a motivational consequence of frustration, it could also be ascribed to previous differential reinforcement for pressing harder under conditions of no reinforcement (e.g., Bolles, 1967).

17.13 SUMMARY

1. Various types of behaviour occurring in conflict situations are discussed.
2. The usual outcome of the simultaneous action of factors for two or more types of behaviour is the suppression of the activities appropriate to all but one. This is called "behavioural inhibition".
3. Sometimes one or both of the conflicting tendencies are expressed in incomplete patterns of behaviour—intention movements, alternation, ambivalent postures, or compromise behaviour.
4. Autonomic activity often occurs in conflict situations.
5. Sometimes, in a conflict situation, an animal shows an activity which appears to the observer to be irrelevant to either of the conflicting tendencies. Such "displacement activities" do not require new principles for their explanation: some examples can be ascribed to autonomic activation, others to "disinhibition".
6. In some conflict situations, the motor patterns appropriate to one of the conflicting tendencies is "redirected" onto another object.
7. Other categories of behaviour occurring in conflict situations include sexual abnormalities, regression, immobility responses, and aggressive behaviour.

section three

THE DEVELOPMENT OF BEHAVIOUR

18

THE STUDY OF THE DEVELOPMENT OF BEHAVIOUR

So far, in studying the immediate causation of behaviour, we have been concerned with an artificially limited span of time. For each unit of behaviour we have searched for events, processes, or conditions, internal or external to the animal, which are present about the same time as the behaviour in question. The identification of causes in this way immediately opens up further questions, for each cause must have a cause. The following chapters, therefore, are concerned with the development of behaviour in the individual.

A descriptive account of the ways in which behaviour changes with age might seem to be a necessary preparation for an analysis of the factors which influence those changes, but in fact really detailed descriptions of the development of behaviour over even a small part of the life span are rather rare.[1] Most research has been concerned with selected problems or aspects of behaviour, and the species used has simply been the most suitable for the task in hand. This concentration on the development of one type of behaviour at a time inevitably carries with it the risk of thinking of each different type as depending on a distinct mechanism, isolated from the mechanisms which underlie other types of behaviour and differentiating independently of them. This danger has been repeatedly stressed by Schneirla (e.g., 1946, 1952, 1956) and his pupils. However narrow the problem may be, it is essential to remember that development involves a nexus of causal relations, with action, reaction, and interaction, both within the organism and between organism and environment, at every stage.

[1] There are, of course, exceptions, among which the studies of Kuo (1932) on the chick embryo, Nice (1943) on the song sparrow (*Melospiza melodia*), Eibl-Eibesfeldt (1951a) on the red squirrel (*Sciurus vulgaris*), Sauer (1954) on the whitethroat (*Sylvia communis*) Messmer and Messmer (1956) on the blackbird (*Turdus merula*), Kruijt (1964) on the Burmese junglefowl (*Gallus gallus*), Hines (1942) and Harlow and Harlow (1965) on the rhesus macaque (*Macaca mulatta*), Gesell (1954) on the human infant, Bolles and Woods (1964) on the rat, and Schneirla, Rosenblatt, and Tobach (1963) on the kitten exemplify diverse approaches.

Having recognized complexity, however, it is also necessary to abstract and simplify in order to analyse the processes involved. Here lies another danger, for until recently most of the questions asked about the development of behaviour have been in the form of dichotomies—is a given piece of behaviour "instinctive" or "intelligent", "innate" or "learned"? That such dichotomies are not only false but sterile has now been pointed out by a number of authors (e.g., Schneirla, 1951, 1952; Hebb, 1953; Lehrman, 1953; Beach, 155; Hinde, 1959b, 1968; Tinbergen, 1963), but the matter merits consideration in some detail.

In the first place, the innate/learnt type of dichotomy can lead to the ignoring of important environmental influences on development. This can occur if it is assumed that the factors influencing the development of behaviour are of two types only—genetic and those associated with learning. Such a view has recently been reasserted by Lorenz (1961, 1965) and Thorpe (1963b), the latter author taking it as a prime aim of ethology "to differentiate between complexity coded in the germ plasm (the truly innate components) and that impressed by the environment (imprinted or otherwise learnt)". Now environmental factors can influence the course of development in many ways, some of which do not come within any generally accepted meaning of learning. To take an extreme example, a given genetic strain of *Drosophila melanogaster* may be capable of full normal flight, or of erratic flight, or incapable of flying at all, depending on the temperature during development (Lát et al., 1961). If all environmental influences on to learning. Similarly, what a rat learns can be influenced by nutrition during development (Lát et al., 1960). If all environmental influences on development are said to operate through learning (as could be inferred from some definitions of learning), then the concept of learning becomes so broad as to be valueless.

A second objection to such "either–or" classifications is that in practice innate behaviour is defined solely in negative terms: a response is said to be innate or unlearned just so long as no learning process (or other environmental influence) has been identified in its ontogeny. Often, indeed, behaviour is labelled as innate on the basis of a "deprivation experiment"—animals which develop in an environment lacking some factor believed to be important are compared with animals living normally. But no series of deprivation experiments can exclude *all* environmental influences. The difficulty here is not merely the theoretical one of proving a null hypothesis: to repeat a truism, all development depends on both nature and nurture. Just as a bird's egg develops only if kept within certain limits of temperature, for any organism there are critical ranges of environmental variables outside which development is distorted or absent. Similarly, there are limits to the environmental conditions under which any pattern of behaviour will develop: in some cases these may be coincident with those in which life itself is possible, so that the behaviour appears after any conceivable deprivation experiment which the animal survives, but usually they are narrower. For this reason, it is sometimes useful to characterize items of behaviour according to their stability or lability under environmental influences (Hinde, 1959a). The form of some movements is so stable that it cannot be modified by any environmental conditions within the viable limits: in particular, modifications cannot be effected by learning (Lorenz, 1961; see pages 19–22). In other cases the form or nature of a movement is labile, varying greatly according to environmental influences, human speech being an extreme example.

Third, dichotomies between learnt and innate behaviour involve the assignment of units of behaviour into one or the other category, rather than an analysis of the factors and processes involved in their development. Since the appearance of any response depends on *both* nature and nurture, attempts to classify responses according to one source or the other are doomed to failure.

If attention is concentrated on *factors or processes*, as opposed to units of behaviour, it is often convenient to distinguish the consequences of "maturation" (tissue growth and differentiation) from those of "experience" (the contributions of stimulation from the developmental medium), although even this is useful only so long as we do not forget that it is only a convenient abstraction (Schneirla, e.g., 1965, 1966). The distinction depends, in fact, on the extent of the universe which is being considered. Processes of maturation by definition occur in, and depend on, the intraorganismic environment: they are not influenced by extraorganismic factors only insofar as the environment relevant to them is maintained within limits by homeostatic mechanisms. Furthermore, the consequences of tissue differentiation may vary with the experience of the individual: even when processes of tissue growth and differentiation have led to the development of an eye, that eye may not be functional until it has been exposed to the light (see page 471). In any case a dichotomous classification of processes still has considerable difficulties, for it implies opposed or incompatible categories, whereas in practice the processes concerned constitute an interacting and reacting system. As we shall see, learning depends upon the genetic constitution of the animal; and behaviour may not be learnt and yet depend on learning processes at an earlier stage in ontogeny (Lorenz, 1961).

In spite of these difficulties, it remains true that the determinants of the characteristics of the organism come from only two sources: the zygote and the environment of the developing individual (Pringle, 1951). Some writers, therefore, have used a dichotomy of "sources of information". For instance, Lorenz (1961, 1965), who regards any environmentally induced modification in the mechanisms determining behaviour as learning (see page 426), emphasizes that an organism acquires "information" from only two sources, its genetic structure and its environment. In his view, therefore, "innate" and "learned" are defined "by the origin of the information which is the prerequisite for adaptedness to any one among the givens of the environment". Thorpe (1963b) emphasizes this point even more strongly, suggesting that it should be possible to determine quantitatively how much of the information necessary to specify an adult characteristic comes from the environment and how much was genetically encoded.

We may consider first Thorpe's implication that "information" can be used in a precise mathematical way to assess the contributions of gene and environment to particular patterns of behaviour. There are at least three difficulties in this. The first is a theoretical one: there is a possibility of interaction between genetic and environmental information. As mentioned above, development depends on interaction between organism and environment. Hebb's (1953) argument, that it is as meaningless to ask how much a given piece of behaviour depends on genetic factors and how much on environmental, as it is to ask how much the area of a field depends on its length and how much on its width, is still cogent. If we turn our attention to the variance found within a population, then it can be argued that the variance of the character under study is the sum of quantities representing phenotypic variance, genotypic variance, and variance in interaction between genotype and environment (Broadhurst and Jinks, 1963; Konishi, 1966): this, however, is another issue.

A second, practical, difficulty concerns the use of information theory in such contexts. To measure the information required to describe a given state or a given pattern of behaviour, we must know the number of states or patterns of behaviour possible. Calculations of the information content of an egg start from knowledge of the proportions of the various atoms present. These, however, are already a selection from those which *might* have been incorporated. Even more so in the adult organism: we can say that to specify a chaffinch's song, say, certain parameters must be defined, but who can assess the number of possible parameters which it might have had? And how can we equate the assumptions made in assessing the information content of the pattern of behaviour with those made in assessing that in the zygote or the environment?

A third difficulty is discussed in some detail by Thorpe himself. Estimates of the information content of the organism indicate an enormous increase from zygote to adult, even in such simple cases as plants which can be grown indefinitely in relatively simple homogeneous media. Neither the hereditary material nor the environment seems to be able to provide the increase in information present in the adult organism. In Thorpe's view, "this seems to push us to a new form of the old vitalist position which assumed that in living organisms purely physical laws no longer operated", though he regards the difficulty as at present unresolved. Thus, in practice, it seems that considerable difficulties beset an information-theory approach to the task of obtaining quantitative measures of the extent to which an adult character is determined by the genes and the extent to which it is determined by the environment.

Returning to Lorenz's more descriptive use of the dichotomy between "sources of information", it is important to bear two points in mind. First, Lorenz uses "learning" in a very broad sense to include "all adaptive modifications of behaviour". Second, Lorenz focuses on a quite different question from more physiologically oriented students of behaviour. He is avowedly not interested in the interplay between organism and environment which is the essence of development. Rather is he concerned primarily with the *adaptedness* of behaviour and with the means whereby this is achieved. For example, when he finds that a stickleback reared in total visual isolation responds more aggressively to a stickleback model with a red belly than to a model without such a red belly (see page 60), he argues that the relevant information (i.e., that it is adaptive to respond aggressively to a stickleback with a red belly) cannot have been environmentally acquired and must have come from an "innate source". [In practice, examples such as this give a much clearer idea of what Lorenz means by "the innate" than statements such as that "the innate" is "what must be in existence before all individual learning in order to make learning possible" (Lorenz, 1965).]

Lorenz's dichotomy throws into relief an important point which is too easily overlooked in studies of development. He points out that, since behaviour is on the whole adaptive (in a biological sense), some explanation is required for the fact that the environmental factors which operate in the development of behaviour act to produce results favourable to the organism, and not deleterious ones. The demonstration that the development of a given adaptive pattern of behaviour depends on specific environmental factors can never be the whole story: we must also ask why the developing organism responds in an adaptive way to those factors, and not in some other way. In particular, we must ask why an animal tends to learn what is good for it. Lorenz reviews evidence that learning abilities are not generalised and diffuse, but limited to certain aspects of behaviour (see pages 583–585); that this limitation is in fact adaptive; and that the limitation is ultimately genetically determined: in his view all learning is performed by mechanisms which contain "phylogenetically acquired information".

Important though this is, Lorenz's thesis runs into a number of difficulties. In the first place, his argument sometimes slides from concern with this dichotomy of "sources of information" to discussion relevant only to a dichotomy of types of behaviour: clearly, if a dichotomy is to be used, one must be clear what is being classified. Second, it depends on the selection of examples. In the case of the stickleback which has no chance to learn what a rival male looks like and yet somehow responds appropriately, it may not advance understanding to ascribe the appropriateness of the response to "innate information", but the nature of the implication is clear. But consider a rodent which, after rearing by its mother, responds to a strange situation in a manner and to an extent which is presumably adaptive. We know that the response can be influenced by experience during rearing (pages 534–538). Is it helpful to describe the amount of stimulation the infant gets in the nest as providing information about the nature of potentially dangerous situations it will experience in adulthood?

Third, Lorenz's dichotomy results in part from the beauty of his own metaphors. From arguments based on cases like that of the stickleback in which the role of learning is limited, he implies that some behaviour depends on a "blueprint" contained in the genome which requires only environmentally provided "building stones" for its realization. But if the nature of the resultant behaviour depends at all on the nature of these "building stones", it cannot be said to be independent of environmental "information".

Fourth, Lorenz's dichotomy leads to a difficulty in the interpretation of deprivation experiments. In the stickleback case (see above), for instance, he argues that if the male is reared in isolation ("if we withhold from our stickleback all information on the colours of its rival"), and it subsequently responds differentially and appropriately, "we can conclude that the information about the colour pattern has been relayed by the genome". Now Lorenz believes that such experiments can justify assertions only about what is *not* learnt. If, as a result of deprivation, the stickleback had failed to respond, he argues that we cannot assert that the response is dependent on experience of other individuals, because deprivation may have involved the withholding of a "building stone" necessary for the realization of the blueprint. Most people would argue just the opposite—namely that the so-called deprivation experiment can only tell us that an environmental influence is important, not that it is not.

Lorenz's dichotomy of "sources of information" thus runs into a number of logical difficulties and is in any case concerned with the adaptedness of behaviour, not with the detailed course of its development. For most students of development, the immediate problem is to disentangle a pattern of changes occurring through time. Each stage in the development of an organism depends on preceding stages, and results from interactions between the organism or part of the organism and its environment. The behaviour of the mature organism is thus not the consequence of a series of straight-line developmental processes leading from gene to adult pattern, modified in some instances by the environment. Rather, we must think in terms of a web of causal relations such that each part may interact with other parts and with the environment at every stage. What then are the relevant and useful questions to ask?

First, given two organisms that differ in behaviour, we can ask if the difference can be traced to a genetic difference, or if it is due to environmental factors. It is in the study of *differences* that quantitative methods are most likely to be fertile, and a dichotomy of *differences* is useful where the other types of dichotomy we have discussed are not (Hinde and Tinbergen, 1958; Jensen, 1961; Hinde, 1968).

Evidence that a difference in behaviour is to be ascribed to genetic differences must come ultimately from the rearing of animals, known to differ genetically, in similar environments (see, e.g., Caspari, 1958; Fuller and Thompson, 1960; Ewing and Ranning, 1967). For instance, adult great tits use their feet to hold particles of food, but adult chaffinches never do so, even when they have been reared from eggs by great tits in a great tit's nest. We can therefore conclude that the differences between the post-hatching environments of the two species are not responsible for the difference, and the operation of genetic factors seems likely. However, such evidence must not be overinterpreted. The occurrence of a behavioural difference between inbred strains or between species is evidence for a genetic influence on the behaviour, but does not imply that environmental factors are not also important. It remains possible that the intra-egg environment is responsible for the above difference between chaffinches and great tits, and further work would be necessary to discover the manner in which this, in its turn, is genetically determined. In addition, we are not entitled to conclude that learning does not enter into the development of the behaviour in the adult great tit—the occurrence of learning may depend on a certain genetic constitution, present in great tits but not in chaffinches. In fact, both observational (Hinde and Tinbergen, 1958) and experimental (Vince, 1964a) evidence indicate that learning does play a part in the development of this skill (see page 443).

Evidence for an environmental influence is the converse of the above: if we take two animals of the same genetic constitution, or two samples from the same strain, and rear them under different conditions, differences in behaviour can be ascribed to those in the environments. If, for instance, we rear some domestic chicks in groups and others in visual isolation from each other, and find that the latter follow a moving model when the former do not, we can associate this difference with that in the rearing conditions (Guiton, 1959). If we can find no difference, we cannot of course conclude that the conditions of rearing have no influence, but only that that range of conditions tested is equally suitable (or unsuitable) as assessed in terms of the criterion used. Nor, from such an experiment, can we conclude much about how the environmental factors operate. Guiton's view that the socially reared birds do not follow the model because they have already learnt to follow each other is only one of a number of possible explanations.

This brings us to the second type of question that can be asked about the development of behaviour—how development occurs. We must ask how the differences in genetic constitution influence the behaviour of the adult, what environmental factors are involved, and how they produce their effects.

Since the organism is an integrated system, any change in its heredity may have diverse consequences. We must thus expect behavioural characters to be influenced by more than one gene substitution, and gene substitutions to influence more than one behavioural character. Although there is considerable variation between characters in this respect, some having a considerable degree of stability and being influenced by few gene changes while others are influenced by the majority of gene substitutions, it is indeed found that most behavioural differences have a polygenic basis (e.g., Keeler and King, 1942). However, the number of genes involved in behaviour differences between strains is often small, and even single gene differences between strains may be associated with marked differences in behaviour. Thus the reduced mating success shown by "yellow" mutant males of *Drosophila melanogaster* is due to courtship of reduced intensity, resulting from the single gene substitution (Bastock, 1956).

Whether the genetic difference involves one or many genes, its effect on behaviour may occur in diverse ways: there may, for instance, be changes in sense organs, effectors, body proportions, or general excitability, as well as in neural organization. For instance differences in foraging behaviour between brown and rufous-sided towhees (*Pipilio*) are related to differences in weight and in the musculature of the legs (Davis, 1957); and differences in the seed preference of European finches to differences in the sizes of their bills (Hinde, 1959a; Kear, 1962; see page 687). Furthermore, the basic difference between two selected strains may not be that implied by any single response measure; for instance Searle (1949) showed that two strains of rats which had been selected for maze-learning ability differed also in "timidity": the low learning ability of the "dull" strain could be a consequence of anxiety in the circumstances of the experiment.

Where the young are dependent on maternal care, the route by which genetic differences produce behavioural ones may be very indirect. Ressler (1963) studied the behaviour of two strains of mice which were reared during infancy by foster parents of the same two strains in all four possible combinations. Offspring of both strains showed more visual exploration at 60 days of age, weighed more and had a higher expectancy of survival, if reared by parents of one strain than if reared by the other. The author points out that the differences could be due to the amount of handling given by parents to young. This, however, is influenced by the strain of the offspring as well as that of the parents: as a result, even this cross-fostering technique cannot completely separate genetic from postnatal parental influences.

In a later paper Ressler (1966) showed that four measures of operant performance in mice could be affected by the strain of mice by which the parents of the experimental animals were reared. Thus a difference in the environment provided for young by the two strains influences the behaviour of the young, and this influence is partially reflected in the parental situation so as to produce a similar influence on the behaviour of the second generation. Thus an environmental influence can be transmitted across a generation. Denenberg and Rosenberg (1967) have shown an effect across two generations: the amount of handling received by the grandmother in infancy influenced the behaviour of her grandpups. In this case the occurrence of free environment experience some time during the mother's early development was necessary for the effects of the grandmother's early handling experience to express itself in the grandpups. Such data indicate that one must not be overhasty in concluding that strain differences are due to genetic factors (Ressler, 1963; Broadhurst, 1961). The complexity of the issues involved is also indicated by the finding that the strain of pups may influence the behaviour of their (foster) parents in contexts other than parent–young interaction (Hudgens et al., 1967).

Just as differences in genetic constitution can affect development in diverse ways, so also can differences in environmental factors. They may, for instance, influence developmental processes directly, produce a change in hormone balance, or provide an opportunity for learning. The student of the development of behaviour must aim at discriminating between these and other possibilities.

In the following chapters some problems concerned with the development of various aspects of behaviour will be considered. While acknowledging the complexity of developmental processes, it is necessary to make an analytical approach, and consider first the development of isolated fragments of the behaviour machinery. Their integration will be considered later. The techniques most used in the study of behaviour have been variants of the deprivation experiment, in which the question asked is: "How does behaviour develop if this or that factor, believed to be important in normal development, is eliminated?". Much of our discussion will therefore be concerned with the extent to which development is stable under changed environmental conditions. Having found that a given experimental factor affects the behaviour under study, we can then ask how it acts.

SUMMARY

1. A dichotomous division of behaviour into "innate" and "learnt" components is not fruitful because
   (*a*) It neglects effects of the environment other than those produced by learning.
   (*b*) The category of "innate" is, in practice, defined solely in negative terms.

(*c*) It involves the classification of units of behaviour rather than the processes upon which development depends.

2. A dichotomy of processes also is of limited value.
3. The possible use of information theory in assessing the contributions of genotype and environment is considered.
4. The most useful questions to ask about the development of behaviour are concerned with differences. Given two organisms that differ in behaviour, is the difference to be ascribed to genetic or experiential factors? And how do those factors exert their effects?

19

DEVELOPMENT: THE FORM AND ORIENTATION OF MOVEMENT PATTERNS

If a pattern of movement differs recognizably between the individuals of a species, we can attempt to relate the differences to variations in either genotype or environment, and thus begin an analysis of the factors controlling its development. Usually, it is environmental variations which are pertinent, and we can then try to assess how the environmental features in question produce their effects. The language a child speaks is determined by the language spoken in its home: we have little hesitation in ruling out any important genetic influence on the differences between the speech of children brought up in homes speaking different languages, and can set about analysing the nature of the "imitation" that occurs. More rarely, individual differences in the form of a movement pattern can be traced to genetic differences, and here, too, we can attempt to trace how the effects are produced: an abnormal gait in mice, for instance, may be due to abnormalities in the effectors or in the middle ear.

If a movement pattern does not show consistent differences between individuals, identification of the factors influencing its development is more difficult. Its constancy of form tells us little except that the range of genotypes and range of environments are adequate for it to develop normally. One possible next step is to assess the effects of depriving the developing organism of environmental influences likely to be important. We can do this by a deliberate deprivation experiment, but some types of environmental influence can be ruled out by direct observation. If, the first time a movement appears in its usual context, it does so in its characteristic form, we can be sure that practice reinforced by the proper functional consequences is not essential. We may consider some examples.

19.1 MOVEMENTS WHOSE FORM IS NOT DETERMINED BY "PRACTICE" OR "EXAMPLE"

The motor movements used in flight by the white butterfly (*Pieris napi*) are already developed when it emerges from the pupa, and such improvement in flying ability as occurs during the first few days is probably related solely to the hardening of the wing cuticle, and not to changes in the movements themselves (Petersen et al., 1956/1957). Practice thus plays little part in the development of the form of the movements. Similarly, young axolotls show swimming movements even before they have left the egg (Coghill, 1929; see also Tracy, 1926); and characteristic prey-catching movements can be elicited from newly metamorphosed specimens of the toad *Xenopus laevis*, a species which is a filter-feeder in the tadpole stage (Eibl-Eibesfeldt, 1962). Young nidicolous birds show the species-characteristic begging movement before they have been fed by their parents: although the upright posture used may be aided by support from the nest structure, the latter is not essential for it (Vince, in press). They also make preening movements before their feathers are fully grown, and show the species-characteristic freezing position on the first occasion that they hear their parents' alarm call (e.g., Nice, 1943; Barraud, 1961). Many of the signal movements (Chap. 27) used by lower vertebrates have been seen in individuals of known history on the first occasion they were exposed to the appropriate stimulus situation (e.g., E. Cullen, 1960); though in gulls (Moynihan, 1959) and goldeneye ducks (Dane and van der Kloot, 1964) such movements are at first more variable than when their signal function is developed.

Although the bulk of the evidence refers to invertebrates and lower vertebrates, similar examples are well known in mammals. Eibl-Eibesfeldt (1951a, c; 1953) found that the movements used in display and brood-rearing by both squirrels (*Sciurus vulgaris*) and hamsters (*Cricetus cricetus*) reared in isolation are normal in form. He also listed sixty-five movement patterns in the desert mouse (*Meriones persicus*), which seem to be independent of experience in their normal functional context. Gauthier-Pilters (1959) recorded that the following motor patterns of the dromedary appear "spontaneously" at the stated intervals in minutes after birth: chewing 10, rolling on the ground 74, jerking of the head 120, yawning 160, tail beating 198, shaking 304, grinding and headshaking 100, sucking 100, kicking 156, urinating 185, rubbing of neck against mother 294. The three phases of rising occurred at 10, 87, and 95 min after birth, and standing upright at 100 min. Although many of these movements may have been made prenatally, and some may even have been functional then, the appearance of the majority must have been without preceding practice in their normal functional context. Many other comparable examples are known.

The human baby has a number of stereotyped movement patterns which appear at birth and can be seen even in infants born prematurely. One example is the lateral head movement which, though not specific to feeding, is used in finding the nipple and sucking. Another is the Moro reflex given on stimulation of the vestibular organ and/or the muscles in the neck, which is almost certainly a phylogenetic remnant of the movement whereby other young primates cling to their mothers (Prechtl, 1965). Smiling appears gradually in the weeks after birth, but it also is not dependent on example, since it occurs even in blind children (Freedman, 1964; Ambrose, 1961).

Turning to some experimental studies, Grohmann (1939) prevented young pigeons from carrying out their incipient flight movements by rearing them in narrow boxes, so that they were partially prevented from using their wings. When the control (normally reared) birds could fly a certain distance, the confined birds were released. Their flight performance was almost as good as that of the controls, indicating that they had not been handicapped by the period during which they had been restrained from making full flying movements.

Another case concerns the development of swimming movements in amphibian larvae (Carmichael, 1926, 1927). The movements first appear in an incipient form while the larva is still in the egg capsule, and gradually improve both before and after hatching. In an experimental group the movement was prevented by rearing the larvae under continuous chloretone anaesthesia from a stage prior to the development of the peripheral nervous system. When a control group, reared under normal conditions, was swimming well, the experimental animals were transferred to fresh water. Carmichael found that, as soon as the effects of the anaesthesia wore off, they were swimming as well as the controls. Although, in repeating this work, Fromme (1941) found some quantitative deficiencies in his anaesthetized groups, there seems no doubt that the main features of the swimming coordination do not depend on practice of any sort.

Surgical interference provides another type of evidence that the development of some fixed action patterns is independent of at least their normal functional consequences. Weiss (1941a, b) interchanged the left and right forelimb rudiments of salamander embryos at a stage when the anteroposterior axes of the limbs were already determined. Normal limbs developed, but they faced backwards instead of forwards. When the nerve connections had been established, the grafted limbs moved just as they would have done if they had been left in their original positions, working to move the animal backwards when the movements of the rest of the body were causing forward locomotion, and vice versa. A year's experience did not change this reversed movement of the grafted legs. Weiss has produced evidence that such motor patterns develop in the same way in the absence of sensory innervation, and that they persist after cord transection down to levels just in front of the tail segments (Weiss, 1950). Comparable results have been obtained in fishes (Sperry and Deupree, 1956; see also Székely, cited on page 45).

Although these experiments are concerned primarily with spinal cord/limb mechanisms, which are known to be relatively unmodifiable, they do serve also to illustrate that many movements not only develop independently of learning, but cannot subsequently be modified by it. Even in mammals there is apparently an early loss of embryonic plasticity, so that rearrangement of central-peripheral relations is subsequently compensated for only with the greatest difficulty, if at all. Monkeys in which the motor nerves to biceps and triceps had been interchanged learnt to inhibit the resultant reversed elbow movement, but made little positive correction (Sperry, 1947).

That the precise form of many species-characteristic movements is independent of the achievement of their normal functional consequences is shown also by the study of hybrids between species which themselves show different movement patterns. Unfortunately, few such cases have been studied in detail, as species showing markedly different movements are usually too distantly related to hybridize. The lovebirds (*Agapornis*), however, provide an interesting case. In this group some species (e.g., *A. roseicollis*) obtain nest material by cutting long strips of paper, bark, or leaves. The strip is grasped at one end in the bill and then tucked into the feathers of the lower back and rump for carrying to the nest. The feathers are raised as the material is tucked in, and subsequently lowered, holding it in place. Other species (e.g., *A. fischeri*) cut material in the same way, but carry it back to the nest in their bills. Hybrids between these two species were studied, from the time when they first started to cut material, by Dilger (1960). Although the hybrids were not successful in carrying material by tucking it into their feathers, they almost invariably tried to do so before carrying it off in the bill. The tucking movements were sometimes incomplete, but more often the movements themselves were normal, the failure being due, for instance, to the bird failing to let go of the strip with its beak after placing it among the feathers, tucking the material into an inappropriate part of the body, or grasping the material in the centre instead of at one end. Thus in many cases the full tucking movement appeared, even though inadequacy in orientation or selection of material prevented it from being successful.

Study of the stereotyped courtship movements of hybrid birds has also provided interesting data. For instance Sharpe and Johnsgard (1966) studied the behaviour of F_2 mallard (*Anas platyrhynchos*) $\times$ pintail (*Anas acuta*) hybrids. Where the displays were similar in the two parental species, they were unchanged in the hybrids; and where the parental species showed displays differing in degree, the hybrids showed intermediate movements. With displays present in only one parent species, the hybrids either performed them in an all-or-none fashion, or produced aberrant versions of them. Lorenz (1958) has shown that some *Anas* hybrids show displays which occur in related species but not in either parental form. (See also, e.g.,

Hinde, 1956c, for finch hybrids; Lade and Thorpe, 1964, for dove hybrids; Hörmann-Heck, 1957, for *Gryllus* sp.)

19.2 PROCESSES OF MATURATION

Where coordination develops with little or no practice, processes of growth and differentiation in the nervous and effector systems, occurring in the homeostatically controlled internal medium, must play a major role in its development. These processes were followed in some detail in Coghill's (1929) classic studies of larval axolotls. The successive stages in the development of swimming movements—reflex flexure, coiling, S-bending—were found to proceed synchronously with the development of new connections in the nervous system, which thus probably mediate them (see also Youngstrom, 1938; Whiting, 1954; Harris and Whiting, 1954; Gideiri, 1966).

This raises the further problem of the processes directing the growth of nervous connections. These are certainly diverse, but a number are well known: for instance, there is the tendency of growing fibres to veer toward organs in a stage of rapid proliferation, and the stimulating effect of ingrowing fibres on dendrite growth within a nuclear field. In the growth of peripheral nerves, mechanical guidance and chemical affinity (discussed in the next chapter) seem to play a leading role (review by Sperry, 1951). Other types of interaction between neurones may also be important: if a mammalian dorsal root is severed, leaving a peripheral field no longer innervated, that field may become innervated by fibres from a neighbouring root: presumably this did not occur before because of some sort of inhibitory interaction. Knowledge of this subject has until recently depended on the study of histological preparations, but tissue cultures of mammalian neurones show them to be in continuous motion: such studies are likely to yield important new data on the factors determining the formation of new connections (review by Doty, 1967).

Another source of evidence comes from studies of function after operative interference. The outgrowth of amphibian spinal axons seems at first to be largely non-selective. Although the termination of a given axon is determined in part by the proximodistal sequence of development within the limb, considerable freedom remains. Furthermore, as we have seen, in experiments involving grafted supernumerary or reversed limbs, normal muscular coordination (though producing abnormal limb movement) is achieved even under conditions in which the axon outgrowth appears to have been highly unusual (see page 45; Wiersma, 1931; Weiss, 1941a, b; Sperry, 1941, 1951, 1963). It has thus been suggested that the central relations of the motor neurones are adjusted to be appropriate to their peripheral connections. On this view the motor neurones must be distinguishable on the basis of the muscles they innervate: just how this occurs is not known. An alternative view is that the motor neurones early on become determined with a preference for a particular muscle. In lower vertebrates every motor neurone branches to many muscle fibres, and each muscle fibre is innervated by many neurones. The motor neurones may interact and compete where they meet, and thus eventually find their own muscle. On this view the normal function of the muscle is restored after a grafting or regeneration experiment because it comes to be innervated by appropriately labelled branches of the appropriate motor axons, the nerve fibres preferentially selecting the pathways which lead them to their proper destination (Mark et al., 1967; Székely, 1968).

Even if the motor neurones are fired according to preestablished patterns, the mechanisms of central coordination pose further problems. Until relatively recently, much of the research in this field was concerned with a controversy as to whether the more complicated patterns of behaviour are built up by the integration of simpler ones; or whether the larger "total" patterns are primary, the simpler forms being "individuated" from them. Coghill's work on *Amblystoma* indicated that the first movements to be shown are simultaneous mass movements of the trunk somites. These result at first in a C flexion, and later in an S flexion in which the body is thrown into lateral undulations which move cranio-caudally, as in swimming (cf. pages 39–42). The limbs move at first as part of these total patterns, but subsequently local patterns differentiate and their components become distinct from the behaviour as a whole. Generalising from these results, and contrary to previous views, Coghill suggested that in all vertebrates local reflexes emerge "as a special feature within a more dominant mechanism of integration of the whole organism". Comparable sequences of development have subsequently been found in a number of lower vertebrates. In teleost fish, movements of the pectoral fins, jaws, and opercula are at first uncoordinated and part of the movements of the trunk: later they become independent and coordinated (Gideiri, 1966; see also Tracy, 1926). Sequences basically similar to those found by Coghill in *Amblystoma* have been described by Youngstrom (1938) in larva of the anuran *Pseudacris triseriata* and by Hughes et al. (1967) in the embryo lizard *Lacerta vivipera*. Even in the chick, movements start with a flexion of the neck (3½ days incubation), while the limbs are still only buds. At 4½ to 6½ days waves starting in the neck spread posteriorly, and the limbs participate in these mass movements. Spontaneous independent limb movements occur only later (Kuo, 1967, and refs. cited; Hamburger, 1964, 1968), though they can be elicited by external stimulation at a time when the spontaneous movements are generalised (Orr and Windle, 1934). As we have seen, in the chick the alternating periods of activity and inactivity originate spontaneously within the central nervous system, and are not substantially affected by either transection of the cord or removal of the sensory input (see page 314; Hamburger et al., 1965; Hamburger et al., 1966; Hamburger and Oppenheim, 1967; see, however, page 445). Comparable rhythms of activity are found in lower vertebrates (refs. in Hughes et al., 1967). Whether the spontaneous and uncoordinated movements of the whole chick embryo give rise to the discrete responses which appear later is an open issue: Hamburger (1968) believes that the coordinated movements used in hatching arise independently.

While Coghill's principle of "individuation" of local movements from general ones is thus well supported in lower vertebrates and perhaps in birds, in mammals discrete reflex responses appear very early—at the same time as, or even earlier than, mass movements (Carmichael, 1934; Barcroft and Barron, 1939; Windle, 1940, 1944). Thus Coghill's generalization cannot be applied without reservation to all vertebrates. Whether there is an absolute discontinuity between lower vertebrates and mammals is, however, still not clear. Hughes (1965) suggests a threefold classification of behavioural development, with the Sauropsida and certain amphibians whose basic behavioural development is entirely embryonic forming an intermediate group. This is based largely on his studies of *Eleutherodactylus*, a toad which on hatching (at a size smaller than a housefly) can jump over a foot. Here the first movement of the legs occurs during trunk wriggles at 7½ days, but a day later they can move independently.

But as the behavioural development of more species is examined, it is likely that any simple system of classification will run into difficulties. It is profitable here to bear in mind Barron's (1950) emphasis on the nature of the response being studied—individuation from a total pattern seems to occur in the development of responses related to the axial musculature, but not in those cases where the appendicular musculature develops independently from the axial system. In any case, the dichotomy between total pattern and local reflex is a theoretical abstraction: the extent and magnitude of the body parts involved in any embryonic or foetal movement are determined by many factors internal and external to the organism (Kuo, 1967). Furthermore Kuo emphasizes that even the concept of the reflex is an abstraction, and that the whole organism is involved to some degree in every response. Finally, generalizations about relations between neuronal development and behaviour across phyla can have only limited validity, since close examination reveals differences even between animals at similar phyletic levels, which are presumably adaptive to different types of embryonic life (e.g., Whiting, 1955; Hughes, 1965; Gideiri, 1966).

It is natural to emphasize neural growth in discussing the development of movement patterns, but the development of the effectors is, of course, of equal importance. This is especially apparent in the development of vocalizations. The threat noises of the elephant seal (*Mirounga angustirastris*) develop from those of the pup, the progressive changes being correlated with an increase in the size of the animal. Similarly, the occurrence of the "clop threat" in this species depends on the development of the male's proboscis (Bartholomew and Collias, 1962). The role of anatomical features in determining the characteristics of sounds used in vocal communication has also been stressed by Rowell (1962) in her study of the agonistic noises of the rhesus macaque.

19.3 THE ROLE OF EXPERIENCE

Although many movement patterns develop in the absence of example and of practice leading to the normal functional consequences, this in itself does not prove that practice plays no role. Many species-characteristic movements have a developmental history involving fragmentary movements and incomplete performance (e.g. Kuo, 1967; Kruijt, 1964; Anthoney, 1968; Vince, in press): feedback from these could influence the course of development (Schneirla, 1956). For example, the development of sniffing behaviour in the rat has been studied in some detail by Welker (1964). In the adult, sniffing involves an integrated and precisely timed sequence of contractions in four groups of muscles which produce (1) bursts of polypnea, (2) recurrent protraction and retraction of the vibrissae, (3) repetitive retraction and protraction of the tip of the snout, and (4) a rapid series of discrete head movements and fixations. Development of this pattern occurs gradually, and during the first 5 days of life only the first of these components is apparent. Synchronous retraction of the vibrissae appears on the fourth day and protraction on the seventh. Repetitive movements of the nose and nostrils appear on the seventh day, and associated head movements on the eighth. At first the polypnea frequencies are only slightly higher than those of normal respiration, but they increase during the next two days, and again from the tenth to eighteenth days. As each type of movement appears it becomes synchronous with those already present. The length of sniffing bouts increases with age, especially between the eighth and tenth days, at which time the sniffing movements become associated with the locomotor components of exploratory behaviour. After about 18 days of age the sniffing pattern changes little. As yet we have no precise information as to the role of the early incomplete performance or of experience in the development of this pattern. Observation alone cannot decide whether the gradual improvement that occurs is a consequence of practice, or merely correlated with it. Although it is known that the coordination is maintained in the absence of olfactory feedback in the adult, it is still possible that proprioceptive and/or environmental factors play a part in its development.

An interesting case is provided by the movements used by some passerine birds in the course of holding an object down with one or both feet. This pattern of behaviour is characteristic of all members of some passerine species (e.g. tits and chickadees of the genus *Parus*) and never seen in others (e.g. chaffinches). Nevertheless, learning plays an essential role in its development in those species where it occurs: the initial attempts of a young tit to hold a mealworm under its feet are clumsy, and are perfected only after some time (Hinde and Tinbergen, 1958). The development of this movement was studied in detail in hand-reared great tits by Vince (1964a). All her hand-reared birds, whether reared with or without objects which could be held with the feet, showed some approximation to the normal patterns as soon as they were given objects to hold: indeed birds reared initially without access to seeds responded to them in this way more readily than birds which had previously not been restricted. But Vince also found that the rates at which the movement patterns became perfected and stabilized varied with the opportunity for practice with suitable objects: this is consistent with the view that learning plays a major part in its detailed development.

Even when the appearance of a new pattern of movement does depend on experience, careful analysis is necessary before the precise nature of the effect can be assessed. For example cockroaches normally clean their antennae with their front legs. If these are amputated, they do not learn to use their middle legs until 8 to 10 days later. However, the learning does not concern the actual use of the middle legs, for if pinned on their backs after removal of the front legs, they are able to use the middle legs immediately. Apparently the learning involves not a manipulative act, but a postural one—standing on three legs with the centre of gravity inside the tripod so formed (Luco, summarized in Bullock, 1966).

When new motor patterns do result from practice, one possibility is that the learning results from a comparison between the feedback obtained and a Sollwert, a partial match being reinforcing. Casual observation of the efforts of a human infant to stand or to walk suggest such a mechanism. That such a process of comparison may occur in the development of some vocal utterances we shall see later; but in most cases it is an unnecessarily complicated hypothesis, and only puts the problem a stage further back, since the development of the Sollwert and the comparator mechanism remain unaccounted for.

The possibility that proprioceptive feedback from incomplete movements may be important in other ways in the development of fixed action patterns has been stressed especially by Schneirla (1952, 1965) and Lehrman (1953), who discuss particularly Kuo's (1932) studies of the development of behaviour in chick embryos. When the embryos are observed through windows placed in the eggshell, changing action patterns can be observed, their form being clearly affected by the shape and relative size of the bodily structures, and by non-organismic factors such as amniotic contractions and movements of the yolk sac. Furthermore the embryo is constantly subject to stimulation both from its own activities and from outside sources. The processes of development thus involve complex interactions between the changing organism and its changing environment. Lehrman speculated particularly on the development of the pecking movement in the chick. Kuo's observations indicated that, as a consequence of early postural changes, the chick's head comes to rest against the thorax and is passively lifted and dropped with the heart beats. Within a few days an active nodding movement appears, which could be a conditioned response to gentle pressure from the thorax each time it is raised. After the sixth day, the head movements occur in relation to a wider range of stimuli, such as pressure from the yolk sac, or contact of the head with the toes. At around this time the beak begins to open and close as the head nods, and after the ninth day fluid enters the mouth as the bill-clapping occurs. On the seventeenth day the head, now much more mobile, occasionally lifts and thrusts forward, and this is followed by beak opening and closing. The elements of pecking thus appear gradually in the egg (see also Gottlieb and Kuo, 1965).

Opinion is divided as to the significance of these observations. Schneirla and Lehrman suggest that early interactions of the embryo with its immediate environment, and the movements which occur, may play an essential role in the development of the post-hatching pecking response. For instance, Schneirla (1965) suggests that the "gustatory and other afferent effects of these actions can function as agents reinforcing a progressive neural control of the head and bill actions by extrinsic stimuli". He also suggests that the reactivity of the head increases first to low intensity contact with soft yielding tissues in the egg, but later its movements come under optic control when "diffusion of impulses from aroused optic centres reaches and activates somatic afferent centres, exciting functions controlled by the latter in embryonic states" (see also pages 503–505).

Such suggestions have been greeted with ridicule by a number of writers (e.g., Lorenz, 1961; Eibl-Eibesfeldt, 1961; Thorpe, 1963a, b) but it is difficult to see why. True, they rest largely on observational evidence and are without experimental support, but the processes which Schneirla suggests are similar to those well known to occur in other contexts. In the first place there is evidence that the activity of an organ, even in the embryo, may help to bring it into a functional state; for instance, Peters et al. (1958) have shown that repeated stimulation facilitates retinal responsiveness in embryo chicks. Second, the importance and ubiquity of interoceptive conditioning, of just the type suggested by Schneirla and Lehrman, has been demonstrated by much recent Russian work (Razran, 1961).

Third, hatching is not a zero-point for the development of behaviour: chicks do "behave" before hatching. Not only do they show spontaneous movements, as demonstrated by Hamburger and his colleagues (see page

440), but they respond to stimuli from their own bodies and from the intra-egg environment (e.g., Kuo, 1967). They also respond to stimuli from the outside world—for instance to the warning call of the cock by remaining quiescent (Baeumer, 1955), and to light by a change in frequency of "beak-clapping" (Oppenheim, 1968). In some species the stimuli produced by the embryos before hatching lead to synchronization in the hatching of the clutch (Vince, 1969). Kuo points out that studies of isolated embryos or part-embryos, in which behaviour is recorded only crudely on kymograph paper, miss much of its complexity and inevitably emphasize spontaneity to the neglect of induced movements. Taking the studies of Kuo and Hamburger together, it is clear that prehatching behaviour involves a complex interplay between chick, its embryonic membranes, and other aspects of its environment, involving both spontaneous and induced movements. However, an influence of environmental factors on behaviour before hatching need not necessarily imply that they influence development. Even when the cerebral lobes of chick embryos are showing spontaneous electrical activity, there may be no transition to the low-voltage fast waves characteristic of attention when the chick responds to stimulation: this suggests that the response involves only lower neural mechanisms (Peters et al., 1965), and thus perhaps has little effect on subsequent behaviour. Thus while it is clear that such environmental factors can influence embryonic behaviour, their influence on development is unproven (see also Gottlieb, 1968a).

A fourth point which must be borne in mind in considering Schneirla's and Lehrman's suggestion is that it is not possible to make a priori assumptions about what sorts of experience will affect behaviour. For instance, Reyniers (1953) showed that young rats delivered by caearean section are unable to urinate until they have received gentle tactile stimulation of the genitalia, as normally provided by the licking of the mother. Such an observation might well not have been predicted.

Finally, there is strong evidence for the human baby that experiences before parturition can influence later behaviour. Prechtl (1965) found a direct correlation between the position of the legs during the last week before delivery and their post-natal posture and pattern of reflex behaviour (Fig. 19.1). A normal infant withdraws his legs to a pinprick on the sole of the foot, but extends them to mild stimulation. Breech babies with extended legs keep their legs in abnormal extension spontaneously, and may extend them after sharp stimulation of the sole. In babies born in a breech presentation with flexed legs, however, the reverse is found—extension to mild stimulation is diminished and the withdrawal reflex is exaggerated. Breech babies with one leg flexed and the other extended showed the patterns corresponding to the leg position. In cases diagnosed as breech presentation in the seventh or eighth month of pregnancy, but turned successfully during the last few weeks and born with a vertex presentation,

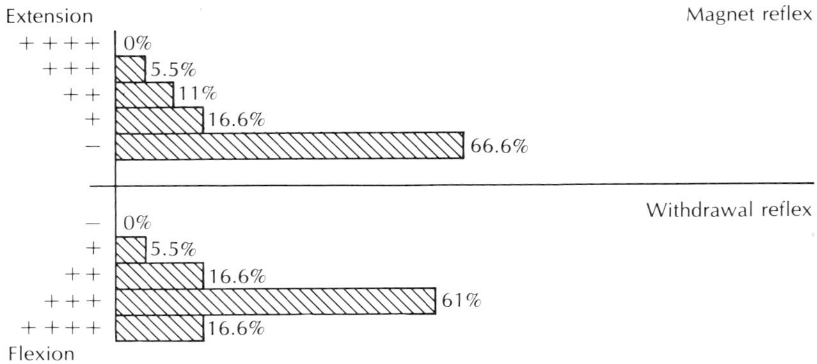

(a) Reflex pattern in 18 newborns after breech-foot presentation

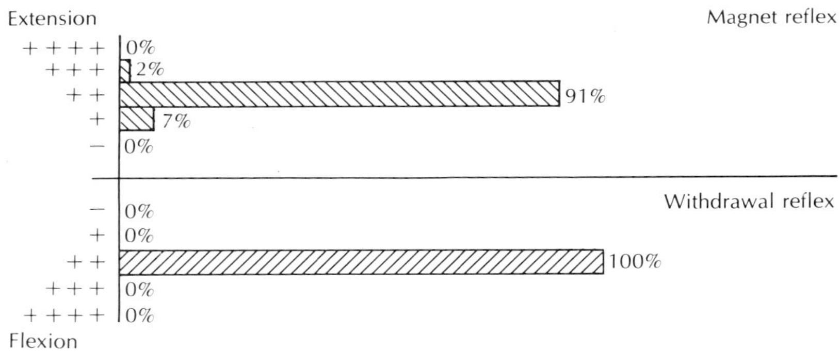

(b) Reflex pattern in 116 newborns after uncomplicated vertex presentation

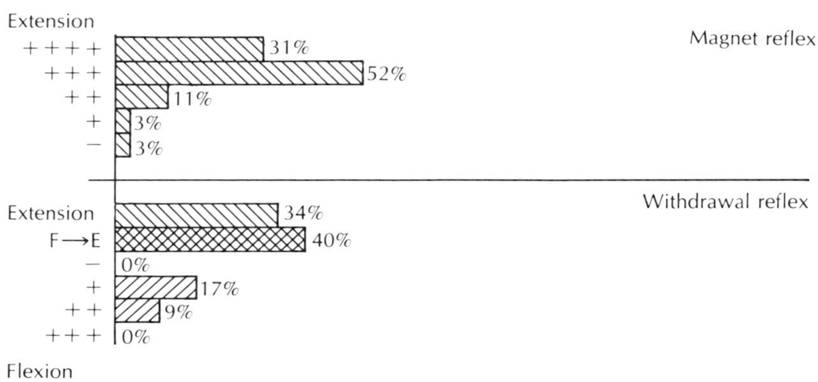

(c) Reflex pattern in 35 newborns after breech presentation with extended legs.

leg posture, mobility and reflex patterns are normal. It thus seems that there must be an influence of the intra-uterine proprioceptive input on post-natal postural and reflex patterns. There is an analogy here to some of Kuo's (e.g., 1967) findings with chick embryos. To cite but one of the examples he gives, only if the embryo chick adopts a position with the legs beneath the body, in which external thrust operates against gravity, do legs develop normally.

In view of these facts, and in the absence of direct experimental evidence, it is safer to consider Schneirla's view that the intra-egg environment and activity may have some effect on the ontogeny of pecking in chicks, than to presume that it does not. The constancy between individuals in the form of the movement, and its apparent environmental stability in development, may thus be consequences of the constancy of the intra-egg environment rather than an indication of a lack of environmental influences on development. Although it seems improbable that the precise coordinations involved in pecking can depend on the relatively unstructured intra-egg environment, as yet we know very little about how far the intra-egg environment could be altered without development being affected.

Lorenz has argued strongly against this view, pointing out that the intra-egg environment is more similar between species than are the behavioural characters supposed by Kuo to depend on them. Thus the suggestion that the embryonic heart beat plays a role in the development of pecking in chicks (see above) does not explain why other species do not peck on hatching, but gape or dabble. This, however, is a different issue: inter-species *differences* may be genetically determined, but lie in differences in susceptibility to commonly experienced environmental influences.

In spite of the controversy over this issue during the past 10 years it remains unresolved: clearly the present need is for an experimental approach.

Turning now to post-embryonic development, the form of a movement can be influenced by the environment in diverse ways. In human infants suffering from rickets the effectors are influenced directly, while in certain human anxiety states a hostile environment may result in the superposition

---

Fig. 19.1 Reflex responses to stimulation of the sole of the foot of newborn babies. (H. F. R. Prechtl, personal communication.)

The magnet reflex is an extension of the lower limb in response to mild stimulation, and withdrawal indicates flexion in response to a pinprick or scratch. The intensity of the response is indicated by the number of crosses, the lengths of the lines indicating the number of infants in each category. *a*, after breech-foot presentation; *b*, after uncomplicated vertex presentation; *c*, after breech presentation with extended legs.

---

of a cringing posture which affects nearly all movement patterns (Barlow, 1959).

When the form of a movement pattern is influenced by learning, the course of that learning may be determined in part by the structure of the species. An example is provided by the behaviour of birds with abnormal beaks. Individuals with gross abnormalities may obtain their food with "scooping movements", or other bizarre patterns apparently quite foreign to the normal repertoire of the species (Fox, 1952; Pomeroy, 1962). Since the development of such unusual patterns clearly involves learning conditioned by the abnormal structures, the development of the corresponding normal patterns may also involve learning similarly conditioned by structure: lack of individual variation in the motor pattern could thus be ascribed, in part, to lack of variation in the relevant effectors (Hinde, 1959a). Of course, the learning of new motor patterns always contains a species characteristic element, in the sense that members of a species will tend to produce movements more similar to each other's than to those of other species. This could be a consequence of similarities in the structure of nervous system or effectors.

In addition, many of the motor patterns characteristic of individual vertebrates are modifications of species-characteristic ones. Thus before take-off most birds show a series of "intention movements" of jumping into the air (Fig. 17.3 and page 400). Andrew (1956d) has described how some buntings, kept in a cage in which the perches were placed too near the roof, showed highly peculiar intention movements of flight: after an initial crouching phase, the body was moved upwards until the bill was vertical, while the legs and neck remained in the same posture as before. These movements, derived from a normal take-off by a change in the association of its components, became extremely stereotyped and persisted when the birds were moved into a larger cage.

As another example, the movement of holding an object under the foot, referred to above, often forms part of the sequence of actions by which tits (*Parus* sp.) are able to pull up a piece of food which is suspended by string from a perch. The bird pulls up a loop of string with its beak and holds it under its foot while it pulls in some more of the string which is still hanging below the perch. However, some species (e.g., greenfinches, *Chloris chloris*) which rarely use their feet in feeding, also rarely use this method to pull up strings, but develop techniques depending on other species-characteristic movements: for instance, many individuals draw the string slowly through their beaks with the mandibulating movement usually used to open seeds. Other patterns used are also probably modifications of preexisting ones (Vince 1961a, b).

A similar building-up of individually characteristic movement patterns from species-characteristic components is seen in the so-called superstitious behaviour of pigeons. If a pigeon is given food on an arbitrarily chosen time schedule, without reference to its behaviour at the time, it begins to repeat certain acts which happen to coincide with the reinforcement. The probability that one of these will coincide with the next reinforcement is thereby increased, and its frequency then increases. Bizarre movements, like turning quickly in a small circle, or tic-like movements of the head, may be learnt in this way (Skinner, 1948; Herrnstein, 1966). Individually characteristic stereotyped movements are especially common in caged animals (e.g., Holzapfel, 1939a) and it is probable that a number of factors contribute to their development.

The development of the skilled movement patterns used by primates raises a number of problems which cannot be discussed here. There is, however, considerable evidence that each skilled movement depends on the elaboration of preexisting patterns and thus, ultimately, on species-characteristic factors (Schiller, 1957). In man feedback is invariably important, comparison of the movement with a "model" or Sollwert of the required movement often playing an essential role in its elaboration (Guillame, cited by Paillard, 1960). This is especially clear when we hear a tune and then try to reproduce it: we first learn the tune, and then adjust our motor output to produce something similar. As we shall see, a similar process is involved in the development of bird song.

The development of tool-using patterns by primates deserves special mention here. Although examples of tool-using behaviour are known in a wide variety of species (e.g., Lack, 1947; Thorpe, 1963a; Hall, 1963; van Lawick-Goodall and van Lawick, 1966), in subprimate forms they are nearly always species-characteristic patterns which, though perfected by practice, can develop independently of example. In primates (e.g., Hall, 1963; Chiang, 1967), and especially in apes (van Lawick-Goodall, 1968), the patterns are much more diverse, and example often plays a role in their development (e.g., Fabry, 1966): the difference from lower forms may of course be one only of degree.

For instance a number of primates have been recorded using objects in agonistic situations (refs. in Hall, 1963), and free-living chimpanzees both use sticks as weapons and throw objects toward adversaries (van Lawick-Goodall, 1968). The waving of sticks forms part of the male chimpanzee's aggressive display, the animal seizing branches and waving them as he runs. The use of sticks as weapons is probably a development of this. Usually the stick is let go, and the movement is merely in the direction of the opponent (see also Kortlandt, 1965). Sometimes, however, immature animals use sticks as weapons: this indicates that ontogenetic development from the adult male display need not occur in each individual.

"Aimed" throwing has been recorded by van Lawick-Goodall in ten individual chimpanzees, but the objective was hit in only five out of forty-four instances: usually the aim was good but the range inadequate. Throwing also probably develops from the male display, during which animals often make movements similar to those used in throwing even though not holding anything, and sometimes let go of objects they are waving about.

Van Lawick-Goodall has also recorded tool-using by chimpanzees in a number of feeding situations. Sticks and stems are used for picking up ants and termites from inside their holes: sometimes the tools are prepared for use and by the stripping off of leaves, etc., and they may be fetched and carried to the feeding place from a distance. This develops gradually in young animals and almost certainly involves imitation: immature chimpanzees probe holes in the presence of their mothers or other sophisticated animals, using similar techniques. They are remarkably persistent, even when unsuccessful because the stalks they are using are too short or the hole unproductive. Sticks are also used as olfactory probes, the chimpanzee pushing the stick into a hole, withdrawing it, and smelling its tip. In addition chimpanzees trying to open a box may poke at it with sticks. Van Lawick-Goodall has also recorded free-living chimpanzees using crushed leaves as a sponge to suck up water from a hole which they could not reach with their lips, and using leaves for wiping their bodies.

That captive chimpanzees can use sticks as a tool to obtain food out of reach behind wire netting, or pile boxes to reach a hanging banana, has been known for many years (Köhler, 1925; see also Döhl, 1966). Previous experience with the objects plays an important part in such examples of "problem-solving", and the role of the species-characteristic investigatory and manipulative movement patterns already in the animal's repertoire has been stressed by Schiller (1957). Imitation of novel movements made by other animals or by humans may also play a role, for chimpanzees are known to be capable of this (Hayes and Hayes, 1952). But the development of tool-using behaviour under experimental conditions could certainly be profitably reinvestigated now that more information on the movement patterns of free-living chimpanzees is available.

Finally, it is worth emphasizing that the development of a movement pattern involves changes not only in the outward form of the movement, but in the mechanisms by which it is controlled. For instance, we saw earlier (pages 52–53) that introduction of a delay into the auditory feedback causes human speech to become incoherent. The speech of 7- to 9-year-old children is more adversely affected by delayed auditory feedback than that of 4- to 6-year-olds, a finding probably related to the increase in control over speech with age (K. U. Smith, 1962; Yates, 1963). The adult type of response to delayed feedback is present from an age of about 12 years (Ratner et al., 1964).

As another example, the extent to which a skilled movement depends on *exteroceptive* stimuli may depend on practice. Thus Poulton (1952) studied the performance of subjects in following a pointer which oscillated in a regular way: after training they could follow with fair accuracy even if the target was momentarily obscured. Broadbent (1958) points out that a consequence of training is that performance of the movement comes to interfere less with other activities: if the pattern of movement is repeated often in the same order, the first part of the stimulus contains almost as much information as the whole sequence, the remaining input becomes redundant, and the neural mechanisms are set free for other tasks.

19.4 AVIAN VOCALIZATIONS

The ontogeny of bird vocalizations raises problems of particular interest. Progress in their analysis has been greatly accelerated by the development of the sound spectrograph, which is an instrument that produces a graph of the frequency of a sound plotted against time, and makes it possible to record the forms of sounds more precisely than those of any other behaviour pattern. Each avian species has a repertoire of sound signals: often there are as many as 15 to 18 types, each appearing in a distinct functional context, with some of these types having considerable variability (Thorpe, 1961b). These signals are characteristic of the species, and interest has been focused in the first instance on whether or not individuals are able to produce them without previous example from other members of their species.

Little work has been done on non-passerine species, but the available evidence suggests that, with a few possible exceptions (e.g., Goethe, 1954), the species-characteristic repertoire is developed independently of example (Lanyon, 1960; Baeumer, 1962; Lade and Thorpe, 1964). The adult calls often develop by elaboration and restriction from those of the chick (e.g., gulls, Moynihan, 1962; domestic chicks, Andrew, 1963a). That some species, such as parrots, can learn by imitation is beyond doubt, but there is as yet no evidence that this ability is used outside captivity, let alone that it is necessary for normal development (Thorpe, 1963a).

Among passerine birds the call notes, which are normally relatively simple sound patterns, also usually develop independently of example (Heinroth, 1926; Messmer and Messmer, 1956; Sauer, 1954, 1955; Lanyon, 1957; Marler, 1956a; Blase, 1960; Thorpe, 1961b). In a few cases, however, certain call notes either fail to develop, or develop abnormally, in birds reared in isolation (Poulsen, 1954; Messmer and Messmer, 1956; Marler, 1956a; Lanyon, 1957). Whether or not the crucial factor in these cases is absence of example is so far unproven. As with song (see below), the observation that call-notes develop normally in the absence of example from other members of the species does not mean that learning plays no part in their development: chaffinches deafened at an early age produce quite abnormal call-notes (Nottebohm, 1967).

The distinction between call-notes and song is not always easy to make. In the Indian hill mynah (*Gracula religiosa*), each bird has a repertoire of three to fifteen loud calls which could functionally be classified as either. All the calls in each bird's repertoire are distinct from those of its mate, but most of them are shared with a few neighbouring individuals of the same sex. These must be acquired by selective imitation of neighbours. Mynahs also have another call-note, used in contexts including alarm, whose precise characteristics are learnt by imitation (Bertram, 1969).

The more complex songs of passerine birds, which usually function in pair-formation or territorial defence during the breeding season, provide material of exceptional interest. Their development has now been studied in a variety of species, and the data obtained are of relevance to a number of basic problems in the study of animal behaviour (e.g., Thorpe, 1961b; Lanyon, 1960; Marler and Hamilton, 1966; chapters in Hinde, 1969). A particularly interesting case is that of the chaffinch, *Fringilla coelebs*, which has been analysed in detail by Thorpe (e.g., 1954, 1958, 1961b; see also Hinde, 1969): song development in this species will therefore be discussed first, and a comparison with other species made later.

The full song of the chaffinch lasts about 2.5 sec, and consists of notes ranging from 2 to 8 kHz. It is divided into three main phrases, of successively lower frequency, followed by a terminal flourish (Fig. 19.3*a*). Individuals usually have a repertoire of one to six songs of this type, differing in their details. This song develops through a "plastic" stage from a rambling subsong which has no definite duration, a wide range of notes, and no terminal flourish (Fig. 19.2). Although the subsong sometimes contains imitations of alien species, these are eliminated in the development to full song. This subsong has itself developed gradually from the fledgling and other calls: it is first heard in late summer, but then ceases until the following spring (Nottebohm, 1967).

Chaffinches hand-reared in isolation from a few days of age develop only a very simple type of song (Fig. 19.3*b*). This is of about the right length and number of notes, but is not divided into phrases and lacks the terminal flourish. This, then, is the song pattern produced when opportunity for imitating other individuals is denied. If such birds are brought up in groups, still in auditory isolation from adult chaffinches, they produce spring songs more elaborate than those of birds isolated individually. This is presumably a consequence of the stimulating effect of countersinging with other individuals. Such songs may be divided into phrases, though the structure of the notes is often abnormal (e.g., Fig. 19.3*c*). The song patterns are usually similar in the members of a group, but differ markedly between

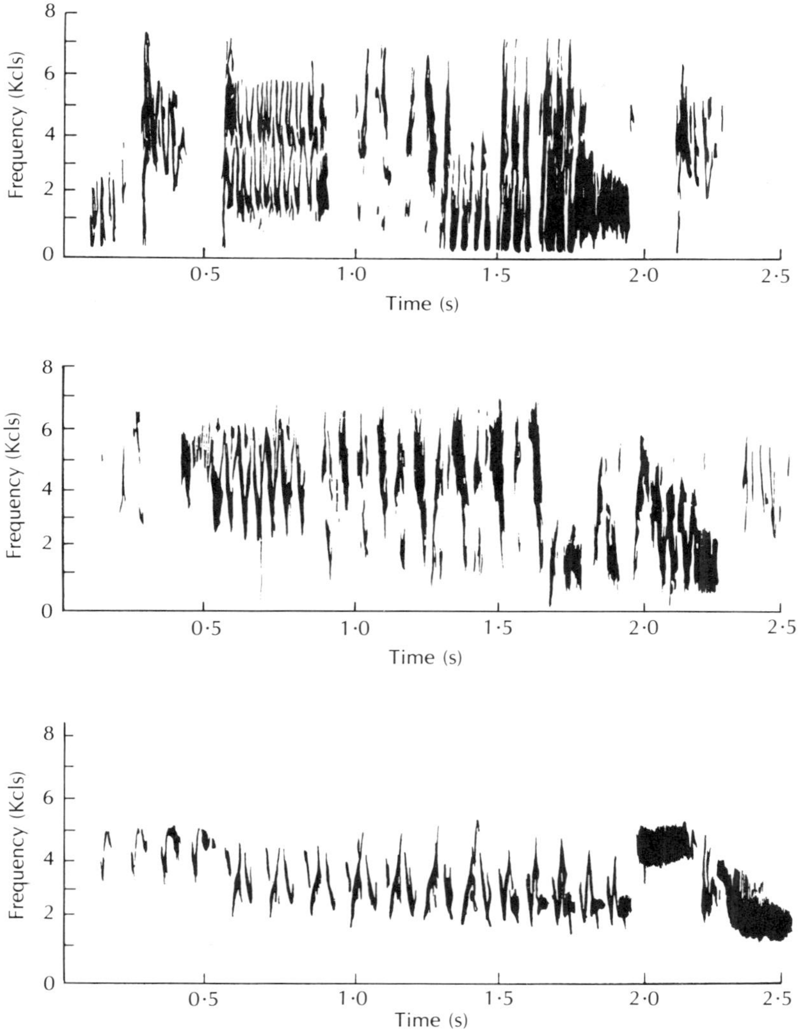

Fig. 19.2 Three stages in the subsong of the chaffinch. The full song of same individual is shown in the upper diagram of Fig. 19.3. (After Thorpe, 1961*b*.)

a, Chirps and rattles having a large range of frequencies; *b*, transition to full song; *c*, spring song, still looser and longer than normal full song.

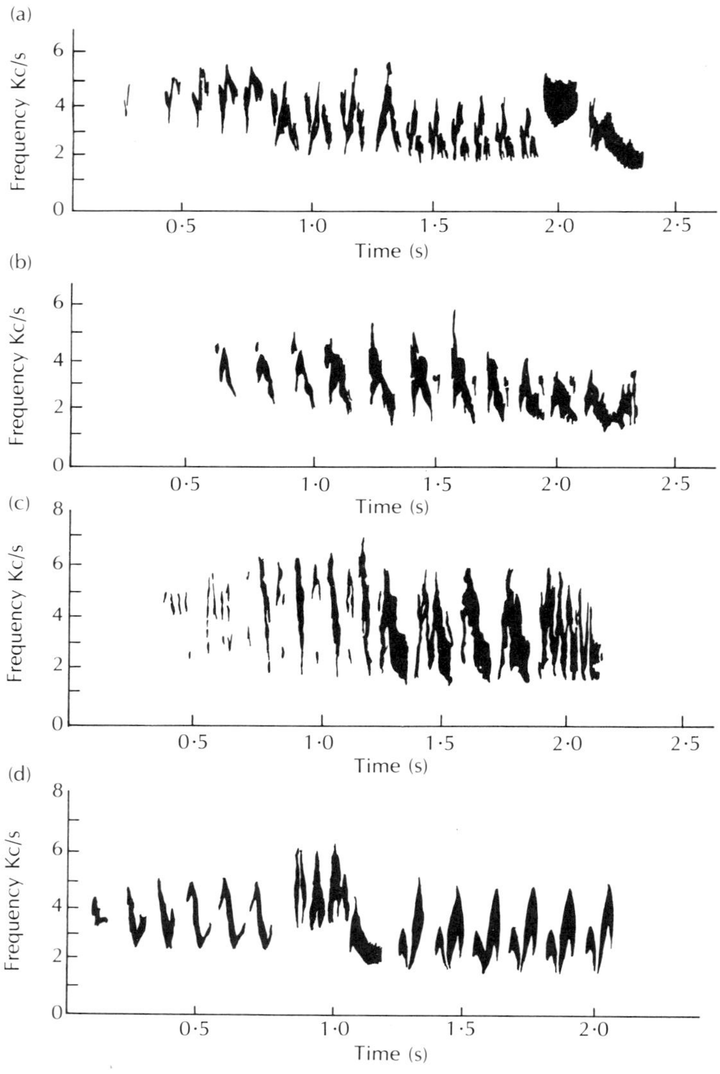

Fig. 19.3 Chaffinch song. (After Thorpe, 1961*b*.)

a, Characteristic normal song; *b*, song of an individual reared in isolation; *c*, song of an individual from a group reared in isolation; *d*, song produced by a bird reared in isolation, after tutoring with a rearticulated chaffinch song with the ending in the middle.

groups. This indicates that learning by imitation is occurring within each group.

Chaffinches caught in their first autumn, and kept subsequently in isolation, develop songs more or less within the normal range. The song is divided into three sections and there is a terminal flourish. The difference between these birds and those reared in total isolation indicates that some learning must occur in the first few months of life, long before the birds themselves begin to sing. If such autumn-caught birds are kept subsequently in groups, they develop songs indistinguishable from the wild ones.

Chaffinches reared in isolation and exposed to tape recordings of chaffinch song during their first autumn and winter subsequently produce nearly normal songs. This again shows that the song pattern eventually produced is influenced by experiences months before the young birds start to sing. However, isolated chaffinches will not learn any song pattern they hear. Using a tape recorder to play back songs at frequent intervals, Thorpe showed that isolated chaffinches did not modify their song as a result of exposure to an artificial chaffinch song in which the notes had a pure tonal quality, and were thus unlike those in normal chaffinch song, but did after exposure to patterns made by modifying chaffinch song; for instance by playing it backwards or with the order of the phrases changed. Figure 19.3*d* shows the song of a chaffinch reared in isolation but exposed to a recording of chaffinch song rearticulated so that the end was in the middle: the imitation was almost perfect. Thus, although chaffinches occasionally incorporate foreign notes into their subsong, these isolated birds will imitate only a limited range of sound patterns in the full song. It is possible that this limitation is in part peripheral and depends on structural features of the syrinx or its nerve and muscle supply, but this explanation is rendered less likely by the wider range of notes which occur in the subsong. Autumn-caught birds, which have heard normal songs during the first few months of life, are even more restricted in the variety of patterns they will learn from the song tutor than are birds isolated from the fifth day. This greater selectivity was presumably acquired during their first few months, before they themselves were singing.

Once full song has developed, no further song patterns are learnt. There is thus a sensitive period for song-learning. A chaffinch castrated before its first breeding season, and subsequently caused to sing at 2 years of age by testosterone, was capable of song-learning. Thus the end of the sensitive period is not merely a matter of age (Nottebohm, 1967).

Since the final song pattern develops gradually from the subsong, and since its form is influenced by experience of hearing song long before the bird itself starts to sing, it seems possible that the effect of experience is to increase the reinforcing value of song. On this view, the development of full song from subsong would be due to the reinforcing value to the bird of hearing itself sing the full song. Two predictions would follow from this. First, deafening the bird after it had experienced song, but before it started to sing itself, would make it impossible for the bird to monitor its own output, and thus would nullify the effects of the early experience. Second, if hearing song acts as a reinforcer for singing, it should also exert a reinforcing effect on other responses made immediately before it is heard.

The first of these predictions has been confirmed by Nottebohm (1967), following earlier experiments of Konishi on other species. Chaffinches, deafened after their songs had developed fully, retained their songs substantially unaltered for at least 2 or 3 years. Chaffinches deafened just before their full songs had become finally stabilized, however, gradually regressed in their singing: the syllables and phrases changed and lost their identity. Other birds were deafened at various earlier stages in their song development. In general the earlier the deafening, the more rudimentary and abnormal the song which developed. Birds deafened after they had started to produce subsong produced songs with syllables resembling the wild type, but birds deafened at 100 days of age, before they had started to produce subsong, subsequently produced only very rudimentary and unstructured songs. Nottebohm's data suggest that sound patterns established before the deafening are probably incorporated in a modified form in the full song (Nottebohm, 1967; Konishi and Nottebohm, 1969).

The second prediction, that song should have the properties of a reinforcer, has been confirmed by Stevenson (1967, 1969). The chaffinches were tested in a soundproof chamber and the effectiveness of song as a reinforcer for perching on that end perch, which had been the less preferred in control runs, was assessed. For autumn-caught testosterone-treated chaffinches song was found to be a reinforcer, but a burst of white noise of similar volume and duration was not (Fig. 19.4). With hand-reared chaffinches not in breeding condition, no reinforcing effect on hearing song was found. Hand-reared testosterone-treated and autumn-caught untreated groups gave intermediate results. Thus the reinforcing effectiveness of song seems to be enhanced both by experience and by androgen levels.

It is possible also to relate the influence of song on song development to its effect in eliciting song. Since chaffinch song-types occur as local dialects, the particular characteristics of each song-type, as well as the more general species-characteristics of chaffinch song, must be influenced by hearing song from other individuals. If songs are played to adult chaffinches in the breeding season, the song-type which is most frequent in each individual's repertoire is also the most effective in eliciting song from that individual. It thus seems probable that earlier experience of a particular pattern both influences the subsequent production of that pattern (or one like it) and enhances the effectiveness of that pattern in eliciting song (Hinde, 1958b; see also Lemon, 1967).[1]

One way of relating all these findings is to suppose that the naive bird has a crude "model", "template", or Sollwert of the species-characteristic song which is improved by experience of that song. On this view sounds which approximate to the model are reinforcing and are also effective in eliciting singing. In both cases, this would suggest, a process of comparison between auditory input and Sollwert is involved. The postulation of a model is of course merely a device for tying together the results of several experiments: the corresponding physiological reality is quite another issue.

Although song-learning is perhaps better understood in the chaffinch than in any other species, particular aspects have been studied in greater detail in other passerines. Some of the principal findings may be summarized as follows:

1. In a number of species individuals reared from the egg or nestling stage in auditory isolation from fellow members of their species sing primitive or aberrant songs (e.g., red-backed shrikes, *Lanius collurio,* Blase, 1960; meadowlark, *Sturnella* sp., Lanyon, 1957; blackbird, *Turdus merula,* Thielcke-Poltz and Thielcke, 1960; white-crowned sparrow, *Zonotrichia leucophrys,* Marler and Tamura, 1964; Oregon junco, *Junco oreganus,* Marler et al., 1962; various other passerines, Vilks, 1966). The only exception amongst passerines seems to be the song sparrow, *Melospiza melodia,* where birds reared by canaries produced songs

[1] At a quite different level of complexity, this finds a parallel in human speech. As we have already seen (page 53), coherent speech depends on the perception by the speaker of the sound of his own voice: if the auditory feedback is delayed, speech is disturbed. We must suppose that the speaker compares the sound of his own voice with some internal Sollwert: if feedback is inappropriate, he attempts to adjust his voice accordingly. M. Halle and K. Stevens have suggested that understanding similarly depends on a process of comparison: the input is accepted and interpreted as a sentence to the extent that it appears to the listener to match some internal representation. However, the number of possible sentences against which the input could be matched is almost infinitely large: there are for instance about 10^{20} possible 20-word sentences in English, and they could not possibly all be stored in the listener's brain. They therefore suggest that the internal representations are generated as they are needed by following the same generative rules that are used in producing speech. On this view, the hearer starts to generate an internal representation on the basis of some hypothesis about the input. The mismatch is reported and used as a basis for the next hypothesis, which will be a closer one, and so on. The successive hypotheses are generated on the basis of the rules the listener uses for generating speech when speaking himself. The rules of language need be assimilated only once—not once by the ear and again by the tongue. Speaking and understanding are thus both manifestations of a single *faculté de langue* (G. A. Miller, 1964).

indistinguishable from the wild type (Mulligan, 1966). Claims of earlier workers that whitethroats, *Sylvia communis* (Sauer, 1954), and blackbirds, *Turdus merula* (Messmer and Messmer, 1956), reared in auditory isolation could produce normal songs are believed by Konishi and Nottebohm (1969) to have been based on inadequate analysis.

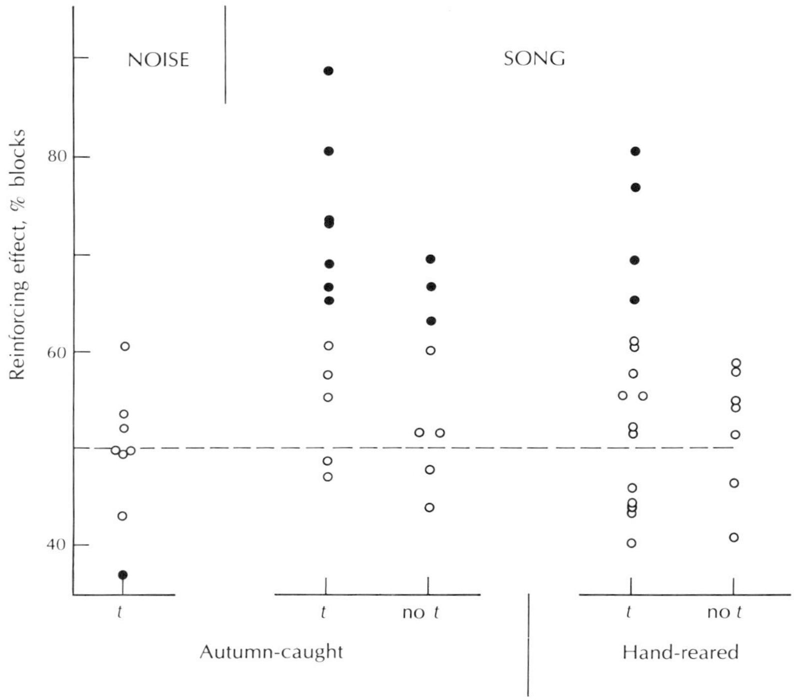

Fig. 19.4 A measure of the reinforcing effect of white noise or normal adult song for first-year male chaffinches. (After Stevenson, 1969.)

Each point represents a different individual, and filled points indicate a significant difference from chance performance (Sign test, $p < 0.05$, one-tailed). Birds were either autumn-caught or hand-reared, and injected or not injected with testosterone (*t*).

2. In a wide range of species song develops gradually, the development involving transition from repeated call-notes and/or subsong, and being associated with increasing androgen levels. The processes have not been traced in detail in any passerine species, but may be comparable to those studied in the chick, where Andrew (1969) has shown that development of various calls and of crowing depends on modifica-

tions to preexisting patterns, and has suggested the existence of a "testosterone mechanism" which, when activated, modifies the nature of the calls given (see page 546).

3. Grouping of hand-raised chaffinches leads to the production of more elaborate song patterns than are found in isolate-reared birds. No such effect occurs in white-crowned sparrows (Marler and Tamura, 1964), but group-reared Arizona juncos *Junco phaeonotus* produce more complex syllables and larger repertoires than isolates—a consequence of group-rearing attributed to "improvisation". Improvisation involving both the rearrangement of phrases and the "invention" of new ones occurs in a number of species (Messmer and Messmer, 1956; refs. in Thorpe, 1961b), and can lead to a surprisingly large repertoire. A detailed description of the improvisation, elaboration, and crystallization occurring in the song of the European blackbird is given by Hall-Craggs (1962), and Tretzel (1967) has described how a local population of this species elaborated and transposed a human whistle. Marler (1959) recorded a robin (*Erithacus rubecula*) which produced fifty-seven different songs, and cites comparable cases in other species. While recombination of elements can occur even in deaf individuals, the role of countersinging in more elaborate forms of improvisation is as yet unknown: labelling the process as "invention" aids little in its understanding, and more detailed studies comparable to that of Hall-Craggs are urgently needed. One process which may play a part is the "drift" described by Andrew (1969) in the calls of testosterone-treated chicks: in a series of calls there is an incomplete tendency to repeat the form of the call just uttered, so that the nature of the call changes gradually through the series.

   However, auditory experience does not always result in an increase in complexity of the song pattern. The Oregon junco *Junco oreganus,* which in nature has a simpler song than the Arizona junco mentioned above, produces more complex songs if reared in isolation than it does in nature. Individuals of this species reared in isolated groups develop songs somewhat more complex than wild birds, but with a higher proportion of normal songs than birds reared in individual isolation. The mutual stimulation amongst the group-reared birds thus serves to bring the course of development back to a normal pathway, rather than to produce further elaboration (Marler, 1967).

4. The range of sound patterns which the chaffinch will imitate appears to be determined by the extent of the resemblance of note-structure to normal chaffinch song. In the white-crowned sparrow, likewise, there is a predisposition to imitate sound patterns resembling the species song (Marler, e.g. 1967). What role earlier experience of the individual's own vocalizations plays in such cases, if any, is not known.

In other species, the range of sound patterns which are imitated is restricted in another way. Bullfinches (*Pyrrhula pyrrhula*, Nicolai, 1956), zebra finches (*Taeniopygia guttata*, Immelmann, e.g. 1969) and parasitic weaver-finches (Viduinae, Nicolai, 1964) learn the song of the individual which reared them: in the first two cases this occurs even if they can hear the normal song of the species. Even here, however, there may also be a predisposition to acquire species-characteristic patterns: zebra finches reared only by females, but able to hear both zebra finch and Bengalese finch (a domesticated form, probably *Lonchura striata*) males, acquired elements only from the former: unlike normally reared zebra finches, however, they acquired elements from several different individual males (Immelmann, 1969). In yet other species there seems to be little restriction on what can be learnt (e.g., meadowlark, *Sturnella* sp., Lanyon, 1957, 1960; mockingbird, *Mimus polyglottus*).

5. In the chaffinch the sensitive period for song-learning extends to the season in which the individual first comes into full song. In some species the ability to mimic persists but is seasonal (e.g. red-backed shrike *Lanius collurio*, Blase, 1960; canary, Poulsen, 1959). In the white-crowned sparrow the sensitive period is much more restricted than in the chaffinch, and is over by the time the birds are 2 to 4 months old. Isolated males exposed to two song dialects, one during their first 2 months and the second between 3 and 10 months, learn only the former (Marler, e.g. 1967). In this species, therefore, the sensitive period is over before the bird itself starts to sing. Konishi and Nottebohm (1969) thus suggest that there may be two critical or sensitive periods—one for the learning of the auditory "template", and the other for the motor learning: in the chaffinch both apparently end together. Even this may be an oversimplification, however, for Immelmann (1969) has evidence that different features of the song of zebra finches may be acquired at different ages. In this species song is first heard at 35 to 40 days of age, and is fully developed at 80 days. After 80 days it can be modified by example only to a very limited extent. The elements of the song can be acquired either before the bird itself starts to sing, or in the early stages of song development. The length, sequence, and rhythm of the song elements, however, are fixed by example during song development (days 40 to 80).
6. The role of auditory feedback has been studied in a number of species by Konishi and Nottebohm (reviewed 1969). Amongst non-passerines, domestic chickens deafened soon after hatching produced virtually normal vocalizations (Konishi, 1963). Konishi points out that deafening does not exclude non-auditory feedback, but Nottebohm showed that when deafening is followed by section of the recurrens branch of the

vagus, which carries afferent fibres from the syrinx, crowing is still unchanged (see also Murphey and Phillips, 1967).

Konishi (1964) reports more detailed experiments with juncos (*Junco oreganus* and *phaeonotus*). The birds were taken from the nest when 5 to 8 days old and kept in auditory isolation until they were deafened by extirpation of the cochlea. The normal song of *J. oreganus* consists of a simple trill, and this appeared in the deafened individuals. The song of *J. phaeonotus* is more complex, usually consisting of three or more parts, each consisting of trills or phrases: this multipartite character was absent in the deafened birds. In both species the form of the syllables tended to be abnormal, and unstable between songs and from song to song, in the deafened individuals. This could be due merely to a role of auditory feedback in maintaining the stability of the syllables (cf. page 456), but Konishi regards this as improbable in view of evidence from other species that the form of the song is unchanged by deafening even though it was originally acquired through the ear. More probably, therefore, the defects were a consequence of songs never having reached the terminal point of "crystallization" into a more rigid form which occurs in normal development. Since juncos raised in acoustic isolation but not deafened develop normal songs, it appears that motor mechanisms producing normal song develop only in the presence of auditory feedback.

In other species, also, normal song development depends on the bird being able to hear itself sing. Black-headed grosbeaks (*Pheucticus melanocephalus*) and American robins (*Turdus migratorius*), for instance, develop even more abnormal songs if they are deafened in early youth than if they are merely brought up in auditory isolation (Konishi, 1965a). Further experiments with white-crowned sparrows showed that this species resembles the chaffinch in that deafening after exposure to the species song but before song development nullifies the effect of the exposure to song, whereas deafening after song development has little effect (Konishi, 1956b). Similarly a song sparrow deafened at 12 weeks did not progress beyond the early stages of song development (Mulligan, 1966).

Konishi and Nottebohm (1969) point out that inter-species differences in the effects of deafening must be interpreted with caution: if all species were to be deafened at comparable stages in their vocal development, more uniform results might be obtained. But it is of interest that the effects of deafening may be drastic even in species like the song sparrow which do not normally copy; monitoring may be important even when imitation of other individuals plays no role. Furthermore the several physical characteristics of songs are not equally affected by deafening. Experiments with deafened birds which did or did not also have the hypoglossal nerve sectioned also suggest that feedback is not a unitary process: different properties of the song may be monitored by different pathways.

In summary, song development demonstrates the impossibility of dichotomising behaviour into categories such as "innate" and "learned" behaviour: its understanding requires the teazing apart of a complex interaction between organism and environment (Chap. 18). In many species, individuals reared in auditory isolation produce abnormal, and usually over-simple, songs. Normal song development is a gradual process, involving progressive transformation and elaboration of more elementary calls, through a mechanism which involves auditory feedback (pages 456 and 460–461). It may depend in part on experience before the individual itself starts to sing, and in part on a gradual adjustment of the motor output: it can thus be regarded as at least a two-stage process (see pages 588–589). Auditory experience may both act as a stimulus to "improvisation" and permit imitation (page 513). To be effective, the experience must occur during a limited "sensitive period" (pages 563–566), and it apparently operates through an effect on the reinforcing properties of song (page 456). It will be apparent that the analysis of song development has far-reaching theoretical implications.

The role of experience in the ontogeny of mammalian vocal signals has so far been little studied, but anecdotal observations of animals reared by hand strongly suggest that the majority of (non-human) mammalian vocalizations develop normally even when the individuals have not heard them from others. Gauthier-Pilters (1959) describes how a young dromedary gave the humming contact note as soon as its head appeared at its mother's vulva, and uttered nearly all the sounds normally observed in dromedaries within the next hour. Learning, however, may play a part in local differences in species-characteristic calls, and certainly plays a part in the "instrumental" noises made by rhesus macaques. These animals, especially the males, shake branches as a challenge: in captivity each individual male develops his own techniques of noise production, such as rattling wire netting or bouncing on wooden boards (Rowell and Hinde, 1962).

19.5 ORIENTATION OF MOVEMENTS

So far we have considered the development of the form of a movement pattern independently of the question of its orientation. In Chap. 3, however, we saw that the orientation of a fixed action pattern may be controlled by stimuli different from those which release the movement itself: it may also depend on different ontogenetic factors. The orientation of most of the movements of invertebrates discussed in Chap. 7 is probably independent of experience, but in the menotaxis involved in the orientation of an ant's track (page 159), the foraging of honey bees (pages 173–175), and the recognition of its hole by a hunting wasp (page 159–160), the mechanism is clearly set by experience. Similarly, in vertebrates some orientation mechanisms are probably independent of previous experience in the particular context: examples are given by the begging of passerine nestlings (p. 146) and the pecking of young herring gulls (page 59–61, see also Chap. 20). Of course, this does not mean that the mechanisms underlying such movements do not have a developmental history involving interaction with the environment.

Frequently, however, the learning of an orientational component plays an important role in the development of functional sequences of behaviour. For instance, Braun (1963) studied the first prey-catching movements of larvae of the fish *Coregonus* and pike (*Esox lucius*) by analysing ciné-films. Many stereotyped components like binocular fixation, approaching, pushing forward and snapping were present, but skill improved later: the precise nature of the improvement is not yet known. Similarly fry of the schooling fish *Menidia* approach the school as soon as they are placed in company with other fish, but require several hours experience before they orient correctly (Shaw, 1960).

Eibl-Eibesfeldt's observations on a number of species indicate that in the species-characteristic activities of mammals it is often only the small units of stereotyped movements that are independent of practice. Thus inexperienced squirrels may bury a nut at the first opportunity: the sequence of digging, depositing the nut, pushing it down with the snout, covering it over and stamping often occurs even when the animal is unable to dig, e.g., in the corner of a room. But the sequence is improved by learning, which affects largely the orientation of the movements: an inexperienced animal may dig a hole, but cover in the wrong place, and sight of the nut then releases covering again.

The movements used by squirrels in opening nuts show similar improvement in the orientation component with practice. These movements include gnawing and splitting with a movement whereby the nut is rotated around different axes. At first furrows are gnawed in many directions round the nut, and only gradually do the squirrels develop the more efficient technique of gnawing a furrow right round. The precise method used varies between individuals (Eibl-Eibesfeldt, 1951a). Similar improvement with practice has been described by Kear (1962) in the dehusking of seeds by finches.

Among other examples cited by Eibl-Eibesfeldt (1963) is the prey-catching of polecats, where shaking, turning over the prey, and so on occur on the first occasion prey is given to a naive animal. The orientation of the bite in the back of the neck, however, is learnt—a learning process in which playing with the litter mates helps. In other small cats, however, the bite is aimed toward the neck independently of experience, though subsequently the predator learns to place it precisely on the nape (Leyhausen, 1965). Play probably also has a role in perfecting the orientation of sexual mounting in mammals, which is often inadequate at first (e.g., mice, McGill, 1962a, b, 1965; rhesus monkeys, Harlow, 1963; personal observation; rats, Folman and Drori, 1965; dogs, Beach, 1968).

It is worth emphasizing that mere observation of improvement with practice is insufficient to establish that the orientation component has been modified by learning. For instance the pecking of young chicks at grain consists of pecking, seizing, and swallowing. Close observation shows that the early attempts are frequently unsuccessful, only about 15 percent of the pecks leading to swallowing (Breed, 1911). If chicks are given their first opportunity to peck at different ages (1 to 5 days), being kept in the dark and fed by hand meanwhile, it is found that the accuracy of striking improves with age. This cannot be due to learning during pecking and must be related to growth: perhaps postural stability and general improvement in neuromuscular coordination are responsible. With every age group, however, there is further improvement with practice, so that birds which have been pecking for some time are more accurate than naive birds of the same age (Fig. 19.5 and Cruze, 1935). Thus the improvement which accompanies practice in the normally reared chick may be a consequence of increased postural stability as well as learning.

When gull chicks peck at their parents' beaks (see pages 59–61) they rotate their heads in such a manner that the beak clasps that of the parent. This develops in wild-reared chicks, but not in those reared in the dark in the laboratory: experience thus has a role (Hailman, 1967).

Experience may play an important part in the ability for orientation and navigation in flying animals. For instance, to allow correctly for the sun's movement across the sky in their oriented movements, bees must have had some experience of the sun's changing position, though this need not extend to one complete cycle across the sky (Lindauer, 1961). Amongst birds, mere orientation sometimes seems to be independent of experience of the sky; even the ability of warblers to orient by the night sky seems not to require previous experience (Sauer, 1957). Navigation and homing, however, involve learning at a number of points. In migratory birds, the area to which the bird returns the next season may be determined during a relatively brief period during the moult (Löhrl, 1959). Pigeons can navigate (i.e., orient toward home when released at a distance) even if they have previously been reared in an aviary from hatching to the time of the release. But this seems to be true only if the birds have been able to see the lower part of the sky; if the rearing aviary is surrounded by a wooden wall, orientation is impaired (Wallraff, 1967). It is possible, however, that this is due not to the absence of learning which would have occurred in the absence of the wall, but rather to disruption of the bird's "sextant" through the rapid angular changes in the bird's horizon as it flew to and fro, due to its abnormal proximity (Matthews, 1968).

Finally, it remains to stress that, in the study of the development of motor patterns and their orientation, an incomplete picture is often obtained if particular movements are considered in isolation. For instance Vince (in press) has described how the orientation of the defaecation response of great tit (*Parus major*) nestlings changes with age. This, however, is correlated with changes in orientation while resting and begging, which in turn are related to the size and shape of the nest. The latter in its turn changes as the young develop in size and strength.

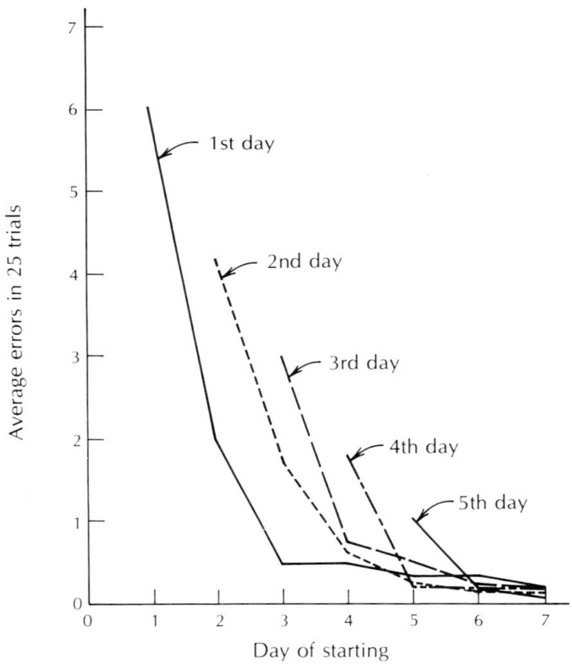

Fig. 19.5 Changes in the accuracy of pecking by chicks. Ordinate, *the number of misses made by chicks allowed to begin practice at the intervals indicated after hatching.* (After Cruze, 1935.)

19.6 SUMMARY

1. Many species-characteristic movements appear in near-perfect form on the first occasion they are elicited in their normal context. Practice reinforced by

the normal functional consequence is thus not essential for their development. This does not show, however, that previous experience is of no significance in the development of the movement: in particular, it is quite possible that intra-egg or intra-uterine experience can influence the form of movement patterns.

2. A number of cases in which the form of movement patterns more or less characteristic of a species are influenced by learning are discussed.
3. Tool-using behaviour in primates is described.
4. Avian vocalizations provide particularly interesting material for study. Although the calls and songs of many species appear in individuals brought up in isolation, in others the song has to be learned. In some of these the learning can best be understood as involving first the acquisition of a Sollwert in early life, and then an adjustment of the motor output to approximate to this pattern.
5. Often improvement in a species-characteristic movement occurs as a result of modification to the orientation component.

20

DEVELOPMENT OF PERCEPTUAL ABILITIES

For information about what an animal perceives we depend on observation of overt behaviour (see pages 56–58). If physically different stimuli always elicit the same group of responses from an animal, we say that it does not discriminate between them. This could be because the ability to discriminate is present but not used in the responses in question. However, when a wrong response has painful consequences, or results in falling off a cliff or in missing a prey, a more probable explanation is that there are inadequacies in the sensory/perceptual apparatus itself (Chaps. 5 and 6). It is with such cases that this section is primarily concerned. To what extent do the abilities to discriminate stimuli depend on experience? Discussion will for the most part be limited to discriminations of spatial relations and shapes, and to the related problem of the control of motor movements by visual inputs.

20.1 ORIENTATION IN SPACE IN SEPIA AND LOWER VERTEBRATES

Perfect performance on the first occasion that an animal makes a response shows that correct orientation is independent at least of previous practice of that response. Thus the cuttlefish *Sepia* catches its prey by first turning toward it and then striking at it with its long tentacles. The strike is made only when the *Sepia* is at the correct distance from its target and, even in a newly hatched *Sepia* feeding for the first time, is nearly always successful (Wells, 1962a). Thus the assessment of the distance of the prey must be more or less independent of previous experience in the prey-catching situation. Presumably the prey-catching mechanism is modified later, for the correct distance for the strike must change as the animal grows (see also von Frisch, 1962).

More detailed studies have been made of spatial orientation in vertebrates. The surface of the retina is connected in a more or less orderly fashion with the visual areas of the brain, mesencephalon or telencephalon according to the phyletic level. In a detailed series of experiments, Sperry (e.g. 1950, 1951) has used various visual responses of salamanders to study how these connections are established. The normal animal strikes in the correct direction when prey objects are presented at various locations with respect to its body axes, and shows an optokinetic following response (see page 164) if a cylinder painted with vertical stripes is rotated round it. If the eyeball is rotated on its optic axis through 180 degrees, leaving the optic nerve intact, it heals in its new position. The retina is then upside down and back to front. As might be expected, the visuomotor responses are then reversed—the prey-catching movements are directed towards the opposite side of the visual field from that in which the prey is located, the optokinetic responses to movement of the visual field around the dorso-ventral or rostrocaudal axes are made in the direction opposite from normal, and so on. These reversals persist indefinitely (Fig. 20.1).[1]

It is also possible to rotate the eye after severance of the optic tract. The optic fibres slowly regenerate and establish connections with the brain, leading to a restoration of vision. As in the previous experiment, all visual responses are then reversed. Thus, although the axons are tangled in an apparently random fashion in the scar region, orderly relations between the retina and the brain have been reestablished. If, in addition, the eyes are transplanted to the opposite orbit, only one of the optic axes need be reversed. On recovery, the animal shows distorted visual responses only in this plane. By these and other experiments, Sperry has shown that, whether the optic axons are growing into the brain for the first time in embryonic development or regenerating from an adult retina, the functional central relations are laid down in an orderly way. Clearly, then, the axons from the retina must be differentiated from each other; each axon must have a qualitative specificity determined by the part of the retina from which it comes. Sperry thus suggests that the orderly patterning of synaptic associations in the brain is determined by biochemical affinities specific for each part of the retina; of the many contacts made by the ingrowing axons with neurones in the central nervous system, only certain specific ones result in the formation of synaptic connections. In a similar way, orderly connections are reestablished after severance of the vestibular nerve, cutaneous nerves, proprioceptive fibres, and so on (Sperry, 1951, 1958, 1964; Franzisket, 1959; Wiemer, 1955).

[1] There seem to be some species differences in the results of such experiments. Szantágolhai (in Gaze and Watson, 1968) found that *Triturus vulgaris* did not relearn the use of a reversed visual field, while *Triturus cristatus* made clumsy attempts to do so. He suggests that the difference is related to a difference in the extent to which eye movements were regained: in the former species the eyes were immobile, but in the latter the animal could receive information about its reversed visual field through conflicting proprioceptive and optic feedback from eye movements.

In the eye, at any rate, the biochemical specificity is not determined from the start of retinal development. If the eye of an embryonic *Amblystoma* is rotated before the late tail-bud stage, vision is erect and normal: the topographical relationship of retina to tectum is not determined until

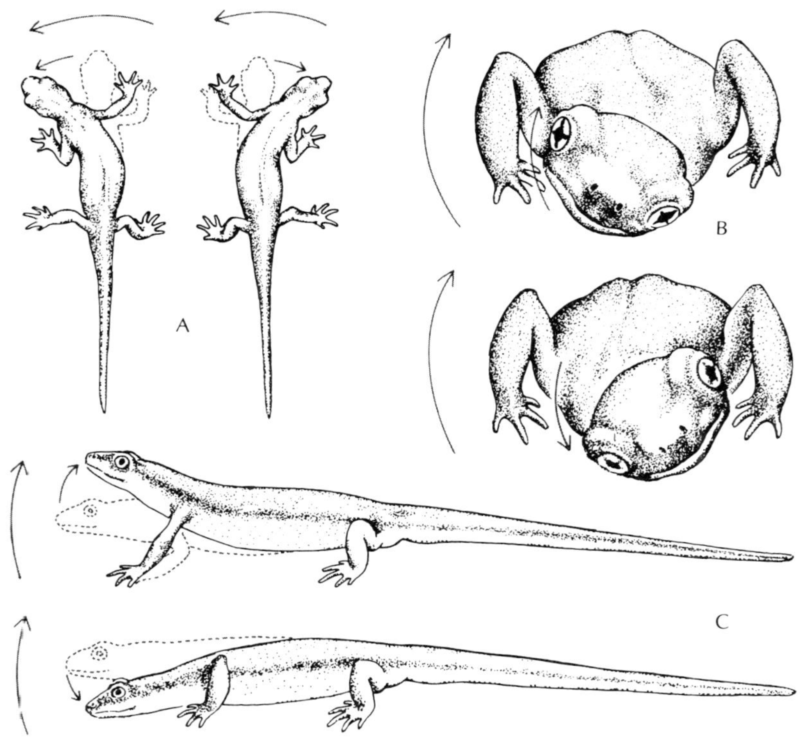

Fig. 20.1 Visuomotor responses of salamanders with normally orientated and inverted eyes. (From Sperry, 1951.)

Concurrent arrows, normal reactions; *arrows in opposite directions,* reversed reactions.

after this stage (Stone, 1948). There is also evidence that the determination occurs independently along the two axes of the eye.

The process of nerve regeneration can be followed in even greater detail by the use of electrophysiological mapping techniques: for instance, the projection of the visual field can be mapped by recording the action potentials evoked in appropriate regions of the central nervous system by a small spot of light in the visual field. In the adult goldfish, Jacobson and

Gaze (1965) found that, when only half of a severed optic nerve was allowed to regenerate into the optic tectum, the resulting projection was similar to that following hemisection of the nerve. When half of the optic tectum was removed and the whole nerve allowed to regenerate, the projection was similar to that which had obtained in the remaining half-tectum before operation. Again it seems that the optic nerve reconnects with precise place-specificity. Similar techniques also provide further evidence that the relations between retina and tectum in amphibians are determined by some form of biochemical specificity. However, some modification to Sperry's original theory was suggested by experiments in which the nasal half of the eyecup of a *Xenopus* embryo was replaced by the temporal half from another embryo, or vice versa, to give animals with "double-nasal" or "double-temporal" eyes. Electrophysiological mapping was carried out 24 months later. In the normal animal the fibres from the temporal part of the retina go to the rostral part of the tectum; those from the nasal to the caudal part. In the double-nasal and double-temporal cases, fibres from each half-retina go to the whole tectum. This seems to imply that any one part of the tectum can form connections with fibres from either half of the retina. Yet the pattern of projection from each half-retina was correct in itself, although it had spread over the whole tectum. A possible explanation is that each ganglion cell is specified by two parameters, derived from gradients along nasotemporal and dorsoventral axes. The tectum would be specified rostrocaudally and lateromedially by similar gradients. The crucial issue, however, would not be absolute values but positions on the gradients. If the gradient is halved by removal of half the eyecup, the remaining half would spread itself out to cover the whole tectum. A point, say, half way along the gradient over the whole eye would be at one end of the gradient on the half-retina, and would thus connect with the corresponding end of the gradient on the tectum. However, still another explanation is possible. It could be that the fibres from each half of the compound eye connect only to their appropriate half of the tectum, the other half of which fails to continue development. Experiments to differentiate the two hypotheses were not conclusive, but they favoured the former hypothesis that the fibres are specified in terms of their rank-order position on a gradient, spacing themselves out over the available tectum in that order. Even this, however, cannot be the whole story, for the degree of specificity varies over the retina, and may increase with time (Gaze et al., 1963, 1965; Gaze and Jacobson, 1963b; see also Sperry, 1968).

In Chap. 5 we saw that there is not merely a point-to-point correspondence between the retina and the optic tectum in the amphibian brain, but that optic fibres of different functional types connect with different layers in the tectum. Lettvin et al. (1959) have confirmed Sperry's finding that after severance of the optic nerve, regeneration results in a restoration of the point-to-point correspondence as indicated by visuomotor performance, and they have also established that the connections are formed with the correct layers. Furthermore, Parriss (1963) has shown that learned shape discriminations are retained after severance and regeneration of the optic nerves in toads.

Sperry (e.g., 1958, 1963) thus proposed that the detailed course of the development of many parts of the nervous system depends on specific chemical affinities between different classes of neurones. Neuronal and synaptic specificity is regarded as a refinement of the cruder processes of induction, embryonic fields, and other basic processes known from the experimental embryology of organ systems; and the neurochemical differentiation of the nervous system which is involved is seen as a refinement of that revealed by the selective action of stains, drugs, poisons, and so on. That such effects are not limited to lower vertebrates is shown by experiments discussed later in this chapter. It must be noted, however, that in addition to the reservation mentioned above, the theory of specific chemical affinities is not of universal applicability: it will not account, for instance, for the regeneration which occurs after rotation of the locust compound eye (Horridge, 1968b).

It is worth adding that even after Sperry's elegant analysis we cannot say that the visual responses of the salamander are independent of the influences of all visual experience; in other amphibians proper retinal functioning is known to depend on exposure to light (Knoll, 1953). It must also be emphasized that though the principle of specific chemical affinities may have wide applicability throughout the vertebrates, the particular manner in which the sensorimotor system develops varies between phyla. Regeneration does not occur to the same extent in higher vertebrates: the adaptability found more in early stages than later, and more in lower vertebrates than higher, parallels the evidence for tissue plasticity in regeneration.

20.2 EXTENT OF THE EFFECTS OF VISUAL DEPRIVATION ON SPACE AND PATTERN PERCEPTION IN BIRDS AND MAMMALS

One way to study the role of experience in the development of perceptual abilities involves rearing animals in the partial or total absence of sensory input. Such work, which was initiated largely by Hebb (1949), has been concerned especially with mammals and birds reared in darkness or without patterned visual stimulation. Studies involving such procedures are not always easy to interpret since acute deprivation may result in inconspicuous but significant changes in the visual system—smaller cells in the visual cortex and lateral geniculate body, reduced thickness and vascularity of the cortex, etc. (Brattgard, 1952; Weiskrantz, 1958; Riesen, 1958, 1961, 1966; Cragg, 1967). The effects of visual deprivation may be apparent in some structures and not others, may affect some functions but not others, and may be limited to a narrow sensitive period (e.g., Fox et al., 1968). They occur in lower vertebrates as well as mammals (Knoll, 1953). Visual deprivation may not merely retard development but may also change its course (e.g., Fantz, 1965). If these difficulties are borne in mind, however, deprivation experiments can provide some useful leads.

That deprivation experiments are sometimes essential is illustrated by a study of the relationship between eye-opening and the appearance of the optokinetic response in deermice (*Peromyscus* spp.; Vestal and King, 1968). In most species and strains studied, the optokinetic response was present on the day of eye-opening. However, a strain which had been selectively bred for early eye-opening did not reach the criterion of optokinetic responding until 2 to 4 days after the eyes opened, which was about the same age as the eyes of the wild strain opened. Thus the study indicates both that visual experience is not necessary for the development of the response and that the extra visual experience of the early strain did not noticeably accelerate the maturation of the response. However, if the origin of the early strain were not known, the time-lag betwen eye-opening and the appearance of the response might suggest a role of visual experience unless dark-reared controls were also studied.

After birds are reared under conditions of sensory or perceptual deprivation, simple reflex responses which do not require localization in space, such as the pupillary reflex, may be unaffected. Mowrer (1936) sewed together the eyelids of pigeon squabs before they had opened. When the lids were opened at 6 weeks, these birds showed normal pupillary responses, but their optokinetic nystagmus, visual startle responses, and ability to avoid objects were impaired. Their behaviour became normal after 3 days in the light. In this case retinal degeneration was at least a possibility: Siegel (1953a, b) found no deficiencies in the optokinetic responses of doves reared for 2 months with their eyes merely covered by translucent hoods.

Space perception in birds has been studied by using the pecking response as an indicator of the bird's ability. Chicks reared in darkness for a few days after hatching show a high proportion of unsuccessful pecks when first given food in the light. However, this is probably due in part to postural instability rather than to perceptual deficiencies. The improvement which normally accompanies practice is due in part to improvement in stability and neuromuscular coordination as well as to learning in a narrow sense (see pages 464–465). If the period in darkness is as long as a fortnight and the birds are fed meanwhile from a spoon, the pecking movement fails to develop at all (Padilla, 1935); but this could be due to the interference of other responses, rather than to sensory/perceptual deficiencies.

Later experiments using the pecking response have produced somewhat contradictory results. Hess (1956b) used a potentially more sophisticated technique than complete visual deprivation. Newly hatched chickens were fitted with hoods, some of which had prisms which displaced the visual field 7 deg to right or left, while in the control animals plane glass was used. In tests of pecking accuracy, the control animals pecked in a pattern centring round the target, while the patterns of the experimentals were displaced laterally by 7 deg. There was little evidence for improvement in lateral accuracy with practice over the first 3 to 4 days of life. However, in an attempted repetition of this experiment in which prisms of a variety of strengths were used, Moray and Jordan (1967) did not obtain a constant error corresponding to the displacement of the prisms, and found that the chicks became progressively less accurate. The conclusions which have been drawn from Hess's experiments have been criticized also on theoretical grounds (Arlinsky, 1967).

In another experiment, Hess studied the role of experience in the use of binocular cues. He allowed chickens normal use of one eye or the other on alternate days, but never both together. Their pecking behaviour was tested when they were 2 to 3 months old and were wearing binocular prisms. The prisms were such that if the chick used binocular cues to estimate the distance of objects at which it was pecking, it would peck short of the object. Since all nine animals tested did in fact peck short, the use of binocular cues for assessing distance seems not to depend upon binocular experience (but see pages 487–488). As always, we must bear in mind the limited nature of such negative conclusions. We still know nothing about the role of other types of experience through the same or other modalities in the development of the chick's powers of visual perception.

That the distance perception of chicks is largely independent of visual experience has also been shown by a quite different method. On the assumption that depth perception is a special case of distance perception, Walk and Gibson (1961) and Walk (1965) confronted the young of various species with an apparent cliff and assessed whether or not they would move in such a way that, had it been real, they would have fallen over it. Their "visual cliff" apparatus consisted of glass raised above the floor, across the centre of which was laid a board. On one side of the board a sheet of patterned material was placed immediately underneath the glass (the "shallow" side), while on the other side a sheet of similar material was placed on the floor (the "deep" side). The animal was placed centrally on the board, and the number of occasions on which it moved out on to each side counted (Fig. 20.2). Chicks tested 24 hours after hatching—and chicks do not normally move far from their mother before this age—invariably left the board on the shallow side, turning away from the deep side. The same result was obtained when the chicks were reared in darkness for 24 hours and tested on exposure to light immediately afterward (Shinkman, 1963).

Of the other species tested on the visual cliff, nearly all were found to avoid the deep side when tested at the age when they normally begin to move about freely—kids and lambs at 1 day, kittens at 4 weeks, human babies at 6 to 14 months. Rats also showed a clear preference for the shallow side, but only if the board was raised so that the glass was out of reach of their vibrissae. Even after being reared for the first 90 days in darkness, rats performed as well as normally reared controls on the visual cliff apparatus. Rhesus monkeys showed a preference for the shallow side from 3 days of age (Rosenblum and Cross, 1963). Only turtles failed to show a near-100 percent preference for the shallow side.

Fig. 20.2 Visual cliff apparatus. Right, chequer-board pattern immediately below the glass; left, some distance below. (From a drawing by Yvette Spencer-Booth.)

In a later study Kear (1967) observed the performance of newly hatched young of a variety of nidifugous birds on a visual cliff. The young of ground-nesting species showed clear preferences for the shallow side of the apparatus, but species normally hatching in trees chose shallow and deep sides about equally. However, the crude scores of the latter gave a false impression: the young ran off the shallow side, but jumped right out on the deep one. This is understandable in terms of the life history of the species concerned: they have as great a need for depth perception as any other, but must jump from the nest to join their mother at the bottom of the tree.

Although a wide variety of species respond to a cliff edge, the cues used differ between them. For instance chicks avoid an optical void even though they are given physical support, but rats, though using visual cues, are indifferent to them if physical support is provided. The nature of the visual cues used has been a matter of some dispute. In chicks, at any rate, binocular vision is not essential for discrimination between "deep" and "shallow" sides (Schiffman and Walk, 1963). The two most probable sensory features on which the depth perception could depend seem to be pattern density, distance decreasing the retinal size and spacing of the pattern elements, and motion parallax, near elements moving more rapidly across the field of vision than distant ones. Walk and Gibson (1961) concluded for both day-old chicks and rats that motion parallax is of primary importance. However, when rats were tested on a visual cliff apparatus with ¼-in. squares on one side and ¾-in. squares on the other, the patterns being placed immediately under the glass in each case, light-reared animals went only onto the large pattern. Dark-reared animals showed no such preference at first, but did so after 24 hours' experience of light. Thus pattern density also can provide a basis for depth discrimination, but the response is apparently acquired by experience (cf. Lashley and Russell, 1934; see also Shinkman, 1962). The position is far from clear, however, for Schiffman and Walk (1963) could find no support for the view that the preference for the larger of two patterns is due to a learned association between largeness and closeness. Furthermore, Shinkman (1963), working with chicks, claimed that cues requiring focussing are critical for discriminating between the deep and the shallow sides, while binocular and motion parallax cues are not. Palen (1965), however, found that chicks showed no preference for sharp as compared with blurred patterns when both were placed immediately under the glass: this is evidence against Shinkman's view, though it must be noted that the birds were unable to bring the blurred patterns into focus at all and so could get no information about distance from them. Other experiments are discussed in detailed reviews by Walk and Gibson (1961) and Walk (1965).

Although chicks with only very brief visual experience show a preference for the shallow side, rearing over the deep side for a period can reduce the extent to which the chicks avoid it (Tallarico and Farrell, 1964). Krames and Carr (1968) confirmed this finding and showed that chicks reared over a solid floor but able to see over a "cliff" through the transparent walls of their cage were less likely to descend on the visual cliff apparatus than chicks reared over a transparent floor. Similarly, although rats normally prefer the shallow side, rats reared above a "deep" pattern come to prefer it to the shallow side, apparently because the latter becomes a novel environment (Kaess and Wilson, 1964; see also Carr and McGuigan, 1965). Experience can also strengthen the normal preference; rats reared in an enriched environment show a stronger preference for the shallow side than do animals reared in laboratory cages (Eichengreen et al., 1966).

In other species experience plays an even greater role. When reared in the dark and tested on the visual cliff apparatus at 27 days, kittens show no side preference: only after a week of visual experience does their performance equal that of normally reared controls (Walk and Gibson, 1961). Held and Hein (1967) showed that kittens can perform correctly on the visual cliff when using only one eye, provided that eye has previously received adequate use during active self-produced locomotion. If the eye has been used only during passive movement, it fails to mediate normal visually guided behaviour. From this observation they argue that the kittens can discriminate deep and shallow sides during self-induced movement, the relationship between movement and displacement of the retinal image being interpreted as distance of the object from the eye. We shall return to this question shortly.

The amount of visual experience necessary for distance perception by kittens has been determined with some accuracy by Munro et al. (in Riesen, 1961), using a different response. When a kitten is held in the hands and lowered toward a surface, the forepaws are extended before bodily contact with the surface occurs. Under normal conditions of rearing this "visual placing" first appears at 22 to 28 days. In this study litters were divided between experimental and control groups, all being allowed normal indoor lighting for the first 18 days of life. Both groups were then placed on a daily schedule of 23 hours darkness and 1 hour light. During this hour in the light, the experimental kittens wore hoods which diffused the light. The controls were tested daily for visual placing. On the day following the appearance of visual placing in a control animal (27 to 29 days) its experimental litter-mates were unhooded and tested during the hour of light. Every one of the fifteen animals tested gave its first visual placing response after 5 hours in the patterned light, even though in some cases the 5 hours was given continuously instead of at 1 hour per day.

The development of visual responsiveness and of visuomotor control thus depends on experience more in kittens than in chickens. In chimpanzees visual deprivation has even more severe effects. Riesen (e.g., 1958) found marked visual deficiencies in a chimpanzee which was reared in darkness except for 90 min light each day. At 7½ months it was given slowly increasing periods in daylight for 10 days. During this period there were frequent episodes of nystagmus and squint, lack of eyeblink to movement in the visual field, and also deficiencies in visual fixation and pursuit. Eyeblinks were first observed on the ninth day of normal exposure to visual stimuli, and fixation and pursuit responses were first seen on the eleventh: the deficiencies subsided during the following weeks. Comparable data were obtained from a number of other animals suffering various degrees of visual deprivation. Of special interest was an animal confined in a supine position during the daily 90-min period when it was allowed to see the room. Although more advanced in these responses than animals reared in diffuse light only or in total darkness, it was less advanced than an animal given an equal period of light but allowed to move around freely (cf. pages 489–496).

Turning to studies specifically of pattern discrimination, experiments on early perceptual deprivation in birds and lower mammals have produced somewhat contradictory results. Pastore (1958, 1959, 1962) reports that ducklings develop abilities for form perception, as well as for size constancy and brightness constancy, independently of visual experience: the learning of visual discriminations is possible for birds only 1 to 2 days old. Siegel (1953a, b) reared doves with translucent goggles and found that they subsequently took rather longer to learn to discriminate shapes than did normally reared controls. In another modality, auditory stimulation appears to be necessary before bird embryos can discriminate the species maternal call (Gottlieb, 1968a). Early rearing in darkness has only a slight effect on visual discrimination in rats (Hebb, 1937a, b), though visual experience is necessary for the development of size constancy (Heller, 1968), and early auditory restriction can produce deficiencies in auditory pattern discrimination (Tees, 1967).

In higher mammals, rearing under conditions of sensory deprivation may result in marked deficiencies in form discrimination though, as we shall see, the extent of the deficiency depends on the test used. Just as cats reared in darkness show impaired performance on the visual cliff apparatus, they also take longer than normally reared animals to discriminate patterns: discrimination of light intensities is, however, not affected (Riesen and Aarons, 1959), and there is little effect on interocular transfer (Meyers and McCleary, 1964; see below). Similar effects have been found in primates (Chow and Nissen, 1955; Riesen, 1958, 1961; Zimmermann and Torrey, 1965; Fantz, 1965). Thus rhesus monkeys deprived of pattern vision from the day of birth to 20 to 60 days of age were similar to newborn monkeys in rate of learning visual discriminations and in various visual responses. Various aspects of visually determined behaviour, including visual cliff performance, improved with time of exposure to patterned light. The rapidity of the improvement of visual acuity and ocular pursuit, and the early appearance of binocular convergence, in the deprived animals after exposure to patterned light indicated that maturational factors were also involved (Wilson and Riesen, 1966). Chimpanzees subjected to visual deprivation also develop pattern and form vision only gradually. The dark-reared chimpanzee referred to above learnt to discriminate horizontal from vertical striations rapidly, but discrimination of human faces took much longer: unlike normally reared infants of similar age, it was not upset by being fed by a stranger during the period after it was first exposed to the light.

However, the effect of rearing under conditions of visual deprivation on ability for form discrimination depends in part on the response whereby the latter is assessed. While Riesen and Mellinger (1956) found that cats with only limited exposure to patterned light took much longer than normally reared controls to learn a visual discrimination when the task required entering one of two alleys, Meyers and McCleary (1964) found a much smaller deficit when the animal was tested in a Pavlovian sling with hind-limb flexion as the required response. This suggests that the effect of visual deprivation revealed in these testing procedures is an ability to orient in space rather than to discriminate forms. We shall return to this point later in this chapter.

The relevance of visual deprivation experiments for the development of form perception in normal animals has recently been questioned by Fantz (1965, 1967) on other grounds. He argues that visual deprivation does not prevent visual development but alters its course, leading not only to neural deterioration but also to oculomotor abnormalities, excessive arousal with novel stimulation, the development of abnormal competing habits, and disturbance of visual preferences. The performance of an animal which has been visually deprived since birth is thus little guide to the abilities of a newborn. Fantz studied the latter directly by using the relative duration of visual fixation on different targets to assess visual preferences: where preferences occurred, they were taken to indicate discrimination. A plan of the apparatus used for infant monkeys is shown in Fig. 20.3. His experiments indicated that infant rhesus monkeys, chimpanzees, and humans can discriminate certain patterns without opportunity for visual learning, and show strong preferences for particular visual targets from birth (see below and Figs. 20.4 and 20.5; Fantz, 1958; see also Kessen, 1967). Although it has been argued that these results could be due in part to intensity discrimination at a retinal level (Sackett, 1963), they certainly demand reexamination of some of the conclusions drawn from visual discrimination experiments on previously deprived subjects.

But in any case, there is no dispute that pattern vision normally improves progressively. It has been known for some time that congenitally blind human subjects whose sight is restored in adulthood show grave disturbances in visual perception which improve only gradually: it is clear, therefore, that many aspects of visual perception depend on visual experience (Senden, 1932; Gregory and Wallace, 1963). There is also evidence that stimulation through other modalities, such as that provided by handling, may influence the development of visual attentiveness (White and Held, 1966). The changes which occur have been followed in detail by Fantz (1965, 1967) using the differential fixation method. In one series of experiments he assessed the stimulus preferences of young rhesus monkeys which had been subjected to various periods of dark-rearing. Testing started toward the end of the period of visual deprivation, when it involved about 5 min of visual experience per day. The stimulus pairs were grouped post hoc into the categories shown in Fig. 20.4. The mean strength of preference for the generally preferred target (shown on the left in Fig. 20.4) is shown in Fig. 20.5; for the non-configurational category the curves represent only the tendency to respond consistently in the same direction. Each point represents an average for all animals for 1 (first week after

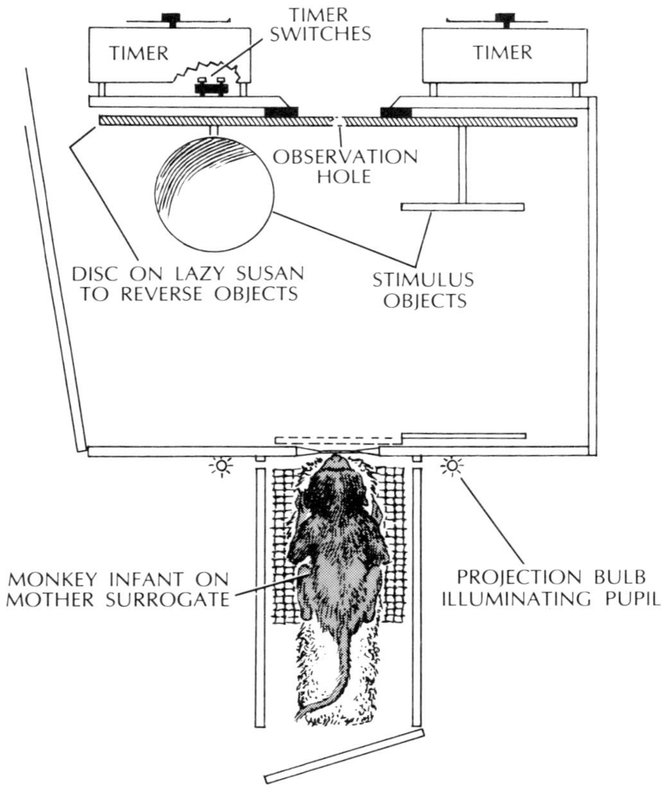

Fig. 20.3 Apparatus for testing visual preferences of infant monkeys. It is shown in plan view, with one of the stimulus pairs (6-in. sphere and circle) in position. (After Fantz, 1965.)

deprivation) or 4 weeks of testing. The subjects which had been deprived for the longest time showed little differential response to configurational variables, though they did to the non-configurational ones. The two less deprived groups showed a preference for patterns over plain surfaces which increased little with experience, but differential fixation of solid as compared

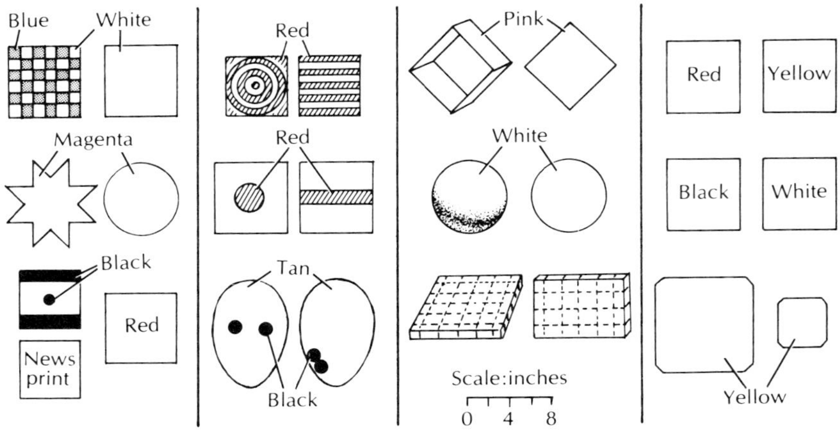

Fig. 20.4 Pairs of targets used in differential-fixation tests of infant rhesus monkeys. The four sections represent the four categories of stimulus variation. Left to right: *patterned versus plain, centred versus uncentred, solid versus plane, non-configurational variables.* (After Fantz, 1965.)

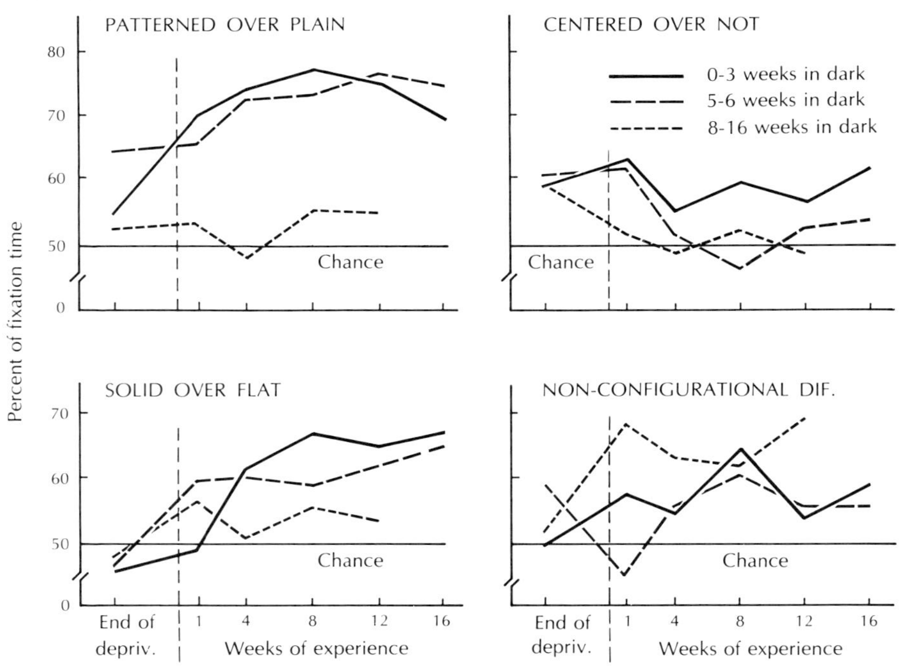

Fig. 20.5 Post-deprivation changes in the visual preferences of infant rhesus monkeys for the stimulus variables shown in Fig. 20.4. (After Fantz, 1965.)

with flat objects increased with time spent in the lighted environment (see also Sackett, 1966).

That the changes which occurred were not solely a function of visual experience was indicated by a detailed study of the stimulus preferences of animals kept in the dark. Preference for patterned as compared with plain objects increased up to the third or fourth week of life. This trend continued if unrestricted visual experience was given before 2 months of age. If it was not, an opposite tendency which appeared at about 1 month continued irreversibly. The long-deprived animals also showed no avoidance of the deep side of a visual cliff apparatus for as long as they could be tested. Thus a period of dark-rearing does not merely delay the development of visual perception but produces changes which may alter the course of, or prevent altogether, normal development, even if unlimited visual experience is provided later.

Fantz (1965) points out that some initial visual selectivity is essential for visual exploration of the environment. Without it, there would be either random eye movements or unvarying visual fixations. Why the visual preferences change with no or very little visual experience is not clear—Fantz speculates that the increase in preference for pattern complexity in dark-reared animals is due to "further maturation of the visual system, a higher optimal level of arousal, and increased neural organization". Given adequate visual experience, the differential exposure to parts of the environment which results from selective visual exploration provides opportunity for learning: what is learnt is determined as much by the nature of the visual preferences as by the environment. Objects, distinguished by high figure-ground contrast, parallax effects, and often movement, tend to be favoured, providing opportunity for their distinguishing features to be differentiated and associated. Such changes which, it may be noted, proceed without differential reinforcement presumably form the basis of "perceptual learning" (see below).

20.3 VARIATIONS IN THE COMPLEXITY OF THE ENVIRONMENT: PERCEPTUAL LEARNING

Another experimental approach to this problem which avoids the disadvantages of drastic sensory deprivation depends on controlling the complexity of the environment. As stressed by Hebb (1949), early experience might be expected to have more beneficial effects, the greater its diversity. A number of workers have therefore assessed the importance of the richness of the early environment for the development of perceptual abilities and the learning which depends on them. In some of these experiments fairly complex tests, like maze learning, have been used. For instance, rats reared in a spacious environment subsequently perform better than rats kept in a restricted environment when tested in a maze: young rats reared for a period in a large space are superior to rats kept at first in a small cage and then placed in a larger cage for a similar period at a later age (Hymovitch, 1952). This question is considered further on pages 561–563.

Of more interest in the present context are studies in which the effects of early experience are assessed by tests of discrimination learning. Such studies have revealed an effect of early visual experience on the ability to discriminate between patterns, which is often ascribed to "perceptual learning" (Gibson and Gibson, 1955). Thorpe (1963a) has stressed its possible importance at all phyletic levels.

Perceptual learning can occur independently of conventional reward. Although rats reared in darkness up to 96 days of age can learn to discriminate between a circle and a triangle (Gibson et al., 1959), early experience of the forms without differential reinforcement appears to increase the ease with which learning occurs (Gibson and Walk, 1956). The effect is not very consistent, and depends closely on the experimental conditions. In particular, experience of forms made from metal cutouts is effective, but when the same forms are merely painted on a rectangular background, the early experience has little influence on subsequent discrimination learning. Gibson et al. (1959) suggested that the difference lies in the "attention-getting" properties of the stimulus, the cutouts having depth at their edges. Gibson also found that prior exposure to only one of the test patterns led to an improvement in discrimination learning. The treatment was at least as effective if the previously exposed pattern was the negative one in the discrimination as if it was the positive (Gibson et al., 1958). The effect thus seems more easily described in terms of a learning of the properties of the stimulus than in terms of the formation of a positive stimulus-response association (but see pages 588–589).

An experiment by Michels et al. (1958) goes further than this in suggesting that experience may involve a degree of abstraction of stimulus properties. They studied the effects of different types of visual experience on the ability of rats to discriminate the magnitude (successively, but in varying order, brightness, area, and height) and form of different stimuli. Animals reared in the dark and given visual experience only during testing showed a gradual improvement on the magnitude problems, but were inferior to the light-reared animals on all tests. The improvement in the magnitude problems, each of which was essentially novel, was interpreted by the authors as indicating that the animals responded not simply to a brightness, size, or light difference, but to differences in stimulus magnitude as such.

The view that perceptual learning involves a learning of the various properties of the stimulus is also supported by experiments in which access to some aspects of the stimulus is controlled. For instance, Meier and

McGee (1959) studied the performance in a tactile-visual discrimination between three-dimensional forms (a cross and a triangle) of rats reared in different ways. Four groups differing in the cages in which they were reared for the first 155 to 162 days of life were used in this study:

1. Visual-tactual experience group. The cage had sides of grey plywood, with a number of three-dimensional figures.
2. Visual experience only. As group 1 but the forms were placed behind plate glass.
3. Minimum experience group. As groups 1 and 2 but free of regular stimulus forms.
4. Control group. The cage had continuous visual access to the laboratory.

In testing, which was started after a few days in standard laboratory cages, group 1 was superior to group 2, and group 2 to group 3. Group 2 was indistinguishable from group 4. These results are explained on the hypothesis that the animal must first learn to differentiate the stimuli, qua stimuli; that is, to differentiate tactual and visual cues related to the edges and angles of the forms. They must then learn which of the objects was associated with food and which was not. Group 1 had already accomplished the first phase at the time of testing and thus had to complete only the second. Group 2 could distinguish the objects only visually and thus had to learn the tactile differentiation in addition to the second phase. Group 3 had to accomplish both phases. Presumably, the visual stimuli from the room permitted group 4 to reach a stage similar to that of group 2. The authors believe that the first stage of learning is independent of reinforcement. It is unfortunate that the high mortality of the animals in this study was not accounted for. The hypothesis that learning in such cases involves two stages is discussed in more detail in Chap. 24.

20.4 THE DEVELOPMENT OF STIMULUS CONTROL

If an animal is trained to respond in the presence of a stimulus in an operant situation, and the stimulus is changed along some dimension, there is usually a decrement in responding. This "generalization gradient", with a peak at the training stimulus, indicates that responding is under the control of the stimulus dimension. A flat gradient (i.e., no change in response rate as the stimulus is changed) indicates lack of "stimulus control". "Stimulus control" thus refers to "the extent to which the value of an antecedent stimulus determines the probability of occurrence of a conditioned response" (Terrace, 1966). The occurrence of a generalization gradient about the training stimulus was a basic assumption of continuity theorists (see pages 585–586) such as Hull (1943) and Spence (e.g., 1956).

Their view was criticized by Lashley and Wade (1946), who argued in effect that generalization was due to a failure to discriminate between training and test stimuli. The ability to discriminate, they believed, depended on experience with the stimulus continuum in question: this allowed the subject to abstract the relevant dimensions of the stimulus.

The reinforcement of an operant response in the presence of a stimulus usually seems to be sufficient to produce stimulus control over the response. However, a number of experiments show the matter not to be quite so simple as this. For example, the role of previous experience has been demonstrated in an experiment by Peterson (1962). Ducklings reared in white light were trained to peck at a key transilluminated by light of 589 mμ, and then given generalization tests with no reinforcement in which the original training stimulus was presented in an irregular series with lights of other wavelengths. A clear gradient of generalization on either side of the training stimulus was found (Fig. 20.6), demonstrating the occurrence of stimulus control. However, ducklings reared from hatching in monochromatic light of 589 mμ and tested in the same way showed a flat generalization gradient, and thus no evidence of stimulus control. Thus the prior rearing in white light was necessary for the development of stimulus control under this experimental procedure. Similar results were obtained by Ganz and Riesen (1962) with monkeys; dark-reared animals initially gave flat generalization gradients after training at only one stimulus value. Later workers, however, did not replicate these conclusions. Malott (1968), working with chickens, and Rudolph et al. (1969), working with quail and chickens, obtained peaked generalization curves even with birds which had had no previous experience of the stimulus continuum. These authors discuss various possible causes for the discrepancy between their findings and Peterson's, but conclude that prior exposure to a continuum is not always necessary for obtaining a peaked generalization curve. The question is thus sub judice.

The further question arises, where prior exposure does affect the production of a peaked generalization curve, by what means does it do so? Peterson (and Terrace, 1966) implied that it was a consequence of differential reinforcement of different points on the continuum before the generalization test, and a number of other experiments support this view. That mere exposure to stimuli along the relevant stimulus dimension before training may be insufficient for the establishment of stimulus control was shown by Jenkins and Harrison (1960) in an experiment using pigeons which were reinforced in the presence of a 1-kHz tone. Only following differential reinforcement (i.e., tone associated with reinforcement, absence of tone with extinction) was stimulus control established. Stimulus control was not established with the tone present throughout training. Similar results have been found with a visual training stimulus when light of a particular intensity filled the entire panel rather than only the area of the response key. Whereas a gradient of responding as a function of stimulus luminance was found following training with a small stimulus size, a flat gradient occurred following training with the entire wall continuously illuminated (Heinemann and Rudolph, 1963). The authors suggest that the large illuminated area was less easily differentiated from the rest of the experimental chamber than the small one, to which responding was directed, and imply that differential reinforcement of responding to the small stimulus had been occurring, although not explicitly programmed by the experimenter.

Another experiment demonstrating the importance of differential reinforcement is due to Heinemann et al. (1968). Pigeons were trained to

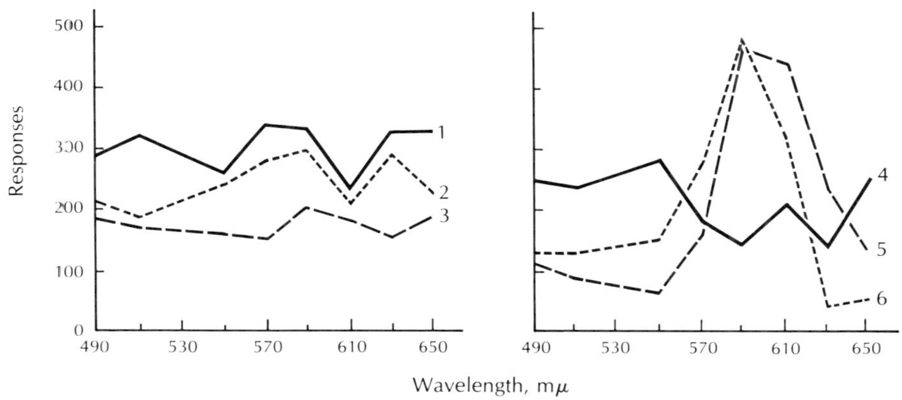

Fig. 20.6 Stimulus generalization gradients for individual ducklings. Birds 1 to 4 were raised in a monochromatic environment of 589 mμ, birds 5 and 6 were not. Training stimulus, 589 mμ. (After Peterson, 1962.)

discriminate between two tones differing in frequency in the presence of light of one colour (*A*), but not in the presence of another (*B*). Generalization curves in the presence of *A* showed control of behaviour by tone frequency. Curves in the presence of *B* were flat.

Again, Baron (1965) trained four groups of pigeons to peck in the presence of a white vertical line on a green background. For one group the negative stimulus was a green background with no line, for a second it was a red background with no line, for a third a red background with a line, and a fourth received non-differential training to the positive stimulus. The groups were then given a generalization test in which the line was presented at various orientations to the vertical. Only the first group, which had learnt the discrimination on the basis of presence or absence of the line, showed a gradient of generalization: this showed a peak at the training orientation. The other groups showed flat generalization gradients. Thus only when the birds were forced to make the discrimination in terms of presence or absence of the line by differential reinforcement did it acquire stimulus control (Baron, 1965, cited by Terrace, 1966).

It is thus clear that differential reinforcement plays at least an important role in the establishment of stimulus control in many of the cases tested. Of course this does not mean that it can be assumed to be necessary for all stimulus dimensions in all animals. However, generalization from the information presently available brings us to this point. In "perceptual learning" exposure to two or more stimuli without differential reinforcement appears to facilitate subsequent differential responding to those stimuli when one is reinforced. In the "establishment of stimulus control", differential reinforcement is given (or presumed to have occurred) during training. In both cases, differential reinforcement is necessary for differential responding, but in perceptual learning it is claimed to be unnecessary in the first stage provided the stimuli are sufficiently "attention getting"—that is, presumably, that they are adequately responded to. A closely related issue is discussed in Chap. 24.

A related problem concerns the generalization of form discrimination. If an animal learns to discriminate one form from another, it will exhibit some degree of generalization to related forms. For instance, a rat trained to respond to a solid triangle will respond also to an outline triangle or a triangle of different size. Some ability to generalize in this way is present in rats and monkeys reared in darkness or diffuse light, though in general stimulus-deprived animals show poorer transposition than experienced ones. Hebb (1949) considered pattern generalization to depend on eye movements, the new stimulus (e.g., an outline triangle) calling forth movements similar to those elicited by the training stimulus (e.g., a solid triangle). However Ganz and Wilson (1967, see refs. cited) found that rhesus monkeys, reared from birth without experience of visual patterns and trained to discriminate a horizontal from a vertical line under conditions in which the stimuli fell only within restricted regions of the retina, subsequently showed generalization to other forms in a manner no different from normally reared monkeys. They suggest that their data can probably be accounted for in terms of the properties of units resembling the simple cells in the visual cortex of the cat, whose selective sensitivity is present when the eyes are first opened (these are discussed on pages 108–115, and 487–489).

20.5 BINOCULAR EQUIVALENCE

The functional equivalence of the two eyes can be assessed by training an animal on a visual discrimination task with one eye covered and then testing it with that eye exposed and the other eye covered. In ducks there is evidence that inter-ocular transfer does not require previous joint experience with both eyes in patterned light (Moltz and Stettner, 1961, 1962): this is presumably related to the complete decussation at the optic chiasma. In higher mammals, which have a corpus callosum, complete and immediate transfer does not occur unless both eyes have had considerable simultaneous exposure to patterned light (e.g., Riesen, 1958). Monocularly deprived cats and chimpanzees show marked deficiencies on the deprived side, even though performance with the other eye is normal: the usual equivalence between the two eyes does not develop, a discrimination habit learnt with one eye remaining specific to that eye. However, pattern-deprived cats, though somewhat retarded in the acquisition of visual discriminations, show little difference in inter-ocular transfer from normally reared ones when tested on a response that does not require visual guidance (Meyers and McCleary, 1964). These authors suggest that previous findings of lack of inter-ocular transfer after visual deprivation reflect a lack of the visuomotor equipotentiality required for the visually guided responses used. The question is, however, complicated. Sechzer (1964,) working with "split-brain" cats in which the corpus callosum is severed, showed that although inter-ocular transfer of a pattern discrimination learned with food approach does not occur, transfer does occur if the discrimination is learned with shock avoidance. This is presumably mediated by an extracallosal subcortical pathway. Further light on this problem is provided by experiments discussed in the next section.

20.6 PHYSIOLOGICAL STUDIES OF PERCEPTUAL DEVELOPMENT IN CATS

New understanding both of the effects of visual deprivation and of the development of relations between the two eyes has recently come from studies of the visual system in kittens. As discussed in Chap. 5, in the normal cat most striate cortex cells are influenced from both eyes, the two receptive fields having similar orientations and retinal locations. The majority of cells are more strongly influenced from the contralateral eye than from the ipsilateral one. If kittens are deprived of patterned visual stimulation and tested when 8 to 16 days old, the responses of cortical cells are similar to those of the adult. Thus the connections responsible for the highly organized behaviour of cells in the striate cortex are present within a few days of birth, and develop to this point independently of patterned visual experience (Hubel and Wiesel, 1963). However, kittens deprived of patterned vision in one eye from birth for 2 to 3 months show defective vision in the deprived eye: correlated with this, most striate cortex cells are driven from the undeprived eye, with normal receptive fields, but very few are driven from the deprived eye. Thus the monocular deprivation must produce a disruption of the system which was functioning normally at birth. Visual experience apparently stabilizes the system against the effects of deprivation. If the kitten is given 1 to 2 months of normal visual experience before monocular deprivation, the severity of the effect is reduced; and in an adult cat even 3 months of monocular deprivation has no effect on the properties of the cortical cells (Wiesel and Hubel, 1963a).

Cells in the lateral geniculate body are less affected by deprivation from birth: most cells show normal receptive fields with an on-centre and off-periphery, or vice versa, but there are some diminution in overall activity and some histological changes, including a reduction in cell area. Again, the changes are less marked or absent if the deprivation is preceded by a period of visual experience (Wiesel and Hubel, 1963b). Little recovery of function occurs after visual deprivation from birth for 3 months (Wiesel and Hubel, 1965a).

Although monocular deprivation from birth disrupts the functioning of the system, binocular deprivation produces effects less severe than expected. Wiesel and Hubel (1965b) sutured the eyes of kittens and tested at 2½ to 4½ months. Although many cortical cells responded abnormally or were unresponsive, there were also many whose responsiveness seemed normal in all respects. Thus the damage caused by deprivation of one eye depends in part on whether or not the other eye is also closed. Presumably, therefore, it is not merely a consequence of disuse, but of abnormal interaction between the two pathways. Additional evidence in support of this view was obtained from kittens given an artificial squint by severance of an extraocular muscle at the time of eye-opening. The animals were reared normally for 3 months to 1 year. When the eyes were then tested separately, no behavioural defects could be detected. Recordings from striate cortex were also normal, except that only about 20 percent of the cells could be driven from both eyes, as compared with 80 percent in the normal cat. Since, as we have seen, most cells are driven from both sides soon after birth, the squint must have caused the cells to shift their ocular dominance so that they came to favour the side which dominated at birth and ultimately lost connection with the other side. Thus lack of synergy between the input from the two eyes is enough to cause disruption of the connections that subserve binocular interaction. A similar result was obtained by covering each eye on alternate days, so that the two eyes never worked together (Hubel and Wiesel, 1965b).

20.7 KINAESTHETIC PERCEPTION IN CHIMPANZEES

In order to assess the effects of deprivation in another sensory modality, Nissen et al. (1951) kept a chimpanzee with its arms and legs enclosed in cardboard cylinders for 30 months, starting at the age of 4 weeks. When the cylinders were finally removed, the animal showed marked tactile-kinaesthetic disabilities in spatial orientation; for instance, only after considerable fumbling and exploration was it able to touch a stimulated point on its skin. Walking and sitting postures were also abnormal, grooming was absent, and when picked up the animal did not cling to the keeper. Its abilities for visual discrimination were normal. Further experiments on this topic are discussed below.

20.8 DEVELOPMENT OF VISUOMOTOR COORDINATION

In many of the experiments discussed so far, perceptual development has been assessed by the measurement of limb movement—for instance, does the animal walk out over a cliff or extend its paws to ward off an approaching surface? They thus involve not merely the processes of perception but also of sensorimotor coordination. An indication that *mere* exposure to the stimuli is not all that is required in such cases is contained in one of Riesen's experiments with chimpanzees mentioned above. An animal which was not permitted to move about in a visual world was deficient by comparison with animals which were. In another experiment Riesen and Aarons (1959) showed that kittens reared in holders but permitted visual stimulation in a patterned environment showed deficiencies in visually guided behaviour compared with normally reared litter mates. Comparable deficits were found in kittens allowed free movement but wearing light-diffusing hoods over their eyes.

Such experiments are open to interpretation in two ways. Riesen (1961), following up Hebb's earlier suggestions, supposed that patterned visual stimulation is important in the development of visually guided behaviour because of the opportunity it offers for developing sensory-sensory associations—that is, associations not involving an overt response—both within and between sensory modalities. On this view, the environment of the confined kittens, though patterned, provided insufficient variation in visual stimulation. Another possibility is that it is not merely the magnitude of the stimulus variation that is important, but whether or not it is self-induced. To investigate this, Held and Hein (1963) attempted to equate the variation in stimulation between the confined and non-confined groups. The groups then differed primarily in the cause of the variation: in one case it was a consequence of the animal's own locomotion, while in the other it was not. This was achieved by using the apparatus shown in Fig. 20.7. The movements of the confined cat *P* about the axis of its own box *d–d* and about the main axis *a–a* were controlled by the movements of the restrained but unconfined cat *A*: visual stimulation was a feedback consequence of its own movement for *A*, but largely the result of equivalent passive movement for the other. The kittens were reared initially in the dark and placed in the apparatus for a daily exposure period beginning between 8 and 12 weeks of age. Subsequently the kittens were tested on a visual cliff apparatus (see pages 473–475) and for visual paw placing (page 476). The actively moving animals were uniformly superior. In a later experiment Held and Hein (1967) showed that this difference in abilities for space perception between kittens reared under conditions of active and passive movement could be induced also between the two eyes of the same kitten. When kittens were reared with one eye open in active locomotion and the other in passive, the "active" eye was found to mediate normal visually guided behaviour but the "passive" eye failed to do so. Hein and

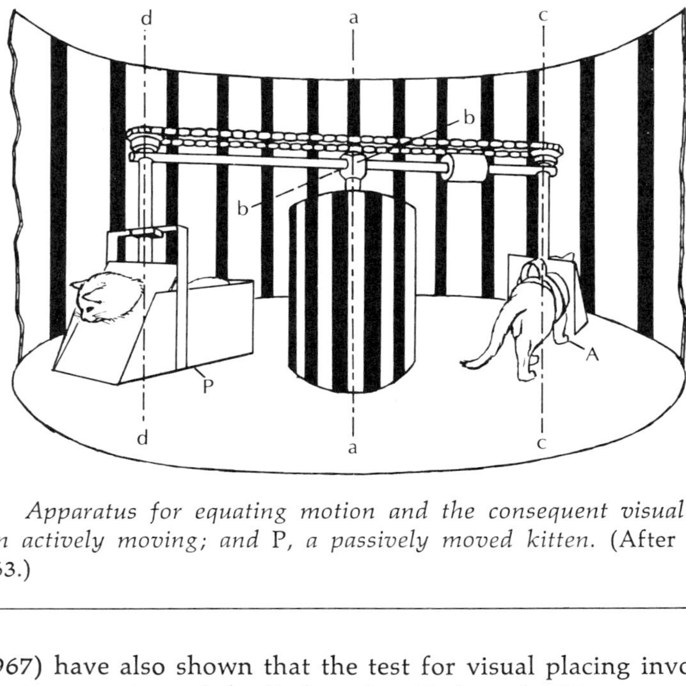

Fig. 20.7 Apparatus for equating motion and the consequent visual feedback for A, *an actively moving; and* P, *a passively moved kitten.* (After Held and Hein, 1963.)

Held (1967) have also shown that the test for visual placing involves two components, extension of the limb and guided reach. Kittens were reared with collars around their necks whenever they were in the light so that they could not see their own limbs. When lowered toward a discontinuous surface, they extended a forelimb, but it often went into the spaces on the surface. Normally reared kittens extended their paws and contacted the surface.

These experiments have been extended to primates. Infant macaques were reared from birth in a chair with broad collars around their necks so that they could not see their limbs (Fig. 20.8). When 35 days old they were allowed to see one hand. At first visually guided reaching with this hand was very poor. Since the animals were able to direct eyes and head and could control movements of the hands with respect to the body, the deficit lay in the integration of the visuomotor control of head movement and the non-visual control of limb movement. When first allowed to see the hand,

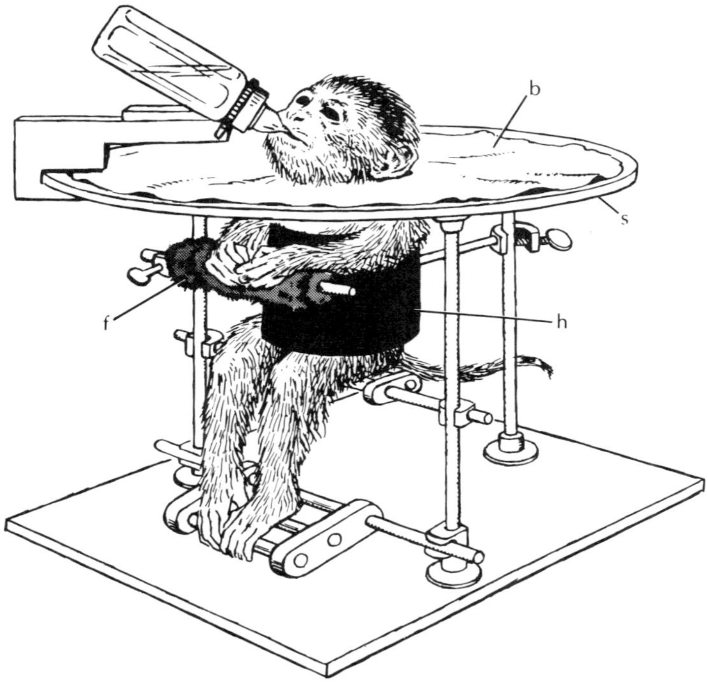

Fig. 20.8 Apparatus for rearing an infant monkey without sight of its limbs. (After Held and Bauer, 1967.)

h, metal cylinder; *f*, fur-covered rod; *s*, plywood body shield; *b*, cloth bib. In addition, a vertical rug-covered cylinder 5 cm in diameter extending from the footrest to the bottom of the hand-shield may be provided.

the monkeys spent much time looking at it, and visuomotor performance gradually improved (Held and Bauer, 1967).

Such results suggest that the role of experience in the development of visual perception cannot be interpreted solely in terms of *S–S* associations, as suggested by Hebb (1949) and Riesen (1958): *S–R–S* sequences seem also to be involved (Held, 1961; see also Meyers, 1964).

This view is consistent with observations on the development of behaviour in human children by Piaget (1952, 1955) and Gesell (1954). For instance, Piaget found that for a child of 3 or 4 months there was no integration of the stimuli presented by an object through different sensory modalities. At 5 to 6 months the visual and tactile aspects of objects began to be coordinated, and Piaget describes the differentiation between "near space" which can be touched and "distant space" which cannot. Throughout this and later stages in the development of visuomotor behaviour, the role of interaction between the child and its environment by movement of eyes, limbs, and body is constantly apparent. White et al. (1964) have shown that children reared with adequate nutritional and physical care, but spending much time lying on their backs staring at a grey ceiling, were retarded in a variety of sensorimotor activities compared with children in a more varied environment. This retardation could be reduced or eliminated by enriching the environment.

At this point we may refer again to the discussion of the role of feedback stimulation in the perception of relative motion (pages 96–101). Even simple organisms behave differently to relative movement of the environment which is caused by movement of their own bodies and that which depends on movement of objects in the environment. To account for this, von Holst and Mittelstaedt (1950) suggested that the consequences of movement (reafferent stimulation) are compared with a representation of the required movement (efference copy or corollary discharge) centrally, further action depending on the results of this comparison (see page 100). But here a further question is raised. How is it that the corollary discharge from the efferent system resulting from a command to, say, move the eyes 10 deg to the right, and the reafference from a shift of 10 deg of the image on the retina, exactly balance each other? Do the growth processes of the organism determine the coding of the messages in each case in such a manner that they do just balance?

That this is not the case in human subjects can be demonstrated quite readily by experimental methods which disturb the reafference. These fall into three main groups (Held, 1961; Teuber, 1960, 1961; Hein and Held, 1962; and refs. cited).

1. *REARRANGEMENT*

If the retinal image is shifted by a wedge prism, there is an apparent displacement of objects in the visual field. This results in various types of sensorimotor discoordination; for instance the subject is unable to put his finger accurately on an object which he sees. With prolonged exposure, however, there is considerable readjustment, and in a matter of hours most of the errors disappear. Readjustment occurs, at any rate to some degree, even when the subject wears spectacles which produce complete inversion of the retinal image (Stratton, 1897; Köhler, 1951). If the prisms are worn for only part of the time, the subject can adapt to both normal and shifted visual fields (Taylor, 1962).

This readjustment depends on a comparison between intended movement (Sollwert) and reafferent stimulation: that is, the subject must make voluntary movements and witness their results. For instance, if subjects wear prisms which displace their visual field laterally, considerable readjustment occurs if the subject is allowed to walk about for an hour, but not if he is pushed about for a similar period in a wheelchair. Similar results are obtained when hand-eye coordination is assessed in subjects wearing prismatic spectacles. Considerable readjustment occurs if they are allowed to move their arms themselves, but not if their arms are moved passively an equivalent amount (Held and Hein, 1958; Held and Bossom, 1961).

2. *DISARRANGEMENT*

In the experiments described above the rearrangement results in a regular relation between the reafference actually received and the reafference which would have been received without experimental interference. Under these conditions readjustment occurs. It is possible, however, to arrange for the discrepancy not to be predictable in this way. Cohen et al. (cited by Held, 1961) used a prism which varied continuously in power, so that the image was displaced over a range of positions, and at any particular time the deviation produced by the prism was unpredictable. The prism produced larger displacements in the horizontal than in the vertical dimension. Under these conditions measures of hand-eye coordination showed no readjustment, with very considerable errors in the horizontal dimension but much smaller ones in the vertical dimension. A similar result is obtained by introducing a delay between the actual and apparent movement of the hand: a delay as short as 0.3 sec prevents adaptation to displacement of the image (Held et al., 1966).

3. *DEPRIVATION*

In these experiments a marked reduction in sensory input is produced. Typically, subjects are confined in soundproof anechoic chambers or in small rooms with a continuous masking noise. The conditions are stabilized, and the subjects may be required to lie down while wearing translucent spectacles and cardboard cuffs to reduce tactile stimulation (e.g., Bexton et al., 1954; Doane et al., 1959; Solomon et al., 1961, and refs. cited). We have already considered some aspects of the results of such experiments (page 216): here we may note that they produced marked perceptual aftereffects which lasted for minutes or hours. It is characteristic of these effects that they are variable and disorderly: not only is there loss of perceptual constancies and dysfunction of colour vision, but shapes are distorted and unstable, with straight edges or lines appearing curved and wavering (Freedman, 1961). The results of deprivation experiments thus contrast with those of rearrangement experiments, for the aftereffects of the latter are relatively regular and predictable. In both cases, however, the time courses of recovery are about the same (Held, 1961).

Such results can be interpreted in terms of a body schema (Head, 1920), an integrated consequence of previous perceptions and memories concerned with the relations between the parts of the subject's body and its relations to the external world (see critical discussion by Oldfield and Zangwill, 1942a, b). We may suppose that the organism's operations with respect to the external world are guided in accordance with such a schema. In rearrangement experiments like those described earlier the new sensory field is isometric with the old one, but shifted along one or more dimensions. Adjustments occur as a result of modifications of the schema to fit the new conditions. Under conditions of deprivation, however, regularities in the sensory input are reduced; the organism has a more or less undifferentiated and homogeneous environment. That the important factor is the absence of regularities in the sensory input, rather than a reduction in the level of input itself, is indicated by the finding that similar perceptual disturbances occur with diffuse light, no light, and randomly flashing lights. Since we cannot suppose the schema to be forgotten during a few hours of deprivation, the degeneration which occurs may well be the consequence of attempts, inevitably unsuccessful, to fit the schema to the new conditions (Freedman, 1961).

Using the language of control systems, Held (1961; Hein and Held, 1962) has produced an elaboration of the von Holst/Mittelstaedt model (Fig. 20.9) which encompasses these experiments. This involves the addition to the reafference model of a "correlation storage unit" where traces of previous combinations of concurrent efferent and reafferent signals are retained. The corelated discharge of the efferent system is supposed to select from the correlation unit the trace combination containing the identical efferent part and to activate the corresponding reafferent trace. This is sent to the comparator for comparison with the current reafferent signal. The outcome of this determines perception and/or further performance.[2]

Under conditions of rearrangement, more than one efferent correlated discharge/reafferent combination will be available for selection from the storage unit by any particular effector discharge. Since readjustment occurs,

[2] There is a certain formal similarity here to Tolman's "means-end-readiness". He writes: "When an *instance* of an *S* is presented, there tends to be released an expectancy . . . that an instance of the kind of performance symbolized by *r*, will lead to an instance of *s* . . ." (Tolman, 1959).

selection is presumably weighted by how recently the combinations were stored. With constant rearrangement, selection of the old combinations gradually ceases. Under conditions of disarrangement or deprivation, however, a number of combinations of equal weight will be available for selection from storage. Thus any one of a number of reafferent signals may be sent to the comparator, whose results will be ambiguous. The model thus accounts for: (1) the possibility of readjustments under conditions of stable rearrangement, (2) the lack of readjustment under unstable or deprived conditions, and (3) the necessity for reafference if readjustment is to occur (see also Held and Freedman, 1963).

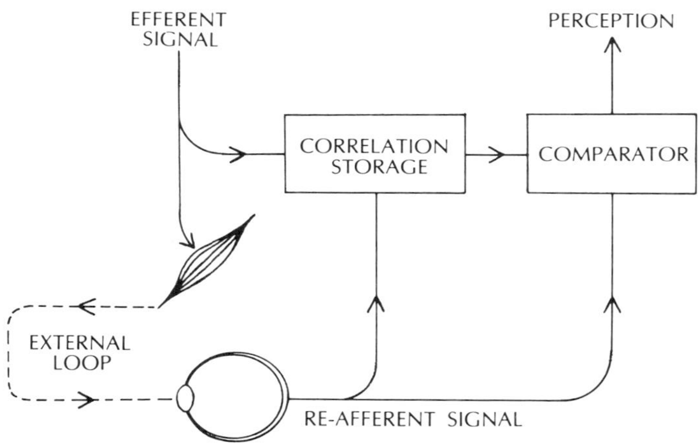

Fig. 20.9 Schematized process assumed to underlie the consequences of neonatal development, rearrangement, disarrangement, and deprivation. (After Hein and Held, 1962.)

Whether adaptation in rearrangement experiments is due solely to recorrelation of movement and feedback is not quite certain. Some authors (e.g., Harris, 1965) have claimed that adaptation to optical displacement produced by a prism results from a change in the "felt position" of the parts of the body relative to each other. This view is based on observations that the altered direction of reaching for visible targets after viewing the optically displaced hand is also present when the subject points at non-visible targets, such as sound sources, or "straight ahead". This is held to indicate that adaptation results from a change in the proprioception mediating the felt position of the reaching arm. On this view, such shifts should all be of equal magnitude. However, Efstathiou et al. (1967) found that subjects showed a greater shift when reaching for a visible target than when reaching for the other hand, and no shift in relocating a remembered position of the adapted arm. They suggest that there are two independent modes of reaching, one based on the matching of arm movements with head orientations toward the target (see page 164), and the other dependent on the felt position of the arm. They suggest that the former is subject to adaptation, while the latter is not and constrains the shifts produced by the former.

Kravitz and Wallach (1966) found adaptation occurring if human subjects could see their immobile forearm through a displacing prism while the arm was resting on a vibrating board. This also suggests that self-produced movements are not essential, and these authors believe that either vibration of the immobile arm or self-produced movements act by enhancing the input from proprioceptors. While such experiments suggest that self-produced movement may not be the only means whereby adaptation to spatial distortion is achieved, Held's experiments indicate that it plays a primary and perhaps essential role (see also Weinstein et al., 1964; Hochberg and Held, 1964).

These experiments have been concerned largely with humans, and we know very little about the extent to which such readjustments are possible in the adults of other higher vertebrates. It may well be that, for most species, there is no need to postulate a correlation storage unit: much of the available information for invertebrates (e.g. Mittelstaedt, 1949; Wells, 1962b) and lower vertebrates (e.g. Sperry, 1943, 1950, 1951) indicates that little readjustment occurs after rearrangement of the sensorimotor systems (pages 467–469). However, some fish show gradual readjustment to disturbances of the sensory inflow used in orientation to gravity and light (von Holst, 1950b), and others show progressive changes in orientation with growth which must be due to a progressive change in the gravity feedback system (Braemer and Braemer, 1958; see also pages 176–178). The effects of rearing mammals under conditions of sensory deprivation indicate that in many species at least sensorimotor coordination is influenced by early experience: adjustment may thus be possible in adult life in the same way as in man.

20.9 THE RELATIONS BETWEEN FORM PERCEPTION AND ORIENTATION IN SPACE

Some of the findings summarized in the preceding sections have recently been related in a manner which should provide a considerable stimulus to studies of the development of perception. We have seen that at any rate some of the physiological mechanisms involved in form perception are present in visually inexperienced animals (Sec. 20.2). A rudimentary type of form vision appears to be present in neonatal primates, and some ability for form discrimination in visually inexperienced cats can be demonstrated provided that they are not required to perform a visually guided response. However, the normal capacity for form discrimination appears to be retarded if the testing procedure requires a visually guided response oriented with respect to the environment (Sec. 20.2). For the development of the visuomotor coordination necessary for such a response, or of most other forms of visuomotor coordination, response-produced visual feedback seems to be essential. Furthermore, the development of visuomotor coordination occurs at a number of levels: sight of the moving limb is necessary for eye-limb coordination, while visual feedback during gross body movement is required for accurate orientation of head and body to visible objects (Secs. 20.2 and 20.8). Again, if only one eye is allowed to see the moving limb, the eyes become dissociated with respect to limb guidance even though each can control head and body orientation (Secs. 20.5, 20.6, and 20.8).

Such considerations led Held (1967/1968) to emphasize the distinction between the analyzing system, which abstracts properties that are invariant over a wide range of sizes and distortions, relational and nonmetrical properties being more important than metrical ones, and the orienting system, which depends on the metrical properties of space and is essential for manipulation and locomotion in space. While development of the orienting system depends on response-produced feedback, that of the analyzing system does not.

However, Held points out that the two systems interact. In rearrangement experiments with human subjects there is an alteration in either visually guided motor responses or in apparent orientations of targets with respect to the body. These amount to alterations in the perceived directions from the body of stimulated points on the retina. It therefore involves also a change in form perception if members of a set of visual directions are displaced in a non-uniform manner. For example, wedge prisms can cause straight lines to appear curved because of differing refraction of the rays of light with different angles of incidence. Experiments show that, while errors of localization produced in this way are subject to almost complete adaptation, apparent curvature and distortion of angle are not. Held suggests that this limited adaptation to shape distortion caused by rearrangement results from an interaction between the orienting system, which adapts, and the analyzing system, which does not.

A dissociation between the ability to identify a stimulus and the ability to orient toward it in space has been shown independently recently by three other investigators. Ingle (1967/1968) found that the equivalence of form for goldfish changed with the direction of the stimuli, suggesting that the neural coding of direction could be based on either shape or position. Schneider (1967/1968) found that the superior colliculus of the hamster is necessary for the sensorimotor control of visuomotor orientation, while discriminative responses depend much more on the visual cortex. The colliculus is necessary only if the discriminative response involves visually guided orientation. Finally Trevarthen (1967/1968) found evidence that vision in primates involves two parallel processes, subserved by anatomically distinct mechanisms, one determining space around the body and the other for the examination of detail in small areas of space. The publication of the findings of these four workers in a single symposium seems likely to open new avenues in the study of perceptual mechanisms and perceptual development.

20.10 INTER–SPECIES DIVERSITY

In conclusion, we see that there are great differences between lower and higher vertebrates in the role of early experience in the development of perceptual abilities. In the lower forms many of the patterns of orientation in visual space can be accounted for in terms of the maturation of nervous connections, though some exposure to light may be necessary for proper retinal functioning. In the higher forms not only does early experience have far-reaching effects, but perception is also adjusted in the light of immediately preceding experience: at any one moment perception must be a consequence of many and diverse effects from the remote and recent past. Presumably, the maturational processes of nerve growth are basically similar in the higher and lower species: just why (causally and functionally) there is this difference in plasticity is not clear.

20.11 SUMMARY

1. The orientation of visuomotor responses in lower vertebrates depends on orderly connections between retina and brain. After severance of the optic tract, rotation of the retina, and subsequent regeneration of the optic pathways, the visuomotor responses of salamanders are appropriate to the point on the retina stimulated and not to the location of the stimulus in space. This indicates that neurones reestablish functional connections similar to those existing before the operation. A mechanism involving specific chemical affinities between neurones has been suggested. Similar findings have been made with other responses and other sensory modalities.
2. Visual experience is not essential for the development of space perception in chicks, as shown by pecking accuracy or performance on the visual cliff. In kittens, and even more in primates, distance perception does depend on visual experience. Some ability for form discrimination is present in visually naive mammals, and improves with experience.
3. The richness of the early environment affects the subsequent performance of rats in form discrimination and maze running. The "perceptual learning" by which early experience influences later ability for pattern discrimination is independent of conventional reward.

4. The role of experience in the development of stimulus control (as shown by a peak in the generalization gradient) is discussed.
5. Binocular equivalence depends on binocular experience in cats and chimpanzees, but not in ducks.
6. The role of binocular vision in the development of the visual system in kittens is discussed.
7. Experience plays a role in the development of kinesthetic perception in chimpanzees.
8. In kittens and monkeys, the experience necessary for the development of visuomotor coordination does not involve merely the development of *S–S* connections, but requires active participation by the subject. The learning is best described as *S–R–S*. Perception of space in human subjects can be upset by a number of techniques which disturb the feedback stimulation received from self-initiated movement. The data from such experiments can be described by a modification of the von Holst-Mittelstaedt reafference model.
9. Some relations between form perception and orientation in space are considered.

21

DEVELOPMENT OF STIMULUS-RESPONSE RELATIONS

Although some responses can be elicited by a very wide range of stimuli and some only by very specific ones, in each case it is possible to delimit the effective ones with some degree of precision (Chap. 4). In vertebrates these stimulus-response relations may be strongly influenced by preceding experience. Since a detailed discussion of the learning processes involved is beyond the scope of this book, attention will be focused primarily on stimulus-response relations in which learning initially plays only a minor part, and on the manner in which those seen in the adult animal develop from them.

21.1 APPROPRIATE RESPONSIVENESS INDEPENDENT OF EXPERIENCE IN THE CONTEXT

In the first place, there are many cases in which stimuli are responded to appropriately the first time that they are encountered. If, for instance, the high-pitched whistle given by adult great tits on the approach of a flying predator is heard by 12-day-old nestling great tits which have been reared by hand away from their kind, they will immediately crouch and freeze (Barraud, 1961): other sound patterns of similar intensity do not elicit this response. A chaffinch 30 days old, which has never seen an owl, will give the mobbing response on the first occasion that an owl is presented to it (Hinde, 1954a).

Similar data are available for stimuli received through other sensory modalities. For example, 2-day-old garter snakes (*Thamnophis sirtalis*) attack chemical extracts of organisms which are normally eaten, but not extracts of other organisms. When different species of snake are compared at the first feeding (i.e., when the snakes are 3 to 7 days old), the differences in responsiveness to extracts of skin substances of small mammals correspond to the species food preferences (Burghardt, 1966, 1967).

Such cases are not limited to very young individuals. To give but one example, the red breast of the male chaffinch is important in intra-species fighting. Females, whose breasts are the natural grey-brown, are dominated by females whose breasts have been dyed red. A similar avoidance of females with dyed breasts is shown even by females reared by hand away from males, so it is not acquired as a result of social experience (Marler, 1955; see also E. Cullen, 1960).

It is, of course, not a complete explanation of the developmental problem to say that, in such cases, the appropriate response is "unlearnt". The observations merely delimit the problem—previous experience in the specific situation is not essential for the appropriate response.

21.2 EXTENT OF INITIAL DIVERSITY OF EFFECTIVE OBJECTS

In some cases not only does the naturally occurring object evoke the appropriate response on the first occasion it is encountered, but experiment reveals some degree of specificity in responsiveness. For instance the young of nidifugous birds usually peck at food objects the first time they encounter them: although they also peck at a wide range of other objects, they are by no means completely catholic in their responsiveness. Hess (1956a) measured the frequency with which young chicks and ducklings would peck at differently coloured chips. Although they had had virtually no previous visual experience, and were not rewarded for pecking, clear preferences for certain of the chips were shown. These were related to the wavelengths of the colours on the chips rather than to their reflectance or purity, the chicks showing a bimodal preference in the orange and blue regions and the ducklings a single sharp peak in the green (Fig. 21.1). In each species, therefore, there was differential responsiveness to colour. Studies of pecking in other avian species suggest that most can be referred to one of three types—either a bimodal preference in red and blue, as with chicks, or a yellow/green preference, as with ducks, or no chromatic preference (Hailman, 1967). We have already seen (pages 484–485) that such differential responsiveness may be influenced by rearing conditions, but the inter-species differences occur when rearing conditions are constant: they are presumably the result of natural selection acting on developmental processes as they occur in the normal species habitat.

In addition to this effect of colour, domestic chicks show a preference for solid objects, pecking more at a solid hemisphere than at a flat disc. This seems to depend on differential surface lighting, but the preference occurs also in chicks reared with light coming from below: thus it does not depend on experience of normal lighting conditions (Dawkins, 1968).

Although the stimulus characters effective for a given response are thus limited, it is important to emphasize that their range is often greater for a naive animal than for an experienced one. The species-characteristic prey-catching response of the newly metamorphosed clawed toad (*Xenopus laevis*) is elicited even by a jet of water directed against the side of the body, or a small moving dot of light (Eibl-Eibesfeldt, 1962). Similarly, while Tinbergen and Kuenen, in their study of begging in young thrushes (page 66), analysed the essentials of the stimulus situation eliciting begging, it is also known that when a young passerine first opens its eyes, it will in fact beg to a very wide range of objects, including a spatula or a pair of forceps. A few days later the response can be elicited only by the object to which the bird is accustomed (see also Schaller and Emlen, 1961). Many of the behaviour patterns used subsequently in reproduction by passerine birds appear erratically in fledglings a few weeks old in response to stimulus situations which would be quite inadequate in the adult (personal observation).

Similarly, some of the responses of the human baby can initially be

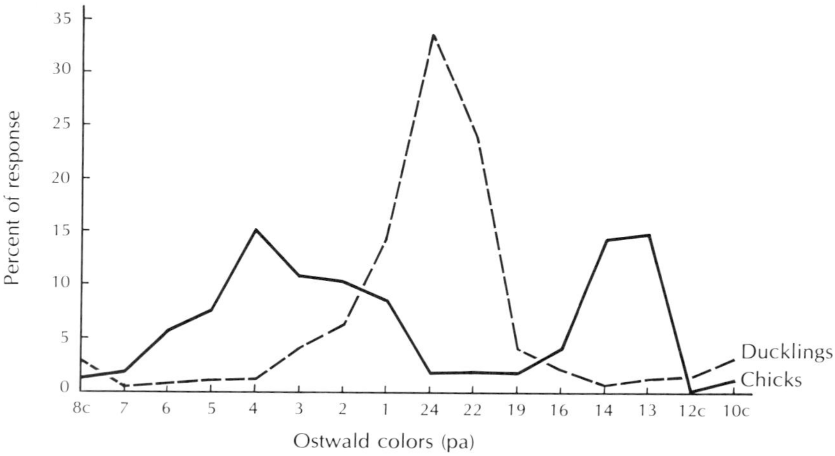

Fig. 21.1 Percentage of total pecks delivered by chicks and ducklings to each of 16 colours. (After Hess, 1956*a*.)

elicited by diverse stimuli. The case of smiling has been studied by Ahrens (1954). After an initial stage in which the baby smiles "spontaneously", there is a period during which two black dots on a white oval background and a wide variety of masks are at least as effective as a human face. Later other features, such as eyebrows, mouth, and so on, become necessary; and, later still, masks become ineffective and only a human face will elicit smiling. By the middle of the first year only one or two familiar faces, such as that of the mother, will elicit smiling, though, when yet older, the range of individuals to whom the baby will smile is again extended (Ambrose, 1960). A comparable example is provided by the sucking behaviour of kittens. Up to 4 days of age a kitten will suckle from any receptive mother, but after 6 days it will suckle only from its own mother, unless it is made extremely hungry. At 20 to 25 days, however, it will once again suckle from strange mothers (Rosenblatt, personal communication).

It is often difficult to determine whether the diversity of stimulus situations eliciting a response in a young animal is due to its inability to distinguish them (i.e., lack of development of the sensory/perceptual apparatus), or to a lack of specificity in the stimuli pertinent to the particular response, or to response to a common feature of stimulus situations divergent in other respects. The phenomenon itself, however, is similar in many ways to that of "stimulus generalization". In conditioning situations an animal which has been trained to respond to one stimulus will also respond to other similar stimuli. The response is likely to be weaker, the greater the difference between training and test stimuli, giving a "gradient of generalization" round the training stimulus (see pages 482–486). The training may involve a variety of discrimination procedures in either classical or instrumental conditioning situations (Cowey, 1968). Although the term "stimulus generalization" is usually reserved for conditioning experiments (e.g., Hilgard and Marquis, 1961), such cases differ operationally from those discussed here only in the initial ineffectiveness of the stimulus.

In line with the possibility that the diversity of stimulus situations effective for particular responses in young animals is apparent only and due to a response to a common stimulus feature which they possess, Schneirla (1959, 1965) has suggested that the essential aspects of the stimulation for many responses of young organisms are quantitative rather than qualitative. They tend to approach, with part or all of the body, sources of stimulation which are quantitatively low, regular, and limited in range of magnitude, and withdraw from those which produce inputs which are high, irregular, and of extensive ranges. On this view, as the intensity of stimulation is increased from zero it first becomes more effective in eliciting a full approach, but, beyond a certain intensity, withdrawal becomes more likely. The important aspect of the stimulus is thus a quantitative one, qualitative differences becoming significant only later.

For instance, some of the data on the stimuli eliciting pecking in young chicks can be interpreted in this way, for many of the stimulus characters (other than colour, see above) important in eliciting pecking seem to involve primarily quantitative features. For example, Rheingold and Hess (1957) showed that brightness of the reflecting surface is an important variable. Some (Engelmann, 1941; Fantz, 1957), but not all (Curtius, 1954), workers find that rounded shapes are more effective in eliciting pecking than other shapes: differential responsiveness is maintained even if pecking experience is unreinforced. Since round shapes are known to be especially conspicuous to a wide range of vertebrates, they may be effective through their quantitative contribution to visual stimulation rather than their particular qualitative characters (Schneirla, 1965). Again, as we saw earlier page 82), Schneirla interprets the various stimulus characters which Tinbergen and Perdeck (1950) found to elicit pecking from herring gull chicks (see pages 59–62) on the same principle, namely, that they owe their effectiveness to the quantitative intensity of visual stimulation they produce. Curtius, *loc. cit.*, however, found that newly hatched lapwings (*Vanellus vanellus*), chicks, and turkeys showed size preferences in their pecking response to models, and that these size preferences were related to their bill sizes, a result perhaps less easy to explain in simple quantitative terms.

Other aspects of chick behaviour can also be interpreted as responses to quantitative aspects of the stimulus. For instance, newly hatched chickens, while responding preferentially to calls of their own species (Gottlieb, 1965, 1968a), are attracted by a fairly wide range of repetitive sounds (Collias and Joos, 1953; see also Kruijt, 1964), as well as by moving objects and patterns (e.g., F. V. Smith, 1962). They are caused to withdraw and give distress calls by a wide range of intense stimuli through various modalities, including temperature (e.g., Collias, 1952; Salzen, 1962; Kruijt, 1964). Schneirla interprets the effectiveness of a dark object moving overhead in eliciting fleeing from nidifugous birds as due not to any qualitative characters of shape (Tinbergen, 1951; Schleidt, 1961; Müller, 1961), but to the sudden change in retinal stimulation (see also page 510 and Melzack et al., 1959; McNiven, 1960).

Quantitative characters seem to be important in a similar way in other species. To cite but one example, in many amphibians such as the toad (*Bufo bufo*), small stimulus objects elicit tongue-flicking, and somewhat larger objects a body lunge as well, but objects larger still induce withdrawal (Eibl-Eibesfeldt, 1951b).

Although Schneirla was concerned primarily with young organisms, similar examples can be found in adults. For example, Beebe-Center et al. (1948) found that rats prefer saccharin to plain water when the concentration is slightly above threshold, but water is preferred to higher concentrations (see also Weiner and Stellar, 1951). Similarly, Warren (1963) found that fluid-deprived mice tended to prefer low concentrations of the alkaloid sucrose octa-acetate (which tastes bitter to man) to water, but avoided higher concentrations.

On the basis of such examples, Schneirla conceives of the mechanisms underlying approach responses in the young organism as depending on a system activated by weak sensory input and involving both central and autonomic nervous systems. The continuing "tonic" activity in these systems forms the basis of normal functioning. When weak stimuli are encountered in pre- and early post-natal development, the system is further but briefly aroused and there is a "phasic" response. This phasic response has the same character as the tonic one and, with repetition, phasic responses become tied to the tonic background. Thus at an early stage the neural mechanisms underlying approach come under the control of low-intensity mechanical and visual impulses. Through tactual-proprioceptive and other feedback mechanisms, the phasic approach responses of head-advancing and turning-toward develop. At hatching or birth the control is extended to optic or other exteroceptors by a diffusion of neural impulses from these newly excited sensory systems, so that approach is elicited by a broad range of extrinsic stimuli which are equivalent in terms of their quantitative neural input. For example, the head-advancing movement of a chick embryo is at first elicited by mild tactile stimulation, but is later controlled visually, in Schneirla's view a result of similar quantitative characteristics of the two types of stimulation. At this stage, through feedback effects, approach movements become phasic, and further experience leads to stimulus differentiation and response specialization.

Schneirla suggests that the withdrawal system, aroused by strong stimulation, does not have a tonic basis since by nature it interrupts the ongoing activity. Arousal of the withdrawal system produces only a phasic response, though this too may become tied to the tonic system as responses come to be elicited by qualitative rather than quantitative features of stimuli. The withdrawal system, aroused by strong stimulation, arises later than the approach system, perhaps because strong stimulation is encountered less frequently.

Just how far Schneirla's generalizations can be taken remains to be seen. His approach poses an immediate methodological problem, since there is no means of comparing the strengths of two stimuli measurable only along different dimensions. In addition, even where quantitative aspects of stimulation seem to be important, it remains true that some stimuli elicit one response, others another. Even in many of the examples which Schneirla cites, qualitative as well as quantitative differences seem to be important. Furthermore, generalization from one sensory modality to another mediated by similarities in quantitative aspects of the stimulation does not seem especially probable and has not yet been adequately substantiated experimentally. But Schneirla's theorizing will certainly initiate an attack on the hitherto-neglected early development of stimulus-response relations, and stimulate the collection of information to supplement the meagre data now available.

21.3 EXTENSION OF THE RANGE OF EFFECTIVE STIMULI

Amongst vertebrates, at any rate, it is a safe assumption that every elicitation of a response is associated with modification to the range of stimuli effective for that response. Many processes are involved in this, and we shall consider first some that lead to an increase in the range of stimuli which are effective.

In the majority of cases, extension of the range of effective stimuli involves the acquisition of eliciting properties by stimuli through association with the initially effective ones. For example, a poor model of an owl elicits a more intense mobbing response from chaffinches which have had previous experience of a more effective one than from experimentally naive birds (Hinde, 1954a); and after a series of presentations involving copulation with dummies of a soliciting female, male chaffinches will mount a normally ineffective male dummy (Hinde, 1959c): in each case the effect is apparently due to conditioning to additional stimuli present in the context of the experiment.

The normal feeding behaviour of the cuttlefish *Sepia* apparently depends on conditioning of this sort. Newly hatched *Sepia* will attack only living *Mysis,* a small crustacean which is probably their normal food in the sea. If they are allowed to attack *Mysis,* however, the latency of attack decreases with successive attacks whether or not the attacks are rewarded with prey, and there is a concomitant increase in the probability of attack on a variety of other moving objects that were at first rarely or never attacked. Thus a cuttlefish which has attacked mysids a few times is much *less* selective than an animal attacking for the first time. The situation is, however, temporary, and the animal later becomes able to learn not to attack patterns which do not result in a food reward (Wells, 1958, 1962a). Similarly, a polecat will attack a stationary rat only after previous experience with a moving one (Eibl-Eibesfeldt, 1963). As a more extreme example, male Siamese fighting fish normally extend their gill membranes in threat as a response to visual stimuli from another male, but this response can be conditioned to an electric shock (Adler and Hogan, 1963) or to a stimulus light (Thompson and Sturm, 1965). Again, Japanese quail (*Corturnix japonica*) can be conditioned to show courting behaviour to the sound of a buzzer (Farris, 1967).

Such cases of extension of the range of effective stimuli by conditioning raise important questions concerned with stimulus selection. When a stimulus elicits a response, there is usually a wide variety of stimuli present which could acquire eliciting properties by association: do all such stimuli become associated, and if not, what determines the selection? In practice, many aspects of the available stimulus complex become associated, so that, after conditioning, the more aspects of the stimulus complex which are changed, the less likely is the response to occur. However, change in some aspects produces a greater decrement than change in others, and so the question becomes one of what causes some aspects of the situation to acquire greater stimulus control over the response than others. One answer is that the aspects of the stimulus situation which acquire control are those to which the animal "pays attention" but, as Terrace (1966) points out, this can be merely a cloak for ignorance.

An important issue here is clearly the nature of the animal's sensory/perceptual equipment: as discussed in Chaps. 4 and 5, this will in part determine selection from the stimuli present. That experience may play a part in the development of that equipment is shown by studies cited in the previous chapter. As might be expected, the nature of the stimuli selected is to some extent characteristic of the species. Most of the evidence on this point comes from a slightly different type of experiment—namely, those in which an animal learns to discriminate between two stimulus complexes and is then tested with individual stimulus components absent: in this way it is possible to assess which characters the animal selected. Such tests show, for instance, that monkeys tend to use primarily cues they can touch (Schuck, 1960), but characteristics of colour also are important, while rats tend to rely on orientation of visual features. Pigeons tend to learn discriminations on the basis of colour in preference to the presence or absence of a line (review by Cowey, 1968). Of course no experiment could demonstrate that any difference, however small, along one dimension was more effective than any difference, however large, in another dimension to which the animal was known to be sensitive: nevertheless, such experiments strongly suggest species differences in the nature of the stimuli selected.

In some cases the more effective stimulus characters are determined also by the nature of the task confronting the animal. For example, dogs were trained on instrumental-conditioning differentiation problems where the discrimination involved either the quality of an auditory stimulus (metronome versus buzzer) or the direction of its source (in front or behind). When the task involved a left leg–right leg differentiation (S_1—R_1, S_2—R_2), they used exclusively directional cues and were unable to learn when only quality cues were available. When confronted with a go–no go differentiation (S_1—R_1, S_2—no R), however, they usually learnt on quality cues (Dobrzecka et al., 1966).

In addition, as Lashley (1938b) found, selectivity may be an individually idiosyncratic matter. This has been demonstrated recently in an experiment by Reynolds (1961a). Two pigeons were reinforced for pecking in the presence of a white triangle on a red background and not reinforced for pecking in the presence of a white circle on a green background. The components of the stimuli were then presented separately under extinction conditions. Since neither bird responded to either the green background or the white circle, both aspects of this complex must have become associated with not responding. One bird, however, pecked only in the presence of the white triangle and the other only in the presence of the red background. Pecking had thus become associated with different aspects of the stimulus situation in the two individuals (Fig. 21.2).

While conditioning is undoubtedly the most common means of ex-

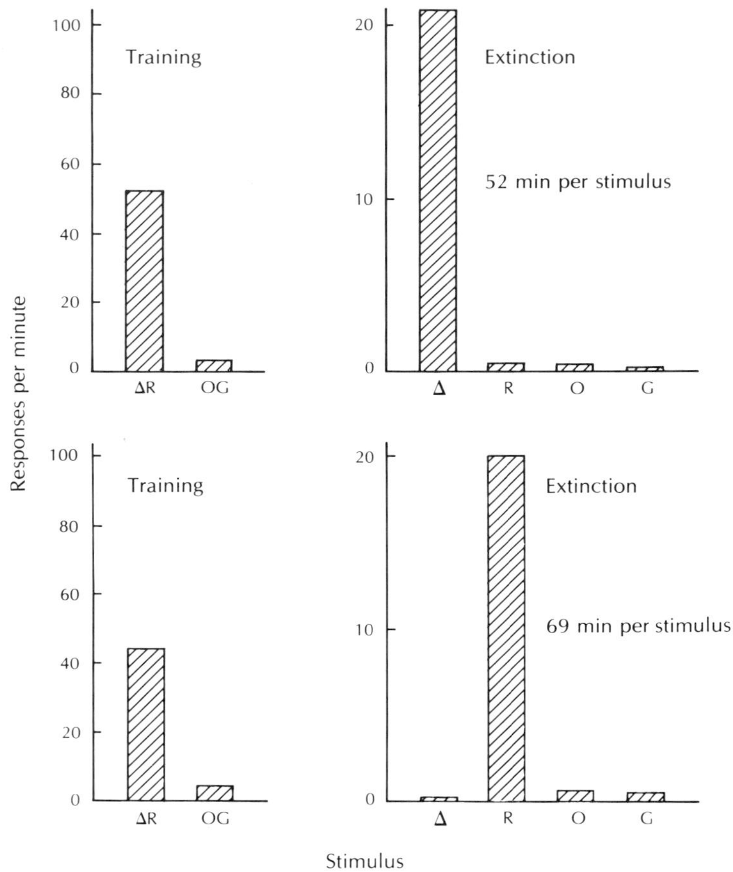

Fig. 21.2 The rate of responding of individual pigeons to a complex stimulus and to the components of the complex stimulus. (After Reynolds, 1961*a*.)

During discrimination training the positive stimulus was a white triangle on a red background and the negative stimulus was a white circle on a green background. During extinction the separate components were presented individually in random alternation.

tending the range of stimulus characters effective for a given response, there are some cases in which it seems not to be involved. In these, elicitation of the response facilitates its subsequent elicitation by stimulus situations which appear to share no characters with that which first elicited it. The process is sometimes referred to as "priming". An example is provided by Noirot's (1964a, b, c; 1965) studies of the maternal behaviour of mice, which have already been mentioned in a number of contexts. Naive adult mice show maternal behaviour (retrieving, licking, etc.) when presented with a 1-day-old living baby, but are much less likely to respond to the weaker stimulus of a 1-day-old drowned baby: there is strong evidence that the difference is due to the ultrasonic cries emitted by the former. However, responsiveness to a drowned baby is markedly increased by previous exposure to a live baby, the effect being at least as great when the interval between the two exposures is several days as when it is 2 min. The increased responsiveness in retrieving does not depend on the performance of the full response during the initial test with the live baby: if the live baby is placed in the nest of an inexperienced female, who thus has no opportunity for retrieving, the female is nevertheless more likely to retrieve a dead baby than animals without previous adult experience of live babies. Furthermore, Noirot (in press) has shown that the auditory (and perhaps olfactory) stimuli from the live young are alone sufficient to enhance subsequent responsiveness to the dead young. In this experiment not only was it impossible for the adult to retrieve, lick, or show the lactation position to the baby in the initial exposure, but there would seem to have been no stimuli, except possibly olfactory, in common between the two situations (see page 308).

21.4 NARROWING OF THE RANGE OF EFFECTIVE STIMULI

Although many species-characteristic responses can initially be elicited by a wide range of stimuli (or by simple stimulus characters provided by a wide range of objects), the range of effective stimuli becomes narrowed in the course of development (cf. James, 1892). Often the process is extremely rapid. Newly hatched nidifugous birds will peck at any spots which contrast with their backgrounds (although they prefer some to others; see above). They quickly learn, however, to deliver the majority of their pecks at potential food objects and to neglect others (e.g., Hailman, 1967). Similarly, the gaping response of young blackbirds is given initially to a very wide range of auditory stimuli, but becomes narrowed by learning so that, under natural circumstances, it is elicited only by the feeding notes of the bird's own mother (Messmer and Messmer, 1956). Tschanz (1959, 1968) has shown that young guillemots (*Uria aalge*) learn to respond selectively to the calls of their own parents during the first few days of life, and Sladen

(1955) has described how individual young adélie penguins (*Pygoscelis adelias*) are called out from among the thousands in the crêche by their parents.

A narrowing in the range of effective stimuli is most likely to occur on the first occasions on which a response is given, but this does not mean that such effects are confined to early development. First-year males of the red-winged blackbird *Agelaius phoeniceus* will mate with models of a wide variety of species, but older males will mate only with models of red-winged blackbird females (Noble and Vogt, 1935).

Dilger (1962) has shown that once some lovebirds (*Agapornis*) have had experience in rearing their own young, they will not rear the young of other species with a differently coloured down. If, however, the eggs of another species are substituted for a female's first clutch, she will subsequently rear only nestlings of the species of her foster young. Myrberg (1964) has evidence that similar principles apply to some cichlid fish, the adults learning to recognize their own young. This conclusion was questioned by Greenberg (1963a, b), who emphasized the importance of the relation between the stage of the host's parental cycle and the developmental stage of the young in fostering experiments, believing that responsiveness of the parents at each stage in the cycle is limited to particular stages of young. Myrberg (1966) has, however, answered this criticism, though he acknowledges that the dominant sensory modality used by the adults in recognizing their young changes from olfaction to vision as the young develop.

Increased selectivity in responsiveness is thus commonplace: what are the processes involved? In many cases, habituation plays an important role. If, as a consequence of frequent exposure to some of the initially adequate stimuli, these cease to be effective, others may remain so. The fleeing response of young nidifugous birds, normally elicited by flying predators, appears initially to be elicited by quite a range of stimuli, though the relationship between size and speed of movement, and also the sudden increase in apparent size which characterizes a swooping model, are important issues (McNiven, 1960; Müller, 1961). Later, shapes which are often seen lose their effectiveness. The earlier view that young turkeys respond differentially to a model shaped like a hawk (Fig. 4.5) and not to one shaped like a goose (e.g., Tinbergen, 1951) was apparently due to preexperimental habituation to geese (Schleidt, 1961).[1]

[1] Two independent groups of investigators have found a differential response to hawk and goose models in mallard ducklings which had not previously been exposed to objects moving overhead. In the case of Melzack et al. (1959) the difference could have been a consequence of preceding tests, but data provided by Green et al. (1966) do indicate that hawk-shaped models produce more locomotor activity. As discussed on pages 348–349, however, there is a difficulty in using only locomotion scores to assess "fear."

Habituation can occur not only because the external stimulus is inadequate to elicit a full response, but because the organism, by virtue of age or condition, at first fails to respond to it. For example, Denenberg et al. (1968) found that a high proportion of rats of a certain strain would kill mice if exposed to them in adulthood. However, if rats of the same strain were reared with mice from 21 to 57 days of age, they did not subsequently kill them. Similar data were obtained earlier for a number of species by Kuo (reviewed 1967): for instance, cats brought up with rats become attached to them and do not attack them.

There is also evidence that habituation can occur when the animal is overexposed to stimuli adequate for a particular response. This sometimes seems to occur, for instance, when animals are exposed to fear-provoking stimuli with no possibility of escape—though this can, of course, also produce other consequences. Another case comes from Noirot's (1964a) studies of the maternal behaviour of mice. When tested with young pups, mice with young of their own responded less strongly than did virgin or non-lactating females: Noirot ascribes this to habituation in the former group.

The obverse of habituation is the reinforcement associated with some stimuli and not others. Chicks continue to peck at stimuli which subsequently release the actions of grasping and swallowing, but after a while cease to peck at those which do not. Young finches (Carduelinae and *Fringilla*) at first show only a limited degree of preference between seeds, but after a while take mostly that kind of seed which they can dehusk most efficiently, obtaining the greatest kernel weight in the least time. Since the different species have beaks of different sizes, they come to prefer different sorts of seeds (Kear, 1962).

Habituation and reinforcement are also important in determining the range of stimuli effective for eliciting sexual behaviour. Hagamen et al. (1963) found that once a male cat had mated with a normally inadequate object such as a rabbit or a teddybear, it would continue to do so with no need even for occasional reinforcement with a real female. In their experiment no cat which had shown three complete mountings on one of the objects ever failed in a subsequent test (see also Valenstein and Goy, 1957). Similarly, male chaffinches which mounted a suboptimal dummy usually continued to do so on successive days, while males which failed to mount showed progressively less interest in the model (Hinde, 1959c). In rats, experienced males prefer the odour of receptive females to that of non-receptive females, though naive males and castrated males do not: the preference is thus learnt. Receptive females prefer the odour of normal males to that of castrates even when naive: in this sex, therefore, the preference does not depend on experience (Carr et al., 1965; Carr et al., 1966).

Sometimes the role of experience in limiting responsiveness is more indirect than the examples cited so far might suggest. For instance the early mounting attempts of Burmese red junglefowl cocks (*Gallus gallus*) are directed exclusively at sitting birds. Since these attempts are always abortive, experience showing that this is the only position possible for successful copulation cannot play a role. Kruijt (1964) suggests that mounting attempts are directed only toward sitting birds because, as a result of experience (see pages 528–530), standing birds elicit aggressive behaviour, which inhibits sexual behaviour. It remains possible, however, that even incomplete mounting has some reinforcing effect (see pages 595–596).

Another issue of importance in limiting responsiveness is the fear induced by strange situations: the range of objects which will elicit begging from a young passerine is reduced in part because those objects with which it does not become familiar subsequently induce avoidance responses. Further examples are given in the next section.

Avoidance of strange situations implies a background of familiarity against which the strangeness is recognized. The role of experience in establishing this has been stressed especially by Hebb (e.g., 1946). Chimpanzees show signs of autonomic disturbance, aggression, and fear when confronted by parts of a chimpanzee's body or by an anaesthetized chimpanzee, and young chimpanzees are disturbed by strangers. Both chimpanzees which have been reared in darkness (see pages 476–481) and congenitally blind human patients who have had their vision restored (Dennis, 1934) initially show no such fear responses to strangers: the responses depend on previous learning. Similarly, kittens, which have been removed from their home cage for a while, hiss and show clear evidence of fear of their mother and litter mates when returned (Rosenblatt et al., 1961). Such cases are by no means confined to higher mammals (e.g., Thorpe, 1963a). For instance, in nidifugous birds fear responses, apart from the freezing responses elicited by the parents' alarm calls, are absent for the first few hours after hatching and appear only later as the environment becomes familiar (Schaller and Emlen, 1962; see also Salzen, 1962).

The influence of previous experience on avoidance behaviour is, however, complex. Fear induced by the unfamiliar depends on previous experience to build up a knowledge of the familiar, but by further experience the unfamiliar by definition ceases to be so. In addition, experience in a diverse environment may result in a reduction in the fear shown to strange objects. Thus 2-year-old chimpanzees reared in captivity are timid in the presence of novel objects and do not investigate or manipulate them to the same extent as do wild-raised chimpanzees of a similar age. The difference is greater, the more severely restricted the conditions under which the laboratory-reared chimpanzees are reared. It cannot be ascribed to deficient arousability, motor ability, or discriminatory ability, and disappears if the animals are exposed to a richer environment (Menzel et al., 1963).

21.5 IMPRINTING

This limitation of the range of stimuli capable of eliciting a response has been studied especially in the context of the following response of nidifugous birds. A newly hatched gosling or duckling will follow practically any moving object it sees in the same way that it would follow its own parent (Spalding, 1873; Heinroth, 1911; Lorenz, 1935). For ducklings, the diversity of effective objects is at least sufficient to include a matchbox and a walking man (Fabricius and Boyd, 1954). This does not mean, of course, that all objects are equally likely to evoke following; some (for instance, round objects) are better than others (e.g., Hess, 1959a; Schaefer and Hess, 1959; Gray, 1961; F. V. Smith, 1962). Within limits, the effectiveness of an object increases with its conspicuousness. Of the models used by Hinde et al. (1956) to elicit following from moorhens (*Gallinula chloropus*) and coots (*Fulica atra*), a large conspicuous canvas hide carried over the shoulders of one of the experimenters was more effective than a life-sized model of a bird. Similarly, Bateson (1964c) found that a significantly higher proportion of day-old chicks approached a conspicuous static pattern than two less conspicuous patterns. But certain stimulus characters can elicit fleeing rather than approach. Young ducklings approach a human being if he sways his body from side to side or walks slowly away (Weidmann, 1958), but flee from him if he moves too vigorously. Such findings are consistent with Schneirla's hypothesis that mild stimulation elicits approach and more intense stimulation avoidance (see pages 503–504).

After a certain period new objects cease to be effective in eliciting the response; the range of adequate objects becomes restricted to ones similar to those already experienced. The learning involved is known as "imprinting", and a considerable amount of effort has been devoted to its analysis. It is conveniently studied in an apparatus of the type shown in Fig. 21.3, where the rotating arm carries a suspended model round the runway. The young bird is placed in the runway and the amount that it follows or avoids the model recorded.

Though there is now no reason for thinking that imprinting is fundamentally different from other forms of learning (e.g., Hinde, 1962b; Thorpe, 1963a; Bateson, 1966 and refs. cited), as Lorenz (1935) and Hess (1959a) believed, it has properties which make it of particular interest. One of these is that the period during which it can occur is limited to a "sensitive period". Some authors have claimed that this is only a few hours in duration: thus Hess (1959a) found a sharp peak 13 to 16 hours after hatching (Fig. 21.4). While later authors have found the period to be less clearly defined than this, the difference is one only of degree: thus domestic chicks will not follow a novel moving object when only a few hours old, nor when they are several days old, but only during the intervening period. The extent of the sensitive period varies, of course, between species: even within one species it may be different for auditory and visual stimuli (Boyd and Fabricius, 1965; Kovach et al., 1968).

The earlier limit to the sensitive period is related to the development of locomotor ability: Hess (1959b) found that the curve of increase in

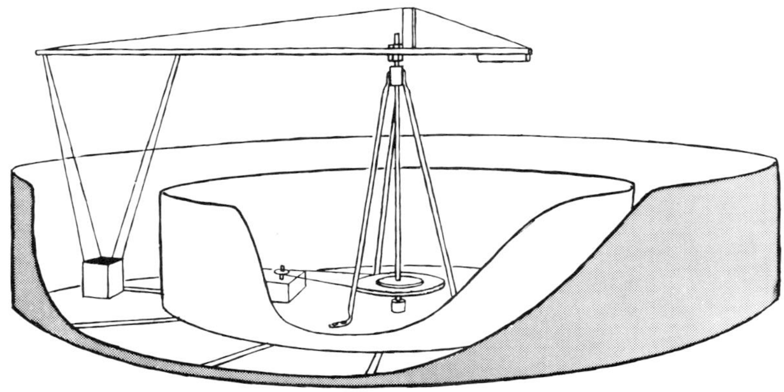

Fig. 21.3 Apparatus used for studying the following response of young chicks. The arm rotates, carrying the box round the circular runway. (From a drawing by P. Bateson.)

Fig. 21.4 The sensitive period for imprinting. The ordinate gives the scores of birds tested at the ages shown in hours along the abscissa. (After Hess, 1959*a*.)

speed of locomotion of chicks with age corresponds closely with the curve showing the onset of the sensitive period (Fig. 21.5). This, of course, does not demonstrate that locomotor ability is itself the operative factor: many perceptual and motivational changes are correlated with it, two possibilities being the developing organization of the retina (Sackett, 1963) and the general excitability of the nervous system (Kovach, 1964). Gottlieb (1961) has shown that the beginning of the sensitive period is related to developmental age, and is little affected by posthatching experience.

There has been more disagreement over the factors limiting the end of the sensitive period. The first suggestion was that the sensitive period is

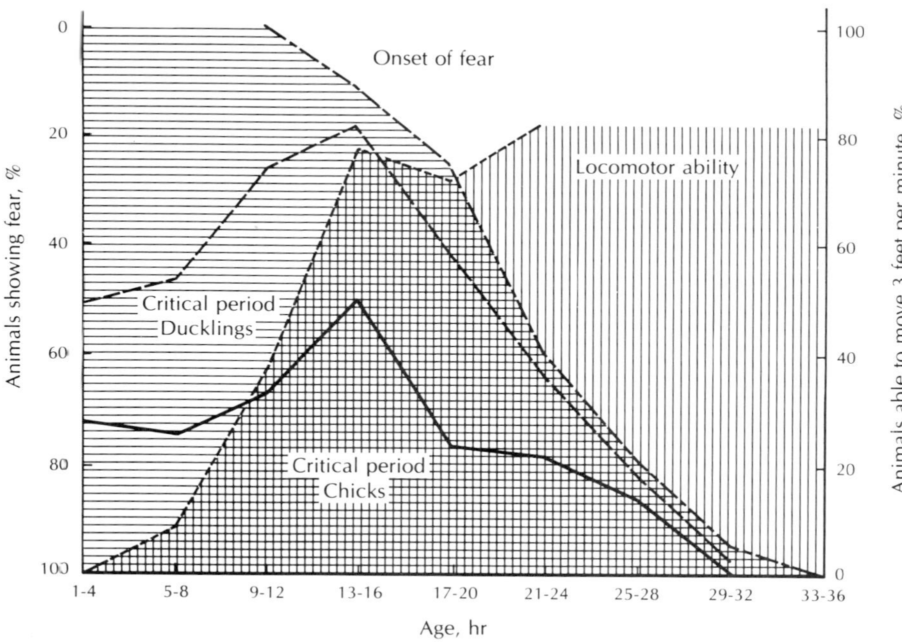

Fig. 21.5 Hypothetical and empirical curves of the critical period of chicks and mallard ducklings. (After Hess, 1959*b*.)

The two shaded areas show the increase with age in the proportion of birds moving more than 3 ft/min in a standard situation, and in the proportion of birds showing fear responses, the latter plotted from above downwards. The cross-hatched area under both lines is the hypothetical sensitive period on the assumption that these two factors are the limiting ones. The empirical period for chicks is shown by the continuous line, and for ducks by the discontinuous one.

limited by a waning of the tendency to follow (Lorenz, 1935; Fabricius, 1951). This carried with it the assumption that, in imprinted individuals, the specific cues learnt during the sensitive period are sufficient to elicit following later. This view was earlier associated with the assumption that the decrease in the tendency to follow was inherently determined, a hypothesis not easily subject to experimental verification. Later authors who postulated a decrease in the tendency to follow (e.g., Jaynes, 1956, 1957, 1958a, b; Weidmann, 1958) either did not specify its cause or ascribed it to experiential factors. However, for coots at least, this theory is inaccurate, since they will generalize their following response to new objects up to at least 60 days of age, provided they are tested in a familiar environment (Hinde et al., 1956). In any case, evidence for such a decrease in the tendency to follow has not yet been produced for any species: the end of the sensitive period is marked by a decrease in the proportion of individuals following, rather than weaker following by those which do (Bateson, 1966).

A second possibility is that the end of the sensitive period is set by the inhibition of following by other types of behaviour which had not previously appeared, fear responses in particular having been suggested (Hinde et al., 1956; Guiton, 1959; Hess, 1959a). In support of this view Hess (1959b) showed that the proportion of experimentally naive birds which would follow a model decreased at the same age as the proportion of birds giving "distress" calls in a standard situation increased (Fig. 21.5). This view left open the problem of the role of experience in the development of the fear responses.

Moltz (1960) extended this hypothesis in an attempt to account not only for the end of the sensitive period, but also for why the young bird follows models at all. He suggested that the stimuli perceived during the first day or two of life, before "anxiety" responses have appeared, acquire anxiety-reducing properties. They are therefore followed later, especially in anxiety-producing situations. In support of this hypothesis, it was shown that habituation to the test run reduces following, and increase in "anxiety" augments it (see also Pitz and Ross, 1961; Stettner and Tilds, 1966; Kovach, 1964: the last author relates the effect to a change in the "general activity of the CNS" rather than to anxiety).

Moltz further suggested that the more attention-catching an object, the more likely it would be to acquire anxiety-reducing properties during the sensitive period: this was consistent with the observations of James (1959, 1960a, b), F. V. Smith (1962), and Moltz (1963) that young chicks will be attracted by and form attachments to flashing lights and to a wide variety of stationary or moving objects. However, Moltz's suggestion that objects are followed because they have acquired anxiety-reducing properties is rendered improbable by cases in which a model is followed within a few moments of its presentation. Moltz (1963) has now modified his views on this point.

A number of experimenters have shown that the end of the sensitive period is not fixed rigidly in terms of developmental age, but depends on the conditions of rearing. Moorhens and coots reared in groups in pens shielded from the general environment hardly follow models at all, though birds reared in groups in more exposed pens follow readily (Hinde et al., 1956). In chicks the sensitive period is extended by rearing them in isolation (Guiton, 1961; Sluckin, 1962) and in ducks by rearing them in diffuse light (Moltz and Stettner, 1961). Since the decline in the tendency to follow the test object at the end of the sensitive period is correlated with an increased tendency to flee from it, this influence of the environment on the length of the sensitive period could be due to an influence on the development of fear responses.

As already discussed (pages 349–350), objects which are strange to an animal are likely to elicit fear responses, though responsiveness to strangeness depends on previous learning of the familiar. It seem probable, therefore, that the sensitive period is one in which the bird learns the details of the environment in which it is reared. Only when this is accomplished are strange objects recognized as such, and only then do they elicit fleeing (Bindra, 1959; Sluckin and Salzen, 1961; Sluckin, 1962; Salzen, 1962; Bateson, 1966; Moltz, 1963). After the sensitive period only enforced proximity to a strange object can result in habituation, and thus a cessation of the attempts to flee.

In a study of the development of fear responses in young chicks, Salzen (1962) showed that drastic alterations in the environment produce fear responses at almost any age. However, chicks reared in isolation, with no experience of moving objects, respond with "pleasure behaviour" to a moving cardboard cylinder even when 7 days old; though chicks which have lived in a social group for 24 hours show fear of the cylinder. This is not easy to understand, but Salzen interprets it on the view that the mobile stimulus can induce positive social behaviour (contentment calls, nestling movements, approach and following) provided it is not responded to as strange; and that it is more likely to be treated as strange if previous experience of moving objects has provided a familiar moving-object percept. This, of course, need imply no absolute distinction between static and mobile objects: we have already seen that chicks may be attracted to static objects, and Thorpe (1945, 1963a) has emphasized that attachment to an environment may be akin to imprinting (page 688).

Further, Bateson (1964a) has shown that the persistence with which chicks avoid a mobile object is reduced if the object is coloured in a pattern similar to that of the static walls of the pen in which they have been reared. This shows that, within a few days after hatching, chicks learn the characteristics of the pens in which they are reared and discriminate objects sufficiently dissimilar from those to which they are accustomed. The ability of the chicks to discriminate was shown to be related to the conspicuousness of the pattern with which the walls of the home-pen were painted: with a more conspicuous (and therefore presumably more easily learnt) home environment, avoidance of differently painted models in the test-pen persisted for longer in each test (see also Matthews and Hemmings, 1963; Klopfer and Hailman, 1964).

This line of argument thus suggests that imprinting consists largely of the development of familiarity with the moving object. Specific rewards such as food or contact with the model are not necessary (e.g., Collins, 1965). Under natural conditions chicks become familiar with at least two types of moving object—the mother and the siblings—and under experimental conditions coots similarly can learn to follow several different objects (Hinde et al., 1956). If familiarity is the issue, then imprinting would seem to have much in common with "perceptual learning" (see pages 481–483 and 586–587), for in both cases the responsiveness to a stimulus is influenced by previous experience of that stimulus independently of its association with any reward (Sluckin and Salzen, 1961; Hinde, 1962b; Sluckin, 1962; Bateson, 1966). Thorpe (1963a) has already emphasized the importance of imprinting as a tool for investigating the genesis of the perceptual side of behaviour.

Bateson (1964b) has produced evidence in support of this view by showing that the learning during the sensitive period can be revealed in contexts other than that of the following response. Thus the performance of chicks in a discrimination problem was augmented when one of the patterns to be discriminated was the same as that on the walls of the home-pen. This is exactly the sort of test used in studies of perceptual learning (see page 482). Similarly Kovach et al. (1966) found that early imprinting experience to visual flicker facilitated learning to approach or avoid the same stimulus a month later. Since the effect was found only in those individuals which approached the stimulus in the training period, they suggested that mere exposure to the stimulus is not sufficient for the early perceptual learning to occur: it may be necessary for the stimulus to become associated with *a* response, though not necessarily with the same response that is used in the subsequent testing whereby the perceptual learning is revealed (see also MacDonald and Solandt, 1966).

The learning of the properties of the object as studied in imprinting experiments should perhaps be regarded as an artificially isolated part of a process of learning about the environment which was started before hatching. On this view, the early distress calls are related to the loss of contact and warmth formerly provided by the egg (Collias, 1952). The response to auditory stimuli from the mother may or may not depend on experience in the egg, but is in any case often fairly unspecific, so that a wide variety of repetitive stimuli are adequate (see below).

There remains the question of why the newly hatched bird approaches conspicuous or moving objects in the first place. Although parental calls may normally play an important role in the laboratory, naive birds will approach silent objects (e.g., Gottlieb, 1965, 1966; Gottlieb and Simner, 1969). At least, however, this is a phenomenon not peculiar to the young of nidifugous birds. As mentioned above, Schneirla (1939, 1965) has emphasized that the young of many species tend to approach stimulation of low intensity and flee from that of high intensity, and this is more or less what happens in the imprinting situation. Thus up to appropriate limits, the more conspicuous an object, the more readily is it followed or approached in an initial exposure during the sensitive period (see page 517).

However, Schneirla's view that chicks approach low levels of visual stimulation because of their quantitative similarity to stimuli encountered before hatching is not easily subject to verification. Furthermore, as Bateson (1966) has emphasized, Schneirla's view implies that the newly hatched chick has a much lower level of behavioural organization than is in fact the case. The chick does not wait passively until exposed to a conspicuous object, but shows behaviour likely to bring it into contact with one. During the few days after hatching, chicks are more active in a grey environment than in one painted in black and white stripes which reflect the same total amount of light; this suggests an inverse relationship between activity and the complexity of the environment (Bateson, 1964d), so that the chick is likely to find stimuli appropriate for imprinting. In harmony with this view is the finding that a conspicuous stimulus of the type used in imprinting experiments may be reinforcing for day-old chicks and ducklings on the first occasion it is encountered: Bateson and Reese (1968) showed that such birds rapidly learned to depress a pedal that switched on a flashing light (Figs. 21.6 and 21.7). Furthermore there is a sensitive period during which the light acts as a reinforcer, for the readiness to learn the response decreases to a minimum about 48 hours after hatching: at about this age there is a corresponding peak in the persistence with which chicks in an imprinting situation avoid the model. The effect of age varies with the rearing conditions, being more marked in socially reared than isolate-reared chicks, and more marked in the latter than in dark-reared birds (Bateson and Reese, 1969). The reinforcing effect could depend on some consequence of approaching the light, such as an increase in its intensity or subtended angle. However, chicks allowed to approach a flashing light did not approach more rapidly in a subsequent test than chicks that had been exposed for a similar period close to the light. So the stimulus onset may be a sufficient reinforcer (Bateson, personal communication).

Thus during the course of the sensitive period, the birds develop a preference for the object with which they are rewarded. Imprinting and operant conditioning thus proceed hand in hand, with imprinting narrowing the range of stimuli with reinforcing properties (Bateson and Reese, 1969). Other workers have shown that the objects on which chicks have been imprinted acquire reinforcing properties as indicated by performance in a *T* maze (Campbell and Pickleman, 1961) or key-pecking situation (Peterson, 1960; Hoffman et al., 1966; Hoffman and Kozma, 1967). Working in a quite different situation, Marx and Knarr (1963) regard their data on the development of a light-contingent bar-pressing response by rats as partial support for the proposition that familiarity per se is a sufficient condition for a stimulus to become reinforcing. However, this latter view cannot be pressed too far, since those stimulus conditions with which an organism is most familiar are those in which it spends most of its time, and which form mere background conditions against which other stimuli are differentiated. The importance of familiarity may, however, increase in anxiety-provoking situations (see page 516).

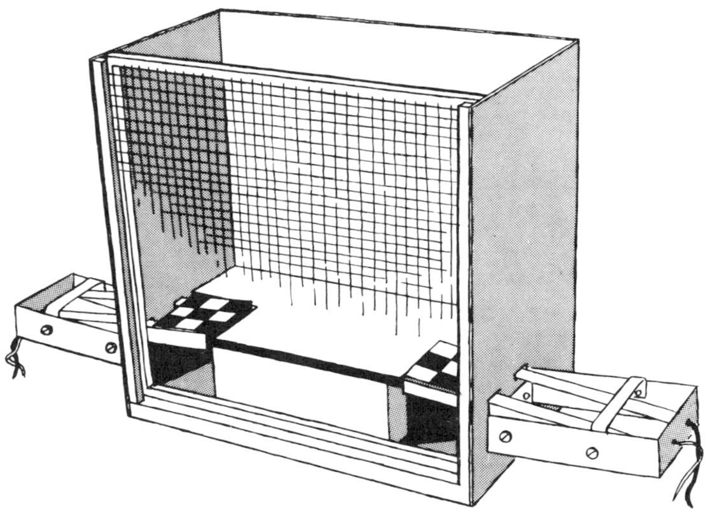

Fig. 21.6 Apparatus used for assessing the reinforcing effectiveness of a flashing light for young chicks. (From a drawing by P. Bateson.)

The light, which was placed outside the box and in front of the glass side, was switched on when the chicks stood on one of the two chequered pedals. The other chequered pedal was used as a control.

The view that, during the sensitive period, conspicuous objects have a reinforcing effectiveness for naive birds which increases with familiarity is consistent with many of the findings of imprinting experiments—for instance, that an imprinted bird shows searching behaviour when the object is absent; and that readiness to follow an object and to respond to it selectively increases with experience. It is also consistent with the facts that reward is unnecessary, that models which evoke fear may later (after habituation and the establishment of familiarity) elicit following, that new models are more likely to elicit following if presented in the same situation as a model on which the bird has previously been imprinted, and so on (Salzen, 1962).

In discussing imprinting of the "following" response, we have so far neglected inter-species differences. In practice, of course, species differ considerably in the length of the sensitive period, in the amount of experience required for imprinting, in the range of objects effective in eliciting following, and so on. An issue of particular interest concerns the relative roles of

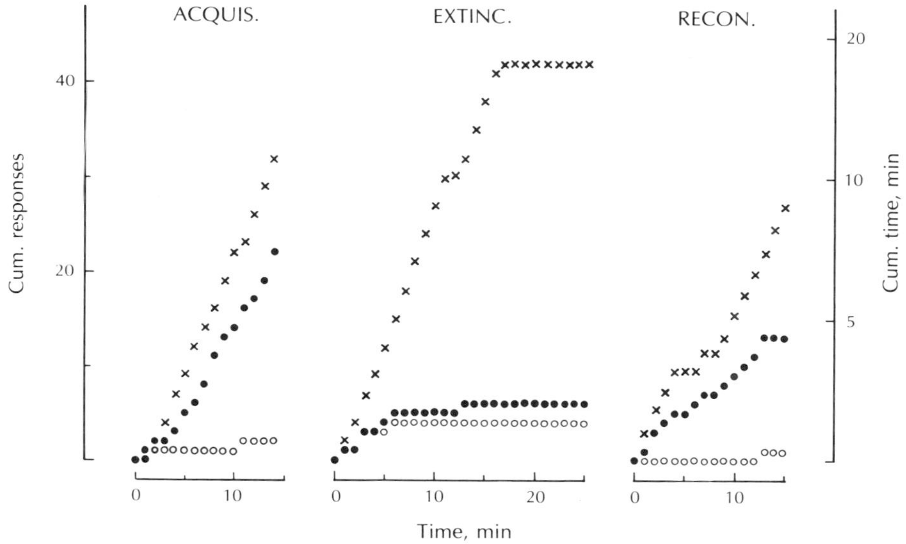

Fig. 21.7 Reinforcing effectiveness of a flashing light for young chicks. (From Bateson and Reese, 1968.)

Cumulative records of the number of presses on the reinforced pedal (*filled circles*) and control pedal (*open circles*), and time spent on the reinforced pedal (×) by a 24-hr-old dark-reared chick during acquisition, extinction, and reconditioning.

auditory and visual stimuli. Domestic chicks are attracted by sounds similar to those which they themselves emitted in the egg (Collias, 1952; Kruijt, 1964), and sound signals play an important part in inducing following (Smith and Bird, 1964a, b; see also Gottlieb, 1965, 1966, 1968b). Fischer (1966) believes that they have at least two roles, serving to produce "arousal" as well as forming a basis for recognition. In most species auditory stimuli can be conditioned as a consequence of exposure to visual characteristics of the object, or vice versa. After training with an object which provides both visual and auditory stimuli, the relative importance of each varies with the developmental age and experience of the individual concerned, and also with the species (Gottlieb and Klopfer, 1962). In ducks which nest on the surface of the ground, the young will approach a wide range of rhythmical sounds, but imprinting to a particular sound occurs only if it is associated with visual stimulation from an appropriate object. In the wood duck (*Aix sponsa*), on the other hand, there is no initial tendency to approach repetitive sounds, but exposure to a sound signal, in the absence of correlated visual stimulation, can produce a subsequent preference for that sound. This is to be understood in relation to the wood duck's nesting behaviour: the nest is in a hole in a tree, and auditory stimuli may be more important than visual ones in governing the ducklings' early responses to its mother (Klopfer, 1959a, b; Gottlieb, 1963).

21.6 ADULT SOCIAL, SEXUAL, AND PARENTAL BEHAVIOUR

Another area in which the stimulus-response relations characteristic of the species are influenced in large part by experience is that of postjuvenile social and reproductive behaviour. Although in many species, for instance, the European cuckoo (*Cuculus canorus*), the stimuli eliciting pair formation and mating behaviour must be determined independently of experience with conspecific individuals, there are many cases in which such experience plays a major role. As Thorpe and Jones (1937) showed, this is the case even among insects. The response to odours made by female ichneumon flies (*Nemeritis canescens*) while laying their eggs is influenced by the larvae on which they were reared. Some adult females which had been reared on wax-moth larvae were attracted to the odour of wax-moth larvae in preference to the odour of the normal host (meal-moth larvae), which was not the case with any of the females reared on the normal host. The extent to which such host-conditioning occurs in other insects appears to vary considerably between species (Monteith, 1962; Thorpe, 1963a). Some light on its nature is provided by an experiment on *Drosophila* which does not involve host selection. The normal aversion to peppermint odour shown by adults is reduced in insects reared on a medium containing peppermint: this is due to habituation, not associative conditioning (Manning, 1967a).

Examples of the effects of early experience on the stimuli eliciting later social or reproductive behaviour are known for nearly all vertebrate groups (see also Chap. 22). To cite but a few examples, adult males of certain cichlid fish normally direct their courtship only toward females, but males reared in isolation initially attempt to mate with both sexes (Noble and Curtis, 1935/1936). In zebra finches (*Poephila guttata*) selection of nest-material and nest-site is influenced by nestling or fledgling experience (Sargent, 1965). In mice, mate selection is determined largely by the female: Mainardi et al. (1965) found that the female's preferences could be strongly influenced by the early olfactory environment, for females reared by perfumed parents showed an increase in preference for perfumed as compared with control males. Social behaviour in the chicken (Baron and Kish, 1960) and rat (Angermeier, 1960) depends on earlier experience. In the latter species the use of tactile receptors in a social situation is responsible for the effectiveness of visual social reinforcement later.

In ungulates, early experience plays an even more important role in "socialization". Lambs reared by hand tend not to graze with the flock when turned out to pasture (Scott, 1945), and early experience has profound effects on their subsequent social behaviour (Hersher et al., 1963; see also Cairns, 1966a, b): in this case early isolation probably affects the behaviour of adults to young as well as the responses of young to adults (see Klopfer and Gamble, 1966, and Klopfer and Klopfer, 1966, on goats).

The effects of early experience have also been extensively studied in guinea pigs (Shipley, 1963), dogs (Scott and Marston, 1950; Scott, 1954; Fox and Stelzner, 1967a, b), and monkeys (Harlow and Harlow, 1965). King (1966) emphasizes the importance of "pleasant" aspects of the mother in the overcoming of fear responses.

Particular interest has centred around the influence of early experience on mate selection in birds. One of the reasons given by Lorenz (1935) for considering imprinting distinct from other forms of learning (see page 513) was that the early experience appeared to have long-term effects, determining not only the objects toward which the young bird directed its filial responses, but also those to which its adult sexual responses were directed. The imprinting thus seemed to have a latent influence on responses not yet in the repertoire of the organism. In addition to the examples given by Lorenz (1935, 1937), the case described in some detail by Räber (1948) may be cited. This was a hand-reared domestic turkey cock which, as an adult, courted most men walking through the park in which it lived, but attacked women. As a consequence of the early imprinting, the turkey behaved as though humans were fellow members of its species. The differentiation between men and women, however, appeared not to be related to the early experience, but to correspond to species-characteristic responsiveness to hanging or flapping accoutrements (e.g., in this case skirts) as indicative of maleness: under normal conditions these are provided by the wattles and drooping wings of a displaying male turkey.

Other examples of adult birds directing sexual responses toward objects similar to those toward which they previously had directed their early filial responses have been cited by a number of authors (ref. in Thorpe, 1963a; Hess, 1959a; Klinghammer, 1967). Many of these accounts have been anecdotal, and care is necessary in generalizing from them. If early experience were indeed all-important in determining sexual responsiveness, it would be reasonable to expect the evidence to be much more abundant. Hens are, in fact, frequently used as foster parents for a wide variety of species, but abnormalities in mate selection among the adults thus reared are relatively rare (Fabricius, 1962). There are, however, some precisely controlled studies which show that the sexual or social responses of *young* chickens or turkeys, induced where necessary by testosterone treatment, are directed preferentially toward objects which the birds had been trained to follow (Schein and Hale, 1959; Guiton, 1961; Bambridge, 1962) or with which they had become familiar (Salzen, 1966). In Salzen's study chicks (7 to 25 days old) showed fear responses to the test object (live chick, dead chick, or hand) unless they had previously been exposed to it. If they had, and if they had been testosterone-treated and the object was appropriate, sexual behaviour appeared. Salzen thus regards experience with the object as affecting later social or sexual behaviour by diminishing fear.

Experiments involving precocious sexual behaviour artificially induced in young birds are, however, not sufficient to show that juvenile experience is in itself sufficient to determine mate selection in adulthood: in normal development adolescent experience may also be important. Under natural conditions the establishment of sexual bonds follows a period of social interactions with other members of the species lasting throughout the whole of adolescence, and it is likely that these have an important influence on mate selection (Hinde, 1962b). For instance, in the bullfinch, juveniles (usually from the same nest) normally form "betrothed" pairs which are subsequently dissolved in favour of new mates by the beginning of the first breeding season. If a bullfinch is reared by hand, in isolation from its kind, it first treats its human keeper as a parent, and then as a substitute for a "betrothed" mate. If, during the autumn or winter, the bird is given opportunity to pair with a member of its own species and the opposite sex, its relation to the human gradually dissolves. If, on the other hand, it is kept in the society only of human beings, it later accepts one of them as its mate and the imprinting becomes irreversible (Nicolai, 1956). Similarly studies by Schütz (1965a) suggest that "sexual imprinting" in ducks depends on experience later than that which determines the imprinting of the following response (see also de Lannoy, 1967).

That adolescent experience is important even in the domesticated species usually used for imprinting experiments is shown by an experiment by Guiton (1962b). Cockerels were reared in visual isolation from each other for 13 weeks and then penned in a group together with females. At 11 weeks, after androgen treatment, they mated with a yellow glove in preference to a stuffed pullet mounted in a crouching position. At 5 months, however, the preference was reversed, though they would still court humans if they had had recent opportunity for courting hens. Thus, in this case, although the early experience certainly influenced adult sexual behaviour, it was largely overruled by the effects of adolescent experience (see also Guiton, 1966).

So far there have been few controlled studies showing that juvenile social experience with an object leads to a preference for that object as a sex partner in adulthood whatever the adolescent experience may be. Schein (1963) compared the sexual behaviour of three turkeys, isolated from other turkeys but exposed to people for their first 32 days of life, with that of three birds kept in groups for some or all of this period. All birds were then kept in social groups until adulthood. At 5 years of age all six birds would court humans in the absence of turkeys or turkeys in the absence of humans, but when offered a choice the first three courted humans and the second three courted turkeys. It was unfortunate that the number of birds available for use in this experiment was so small. Another study which comes near to satisfying the requirement of controlled adolescent experience is that by Warriner et al. (1963), in which it was shown that male (but not female) pigeons tended to mate with birds of the same colour as the adults that reared them. Even here, however, the young were left with their (foster) parents until 40 days old, but were thereafter confined in visual isolation until the mating tests. Their parents were thus the only adult birds they had seen previously.

In a similar type of experiment Immelmann (1965) placed the eggs of zebra finches (*Taeniopygia guttata*) in the nests of society finches (*Lonchura striata*). The young zebra finches were removed from their foster parents when 33 to 66 days old and housed individually until they reached sexual maturity at about 4 months of age. In a choice situation zebra finch males then courted society finch females and ignored or attacked females of their own species. Zebra finch females, however, responded to the courtship of zebra finch males rather than to society finch males. The difference between the sexes is presumably related to the sexual dimorphism in colour pattern.

Some of the most detailed data available come from Klinghammer's (e.g., 1967) studies of mourning doves (*Zenaidura macroura*) and related species. He found that squabs removed from the nest before 8 days of age and reared by hand remained permanently tame to humans: squabs removed much later than that subsequently showed fear of humans. If the former were hand-reared in isolation from other birds and kept isolated until sexual maturity, they would then pair with humans. However, over the course of several breeding seasons during which they were exposed to mourning doves, such birds would gradually come to take mourning doves as mates. If hand-reared with a partner, they would always pair with doves. This effect of exposure to other doves also operated in birds hand-reared in isolation but placed in a community cage from 52 days of age. Thus the nature of the sex partner which will later be chosen is not determined once and for all in early infancy: experiences throughout development, and indeed throughout life, may affect it. Further studies in which experience at different ages is carefully controlled are urgently needed, especially as inter-strain and inter-species differences in the role of experience may be considerable (Guiton, 1962a).

Thus we see that experimental evidence for the unique importance of early experience in determining later choice of a sex partner, as was claimed by Lorenz, is so far lacking, and that in most species pair formation occurs after, and is influenced by, an extended period of social experience. The most economical hypothesis is that the influence of previous experience on later choice of a sex partner is in part permissive, in that a familiar object does not elicit fear responses, and in part a special case of greater readiness to respond to a familiar stimulus. Bateson (1964b) has already shown that early experience influences not only the "following" response of domestic chicks but also discrimination learning for a food reward, and there seems to be no reason why sexual behaviour should not also be influenced in this way (see pages 547–548). Thus experience in early development is merely one of the factors determining choice of a sex partner, and its effects are not irreversible. In a free environment, however, very early experience may be of special importance because "taming" is a *sine qua non:* if objects or individuals are still feared in adolescence they may be always avoided and thus can never become familiar (Bateson, 1966; Salzen, 1966; Klinghammer, 1967).

Finally it must be emphasized that the relative effectiveness of earlier experience, and of a predisposition to pair with birds of the same species which does not depend on experience with them, varies greatly between species. As noted above, even within one species the sexes may differ in the extent to which social experience influences later choice of a sex partner. To cite another example, mallard drakes reared together in groups of five to ten individuals subsequently pair with drakes. Both partners in such pairs behave as males, and complete copulation never occurs: the pair is nevertheless stable, even in the presence of females. Females, however, never become homosexual when reared in this way (Schütz, 1965b). This difference between the influence of experience on males and females is presumably related to the sexual dimorphism of ducks: females of the only monomorphic species tested did pair with the species with which they were reared.

21.7 SUMMARY

1. Many stimuli are responded to appropriately the first time they are encountered. Usually, however, the range of effective stimulus situations is greater for a naive animal than for an experienced one.
2. This lack of specificity in the stimuli eliciting the responses of very young

animals may be due to the essential aspects of the stimulation being quantitative rather than qualitative. Schneirla has suggested that pre- and early post-natal animals tend to approach sources of weak stimulation and withdraw from strong ones.

3. Experience may lead to a widening of the range of stimuli effective for a given response. Some such cases involve "priming", but the majority depend on learning by association. Selectivity in association is discussed.
4. Experience may also lead to a narrowing in the range of effective stimuli. Habituation and reinforcement play important roles. Responsiveness is also limited by the fear induced by strange situations. Here experience plays a dual role, because fear of the unfamiliar depends on previous experience to build up a knowledge of the familiar, but further experience renders the unfamiliar familiar.
5. The learning process whereby the range of stimuli eliciting the following response of nidifugous birds becomes narrowed has been termed "imprinting", but this is not a special form of learning. It is limited to a sensitive period: this period comes to an end, in part, because the bird begins to flee from strange objects.
6. In many species experience plays an important part in determining the objects toward which adult social, sexual, and parental behaviour is directed, but the earlier claim that juvenile imprinting determines the object of the sexual responses of adult nidifugous birds, independently of adolescent experience, has not yet been substantiated.

22

DEVELOPMENTAL ASPECTS OF MOTIVATION

22.1. THE NATURE OF THE PROBLEMS

The activities of young animals are often not related to the motivational contexts in which they will appear in adulthood (e.g., Holzapfel, 1939b, 1949). For example, when a fledgling great tit first starts to peck at small objects that contrast with the background, it is most likely to do so when it is not hungry: when it is hungry it begs from its parents. A week or two later it shows fragments of reproductive behaviour—snatches of subsong, nest-building, and copulatory behaviour—which seem unrelated to the motivational factors which control such activities in the adult. Similarly the play activities of young animals—fighting, hunting, nesting—are largely independent of those motivational factors which will later be associated with them. Indeed the same movement pattern may form part of different functional and causal systems in young and adult. An example is provided by the Cichlid fish *Etroplus maculatus*. The young of this species frequently swim to the parent and skim or bounce against its side. This "glancing" is a movement of ingestion, the young taking in mucus from the side of the parent. It is always directed toward a much larger individual. In adulthood the same movement forms part of the intra-pair behaviour and is directed toward individuals of the same size (Ward and Barlow, 1967). Cases such as this show how functional categories, such as eating, fighting, and sexual behaviour, can be even more misleading in the study of development than in that of immediate causation.

The development of aggressive behaviour in the Burmese red junglefowl (*Gallus gallus*) illustrates the diverse stages through which the motivation of species-characteristic patterns may pass in development. When chicks are about 1 week old the hopping which occurs in the course of general locomotion sometimes results in clashes between individuals. After about a week this hopping becomes oriented toward other individuals, and clashes become more frequent, the birds bumping their breasts against each other. Frontal threatening, leaping, aggressive pecking, and kicking appear, respectively, at 8 to 9, 9 to 12, 10 to 20, and about 21 days after hatching.

At this stage the factors controlling fighting seem to be similar to those controlling locomotion; for instance fighting is more common in active chicks, and both fighting and locomotion are increased by a period of confinement. In addition there is evidence that the factors controlling feeding also affect aggression: not only does the motor pattern of aggressive pecking resemble that of feeding, but aggressive pecking is more common in the fighting of hungry chicks than in that of satiated ones, even though fights themselves are no more frequent.

After about 3 weeks fighting becomes more effective, and aggressive pecking occurs more often. Probably as a result of the strong tactile stimulation received during encounters, other chicks come to elicit escape responses as well as attack. The fights involve ambivalence between attack and escape (see Chap. 16) and are accompanied by a variety of apparently irrelevant movements, such as head-shaking and preening (see pages 411–413).

After about 4 weeks fighting starts to be elicited by the proximity of another individual. It is also especially likely to occur after alarming stimulation and after the performance of comfort and feeding movements. By 50 to 80 days the associated internal states are apparently more enduring, being present both before and after the fights, as shown by the appearance of new ambivalent patterns like the "side-display" and "waltzing". Further development entails changes in the relative preponderance of the tendencies to attack, to flee from, and to mate with other individuals, and aggressive behaviour becomes more independent from those factors with which it was associated in the chick: for instance after 6 months the relation to locomotor activity is no longer discernible (Kruijt, 1964). Thus the behaviour develops through integration and interaction between causal factors and movement patterns which were initially independent of each other.

In this development social experience plays a complex role. With young (domestic) chicks, aggressive behaviour is shown toward strange moving objects before it is shown toward social companions, and it seems probable that habituation delays the response to the latter (Evans, 1967). In junglefowl, also, social experience can have a negative effect: week-old chicks reared in isolation show more advanced fighting behaviour than do socially reared birds. However, social experience also plays a positive role in acquainting chicks with the stimuli appropriate for eliciting fighting: chicks raised in isolation come to fight their own tails, perhaps because these are the only moving objects in close proximity to them (Kruijt, 1964; see Andrew, 1964b, for additional data on domestic chicks).

Such developmental changes depend in part on maturational processes, including changes within the central nervous system itself. The gradually increasing involvement of the more rostral parts of the central nervous system in the responses of chick embryos to stimulation (Peters et al.,

1965), or in the micturition of rabbits (Bradley and Wright, 1966), must depend on developmental changes within the nervous system. Even these, however, may not be independent of environmental influences, and to understand developmental changes it is necessary to teaze apart the interaction between organism and environment which continues throughout development. Progress to this end can sometimes be achieved by experimental manipulation. For instance, the killing of rodents by cats is normally the consequence of an almost inevitable series of experiences which elicit a series of species-characteristic responses—attention to a moving object, pursuit, biting on contact, the taste of blood, and so on. Nevertheless, according to their early experience, cats may kill rodents, ignore them, or play with them (Kuo, e.g., 1967): by appropriate alteration of experience the course of development can be changed. Such cases pose a host of developmental problems. What precisely are the processes whereby activities like eating, fleeing, fighting, and mating come to appear in the appropriate motivational contexts and in the presence of the appropriate stimuli? As yet we have only scattered facts which cannot be pieced together into a coherent story. Indeed even descriptive data about what happens in the development of functional sequences of behaviour are scarce (page 425), and the few detailed studies which are available serve to emphasize that no easy assumptions can be made about which characteristics of a response are influenced by any particular type of experience and which are not. For example, during the period 105 to 138 days of age, various characteristics of sex behaviour of male rats change (e.g., the frequency of ejaculations per hour increases, number of intromissions before ejaculation decreases, postejaculatory latency decreases). The last of these is influenced by mating experience, but the others are a function of age and not of mating experience (Larsson, 1959a).

Some of the processes with which we are concerned in the study of development may differ in kind from those known to occur in adults. Schneirla (cited on page 503–505) has made some suggestions of this sort, and we shall meet others (e.g., pages 534–540). It would seem sensible, however, to see how far the processes which have already been studied in adults can explain the developmental changes we observe in young animals. For example, the use of operant techniques has shown that in adults the frequency of species-characteristic patterns, such as the calls of budgerigars (Ginsburg, 1960) and various mammals (Myers et al., 1965; Lal, 1967; Burnstein and Wolff, 1967, and refs. cited) and the preening of pigeons (Hogan, 1964), can be increased by appropriate positive reinforcement. Furthermore stimuli known to elicit species-characteristic behaviour patterns can act as reinforcers for an operant response: the reinforcing properties may be either positive, as with the visual presentation of an adversary for Siamese fighting fish *Betta splendens* or fighting cocks (Thompson, 1963,

1964; see also Thompson, 1969a; Hogan, 1967; Goldstein, 1967), song for a male chaffinch *Fringilla coelebs* (Stevenson, 1967, 1969), and the opportunity to kill a mouse for mouse-killing rats (Myer and White, 1965); or negative, as with an alarm call of the chaffinch (Thompson, 1969b). As yet, however, there have been few comparable investigations with young animals (but see Bateson and Reese, cited on page 519), and the role of similar processes in young animals is in urgent need of investigation.

In the following pages we shall confine our attention to a few functional categories of behaviour in order to illustrate the nature of the problems which arise.

22.2 AVOIDANCE, ANXIETY, AND EMOTIONALITY

We shall consider here the complex of behaviour which includes fear and escape responses in strange situations or from rivals or predators, "anxiety" or "emotionality" in a strange situation, and so on. That these are grouped together carries no implications as to their precise developmental relationships with each other (see pages 348–350). Indeed, there are problems here whose complexity has only recently become apparent, for even closely related responses may differ in developmental history. For instance, junglefowl chicks show escape responses to intense tactile and auditory stimuli immediately after hatching, but visually elicited escape responses develop only later. Furthermore, the escape responses elicited through the three modalities differ in nature; for instance, fast silent running is not elicited by tactile stimuli, and at most rarely by visual ones; and running in response to sounds, unlike the others, is often not directed away from the source of stimulation (Kruijt, 1964).

Although detailed studies are scarce, it is clear that fear and withdrawal responses develop gradually both before and after birth or hatching. Domestic chicks show distress reactions to cooling before hatching and will cease movement in response to intense stimuli but, apart from reflex responses like blinking, are unresponsive at hatching to stimuli which will later evoke fear or avoidance behaviour. Responses to loud noises and to visual and painful stimuli increase to a plateau a few days after hatching (e.g., Phillips and Siegel, 1966). Similar changes occur in other nidifugous birds (e.g., Salzen, 1962; Kruijt, 1964), though their precise time course is difficult to trace owing to the complexity of fear responses and the influence of environmental factors on their development (see pages 532–534).

In mammals, also, distress reactions appear at or shortly after birth, when they can be elicited by pain, low temperature, and so on. Fear of visual novelty appears much later—at about 2 weeks in dogs, 3 weeks in monkeys, and 6 months in humans. In both mammals and birds stimuli from the mother can reduce fear responses in the young: in general, tactile stimulation is necessary in the early stages, but later visual stimuli alone suffice (Bronson, 1968a, b).

As mentioned previously in several other contexts, both orientation and avoidance responses may be elicited by a discrepancy between the perceived stimulus situation and that which is familiar. Habituation thus plays a major role in their development (e.g., Salzen, 1962; Phillips and Siegel, 1966). It has even been claimed that chicks receiving photic stimulation before hatching show less avoidance behaviour immediately after hatching than chicks incubated in darkness, and that the difference could be due to adaptation to the general overall level of stimulation (Dimond, 1968).

Habituation, in turn, may have profound and far-reaching effects, which are revealed when opportunity for habituation is denied. Birds reared in partial or total isolation, or in a restricted environment, are not only more afraid of others of their kind than are normally reared individuals (e.g., Katz, 1937), but are more prone to give avoidance responses to strange objects or situations (e.g., Hinde et al., 1956; Schaller and Emlen, 1961; Salzen, 1962). They also show quite different social behaviour: the rate of distress calling by 3-day-old domestic chicks placed alone in a strange situation is decreased by the presence of a mirror in group-reared birds, but increased in chicks reared in isolation (Kaufman and Hinde, 1961; see also Marley and Morse, 1966). Comparable data are available for puppies (Fredericson, 1952; see also below) and kittens (Rosenblatt et al., 1961). Rather surprisingly (though see pages 541–543), the greater tendency of animals reared in isolation to show fear responses may lead later to unusual aggressiveness. Kruijt (1964) found that junglefowl reared in visual isolation from conspecific individuals went through a stage in which they showed excessive escape behaviour, but later the males became very aggressive toward their tails and toward humans.

Similar effects are well known in primates. Chimpanzees reared in a moderately restricted environment show marked avoidance of novel objects, especially at about 2 years of age (Menzel, 1964b). Rhesus monkeys reared in total isolation for 6 months or a year show terror when first released, and profound disturbances in many aspects of behaviour. They never develop normal social behaviour (Harlow and Harlow, 1965). Whether the restricted conditions of early rearing were solely responsible for these effects is, however, not clear: in most such studies the effects of restriction have not been separated from those of early maternal deprivation, and Menzel (e.g., 1963) believed that the latter was responsible for some of the symptoms (see also page 650). However, the restriction certainly has some effect: Mason and Sponholz (1963) found that rhesus monkeys brought up in total isolation showed more intense fear responses than animals allowed to see other individuals: since both groups were maternally deprived, the greater impoverishment in the environment of the former group was presumably responsible for the difference. Probably both factors play a role, though by different mechanisms: among other things the mother ameliorates the effect of strange stimuli and thus influences the nature and amount of stimulation received by the infant. The two factors also interact, for there is evidence suggesting that maternal deprivation in mammals may be associated with a permanently heightened responsiveness to visual novelty (review by Bronson, 1968a, b).

In addition to habituation, conditioning plays an important role in the development of fear and avoidance behaviour. Of special importance in the early development of many higher vertebrates is conditioning to stimuli associated with the alarm signals of the parent or other adult. It is perhaps for this reason that wild-reared monkeys respond much more vigorously to snake-like objects than do laboratory-reared ones (Joslin et al., 1964). Association with stimuli whose efficacy in eliciting fear responses is independent of previous experience of them has already been mentioned (see page 506). In addition, social experiences may play a vital role: if young mice are exposed to attack by more aggressive individuals, they become more likely to show withdrawal responses in fighting situations (Kahn, 1951; see also Bevan et al., 1960, and refs. on pages 541–543).

Using "fear" as an intervening variable in this context, Miller has shown that responses which reduce fear are easily learned. For instance, rats will learn to rotate a wheel or press a bar in order to escape from a compartment in which they had previously been shocked. Assessing the usefulness of a drive-reduction hypothesis (see page 594) in accounting for this, Miller (e.g., 1948b) suggests that stimuli from the compartment elicit fear as a result of previous conditioning. The reduction in fear produced by escape reinforces the wheel-turning or bar-pressing (see also Brown and Jacobs, 1949; Kalish, 1954). Internal fear or anxiety responses have been similarly postulated to explain the acquisition of many types of behaviour ranging from the tics of neurotics to complex higher human activities (Brown, 1961; Brown and Farber, 1951). However, attempts to anchor the intervening variable, fear, to particular autonomic states have not met with great success (e.g., Sharpless, 1961).

A further point about the long-term effects of avoidance responses must be made. Stimuli eliciting avoidance responses may produce not only the immediate behavioural response, but also a complex of internal changes, involving the autonomic system, which are correlated with relatively long-lasting influences on behaviour. When the stimuli in question are painful, the internal changes may include stomach lesions as well as changes in heart rate, etc.: their severity may depend on the predictability and avoidability of the stimuli (e.g., Miller, 1969; see also page 405–406).

The complex and far-reaching effects of severe restriction during rearing are dramatically illustrated in a study by Melzack and Scott (1957).

Dogs were reared to maturity under conditions which severely restricted their sensory experience and screened them from noxious stimulation. Such animals learned to avoid various types of painful stimulation (nose-burning, electric shock, etc.) much more slowly than did controls reared under more normal conditions. At first the restricted animals showed no avoidance responses at all, walking repeatedly into lighted matches, striking their heads against water pipes, and so on. Their behaviour on release from confinement was also characterized by a very high level of behavioural arousal and by a failure to attend selectively to environmental stimuli. They also showed poor performance on a delayed response test and were submissive to normally reared animals. Melzack (1965) interprets their behaviour as due to an inadequate filtering of sensory input, so that excessive central nervous arousal was produced. Evidence in support of this view was obtained from a study of severely restricted beagles which not only had severe behavioural deficits, but on release from confinement showed more marked EEG changes than the control dogs, indicating higher central nervous arousal. Melzack emphasizes that comparable effects are obtained when dogs reared in a free environment are placed in a cage. The restricted animals thus differ only in degree from normally reared ones: in both cases previous experience is inadequate to enable the animal to cope with the radical environmental change.

The persistence of the behavioural disturbance shown by dogs and monkeys on release from severe sensory restriction seems to demand principles in addition to those discussed so far. One possibility, suggested by Fuller and Clark (1966a, b), is that the initial emotional reaction to release from restriction leads to responses, such as withdrawal to a corner, which are incompatible with adjustment to the new environment and which may be reinforced. In support of this view, they found that under certain conditions chlorpromazine could aid post-isolation socialization, and that experience with objects or a puppy during isolation did not. Though the process they suggest could play a role, their data scarcely support the importance they attach to it.

The effects discussed so far mostly come within a broad but generally acceptable definition of learning. Recent studies of the development of behaviour in mammals, however, have shown that apparently trivial aspects of early experience may have profound effects on later behaviour which, though revealed in learning tests, are probably mediated in a quite different way.

Bernstein (1952) found that the mere handling of juvenile rats improved their subsequent performance in a *T* maze. Hunt and Otis (1955) assessed the effects of various treatments given in infancy—daily handling alone or with food deprivation, daily electric shock, etc.—on the readiness of adult rats to emerge from their cages on to an open runway when hungry: a control group of animals was not handled. While no differences among the experimental groups were found, the control (non-handled) group emerged from the cages less readily than did any of the experimental groups. It thus seemed that the noxious stimulation in infancy produced a reduction in "emotionality" among the adults.

In subsequent experiments with rats, Levine (e.g., Levine et al., 1956; Levine, 1962a, b) showed that "stimulation in infancy results in the capacity for the organism to respond more effectively when confronted with novel situations, or, in other words, to exhibit a diminished emotional response to such novel stimuli. . . ." In the early experiments the stimuli used were handling for a few minutes a day, or mild electric shock, administered from birth to weaning (about 21 days). Response measures included the amount of defaecation and freezing shown in an open field test, the amount of water drunk after a period of deprivation, and so on. Later, other behavioural indices, as well as measures such as length of survival with total food and water deprivation (an unnecessary, cruel procedure), or survival with known pathogenic agents, were used by other investigators. The most effective age for the stimulation lies during the first weeks of life, though the optimum period varies considerably with such factors as the intensity and conditions of stimulation, the dependent variable used, and the animal's experience between stimulation and testing: because of the influence of such factors it is of doubtful utility to speak of a critical period for stimulation (Denenberg and Bell, 1960; Meyers, 1962; Spence and Maher, 1962; Denenberg, 1968). Although there have been some reservations (Broadhurst, 1961; McMichael, 1961), these results have now gained general acceptance.

Many of the results obtained have been interpreted in terms of an effect of the early experience on "emotionality", a concept not always clearly defined but usually assessed in terms of defaecation and lack of mobility in an open field. As often used, the concept is, in fact, a loose one. The various measures by which it has been assessed are not always highly correlated (e.g., Paré, 1964), and they may be affected differentially by the same changes in the early environment (Denenberg, Schell et al., 1966; DeNelsky and Denenberg, 1967a, b). Furthermore the relationship of the conventional measures of emotionality in the open-field situation, defaecation and activity, to heart rate, which activation theorists (see page 214) would expect to be positively correlated with emotionality, is in fact the opposite to that expected (Snowdon et al., 1964). Since in addition developmentally controlled studies show that defaecation may increase or decrease in a disturbing situation according to the subject's own developmental history (Tobach and Schneirla, 1962), considerable doubt has been cast on the usefulness of the concept of emotionality (see also Furchtgott and Cureton, 1964). The difficulties have arisen, however, in part from the loose way in which it has been used. Whimbey and Denenberg (1967) used rats in a study involving four experimental variables (handling of mothers, handling of pups, housing conditions before and after weaning) and an extensive battery of tests in adulthood. When the data were subjected to factor analysis (see pages 377–380), defaecation in the open field loaded on one factor (called the "emotionality" factor), while activity loaded on both the "emotionality" and "exploration" factors. Furthermore, activity on the first day in the open field correlated positively with defaecation, but on subsequent days the correlation was negative. Measures of "activity" are in fact misleading unless coupled with knowledge of the nature of the activity. The Whimbey and Denenberg study has thus placed the concept of emotionality, as used in this context, on a much firmer basis than previously.

What are the stimulus conditions necessary in infancy to affect adult emotional behaviour? In a number of experiments with rodents (e.g., Denenberg and Whimbey, 1963a) it was shown that a few minutes' handling each day during infancy produced effects indistinguishable from those of noxious electric shock. Other studies suggest that the resultant temperature change, rather than handling itself, may be the crucial variable (Hutchings, 1963; Schaefer, 1963). Yet other studies show that mere shaking of the cage, or moving it about the laboratory, is as effective as handling in reducing "emotionality". The effectiveness of such diverse treaments led to the view that quality and quantity of stimulation are unimportant. However, Denenberg, Schell et al. (1966) found that *lack* of noise in the rearing room could produce effects similar to those of handling during infancy: the weaning weights were smaller in handled as compared with non-handled animals, but larger in rats reared in a noisy as compared with a quiet room. Similarly handled rats were more active and noise-reared rats less active in an open field than their appropriate controls. Furthermore, very intense stimulation in infancy leads to an increase rather than a decrease in subsequent emotionality (Lindzey et al., 1963). Bell and Denenberg (1963), working with infant mice and studying the effects of infantile shock on avoidance learning and performance in an open field, showed that an intermediate amount of stimulation in infancy leads to the best performance on certain learning tasks in adulthood. The optimum amount of shock administered in infancy, however, varies with the age at which it is administered, so that there seem to be different sensitive periods depending on the intensity of stimulation and other variables (see also McMichael, 1966).

Experiments on the relation between early stimulation and adult emotionality or performance thus do not lend themselves to easy generalizations. Denenberg (1964) has suggested that the amount of stimulation in infancy reduces subsequent emotionality in a monotonic fashion, but is related in an inverted U fashion to adult performance in tasks involving noxious stimulation and of moderate difficulty. For tasks which are easy or difficult he suggests that the relationship between intensity of infantile stimulation and adult performance is monotonic, though opposite in slope (Fig. 22.1). Goldman (1965; in press), however, regards Denenberg's monotonicity hypothesis as holding only over a limited range: in her view emotionality is a curvilinear function of stimulation in infancy. This would account also for the experiments in which rats given electric shocks in infancy were as or more emotional than non-handled controls (see also Ader, 1966; Henderson, 1966; Newell, 1967). On this view it is possible to account for the inverted U relationship between infant stimulation and adult learning performance directly, the only assumption being that high levels of emotionality interfere with performance on aversively motivated tasks.

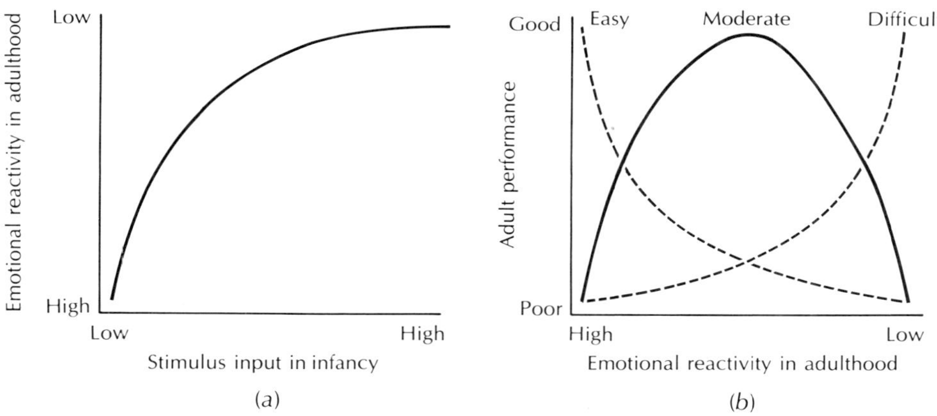

Fig. 22.1 a, *Theoretical relationship between stimulus input in infancy to emotionality in adulthood;* b, *theoretical relationship between performance in adulthood and emotionality level for tasks of varying degrees of difficulty.* (After Denenberg, 1964.)

The consequences of infantile stimulation in rats are extremely widespread. In appropriately stimulated animals the eyes open earlier, hair growth is more rapid, weight at weaning may be greater or less than controls (depending upon the stimulation and the strain of animal used), weight in adulthood is greater, there is an earlier maturation of the hypothalamic-hypophyseal-adrenal system, a heavier subcortical brain, and earlier myelinization of the central nervous system, than in non-stimulated controls (Levine and Alpert, 1959; Levine, 1962b, c; Tapp and Markowitz, 1963). The effects on brain development are complex: Altman et al. (1968) found that handling produced "infantilization", but postulate that this may lead to more adaptive consequences in adulthood. Furthermore, many aspects of behaviour are affected, including the appearance of complex patterns, perception, exploratory behaviour, and learning (Denenberg and Morton, 1962b; McMichael, 1966; Gard et al., 1967; DeNelsky and Denenberg, 1967). In human infants visual attentiveness appears to be influenced by previous non-specific stimulation (White and Held, 1966).

Such findings have prompted suggestions that the diverse types of early stimulation used in these studies act by producing a common mediating effect which then produces diverse consequences. Changes in the cerebral cortex (Bovard, 1958) and in nervous function in general (Lát, 1963) have been proposed. However, since infantile stimulation affects responsiveness in novel or fear-producing situations, the response is likely to involve the adrenal-pituitary axis. That the neonatal rat responds with rises in adrenal and plasma corticosterone to a variety of stressors, including the standard handling procedures used in so many of these experiments, has been established by a number of experimenters (e.g., Denenberg, Brumaghim et al., 1967). Using adult rats, Levine (1957, 1962b) compared the adrenocortical response to stressful situations of rats which had or had not been experimentally stimulated as infants. In general, the results confirmed the view that the non-stimulated animals are more reactive to the unpleasant situations used, for when measured 24 hours later their adrenals were heavier. Paradoxically, however, the stimulated animals showed as adults a more rapid rise in circulating corticosteroids after electric shock, and a greater output in the first 15 min after the shock (see also Haltmeyer et al., 1967). Thus it is the pattern of response to stressful situations which seems to be affected by stimulation in infancy. The non-stimulated animal's response to unpleasant situations seems maladaptive in that it is less vigorous to painful stimuli like electric shock than that of a stimulated animal, but greater to relatively non-noxious situations like the open field.

Levine and Mullins (1966) suggested that handling, by causing variations in the concentration of adrenal steroids in the infant, thereby modifies the control system governing their secretion, so that concentration can vary in a graded manner in the adult. Denenberg and Haltmeyer (1967) found a linear decrease in plasma corticosterone at weaning as a function of the number of days on which rats were handled—a result in conformity with Denenberg's suggestion of a monotonic relationship between amount of handling and emotionality (see above). Other workers have not found a significant relationship between open-field behaviour and corticosterone (Ader et al., 1967), so the relationship between adrenal activity and behavioural emotionality cannot be regarded as certainly established: however, some of the discrepancies undoubtedly stem from differences in techniques and from the misleading nature of straight activity scores (see above).

The effects of infantile stimulation on subsequent emotionality could be due to disturbance of the mother rather than to a direct effect on the infant: most experiments have confounded the two. It is known that even during pregnancy, maternal influences may affect the subsequent behaviour of the offspring. De Fries et al. (1967a, b) transplanted ovaries from two mouse strains into F_1 hybrid recipients, and thus obtained inbred offspring carried by either hybrid or inbred mothers. Differences in the maternal environment had a considerable effect on body weight and a small one on open-field behaviour. The consequences of subjecting mothers to stressful experiences during pregnancy may be more marked than this and can lead to greater emotionality in the offspring (Thompson, 1957; Doyle and Pratt Yule, 1959; Hockman, 1961; Thompson et al., 1962; De Fries et al., 1967a, b).

Maternal characteristics may also exert a postnatal effect on subsequent behavioural development. For instance, Ressler (1963) showed that genotype-correlated characteristics of a foster parent exert a postnatal influence on visual exploration, body weight, and viability in two strains of mice.

The experiential effects resulting from stimulation at any one age, pre- or post-natally, may interact with those at others. Denenberg and Morton (1962a) therefore assessed the effect on adult emotionality of a variety of environmental and social variables applied during infancy. They found that the adverse effects on adult emotional behaviour of lack of early stimulation can be offset by providing a fairly diverse environment after weaning, but are accentuated by rearing the animals in company with stimulated, non-emotional controls. Furthermore, the more emotional mother rats (like human mothers, e.g., Ottinger and Simmons, 1964) tended to rear more emotional offspring. Experiments in which infants were switched between mothers showed that while the effect was not a genetic one, the mother's influence was exerted both pre- and post-partum (see also Reading, 1966). In fact, the effect is also reciprocal, the mother's emotionality being affected by that of her young (see also Barnett and Burn, 1967). It is thus clear that although stimulation as applied through daily handling by an experimenter is effective in reducing adult emotionality primarily during a few days after birth, emotionality can be affected by other environmental factors before and after this period.

The effect of infantile stimulation can even carry across generations. Denenberg and Whimbey (1963a, b) found that the experiences which a mother rat received in infancy had an effect on her offspring's body weight at weaning and open-field behaviour in adulthood. The effects were mediated both through the prenatal mother–foetus relationship and in the course of postnatal mother–infant interaction. Similarly Ressler (1966) found that the operant behaviour of mice was influenced by the strain of foster grandparents, and Levine (1967) showed that handling of the mother could produce effects on the adrenocortical response to stress of weanling rats similar to those produced by handling of the subjects themselves in infancy.

Most of the experiments on the effects of infantile handling have been carried out with rats or mice. Comparable effects are known in other species; for instance dogs and cats (Meier, 1961; Fox and Stelzner, 1966, 1967; and refs. in Levine, 1962a). However, great care must be exercised in generalizing from one strain or species to another. In one experiment, for instance, adult mice stimulated in infancy showed *less* rather than more resistance to total food and water deprivation than unstimulated controls (Denenberg and Karas, 1959). King and Eleftheriou (1959) subjected two species of deermouse (*Peromyscus*) to infantile stimulation. Adult avoidance learning in a Skinner box situation was improved in one species as a consequence of the infantile stimulation, but impaired in the other. Similar genetic influences on the effects of infantile stimulation have been found by other workers (review by Levine, 1962a). In some cases genetic differences in the effects of infantile stimulation may be related to correlated differences in rates of development. Further studies are listed in a comprehensive review by Bell (1965). In spite of the inter-species differences, these studies indicate new ways to examine the effects of the nature of the mother–infant relationship, and of maternal deprivation, on subsequent behaviour characteristics in man (Bowlby, 1960) and monkeys (Harlow, 1962; Spencer-Booth et al., 1965).

It will be apparent that experiences with both the inanimate and social environments may affect subsequent fear and avoidance behaviour and emotionality by a number of mechanisms. These effects will have further ramifying consequences on nearly all other aspects of behaviour, and it is necessary to bear them constantly in mind when considering the effects of environmental manipulations on development.

22.3 EARLY EXPERIENCE AND PHASE DIFFERENCES IN LOCUSTS

A more extreme example of a general influence of early experience is given by the solitary and migratory phases of the locust (*Schistocerca gregaria*). These differ in colour, morphology, and behaviour, and were once classified as separate species. The difference is, however, determined in large part by early experience. Locust hoppers brought up in the laboratory in groups develop into the migratory type, whereas those brought up singly develop into the solitary phase.

Under natural circumstances, the hoppers are likely to become aggregated, in part through their tendency to climb up plants to bask in the sun, and in part through a visual attraction to other hoppers (see page 99). If the population is sufficiently dense, the crowding thereby produced leads to a change in behaviour: the hoppers become more active, and their response to the close proximity of another individual changes from withdrawal to further approach, tactile investigation, and so on. Once the tendency to group has occurred, it cannot be completely reversed. There is thus positive feedback, and the crowding may become sufficient for the hoppers to develop into the migratory phase. Furthermore the increased activity leads to the swarms "marching".

Although the course of development toward solitary or migratory phases is thus environmentally determined, species differ in the extent to which crowding produces a tendency to group. Furthermore there is some evidence that crowding affects the morphology and behaviour of the offspring: cytoplasmic inheritance seems to be involved. Thus in some but not all species the social conditions of early rearing produce a radical change in the direction of development which may affect even the next generation (Faure, 1932; Ellis, 1953, 1962, 1963, 1964, and refs. cited).

22.4 AGGRESSIVE BEHAVIOUR

In the introduction to this chapter Kruijt's (1964) study of the development of aggressive behaviour in junglefowl was quoted to exemplify the complexity which may lie in the developmental histories of apparently simple adult activities. The manner in which the development of aggressive behaviour can be influenced by experiential factors is clearly a matter of rather fundamental importance, but it has not yet received the attention it deserves. There are, however, abundant indications of the inadequacy of Lorenz's (1966) implication that experiential influences on aggressiveness are of minor importance in most vertebrates.

In the first place, the type of maternal care may influence subsequent aggressiveness. Hudgens et al. (1967) found that mice reared from 4 days by rat mothers fought less than mouse-reared mice. How much this was consequence or correlate of other differences observed (e.g., weight, activity) has not yet been determined. It may, of course, have resulted from a difference in early general stimulation, as discussed in Sec. 22.2 (see also Denenberg et al., 1966; Hudgens et al., 1968).

Rearing in social isolation also affects later fighting behaviour, though the factors determining the nature and extent of the effect are far from clear. In rodents, some investigators have reported that animals reared in isolation were less aggressive than controls (e.g., King and Gurney, 1954; Hutchinson et al., 1965). Levine et al. (1965a, b), however, found that mice reared in isolation since weaning were more aggressive than controls, the extent of the effect differing markedly between strains. Sigg et al. (1966) obtained a similar result, but found that isolation had no effect when animals were hypophysectomized or gonadectomized (see also Ginsburg and Allee, 1942; Uyeno and White, 1967). Cats (Seitz, 1954; Rosenblatt, personal communication) are also apparently more aggressive if reared in isolation, and in deermice (*Peromyscus*) the direction of the effect varies with the species (Rosen and Hart, 1963). Kuo (1967) regards partial isolation as essential for training birds and dogs as "good fighters".

These discrepancies are probably due in part to differences between species and techniques. But it is likely also that the extreme conditions of rearing in social isolation would affect fighting behaviour by a number of processes, the overall result depending on which predominates. In the first place, a fellow member of the species will be a novel object to an individual reared in isolation and therefore likely to elicit fear responses. Second, in social species early experience may influence the extent to which aggression is inhibited by fear or other types of behaviour. Rhesus monkeys reared in isolation may persist in responding aggressively to individuals much larger and more powerful than they are in spite of severe punishment, being apparently unable to learn to curb their aggression. Third, in the Burmese red junglefowl, individuals reared in total social isolation not only show abnormally intense escape and aggressive behaviour, but also inadequate integration between them: the changeover from one to the other occurs erratically and is not accompanied by ambivalent behaviour (Chap. 16; see also Middlebush et al., 1968).

Experience during fighting also has a marked effect on subsequent aggressive behaviour. Mice subjected to defeat in youth are likely to be subordinate in adulthood (Kahn, 1951). Such effects of early experience are often long-lasting, but comparable changes occur even in adult animals (e.g., Ginsburg and Allee, 1942; Scott and Fredericson, 1951). Lagerspetz (1964) found that defeats decreased the aggressiveness of both aggressive and non-aggressive mice, while victories increased the aggressiveness only of initially aggressive ones: the effects disappeared in the course of a few weeks' isolation. A review of the effects of submissive experience on dominance is given by Thines and Heuts (1968).

Where fighting takes place with respect to an object, such as food, it is to be expected that success or failure in obtaining the object will influence subsequent fighting behaviour. Kuo (1967) reports long-term experiments in which dogs were trained as fighters, a large part of the training depending on encounters with weaker individuals at a food bowl. In more controlled situations it has been shown that aggression can be reinforced by both food and water (Ulrich et al., 1963, and Azrin and Hutchinson, 1967, in rats; Reynolds et al., 1963, in pigeons) and intra-cranial electrical stimulation (Stachnik et al., 1966).

Furthermore, studies reviewed in Sec. 15.5 show clearly that the performance of aggressive behaviour, or the appearance of stimuli eliciting it, or stimuli previously associated with aggression-eliciting stimuli, can have reinforcing properties. Thus, fighting may give rise to fighting.

Finally, as we have already seen, the stimuli which elicit aggressive behaviour can be influenced by conditioning (e.g., Thompson et al., 1965; Creer et al., 1966; Vernon and Ulrich, 1966). It seems possible that experience may affect the stimuli which elicit and direct aggression independently, since monkeys reared in isolation often come to bite their own arms and may persist in doing so while later living in groups (personal observation).

It remains true, of course, that even individuals kept in isolation for long periods after birth may still show not only aggression but also the consequent social adjustments (e.g., Ratner, 1965). But the studies cited show that aggressive behaviour may be markedly influenced by experience, especially early experience. The implications of this for our own species are of course considerable. A possible developmental sequence which has been suggested for the human case is as follows. The anger response is first elicited by frustrating situations more or less independently of previous experience in them. With the development of motor control and perceptual sensitivity, this becomes more and more directed toward the source of frustration. On occasion it will then have the effect of causing injury and/or otherwise removing the frustration and thus be reinforced. The association of the perception of pain in others with the removal of the source of frustration leads to the acquisition by the former of secondary reward value (review and critique by Feshbach, 1964). Comparative evidence, as we have seen, suggests that a primary reinforcing effect of fighting is also likely. In any case, subsequently, the eliciting situations, the reinforcing situations, and the nature and intensity of aggression may all be influenced by experience during fighting, experience of aggressive behaviour by others, and experience not specifically related to fighting.

22.5 SEXUAL BEHAVIOUR IN MAMMALS AND BIRDS

The development of sexual behaviour involves both hormonal action on the mechanisms of mating behaviour and experiential factors. With regard to the former, the most detailed studies have used guinea pigs and rats (e.g., W. C. Young, 1961b), where processes of at least two types are important:

1. *HORMONES FROM FOETUS OR MOTHER HAVE A PERINATAL ORGANIZING ACTION ON THE TISSUES MEDIATING MALE SEXUAL BEHAVIOUR*

For instance, if pregnant guinea pigs are given testosterone propionate, the genitalia of the female offspring resemble those of males: if the young are ovealiectomised and treated with gonadal hormones, their behaviour is much more male-like than that of similarly treated controls whose mothers did not receive testosterone (e.g., Young, 1961b, 1965; Young et al., 1964; Gerall, 1966).

In the rat the period of susceptibility is primarily postnatal rather than prenatal (Fig. 22.2): the data indicate that male and female rats are born with a nervous system undifferentiated behaviourally though potentially showing female cyclicity, and that this is normally differentiated in the male

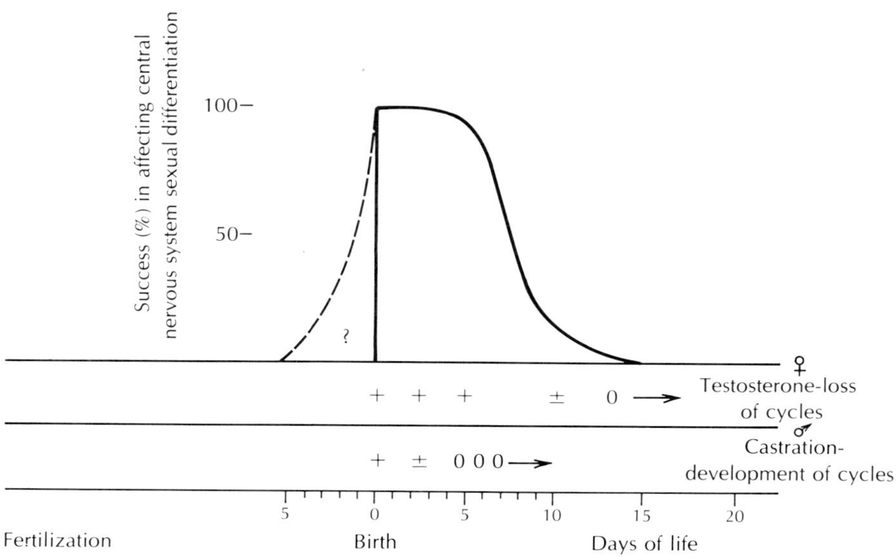

Fig. 22.2 Summary of various data concerning the period during which the central nervous system of the rat becomes sexually differentiated. (From Levine and Mullins, 1966, after Harris, 1964.)

The upper line of symbols (+, effective; ±, possibly effective; 0, ineffective) shows the days after birth when administration of testosterone to the female rat evokes masculinization of the nervous system. The lower line of symbols shows the days when castration of the newborn male rat allows retention of a female-type nervous system. The curve shows success in affecting central nervous system sexual differentiation.

by the action of testosterone (Harris and Levine, 1965). Thus female rats treated with testosterone at 4 days of age show a complete loss of oestrous cycles and of sexual behaviour in adulthood. Even when ovariectomised and given oestrogen-progesterone therapy they remain sexually unresponsive. However, if ovariectomised and given testosterone when they reach adulthood, they show the complete masculine mating pattern (see also Whalen and Robertson, 1968). Treatment of 4-day-old females with low oestrogen doses has only a slight effect on later sexual behaviour. Treatment of 4-day-old males with testosterone has no effect on their sexual behaviour in adulthood, while oestrogen produces some loss which cannot be compensated by testosterone therapy: this is ascribed to a partial functional castration at the crucial period, though inadequate penile development may also be involved. Neonatally castrated males can show ovarian cycles and female sexual behaviour in adulthood, though their lordotic responses after oestrogen treatment are inferior to similarly treated spayed females, and their mounting rates after androgen treatment are higher: the latter results suggest that the sensitive period may be partially prenatal (A. A. Gerall et al., 1967; see also Gerall and Ward, 1966; Larsson, 1967a, b).

The organizational influences of the androgenic secretions of the foetus and the mother are not limited to the brain. Quite apart from their effects on the external genitalia, which in the rat are determined before the mechanisms of hypothalamic control (Saunders, 1968), spinal mechanisms also seem to be affected. Male rats castrated when 4 days old and given testosterone in adulthood were sexually motivated but unable to complete the mating sequence. After spinal transection they showed impaired genital responses. Since males castrated at 10 days showed no such deficit, the neonatal androgen must have an organizing influence on the spinal reflexes of copulation (Hart, 1968c). This probably involves the spinal mechanisms themselves, though more peripheral effects have not been ruled out.

While the general principle that gonadal hormones administered in infancy have a marked effect on the development of reproductive behaviour is well established, the process is not simple. In particular, the effects of the different steroids is not so specific as was first thought, larger doses of oestrogen than those used by Harris and Levine producing results similar to those produced by testosterone (Levine and Mullins, 1966).

Other workers have shown that the effects of such perinatal hormones are not limited to behaviour related to copulation: other aspects of behaviour in which the sexes normally differ may also be affected (Goy, 1968; Gray et al., 1965; Swanson, 1967; Edwards, 1968; Bronson and Desjardins, 1968).

2. *AT PUBERTY THE INCREASE IN ENDOGENOUS GONADAL HORMONES PRODUCES A READINESS TO SHOW SEXUAL BEHAVIOUR*

This is a result of action by the hormones on mechanisms which may have become partially organized even in animals castrated during the prepubertal period (Beach and Holz, 1946; Riss et al., 1955; Rosenblatt and Aronson, 1958a, b). In male hamsters some aspects of the sexual pattern develop even in animals castrated and adrenalectomised before puberty, indicating that such fragmentary sexual patterns as castrates show is not due to androgens from the adrenal cortex (Warren and Aronson, 1957). Nevertheless quantitative examination of the sexual behaviour of rats shows that the mating pattern after standard hormone treatment differs between animals castrated pre- and post-pubertally—the number of intromissions preceding ejaculation is greater in animals castrated before puberty. Since differences in sexual experience cannot account for this effect, Larsson (1967) suggests that at puberty the testicular secretions produce changes in responsiveness to hormones in the tissues mediating sexual behaviour. Similar processes in normal development could account for some of the known age changes in copulatory performance (see page 530). That such changes in central responsiveness are reversible is suggested by Hutchison's (1969) finding that the effectiveness of hormones in inducing courtship behaviour in ring doves (*Streptopelia risoria*) decreases with time since gonadectomy.

Although sex hormones are usually considered to have separate effects on particular items or categories of behaviour, it has been suggested that diverse effects may be produced by a single type of change. Andrew's (1963a, 1969) detailed analysis of the influence of testosterone on the calls of domestic chicks is of special interest here. Injections of testosterone during the first week or two of life lead to the appearance of a number of new patterns, each apparently the result of a separate effect of testosterone. However, Andrew suggests that these changes, though complex, are all generated by a simple change in a single functional system. This view arises from the finding that most calls of the untreated chick, normally considered as consisting of a number of discrete types, in fact form a continuous series differing from each other only in degree. Andrew argues that these calls of the untreated chick also have a common causal basis, occurring when stimuli which are inherently conspicuous, or have acquired significance through learning, are being scanned. The new calls which testosterone seems to produce are in fact each a distortion of some part of the normal series, involving prolongation and, often, trilling. The call produced is thus a disturbance in these directions of the call which would otherwise have been given. A complication is introduced by the fact that any particular testosterone call may be developed from different parts of the juvenile series in different individuals: the classification of the calls developed depends not only on their form, but also on the behaviour and situation associated with them. Andrew ascribes the action of testosterone to the induced activity of a "testosterone mechanism", activated by situations similar to those controlling normal vocalization, but having a higher threshold: whether the postulation of such a mechanism helps his thesis is an open issue. He does, however, produce considerable circumstantial evidence that the processes which he has studied in chicks resemble those occurring when song and other testosterone-dependent vocalizations develop in adult birds.

Although gonadal hormones are of fundamental importance in the control of sexual behaviour in most adult vertebrates, young animals often show mating patterns when their gonads are showing at most slight endocrine activity (see page 528). In chicks of the red junglefowl, Kruijt (1964) believes that copulatory activity occurs when the tendencies for aggressive and fleeing behaviour are low and balanced, a conclusion supported, for instance, by the observation that it depends strongly on the dominance rank of the male and the size of the group, and is directed to lower-ranked birds.

Often the appearance of sexual patterns in young animals is extremely spasmodic, occurring only rarely and in a small proportion of individuals (e.g., Barraud, 1961). In other cases, such as the penile erection of male human infants, it is quite regular. However, a semantic difficulty arises here over the applicability of the label "sexual" in such cases. For example, the lordosis response of guinea pigs appears in the neonate, but it is then part of the excretory pattern and unrelated to sexual behaviour both functionally and in its independence of ovarian hormones (Beach, 1965).

In addition to hormones, experiential factors play an essential role in the development of sexual behaviour. Indeed their role is often underestimated, since factors which appear at first sight irrelevant may exert a profound influence. For instance, platyfish reared in aquaria of clear glass show sexual behaviour in adulthood, but individuals reared in aquaria of frosted glass do not (Shaw, 1962).

The most important experiential factors, however, are social ones. In birds, males reared in social isolation may fail to mate with females because they show more fear than group-reared controls. When sexually mature male chaffinches, hand-reared in isolation from others of their kind, were presented with females stuffed and mounted in the soliciting posture, they showed some courtship but would not approach the model. The same model elicited copulation from a high proportion of normally reared birds. Under ordinary circumstances chaffinches become habituated to the proximity of other individuals, but to the isolate the soliciting female was a strange object, and therefore elicited an unusually strong tendency to flee as well as sexual behaviour (Hinde, 1959c). Fisher and Hale (1956/1957) similarly found that male domestic fowl reared in social isolation courted female chickens but would not mount a crouching bird: this is in marked contrast to the behaviour of normally reared cocks, for whom crouching readily elicits mounting attempts. After experience with live females some of these males attempted to copulate by force with standing females rather than with crouching ones. Kruijt (1962) also found that junglefowl cocks reared in isolation for 10 months were unable to perform normal copulatory behaviour, and suggested that this was due to deficiency in the integration of aggressive and escape behaviour in early life: the agonistic behaviour was abnormal from quite an early age, the birds showing fits of very strong escape behaviour alternating with periods of extreme aggressiveness. Normal mating behaviour was shown by cocks reared in isolation but placed with females before they were 10 months old. In group-reared birds, copulatory behaviour comes to be elicited solely by females and copulatory responses to inappropriate objects disappears as a result of experience (Kruijt, 1964).

In rats, the evidence about the role of early social experience on later sexual behaviour is rather contradictory. Beach (1942a, 1958a) found that male rats reared in isolation were more responsive to females and more likely to copulate than males reared with females: this was attributed in part to the greater excitement engendered in the isolates by contact with a second animal. Kagan and Beach (1953) also found that males kept in isolation from the 37th to 100th day of life were more likely to show the complete copulatory pattern in a mating test than were animals which had had weekly contact with a male or a female during the same period of isolation. They attribute this difference to patterns of social behaviour formed by the latter groups which partially reduced the sexual responses.

Zimbardo (1958), however, obtained different results. Male rats reared in part-time cohabitation showed better performance in mating tests than males reared in isolation. H. D. Gerall et al. (1967) also found that rearing in social isolation disrupts mating ability in male rats. Of various procedures used, only cohabitation with a female for 3 weeks or more resulted in ejaculation, and even then many animals were very unskillful (see also Folman and Drori, 1965). The deficit was regarded as due both to lack of practice and to inadequate responsiveness to the stimuli mediating sexual behaviour. That responsiveness to the chemical stimuli mediating sexual behaviour does depend on experience in male rats has in fact been established (Carr et al., 1966; see page 511). That these are not the only means whereby social experience affects mating behaviour is suggested by a study by Drori and Folman (1967). Male rats kept in unisexual groups for 16 months, although then able to impregnate females when caged with them, were markedly inferior in copulatory performance as compared with experienced males. They improved gradually over the next few months, though at 20 months mean penis weight was smaller than in the control group.

Rearing in isolation has also been shown to affect mating performance in guinea pigs, where experiments by Valenstein et al. (1955), with several genetic strains, showed that males reared in a social situation scored better in sex tests than males reared in isolation. Moreover, male cage mates were adequate for the organization of sexual behaviour in males: males reared with spayed females were inferior to males reared with either intact females or with other males, apparently because of the lack of sexual stimulation these females provided (Valenstein and Goy, 1957). The social experience produced its effect before the animals were 25 days old (Young, 1961b). That social experience through the whole of this period is not necessary is indicated by Dieterlen's (1959) finding that individuals isolated when 4 to 10 days old respond appropriately to olfactory stimuli from other individuals, and mate normally even when completely inexperienced.

Gerall (1963) has obtained data on the sexual behaviour of the guinea pig not wholly consistent with those of Valenstein and Goy, *loc. cit.* His data indicated that practice in mounting behaviour is not an important issue, for guinea pigs separated from the group at 10 days only by a wide mesh screen, which permitted visual, olfactory, and some tactual contact but no mounting behaviour, showed no deficiency. Males isolated from 10 days of age were more likely to mate than males isolated from 2 days: the age at which isolation occurs is thus a crucial variable (H. D. Gerall, cited by Harper, 1968). Gerall (1963) suggests that the sexual inadequacies of the isolates were due to interference by other types of behaviour (e.g., play responses, see above) and to inadequate responsiveness to the relevant stimuli from the female. However Harper (1968) found that males isolated from birth and tested with stimulus animals at 80 to 90 days of age "were more playful than the controls", but "their behaviour was not considered aberrant". Furthermore when first placed with a receptive or potent member of the opposite sex, the animals reared in isolation showed normal copulatory behaviour. He concluded that, though social experience plays a role in the development of sexual behaviour, it is not essential.

These experiments indicate that experience with companions plays an important part in the organization of the sexual behaviour of rodents. Since incipient mating behaviour may occur at a very early age, the experience may operate in part through effects of incomplete sexual behaviour (McGill, 1962a), though this is not the only issue. Experience gained in early (pre-adolescent) mounting attempts has an important influence on later sexual behaviour also in the higher mammalian groups (Whalen, 1961, 1963a; Beach, 1968b): in chimpanzees and rhesus monkeys learning during mating attempts is essential for the organization of the components of sexual behaviour into a functional system (Nissen, 1954; Harlow, 1962; Harlow and Harlow, 1965; Mason, 1965).

Experience during puberty or early adulthood also plays an essential role in the organization of the mating pattern. Rosenblatt and Aronson (1958a) castrated thirteen male kittens at the age of 4 months. As adults they showed little sexual behaviour when tested with oestrous females. Eleven of these were subsequently treated with testosterone propionate. Six were given sexual experience during the period of hormone administration, four showing the complete sexual pattern. The other five were kept isolated from females during the period of hormone administration, but tested subsequently. Three showed no increase in sexual performance over the preandrogen level: the other two showed some increase during the period of testing, suggesting that experience in the tests soon after hormone withdrawal was important. The organization of sexual behaviour thus requires appropriate environmental stimulation while the animal is in an adequate hormonal condition. Rosenblatt and Aronson (1958b) also showed that the sexual behaviour of cats persists after castration only if the male has had some sexual experience before castration, a result again pointing to the necessity for a combined action of hormonal state and external stimulation for the organization of the mating pattern. Once this organization has occurred, however, mating behaviour may continue even with very low hormone levels in this species (see also Rosenblatt, 1965). In beagles, however, Hart (1968b) found no evidence that inexperienced animals were inferior to experienced ones in post-castration sexual performance. Mating experience has been shown to affect copulatory behaviour in hamsters (Bunnell and Kimmel, 1965).

The role of learning during mating appears to be greater in male than female mammals (Beach, 1947, 1948). For instance experienced male rats prefer the odour of receptive females to that of non-receptive females, but naive and castrated males do not. By contrast experienced females, whether receptive or not, and naive receptive females prefer the odour of normal males to that of castrates, though naive non-receptive females do not. Thus the male's preference depends on mating experience, while the female's does not (Carr et al., 1965). This view is in harmony with the greater effect of cortical ablation on the sexual behaviour of male rats as compared with females (Beach, 1940, 1944; see page 239).

Beach also pointed out that the extent to which sexual behaviour may become independent of its normal hormonal background is greater in the higher mammals. While this view is almost certainly correct, a loophole in the supporting evidence has been pointed out by McGill and Tucker (1944). Studying the persistence of sexual behaviour after castration in mice, they found the ejaculatory reflex was retained far longer in genetically heterozygous individuals than in homozygous ones. Generalizations about the relations between the hormonal dependence of sexual behaviour and phylogenetic status have depended in part on data from inbred rodents and mongrel cats and dogs, and the possible influence of the degree of heterozygosity in the strains used has been overlooked.

The preceding discussion has been concerned primarily with copulatory behaviour itself. We have seen, however, that in many species this is accompanied or preceded by courtship displays, which are themselves often associated with conflict between incompatible types of behaviour, including attack and escape. The ontogeny of the ambivalent courtship displays have so far been studied in detail in only one species, the red junglefowl (*Gallus gallus*) and involve complex interactions between sexual, attack, and escape behaviour much influenced by environmental factors (see, e.g., pages 528–532, 541–542; Kruijt, 1964).

In conclusion, it will be apparent that social experience influences sexual behaviour in a number of ways, affecting copulatory readiness, responsiveness to the appropriate stimuli, and mechanical skill in mating, as well as other types of social behaviour which may interact with sexual behaviour. As Rosenblatt (1965) stresses, in higher vertebrates sexual behaviour grows out of affective reactions between individuals and is rooted in a general background of social responses developed earlier in life.

22.6 FEEDING

Evidence from a number of sources shows that the early feeding movements of neonatal mammals are relatively independent of motivational factors associated with food deprivation. Thus the amount the neonate sucks often shows considerable independence from the amount of food it obtains by sucking: puppies fed from bottles with nipples having large holes, which thus satisfied their nutritional needs easily, sucked more at objects outside the feeding situation than did puppies fed through small holes (refs. in Beach and Jaynes, 1954; Ross, 1950; James, 1957). Comparable data have been obtained with human infants (Ploog, cited by Eibl-Eibesfeldt, 1967). Indeed Bridger (1962) found that human newborns sucked more when satiated and experimentally aroused than when food-deprived. Koepke and Pribram (1967) reached a similar conclusion in a rather different way. Two groups of kittens were maintained by a stomach-loading technique, one of them being allowed to suckle for 6 hours a day from a lactating female and the other from an anaesthetised non-lactating female. There was practically no difference in the amount of time spent sucking between the two groups over the first 3 weeks of life. Subsequently the group with the non-lactating female sucked less: this suggests that milk ingestion has a reinforcing effect important for the maintenance of sucking.

There have been some differences of opinion as to the effect of consummatory stimuli from the stomach on sucking behaviour. The studies just cited suggest that any such effect is small. However, Satinoff and Stanley (1963) assessed the effect of stomach loading by a tube on the amount of nutritional sucking behaviour shown by puppies. Even with 1-day-old puppies the non-loaded animals spent more time sucking and drank more milk than did the preloaded ones, and the effect varied little with age over the 11 days of testing. These authors regard the negative results obtained in earlier studies as due to the less sensitive procedures used or to insufficient preloading (see Stanley and Bacon, 1963). James (1963b), however, regards the positive results obtained by these authors as due to overloading beyond a normal day's nutritional intake: "rather large" injections of gruel into the stomach of puppies deprived from $3\frac{1}{2}$ to 4 hours did not terminate the sucking response (James and Rollins, 1965). Comparable data are available for lambs (Alexander and Williams, 1966).

In the study of kittens by Koepke and Pribram (1967) cited above, non-nutritive sucking on an anesthetised non-lactating female was sufficient to maintain sucking for three weeks. Similarly Kovach and Kling (1967) found that after complete sucking deprivation and individual isolation for 23 days kittens were unable to initiate sucking from the mother, but if raised in groups by forced feeding they sucked on each other's fur and genital areas and were subsequently able to initiate sucking on the mother. The study of Alexander and Williams (1966) shows that denial of the opportunity to suck or receive nourishment leads to a rapid decline in sucking in lambs. Newborn Merino lambs were prevented from sucking by a cover over the maternal udder: the time spent sucking declined from about 30 percent during the first 2 to 3 hours of life to less than 5 percent by 12 hours after birth. If the udder was uncovered at 6 to 24 hours after birth, fewer lambs sucked successfully than in control groups. Taken together, these studies indicate that milk ingestion is at first of little importance for the maintenance of the sucking response, but the opportunity to suck is. A similar phenomenon is well known in human infants (Gunther, 1955). Once sucking has been initiated, the young animal learns the characteristics of the milk source, and after a while it can be transferred to another only with difficulty (e.g., Rosenblatt et al., 1961; Kovach and Kling, 1967; see Chap. 26).

In addition to sucking, other stereotyped movement patterns used in feeding may at first show considerable independence from the motivational factors which will later control them. For instance the killing bite of kittens appears at first in play and is not sufficiently vigorous to be effective. Playing with the prey seems to excite the kitten to the point where it bites effectively, but competition from mother or siblings can fill the same role (Leyhausen, 1965).

The evidence thus suggests that, in very young mammals, visceral factors may not control feeding behaviour to the extent that they do later. The means whereby they acquire this control are far from clear: for instance, "air drinking" (see page 253) is reinforcing even for 18-day-old rats reared without access to water bottles (Riccio et al., 1967).

The control of feeding behaviour in the rat develops rather rapidly: eating starts at about 14 days, and by 21 days the animal is weaned and regulates its diet with competence. Teitelbaum has, however, studied what is presumably the normal sequence in rats whose development was slowed by thyroidectomy soon after birth. When weaned at the normal age, they weighed 10 to 45 g, instead of the usual 55 g. Those weighing about 15 g sucked reflexly, but were completely aphagic and adipsic. The 25-g animals would lick and nibble at wet and palatable food, but were anorexic and adipsic, and usually died. A 35-g animal would regulate its caloric intake on a liquid diet, but would not take dry food and water. A 45-g animal would eat normally, but its drinking would be limited to mealtimes, and it would be very finicky. It will be noted that the types of behaviour shown by animals of successively higher body weights at weaning parallel those occurring during recovery from lateral hypothalamic lesions (Teitelbaum, Cheng, and Rozin in Valenstein, 1968).

The relationship between a given degree of food deprivation and the amount of feeding behaviour shown is, however, affected by experience (see Lawrence and Mason, 1955; Ghent, 1957; Dufort, 1964; review by Bolles, 1967). Furthermore, infantile food deprivation can influence the rate of eating in adulthood. Marx (1952) restricted recently weaned rats to two brief feeding periods per day. They were then fed *ad lib.* for 30 weeks. On subsequent testing for rate of eating after deprivation the experimental animals fed more rapidly than did control animals who had had food *ad lib.* throughout (see also Mandler, 1958; Renner, 1967; and Elliott and King, 1960, for puppies).

Further problems are raised by the relations between the acceptability of a food substance and its reinforcing value. For instance, saccharin, although non-nutritive, is well known to have a reinforcing effect in some species (Sheffield and Roby, 1950). There is, however, evidence that the effect of saccharin is greater if it is experienced in the presence of need reduction. Rats preloaded by stomach tube with sugar and then allowed to drink saccharin showed an increase from day to day in the amount ingested, while rats preloaded only with water did not (Capretta, 1962). Such a result is, of course, consistent with the evidence for the multiple nature of the consummatory stimuli for eating, as discussed in Chap. 10. The developmental processes governing the selection of dietary constituents are as yet little understood, but it seems that many aspects of the regulatory processes are independent of previous learning. For instance inexperienced sodium-deficient rats select sodium salts within seconds of first tasting them; and it seems that experience of a dietary constituent by a non-deficient animal can lead to its subsequent selection (see, e.g., Krieckhaus and Wolf, 1968, and refs. cited).

The role of visceral factors in the control of eating becomes relatively smaller with age, as external influences become increasingly important (Anderson, 1941). One of the most intangible factors here is the time of day. Although the activity of very young rats is more or less continuous, a circadian rhythm (see Chap. 14) is apparent after about 14 days (Richter,

1927). Such a circadian rhythm does not necessarily depend on external timers: mice kept for two generations under constant light and temperature showed a circadian rhythm of feeding behaviour, though this was out of phase between individuals (Wiepkema et al., 1966; see also Bolles and Stokes, 1965). Beyond this, a tendency to feed during particular parts of the cycle may be established by experience. A rat does not run well in a maze until it has been on a regular feeding schedule for some time: only when it has had repeated experience of hunger followed by eating will it work well for food (Verplanck and Hayes, 1953; Reid and Finger, 1955; Birch et al., 1958).

Quite apart from temporal effects, experience of the feeding situation has a marked effect on food intake (e.g., Moll, 1964). This may be in part a consequence of the extinction of exploratory behaviour: the effect is much greater in wild rats (Barnett, 1963).

The development of hoarding behaviour in rats is closely related to that of feeding and shows similar complexities. Some of these may be mentioned here. Wolfe (1939) reported that rats hoarded more if they were reared on pellets than if reared on powdered food, and Rosenblatt (1950) found that rats reared on a liquid diet carried pellets to the home cage but left them scattered over the floor, though normally reared rats placed them in piles. Wolfe, *loc. cit.*, also found that rats hoarded more if they were deprived of food prepuberally than if they were not. This effect does not occur if the animals are deprived for a period starting 12 days after weaning, and is apparent only if the hoarding tests are preceded by a period of deprivation (Hunt, 1941; Hunt et al., 1947). Comparisons of hoarding behaviour in adulthood between animals from small and large litters shows that the latter show greater food-hoarding activity: this may be related to greater food deprivation or competition at an earlier stage (Seitz, 1954). Even the effects of infantile deprivation are more complex than would appear from the studies just cited, for they are more marked in males than females, and infantile food deprivation affects the hoarding of food by adults but water deprivation does not affect the hoarding of water-soaked cotton (McKelvey and Marx, 1951). Infantile food deprivation may affect adult hoarding in part because it results in a higher rate of eating, allowing more time for hoarding pellets in the test (Marx, 1952).

A relationship between the developmental factors influencing feeding and hoarding behaviour is also indicated by neurophysiological data. Electrical stimulation in those hypothalamic areas which promote feeding can elicit intense hoarding behaviour in satiated rats: the behaviour is similar to that produced by long-term food shortage (Herberg and Blundell, 1967).

Although the development of feeding behaviour has been studied more closely in mammals than in any other vertebrates, comparable principles certainly apply to birds. Thus pecking in newly fledged passerine birds is most likely to occur when the birds are not hungry: when they are hungry, they beg from their parents. Even in chicks, pecking at first shows little relationship to food deprivation (Hogan, 1965). The stimuli eliciting pecking are at first generalized, but objects with the properties of roundness and brightness are preferred (see pages 503–504). Subsequently preferences are channelled by the reinforcing properties of the objects pecked at (Fantz, 1967), but the way in which these operate is complex. Sterritt and Smith (1965) tested chicks in a Y maze with a variety of incentives and found that neither the delivery of nutrients directly into the crop by a plastic tube nor the elicitation of pecking were reinforcing alone. However, all incentives which elicited pecking at the time the nutrients were delivered were reinforcing. They thus concluded that the internal effects of the delivery of nutrients and feedback stimuli from pecking were both reinforcing, but only in interaction with one another. Furthermore a number of environmental factors, including the level of illumination and the presence of social companions, influence the development of adequate feeding behaviour (May and Dorr, 1968). Finally, hoarding is known in a number of bird species, and in the raven (*Corvus corax*) it is known to increase with food deprivation (Gwinner, 1965).

22.7 MATERNAL BEHAVIOUR

Motivational aspects of the development of maternal behaviour will not be considered in any detail here, but two controversial studies on the rat must be mentioned. Birch (1956) prevented pregnant rats from licking themselves by placing wide rubber collars round their necks, and found that most animals treated in this way did not attend to their young at birth or establish effective maternal behaviour. He therefore supposed that the preparturient grooming behaviour of the female played a role in the establishment of maternal behaviour. Subsequent detailed data by Roth and Rosenblatt (1966) show that self-licking does in fact become concentrated along the nipple lines and in the anogenital region as parturition approaches and promotes development of the mammary glands (see also Steinberg and Bindra, 1962; Rosenblatt and Lehrman, 1963). However, Coomans (in Eibl-Eibesfeldt, 1958) obtained quite different results from Birch, and concluded that such deficiencies in maternal behaviour as he found could be accounted for in terms of mechanical interference. As Rosenblatt and Lehrman (1963) point out, the strains of rats and methods used differ in the two cases, so that attempts to reconcile the two studies are urgently needed. One such, by Christopherson and Wagman (1965), did not confirm Birch's result.

A parallel case is given by Riess's (1950) finding that female rats, reared in cages containing no food or other objects which could be picked up and carried about, failed subsequently to build nests, or even to establish a proper nest site in a test cage. He concluded that, in the course of carrying food about, rats learn to pile up warm nest material in one corner of the cage. Eibl-Eibesfeldt (1961), however, produced evidence that the failure of Riess's rats to build nests was due to a deficiency in the experimental conditions. In a new environment rats show mainly exploratory behaviour, and thus do not form the attachment to a particular part of the cage necessary for nest-building. Wehmer (1965) could find no differences between rats reared in cages containing movable objects, immovable objects, or no objects. Although the last group showed a delay in preparturient nest construction, no postparturient differences were found. His result favours Eibl-Eibesfeldt's view.

22.8 STEREOTYPES—FUNCTIONAL AUTONOMY

It often seems as though with repetition some activities become progressively more independent of the external stimuli and/or internal conditions with which they were associated initially. In captivity animals often develop behaviour stereotypes or tics which are repeated monotonously. They may walk repetitively to and fro, or show movements resembling incomplete locomotor or feeding patterns (Holzapfel, 1939a; Hediger, 1950). These movements often consist of the early stages of a behaviour sequence which has become fixed and modified to a varying degree. Thus the flying to and fro of a bird in an aviary develops from attempts to fly out of the aviary, though it may be so stereotyped that the bird flies on a constant track from side to side, perching in the same place each time. Such behaviour often becomes more or less divorced from the conditions which initially evoked it. The stereotyped flying to and fro may occur in the absence of any of the stimuli which normally elicit fleeing: apparently the bird shows this behaviour in the absence of fear whenever there is no overriding tendency to behave in other ways. A comparable case has been studied experimentally by Datel and Seward (1952). These authors assessed the amount of ear-scratching shown by two groups of rats—one group in which the ears were treated with collodion for 5 days and a control group not so treated. The collodion caused a large increase in ear-scratching, some of which persisted for at least 16 days after the last application. Since the difference between the amount of scratching shown by experimental and control groups did not diminish during these 16 days, it seems improbable that the greater amount of scratching shown by the experimentals was merely due to diminishing after-effects either of the collodion or of the excessive scratching during the collodion period on the tissues themselves (see also Earl, 1957).

Since such responses persist even though the factors which originally gave rise to them are no longer present, there has, presumably, been some change in their motivational basis. This has been described as the acquisition of "functional autonomy", though this is only a descriptive label (Woodworth, 1918; Allport, 1937). Bindra (1959) has pointed out that autonomy is never complete, but occurs with respect to a limited set of conditions, and can be understood to some extent at least in terms of processes known to be important in other contexts. First, cues frequently associated with the response may become substitute cues, capable of eliciting the response in the absence of those factors which were necessary initially. Second, as a result of repeated performance, fewer of the original cues are necessary to elicit the response—individual cues become increasingly effective. Third, Bindra suggests that activities acquired on the basis of one reinforcer are maintained in its absence because something else comes to reinforce the same activity—possibly the performance of the activity itself may have reinforcing value. There is, of course, a danger of circular argument here.

The general principle that the performance of a response can lead to an increase in the probability that the response will be given on subsequent occasions may be of wide applicability. Male grasshoppers reared in isolation, while able to sing the various songs of their species quite normally, do so very infrequently until placed in a singing group. The frequency of stridulation then increases markedly and permanently (Haskell, 1957).

Since stereotypes are especially liable to develop in monotonous environments, one possible explanation for their persistence is that they provide a source of stimulation, exteroceptive and proprioceptive. Amongst chimpanzees, stereotypes are frequent amongst individuals reared in impoverished laboratory conditions, but absent from wild-caught individuals (Davenport and Menzel, 1963): a similar generalization applies to rhesus monkeys. Whether environmental restriction or rearing apart from the mother is the basic issue here is not known: in a recent review Bronson (1968a, b) places the emphasis on the latter. The effects of cage conditions are apparently seen only in young chimpanzees, for 6 months' restriction of 1½-year-old animals does not produce stereotypes (Davenport et al., 1966): in other mammals, however, caging can induce stereotypes in older individuals (Loizos, 1967). In rhesus monkeys, stereotypes become less common the larger the cage in which the animal is kept (Draper and Bernstein, 1963). Monotony of the environment may also be involved in the rocking and other stereotyped movements shown by children, especially when in hospital: in mental defectives such movements are more frequent in blind than in seeing individuals (Berkson and Davenport, 1962).

The conditions conducive to the development of stereotypes are not necessarily the same as those leading to their appearance in animals which have already acquired them. In canaries the frequency of stereotypes is reduced by opportunity for interaction with other individuals (Sargent and Keiper, 1966), and in other species they are often immediately (as well as in the long-term) associated with "boredom": they are also often elicited in situations of novel stimulation or high anxiety (Nissen, 1956; Berkson et al., 1963; Hutt and Hutt, 1965, 1968). Berkson and Mason (1964a, b) report that rocking and swaying, but not repetitive head movements or other non-repetitive stereotyped acts, are increased by white noise levels and food deprivation. In voles a turning movement, once acquired, may be elicited by practically any sort of environmental variable (Fentress, 1965). Suggested functions are the reduction of tension or anxiety (Hines, 1942) and the modulation of excessive stimulation from other sources (de Lissavoy, 1963).

22.9 SUMMARY

1. The activities of young animals are often not related to the motivational factors associated with them in adulthood; and the motivational bases of apparently simple activities often have a complex developmental history.
2. Fear and withdrawal behaviour are profoundly influenced by experience through a variety of mechanisms. In addition, non-specific stimulation in infancy, such as that produced by handling or mild electric shock, may have profound influences on a wide variety of juvenile and adult characteristics, and especially on "emotionality" and other related aspects of behaviour.
3. The effects of experience on fighting behaviour are discussed.
4. Motivational aspects of sexual behaviour are influenced by hormones at at least two stages in development. Sexual behaviour is also much affected by the social conditions of rearing.
5. The relations between the feeding movements of young animals and the consequences of food deprivation and ingestion are discussed. Some ways in which experience may affect the feeding behaviour of adult animals are mentioned. Some similar effects are found with hoarding behaviour.
6. Some problems concerning the development of maternal behaviour in rats, and the formation of stereotypes, are mentioned briefly.

23

DEVELOPMENTAL ASPECTS OF LEARNING

23.1 THE OCCURRENCE OF AGE CHANGES IN LEARNING ABILITY

Although it is now well established that simple learning can occur at or before birth or hatching (e.g., in chicks, Schneirla, 1966; rats, Thoman et al., 1968; human neonates, Lipsitt, 1967), it also appears that learning ability improves with age. For instance, adult cuttlefish given prey-objects which are inaccessible inside glass tubes learn more rapidly than do young animals not to attack them: it seems to be impossible to train *Sepia* of less than 3 or 4 weeks old not to attack their prey under conditions in which adults learn rapidly (Wells, 1962a, b). In chicks, the ability to acquire an active response (moving from a cold place to a warmer place where there is some food) is present on the day of hatching, but a passive avoidance response (absence of approach) cannot be acquired until the third day (Peters and Isaacson, 1963; Fischer and Campbell, 1964).

While such experiments demonstrate differences in speed of learning under the conditions used at each age, their significance for changes in learning ability is not so clear. Not the least difficulty here is that of equating motivational conditions across ages. For instance some studies of age changes in young chicks involved thermoregulation, which changes markedly with age (see also Zolman and Martin, 1967; Zolman and Becker, 1968). The difficulty is particularly acute with feeding responses: young animals have a higher surface/mass ratio, higher activity level, and higher metabolic rate than older ones, so that it is nearly impossible to equate amount of food deprivation between young and adults (Campbell, 1967). However, in a search for a way to equate motivation and reinforcement across ages, Campbell (*loc. cit.*) found that, at moderate shock intensities, the change in shock intensity necessary to produce either avoidance of or preference for a specified level of shock by rats is more or less independent of age. Using shock avoidance, he was then able to show that there is little change in speed of learning between 25 and 175 days, though retention is better in the older animals (see also Campbell and Campbell, 1962). Similar age relations were found in a study of avoidance conditioning by Kirby

(1963) and in a discrimination problem with food reinforcement (Campbell, 1967). Henderson (1965) likewise found no age differences in the acquisition of a fear response by mice, though there are differences in the rate of extinction (Denenberg and Kline, 1958).

Such results indicate that there is little change in speed of acquisition in simple learning tasks, though retention develops with age. Campbell (1967) stresses that the latter finding does not mean that early learning experiences are not as important as adult ones. For one thing, normal development involves the acquisition of sequentially dependent responses, so that failure to learn one may adversely affect the subsequent acquisition of others. For another, occasional experiences may serve to reinforce and retain early conditioned fear or avoidance responses. He emphasizes that the generality of his conclusions for different species, different ages, and different responses remains to be assessed.

In addition to the problem of equating motivation and reinforcement across ages, another source of difficulty has been the presumption that learning abilities change in a simple manner with age. In discrimination learning this is not always the case. For instance Cane and Vince (1968) found age changes which were far from simple in the performance of quail (*Colinus virginianus* and *Excelfactoria chinensis*) on discrimination tasks. In learning to discriminate between white and grey boxes, the former species gave a performance curve which declined from 0 to 5–6 weeks and then rose. This suggests that performance is governed by the interplay of a number of different factors.

With more complex tasks, learning ability certainly does change markedly with age. Detailed data on a variety of tasks are available for rhesus monkeys (Zimmerman and Torrey, 1965). Classical conditioning and simple discrimination learning are possible very early in life, and do not change radically after the first few months of life in laboratory-reared animals: for instance brightness and colour discrimination learning improves little after 20 to 30 days of age, but size, form, and depth discrimination improves at least up to 90 or 120 days of age. In contrast to these relatively simple tasks, learning set formation (see pages 661–662) is not apparent in 90- and 120-day-old animals, and 1-year-olds are inferior to adults. There is a similar marked improvement over a considerable age span on such tasks as delayed response learning and patterned string problems (see also Mason and Harlow, 1958; Harlow, 1959a, b; Harlow and Harlow, 1962, 1965). The age changes therefore depend in part on the nature of the problem.

23.2 THE BASES OF AGE CHANGES IN LEARNING ABILITY

Insofar as performance in tests of learning or retention does change with age, what factors could influence the changes? In the early stages, learning may be limited by growth and differentiation in the nervous system or the effectors. The development of behaviour in kittens, and the ease with which conditioned responses can be established in puppies, is paralleled by the progress of myelinization (Langworthy, 1929; Harman, cited by Scott, 1954). In the human infant learning to walk or control micturition clearly requires a certain degree of muscular development. The poor ability of *Sepia* to learn not to attack prey during the first few weeks of life is probably related to the late development of the vertical lobe of the brain (Wells, 1962b).

In other cases the ease of learning may be affected by the development of perceptual abilities as discussed earlier (Chap. 20), or by changes in the response repertoire. Tinbergen (1942) records that immature Eskimo dogs, which do not defend territories, seem incapable of learning the territories of others: but when they begin to show territorial behaviour, they learn to avoid the territories of others very quickly.

Another issue is the changes which occur with age in inquisitiveness and persistence. For instance, rats show increasing exploratory behaviour in a Y maze from 50 to 116 days of age (Williams et al., 1966), but a decrease in exploratory sniffing in a standard situation from 15 to 24 months (Goodrich, 1966). Welker (1956) found that young chimpanzees (1 to 2 years) are more cautious in approaching or touching small stationary objects than are older ones. The initial responsiveness of 3- to 4-year-olds was similar to that of 7- to 8-year-olds, but in the latter case it waned more quickly.

Vince (1961a) found changes of this sort to have an important influence on the ability of passerine birds to pull up food suspended from a perch by a piece of string. Juveniles were more successful than adults, in part because of the greater time they spend responding to the bait in each test, and in part because of the greater variety of methods they tried. Vince (1960) showed that age changes in learning ability cannot be interpreted in terms of a unitary learning capacity, but must be analysed in terms of a number of characteristics, each of which has its own developmental history. In these birds both responsiveness to new stimuli and the stability of unreinforced responses rise to a peak and then fall, but at different rates (Fig. 23.1; Vince, 1961a, b; see Fox and Spencer, 1967, for a similar approach to age changes in delayed response learning in dogs). In birds (Vince, *loc. cit.*), rats (Sackett, 1967; DeNelsky and Denenberg, 1967a, b), and primates (Menzel et al., 1963), responsiveness to novel stimuli is much influenced by the conditions of rearing.

Stimulated by Hebb (1949), a number of workers have studied the effects of rearing conditions on various aspects of later learning ability (cf. pages 481–482). Although differences may not be found with simple tasks, such as the discrimination of horizontal and vertical striations by rats (Woodruff and Slövak, 1965), they may be marked with more complex ones. Rats and dogs reared unrestricted in a rich environment perform better on maze and other learning tests than those reared under restricted conditions (e.g., Bingham and Griffiths, 1952; Forgays and Read, 1962; Hymovitch, 1952; Thompson and Heron, 1954; Woods, 1959; Nyman, 1967). No doubt several factors are involved. One is the greater novelty of the test situation for the restricted animals, resulting in a greater amount of fear and exploratory behaviour (Zimbardo and Montgomery, 1957; Woods, 1959). Retardation in the development of perceptual abilities produced by the restricted en-

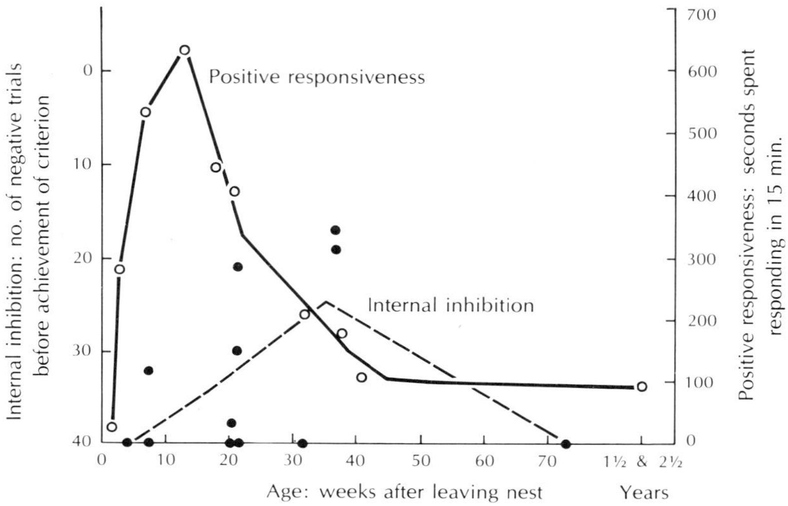

Fig. 23.1 Changes in positive responsiveness and negative responsiveness of hand-reared great tits to small coloured objects. (From Vince, 1960.)

vironment is probably also an issue. This is suggested by the finding that the age at which access to the richer environment is permitted is important: rats reared in a large space at an early age are superior in maze performance to rats placed in the large space later. Forgays and Read (1962) similarly found that 3 weeks' experience in a more complex environment given just after weaning was more effective in enhancing subsequent maze performance than when given either earlier or later: it is, of course, hardly surprising that the group placed in the complex environment for the first 21 days of life did not profit much from it, since during much of this time they would not have been mobile or visual. Nyman (1967), however, studying the effects of exposure to a free environment on spatial discrimination, found a rather later optimum: the free environment had a greater effect at 50 to 60 days than at 30 to 40 or 70 to 80 days.

Although experience at a relatively early age is thus most beneficial, Woods (1959) found that the detrimental effect of early sensory and motor deprivation was reduced by subsequent exposure to an enriched environment. Moreover the relation between the nature of the early experience and the subsequent test conditions is an important issue. Walk (1958) compared the performance on an elevated maze of two groups of rats: one had been reared in a large cage with opportunity to see and have commerce with objects placed round it (visual-motor group), while the other had been restricted to the centre portion of a similar cage by transparent plastic sides (visual group). Tested in the light, the former group was superior, but no differences were found when tests were carried out in the dark. (This experiment replicated an earlier one by Forgus (1955), but obtained different results: see also Forgus (1958).) Other relevant studies are cited in Chap. 20.

It has also been claimed that there is an influence of age and experience on the effects of brain lesions on learning. Tsang (1937) found that rats operated on at a young age were less affected than rats given similar lesions when older, though his results were not confirmed by Wilcott (1967). A number of studies, however, show that rearing conditions influence the deficit resulting from brain lesions. A large anterior cortical lesion has a greater effect on the performance of cage-reared rats in a learning test than a small posterior lesion, as predicted from the law of mass action. With rats reared in a free environment, on the other hand, the smaller posterior lesion produces the greater deficit (Smith, 1959; see also Schwartz, 1964). More recently, rats reared in isolation have been shown to differ in a number of aspects of brain chemistry and morphology from those reared in a more complex environment (Rosenzweig et al., 1962; Diamond et al., 1966; Rosenzweig and Leiman, 1968).

Yet another aspect of early experience which has been shown to affect later learning ability is nutrition. It is unfortunate that this basic factor has so far been so little studied (but see, e.g., Lát, 1967b).

23.3 SENSITIVE PERIODS FOR LEARNING

The concept of sensitive periods during which organisms are especially susceptible to particular environmental influences is of wide applicability, being useful to embryologists, clinicians (e.g., Kalter and Warkang, 1959), psychologists, and many others. The utility of the concept of a sensitive period for the effects of environmental stimulation on subsequent emotional behaviour and other aspects of development was mentioned in the previous chapter. Our present concern is with the particular case of sensitive periods for learning. We have already discussed two examples: the imprinting of the "following" response of young nidifugous birds is normally limited to the first few days of life (pages 513–522), and some species of song birds cannot add new songs to their repertoire after a certain age or stage (pages

451–462). As another example, Hess (1964) found that when the response of pecking at particular stimulus configurations was reinforced with food in 3-day-old chicks, it was not readily subject to extinction: if older or younger chicks were given similar treatment, they readily reverted to their initial pecking preferences when reinforcement was removed.

In general, it is a useful working assumption that no particular case of learning would occur with equal facility at all stages of the life cycle. In cases where a learnt modification to behaviour occurs most readily in a fairly restricted stage of development, we can refer to that stage as a sensitive period. Viewed in this way, the sensitive period refers not to a sharply defined period during which the learning can occur and outside which it cannot, but to the result of a gradual change in the ease or probability of learning. Furthermore, the period during which any specified type of learning can occur most readily is not rigidly determined but may itself be influenced by experience. The term "sensitive period" is thus to be preferred to "critical period", sometimes used in this context.

The latter term has in any case been used in a quite different sense (e.g., Scott and Marston, 1950; Scott, 1958, 1962; Williams and Scott, 1953) to refer to periods of an organism's life in which diverse aspects of its behaviour are particularly susceptible to modification. For instance Fox (1962) defines "critical period" as "the time when early experience has the greatest effect on later behaviour". This is perhaps useful when interest centres on correlations between dramatic changes in the organism's behaviour and changes in the central nervous system. But it can also be misleading, for the various types of behaviour in an organism's repertoire are susceptible to modification at different stages. The sucking behaviour of human babies is much influenced by what occurs on the first few occasions they are given the breast: whatever is learnt then is resistant to subsequent modification (Gunther, 1955). The learning of the characteristics of the mother's face which elicit smiling occurs somewhat later (Ambrose, 1960), and that involved for instance in toilet training, reading, and many aspects of social and sexual behaviour progressively later still (e.g., Gesell, 1954; Piaget, 1952; Bowlby, 1952). Furthermore, sensitive periods are not characteristic solely of young animals. In mother goats and sheep there is a brief sensitive period after parturition during which the characteristics of the young are learnt: any infants which approach after that time are repelled (Collias, 1956; Blauvelt, 1956; Klopfer and Gamble, 1966). Goats separated from their young for an hour after birth fail to show normal individual-specific maternal behaviour (Hersher et al., 1958), though a dam will sometimes accept a kid after this sensitive period if forcibly confined with it (Hersher et al., 1963). Thus, although the general learning abilities of an organism may change with age (Harlow and Harlow, 1962; Welford,

1962), the concept of sensitive period is most usefully applied to specified items of behaviour, and not as a global term to the organism as a whole.

Scott's use of the concept of sensitive period has also been criticized on rather different grounds by Schneirla and Rosenblatt (1963). These workers found that the social approach made by a kitten to its mother before feeding undergoes a course of development in the litter situation which is different from that found in isolated kittens fed from an artificial mother. Isolation of kittens for specified periods at different ages showed that the isolates behaved differently from controls who remained with the litter, at whatever age the isolation occurred: experiences in the litter situation must be continuously modifying the approach behaviour (Rosenblatt et al., 1961). Such a finding is not compatible with Scott's picture of critical stages in which learning occurs, but is compatible with the view that the particular characteristics of the natural approach behaviour are shaped by a continuous interaction between organism and environment which changes in nature as development progresses.

Even when used in this way, the term "sensitive period" may involve a considerable simplification. We have already seen how the responsiveness of great tits to small brightly coloured objects at first increases with age, reaches a maximum at about 15 weeks, and then declines. These changes are influenced by experience (Vince, 1960): the age at which the learning involved in this response occurs most readily also varies with experience. Sensitive periods therefore do not occur at fixed stages in an animal's development, and it is necessary in every case to search for those factors which change with age and influence the ease with which learning can occur. This may involve a detailed analysis of the ontogeny of behaviour, and a study of the interactions between the organism and its environment at each moment in its development (Schneirla, 1956).

Another complication is suggested by an experiment by Denenberg and Bell (1960), though it is concerned not with a sensitive period for learning but with a sensitive period to an environmental influence which affected later learning. They studied the effects of stimulating mice at 2 to 3 days, 8 to 9 days, and 15 to 16 days, with various strengths of electric current, on avoidance learning in adulthood. The age at which the stimulation was administered was clearly important, but the age at which the stimulation had the most marked effect varied with the current strength. Studies such as this (see also, e.g., Denenberg, 1962, 1964, 1968) suggest that even to consider each type of behaviour to have a sensitive period for each kind of environmental influence is an oversimplification: nevertheless "sensitive period" remains useful as a descriptive term.

The factors which limit sensitive periods for learning are as diverse as the types of behaviour involved. Some factors which may set the earlier limit to a sensitive period were mentioned on page 514–515: of more interest are those which bring it to an end. One is a change in the response repertoire of the organism. Razran (1935) has suggested that certain types of simple conditioning experiments on children become increasingly difficult with age because of the increasing self-consciousness and sophistication of the subjects.

We have already seen that the learning involved in the "following" response of nidifugous birds eventually becomes impossible because of the overriding fear of the strange object or environment used to elicit the response (pages 517–522). Here the sensitive period for learning is in fact brought to an end by the learning process itself—learning involves building up a schema of the familiar, and it is deviations from this which elicit fear (see Bateson, 1966, cited on pages 517–518). Similar processes may limit the sensitive period for the contact response of young mouth-breeding cichlids (Bauer, 1968). On the response side, the ability of a chaffinch to learn new songs seems to be brought to an end by the learning of the songs used in its first breeding season (Thorpe, 1958), and the ability of chicks reared in darkness for a fortnight to learn to peck seems to be prevented by the intrusion of previously learnt incompatible responses (Padilla, 1935).

The problem of sensitive periods has recently been reviewed elsewhere (Thorpe, 1961a; Hinde, 1962a), and the examples cited here are intended to emphasize one point only, namely, that the problem of sensitive periods for learning is the problem of the ontogeny of behaviour itself. Only when the changing behaviour and abilities of the organism have been analysed in terms of its interaction with the environment at each stage can the factors limiting sensitive periods be understood. The problems involved are important not only from a theoretical viewpoint. Since development depends on an interplay between organism and environment, disturbance in the latter (and we may here include all forms of sensory input, proprioceptive as well as exteroceptive) may have far-reaching repercussions. Often there are regulating mechanisms to minimize such effects, but an environmental deficiency during a sensitive period may be difficult to correct. The disabilities of deaf children are likely to be cumulative as communication by language becomes more important, affecting diverse aspects of behaviour and personality, but the effects can be minimized if appropriate measures to establish such communication as circumstances permit are taken sufficiently early (Lewis, 1963).

23.4 SUMMARY

1. Present evidence indicates that differences with age in performance on simple learning tasks are small if motivation and reinforcement are controlled, though retention does improve with age. Performance on more complex tasks does change with age.
2. The bases for age changes in learning are discussed. These include the development of the nervous system and effectors and the effects of rearing on perceptual and motor development.
3. When learning does occur most readily at a particular stage of the life cycle, we can refer to that stage as a sensitive period. Sensitive periods are usually not sharply defined, and some limiting factors are discussed.

24

SOME ASPECTS OF LEARNING

Learning, perhaps the most interesting and intensively studied group of processes involved in the development of behaviour, will not be discussed at length. Rather we shall concentrate on a few controversial issues which merit reconsideration in the light of contributions from workers outside the mainstream of learning theory, and which are particularly relevant to other problems we have been considering.

An extensive review of learning in submammalian species has been given by Thorpe (1963a); and the intensive work on learning in pigeons, rats, and monkeys is summarized in many textbooks. Estes et al. (1954), Hilgard (1956), Koch (1959), Hilgard and Marquis (1961), and Bullock (1966) may be consulted for summaries and critical appraisals of the theoretical views which have arisen from this work. Recent findings concerning the cellular basis of learning are reviewed by Kandel and Spencer (1968).

24.1 WHAT IS LEARNING?

While there is usually agreement as to whether any particular change in behavioural organization is an example of learning, there is no generally accepted definition. Differences between definitions largely reflect the preoccupations of the authors. The psychologist uses the term with reference to a relatively permanent change in behaviour, the physiologist to an inferred change in neural substrate. The biologist may limit its use to "adaptive" change, while the psychopathologist uses it to include the acquisition of tics and motor stereotypes. Most definitions, however, utilize exclusion—learning refers to changes which cannot be understood in terms of maturational growth processes in the nervous system, fatigue, or sensory adaptation. Definition by exclusion is seldom wholly satisfactory, and here this is especially the case. For instance, maturation is difficult to define except either by exclusion of learning, in which case the argument becomes circular, or in terms of growth processes which may not be separable from the as yet unknown processes on which learning depends.

A second difficulty is that, although it is rarely stated explicitly, learning is usually taken to refer to changes in the central nervous system. Many behavioural adjustments involve both intra-neural and extraneural changes which interact with each other, so that a change in behaviour produced by experience may entail both. Furthermore, while with higher animals it is convenient to restrict learning to cases believed to involve change in the nervous system, the question of whether modifications in behaviour shown by a single-celled animal are the result of learning then becomes partly a matter of definition. While it is certainly important not to entangle our definition of learning with any provisional models we may have as to how it occurs (Pantin, 1964), if we stretch our definition too widely, learning ceases to be a valuable concept (Hinde, 1968).

Yet another difficulty is that learning is usually taken to refer only to relatively permanent effects on behaviour. We have seen, however, that it is sometimes difficult to separate the effects of exposure to a stimulus according to their degree of permanence (Chap. 13). The waning of the owl-mobbing response of chaffinches, for example, depends on a number of different processes which recover at different rates (pages 298–305). In addition, permanent changes may depend on short-term effects (see below).

In practice, this difficulty of definition is unimportant, except in one respect. When it is glossed over, there is inevitably a tendency to regard learning as homogeneous, a single category of changes requiring but one explanation and to be understood in terms of a single set of laws. As long as learning is defined by exclusion, the onus of proof rests with those who hold it to be a homogeneous category.

24.2 HOW MANY KINDS OF LEARNING?

In discussing our own behaviour, the processes for which we use the term "learning" are diverse: we talk about learning to ride a bicycle, to judge distances, to recite a poem, to translate French, to find our way home. At the behavioural level of analysis, at any rate, these are pretty diverse. Since different animals achieve the same results in different ways, the diversity of learning processes is likely to be even greater when cross-species generalizations are involved. Indeed, it is even possible for distinct memory systems to arise in the same animal, for in *Octopus* visual and tactile memories depend on anatomically distinct parts of the brain (J. Z. Young, 1965). Nevertheless, much effort has been made to relate all examples of learning to one or two basic types. The problem here is partly one of the level of analysis. All learning, however broadly defined in terms of behaviour, could depend on similar processes at a subcellular level. But since we know little about these processes, and since learning is in any case usually defined behaviourally, it is necessary to assess the degree of diversity of learning as studied at the behavioural level.

1. Short-term versus Long-term Storage

In the first place, a distinction is often drawn between short-term and long-term or permanent storage processes. This distinction is based partly on the observation that a person who has been concussed loses his memory for events which occurred just before the accident, but not for earlier ones. This has been paralleled in the laboratory. Rats were given electroconvulsive shock at various intervals after each training trial on an avoidance conditioning task: those animals which received the ECS within a few minutes of the trial subsequently performed less well than those which received it later (Duncan, 1949; Gerard, 1961; McGaugh, 1965; Chorover and Schiller, 1966). In addition to ECS, similar effects have been found after a number of other treatments, such as certain anaesthetics and "spreading depression" of cortical activity.

Such evidence has been used to support the view that the consequences of an event first affect a short-term storage system, thus producing a "short-term memory", which is susceptible to disruptive treatments. Only later is the experience stored in the long-term memory, which has much greater stability. The ECS or other treatment is held to disrupt the short-term memory, thus preventing permanent storage.

The short-term storage is held to be distinct from the very short-term preperceptual trace which decays in about ½ sec: a number of studies on human subjects have shown that material must be responded to during its presentation or within this ½-sec period afterwards if it is to be subsequently available for recall (e.g., Averbach and Sperling, 1961; Mackworth, 1962). Short-term storage is also held to be distinct from long-term storage. For instance, it is said to involve activity in neural mechanisms which decays autonomously with time, and to be susceptible to interruption by mechanical, electroconvulsive, and other stimuli, as discussed above. Long-term storage, on the other hand, is believed either to be permanent or near-permanent, with the stored associations subject to only slow decay by disuse. It is usually held that long-term storage requires the preceding short-term neural activity for its consolidation (Hebb, 1949). Broadbent (1958) has emphasized that the mechanisms of short-term storage have a limited capacity which can be overloaded if the rate of presentation of information exceeds a certain level, while the long-term storage is effectively infinitely expandable. In addition, Broadbent (e.g., 1966) has argued that short-term memory can be affected by anything which distracts attention, while the effect on long-term memory of material intervening between training and recall depends on the nature of that material.

The evidence on which the distinction between short-term and long-term storage depends has been drawn primarily from studies of rat and human subjects. That it may be applied even outside the vertebrates is suggested by some experiments with *Octopus*. A blinded *Octopus* soon learns to reject inedible objects by means of tactual cues. If the object is presented in a rapid series of tests to one arm, it comes to be rejected by that arm but not by other arms. If the tests are spaced, however, the effect spreads to other arms, none of which will accept the object. Wells (1959) thus suggests that the changes involved in learning related to one of the arms take some time to spread through the supraoesophageal lobes of the brain.

Although a distinction between short-term and long-term memory traces has thus been valuable in a wide variety of contexts, a number of difficulties arise. First, the interval after initial learning at which the treatment is still disruptive varies greatly between studies. This is partly a function of the duration and strength of the ECS: for instance Alpern and McGaugh (1968) found that 0.2 sec of shock impaired memory only when given immediately after training, while 0.8 sec was effective 3 hours afterwards. It may also be a function of the nature of the treatment: carbon dioxide is said to be effective for longer than ECS (Paolino et al., 1968; see also Kincaid, 1968). This suggests that the different treatments may act in different ways and makes it less easy to postulate a single process of transference from one memory store to another.

Second, the mode of action of the ECS is open to question. It has been suggested, for instance, that it acts by producing conflict (Coons and Miller, 1960), or competing conditioned responses, or Pavlovian inhibition (Lewis and Adams, 1963; Lewis and Maher, 1966). While there is strong evidence that it is the shock, and not the convulsion, which is responsible for the amnesia (Lee-Teng, 1967; Jarvik and Kopp, 1967), it seems probable that the ECS produces amnesia directly for events up to 10 sec previously, while the longer-term effects are due to other consequences of repeated ECS treatment (Chorover and Schiller, 1965; Lewis et al., 1968).

Third, there is so far little clear evidence that the disruptive treatment affects storage: it could equally well impair recall. As Weiskrantz (1968b) points out, there is a need here for further studies in which the interval between treatment and testing for retention is varied. If retention increases with this interval, this would constitute strong evidence for thinking that it is the availability of the memory trace which is affected.

Finally, a number of studies, particularly some concerned with the storage of verbal material by adult human subjects, suggest a continuity between short-term and long-term storage (Melton, 1963), or that consolidation takes place in several phases (Barondes and Cohen, 1966).

The status of the concept of short-term memory is thus still an open issue.

2. Classical and Instrumental (e.g., Operant) Conditioning

Most theorists have been concerned with only one or two animal species, and some have worked with the assumption that all learning is of one type. Watson (1919) and Guthrie (1935), for instance, both used a single principle of conditioning or associative learning which is related to the Pavlovian conditioning experiment. Hull (1943, 1952) likewise made reference to only a single type of learning. Nevertheless by introducing intervening variables he was able to give descriptions of a suprising wealth of behavioural phenomena.

There are, however, two different types of situations in which learning is commonly studied, and in association with these two basic forms of learning are often postulated (e.g., Maier and Schneirla, 1935; Schlosberg, 1937; Konorski, 1948). As an example, the distinction between classical and operant conditioning used by Skinner (e.g., 1938, 1966) may be considered. In the first place, Skinner recognized two kinds of responses: those which are immediately elicited by known stimuli (respondents) and those which, prior to conditioning, are not necessarily correlated with any identifiable stimulus (operants). The knee jerk to a tap on the knee is a respondent; the pressing of a lever an operant. Skinner relates this distinction between two types of response to different kinds of conditioning.

Respondents are said to be susceptible to classical or Pavlovian conditioning. For example, food presented to a hungry animal will elicit salivation. Conditioning then depends on a temporal correlation between a stimulus which originally does not elicit salivation (e.g., a tone) and the unconditioned stimulus (food). With repeated presentations of tone followed by food, the tone alone will come to elicit the response of salivation. It will be noted that the choice of response is determined by the choice of unconditioned stimulus. The paradigm is as follows:

$$S^2_{\text{(tone)}} \quad \rightarrow r_{\text{(responses not measured)}}$$

$$S^1_{\text{(food)}} \quad \rightarrow R_{\text{(salivation)}}$$

Responses not correlated with an eliciting stimulus are held to be susceptible to operant conditioning. Here the choice of response is not determined by an eliciting stimulus, but rather by arranging a temporal sequence between a response which the animal performs from time to time in the situation and a reinforcing stimulus. Say, for example, we want to condition lever pressing in the presence of a tone by a hungry animal. To begin with, neither the tone nor food elicit lever pressing, but once a lever press while the tone is on is followed by a food presentation, the food increases the probability that the animal will press again. The following stimulus-response chain will be established:

$$S^2_{(\text{tone})} \text{—}R_{(\text{lever press})} \longrightarrow S^1_{(\text{food})}$$

Although, as conditioning proceeds, the tone will set the occasion for a bar press in a temporal sequence similar to the tone's elicitation of salivation, the operations which went into building up the two relations are quite different. Since the operant response acts upon the environment, operant conditioning is subsumed under the heading "instrumental conditioning". It need not involve a freely repeatable response like lever-pressing: the response can be any behaviour which produces a reward, such as running a maze. It will be noted that the distinction between classical and operant conditioning is made primarily in terms of procedure: the extent to which the underlying processes differ is an open issue (Hilgard and Marquis, 1961; Rescorla and Solomon, 1967; see also pages 579–580).

One of the earliest examples of instrumental conditioning was provided by Grindley's (1932) work with guinea pigs. The animal was placed in the experimental situation, which included an auditory signal. Whenever it made a certain movement (for instance, turning of the head), it was rewarded with food. Head movements then became more frequent in the experimental situation. Thus the stimuli indicative of the experimental situation set the occasion for the response, but did not elicit it in the same manner as, say, light elicits pupillary contraction.

The operant situation devised by Skinner (1938) contains a mechanism which the animal can operate by a suitable response, usually pressing a lever in the case of a rat or pecking a disc in the case of a pigeon. A suitable reinforcer is delivered to the subject either with every response, or on a previously arranged schedule, e.g., every *n*th response, or on the first response after an interval of *n* seconds from a previous response or reinforcement. This "schedule of reinforcement" (see Ferster and Skinner, 1957) is of crucial importance in determining the subsequent rate and pattern of responding (Fig. 25.8). Withholding reinforcement results in a decrease in rate of responding (i.e., extinction). The rate and pattern of responding in extinction depends largely on the previous schedule of reinforcement: in general, if reinforcement presentations are spread out in training in any manner which makes the change to an extinction condition difficult to discriminate, resistance to extinction is high. However, it must be added that the generality of the view that the amount of responding in extinction is directly related to the similarity between stimulus conditions in training and extinction is not established (e.g., Kelleher, 1966).

While the above distinction between classical conditioning and operant conditioning serves to clarify procedural differences, it should not imply that either type of conditioning necessarily occurs in isolation from the other. That is, in many of the situations in which classical conditioning occurs, the response studied (e.g., salivation) is accompanied by skeletal responses which may be conditioned according to the operant paradigm: where they seem to be absent, the conditioning may be extremely difficult, though it has been achieved (Solomon and Turner, 1962; Yehle et al., 1967). Conversely, in many situations in which operant conditioning occurs, the operant is closely followed by a classically conditioned response. That an operant response can nevertheless be relatively independent of a classically conditioned response was demonstrated by Ellison and Konorski (1964). A schedule was arranged such that a dog had to respond repeatedly to one stimulus in order to obtain a second stimulus, which was then followed by food: the first stimulus came to evoke the instrumental response without salivation, and the second stimulus salivation. Instrumental and classical conditioning were thus separated.

A second point arising from the classical versus operant distinction concerns their relation to visceral and skeletal responses. It has been held that the classical procedure applies to visceral responses and the operant to skeletal responses. If visceral responses could be modified only by classical conditioning, they could be learned and maintained only with stimuli which have the previous ability to elicit the same visceral response. If they can be learned as instrumental responses, on the other hand, they could be acquired and maintained in a variety of ways. The issue is thus one of considerable importance. Miller (e.g., 1961c) has in fact repeatedly challenged the view that visceral responses are not susceptible to instrumental conditioning, and a number of recent experiments from his laboratory support his claim. For instance, thirsty dogs rewarded by water for bursts of spontaneous salivation showed a progressive increase in salivation, while other dogs rewarded for periods without salivation showed progressive decreases (Fig. 24.1): correlational data indicated that it was improbable that any skeletal response could have acted as an intermediary (Miller and Carmona, 1967). Even more conclusive evidence came from a study using heart rate rather than salivation. Rats were trained to press a bar for intra-cranial electrical stimulation (see pages 220–221 and 277–278). They then received the ICS for either relatively fast or relatively slow heart rates, and showed cardiac acceleration or deceleration, respectively. Since the experiments were carried out under curare, the possibility of mediation of the response by the skeletal musculature was ruled out (Trowill, 1967). In this experiment the changes in heart rate were fairly small, but Miller and DiCara (1967) showed that, by requiring the rats to meet progressively more difficult criteria for reward, it was possible to produce changes of 20 percent in the heart rate (Fig. 24.1). The rats also learned to respond discriminatively to cues that heart-rate changes would be rewarded. In subsequent experiments Miller (1969) and his coworkers (see also DiCara and Miller, 1968) have shown that instrumental learning of a wide variety of visceral changes is possible—heart rate, intestinal contractions, blood pressure, urine formation, and EEG voltage. In some cases at least the learning will transfer from the curarized to the non-curarized state.

Furthermore, from an embryological and phylogenetic point of view, the distinction between skeletal and visceral musculature is not as clear as

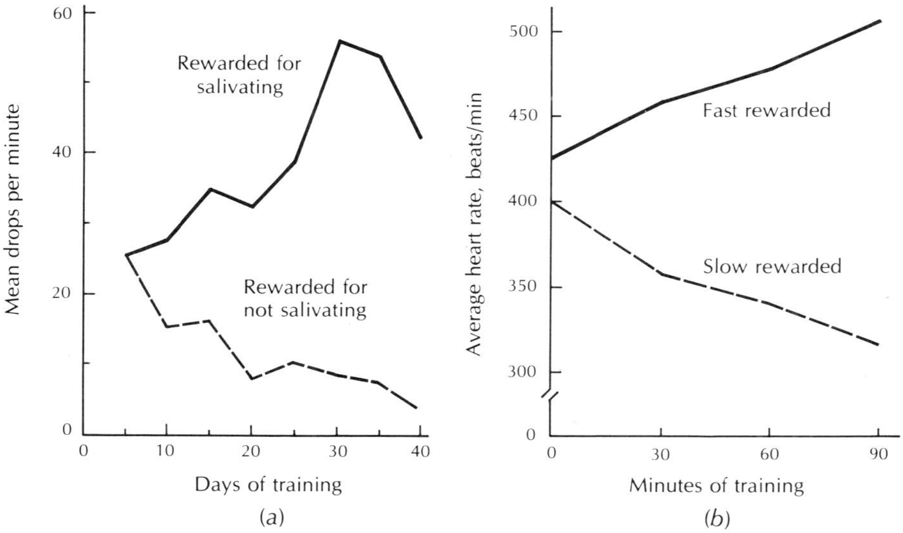

Fig. 24 1 a, Mean curves of instrumental learning by thirsty dogs rewarded with water for increases or decreases in spontaneous salivation. (From Miller and Carmona, 1967.) *b, Mean curves of instrumental learning by curarized rats rewarded with stimulation of the median forebrain bundle for increases or decreases in heart rate.* (From Miller and DeCara, 1967.)

it first appears (e.g., Goodrich, 1930): for instance, much of the gill musculature of fishes, though resembling skeletal muscle in structure, develops from the lateral plate mesoderm, like visceral musculature. On embryological grounds it thus seems improbable that their control should involve distinct learning mechanisms.

Although the distinction between operant and respondent behaviour has been useful in the laboratory, there is still much disagreement as to the extent of the difference between classical and instrumental conditioning.[1]

[1] See Symposium 4, 18th Int. Cong. Psychol., Moscow.

In any case, it seems unlikely that such a clear dichotomy exists in nature. Whereas in most laboratory experiments on operant conditioning there is initially no relation between the cue stimulus and the operant response, in nature, as we have seen, there often is a relationship between eliciting stimulus and response. For example, young nidifugous birds peck at a wide variety of objects which contrast with the background. Although pecking may not lead immediately to the ingestion of edible objects, the activity continues for a while. This is functionally important since it permits a discrimination between food and non-food objects to develop.

As stressed in Chap. 22, we still know little about the development and maintenance of such responses of young animals, but a study of licking in adult rats has been revealing. Rats normally lick in rhythmic bursts, a number of licks being elicited by a single drop of fluid. The licking rate within the burst seems to be independent of deprivation and of the sucrose concentration of the fluid. However, it increases with drop size, decreases with repetition, and can be classically conditioned. Such considerations suggest that the licking rate within a burst is best regarded as elicited behaviour with the characteristics of a phasic reflex. By contrast the overall licking rate (including both the bursts and the pauses between them) is affected by deprivation, sucrose concentration, and schedule of reinforcement. Thus these characteristics of drinking behaviour, as well as the initial approach to the drinking tube, are under close control of motivational and postingestational factors and follow rules similar to more conventional operant responses. It is thus apparent that some insight into the control of licking can be gained by regarding it as a mixture of respondent and operant behaviour. Even then, however, there are some anomalies: for instance licking rate decreases with increases in the size of a fixed ratio schedule, unlike the more usually studied pecking or bar-pressing (Hulse, 1967; Hulse and Suter, 1968). Further laboratory work with such types of stimuli and responses would be most profitable, both for the learning theorist and the ethologist.

3. Tolman's Capacity Laws

The theorists mentioned so far have been concerned in the first instance with a limited range of animal types. From the point of view of economy, their attempts to interpret diverse instances of learning in terms of few principles have been very successful. When a wider range of animals is considered, however, the learning phenomena become more obviously heterogeneous. There is a virtue in not losing sight of this heterogeneity. Tolman (1932, 1959) met this point not by reference to a large number of types of learning, but by relating performance in more-or-less standard learning situations to a series of capacity laws. These may be listed as follows (Hilgard, 1956):

(*a*) Formal means-end-capacities, that is, the ability to form an expectancy of a stimulus and act in accordance with it.
(*b*) Discriminative and manipulative capacities.
(*c*) Retentivity.
(*d*) Means-end-capacities needed for alternative routes, detours, etc.
(*e*) Ideational capacities, permitting, for instance, the comparison of alternatives.
(*f*) Creative instability.

Tolman thus took three types of learning experiments—classical conditioning; trial-and-error learning, as studied in a maze; and inventive learning. He then interpreted each of these in terms of the capacity laws, using, in addition, other laws relating to the nature of the material and the manner of its presentation. The simpler cases of learning require fewer laws than do the more complex ones: thus, classical conditioning requires only capacity laws (*a*), (*b*), and (*c*). Trial-and-error learning requires also (*d*), though (*e*) enters into the more complex cases. Inventive learning requires also creative instability (*f*).

4. Thorpe's Classification of Learning Types

Thorpe (1956, 1963a), who has provided the most wide-ranging review of animal learning, has met this problem of diversity in another way. Using a descriptive classification of learning types, he accepts that the underlying mechanisms may be different in different species, and thus leaves open the extent to which examples of the same type drawn from different phyla depend on similar mechanisms. His types are as follows.

(*a*) *HABITUATION*

This is defined as "the relatively persistent waning of a response as a result of repeated stimulation which is not followed by any kind of reinforcement". Habituation is further characterized as being specific to a particular stimulus: its relative permanence also distinguishes it from fatigue and sensory adaptation. A classical example is given by Humphrey's (1933) work on the snail's (*Helix albolabris*) withdrawal of its tentacles in response to mechanical shock. With repetition of the stimulus, the response gradually wanes. Comparable examples can readily be found in every phylum of the animal kingdom.

Three points about Thorpe's definition must be made. First, the phrase "relatively persistent", though ambiguous, is clearly intended to rule out the short-term waning phenomena discussed in Chap. 13. It may be noted that the term "habituation" is used by physiologists to refer to such short-term effects: in any case, the distinction between short- and long-term phenomena is not always easy to draw. Second, the reference to "repeated stimulation" should not be taken as excluding continuous stimulation. Third, the reference to the absence of reinforcement begs one of the most interesting questions about habituation: we shall return to this later.

Insofar as both involve a waning in response strength in a particular situation, there are resemblances between habituation and the extinction of an operant response. The distinction between them seems to depend primarily on the manner in which the response was acquired in the first instance, and it is perhaps for this reason that habituation was ignored by the earlier theorists. It must be noted, however, that extinction involves the removal of a reinforcing stimulus, while habituation is sometimes applied to cases of response decrement in which any absent reinforcement has not been identified.

Thorpe points out that, at any rate in the simpler cases, habituation involves merely a decrease in intensity or number of responses, and he thus regards it as the simplest form of learning. A waning of responsiveness with the formal properties of habituation, and lasting for days, can be found even with a reflex response in spinal rats (Griffin and Pearson, 1967). However, two types of evidence suggest that some cases of habituation may not be as simple as they appear. The first comes from Sokolov's (1960) work on the extinction of the orientation reflex in dogs discussed in Chap. 6. This comes within the above definition of habituation, but seems to imply some sort of central reorganization, rather than a mere decrease in response strength. Whether such results could also be obtained with animals at a simpler level of organization, such as snails, remains to be seen: it is at least likely that the mechanisms underlying habituation differ considerably up the phyletic scale. In any case, the interpretation of Sokolov's experiments is not clear (see Chap. 6).

The second line of evidence has also been mentioned earlier. Studies of the mobbing behaviour of the chaffinch show that response to a stimulus is accompanied by complex changes which may lead to both an increase and a decrease in response strength on subsequent occasions, and which decay at different rates (pages 298–305). If habituation is defined in terms of the waning of the response, it is the resultant of these processes, and thus far from simple. If, on the other hand, habituation is held to refer only to the relatively permanent decremental processes, it cannot be measured directly.

These points in no way diminish the importance of the previously rather neglected phenomena grouped under the rubric of habituation. But they should remind us that the phenomena are less simple than they appear, and may depend on mechanisms differing greatly in complexity. The relations between behavioural habituation and short-term neural activity are discussed in various contributions to Horn and Hinde (in press).

(b) *CLASSICAL CONDITIONING*

Conditioning can apparently occur simply as a consequence of simultaneous stimulation of two cortical areas (Doty and Guirgea, 1961; but see Wagner et al., 1967). Furthermore, it seems not to require performance of the unconditioned response: for example, dogs will salivate to a conditioned stimulus even if, during the training trials, salivation to the unconditioned stimulus was blocked by atropine (Finch, 1938); and a fear response can be conditioned under curare (Solomon and Turner, 1961). In any case Thorpe, like other writers, regards classical conditioning, as investigated by Pavlov, as an artificially isolated part of the learning process, and feels that, as usually conceived, it is unlikely to form the fundamental unit out of which more complicated learning processes are constructed. Thus he emphasizes that the reflex is in any case an idealized concept; that stimuli are never in fact simple, for higher animals always respond to the situation as a whole with the stimulus presented by the experimenter differentiated against it; and that the conditioned response is always in fact "anticipatory", the animal behaving as if "expecting" the unconditioned stimulus rather than just attaching the unconditioned response to the new stimulus.

The conventional view of the conditioned reflex is, in fact, a simplification. It may involve more than the association of a new stimulus with the unconditioned response, for the conditioned response is seldom the same as the unconditioned one, and often is only part of it (Hilgard and Marquis, 1961). If the restrictions on the animal's freedom of movement usually imposed in conditioning experiments are removed, the conditioned response is accompanied by behaviour which appears to the observer to be appetitive to the arrival of the unconditioned stimulus.

Finally, conditioning as usually defined may cover changes in behaviour of markedly differing degrees of complexity. It is hard to avoid the view that conditioning in annelid worms involves little other than a change in stimulus-response relations. Equally, however, it seems likely that the conditioning of salivation in man is more than a mere switching of connections, and involves a much more far-reaching reorganization of behaviour. Like habituation, therefore, classical conditioning, as defined behaviourally, covers changes which differ greatly in complexity (see also Zener, 1937; Hilgard and Marquis, 1961).

(c) *TRIAL–AND–ERROR LEARNING*

This is defined by Thorpe as "the development of an association, as the result of reinforcement during appetitive behaviour, between a stimulus or a situation and an independent motor action as an item in that behaviour when both stimulus and motor action precede the reinforcement and the motor action is not the inevitable inherited response to the reinforcement" (Thorpe, 1956, 1963a). Learning to run mazes and open problem boxes provide classic examples. In practice, Thorpe regards trial-and-error learning as always involving both classical conditioning and what he calls "instrumental" conditioning. He holds that pure instrumental learning would be as much an isolated part of the learning process as he already believes classical conditioning to be.

Behaviour which comes within the category of trial-and-error learning varies markedly in complexity. Changes which meet most of the criteria of trial-and-error conditioning have been found in a headless cockroach and even in an isolated small section of a cockroach containing a ventral ganglion, a pair of legs, and their associated muscles. The preparation was mounted over water in such a way that if the leg extended, a circuit was completed and an electric shock administered. After a few minutes the leg remained flexed, and the periods of flexion increased with successive stimulations. Yoked controls which received the same amount of shock, but at random with respect to leg position, showed no such changes (Horridge, 1963; Pritchatt, 1968). The effect was retained for several hours, but beyond that the permanence of the learning could not be assessed. In an extension of this experiment, Hoyle (1965) showed that control of the leg-raising muscle could depend only on the frequency of efferent impulses to that muscle. This was achieved by giving electric shocks to the efferent nerve when the impulse frequency in it was above or below certain limits. Thus the efferent frequency must be measurable by the insect, which must also be able to associate particular values with shock avoidance. This may play an important part in long-term postural adjustments (Horridge, 1968a).

Even in whole animals, the crude criteria of trial-and-error learning may embrace phenomena of markedly different complexity. Thus Schneirla (1959) has emphasized the fundamental differences between the learning involved when an insect runs a maze and that occurring in a rat in a similar situation. The insect first masters the individual choice points and later integrates the habits thus formed in a stereotyped way, while the rat starts its local learning and its overall integration together on the first run. Furthermore, once the maze is learned, the insects have only a limited ability to learn to run it in the reverse direction or under changed conditions, while rats can do so readily (see also Weiss and Schneirla, 1967).

Indeed, just as with classical conditioning, there is increasing evidence that generalizations about trial-and-error learning may lead to oversimplification. In many cases the learning cannot be described in terms of new S–R connections but requires the postulation of quite complex intermediaries: this is discussed later (pages 585–594).

(*d*) *LATENT LEARNING*

Thorpe defines this as "the association of indifferent stimuli or situations without patent reward". In most maze-learning experiments, a reward is present in the goal-box throughout training. If rats are allowed daily access to a maze which has no reward in the goal-box, the number of errors they make falls only gradually by comparison with animals that are rewarded. If, after a number of days, the previously unrewarded animals are given a reward in the goal-box, their performance improves very rapidly to a level similar to that shown by animals which have been rewarded throughout. Thus the initially unrewarded group must have learnt something during their early trials—their learning was "latent" and did not show up in performance until activated by the reward. This has been described by saying that the rat initially learns "what leads to what", in the same way as it learns the relations between objects when exploring a novel spatial array (see page 586). In more general terms this can be described by saying that the rat learns conditional probabilities of association between stimuli.

This classical latent learning experiment was the forerunner of many others which have been reviewed in detail by Thistlethwaite (1951) and MacCorquodale and Meehl (1954). These include studies in which the experimenter provided a reward, but one irrelevant to the animal's motivational state as controlled in the experiment. For instance, rats which are hungry but not thirsty are given forced runs through a *T* maze leading to food in one arm and water in the other. If subsequently tested when thirsty, they will tend to choose the water side: learning thus took place under an irrelevant motivation. After a critical review of the literature, MacCorquodale and Meehl (*loc. cit.*) concluded that, although some experiments had given negative results, the majority of experiments on latent learning appear to demonstrate its occurrence.

The essential element in Thorpe's definition of latent learning is absence of reward. This in itself, however, cannot serve to differentiate it from trial-and-error learning. As we shall see, some theorists do not regard reinforcement as essential for trial-and-error learning, so the distinction from latent learning disappears. Even for reinforcement theorists there need be no distinction: Hull (1952) came to regard habit strength as acquired independently of the magnitude of the reinforcement, so any "latent" learning could occur in relation to some postulated minimal reinforcement, such as return to the home cage at the end of the maze: it is thus not separable from trial-and-error (but see Watson, 1961).

It will be argued later that the relationship between latent learning and trial-and-error learning is indeed close. In higher organisms, at least, the latter could involve two processes: a learning of the characteristics of the situation, and a translation of this into performance. Latent learning involves only the former.

(*e*) *INSIGHT LEARNING*

This involves "the production of a new adaptive response as a result of insight", where insight is defined as "the apprehension of relations". The controversy as to whether insight learning differs in kind or only in degree from trial-and-error learning is an old one. The argument has been prolonged largely because insight learning was held by some to involve "ideation", while trial-and-error learning could perhaps be explained in purely connectionist terms. We shall discuss this problem again in the next section.

Thorpe includes here imprinting: as discussed previously (pages 513–522), the apparently peculiar characteristics of imprinting depend more on the context than on the learning process itself. He includes also "imitation," but is careful to point out that a number of behaviourally diverse phenomena are often subsumed under this label:

(1) *Social Facilitation.* This refers to the performance of a pattern of behaviour already in an individual's repertoire, as a consequence of the performance of the same behaviour by other individuals. Yawning is a common example in man. Since the effect is immediate, learning need not be involved, though of course the propensity to show social facilitation may depend on learning.

(2) *Local Enhancement.* This refers to an increased tendency to respond to part of the environment as a consequence of the response of another individual to it. For instance, the spread of the habit of opening milk bottles through a local population of tits (*Parus* spp.) probably depends in part on this (Hinde and Fisher, 1951). A more carefully analysed case is provided by Klopfer (1959b): a greenfinch (*Chloris chloris*) which has learnt to discriminate between a palatable and unpalatable food source starts to make errors when it sees an untrained companion doing so. The reverse also occurs, one bird learning to avoid a potentially dangerous situation by observing the behaviour of experienced birds (Klopfer, 1957), though this pro-

cedure is not always successful with stimuli which are only mildly aversive (Sexton and Fitch, 1967).

(3) *True Imitation.* Thorpe reserves this term for the "copying of an otherwise improbable act or utterance". Except for vocal imitation in birds (see pages 451–462), he believes this is limited to the primates (see, e.g., Hayes and Hayes, 1952; Kellogg, 1968).

These categories provide a useful means for the study of imitation, though many cases are difficult to assign to one or the other. For instance, Dawson and Foss (1965) found that a budgerigar (*Melopsittacus*) learning to take a lid off a pot tends to use the same movement as a more experienced bird which it has been allowed to observe: this however hardly fits into Thorpe's category of true imitation since the movement patterns were not novel. Again, van Lawick-Goodalls' (1968) description of the manner in which young chimpanzees learn to use stalks to probe for termites (see page 450) in the presence of their mothers suggests that all three categories of imitation are involved.

In conclusion, two points about this classification of learning types, both emphasized by Thorpe himself, must be repeated here. First, the categories intergrade. Second, each contains examples of greatly differing complexity. One of the reasons for these points is worth noting: each of Thorpe's categories may involve perceptual learning (pages 481, 586), and the extent and importance of this varies between responses and between phyla.

24.3 SOME CONSTRAINTS ON LEARNING

At a number of points in the preceding chapters it has been implied that there are constraints on what an animal will learn which are not solely ones of learning capacity. This is a point which has often been overlooked just because, for obvious reasons, the properties of learning are usually studied in situations devised to ensure that learning will occur. It is thus worthwhile to summarize briefly some of the ways in which an animal's learning ability may be restricted.

First, there are restrictions on the stimuli to which the animal responds, and which can therefore become conditioned. These depend in part on limitations of the sensory/perceptual mechanisms (Chaps. 4 to 6). Amongst the stimuli to which it can respond, it is more likely to respond to some than others; and it learns more about some than others (see Chaps. 20 and 21). But even amongst stimuli to which the organism is responsive in some contexts, learning ability may be restricted in others. For instance *Octopus* compensates appropriately for the weight of objects it picks up, but is unable to distinguish objects differing only in weight (see page 77). Furthermore, as we have seen in Chap. 21, there are species-, individual-, and situation-specific differences in the stimuli most likely to become conditioned or otherwise responded to. To cite three further examples, the chaffinch selects from the sounds it hears those characteristic of the species-song to reproduce in its full song (Chap. 19), and the digger wasp *Bembix rostrata* has a much greater tendency to learn characteristics of the environment in the vicinity of its nest-hole when it leaves the hole than just before it enters (van Iersel and van den Assem, 1964). Similarly bees can learn to orient by complex shapes when steering a course to and from the hive, but not when they are signals to indicate presence or absence of food on a tray (page 95).

Amongst the stimuli which can readily influence the organism's behaviour, responsiveness wanes more rapidly to some than to others (see Chap. 13). In such cases, one must assume that the rate at which particular responses habituate has been determined by natural selection. Even more important, and often lost sight of, is the fact that the situations which are reinforcing must also have been established by natural selection (Lorenz, 1966). The fact that food, water, or warmth is reinforcing is not an intrinsic property of the stimuli, but of the animal. That many reinforcers are species-specific becomes much more apparent when learning is studied in situations other than the conventional laboratory ones, and when reinforcing properties of stimuli other than food and water are investigated (Stevenson, 1969; and pages 597–599).

Indeed, species may differ in such a way that one learns in a context in which the other does not, even though there is no evidence of a difference in general learning capacity. An example is provided by the orientation behaviour of sunfish (*Centrachidae*), which are found only in the northern hemisphere, and the Cichlid *Aequidens portalegrensis*, whose range spans the equator. The former, after rearing in artificial light, can orient by the sun at their first exposure, provided the sun moves from left to right, but they cannot adjust to a sun moving as it would in the southern hemisphere. The Cichlid, by contrast, can orient by either a northern or a southern sun, depending only on which way the sun was moving when it first saw it (Braemer and Schwassmann, quoted by Lorenz, 1966).

Again, Egyptian vultures (*Neophron percnopterus*) are known to carry a stone to an ostrich egg and throw it repeatedly at the egg until it is broken (van Lawick-Goodall and van Lawick, 1966). No other related species of vulture has ever been observed to throw stones, though they obviously have the effector equipment for doing so. It is extremely unlikely that, even in an experimental situation, stone-throwing in these other species could be conditioned. A more well-known example is the impossibility of teaching chimpanzees to make full use of their vocal capacities (Hayes and Hayes, 1952; Kellogg, 1968).

Another case worth consideration here concerns a difference between sexes in what will be learned. It is well established that the opportunity to copulate is reinforcing for male rats: they will learn an instrumental response which leads them to a receptive female. Oestrous female rats, however, run no faster to sexually active males than to passive ones. From this it has been concluded that "mating does not reinforce instrumental behaviour in the female rat" (Bolles, 1968). But it remains at least possible that, had cessation of locomotion or even lordosis been the response on which mating had been contingent, a reinforcing effect would have been found.

Another case in which the course of conditioning was affected by the operant response chosen is cited by Sevenster (1968). Sticklebacks, reinforced by the opportunity to court a female, soon come to show a high frequency of swimming through a loop. The response of biting at a stick, however, reaches only moderate frequencies when reinforced in this way: during extinction, high response rates occur. Sevenster suggests that the low response rates during reinforced sessions were due to inhibitory aftereffects of courtship on the aggressive response of biting. When the reinforcer was the opportunity to behave aggressively to another male, high response rates were obtained.

Other restraints may be a consequence of maturation. For example, about 14 days after hatching, chaffinches hop over the ground, occasionally pecking at it. However, even in the midst of food, they will starve unless fed by a parent or fosterer. Training is impossible, since if food is gradually lowered towards the ground, the birds stop looking at it: while if it is held in front of them, they only gape (Stevenson, personal communication).

Finally, in the natural situation there may normally be environmental restrictions on what is learnt: nidifugous birds become imprinted on their parents rather than on other conspicuous objects because their parents are present and available.

Thus what an animal can learn is limited in a number of ways. Analysis of the constraints is an essential step in identifying the conditions necessary for learning to occur.

24.4 COGNITIVE AND S–R THEORY

In higher animals, at any rate, learning rarely involves a mere shifting of relationships between external stimuli and responses. Many theorists have therefore postulated events between stimulus and response, but there has been considerable disagreement as to their nature. Behaviouristically oriented psychologists have regarded them as covert stimulus-response sequences and have relied especially on kinesthetic stimuli as integrators of behaviour. "Cognitive" theorists, on the other hand, lay emphasis on the restructuring of central processes without regarding these central processes as basically similar to stimulus-response sequences. To put it at its simplest, a stimulus-response explanation of the learning of a maze would be given in terms of learning which responses must be made to which stimuli, while a cognitive explanation might be in terms of the learning of a "cognitive-map" of the maze such that the animal acquires a knowledge of "what leads to what", or of the probability that this stimulus will be followed by that. In practice, the results of most learning experiments can be interpreted in terms of either view, theorists of both types displaying remarkable ingenuity in accommodating the diverse data now available (Osgood, 1953).

The controversy is an old one, and all its ramifications will not be pursued here. It is worthwhile, however, to consider some experiments bearing on the nature of the central changes which occur in various types of learning situation.

1. Sensory Preconditioning

An early experiment provides a convenient starting point. An auditory stimulus was given to dogs at intervals, followed in each case by a visual one. Any learning which occurred was at least not apparent. Subsequently a response was conditioned to one of the stimuli. It was then found that the response could also be elicited by the other stimulus (Brogden, 1939). The early presentations must therefore have led to some association between the stimuli. Although some of the early studies on this "sensory preconditioning" were inadequately controlled, the occurrence of learning under these circumstances has since been well established (review by Seidel, 1959). While such experiments appear to demonstrate the formation of direct associations between stimuli, they pose no real problem for the S–R theorist: covert S–R sequences could serve as intermediaries. Indeed Parks (1968) has shown that the frequent repetition of two stimuli hinders the development of a learned association between them, and argues that this supports the view that the orientation response acts as a mediator in sensory preconditioning.

2. Perceptual Learning

More recently a number of experiments have shown that variations in perceptual experience during development, although producing no effects apparent in behaviour at the time, may have consequences which can be revealed in subsequent testing. For instance, we have already considered evidence indicating that rats, if given early experience of a circle and a triangle without differential reinforcement, will subsequently learn to discriminate between them more readily than do rats without such experience (see pages 481–483). Such processes, by which an organism's readiness or ability to respond to variables of physical stimulation are improved, are often referred to as "perceptual learning" (Gibson and Gibson, 1955). The importance of perceptual learning in a wide range of species has recently been stressed by Thorpe (1956, 1963a). He argues that even the simplest stimulus is a relational one, so that perception always involves relational properties. In his view species which require perceptual experience (see pages 467–499) are using it to build up, by various learning processes, the ability to respond to the significant patterns presented by the environment.

The concept of perceptual learning has not gone without criticism. In the first place, attempts to define perceptual learning have been inadequate (Bevan, 1961) and unoperational. The occurrence of perceptual learning can be detected only through a change in behaviour: whether or not it differs from learning in a classical or instrumental conditioning situation in any absolute sense is not yet clear. One suggested criterion is that, as perceptual learning proceeds, the organism responds more and more selectively to the particular type of stimulus, the implication being that in conditioning, by contrast, the animal generalizes progressively more readily to other related stimuli. However, no simple statement can be made as to whether continued training in the conditioning situation will increase or decrease the slope of a generalization gradient (e.g., Hilgard and Marquis, 1961). In any case, a conditioning situation which does not involve perceptual learning, and yet is operationally similar to the situations in which perceptual learning is studied, is hard to find.

A second difficulty concerns the claim that perceptual learning does not depend on reinforcement. For instance, Kerpelman (1965) found that preexposure of rats to stimuli had less effect on their subsequent performance in a discrimination learning situation if, during the preexposure period, the stimuli were removed for 2 hours daily around the feeding period. He thus suggests that differential reinforcement plays a role. However, as discussed on pages 483–486, the establishment of stimulus control appears to depend on differential reinforcement: so also does detection of perceptual learning, and the difference may be more apparent than real.

Be that as it may, the experiments on perceptual learning show that experience may produce changes which are not immediately revealed by changes in behaviour. This does not *necessarily* mean that the learning is not mediated by a response. We have already seen that, in higher mammals, the learning of the characteristics of visuomotor space involves comparison of the feedback stimulation received as a result of a response with an expected feedback (pages 489–499): learning is S–R–S rather than just S–S. Whether perceptual learning *always* requires mediation by a response (e.g., eye movements) is not known.

3. Two-stage Discrimination Learning

Changes in central organization which are not immediately revealed in behaviour may accompany learning as studied in many conventional contexts. Lawrence (1949, 1950) showed that an acquired ability to discriminate between two stimuli had some measure of independence from the particular response which the animal was required to make to them. For instance, when rats learnt to discriminate between cues in one situation, the acquired distinctiveness of the cues was transferred to new situations even when the significance of the cues was reversed. Lawrence interpreted his results in terms of a hypothetical mediating response which made the cues more distinctive (see also Estes, 1942, 1948; Schoenfeld et al., 1950).

More recently Sutherland (1959) has suggested that discrimination learning is a two-stage process: the animal first learns to respond to the quality by which the two stimulus situations are maximally differentiated—for instance it learns that the selection must be made in terms of brightness, or of shape—and then it learns to attach particular responses to the characters in question. Reward on any one trial is presumed to increase the probability that the animal will pay attention to the same character on later trials. (We have already noted that the term "attention" can be used as a cloak for ignorance, but it is used here to indicate that the animal's behaviour is being determined by one stimulus character rather than another, and is preferred to the terminology sometimes used in this context in which animals are said to "switch in analysers", each of which is supposed to analyse one aspect of the stimulus situation.)

The hypothesis that discrimination learning involves two stages permits statements about whether any particular training procedure strengthens one stage in relation to the other, and thereby encompasses results which would otherwise pose difficulties. For instance, if rats or *Octopus* are trained in a discrimination, and the cues are then reversed, they learn the new problem faster if they have been overtrained on the original task. On the present hypothesis, the first stage (learning the types of cues which are relevant) continues during the overtraining period, though this cannot be revealed in performance: when the cues are reversed, this learning is still relevant, and thus the overtrained animals are better equipped for the reversal (e.g., Mackintosh, 1964b).

It would be predicted that this overtraining reversal effect would be most marked when many irrelevant cues are present in training, and/or when the relevant cue is not conspicuous. In such cases the overtrained animals would continue to pay attention to the relevant one after reversal, while the non-overtrained animals would be more prone to switch attention to irrelevant cues. In a series of reversals the rate of improvement was indeed found to depend on the number of irrelevant cues (e.g., Mackintosh,

1964b). In *Octopus* also, reversal learning occurs more readily after overtraining on the initial discrimination if irrelevant cues are present (Mackintosh and Mackintosh, 1964).

Mackintosh (1963) likewise interpreted the extinction performances of overtrained and non-overtrained rats on the assumption that discrimination learning involves two stages, and that with non-overtrained animals extinction affects mainly the first stage (isolation of the relevant stimulus dimension), while with overtrained animals it affects the second (attachment of responses to stimuli). Other effects which can be interpreted in terms of this hypothesis are the deleterious effect of overtraining on a shift which does not involve reversal, certain properties of transfer of training along a continuum, and the partial reinforcement effect (see Sutherland et al., 1963, and reviews by Sutherland, 1964, 1966; Mackintosh, 1964b, 1965a, b).

There are, however, some experiments which appear to conflict with this hypothesis. First, there is some question as to the generality of the finding that overtrained animals perform better in reversal learning. Although a positive effect has been found in a number of experiments on rats and *Octopus*, and also on pigeons (Williams, 1967a), it is not invariably found even in rats (Lukaszewska, 1968); overtraining was found to *retard* reversal learning in experimentally sophisticated cats and monkeys (Beck et al., 1966), chicks (Mackintosh, 1965b), and fish (*Macropodus opercularis*, Warren, 1960; goldfish, Mackintosh et al., 1966). Indeed Hirayoshi and Warren (1967) did not find that overtraining facilitated reversal learning in an experiment using naive kittens and involving just those conditions (i.e., multiple irrelevant cues, relevant cues initially low in attention values) which Mackintosh had held to favour the effect.

However, Mackintosh et al. (1966) have suggested that such results may be due to the absence of a differential effect of the overtraining on the two processes involved in the learning. In support of this they have produced evidence both between and within species that when overtraining facilitates reversal, non-overtrained animals extinguish the first stage of learning faster than the response attachments, while when overtraining does not facilitate reversal, both stages extinguish at similar rates however extensive the overtraining. Furthermore, Mackintosh (1969) has confirmed that overtraining facilitates reversal learning with difficult discriminations and large rewards.

A second difficulty is presented by the finding that animals learn discriminations faster if two cues are present and relevant than when only one is relevant (e.g., Warren, 1953). It is usually assumed that the animals learn about each cue as fast as they would if it were presented alone. This is in conflict with Sutherland's theory in that he had supposed that the more an animal attended to and learnt about one cue, the less it could learn about others: the animal would therefore not learn so fast with two cues presented together as with one (e.g., Sutherland and Mackintosh, 1964). Support for Sutherland's view comes from the finding that if rats are trained with two relevant cues for a fixed number of trials there is a negative correlation between the amount individual rats learn about each cue (Sutherland and Holgate, 1966). A further prediction from Sutherland's hypothesis, that such rats would learn less about one of the cues than rats trained with only one cue relevant, received only partial support. When the animals were trained with two relevant visual cues they learnt more about one of them than did rats trained with only one relevant cue. However, if one cue was visual and one tactile, the group trained with both relevant learnt less about a single cue than did subjects trained with only one relevant (Sutherland and Andelman, 1967). In any case, this is a detail of the theory, not concerned with the basic view that discrimination learning involves two stages.

A third difficulty is that rats overtrained on a position discrimination do not learn the reversal faster if overtrained. Mackintosh (1965c) argues that this difference from visual discriminations is because irrelevant cues are usually present in the latter, but not in the former. In addition, he suggests that, with overtraining, control may be gradually transferred from external to proprioceptive control: reversal cannot occur until external control has been reestablished (see also Lovejoy, 1966).

Finally an experiment by Siegel (1967), while confirming some of the data on which the overtraining reversal effect depends, suggested that a two-stage learning theory was not wholly adequate. Rats were trained or overtrained on a discrimination in which two sets of cues (black versus white and textural) were present. They were then tested on either a simultaneous or a successive discrimination in which either the cues relevant in the initial training were reversed (intra-dimensional shift), or the other cues were relevant (extra-dimensional shift). Overtraining facilitated performance on the simultaneous intra-dimensional shift, as expected. However, it did not adversely affect the simultaneous extra-dimensional shift, as might be expected if the initial training involved learning the relevant stimulus dimension. Furthermore, while the successive intra-dimensional shift was learnt faster than the successive extra-dimensional, overtraining did not facilitate further the successive intra-dimensional shift or further hinder the successive extra-dimensional shift. Siegel watched the actual behaviour of his animals closely, and suggests that some of the anomalous results can be understood in terms of a necessity for the animals to overcome initial position preferences. His interesting data, however, supplement rather than disprove the two-stage learning viewpoint.

In summary, the two-stage view of discrimination learning has considerable support, and some of the possible objections can be met. While attempts at formalization (e.g., Lovejoy, 1966) are premature, the evidence is strongly against the interpretation of discrimination learning in simple stimulus-response terms, at any rate in rats and octopuses.

4. Place Learning

Evidence that some aspects of maze running can be understood more easily in cognitive than stimulus-response terms is provided by experiments on "place-learning". The cruder stimulus-response theories would suggest that an animal which has learnt a maze moves from start to goal according to a fixed sequence of movements. However, in some experiments interference with the movement sequence does not prevent attainment of the goal. Rats which had learnt to walk through a maze subsequently demonstrated their learning when severe cerebellar damage caused gross abnormalities in their locomotor pattern (Lashley and Ball, 1929).

If an elevated maze is arranged in such a way that rats have to learn either a spatial habit (i.e., always to go to the same place, irrespective of whether this involves turning to right or left) or a response (e.g., turning left), the former is the more easily learnt if cues from outside the maze are readily available (Tolman et al., 1946, 1947). If such cues are not available, however, response learning is more important (e.g., Blodgett and McCutchan, 1948). Maze-running may thus involve either the learning of spatial habits or that of responses or both: the relative importance of each varies with the cues available, and also with the stage in training reached (Kendler and Gasser, 1948). Other experiments have investigated the extent to which rats could make use of alternative paths when the main route to the goal was blocked: on the whole they favoured the place-learning hypothesis, though they also indicated that acquisition of the "cognitive map" is surprisingly influenced by such variables as "alley width", and that the map does not correspond in any very direct way with the maze (e.g., Keller and Hill, 1936; Ritchie, 1948; see also Deese, 1951). Seldom cited in this context, the work on the abilities of birds to navigate (see Chap. 7) would also seem to stretch any theory couched in simple stimulus-response terms beyond the limits of credibility. Another line of evidence drawn from experiments on maze-learning is that, in certain types of mazes, rats appear to adopt systematic modes of solution which are tried in turn until success is achieved (Krechevsky, 1937). Such a finding is again more easily interpreted in cognitive terms.

From these experiments, it is reasonable to conclude that many of the phenomena of learning experiments cannot economically be described in simple stimulus-response terms. The theorist who would attempt to do so must concede that the covert responses have behavioural laws different from those governing overt responses: such a conclusion results in a loss of the logic of the stimulus-response approach. The evidence quoted has been drawn primarily from laboratory experiments on rats: we may now refer to some work on other species, mostly already discussed in earlier pages, which seems to demand explanation in terms of a central change that can be initially independent of any change in overt behaviour and cannot easily be interpreted in terms of covert S–R sequences.

(*a*) *Orientation.* Many orientational responses of insects involve either movement bearing a similar relationship to two sorts of stimulation (e.g., light and gravity) or movement whose relationship to one sort of stimulation varies with time. Such cases can be understood more easily in terms of a control system with variable Sollwert than in terms of multiple S–R mechanisms, each serving a different orientation, together with a mechanism to select between them (see Chap. 7). Although much simpler than the other examples discussed here, these cases of orientation in insects are of special interest in indicating the existence of a type of mechanism which both S–R and cognitive theorists might own.

(*b*) *Search Image, Reward Expectancy, and Selective Attention.* Much appetitive behaviour involves a scanning of the environment until a particular set of stimuli are encountered, and can be understood on the assumption that the animal is using a "search image". In other words, searching involves not only a change in behaviour, but also a change in responsiveness not immediately apparent in behaviour (see Chap. 5). Similarly, as we have seen, the phenomena of selective attention cannot be understood in terms of a passive and peripheral filter, but must involve central mechanisms sensitive to subtle categories of stimulation (see Chap. 6).

(*c*) *Habituation.* As discussed on pages 139–140, some of Sokolov's experiments on the extinction of the orientation reflex seem to be explicable only in terms of some central restructuring. Sokolov referred to the process as the formation of a "neuronal model", a pseudo-physiological term which has the merit of implying a direction for future research, but does not explain the phenomena any more than does description in terms of a "cognitive schema" (see pages 131–143 and discussion by Vinogradova and Horn in Horn and Hinde, in press).

(*d*) *Song-learning.* As discussed in Chap. 19, the development of the species-characteristic song by chaffinches seems to involve first the acquisition of a Sollwert and then the adjustment of the motor output until correspondence with it is achieved.

(*e*) *Imprinting*. The learning associated with the following response of nidifugous birds was discussed in Chap. 21. It involves a learning of the characteristics of the environment which has consequences on the birds' subsequent responsiveness to stimuli in other contexts. Bateson (1964a) has shown that the same early treatment which determines the effectiveness of stimuli for eliciting the following response in an imprinting experiment also influences the speed of learning when the same stimuli are used in a discrimination experiment, where the response is quite different. Clearly the learning is more easily described as involving first a learning of the characteristics of the object, rather than in crude S–R terms.

(*f*) *Conceptualization*. The extent to which subhuman animals are capable of concept formation has long been a matter for speculation, but has only recently been subject to experimental study. Herrnstein and Loveland (1964) trained pigeons to respond to the presence or absence of human beings in photographs. The people might be anywhere on the photograph, and were clothed or nude, adults or children, in varied postures, black, white, or yellow, in fact so diverse that simple stimulus characterization seems ruled out. These experiments thus suggest very considerable powers of conceptualization in pigeons. (See also Köhler, 1955; Millenson, 1962.)

A somewhat similar approach has been used by Lehr (1967). He trained two monkeys to select three different insect pictures of the same colours instead of pictures of withered leaves, fruit, branches, etc., of similar area and colour. In later tests they showed significant preferences for insect pictures over those of other natural objects. Similar results were obtained by training on flower pictures as compared with other plant organs. However, the generalization appeared to depend on a few criteria common to all flower patterns: this suggests a mechanism which some would feel hardly to merit the label of "conceptualization."

Faced with such facts, it is clear that learning which appears to involve only changes in stimulus-response relationships may, in fact, involve a central reorganization not immediately evident in behaviour. This does not necessarily mean that all learning is to be interpreted in this way. Any student of lower animals, familiar with the rigid forms of behaviour which they often show, must believe not only that for some examples of learning descriptions in stimulus-response terms are adequate, but also that anything else would be erroneous. The present value of an S–R approach thus lies not in its theoretical implications, but in its emphasis on investigating the stimuli which ultimately control the responses measured.

24.5 THE ROLE OF REINFORCEMENT IN THE FORMATION OF STIMULUS–RESPONSE RELATIONS

Another dilemma which has divided learning theorists concerns the role of reinforcement in learning. Some hold that the only requirement for stimuli and responses to become associated is a certain temporal relationship between them—this is learning by "contiguity". Since connections are not established between all stimuli and all responses which occur in association, it is necessary to explain why some things are learnt and not others. One way in which contiguity theorists have met this difficulty is by suggesting that the reward which appears after the response takes the animal out of the situation, and thus prevents the stimuli which were present just before the reward was obtained from becoming associated with any further responses (e.g., Guthrie, 1935). Estes (1959) proposes a stimulus sampling, so that only the stimuli being sampled at the moment the response is made get attached to the response. Yet another hypothesis is that the motivational state determines which stimuli the organism will attend to, and thus which become associated with each other or with responses (e.g., Tolman, 1959): in this case the reward acts by confirming expectancies about the nature of the environment.

Other workers, however, have held that such a close temporal association between stimuli and responses is not a sufficient condition for learning to occur. It is also necessary, they maintain, for the response to be followed by some "reinforcing" event. For instance, in operant conditioning a reinforcer can be defined as an event which increases the frequency of a response when the event is contingent upon a response of that class. If control of behaviour is the sole aim, such an operational definition may be all that is needed: it is an empirical matter to determine which events are reinforcers for a particular response and which are not (e.g., Skinner, 1938). However, certain generalizations beyond the definition can be made and then subjected to empirical test. For instance, one sort of generalization is that a stimulus which has been demonstrated to be a reinforcer in one situation will also serve as a reinforcer in another: the limits of this generalization, in terms of responses and species, can then be established. A related approach, aimed at an understanding of the nature of reinforcement, is to seek generalizations about those events which lead to reinforcing effects and those which do not. The theories which have been most productive here are those ascribing reinforcement to the satisfaction of a biological need or drive (drive reduction), to the removal of stimuli associated with such a drive (drive-stimulus reduction), or to the appearance of stimuli previously associated with them (secondary reinforcement) (see, e.g., Hull, 1952; Miller and Dollard, 1941). Over the years it has become increasingly clear that such theories, while retaining a certain heuristic value (e.g., Miller, 1959), cannot be stretched to cover all cases. Although there is no satisfactory alternative, it is worthwhile to survey some of the evidence which leads to this view. In so doing, we shall accept that "drive" is a useful intervening variable in this context (see Chap. 8).

In the first place, drive-reduction theory in its original form implied that responses were learned by reduction in the factors which gave rise to them. However in some cases, at least, consummatory stimuli can act as reinforcers even though the initiating factors are still present. For instance, the ingestion of liquid made sweet with saccharin both reduces subsequent ingestion and serves as a reinforcer, even though saccharin is non-nutritive (Sheffield and Roby, 1950; Miller and Kessen, 1952).

Even though saccharin does not reduce the need for food, it does reduce the intake of food. This is in harmony with the view that reinforcing effects are correlated not with need reduction or the removal of instigating factors, but with a reduction in "drive" as indicated by the readiness with which the animal shows the behaviour in question. A number of experiments do in fact show a correlation between the effectiveness of a stimulus in bringing the behaviour to an end and its effectiveness as a reinforcer. For instance milk injected directly into the stomach not only inhibits eating less than milk drunk by mouth, but is also a less effective reinforcer for bar-pressing. Saline injected into the stomach is less effective than either in both respects (Kohn, 1951; Berkun et al., 1952; Miller and Kessen, 1952). Whether the consummatory and reinforcing consequences of stimulation are mediated in the same way in these cases is an open issue: there is some evidence for a difference in temporal characteristics, the reinforcing effect of stomach loading being slower than the satiating effect (see also Hammer, 1969).

However reinforcement is not necessarily associated with drive reduction. Sheffield et al. (1951) found that intromission served as a reinforcer for male rats, even though ejaculation was not permitted. Intromission is likely to enhance the tendency to show sexual behaviour rather than to diminish it, and thus reinforcement seems to be associated with an increase rather than a decrease in drive. This argument remains valid even though later workers have found that intromission with ejaculation is more effective as a reinforcer than intromission alone (Kagan, 1955; see also Whalen, 1961). In any case the relative reinforcing effectiveness of a non-receptive female, intromission, and ejaculation depend on experience. Ware (1968) found that these situations did not differentially influence the mount latency of sexually naive male rats, but after experience ejaculation was most effective. However recent knowledge makes this case difficult to interpret in terms of simple drive-reduction theory, since the consequences of both intromission and ejaculation on subsequent responsiveness are complex (see pages 325–332). Indeed it raises doubts as to the usefulness of unitary drive concepts in this context. Both eating and sexual behaviour come to an end as a result of a number of different feedback mechanisms (Chaps. 10 and 15), and the several different consequences of responding could play a role in reinforcement (e.g., Smith and Duffy, 1955, 1957); indeed reinforcing effectiveness may be influenced by interaction between them (Sterritt and Smith, 1965).

Although some of these cases imply that performance of the consummatory response (e.g., eating) is not necessary for a reinforcing effect to be obtained (see also, e.g., Coppock and Chambers, 1954, cited on page 252), others indicate that consummatory behaviour (or its stimulus consequences) can be reinforcing. There are indeed a number of cases in which the mere performance of a response, without apparent drive reduction, can be reinforcing. For instance, King and Weisman (1965) found that *Peromyscus* would bar-press for sand, which provided them with an opportunity to dig; and the opportunity to gnaw on an object is reinforcing for rats caused to gnaw by hypothalamic stimulation (Roberts and Carey, 1965). Even the opportunity to play is reinforcing for primates (Mason, 1965). An extreme example of this viewpoint is Premack's (1959) "prepotent response" theory that any dominant response is capable of acting as a reinforcer for a preceding weaker response. For instance, thirsty rats will run more in a wheel if they are given the opportunity to lick a waterspout contingent upon wheel-running, and they will lick the waterspout more if the running wheel is available only after they have licked the waterspout. Or again, children increased their rate of eating candy when the operation of a pinball machine, a response with an initially higher rate than candy eating, was made contingent upon it. On this view, therefore, acts need not be consummatory (i.e., coming at the end of a chain) to be reinforcing. However, there are obvious problems in deciding which of two very different responses is dominant when they cannot be measured on the same scale. Furthermore it is still possible for the drive theorist to postulate a drive which is reduced by the response in question—in this case, for instance, a pinball machine drive (Hilgard and Marquis, 1961).

Usually it is not possible to separate the reinforcing effectiveness of the performance of a response from that of the stimuli which elicited it. Thompson (1963), for instance, found the presentation of its mirror image to be reinforcing for a Siamese fighting fish *Betta splendens:* a moving model and a stationary model were less effective as reinforcers than the mirror, and also less effective in eliciting display (see also Hogan, 1967). He thus suggests that the stimulus situation most effective for releasing the display is also the best reinforcer, though he notes (1969a) that the performance of the response could be the crucial issue. Some stimuli, however, are effective even though they elicit at most an orientation response: thus Stevenson (1967, 1969) showed that the species song was reinforcing for testosterone-treated male chaffinches even though they showed no more than an attentive response to it (see page 456). In such cases the effectiveness of the stimulus as a reinforcer seems to be related to its adaptive significance to the species concerned. Thus in her studies with chaffinches Stevenson (*loc. cit.*) found that song was ineffective to males in winter condition, and that a burst of white noise similar in intensity to the song was not reinforcing even for testosterone-treated birds. Similarly Butler (e.g., 1953, 1964) found that visual stimulation by other monkeys was a more effective reinforcer for rhesus macaques than that provided by dogs, toy trains, etc.

Nevertheless, some stimuli or stimulus changes which appear to have little relevance to the natural situation can act as reinforcers (review by Kish, 1966). A number of investigations have shown that rats will learn to press a lever if an increase in illumination follows (e.g., Girdner, 1953; Kish, 1955; Kling et al., 1956; Hurwitz and De, 1958; McCall, 1966; Tapp and Simpson, 1966). Comparable effects have been found in monkeys (Fox, 1962). Other studies have shown the opposite effect, the animals learning to press a lever to decrease the illumination (Keller, 1941; Flynn and Jerome, 1952; C. L. Roberts et al., 1958; Barry and Symes, 1963): in some of these studies the rats were found to work to turn the light both on and off. Not all experiments produce a positive result. Among the factors conducive to bar-pressing for light are the intensity of the light, moderate intensities being most effective: the optimum is influenced by age and past experience (Levin and Forgays, 1959; Lockard, 1963a; Roberts et al., 1958; D'Amato and Jagoda, 1962; Kish and Baron, 1962), and the response rate may be affected by the state of hunger or thirst (Clayton, 1958; Forgays and Levin, 1958; Davis, 1958; Kiernan, 1965).

The reinforcing effect of stimulus onset and offset could be related to the property of stimulus novelty (Berlyne, 1960) and/or complexity (Barnes and Baron, 1961). Barry and Symes (1963) found that bar-pressing by rats was reinforcing whether the light was turned on or off, and Premack et al. (1957) showed that the rate at which rats would press a bar for light reinforcement increased with the length of a preceding period in total darkness (see also C. L. Roberts et al., 1958). However, the precise role of novelty in this context is not clear. Berlyne et al. (1966) found that for rats maintained in a noisy room, familiar stimuli (light changes or buzzer sounds) were reinforcing, while for rats maintained in a quiet room the opposite was the case. They therefore suggest that reward value is an inverted U function of arousal (see pages 218–220) value, which increases with the novelty of the stimulus and the arousal level of the subject. However, while novelty is certainly a property of many reinforcing situations (see below), the persistence of bar-pressing over days rather counts against its importance here: Kavanau (1967) found that deermice able to control the light intensity in their cages (see below) continued to do so for months.

A related possibility is that stimulus change is itself reinforcing (Forgays and Levin, 1959). On this view, bar-pressing rate should be an increasing function of the amount of light change in either direction, at least up to a point. Although Lockard (1963b) rejected this view, later experiments by McCall (1965, 1966) gave strong support to it, bar-pressing rate being an increasing function of the amount of change rather than of particular light intensities. Rearing conditions and light deprivation (Sackett, 1965) influence the amount of change which is acceptable. McCall emphasizes that the effect is superimposed on other preferences, and stresses the rat's nocturnal nature. In a later experiment designed to compare the relative roles of stimulus change and stimulus preference, Lockard (1966) found the former played a role only early in testing, while the latter increasingly predominated.

Yet another possibility in this confused area of research is that the reinforcing effect of the light is partially due to the opportunity it provides for visual exploration. Robinson (1961) argued that, on this hypothesis, increase in light intensity in an already lighted environment should be less reinforcing than a similar increase from preceding darkness, and that light decrease should be more reinforcing than light offset. His results were in accordance with these predictions (see also Eacker, 1967).

The relevance of such experiments to the natural life of the animals has so far rarely been considered: as Stevenson (1969) points out, the effectiveness of a stimulus change as a reinforcer is likely to be related to its adaptive significance (in terms of natural selection) to the animal concerned. In fact the alternation between total darkness and bright light often used in studies of sensory reinforcement is quite unsuitable for nocturnal animals (Kavanau, 1967), and the relation between the experimental sessions and the animals' circadian rhythms is usually neglected. When deermice *Peromyscus maniculatus* were allowed to control their illumination with two levers, a press on one increasing and on the other decreasing the illumination by one out of ten steps, they maintained circadian rhythms in the absence of external cues, and selected light intensities similar to those they would experience at corresponding phases of activity in the wild. A circadian periodicity in need for stimulus change has been found also in rats (Thor and Hoats, 1968).

Turning to another sensory modality, Barnes and Kish (1961) found that sounds of fairly low frequency and intensity were reinforcing for mice, but the effect was only small. Once more, since the auditory sensitivity of mice extends far into the ultrasonic range, stimuli more relevant to the animals than those used in this study might have been more effective (Stevenson, 1969). Again, kinesthetic feedback may be involved in the reinforcing effectiveness of wheel-running (see page 258), but mere activity is not the only issue. Kavanau (1966) found that deermice (*Peromyscus maniculatus*) preferred to run in a large round wheel containing hurdles or in a square "wheel" than in plain round wheels. He therefore suggests that one of the reinforcing aspects of wheel-running is the requirement for split-second timing, coordination of movements, and quick reflex actions. Most of the house mice tested preferred a small plain round wheel. Kavanau (1967) also found that the wheel-running of deermice tends to be oriented with respect to the light source and changes in both amount and character with aspects of the light regime. The reinforcing nature of wheel-running, it may be noted, varies with the internal state. Mice provided with a motor-driven wheel which they could turn on or off with levers would turn it on if the experimenter turned it off when they had turned it on, and vice versa.

It will be apparent from the examples cited that the diversity of events which are effective as reinforcers poses grave problems for a drive-reduction or drive–stimulus-reduction theory of reinforcement. Learning may be associated with the performance of a response or with a stimulus change, and the latter may involve not only decreases but also increases in an inferred drive. Furthermore, although the effectiveness of events as reinforcers is related to their biological relevance, no theory which relates reinforcement to the performance of specific consummatory acts (e.g., Glickman and Schiff, 1967) would seem able to cope with the facts of sensory reinforcement. In fact the diversity of the above results defy generalizations about the nature and effects of sensory reinforcers valid for all responses of all species in all conditions.

A case of special interest is provided by exploratory behaviour. As we have seen, this cannot be explained in terms of the traditional biological drives like hunger, and is not merely an aspect of general activity (pages 351–355). However the opportunity to explore a novel environment is reinforcing in the sense that a response which leads to such an opportunity tends to be repeated (see references on pages 345–355). Berlyne (e.g., Berlyne and Slater, 1957) has suggested that some of these results could be accounted for in terms of a reinforcing effect not of the novelty of the stimuli which the rat explores, but of the spaciousness or complexity of the environment it encounters: the difference, however, seems to be only one of degree (see also Halliday, 1968). Furthermore, in some experiments novelty has been shown to be reinforcing when complexity was controlled. Rats show a smaller latency before running to a novel card than before running to a familiar one (Stretch, 1960; see also Chapman and Levy, 1957; Franken, 1967). Working with hamsters, Schneider and Gross (1965) found that a constantly changing series of stimulus objects retained their reinforcing value, but similar objects lacking in novelty did not.

Attempts to explain the reinforcing effects of novel stimulation run into the following difficulties. If an "exploratory drive" elicited by novel stimulation is postulated, an animal which performs an instrumental response on the basis of subsequent access to novel stimulation must be doing so in the absence of the drive. Furthermore, if the instrumental behaviour is reinforced by the same stimuli which produce the drive, learning must depend on an increase rather than a decrease in drive. We have already noted the suggestion that drive induction forms the basis of the reinforcing effects of incomplete copulation (page 595), and the same has been proposed for the case of exploratory behaviour (Montgomery, 1954). But with such a view drive reduction loses much of its value as a unifying concept: the chief merit of the drive-reduction hypothesis was that it seemed to permit prediction of when learning would occur (e.g., Watson, 1961).

The alternative view, that lack of novel stimulation initiates the exploratory drive, has been expounded especially by Fowler (1967). His thesis involves a distinction between the consummatory behaviour of peering, examining, sniffing at, and otherwise investigating the novel situation, and the instrumental behaviour which precedes this. Emphasizing the "satiation" or "boredom" effect of unchanging stimulation, he proposes that the exploratory drive depends on the length of deprivation of a change in stimulation; and that the magnitude of the incentive or reinforcement, involved in the learning of the instrumental behaviour leading to exploration, depends on the magnitude of the stimulus change which occurs. Stimulus-change deprivation and stimulus-change reinforcement thus take the place of the more conventional food deprivation and food reinforcement. This view is supported by runway experiments in which these independent variables were manipulated.

Watson (1961) has argued that, if exploration is supposed to depend on a drive which increases with deprivation from stimulus change, an animal allowed to explore a novel environment which is unlimited in extent would show a gradual decrease in exploratory behaviour, owing to decrease in the drive. He believed such evidence as was available to be against this view. The argument is, however, not a very forceful one since it could be held that the incentive motivation acquired through exploring more than compensated for the decrease in drive.

Just because so much of the controversy about the role of reinforcement in learning has arisen from attempts to formulate laws of learning at the behavioural level applicable to all learning in all situations, it is worthwhile to push the argument one stage further. We have seen that the situations which bring behaviour to an end differ according to the type of response considered. In vertebrates feeding behaviour ceases after the advent of a variety of consummatory stimuli. Sexual behaviour also depends on a change in the stimulus situation. Fear behaviour ceases with the removal of an external stimulus. Now exploratory behaviour, the orientation reflex, and so on, cease without any external change, leading one to postulate a change in internal organization. In feeding and sexual behaviour the consequences of responses which bring the behaviour to an end also make the preceding instrumental responses more probable on subsequent occasions. If the same is true of exploratory behaviour, then its reinforcement must come from internal changes. On this view, then, the investigatory actions of exploratory behaviour are elicited by novel stimuli and come to an end when those stimuli cease to be novel as a result of certain central changes.

Returning to the more general problem, a common feature of all reinforcers [apart from intra-cranial electrical stimulation (see below)] is that the animal pays attention to them, or that they elicit an orientation response. It is significant that, while the mere exposure of rats to various shapes facilitates their subsequent ability to learn to discriminate between them, the exposure is more effective, the more conspicuous the shapes (page 482). Again, mere exposure of the young of some passerine birds to song patterns modifies their subsequent pattern of singing, but the effective song patterns are those which resemble the species-characteristic song or come from a particular individual (pages 459–460), and these are presumably those most effective in eliciting an orientation response. Indeed it was at one time suggested that the orientation response in itself constitutes reinforcement. This, however, does not get us very far since the argument becomes circular: virtually all stimuli which affect the animal elicit some degree of orientation response, at least initially.

Berlyne (1967) prefers the hypothesis, not of course incompatible with the last, that reinforcement results from changes in arousal. Many observations suggest that a reduction in arousal may be important. For instance, many rewards (e.g., food, copulation) are followed by quiescence, and the termination of many conditions which are arousal-increasing (e.g., pain, fear) is reinforcing. To account for the fact that some arousal-increasing situations seem to be reinforcing (e.g., novel stimuli, exploratory situations), Berlyne earlier (1960) argued that while in general a decrease in arousal is reinforcing, a moderate increase is not necessarily punishing and may be followed by a decrease which is reinforcing (for instance when the novel stimuli become familiar as a result of exploration). He also suggested that, in such situations, arousal may be increased by anticipation of the unknown and subsequently reduced by exploration.

More recently Berlyne has come to the view that reinforcement can also be associated with an increase in arousal. He points out, for instance, that food and copulation have arousal-increasing as well as arousal-reducing properties. Since marked arousal increase is clearly aversive, Berylne is forced to argue that it is the degree of arousal increase which is crucial, and that the higher the level of arousal, the less rewarding is a particular arousal increase. Nevertheless he stresses that reinforcement must work through "something positive, something starting or increasing in intensity, rather than something ceasing or declining". He thus suggests that, with increase in arousal potential, first reward and then aversion effects increase to a maximum. Their algebraic sum thus reaches a positive peak with moderate changes, and then decreases and becomes negative, the position of the peak probably depending on arousal level.

Two points about these suggestions must be made. First, there are general difficulties with the concept of arousal which have already been discussed in Chap. 9. Although Berlyne himself now regards arousal as little more than a first factor "reflected by the common variance of many response measures", he seems to be reifying it much further than this in his analysis of reinforcement. Second, his hypothesis is partially circular and seems to be in danger of being incontrovertible.

In parenthesis here we must add that studies of intra-cranial stimulation (see pages 219–221), apparently such a bright hope a decade ago, have not provided the insight into reinforcement that once seemed possible. While it is clear that it can act as an effective reinforcer for rats (and other species, Andrew, 1967; MacPhail, 1967) in an operant situation, the mode of action of such an unnatural source of stimulation is far from clear. Although a number of brain sites in which self-stimulation occurs are also sites at which stimulation produces forward approach (Valenstein, 1964) or activities which are themselves reinforcing (e.g., feeding, copulation; see page 595), the generality of such correlations is far from clear: certainly some sites yield moderate self-stimulation and moderate escape (Olds and Peretz, 1960). In any case ICS may act in part by affecting sensory input (Valenstein, 1964). Deutsch (1960; Deutsch and Deutsch, 1966) has suggested that ICS has separable reinforcing and motivating properties: for instance, running speed reinforced by ICS decreases immediately if the inter-trial interval is lengthened, suggesting the rapid decay of a motivating effect (e.g., Gallistel, 1966), and certain of the stimulus parameters which influence response rate do not affect discrimination learning, while others do (Keesey, 1966). However, Scott (1967) found ICS to be effective as a reinforcer even with well-spaced trials.

So where does this leave us? It is clear that easy generalizations about the nature of reinforcing events are beyond our grasp. Drive-reduction theories have provided a valuable spur to research in the past but cannot be universally applied, and yet there seems no satisfactory alternative. Perhaps the search for a unifying concept of reinforcement is premature. At any rate a more urgent goal would seem to be to specify in behavioural terms which situations are reinforcing for a wider variety of types of behaviour than have been studied hitherto.

24.6 SUMMARY

1. Certain difficulties in defining learning are discussed.
2. Attempts to classify types of learning include short- versus long-term storage, classical versus instrumental conditioning and Tolman's capacity laws. Thorpe offers a wider classification in terms of habituation, classical conditioning, trial-and-error learning, latent learning, and insight learning. The overlap between categories as well as the range of complexity within categories must be emphasized.
3. Just what an animal can learn is restricted in a number of ways. Limits are imposed by sensory/perceptual mechanisms, the ability to perform particular responses, species and individual characteristics, maturation, and environmental influences.
4. Learning situations which pose problems for a strict S–R theorist also have parallels in more natural situations. Sensory preconditioning, perceptual learning, two-stage discrimination learning, and place versus response learning are discussed in this context.
5. The role of reinforcement in learning involves not only the strict operational definition of reinforcement, but also generalizations made from it and inferences about possible mechanisms of reinforcement. Sensory reinforcement and exploration are considered with reference to theories of drive reduction, drive induction, and arousal.

25

INTEGRATION

The behaviour shown by an animal can be labelled in terms of a physical description of the movements involved or in terms of their consequences (Chap. 2). In the repertoire of any organism the number of discrete types of behaviour, as differentiated in one or other of these ways, is usually large. They do not, however, occur at random, but in integrated sequences which, for the most part, lead to biologically adaptive consequences. After a stickleback has made a zigzag dance to a female, it will often swim toward the nest, and this is likely to be followed by other courtship activities which lead to fertilization of the eggs: when a great tit sings, it is especially likely to show functionally related activities, such as flying from tree to tree on its territory and attacking intruders. The aim of this chapter is to examine the ways in which discrete items of behaviour are integrated into functional sequences.

The initial data here concern the extent to which different activities are temporally correlated with each other: there are in essence two procedures by which this can be assessed. One method is to record the frequency with which activity *A* is preceded or followed by activities *B*, *C*, *D*, etc: if the frequency exceeds the chance level, the two activities are said to show sequential association. The data can be subjected to factor analysis, so that it is possible to assess the extent to which correlations between a large number of activities can be ascribed to a small number of common factors (see Chap. 16).

Alternatively, certain items from an animal's repertoire can be recorded over standard periods of time, and the relations between their frequencies examined. The relations may be monotonic, the frequency of one activity increasing or decreasing as that of the other increases, or non-monotonic, perhaps with a U-shaped relationship between the two frequencies. The lengths of the time units in such analyses are of crucial importance. The use of day-long observation sessions to count how often a great tit sings, examines a nest-hole, and threatens other great tits with the head-up threat posture, would indicate that all these activities are positively correlated—all are more frequent in the spring. If, however, their frequency was examined during 1-min periods taken at intervals during the first few hours after dawn in early April, negative correlations would appear.

If such methods show that two activities are positively or negatively correlated to a significant extent, we must search for relationships, direct or indirect, between the factors or mechanisms controlling them. But the limitations of deductions made in this way must be recognized. Apart from the difficulty of the time units employed, it is important to recognize that the correlations are derived from behavioural data, and inferences about underlying physiological mechanisms must be made with caution. A correlation between two activities *A* and *B* could arise in a number of ways, and we must now examine some of these. We shall do this in terms of a number of overlapping principles, several of which may operate in any one case (see Hinde and Stevenson, 1969b).

25.1 COMMON CAUSAL FACTORS

In Chap. 3 we saw how more and more of the different components of a fixed action pattern may appear as the causal factors specific to that pattern become more intense. The components all depend on the same causal factors (this is our definition of a fixed action pattern), but differ in the threshold levels at which they appear.

In the same way, but on a longer time scale, diverse activities may share causal factors. A particular experimental manipulation—it may involve the presentation of an external stimulus or an alteration to the internal state—alters the probability of some activities but has relatively little effect on others. For instance, administration of male sex hormone to a male chaffinch influences whether it will show courtship to a female, the extent and the nature of its aggressive response to a male, the frequency with which it sings, and so on: androgens act as a causal factor common to many different activities, though, of course, all are not affected equally. Similarly, deprivation of food may influence the intensity of a rat's response to food when it is presented, the speed with which it will run down a runway at the end of which it has previously found food, the amount of time it spends searching its cage, and so on. The same is also the case with external factors—stimuli from a baby mouse increase the probability of retrieving, nest-building, licking, and adoption of the lactation position in an adult (Beniest-Noirot, 1958). Similarly, stimuli from the eggs of the herring gull influence a number of different activities concerned with incubation (Baerends, 1959b).

The multiplicity of consequences produced by an apparently simple change in an independent variable can arise in a number of ways:

1. The change in the independent variable may not be as simple as it appears, for it may entail separate but correlated changes which influence different responses. Such a situation arises when the stimuli for two responses consist of separate features of a complex situation (e.g., Chaps. 6 and 16). The various stereotyped movements used by canaries in nest-building appear in temporal association partly because they are all elicited by stimuli encountered when the bird is in the nest, though the precise stimulus for each one is different from that for the others. The zigzag dance which a male stickleback shows to a female represents alternate aggressive advances and sexual leading to the nest, for a pregnant female has both some of the characteristics of an intruder which normally elicit aggressive responses and the swollen belly and upright posture which elicit sexual responses (Fig. 25.1 and van Iersel, in Sevenster, 1961).
2. A common factor could act more or less independently on the mechanisms controlling two types of behaviour. Thus when a hormone influences more than one type of behaviour, it may do so by action on different brain sites rather than (as has previously been suggested) through a superordinated mechanism, or by influencing both central and peripheral mechanisms, as with the sexual and nest-building behaviour of female canaries (pages 633–640). Since the action of one hormone on one type of behaviour may be mediated through more than one mechanism (see pages 230–239), it is even more likely that effects on two activities are.
3. A single factor could act on a mechanism which, in turn, influences the occurrences of more than one response. The appearance of a predator may evoke autonomic activity and thus a number of different responses which all depend on the autonomic system.
4. The independent variable could exert an inhibitory effect on some activities leading to an apparent association between activities not so affected. Thus many activities which an animal shows at night are correlated just because they are the only activities not inhibited by darkness. Similarly, activities may be associated because they share a low priority among the organism's activities and can occur only in the absence of response tendencies of greater priority. This may be one reason why different "toilet" or "comfort" movements tend to be associated together: they rarely interrupt other types of behaviour already in progress but are themselves easily interrupted.

Whatever the mechanisms by which a common causal factor produces its effects, a difference in threshold between the various activities is usually to be observed. Thus with increase in the quantity of oestrogens injected into female canaries they first hop about over nest material and occasionally

Fig. 25.1 Schematic representation of the sexual behaviour of the three-spined stickleback. (After Tinbergen, 1951.)

The female, with a swollen belly (*top right*) enters the territory of the male, swimming in a special posture. The male courts her, and when she responds he leads her to the nest. If she follows, he adopts a special posture by the nest entrance. When she enters (*inset*), a trembling movement by the male elicits spawning. Subsequently the male enters the nest and fertilizes the eggs (*not shown*).

pick it up, then pick it up and carry it about inconsequentially until they drop it, and finally gather it, carry it to a nest site, and place it there. Similarly, when a hawk flies over, a passerine bird may respond with anything from a momentary glance to a dash to cover, depending in part on the intensity (size, proximity, etc.) of the stimulus. In such cases the early phases of the sequence consist of low-threshold activities, requiring a lower level of motivational factors, or a weaker eliciting stimulus, than the behaviour which comes later.

While common causal factors may be responsible for temporal associations between different activities, they can influence the ordering of activities within a sequence only to the extent that the latter differ in threshold. As an example of such an effect, the preening movements shown by terns after bathing can be divided into three groups, of which the principal constituents are (1) head-shaking and preening of breast and back, (2) shaking and preening of shoulder and wing, and (3) preening of wing pinions and tail. The occurrence of these groups of movements is correlated with the completeness and frequency of preening movements. Since the differences cannot be accounted for by differences in external stimuli for preening, van Iersel and Bol (1958) suggest that they are due to differences in internal factors for preening. On this view the three groups of activities have different thresholds: changes in the nature and intensity of preening movements through a bout are due to fluctuations of internal factors producing changes in the proportion of high-threshold movements. Similar suggestions relating to the patterning of chain responses have been made by Bastock and Manning (1955) for the courtship of *Drosophila melanogaster* and by Tugendhat Gardner (1964) for the hunting of salticid spiders. But the sequential patterning produced in this way is unlikely to be more than crude.

To some extent, the relations between the causal factors controlling the behaviour of an animal can be pictured in a hierarchical way. For example, the reproductive behaviour of male sticklebacks can be classified as shown in Table 25.1 (Tinbergen, 1951). Other types of behaviour, such as feeding, could be classified in a similar way. Such an hierarchical classification can also be regarded as a hierarchical arrangement of causal factors. Thus all reproductive activities are influenced by certain hormones, but at the next lower level, aggressive behaviour and sexual behaviour are elicited by a rival male and a female, respectively. Each type of aggressive behaviour similarly has in addition certain factors specific to it. This argument has been taken one stage further with the suggestion that each level in the hierarchy represents a neural mechanism, that the causal factors act on those neural mechanisms, and that each neural mechanism influences the state of those subordinated to it. Tinbergen (1951) elaborated a scheme of this type, but its dependence on an energy model of motivation (see page

TABLE 25.1. Basis of a classificatory system for the reproductive activities of the male three-spined stickleback. Modified from Tinbergen (1951).

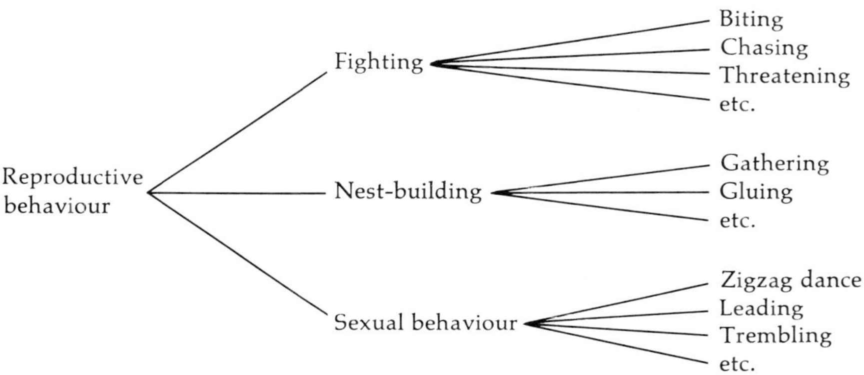

201) reduced its utility. While the abundant evidence that localized central stimulation can produce integrated and functional sequences involving several types of behaviour lends partial support to the view that the mechanisms underlying behaviour could be hierarchically organized, satisfactory physiological confirmation for the higher levels of such a hierarchy of mechanisms is so far lacking. At the lower levels, however, the structural basis for hierarchical organization is known in some cases. For instance, insect locomotion depends successively on segmental mechanisms, intersegmental connections, and influences from the sub- and supra-oesophageal ganglia, each having a more general influence (e.g., Weiss, 1950; Wiersma, 1962; Roeder, 1967; see also Chap. 12).

25.2 CHAIN RESPONSES

In a chain response, each movement or activity brings an animal into a situation where the next is evoked. The principle is well known at the reflex level: the alternating movements involved in walking, nystagmus, and so on depend in large part on the first phase of the movement inducing the second. A case discussed in some detail by Sherrington is that of the chewing movements of a decerebrate cat. If its jaw is touched with a piece of meat, it drops sharply. The muscles which close the jaw are thereby stretched, their proprioceptors stimulated, and their motor neurones caused to fire. As the jaw closes through the meat, the pressure causes stimulation of receptors in the periodontium. The muscles which close the jaw reflexly cease to contract. The jaw-opening muscles, already tensed, can then open the jaw. The proprioceptors of the closing muscles are thereby stimulated again, and the process is repeated (Sherrington, 1917). Of course, the intact cat's chewing movements are not solely dependent on this simple mechanism: data from other responses (see Chap. 3) would suggest an interaction between successive stimuli and central coordinating mechanisms. Nevertheless, with more complex activities, the proprioceptive input from one response may play a part in eliciting another (Lind, 1959).

More obvious, perhaps, are cases in which the stimuli for the later responses are external stimuli which the animal encounters as a consequence of its earlier behaviour. In such cases the chain nature of the sequence may be revealed when sudden breaks are correlated with the absence of a particular external stimulus, or when the presentation of a limited stimulus situation elicits only some of the links in the chain. The hunting behaviour of the wasp *Philanthus triangulum,* which preys on bees, provides an example (Tinbergen, 1935, 1951). When hunting, it flies from flower to flower in search of a bee. It is responsive to visual stimuli from any moving object of about the right size, but at this stage indifferent to the scent of bees. When the visual stimuli are perceived, it turns toward the prey and hovers 10 to 15 cm leeward of it. The wasp is now sensitive to bee-scent: dummies without it are at once abandoned, but the wasp seizes with a sudden leap ones that smell appropriately. The further responses, including the stinging of the bee, depend on additional as yet unanalysed stimuli which are not provided by simple dummies. Another example is provided by the behaviour of the crab *Ocypode saratan* in constructing its sand pyramid: each activity depends on stimuli encountered as a consequence of the previous one (Linsenmair, 1967).

Such chains are often of great importance in integrating the behaviour of two individuals of the same species. Thus the zigzag dance of the male three-spined stickleback (Fig. 25.1) is elicited by certain stimuli from the female (page 59). When the female swims toward him, he turns and leads her to the nest. She follows, and this stimulates the male to point his head at the nest entrance. His behaviour causes the female to enter the nest. Stimuli from the female elicit "trembling" from the male (see inset Fig. 25.1), which induces the female to spawn. The male is then stimulated to fertilize the eggs (Tinbergen, 1951), and stimuli from the eggs produce a decrease in his tendency to show further courtship behaviour (Sevenster-Bol, 1962).

Many other cases of social behaviour in which two individuals respond reciprocally depend in part on a chain sequence (e.g., courtship of queen butterfly *Danaus gillippus,* Brower et al., 1965; courtship of salamander, Joly, 1966; mating of cockroach, Roth and Barth, 1967; courtship of *Drosophila,* R. G. B. Brown, 1965; copulation of turkeys, Schein and Hale, 1965). In most cases (including that of the stickleback), however, the chain is far from rigid: links may be missed, or the behaviour may continually revert to an earlier stage in the sequence even though the stimuli for the next stage are available. One example of this has already been seen in the copulation of greenfinches (Fig. 16.5). As another, Fig. 25.2 shows the sequences of courtship activities observed by Baerends et al. (1955) in *Lebistes reticulatus,* the thickness of the lines indicating the frequency with which each transition was observed (see also, e.g., Aronson, 1949; Hinde, 1953a, 1954c; Morris, 1958a, b; Dane and van der Kloot, 1964). The variability in such sequences implies that the sequence is largely controlled by internal factors as well as by the external stimuli: a given act by one individual may thus produce any of several (but usually not all) of the responses from the other. Thus Fig. 25.3 shows first the "ideal" sequence of responses in the courtship of the fish *Badis badis,* and underneath the extent of "stimulus overlap" (Barlow, 1962b).

Indeed most instances in which responses appear in a regular sequence do not depend solely on successive stimulus situations each arising as a consequence of the previous response, but also on common causal factors for the various responses. Often, as we have seen, the later activities in the sequence require a higher intensity of these factors than do the earlier ones: if they are inadequate, behaviour reverts to an earlier point in the sequence. This is undoubtedly the case with the stickleback's sexual behaviour, where all the activities of each sex depend both on common endocrine factors and on a short-term state of heightened responsiveness, as well as on the external stimuli. Similar principles apply to the courtship of *Drosophila melanogaster* (Bastock and Manning, 1955), the feeding of sticklebacks (Tugendhat, 1960a), the fighting of ants (Wallis, 1962), and the hunting of salticid spiders (Tugendhat Gardner, 1964). A similar conclusion was reached by Simpson (1968) from his analysis of the threat behaviour of the Siamese fighting fish (*Betta splendens*). Although the frequencies with which a fish gives various display movements depend on the movements made by its rival in the preceding few seconds, a particular move of the rival is not a necessary condition for the occurrence of one of the fish's own display movements. In this study of this complex case (see pages 392–393) Simpson developed a methodology likely to be valuable in the analysis of interactions which depend on the chain-response principle only to the extent that a given response by one individual increases the probability of another response in its partner.

It is worthwhile to compare the naturally occurring chains considered so far with the "operant chains" which can be built up in the laboratory. These can be quite complicated: for instance, Pierrel and Sherman (1963) trained food-deprived rats to successively climb spiral stairs, push down and cross a drawbridge, climb a ladder, pull in and pedal a car through a tunnel, climb stairs, run through a tube, and take an elevator down to an area where finally, when a buzzer sounded, pressing a lever produced food. In spite of this complexity, the operations necessary to establish such a

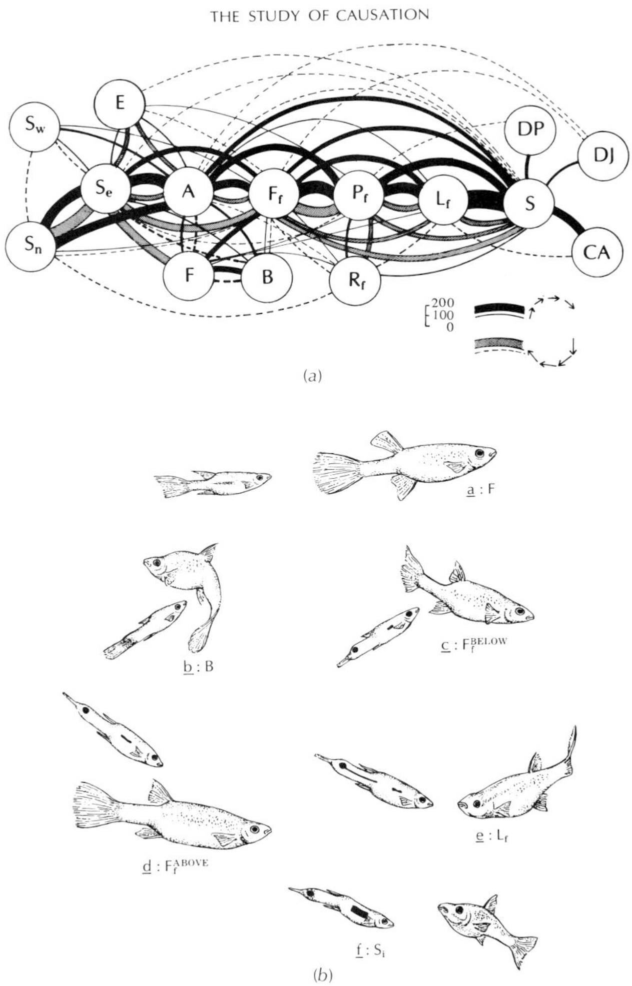

Fig. 25.2 a, The sequence of courting activities shown by male guppies; b, preliminary courting activities; c, advanced courting activities, with variations of the sigmoid posture, S; d, copulation attempt. (From Baerends, Brouwer, and Waterbolk, 1955.)

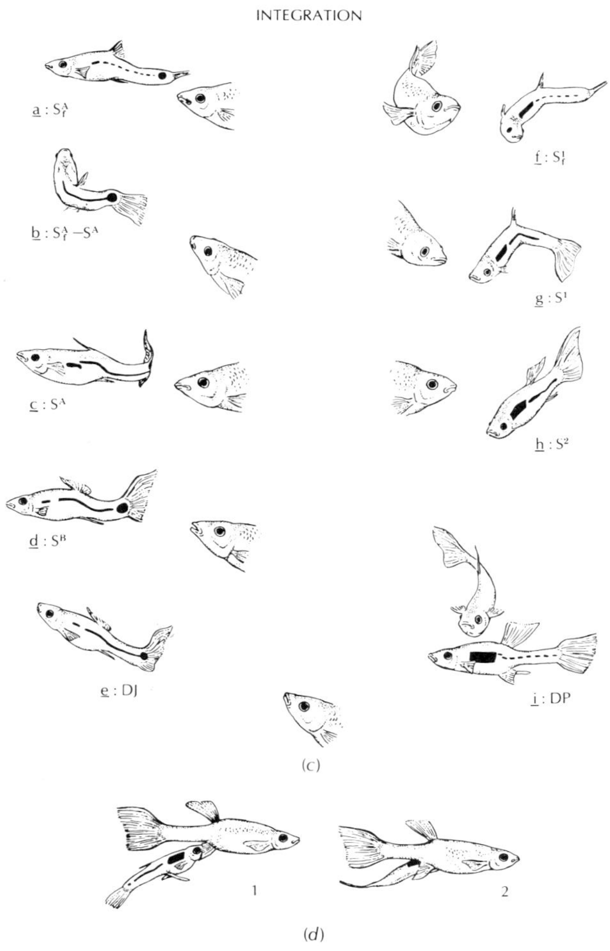

The thickness of the lines in *a* indicates the frequency with which one activity is followed by the one to which the line leads. Black lines should be read from left to right, hatched bands and dotted lines from right to left. The activities are illustrated in *b*, *c* and *d*, and Fig. 3.3. The abbreviations are as follows: *A*, approach; *B*, biting; *CA*, copulation attempt; *DJ*, display jump; *DP*, display posturing; *E*, evading; *F* and F_f, following; L_f, luring; P_f, posturing; R_f, retreating; *S*, sigmoid posture; *Se*, searching; *Sn*, snapping; *Sw*, swimming about. The male's colouration changes according to his internal state.

sequence are limited in number. First approximations to the final response (in this case, lever pressing) are reinforced (in this case, with food). The response is "shaped" by reinforcing only responses closer and closer to that required. Then a discriminative stimulus for that response is established by

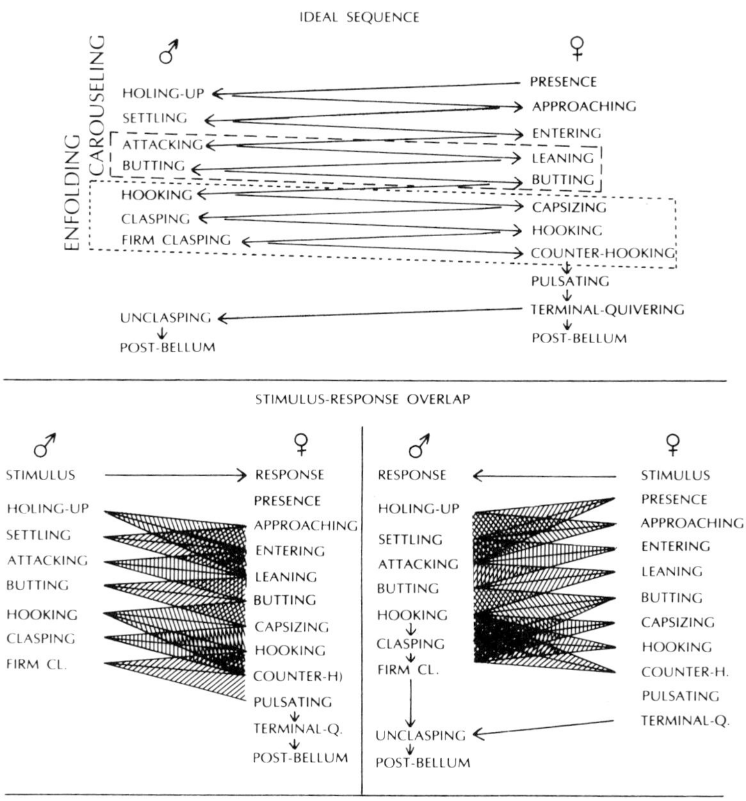

Fig. 25.3 The courtship behaviour of the fish Badis badis. (After Barlow, 1962*b*.)

Upper diagram, a hypothetical scheme in which the sequence of behaviour shown by male and female is regarded as a chain response; *lower diagrams*, the actual stimulus response relations which occur.

reinforcing the response only in the presence of that stimulus. This discriminative stimulus for the ultimate response can then be used as a conditioned reinforcing stimulus for the penultimate response, and so on back along the sequence. The effectiveness of such conditioned reinforcers depends on the ultimate "primary" reinforcement at the end of the chain (review by Kelleher, 1966). Some of the responses may need a lot of shaping, but others, such as running through a tube, may be part of the rat's normal repertoire: the difference is, of course, one only of degree.

Unlike such operant chains, in many naturally occurring sequences the performance of the early stages does not depend on prior performance of the later ones. The nest-building behaviour of canaries provides an example (Hinde, 1958a). In the breeding season the female shows chains of behaviour consisting of picking up and gathering material, carrying it to a nest site, and building it into the structure of the nest. Canaries reared from a few days of age without access to nest material, and tested in full breeding condition 10 months later, respond to nest material appropriately with a very short latency and go through the full sequence of building behaviour. If given a choice of material, they select appropriately, and subsequently build normal nests. Furthermore, at the beginning of the breeding season, when hormone levels are low, normally reared birds show the early phases of the sequence repetitively even though they never lead to sitting/placing. If birds are kept without a nest site, they will continue right through the breeding season to pick up and carry material inconsequentially, even though again it never leads to sitting/placing. In all these cases, performance of the early phases of the chain does not depend on previous performance of the later ones. If the sequence were an operant chain solely dependent on some form of primary reinforcement resulting from the behaviour of nest-building movements, it would surely extinguish. The frequency of the behaviour depends much more closely on the endocrine state than on reinforcement.

Nevertheless, learning clearly plays an important role in the integration of the sequences. Birds which are deprived of nest material develop substitute habits. Some individuals pull out their own feathers, carry these to the nest, and place them there. This may lead to the whole breast becoming denuded of feathers. Others, instead of actually plucking their own feathers, pick one up without detaching it from the skin, fly to the nest pan holding on to it, and go through the motions of placing it and building there. Such behaviour usually becomes very stereotyped—some females consistently carry a breast feather, others a tail feather, and each individual habitually flies to a particular point in the cage in order to pick up its own feather. Furthermore, whereas undeprived birds carry material from the floor of the cage to the nest-bowl but practically never in the reverse direction, individuals deprived of material most of the time carry both to and from the nest if occasionally given access to material (Hinde, 1958a).

Such observations suggest that habits which enable parts of the response sequence to be performed are learned, even though no nest is constructed. The possibility that the movements of sitting/placing in the nest are reinforcing, and that this acts retroactively through the sequence to establish the earlier phases, does not account for the performance of incomplete sequences when motivation is low, or in the absence of a nest site. Nor does it account for the manner in which undeprived birds carry only to the nest site and not away from it. Another possibility is that the performance of each phase of nest-building behaviour is self-reinforcing. Such a view gains some support from the studies, cited on page 596, showing that the performance of species-characteristic responses or the stimulation associated with them can be reinforcing. The difficulty with such a line of thought, however, is that without adequate support it leads too readily to an explanation which will explain anything.

Although the establishment and maintenance of operant chains depend on reinforcement, in performance they resemble the naturally occurring ones. Like them, they are far from rigid, for the probability of an operant response to a discriminative stimulus is not fixed but varies with internal and external factors: if some characteristic of the stimulus is changed, responding does not necessarily cease, but occurs less frequently than to the original training stimulus (e.g., Terrace, 1966). Thus in both naturally occurring and operant sequences the relations between stimulus changes and responses are probabilistic. A precise comparison between the two cases will not be possible until more is known about the extent to which S–R dependency is related to species characteristics, past developmental history, internal factors, and/or reinforcement in both cases.

25.3 INHIBITORY RELATIONS BETWEEN ACTIVITIES

At any one moment, causal factors for many types of behaviour are likely to be present. A male stickleback may, for instance, be on its territory, in reproductive condition, in the presence of food, a ripe female, and nest material. Yet animals usually do only one thing at a time: at some stage the causal factors for the other activities fail to produce their effects. We can describe this by saying that there are inhibitory relations between the activities concerned. This is a statement in terms of behaviour and carries no implications about whether the inhibition occurs by direct negative effects in the central nervous system, or by competition for receptors or effectors. Various mechanisms involved have already been discussed in the contexts of selective attention (Chaps. 5 and 6) and responses in conflict situations (Chaps. 16 and 17; see especially pages 396–400).

25.4 EFFECTS OF ONE RESPONSE ON ANOTHER, NOT RELATED TO CHANGES IN THE EXTERNAL STIMULUS SITUATION

As discussed in Chap. 13, each time an animal is exposed to a stimulus which elicits a particular response, the probability of the same response being given again is affected. The effect may be either positive or negative, leading to either an increase or a decrease in response strength. If a number of activities are linked either positively (e.g., by chaining) or negatively (e.g., by inhibitory relations), consequences (internal or external to the animal) of the performance of one of them may also have repercussions on the probability of others.

An example is provided again by the nest-building of female canaries. As we have seen, the activities involved can be subdivided into those involving gathering material, carrying it to the nest site, and building it into the nest. Their integration depends in part on motivational factors which they share (endocrine factors) and a more short-term central state revealed in a heightened probability of any of the nest-building activities being shown. Changeover from one activity to the next depends in part on stimuli encountered during the course of the first: the female cannot "carry" unless she has gathered material, and does not usually visit the nest unless she is carrying. The presence of the adequate stimulus situation, however, is not in itself sufficient to bring about a changeover; the female may carry material to and fro between perches for a long time before visiting the nest. One factor here is that the three groups of activities require progressively higher levels of the motivational factors: for instance gathering is less likely to lead to carrying, or carrying to sitting or building, early in the nesting season than later (see pages 607–608). It is probably for this reason that bout lengths are shorter, the more vigorous the nest-building behaviour (Fig. 25.4), and thus, presumably, the stronger the motivational factors. However this is not the only mechanism involved. Examination of the bout lengths of the several activities indicates that interruption of the activity in progress is an important issue. Long bouts of gathering tend to be followed by long bouts of carrying and of sitting/placing. But for a given length of gathering bout, carrying bouts followed by sitting/placing are shorter than carrying bouts followed by a temporary cessation of all nest-building behaviour (Fig. 25.5). In other words, the bout lengths tend to be longer if they are not immediately followed by the next link in the chain—that is, by implication, if they are not interrupted by it. The ease with which they can be interrupted may depend on decremental effects consequent upon their performance. If such effects accumulate faster than they dissipate, they may be instrumental in permitting the next activity (Hinde, 1958a).

The relations between the successive activities in the courtship of the Glandulocaudine fish *Corynopoma rissei* may have a similar basis. These are, to a considerable degree, independent of changes in external stimuli but, once the sequence has started, the probability of any particular action varies with the immediately preceding one. The dependence of the second activity on the first decreases as the time interval between them increases; and the time interval at which the second is statistically independent of the first varies between the different activities (Nelson, 1964a). Here, however, the influence of the performance of one response on that of the next seems to be more than merely permissive, as was suggested for canary nest building: more direct positive or negative after-effects may be involved.

We have already seen how the performance of one activity may be associated with a simultaneous inhibitory effect on that of others (see pages 396–400). Consideration of cases in which the inhibitory effect is produced by an external stimulus shows how either positive or negative after-effects of one activity on another can come about. At a simpler level of integration, the undulatory movement of the dorsal fin of a sea horse can be inhibited by peripheral stimulation: when the latter ceases, the movements of the dorsal

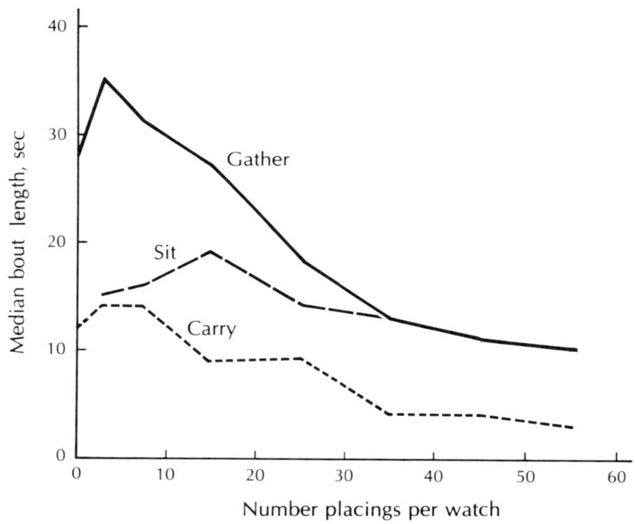

Fig. 25.4 Canary nest building. Length of bouts of gathering, carrying, or sitting, as a function of the number of times material is placed in the nest. (After Hinde, 1958*a*.)

fin are resumed, and for a while have an increased amplitude (von Holst, 1936). Earlier, Sherrington (e.g., 1906), from a study of spinal reflexes, had suggested that, after the withdrawal of inhibition, the excitation of previously inhibited units is likely to develop ("successive induction"). The opposite effect also occurs: the swimming movements made by a medullary goldfish are increased when a current of water is directed against the body, but subsequently fall below the normal level.

Kennedy's studies of the relations between "flying" and "settling" in *Aphis fabae* show phenomena which are similar, at least at the behavioural level of analysis. This insect flies for a while after becoming adult, and then settles on a leaf to feed. Kennedy (1958, 1965, 1966; Kennedy and Booth, 1963, 1964) has shown that elicitation of each of these activities has an inhibitory effect on the other. When one activity ceases, however, the previously inhibited activity may appear with either an enhanced (antagonistic induction) or a reduced (antagonistic depression) strength (Fig. 25.6). Which of these effects occurs depends on the precise conditions. The nature

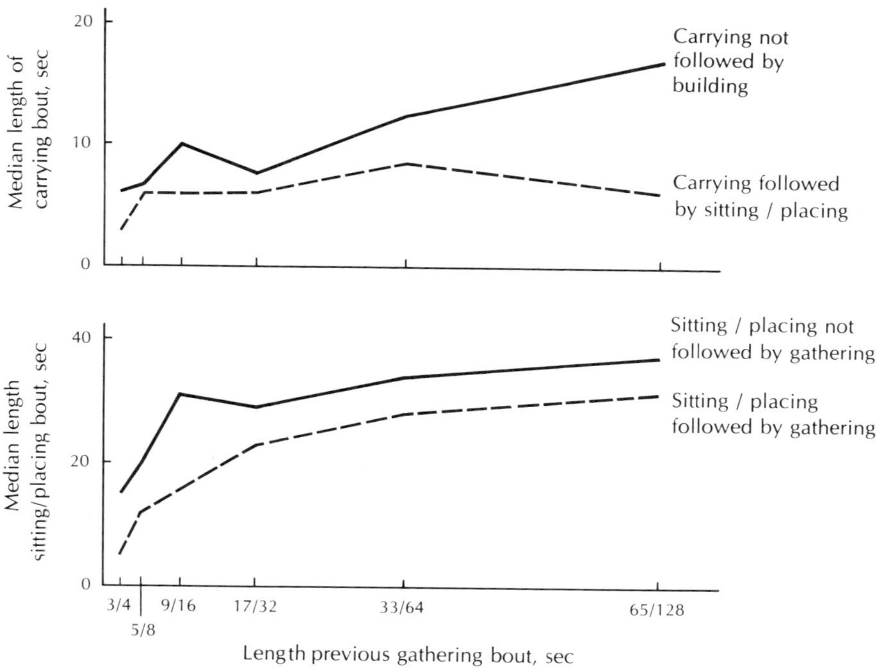

Fig. 25.5 Canary nest building. Length of carrying bouts and of sitting/placing bouts as a function of the length of the previous gathering bout. (After Hinde and Stevenson, in press.)

of the after-effect of flight on settling, for instance, depends on the nature of the surface, the strength of the settling response at previous landings, and the length of time the insect has been flying.

There is evidence that it is not the performance of one activity which has the inductive and depressive after-effects on the other, but the stimuli which elicit it. Thus surfaces differ in their effectiveness in eliciting settling. Since weak and strong stimuli for settling have typically different after-effects on flight even when the amount of settling is held constant, it must be these stimuli which subsequently affect flight and not the settling activity itself. Kennedy suggests that the stimulus, in eliciting one activity, simultaneously inhibits the antagonist: this inhibition then produces the after-effects.

Kennedy's interpretation of these results in terms similar to those of Sherringtonian successive induction and inhibition does not imply a similarity of mechanism; he himself points out that the time scale and level of complexity are quite different from the phenomena which Sherrington studied. What is important is the reciprocal influence of each of these activities on the other, and the demonstration that flight is not appetitive to settling and feeding in the sense that both are caused by the same factors, for it is in part flying (or the stimuli eliciting it) which creates a state in which the insect is ready to feed.

Another example, though the author does not choose to interpret it in terms of successive induction, concerns the "rocking" movement made by

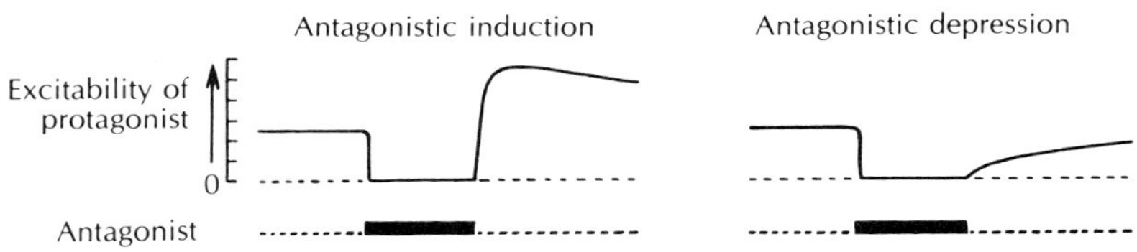

Fig. 25.6 Schematic diagram of "antagonistic induction" and "antagonistic depression". (From Kennedy, 1964.)

Hemileucid moths on assuming the resting position after preceding activity (Bastock and Blest, 1958; Blest, 1958, 1959). The number of "rocks" given increases with the length of a preceding flight according to a simple linear relationship, the relationship changing somewhat with age from eclosion. Since the duration of flight controls the subsequent strength of the rocking response, it must be registered in some way. This registration appears not to depend on exteroceptive or proprioceptive feedback resulting from flight, or to change in metabolic reserves. In the absence of plausible alternatives, it thus seems that the control of rocking depends directly on activation of the central nervous mechanisms involved during flight, and not on its peripheral consequences.

It is possible that a similar description could be applied to the sudden increase in aggressive or fleeing behaviour which is seen immediately after copulation in many vertebrates. As discussed in Chap. 16, the preceding courtship involves tendencies for a variety of incompatible types of behaviour, such as attack, fleeing, and behaving sexually. At copulation the first two are suppressed, but may find expression as soon as it is over.

Similar "induction" may also be observed in conditioning experiments. In the case of classical conditioning (page 572), a response such as salivation, which after training has come to be elicited by a positive conditioned stimulus, may be enhanced if that stimulus is preceded by negative stimulus which temporarily inhibits the response (Pavlov, 1955). A similar effect can be produced in an operant situation. First responding is reinforced in the presence of either of two stimuli which are presented in succession. If responding to only one of the stimuli is then reinforced, the rate of responding to the other (now the "negative" stimulus) decreases. Furthermore, the rate of responding to the positive stimulus increases over its level before the discrimination training. Since the change in rate to one stimulus is in a direction away from the rate to the other stimulus, the change is called "contrast" (Reynolds, 1961b). That an initial period of making a response to the negative stimulus is essential for the increase to the positive stimulus can be demonstrated by introducing the negative stimulus gradually, so that responding to it never occurs. There is no increase in responding to the positive stimulus as a result of such "errorless" discrimination training (Terrace, 1966).

25.5 CONSUMMATORY STIMULI AND/OR REINFORCEMENT

A special case of common causal factors (see Sec. 25.1) is provided by stimuli which bring a sequence to an end. Insofar as such "consummatory stimuli" affect a number of different patterns of behaviour, they play a role in their temporal integration. A number of examples, including consummatory stimuli for courtship in the three-spined stickleback and for eating and drinking in mammals, were discussed in Chap. 10. Another is provided by the nest-building behaviour of domesticated canaries. Here stimuli from the nest which the female herself has built produce a decrease in nest-building behaviour, and thus play a part in producing the reduction in all nest-building activities which occurs around the time that the eggs are laid (Fig. 25.7).

In this case the female does not cease building as soon as she comes into contact with a completed nest: rather, repeated stimulation produces a gradual reduction in the intensity of her building. This is in fact often the case: the action of consummatory stimuli is rarely all-or-none. Furthermore, the stimuli consummatory to a sequence do not necessarily decrease the frequency of all the units in that sequence. In the case of the canary nest-building they probably do: stimuli from the nest influence the subsequent probability of gathering the material, of carrying it to the nest site, and of sitting-building it into the structure. In such a case the consummatory stimuli play an obvious role in the observed temporal correlations between the units. In the male bitterling (*Rhodeus amarus*), however, a rather more complicated situation seems to occur. In this fish the female lays her eggs near a fresh-water mussel to whose gills they adhere. The smell of the freshly laid eggs produces an increase in the aggressiveness of the male and also an increase in "skimming"—a movement much used in the early phases of courtship but also associated with sperm ejection after the female has laid her eggs. Other sexual movements, however, are decreased by the smell of the eggs, apparently as a consequence of the increase in aggressiveness. Analysis shows that the extent to which the various sexual movements are inhibited varies, with most movements being markedly affected but skimming not at all. Thus the smell of the eggs serves as a consummatory stimulus for most sexual activities (though acting through the increase in aggressiveness it induces), but not for skimming (Wiepkema, 1961). The effect of the smell of eggs on the temporal correlations between the various sexual activities of the bitterling is thus rather different from that of the nest on the nest-building activities of the canary.

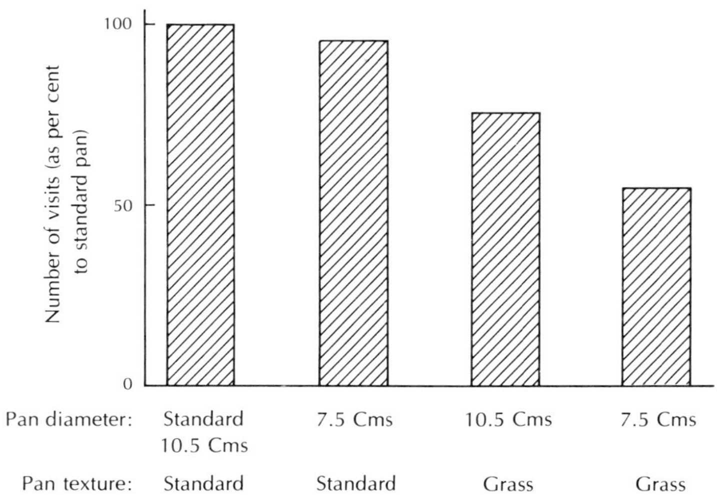

Fig. 25.7 Effect of substituting pans differing from the standard pan in diameter or texture on building behaviour of female canaries. Columns indicate number of visits to substituted pan as a percentage of the number of visits to the standard pan on adjacent days. (After Hinde, 1965.)

In addition to bringing a sequence of behaviour to an end, such consummatory stimuli may also influence the probability that a particular response will recur, and thus lead to the occurrence of a sequence of responses.

As we have seen (Chap. 24), if a response of a food-deprived animal is followed by a brief presentation of food, then the frequency with which that response recurs may be increased: it will decrease again if food is no longer presented (page 573). Such reinforcement affects not only the maintenance of the operant responses, but also their patterning in time. How the occurrences of the response are distributed over time depends largely on the "schedule of reinforcement", which specifies how many operant responses must be made to obtain reinforcement, and usually involves a temporal requirement as well. On a variable interval schedule, for instance, a pigeon is reinforced for pecking a disc at an interval after the previous food presentation which varies in a random fashion from one occasion to the next: a rather steady rate of responding results. If, on the other hand, food is given on a fixed interval schedule, that is, for a response made a fixed t or more seconds from the preceding reinforcement, then the bird will peck the disc at a rate which is at first low but increases up to the time when the next reinforcement becomes available, resulting in a characteristic "scalloped" pattern (Ferster and Skinner, 1957; Fig 25.8).

The precise manner in which such temporal patterns are controlled is, however, still not known. It has been suggested that the sequence of responses depends on "chaining", with each response acting as a dicriminative stimulus for the next. But an explanation solely in terms of the consequences of responding cannot account for a regular rate on a variable-interval schedule and a scalloped pattern on a fixed-interval schedule. Furthermore, when moment-to-moment control by response-produced stimuli is disrupted, for instance by black-outs or by a stimulus in whose presence responding is never reinforced, the scallops nevertheless remain over the normal part of each interval (Blough and Millward, 1965; Kelleher, 1966). A sequence of fixed-interval responding must therefore involve some additional temporal reference, possibly related to previous reinforcement (Dews, 1962). Such a temporal reference could be based on the inter-response times which have actually occurred in the situation. For instance, on a fixed interval schedule reinforcement comes when the sum of inter-response times since the last reinforcement equals or exceeds the value of the interval: response rate could thus depend in part on the increasing probability of reinforcement as inter-response times accumulate (Morse, 1966). However, a fixed interval schedule, which specifies merely that a response should be made some time after the interval has elapsed, still leaves many characteristics of the response sequence uncontrolled. Indeed while responses to most conventional schedules can be loosely described in terms such as "scallops" or "steady rates", more precise measures, such as the rate at any instant or the actual inter-response times, vary among subjects and even within subjects. In other words, such schedules do not control important factors which affect responding. To the extent that each operant response depends on a preceding response, whether it be the operant response or some other unrecorded response, a schedule which leaves such dependencies unspecified loses control over response emission (Blough, 1966).

The extent to which reinforcement can control responding in a restricted situation makes it likely that it plays an important role in the integration of naturally occurring sequences. However, as we have seen with canary nest-building, not only reinforcement but also common causal factors, chaining, and other effects of performance may also be involved. The precise roles of these different processes in the integration of naturally occurring sequences remain to be investigated.

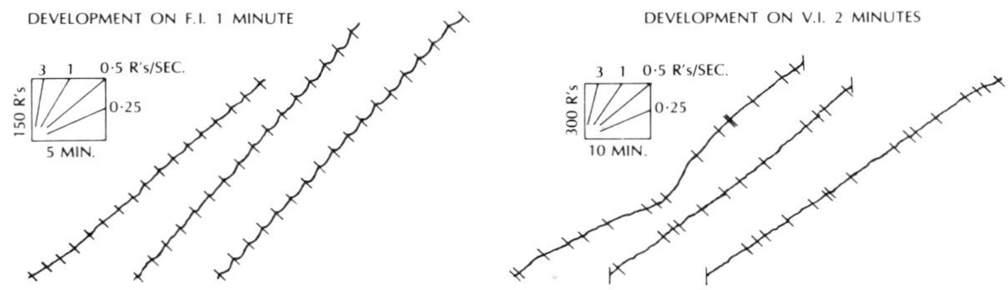

Fig. 25.8 Cumulative records of responses obtained on (left) *a fixed-interval schedule, with reinforcement available at the end of each minute; and* (right) *a variable interval schedule, with reinforcement available on average every 2 min. The scales for ordinate (cumulative number of responses) and abscissa (time) are shown by the inset. The short lines represent the occurrence of reinforcement.* (From a drawing by Elizabeth Steel.)

25.6 GOALS

Many sequences of behaviour appear to involve a series of responses which are selected or directed in such a manner as to bring the animal to a goal situation. In such cases the various responses which it makes before the goal is reached may be said to be integrated by the goal. The precise interpretation of such cases, however, is a subject on which extreme views have been expressed. When the "purposive" drives of the early twentieth century were countered by the reflexes and tropisms of the pioneer behaviourists, the use of words like "goal" or "directiveness" came to be regarded by many as disreputable. This was due in part to the fact that such terms can carry meaning in addition to that necessary for them to fulfill their descriptive role. But now that it is recognized that even machines with negative feedback can be regarded as teleological mechanisms (Rosenblueth et al., 1943), it has become more respectable to interpret animal behaviour as goal-seeking. And since Craik's (1943) suggestion that thinking can be regarded as a form of symbolism of the same kind as that which goes on in a computer, it has become less difficult to postulate in animals a "Schema", "cognitive map", "neuronal model", or Sollwert with which the external situation can be compared. However, we must ask where such concepts are necessary. Also we must not forget that animals are diverse: they not only achieve the same thing in different ways, but also have capabilities which differ markedly from phylum to phylum and species to species. We must therefore ask just how broad are the generalizations which can be made here: in seeking to analyse the causal mechanisms which underlie adaptive behaviour, should we regard all animal behaviour as "directive" (Russell, 1945), and can we discriminate between degrees of directiveness?

Discussion here must be concerned in the first instance with questions of terminology. The concept of a goal refers always to a situation which brings a sequence of behaviour to an end. However, the converse is not true, for sequences of behaviour may be brought to an end in ways that cannot usefully be described as involving the achievement of a goal. We have, for instance, already considered the roles of changes in responsiveness associated with performance (Chap. 13), habituation (Chaps. 6 and 13), removal or change in the eliciting stimuli (e.g., pages 332–350), and extinction (Chap. 24). If the label "goal-directed" is to be useful in categorizing behaviour, it is unlikely to be so if we define it so broadly that it includes such cases. Even many cases where a behaviour sequence is brought to an end by consummatory stimuli (Chap. 10) hardly merit the label, for the role of the consummatory stimuli may be merely inhibitory (Hinde and Stevenson, 1969a).

A common factor in definitions of goal-directed behaviour is the implication that the animal has some kind of internal correlate of the goal situation (e.g., Schneirla, 1949; Thorpe, 1963a). More operationally, this could imply that goal-directed responses are those which are governed by discrepancy between the present situation and a goal situation: the behaviour ceases when the two become the same. A number of such cases have already been considered in preceding chapters—for instance the corrections made by a singer attempting to maintain a note of constant pitch (Chap. 3) and the optomotor and orientation behaviour of insects (Chaps. 5 and 7). At first sight we have here a simple criterion for goal-directed behaviour: whereas behaviour which is merely elicited by a stimulus increases in frequency or intensity the nearer the stimulus approaches a certain value, behaviour directed toward a goal might be expected to decrease as the incident stimulus situation approximates that of the goal. However, there are several difficulties. First, as discussed on pages 131–144, the step from behaviour which appears to be governed by discrepancy between the present stimulus situation and Sollwert to the implication of a comparator mechanism is one that requires great caution. Second, cases such as those cited above are vastly simpler than most behaviour which is more usually described as goal-directed, such as maze-running. When we consider the latter, things are not so simple. Rats often run fastest as a goal is approached (Miller, 1959), and human subjects may work hardest as a task nears completion (Welford, 1962). In such cases response strength bears no constant relation to the discrepancy, and we thus have no simple criterion for goal-directed behaviour.

Thorpe (1963a) gives additional criteria by which a set of actions may be defined as purposive or goal-directed: that there should be a large but not unlimited series of alternatives for action, and that there should be some ability to choose the more expedient or simpler of two or more courses of action. The intention behind the latter is clear: it refers to the ability, for instance, to take a short-cut in a maze rather than following a set routine (e.g., Osgood, 1953). As a general criterion, however, it is not very useful, for a more or less expedient course can hardly be said to exist for an animal which never uses it. Thorpe does not specify the criteria for the existence of alternative courses of action: Does it imply that the effectors for alternative courses are available, or the central nervous mechanisms, or the ability to develop new courses?

Furthermore, the occurrence of alternative actions before a consummatory situation is reached is certainly not itself adequate grounds for implying "goal-direction". For instance, Jennings' classic observations on *Stentor*, the protozoan which may "try" several different courses of action when exposed to a stream of carmine particles until it avoids them, was cited by Russell (1945) as implying "a power of varying behaviour according to the result of previous action, a power which is essentially a psychological one, a power of relating events and acting in accordance with the situation". This, however, is an overinterpretation: each type of behaviour is stereotyped, and its cessation could be due merely to inhibitory effects consequent upon performance (Chap. 13), rather than to response to error signals.

Thorpe (1963a) also uses complexity and variability in behaviour as evidence for goal direction. As one example, he cites accounts of nest-repair behaviour in wasps and other invertebrates which demonstrate a surprising flexibility in both the types of behaviour shown and the manner in which it is organized into sequences. As yet, however, there has been no analysis of the precise stimuli eliciting the several responses, and thus no conclusive proof that (for instance) the response of a wasp to a hole in its nest is a response to a discrepancy between mutilated and complete nest, rather than to the edges of the hole.

A second case discussed in detail by Thorpe (1963a) is the nest-building of the long-tailed tit (*Aegithalos caudatus*). Citing a descriptive account by Tinbergen, Thorpe shows that this involves thirteen motor patterns and responsiveness to eighteen different releasers. He suggests that such complexity provides evidence of goal direction, arguing that it must indicate that the bird has some "conception" of what the completed nest will look like, and that "the addition of a piece of moss or lichen here and there will be a 'help' towards the ideal pattern, and that other pieces here and there would detract from it". There are in fact three points here. First, evidence from other avian species does indeed show that the intensity of building behaviour is influenced by stimuli from the nest (pages 620–621): by itself, however, this does not necessarily imply comparison with a Sollwert. Second, it is at least likely that the orientation of particular building movements is influenced by irregularities, gaps, or excrescences in the existing structure; but even if this can usefully be said to involve comparison with a Sollwert, it is at most a matter of the elicitation of the individual movements, and quite a different issue from comparison between present structure and completed nest. Crook (1964b), studying the nest-building of weaver-birds, found no evidence for a process of comparison between the stimuli provided by the nest and a goal situation. When nests were mutilated, the repair behaviour was essentially a stereotyped repetition of earlier phases of building, and was successful only so long as the orientation of the nest was not changed. If the orientation was changed, a new initial ring was built, and from that a new nest which did not conform to the old.

A third point is that Thorpe, in his discussion of this and other cases of nest-building, suggested that error signals contribute to integration in another way. Not only are movements elicited and orientated by discrepancies between the present structure and the completed nest, but movements which reduce that discrepancy, even though not achieving the goal immediately, are more likely to be repeated. This would bring nest-building into line with a number of cases of the development of motor patterns (see Chap. 19), but the evidence for it is as yet meagre. Such evidence as is available indicates that the efficiency and orientation of the movements depend more on the level of motivational factors than on experience (see page 615).

A third case, cited by Russell in the same context, is of quite a different order. As discussed in Chap. 24, if rats are subjected to spinal or cerebellar operations so as to interfere with their motor coordination, they may nevertheless use quite novel movements to make errorless runs through a maze (Lashley and Ball, 1929). The essential point here is that the new movements are not stereotyped, but selected from variable patterns in such a manner as to bring the animal nearer the goal. Furthermore, the new patterns are "directly and efficiently substituted without any random activity".

Contrasting these examples we may draw two conclusions. First, persistence with varied action cannot be taken as evidence of "goal direction" in any useful sense unless the actions are selected from a large repertoire of stereotyped movements, or from variable movements, in such a way that they bring the animal nearer a goal. Second, the most sophisticated goal-directiveness implies selection of the most economical course from starting point to goal, and thus, for any given starting point, lack of variability. When a monkey given a mechanical puzzle bites, bangs, pushes, pokes, jiggles and jogs it in an apparently infinitely variable fashion, the behaviour is goal-directed only in the sense that solution of the puzzle will bring the behaviour to an end, and not in the sense that it is guided by error signals. Far from being an indication of goal-directiveness, variability in the types of behaviour shown may indicate the opposite, for it implies little selection from all possible activities. The apparent variability of efficient goal-directed behaviour lies only in the large number of starting points from which the goal can be reached.

When it is useful to describe behaviour as goal-directed, the whole sequence often involves not merely one final goal, but the achievement of a succession of subgoals. This can be illustrated by Deutsch's (1960) model of maze running by rate. Deutsch argues that the learning of a maze involves not the learning of a series of responses (see page 615), but of a succession of subgoals which, if achieved in turn, lead to the ultimate goal of the food box. Here we are concerned primarily not with the acquisition of the behaviour, but with its control once it has been acquired. To account for this Deutsch has produced a conceptual scheme, based on a feedback model of motivation and utilizing also the principles of common causal factors, chain responses and inhibitory relations, which accounts for many of the facts of maze running. His motivational model is shown in Fig. 25.9 and indicates the role played by feedback. Deutsch further supposes that, underlying the performance of a complex task (like the correct running of a maze) is a series of units each consisting of an analyser (or sensory/perceptual mechanism capable of selective responsiveness), a link, and a motor unit. The links are connected together in such a way that when the primary link (referred to in Fig. 25.9 as the "central structure") is activated by a change in the medium, all the other links are initially also activated (i.e., principle of common causal factors). As soon as an analyser is stimulated, excitation is no longer passed to units more remote from the primary link, but is directed to the motor mechanism (Fig. 25.10). That unit then acts as a goal-seeking mechanism, moving so as to optimize its own stimulation. When this is achieved, the next analyser is stimulated (chain-response principle), and stimulation through it is optimized by the activity in its motor mechanism. The process continues until the activity in the primary link is inhibited by the appropriate consummatory stimuli. With additional assumptions about changes in resistance in the connections consequent upon performance, Deutsch's scheme describes a mechanism which will learn to run through a series of goals in turn, and stop when it gets to the end. It will avoid blind alleys, and take short-cuts, but cannot make detours. The ability to take short-cuts is of particular interest as it implies that the machine will, if opportunity arises, pursue goals nearer (in some sense, spatial or otherwise) to the final consummatory one, instead of

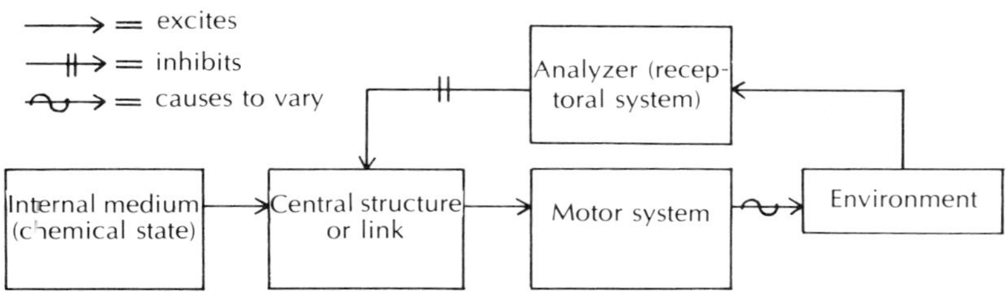

Fig. 25.9 Deutsch's feedback model of cotivation. (After Deutsch, 1960.)

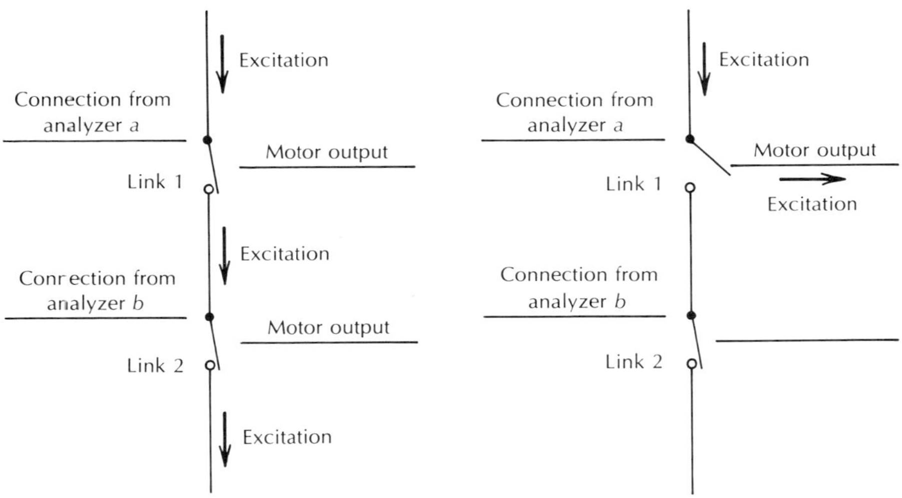

Fig. 25.10 Deutsch's scheme for the mechanisms underlying goal-directed behaviour. (After Deutsch, 1960.)

Left, neither analyser is stimulated and excitation is conducted through the links; *right,* stimulation of *analyser a* switches the excitation to the corresponding *motor output,* cutting off *link 2.*

going through its routine: in other words, selection from among the possible goals is not random, nor dictated entirely by the environment. It is thus goal-seeking, both in the sense that subgoals are sought by each motor mechanism in turn, and in the sense that some ability to select appropriate subgoals is present. Such schemes may be useful in displaying the requirements for "explanations" (at any analytical level) of maze running. How much they will help in ultimate understanding is a matter of opinion.

At this point it becomes important to emphasize that behaviour may be goal-directed at one level of integration but not at another. In the early stages of nest-building, the movements used by a weaver-bird in stitching may be oriented and coordinated to produce a perfect stitch, but the stitches are not oriented in such a way that a nest is produced. The lower levels may be goal-directed, but there is no direction of behaviour toward the completion of a nest. Eliot Howard's classic field studies of bird behaviour provide many other examples of the manner in which species-characteristic movements are performed in an incomplete, disoriented, and functionless manner when motivation is low. Even when motivation is at full strength, biological consequences may be achieved without the operation of teleological mechanisms guiding behaviour towards them: in nest-building the goal-directed mechanisms controlling the part-activities may be integrated by stimuli from the growing structure to produce a finished nest (see Figs. 15.2 and 15.3).

Such cases can be accommodated in the scheme devised by Miller et al. (1960) to embrace the goal-directedness of human behaviour. These authors suggest that the basic unit of behavioural organization (the TOTE unit) depends on negative feedback as illustrated in Fig. 25.11*a*. The test unit serves to compare the input with some criterion established in the organism, the response occurring as a result of incongruity. After the response, a further test may or may not reveal congruity. If it does, action ceases. The sequence is thus Test-Operate-Test-Exit.

Such a scheme can be applied at many levels, from the reflex, where the diagram may represent nervous impulses passing along the neurones, to more complex behaviour, where it may be information or control. Such TOTE units, they suggest, may be arranged in an hierarchical manner, which they illustrate by the mechanism which might underlie hammering a nail (Fig. 25.11 *b*). Hammering has two phases—lifting the hammer and striking the nail. It must continue until the head of the nail is driven down flush. The sequence of events is thus as follows:

Test nail. (Head sticks up.) Test hammer. (Hammer is down.) Lift hammer. Test hammer. (Hammer is up.) Test hammer. (Hammer is up.) Strike nail. Test hammer. (Hammer is down.) Test nail. (Head sticks up.) . . . Test nail. (Head flush.) Control transferred elsewhere.

We need not dwell here on the many ways in which such a model can be used to clarify the structure of human behaviour. Important for our present purpose is the emphasis laid on hierarchical organization, for this permits the possibility that behaviour can be goal directed (in some sense) at one level but not at another. Miller et al.'s insistence that even the simplest reflex is based on a TOTE structure is both not necessary and misleading. Indeed, it is a major task for comparative psychologists to specify at just which levels and which stages such a model is applicable, and at which it is not (cf. Chap. 3).

In conclusion, we see that any definition of goal-directed behaviour

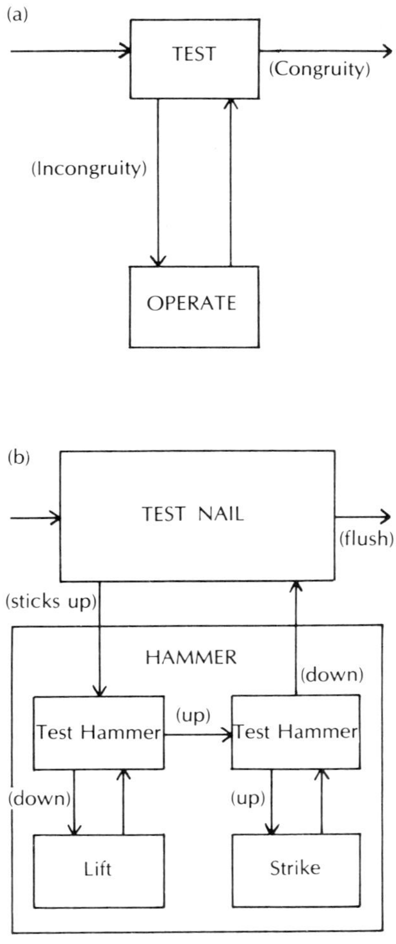

Fig. 25.11 a, *The TOTE (Test-Operate-Test-Exit) unit;* b, *a hierarchy of TOTE units which describes the hammering of nails.* (After Miller et al., 1960.)

raises difficulties. While there is a sense in which it is a useful category, distinguishing sequences of behaviour which seem to involve principles of integration additional to those discussed earlier in this chapter, it is hard to pin down precise criteria. Clearly, to be useful the term "goal" must be applied to situations which bring behaviour sequences to an end. It must be further limited to cases in which the behaviour involves response to a discrepancy between present and goal stimulus situations: this clearly implies feedback. However, it must be remembered that such a discrepancy is not always easily defined and that it can affect behaviour in a number of ways: it may determine the intensity of the behaviour, or its direction, or its nature, or even the subgoal toward which it is directed. The extent of the variability of the behaviour, while throwing some light on the nature of the mechanisms involved, is not itself useful as a criterion. In view of such difficulties in definition, it is clear that we are a long way from being able to use goal-directiveness as anything more than a crude descriptive concept. Nevertheless the use of a control systems approach, even in a qualitative way, shows that there need be nothing mystical about goal-directiveness: when used in this restricted way, it is susceptible to causal analysis.

25.7 SUMMARY

1. The mechanisms whereby the various types of behaviour in an organism's repertoire are organized into functional sequences are discussed.
2. The sharing of causal factors by a number of types of behaviour may assist in their functional integration.
3. In a chain response, each movement or activity brings an animal into a situation where the next is evoked. Usually such chains are far from rigid. The differences between naturally occurring chains and those established in the laboratory are discussed.
4. Inhibitory relations between activities, as discussed in Chap. 17, play an essential role in integration.
5. The performance of any one response in a sequence may affect the probability of others. Such effects can be described as "induction" and "depression".
6. Behavioural sequences may be integrated through the consummatory situations which bring them to an end and which may have reinforcing effects influencing both the frequency and temporal patterning of responses.
7. "Goal-directedness" has been used to refer to mechanisms of greatly differing complexity and has little explanatory value. Although useful in focussing attention on certain types of behaviour, it is difficult to set up general criteria for "goal-directedness" which can be satisfactorily applied to a variety of situations.

26

INTERACTION IN DEVELOPMENT

Earlier we noted that studies of development usually involve the isolation of particular aspects of behaviour, whereas in practice the development of each aspect depends on the development of others in the course of a complex interaction between organism and environment. In this interaction, it is not only the organism which changes. The environment is not static, and may itself be affected by the organism: this is, of course, especially the case with the social environment. Thus, before we can come to grips with the development of the behaviour of the whole organism, we must somehow teaze apart this interaction between the changing organism and the changing environment. In this chapter we shall consider three cases where this task is being attempted. The first concerns the seasonal development of breeding behaviour in a bird: here the interaction is mainly (though not entirely) with the physical environment. The other two involve the development of young mammals, where the mother constitutes the most important element in the environment.

26.1 REPRODUCTIVE DEVELOPMENT IN FEMALE CANARIES

To reproduce successfully, a female canary must show an overlapping sequence of types of behaviour—pair formation, courtship, and copulation; nest-site selection and nest-building; egg-laying, incubation, and feeding the young. Each of these activities must appear at the right time in relation to the others, at the right time with respect to the physical and social environment, and of course at the right time with respect to egg-laying. The correct timing is achieved by a complex interaction between the female and her environment. It is convenient to consider first two aspects of this which have been mentioned previously in other contexts, and then to consider in broader outline the integration of the whole.

First, the interaction with the mate. The canary is closely related to the greenfinch and more distantly to the chaffinch, and its courtship follows a sequence similar to that already described for those species (Chap. 16). At pair formation the male is initially hostile to all other individuals. Receptive females are submissive to his threats, and this causes the male to change from mainly aggressive behaviour to courtship. Courtship can be described as involving conflicting tendencies to attack, flee from, and mate with the female. As a result of stimulation by the male, the female's behaviour gradually changes, producing a further decrease in the aggressiveness of the male. At this stage his courtship can be understood in terms of conflicting tendencies to flee from and mate with the female. Gradually the latter predominates and mating occurs, though many attempts are unsuccessful (compare Fig. 16.5). This brief account is of course only a sketch of a complicated interaction lasting days, weeks, or even months (Hinde, 1955/1956).

The second aspect is the sequence of activities involved in nest-building. As we have seen (Chap. 25), this can be analysed in the first instance into three groups of activities: gathering, carrying, and sitting/placing. The frequency with which each of these occurs can be influenced by the same endocrine factors, and since they show correlated short-term fluctuations, they presumably also have other short-term factors in common. The changeover from each activity to the next depends in part on response-produced external stimuli, so the chain-response principle is involved. The changeover also depends on the different thresholds of the several activities and perhaps on decremental effects consequent upon performance. Since the canary does not normally gather material while sitting in the partially built nest, though material suitable for carrying is obviously available there, there must also be inhibitory relations between the different activities. We have seen that the role of experience in the integration of the sequence is an open issue (pages 615–616).

As discussed in Chap. 25, many of the changes in nest-building behaviour which occur as the nest is built do not demand the postulation of any reinforcing effect of approximations to the finished structure. Stimuli from the partially completed nest do, however, have a number of other effects on nest-building behaviour (Hinde, 1958a). Substitution of an artificial nest of smaller diameter than the nest-pan in which the female is building results in:

1. A change in the frequency of certain of the nest-building movements which the female makes as she is sitting in the nest cup. For instance "scrabbling" with the feet is more frequent in a small nest cup than in a large one.
2. The structure of the nest also influences the orientation of these nest-building movements, as illustrated more dramatically by the weaver-birds (Figs. 15.2 and 15.3).

3. Increased selection of lining material for the nest. Under laboratory conditions, if grass and feathers are provided, grass is used for the main cup and feathers for the lining. At any stage in nest-building, substitution of a nest cup of smaller diameter results in an increase in the proportion of feathers carried. Figure 26.1 shows data for two individual females which were given a small-sized nest cup on alternate days and a normal one on the intervening days. The abscissa is the date with respect to the laying of the first egg; the ordinate is an index of the proportion of feathers carried. The lines fitted to the data for the small cup (the filled circles) show a higher proportion of feathers than those fitted to the data for the large cup (the open circles) (see also Hinde and Steel, 1962).

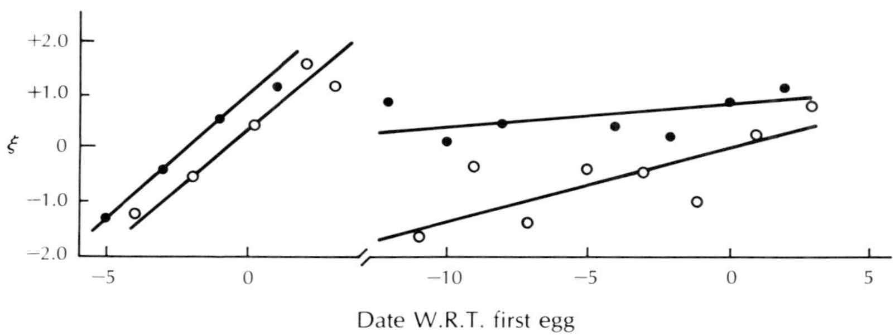

Fig. 26.1 Selection of nest material by female canaries as a function of type of nest-pan and date with respect to egg-laying. Feathers and grass were provided. (After Hinde, 1958.)

Ordinate, a measure of the proportion of feathers selected; *abscissa,* date with respect to the first egg. *Open circles,* observations with standard pan; *closed circles,* observations with smaller pan. Data for two individual females are shown.

4. A decrease in the total amount of building behaviour. This normally occurs around the time of egg-laying (Fig. 25.7).

It will be apparent that the course of nest-building behaviour is determined by an interaction between the female and the nest structure which (in normal circumstances) she creates.

Having discussed the interactions between the female and her mate and between the female and the nest structure, we may now consider other aspects of her reproductive development over the season as a whole (reviewed by Hinde, 1965; Hinde and Steel, 1966; Hinde, 1967a). These are summarized in Fig. 26.2. The female's reproductive development early in the

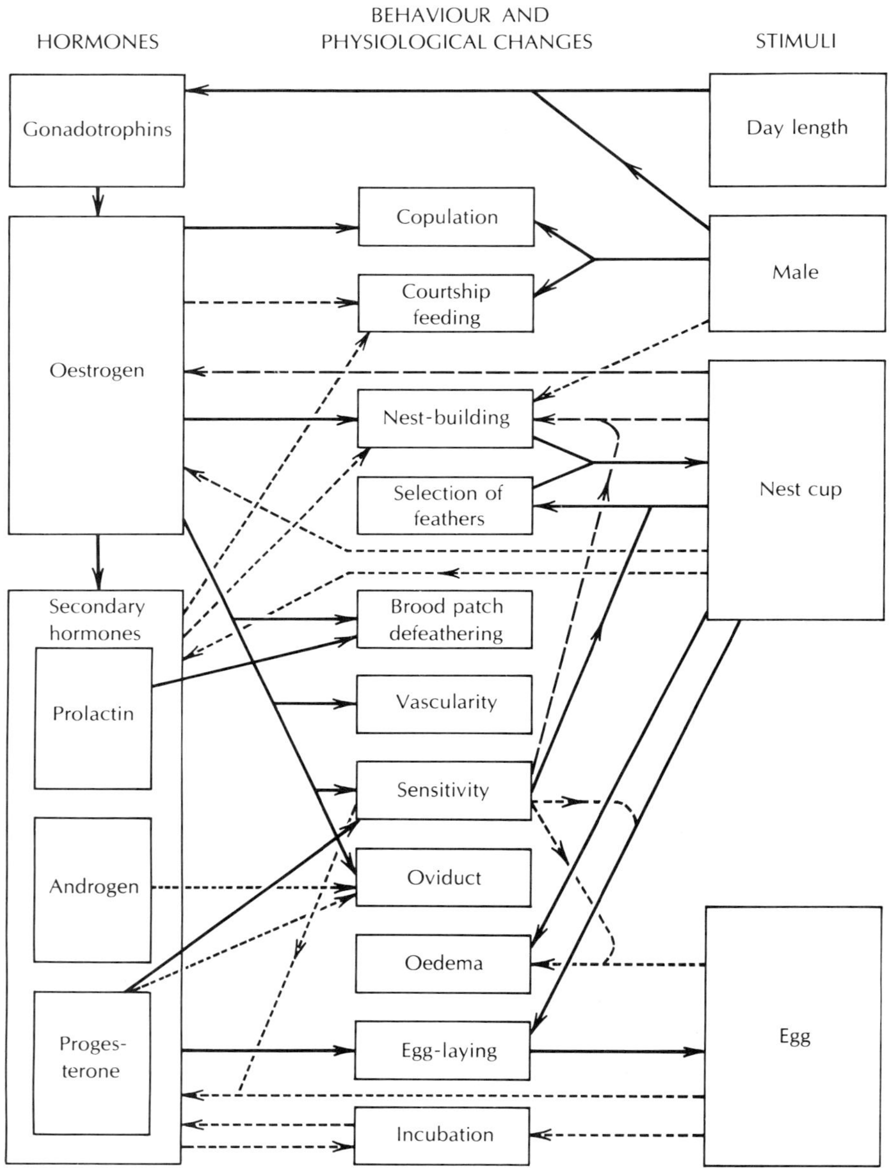

Fig. 26.2 Relations between external stimuli, internal hormonal conditions, and reproductive development and behaviour of female canaries. (Modified after Hinde, 1965.)

Continuous lines, experimentally established positive effects; *discontinuous lines*, negative effects; *short dashed lines*, probable relationships not yet established with certainty. Where the extent to which an effect may be indirect has not yet been established, only the direct effect is shown.

natural breeding season is accelerated by increases in daylength. Comparable changes can be induced by injections of gonadotrophins, so the daylength presumably acts by a pituitary mechanism. The female's reproductive development is also augmented by stimulation from the male: females kept without a male show a slower reproductive development than females kept with one (Fig. 26.3).

As a result of the hypothalamic-pituitary-endocrine changes thus induced, the female starts to nest-build, participates in courtship feeding, and mates with the male. Nest-building occurs especially in the week or two before egg-laying (Fig. 26.4*a* and *b*), but the female's reproductive development up to this point is often not smooth: she may show cyclical changes in which nest-building activity rises to a peak and dies away without being followed by egg-laying (Fig. 26.4*c*–*e*).

During the weeks before egg-laying the female develops a brood-patch: her ventral surfaces become defeathered, vascular, and later oedematous. They also become increasingly sensitive to tactile stimulation (Fig. 10.1).

Copulation and nest-building reach their peak simultaneously a few days before egg-laying. Nest-building activity coincides with a period of rapid ovarian development and can be induced by large doses of exogenous oestrogen, but its precise hormonal basis is not yet clear. The cyclical changes in reproductive development (Fig. 26.4*c*–*e*) may be correlated with the maturation of successive groups of ovarian follicles which cease to develop and become atretic before ovulation. The several aspects of brood-patch development, and that of the oviduct, can be induced by exogenous injections of oestrogen in combination with prolactin, androgen, and/or progesterone (Hutchison et al., 1967, 1968a). The secretion of these secondary hormones is a consequence of either gonadotrophin secretion, oestrogen secretion, or both.

The female receives stimulation through her brood-patch from the nest cup which she herself has built: the effectiveness of a given nest is thus augmented by her growing sensitivity to tactile stimulation (Fig. 10.1). It is probably largely for this reason that the proportion of feathers she carries to the nest increases as egg-laying approaches even when the external stimulation remains constant (Fig. 26.1). In addition to its effect on nest-building behaviour, stimulation from the nest (or from nest-building activities) also accelerates further reproductive development: females not allowed to complete nests lay later than those which are (Fig. 26.4). The influences of the male on the later aspects of reproductive development are in fact probably exerted in part through the nest which he stimulates the female to build.

The causal bases of incubation and feeding the young have not yet been studied in this species. It will be apparent, however, that the control of reproduction does not depend on a series of independent causal chains, but on a complex nexus of events both inside and outside the organism,

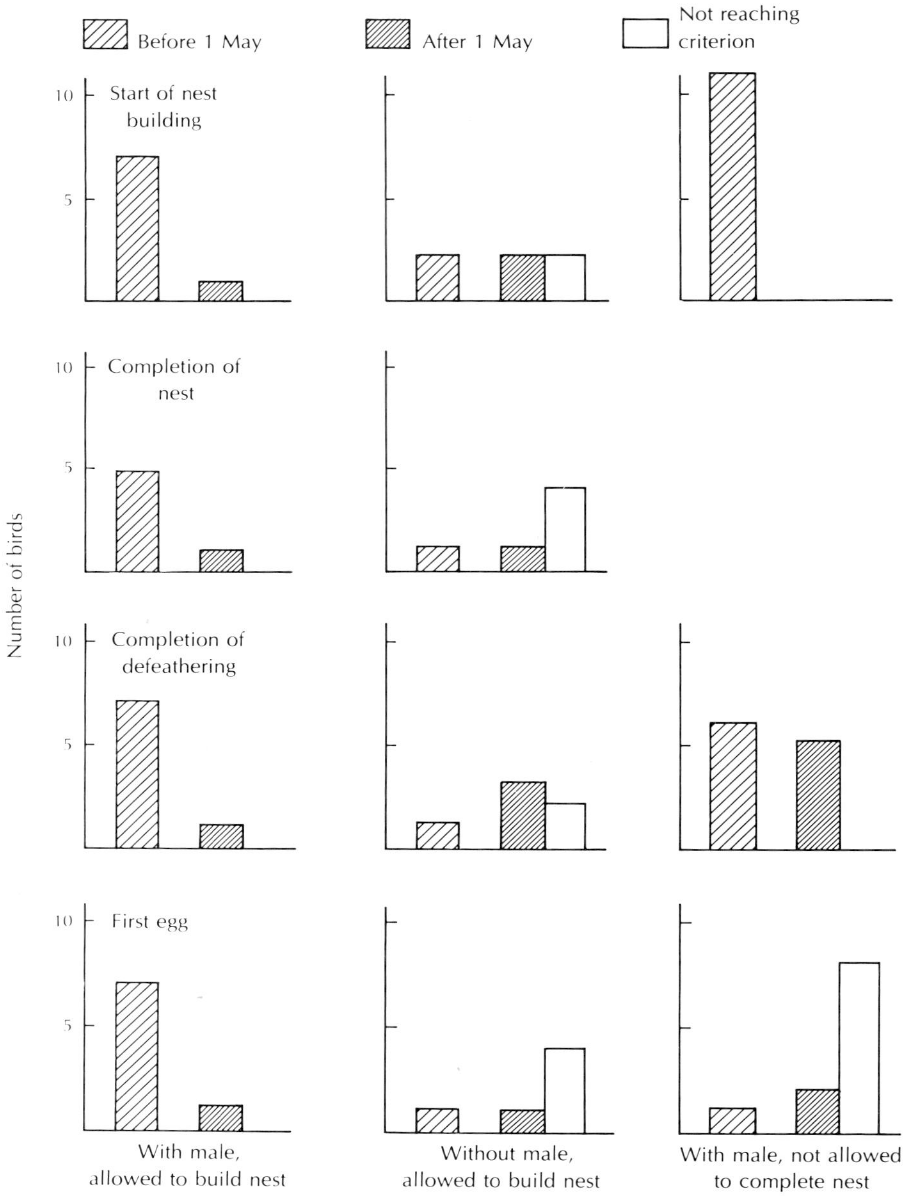

Fig. 26.3 The effect of depriving female canaries of a male and/or the opportunity to build a nest on various aspects of reproductive development. (After Hinde and Steel, 1966.)

The columns show the number of females, kept under the conditions specified, reaching the criterion by May 1, or after May 1, or failing to reach it during the breeding season.

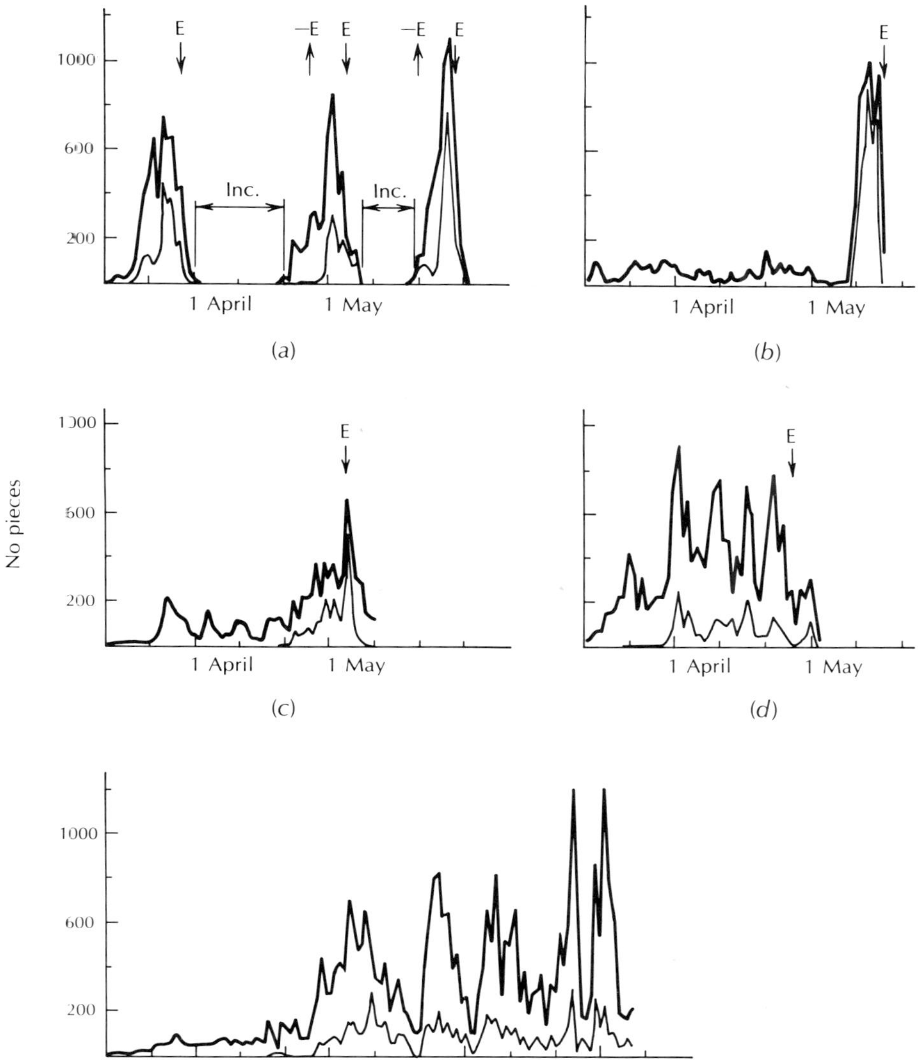

Fig. 26.4 Changes in nest-building activity through the season of five individual unpaired female canaries. The building activity was assessed in terms of the number of pieces of grass removed from a dispenser (thick lines) *or placed in the nest-pan* (thin lines) *daily.* (After Hinde, 1967.)

E indicates laying of first egg of clutch; –*E* indicates egg was removed by the experimenter. *a*, A female which laid three times, incubating between; *b*, a female which showed little building until just before she laid late in the season; *c,d*, females which showed roughly cyclical fluctuations in building before laying; *e*, a female which showed fluctuating building but did not lay.

involving both positive and negative feedback loops. The interactions represented in Fig. 26.2 undoubtedly simplify the complexity which exists in nature, but the general principle that avian reproductive development depends on systems of this type is confirmed by Lehrman's (e.g., 1965) detailed studies of the ring dove (*Streptopelia risoria*), and by data from other species (see reviews by Lehrman, 1961, 1962; Farner, 1966).

26.2 MOTHER–INFANT INTERACTION IN CATS

The interaction between cat and kittens has been the subject of detailed study by a number of investigators (e.g., Schneirla and Rosenblatt, 1961; Rosenblatt et al., 1961; Schneirla et al., 1963, and refs. cited; Ewer, 1959, 1961). Their work shows that the development of behaviour in the kitten depends on a continual and progressive interchange between cat and kittens which begins at or before parturition. Components of this interaction are the hormonal condition and related behavioural changes in the mother, and the motivational, perceptual, and motor development of the kittens; related changes in both mother and young are based upon experience in the litter situation. Unfortunately, it is not possible here to do more than hint at the intricate mosaic of factors which has been revealed by careful study under controlled conditions.

Parturition itself involves both internal changes in the female (hormonal and muscular) and a change in her environment (appearance of neonate and birth fluids). The female exhibits a number of different types of behaviour (self-licking, licking newborn or floor, eating after-birth, etc.) which result from an interplay between the stimulating effects of endogenous events (e.g., uterine contractions and the passage of the foetus) and of exogenous stimuli resulting from the by-products of these processes, such as the birth fluids, the neonate, and the placenta. During parturition, intervals of intense activity, which facilitate delivery, are interspersed with periods of exhaustion and rest, during which the kittens may initiate nursing. Throughout, the female directs much of her behaviour (for instance licking) towards the posterior part of her body; this may be induced by changes in cutaneous sensitivity and contribute towards parturition.

Parturition is followed by a resting period, during which licking and other responses of the female to the neonate intensify the bond between them. Nursing usually begins within the first hour after birth. It is usually first initiated by the female's enclosing response: she approaches one or more of the kittens, lies on her side, and draws her front legs ventroposteriorly and her back legs ventroanteriorly, thus enclosing the kittens. The kittens are guided to the nipples by their responses to the shape of the mother's body, but at first they gain them only after much nuzzling and fumbling. From the first day onward their suckling behaviour is much influenced by experience: individual methods of suckling, often involving specific nipple preferences, are soon developed (see also McBride, 1963; Ewer, 1961). The various types of adjustment between mother and infant in the feeding situation are affected of course by the perceptual-motor development of the kitten.

The length of time which the female spends nursing the kittens depends at all ages on the number of kittens in the litter. It decreases as the kittens get older (Fig. 26.6). For about the first 20 days, nearly all nursings are initiated by the mother (Fig. 26.5). Between days 20 to 30, feeding is initiated by either kittens or mother, the changes paralleling the perceptual development of the kittens (some aspects of which were discussed in Chap. 20), increased efficiency in feeding and orientation, and changes in their mobility, vigour, and so on. The feeding process provides a focus around which much of the learning involved in the establishment of the social relationship occurs, but during this period other responses (e.g., "play") appear. During the next period the initiation of feeding depends more and more upon the initiative of the kittens as they follow the mother about. The female actively repels and avoids the kittens, and the feeding periods become progressively shorter.

During the early days after birth, relations between mother and kitten are strongly influenced by the latter's attachment and orientation to the "home-corner", a region of the cage which has been saturated chemically by the female. As early as a few hours after birth, odours from this site, as well as contacts with the female, have a quieting effect on the neonate. Initially the kitten orients to this corner on a tactile chemical basis, but later (about 5 days) visual orientation to the home region, and later still to the entire cage, is achieved. Early on the mother also plays a large role in retrieving the young if they stray from the home-corner, responding primarily to their vocalizations, and at this stage the mother plays a major role in keeping the litter together. As the kittens grow, the behavioural relationship between mother and kittens changes. Not only do the kittens play a progressively greater role in initiating feeding, but the mother actively withdraws from them for much of the time. That this withdrawal is not due solely to endogenous changes in the mother is indicated by the fact that it is less marked in one-kitten litters than in larger ones.

By temporarily removing kittens from mother and litter mates and subsequently returning them to the litter situation, Schneirla and Rosenblatt (1961) demonstrated "the necessity of a continued behavioural and functional interchange with female and litter mates if the kitten is to develop an adequate suckling adjustment typical of its age group. Psychological processes concerning perceptual and behavioural organization are required in which organic factors underlying reciprocal stimulation play a basic, inextricable role". It is thus apparent that an understanding of the development of social behaviour can be obtained only from a study of the interaction between the kitten, its mother, and its litter mates through each

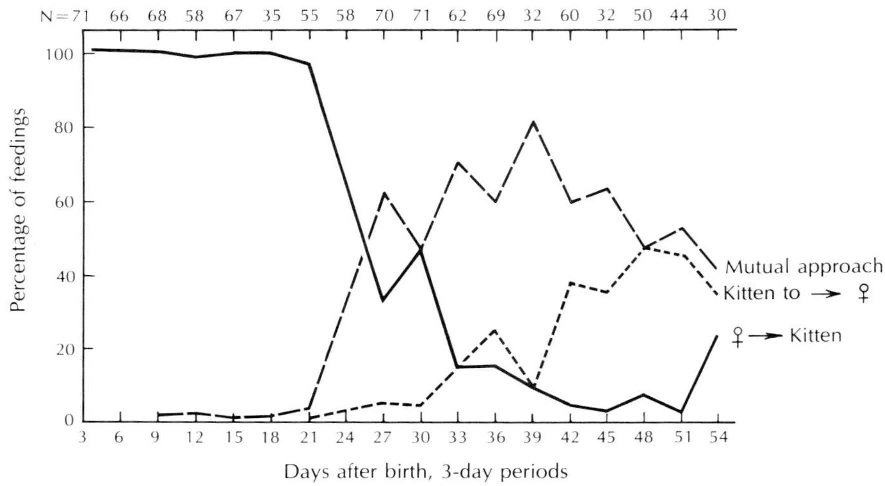

Fig. 26.5 Initiation of feeding in three litters of cats, based on daily 2-hr periods of observation. N, *the number of feedings observed in each 3-day period.* (From Schneirla and Rosenblatt, 1961.)

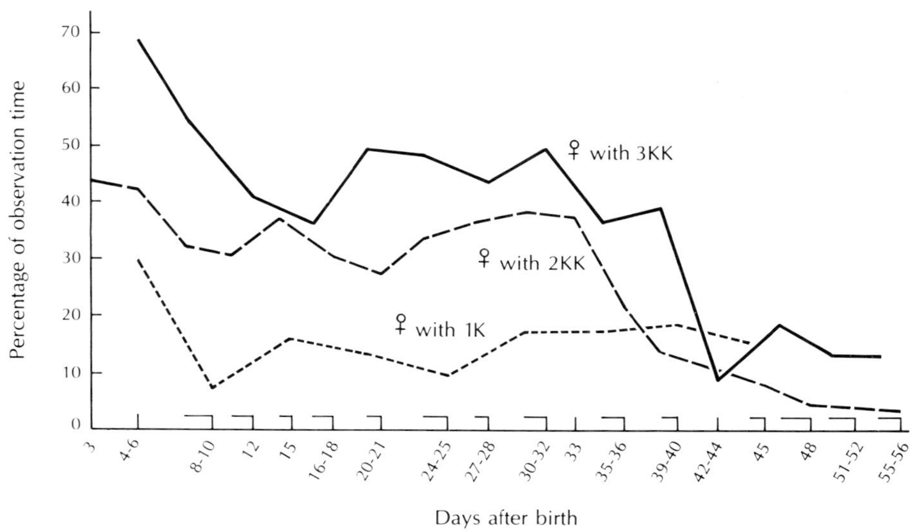

Fig. 26.6 Time spent by three cats in nursing their kittens, based on daily 2-hr observation periods. (From Schneirla and Rosenblatt, 1961, after Schneirla, Rosenblatt, and Tobach, 1963.)

stage of development. A comparable analysis of the maternal behaviour of the rat, in which the role of mother–infant interaction in influencing the mother's changing responsiveness to the pups is emphasized, is given by Rosenblatt and Lehrman (1963 and refs. cited).

26.3 MOTHER–INFANT INTERACTION IN RHESUS MONKEYS

Because of their phylogenetic relationship to man, particular attention has been paid to the development of behaviour in primates. Descriptive data are now available for a considerable number of species, but for considerations of space the present discussion is confined to studies of captive rhesus monkeys as reviewed by Harlow and Harlow (1965; see also Hansen, 1966) and Hinde and Spencer-Booth (1968). Harlow and his collaborators were concerned mostly with monkeys living in rather small cages which could be observed closely and kept under fairly strict experimental control. Hinde and Spencer-Booth studied captive groups, each consisting of a male, three or four females, and their young: these were set up to provide the animals with a rather richer environment than a small laboratory cage, in which control and recording more accurate than field conditions permit would still be possible. The data obtained were qualitatively similar to those derived from field studies (e.g., Southwick et al., 1965).

The infant rhesus monkey spends much of its early life attached to its mother in a ventroventral position. At this stage two groups of reflexes are of major importance. The first of these is concerned with feeding, and includes the rooting reflex, and probably also locomotor reflexes which serve to bring the infant near the nipple, such as forced upward climbing. The second group includes all those patterns which assist the infant in establishing and maintaining itself in the ventroventral position. The reflex behaviour of the infant would, however, be quite ineffective if it were not reciprocated by appropriate passive and active behaviour on the part of the mother. This includes cradling, nursing, grooming, restraining, and retrieving. That these responses depend on the mother's own experience is shown by the Harlows' finding that mothers which had been reared on wire mother-surrogates (see below) did not themselves show adequate maternal behaviour.

The infant rhesus appear to learn to recognize their own mothers in the first week or two of life: in an experimental situation learned discriminations are known to be possible from 10 days (Zimmermann and Torrey, 1965). Perceptual abilities and manipulative skills improve during the early weeks (see pages 477–481 and 489–499): eye-hand-mouth coordination permits some self-feeding by about 3 weeks, and the infant becomes capable of running on vertical wire-netting by about 6 weeks (Hinde et al., 1964).

The infants first leave their mothers when about a week old. Their growing independence is illustrated in Fig. 26.7, which is based on observa-

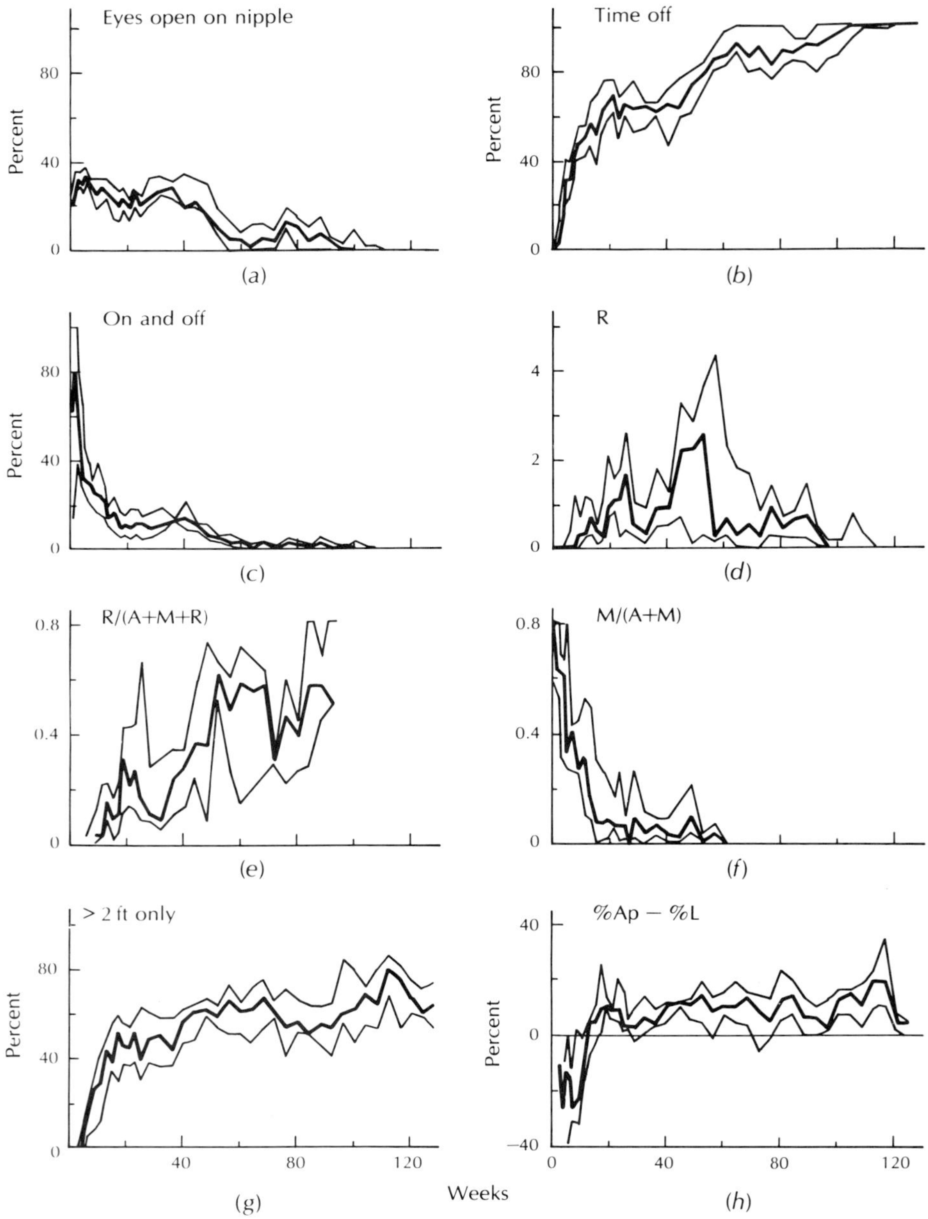

Fig. 26.7 Age changes in rhesus monkey mother-infant interaction. (From Hinde and Spencer-Booth, 1968, by permission of the Royal Society.)

a, Number of ½ min in which awake on nipple; *b*, number of ½ min in which recorded off mother; *c*, number of ½ min in which recorded both on and off mother as tions of the small captive groups made during the morning hours. The time spent off the mother increases rapidly from the second to about the twentieth week, and again just before the end of the first year (Fig. 26.7*b*). There are corresponding falls in the times spent on the nipple (Fig. 26.7*a*) and asleep. The decrease in the number of half-minutes in which the infant was recorded both on and off its mother, expressed as a percentage of the number in which it was off her (Fig. 26.7*c*), shows that the periods off the mother were increasing in length.

Although the infant continues to spend a fifth of the daylight hours awake and attached to the nipple for most of its first year, only a small proportion of this time is spent sucking. The nipple serves as an additional point of suspension from its active mother, and as a "pacifier". That the nourishment which the mother supplies represents only one aspect of her attractiveness for the infant was demonstrated by Harlow and Zimmermann (1959). Infant rhesus were removed from their natural mothers and reared on surrogate mothers. In one experiment each infant had two surrogates, one consisting primarily of a wire cylinder and the other covered with cloth. Half of the infants were fed from the wire mother and half from the cloth mother. Irrespective of which mother they were fed from, the infants spent much more time on the cloth mother than on the wire one (Fig. 26.8). In other experiments it was shown that the opportunity to cling to the cloth mother, but not the wire mother, allayed the infant's anxiety in a strange and fear-provoking situation. Evidently the "contact comfort" supplied by the mother is an important issue.

As the infant develops, its relationship with its mother gradually changes. To understand the dynamics of these changes, it is necessary to teaze apart the roles of mother and infant in their interaction. For instance, is the growing independence of the infant primarily a result of changes in the infant's behaviour, of changes in the mother's behaviour, or both? A

---

a percentage of the number in which they were off at all; *d*, number of times infants were rejected by their mothers; *e*, ratio of number of times infants were rejected to total number of time infants attempted to gain nipple or were picked up by their mothers; *f*, ratio of number of times infants were picked up by their mothers on mother's initiative, to total number of times they were accepted or picked up by them; *g*, number of ½ min which infants spent wholly more than 2 ft from their mothers as a percentage of the number in which they were off at all; *h*, difference between percentage of approaches (from more than 2 ft to less than 2 ft) which were due to movement by infants, and percentage of leavings (from less than 2 ft to more than 2 ft) which were due to infants.

In *a*, *b*, and *d*, data are expressed as a percentage of ½ min watched. In each case the thick line represents the median and the thinner lines the limits of the interquartile range.

convenient starting point is to consider how far observed changes in the mother–infant relationship can be understood as effects of one or more of four possible simple types of change in the participants—namely, an increase or decrease in the tendencies of mother or infant to respond positively (i.e., other than by avoidance or aggression) to the other. The hypothesis that changes in mother–infant interaction depend on only four possible types of basic change will of course prove to be an oversimplification, but it provides a convenient starting point.

The method is illustrated in Table 26.1, where the predicted effects of these changes on four representative measures are shown. The measures are the percentage of time the infant spends off its mother; the number

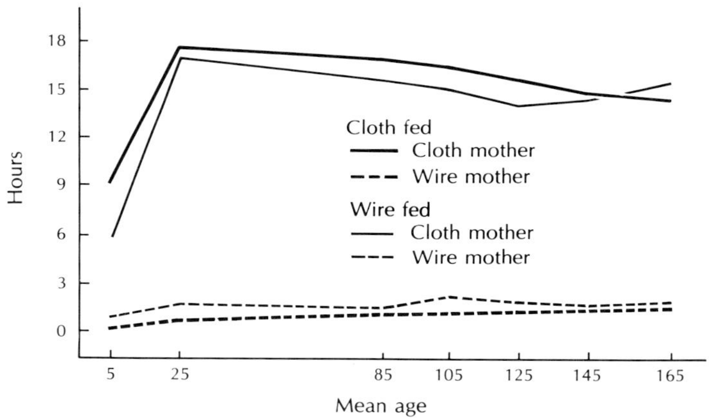

Fig. 26.8 Mean number of hours per day spent by infant rhesus monkeys in contact with cloth and wire mother surrogates. (From Harlow and Zimmermann, 1959.)

of times its attempts to gain the nipple which are rejected by the mother (R) divided by the number of times it gains ($A + M$) or unsuccessfully attempts to gain (R) the nipple; the proportion of half-minutes off the mother which it spends wholly more than 2 ft from her ($>$ 2 ft only: 2 ft is about the distance the mother can reach); and a measure ($\% Ap - \% L$) of the infant's role in maintaining proximity to the mother. The last is calculated from the difference between the percentage of occasions in which the distance between infant and mother decreases from more than 2 ft to less than 2 ft (Ap) which are due to movement by the infant, and the percentage of occasions on which it increases (L) similarly (Ap and L stand for approaches and leavings).

TABLE 26.1. The predicted directions of the consequences on certain measures of mother–infant interaction of the four basic changes indicated on the left. The direction of the arrow indicates whether the individual under which it lies shows an increased or decreased tendency to approach the other. For explanation, see text.

		Time Off	$R/(A + M + R)$	> 2 *ft only*	$\% Ap - \% L$
infant ⟵	mother	+	−	+	−
infant ⟶	mother	−	+	−	+
infant	mother ⟶	+	+	+	+
infant	mother ⟵	−	−	−	−

It will be seen that time off and time more than 2 ft from the mother, although not of mathematical necessity correlated with each other, would be affected in similar directions by all the postulated changes. Likewise the relative frequency of rejections ($R/A + M + R$) and the infant's initiative in maintaining proximity should be similarly correlated. Thus the extent to which such pairs of measures are in fact correlated provides some indication of the adequacy of the four postulated changes in explaining the observed phenomena. It will also be observed that an increase in time off the mother can be ascribed primarily to the infant if associated with a decrease in the relative frequency of rejections, and to the mother if associated with an increase. Similarly an increase in time at a distance from the mother is due to the infant if associated with a decrease in the infant's initiative in maintaining proximity, and vice versa [see Hinde (1969) for further details].

The application of the method is illustrated by the hypothetical data in Fig. 26.9. If the open circles represent mean scores of particular measures for all infants, we can calculate correlations between the measures over each of the age ranges delimited by the arrows. It will be apparent from Fig. 26.7 that the time off the mother (Fig. 26.7*b*) is positively correlated with time more than 2 ft from her (Fig. 26.7*g*), and relative frequency of rejections (Fig. 26.7*e*) with $\% Ap - \% L$. This justifies further use of this approach. The figure also shows that the relative frequency of rejections (Fig. 26.7*e*) is on the whole positively correlated with the time off the mother (Fig. 26.7*b*); and the time at a distance from the mother is positively correlated with the infant's initiative in maintaining proximity ($\% Ap - \% L$, Fig. 26.7*h*). Thus the increasing independence of the infant with age is due immediately to a change in the mother's behaviour rather than to one in the infant's. The proportion of occasions on which the infant went on the nipple

$(A + M)$ and the initiative was the mother's (M) decreases as the infant grows older (Fig. 26.7*f*): this also is in harmony with the view that the increasing independence of the infant is due in large part to a change in the mother's behaviour. This decrease in maternal attentiveness is shown also by other measures: for instance the frequency with which the mother restrains her infant decreases with its age, becoming zero at 40 to 50 weeks, and the frequency with which she behaves aggressively to it increases. The absolute frequency of rejections (R) rises to a peak but then, unlike the relative frequency $(R/A + M + R)$, falls off (Fig. 26.7*d*): this fall must be ascribed to decreased persistence by the infant.

This emphasis on the mother's role in permitting or promoting the independence of the infant does not of course imply that the changes in

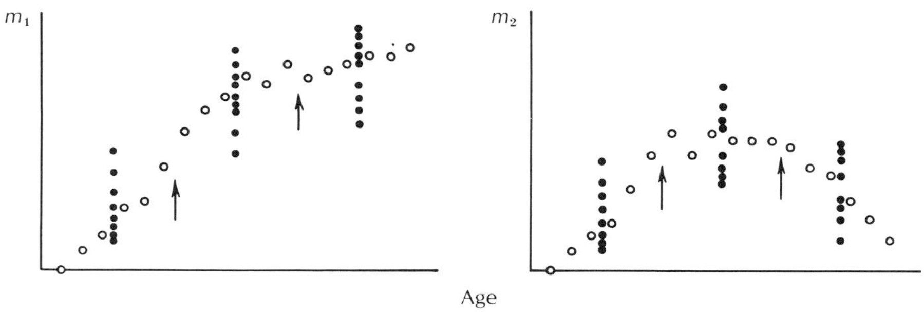

Fig. 26.9 Relationship between age changes in two hypothetical measures, m_1 and m_2. (From Hinde and Spencer-Booth, 1968, by permission of the Royal Society.)

Open circles, medians for all animals at each age; *black circles*, mean for individual animals over age ranges whose limits are marked by arrows.

her behaviour arise endogenously: they may be initiated by changes in the infant's behaviour, such as its increasing demand for milk or locomotor activity. These are consequences of its development, which in turn depends on the mother. The changes in the relationship thus depend on complex interactions, but this analysis shows that the mother is not merely a guardian and passive provider of the infant's needs: she also plays an active part in promoting its independence.

Figure 26.9 shows how a similar analysis of the nature of individual differences can be carried out. The black circles represent hypothetical individual means over the time periods delimited by the arrows: the rank-order correlations between the two measures can be assessed. Analysis shows that, although most differences in mother–infant interaction are due to both partners, maternal differences are of primary importance in the crucial period between 7 and 12 weeks, and become less so.

As the infant comes to spend more time off its mother, it investigates and explores its environment more and plays more with its peers. The latter consists primarily of "approach-withdrawal" and "rough-and-tumble" play. The development of peer-peer relations has been studied by Harlow and his colleagues in situations like that shown in Fig. 26.10. The mothers are confined in the living cages, which can be connected to the playpen units by openings of a size which permits passage only of the infants. The amount of play behaviour can be recorded during daily sessions in which the infants are allowed access to each other. Harlow and his colleagues also

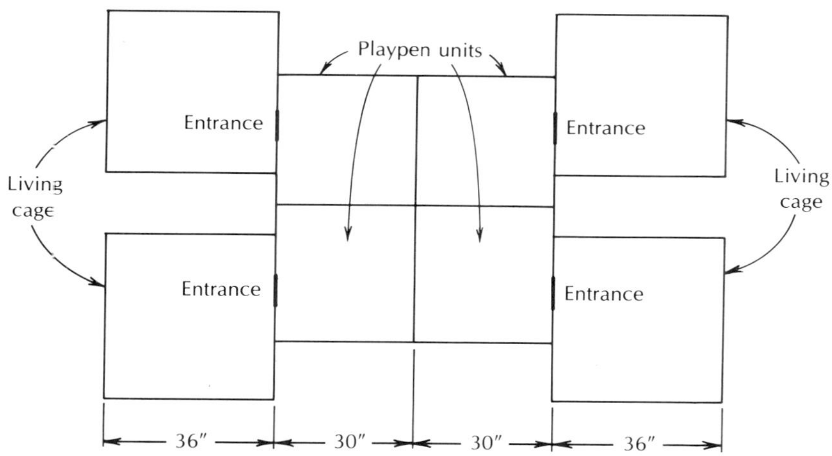

Fig. 26.10 Floor plan of playpen test situation used for testing effects of early social environment on development of young rhesus monkeys. (From Harlow and Harlow, 1965.)

used more complex "playroom" conditions, in which infants without their mothers were studied in a more complex physical environment. Under such conditions both types of play increase up to an age of 2 to 3 months. When infants are living in social groups, play behaviour remains frequent until the end of the second year. Males show a greater tendency to initiate rough-and-tumble play than do females, and the amount of play behaviour is also markedly influenced by experience. Infants brought up in total social isolation for 6 months show much less play and also less social threat when tested in a playroom situation over a 32-week period, than do infants reared in bare wire cages where they can see and hear, but not touch, other infants (Fig. 26.11*a*,*b*). More subtle effects of experience were apparent in larger groups living under more nearly natural conditions. For instance, age changes in the nature of the social play suggest that the younger infants gradually learn to avoid the play-attacks of their peers.

Sex differences become apparent quite early on. Males not only initiate more rough-and-tumble play, but also show more social threat (Fig. 26.12), branch-shaking, and mounting behaviour from an early age.

The course of the infant's behavioural development is highly influenced by its interaction with its mother and its peers. To take the extreme case,

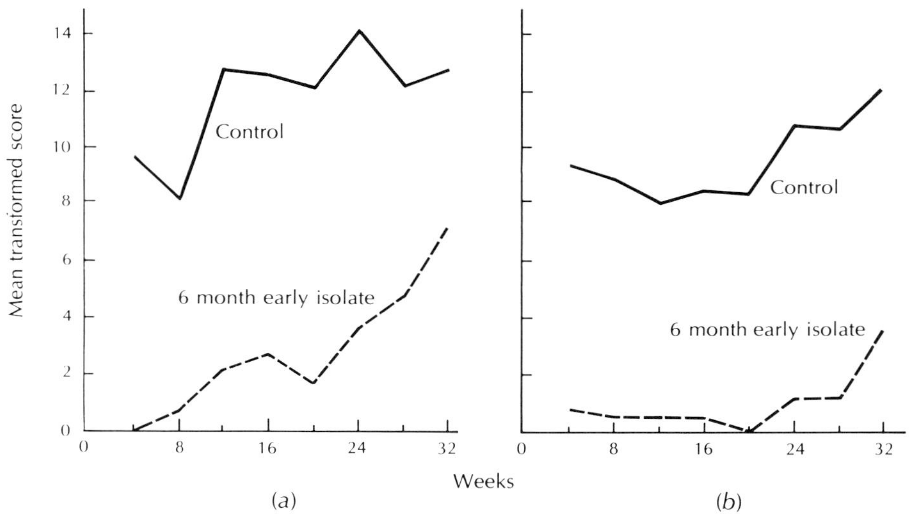

Fig. 26.11 a, Play responses; b, social threat responses by 6-month semi-isolated controls and 6-month (early) total isolate rhesus monkeys. (After Harlow and Harlow, 1965.)

infants reared in bare wire cages are subsequently usually (though apparently not always, Meier, 1965) incapable of normal adult sexual behaviour, and those females which do become pregnant show quite inadequate maternal behaviour. Under more normal conditions, peers attract the infants to play away from their mothers, while the mothers control the infants' interaction with their peers and may interfere if play becomes too rough. Thus interactions with mother and with peers influence each other mutually.

If an infant's mother is removed for a few days, the infant becomes depressed, showing a reduction in play behaviour and general activity. When the mother is returned, most infants are at first intensely clinging: a complex interaction ensues in which the mother first accedes to the infant's demands and then rejects them, until finally a more harmonious relationship is reestablished. Figure 26.13 shows the median time spent off the mother by a group of control animals (thick line) and by four animals whose mothers were removed for 6 days at 30 to 32 weeks. It will be seen that two of the animals were quite markedly affected on this measure for some months after the separation experience. The full effects of such a period of separation can only be assessed in terms of a number of measures of mother–infant interaction and infant behaviour: present evidence indicates that they may be considerable. One of the most important variables affecting the severity of the effects seems to be the nature of the mother–infant

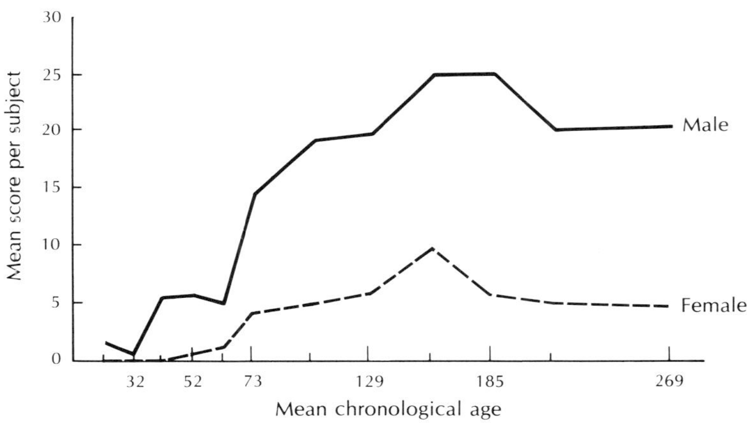

Fig. 26.12 Development of threat response by male and female rhesus monkeys. (From Harlow and Harlow, 1965, after Harlow, 1962.)

relationship beforehand (see Spencer-Booth and Hinde, 1967; Seay and Harlow, 1965; Kaufman and Rosenblum, 1967).

Under natural conditions, mother and infant live in the complex social structure of a troop. Other animals in the troop may display considerable interest in the infant, looking at, touching, grooming, or holding it, playing with it, behaving aggressively or sexually towards it. In laboratory social groups, investigatory and maternal behaviour is directed especially towards infants under 12 weeks old, and subsequently decreases: females 2 years older than the infant are more prone to show such behaviour than females of any other demographic category (Spencer-Booth, 1968).

The mothers usually resent such attentions and protect their young either by restricting their movements, removing them, or threatening the other monkeys. That the attentions of the social companions may have a marked effect on the mother–infant relationship was shown by a comparison between mother–infant pairs living in a social environment with mother–infant pairs living in social isolation. In the latter the infants spent more time off their mothers and went to a distance from them in more half-minutes, than in the former (Fig. 26.14). Comparison with the data on the relative frequency of rejections and with that on the infant's initiative in maintaining proximity (cf. Table 26.1) shows that the difference was primarily due to the greater permissiveness of the mothers in the isolate mother–infant pairs. Indeed it will be seen that, in the isolate animals, the

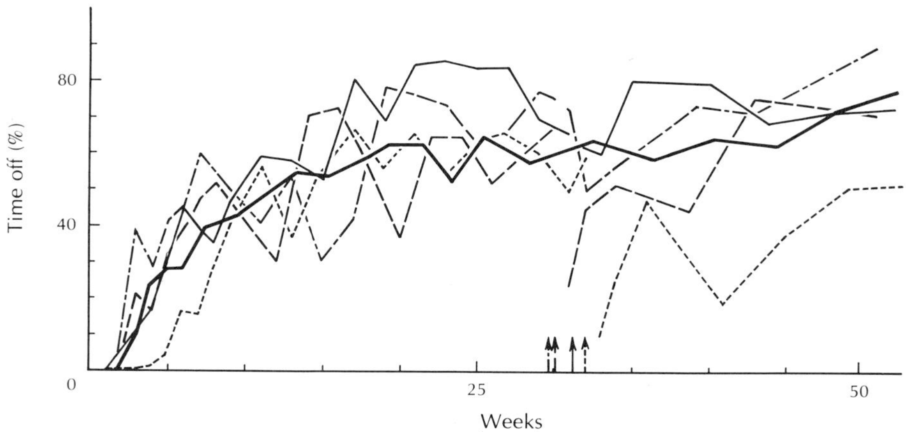

Fig. 26.13 Percentage of ½ min in which infants were recorded off their mothers from 0 to 52 weeks old. (From Hinde and Spencer-Booth, 1968, by permission of the Royal Society.)

Thick line, the median for 8 control animals; each of the *other lines,* an individual whose mother was removed for 6 days at the age indicated by the arrow.

infants were primarily responsible for maintaining proximity between mother and infant from a very early age (i.e., $\%Ap - \%L$ positive). However, although the infants went away from their mothers more often, they spent fewer whole half-minutes there. This difference is probably to be ascribed to the absence of peers: when the isolate infants left their mothers, they soon returned because she was the only social companion available.

In the rhesus monkey the male plays a relatively small direct role in the infant's development, but he influences it indirectly in a number of ways. For instance, not only does he take a major share in the protection

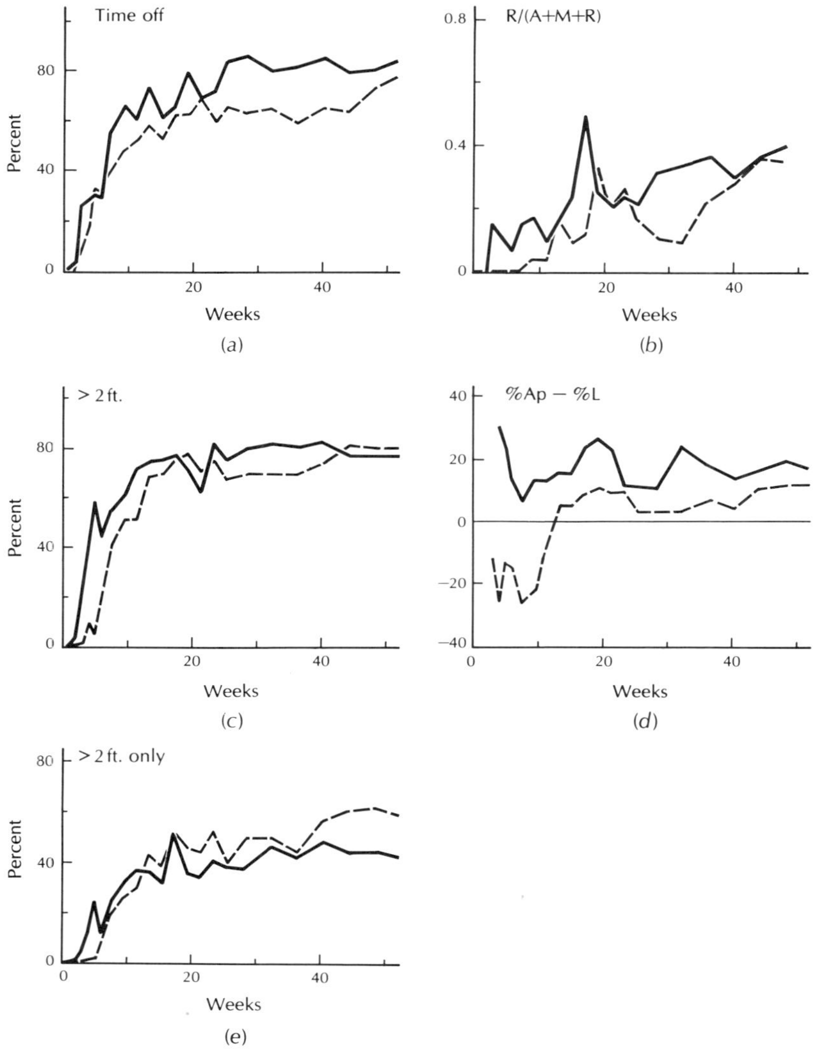

Fig. 26.14 Comparison of mother–infant interaction in isolate and group-living mother–infant pairs. (From Hinde and Spencer-Booth, 1968, by permission of the Royal Society.)

a, Number of ½ min in which recorded off mother as a percentage of ½ min watched; *b*, ratio of number of times rejected by mother to total number of times infants attempted to gain nipple or were picked up by mother; *c*, number of ½ min in which infants went more than 2 ft from their mothers as a percentage of ½ min watched; *d*, difference between percentage of approaches (from more than 2 ft to less than 2 ft) which were due to movement by infants, and percentage of leavings (from less than 2 ft to more than 2 ft) which were due to infants; *e*, number of ½ min which infants spent wholly more than 2 ft from their mothers as a percentage of the number in which they were off at all. Medians only shown. *Continuous lines*, isolates; *discontinuous lines*, group-living animals.

of the troop, but also he settles squabbles between the females, often chasing and mock-biting the instigator.

The data summarized in the preceding paragraphs represent only a small part of the information now available on the development of behaviour in rhesus monkeys, let alone on subhuman primates in general. It shows, however, the way in which development depends on a complex and continuous interaction between infant, mother, social companions, and other environmental factors. At or soon after birth the mother has a repertoire of maternal behaviour patterns which interlock with and depend on the primarily reflex patterns of the infant. Through commerce with its environment, the infant's perceptual, manipulative, and locomotor powers develop. As it grows, it makes greater demands on its mother and interferes more with her own maintenance and comfort activities. The care and solicitude which she shows to the newborn diminish, though she continues to care for him. The nature of the mother–infant relationship gradually changes, with the mother playing an important role in permitting and promoting the infant's independence. The infant is attracted away from the mother by its peers: these and the maternal responsiveness to the infant shown by more mature social companions have a marked influence on the mother–infant bond. In the ordinary course of events the bond gradually weakens, but sudden change, such as that produced by a period of separation, may change the course of development in a manner which is not fully reversible. Some data on other subhuman primates is reviewed by Hinde (in press).

26.4 SUMMARY

Development involves a continuous interaction between the changing organism and its changing environment. Before development can be understood, factors involved in this interaction must be teazed apart. Studies of the development of reproductive behaviour in female canaries, and of the development of behaviour in kittens and infant rhesus monkeys, illustrate the nature of interaction in development.

section four

EVOLUTION

27

EVOLUTION AND BEHAVIOUR

The increased interest in the study of animal behaviour which occurred around the turn of the century was due in no small part to the stimulus provided by the theory of evolution by natural selection. Darwinism immediately posed questions like: "Are the differences in behaviour between animal species and, more especially, between animals and man, of kind or of degree?", or "To what extent can the principles used to explain human behaviour be applied to animals, and vice versa?" Before such questions could be answered, further knowledge about the causation and development of behaviour was required: this raised problems which have continued to occupy the major effort in psychological research ever since. Attempts to specify the precise similarities and differences between the behavioural capacities of animals continued, however, and work on this problem has recently been considerably accelerated (Schneirla, 1952).

Such studies have been concerned largely with macroevolution, since closely related species differ little in their abilities: [1] the comparisons made have thus been mostly cross-phyletic. Research on microevolution and behaviour has been concerned with questions which, though rather different, also spring from the theory of evolution by natural selection; for instance, "Can behavioural characters be used in the study of systematic relations between species?" and "What has been the detailed course of the evolution of those characters?" These issues were raised early in the century by such men as Heinroth and Whitman, and have recently been pursued actively by Lorenz, Tinbergen, and others influenced by them.

If behaviour has evolved under the influence of natural selection, then the differences in behaviour between closely related species must have survival value. Why is it advantageous for this species to be social and not for that? Why does this species have one sort of courtship display,

[1] "Abilities" here is not intended to include sensory and motor capacities, which sometimes differ considerably even in closely related species.

while its close relative has a superficially quite different one? This question of the adaptedness of behaviour has been the subject of speculation for some time, but only recently has it become a subject for experimental study. Further problems concern the role played by behaviour in evolution: for instance, we must know how behaviour contributes to speciation, and how those behavioural differences between populations which are important in speciation arise.

For a full understanding of behaviour, these groups of problems are as important as the questions of immediate causation and development considered so far in this book. Here they will be merely outlined, the aim being to introduce the problems and show their connection with those previously discussed rather than to provide an extensive review.

27.1 QUALITATIVE AND QUANTITATIVE DIFFERENCES BETWEEN BEHAVIOURAL CHARACTERISTICS AT DIFFERENT EVOLUTIONARY LEVELS

That animals with different degrees and types of structural complexity show corresponding differences in the complexity of their behaviour is immediately apparent. To specify the precise nature of these differences, however, is a task of great difficulty. From the evolutionary point of view, interest must centre on how each species functions as a whole in the environment in which it lives. But for practical reasons comparison must be confined to particular aspects. Furthermore, comparisons between the abilities of animals require standard tests suitable for those abilities: since each species is adapted to meet different situations, any one test is rarely equally suitable for many animals (Tinbergen, 1951). Even if a species does appear to be deficient in some faculty, as defined by a particular type of test, we must still refer back to the natural situation to assess the extent to which this is compensated for by the development of other faculties.

In spite of these difficulties, the problem is an important one, since the student of behaviour seeks to make generalizations. A knowledge of differences in the organization of behaviour between phyletic levels is necessary not only to define the limits of applicability of these generalizations, but to indicate the language in which they should be couched (Schneirla, 1949).

We may take the evolution of learning abilities as an example. This is a problem of peculiar interest, but cross-phyletic comparisons run into many difficulties. Few tests are equally suitable over a wide range of species, for differences in test scores may reflect differences in sensory, perceptual, or motor functioning, rather than in learning ability itself. The difficulties of equating motivation and reinforcement which arise when individuals from the same species but of different ages are compared (see pages 559–560) are virtually insurmountable when the comparison is between species. Furthermore, learning processes are themselves diverse, so that different species may solve the same problem in different ways.

However, a few learning situations that have been used over a wide range of animal types may be mentioned. Habituation (see pages 577–578) to stimuli not associated with consequences significant to the organism occurs in every phylum, and its general characteristics are at least superficially similar in all. But since the underlying complexity which has been demonstrated in mammals (see pages 131–143) has not so far been shown to occur at levels lower than birds, it remains possible, and indeed probable, that the processes involved are in fact very different at different phyletic levels. Detailed analyses of habituation in a wider range of species are necessary before this point can be settled.

Classical conditioning experiments have been performed on a great variety of species, and at first sight there are few obvious differences in the rate at which conditioning occurs (e.g., Razran, 1933). However, Voronin (1962), reviewing experiments on learning in a number of vertebrate species, points out that while the number of trials required for conditioned reflex formation differs little between goldfish and chimpanzee, the rate at which extinction and differentiation occurs is slower in the lower species. Voronin's review thus demonstrates clearly that, within the vertebrates, there are great differences in some aspects of this type of learning ability and little in others: analysis must therefore precede comparison.

Discrimination learning tests have been performed on a wide range of animals, but the results often fail to differentiate widely separated species. This is due in large part to the impossibility of equating task difficulty in different species. Members of several phyla have also been tested in roughly comparable maze-learning situations. Warren (1957), reviewing a number of studies on the basis of crude learning scores, found an increase in the rate of learning and the complexity of problems which could be learnt among invertebrates, from worms to ants. A similar increase in learning ability occurs in vertebrates, but "among the vertebrates, only mammals are clearly superior to insects with respect to maze learning". Again, however, superficial comparisons can be misleading. In a more sophisticated comparison of maze-learning in ants and rats, Schneirla (1959) showed that there are marked qualitative differences in the ways in which a maze is learnt. A precise analysis of the learning process is thus necessary before comparisons can be significant.

Similar difficulties inevitably beset any attempt to use a standard technique for comparisons across a broad range of phyla. Greater hope lies in comparisons within a more restricted range—for instance within the Mammalia—though even here the difficulties of standardization are enormous. A common approach has been to examine rather more complex problems than those mentioned in the preceding paragraphs: this would be more likely to show up differences and might also make the standardization of stimuli and responses relatively less important. For instance, instead of requiring an animal to learn to respond to one object and not to another, it may have to learn to pick the odd object out of three (e.g., Strong and Hedges, 1966).

A technique which has been held to be useful here depends on reversal learning—the subject is required to learn a discrimination to a given criterion, and then the cues are reversed and the time taken for relearning measured. Rajalakshmi and Jeeves (1965) assessed performance as the

$$\text{Reversal index} = \frac{\text{trials (or errors) to criterion on reversal learning}}{\text{trials (or errors) to criterion on original discrimination learning}}$$

and claimed that, within any one species and kind of task, the index was relatively independent of task difficulty. It did, however, vary with the nature of the task—for instance, with whether form or position discrimination was used. They also found that, in all those species for which data were available (rat, cat, monkey, and human), the reversal index decreased with age. When comparable studies of animals of similar maturity were compared, the index decreased over these four species from rat (2.77 to 3.31) to man (0.32) (Rajalakshmi and Jeeves, 1965). However, the value of this index is questionable. Warren (1967), using cats and monkeys, found that it did not vary systematically among groups differing in age, species, previous testing experience, task difficulty, and so on, and that values for the same animals in different apparatuses were not correlated. The value of the reversal index was also related to the number of errors in the initial training. Furthermore Sheldon (1968) points out that the number of trials on the original learning may not accurately reflect the difficulty of the task: for instance, the animal may respond initially to inappropriate aspects of the situation. This would yield a spuriously low reversal index. Cross-species comparisons must thus be treated with caution.

Another approach involves attempts to assess the effects of training on one problem on the readiness with which an animal learns another related problem. If the ability to transfer from one problem to the next is independent of the difficulty of the individual problems, it might provide a basis for cross-species comparisons which had some degree of independence of the individual problems. Examination of the rate of improvement in a succession of discrimination reversals (see previous paragraph) has been tried here with a few species, and the evidence suggests that error reduction across successive reversals increases with phylogenetic status (e.g., Warren, 1965b; Gossette et al., 1966; Stettner et al., 1967; Capretta and Rea, 1967). The cross-species comparisons are, however, open to question, since there was little standardization of procedure (Sheldon, 1968).

A more promising technique involves testing for the ability to form a "learning set". Typically, an animal is given six trials on an object-discrimination problem involving a single pair of objects. Regardless of the animal's performance, a new pair of objects is then presented for six trials, and so on until several hundred pairs of objects have been used. With the first trials with each pair of objects, when chance alone can operate, 50 percent will on average be correct. In the early discriminations, trial 2 scores are little more successful than trial 1. With successive problems, however, trial 2 scores improve, and after two or three hundred problems rhesus monkeys may be obtaining 80 to 90 percent correct responses on the second trials of each discrimination problem. If improvement between problems were independent of learning within problems, the rate of improvement of trial 2 scores could be used for cross-species comparisons. Thus squirrel monkeys and marmosets improve considerably less rapidly than rhesus monkeys (Fig. 27.1; Harlow, 1959b). Comparison with subprimate forms must carry reservations about differences in the suitability of the testing conditions, but rats, cats, and raccoons appear to be markedly inferior to primates (Koronakos and Arnold, 1957). The performance of pigeons and young chickens is comparable to that of cats (Zeigler, 1961; Plotnik and Tallarico, 1966). Although the presumption that learning set performance is independent of the difficulty of the individual problems is open to question (Sheldon, 1968), learning set experiments provide one of the most nearly satisfactory sources of evidence for differences in learning ability between primates and other mammals (e.g., Warren, 1965a).

These are, of course, not the only techniques available for comparisons of learning ability, but they are sufficient to demonstrate that direct comparisons of crude learning scores are of restricted value: they must be accompanied by analyses of the learning process. Largely owing to the work of Schneirla (e.g., 1949), and more recently Bitterman (e.g., 1960), the importance of making comparisons only after an appropriate analysis is now becoming widely recognized. Bitterman, for example, found that while responses learned by rats under continuous reinforcement extinguished more rapidly than responses learnt under partial reinforcement with number of training trials equated, no such effect occurred in certain fish. This led him and others to explore the factors determining resistance to extinction and thus to make more meaningful cross-species comparisons (Gonzalez et al., 1965; Gonzalez and Bitterman, 1967; and refs. cited.) A valuable review and evaluation of many of the methods available for inter-species comparisons of learning ability is given by Sheldon (1968; see also Weiskrantz, 1968, p. 168).

If the student of comparative learning could be prepared to forgo neat quantitative assessments of inter-species differences, deeper understanding could perhaps be gained by looking not at the speed with which they learn particular tasks, but rather at the range of tasks they can master and the nature of the errors they make. Such an approach could, for instance, be developed by applying Piaget's (1952, 1955) methods for studying ontogenetic development to phylogenetic differences. But whatever methods are used, it will always be necessary to relate test data to the ecology of the species concerned. One example of this is provided by Vince's (e.g., 1961b) studies of the ability of passerine birds to pull up food suspended on the end of a string: both the method used by and the success of the species she tested were related to the extent to which the species used their feet to hold food-objects in nature (see page 448). Another case is provided by Jolly's (1964a, b) studies of learning in prosimians. The species she studied fell into three groups in their approach to objects. Some stared at the object for a long time and then struck it with the hand. Others gave shorter bouts of

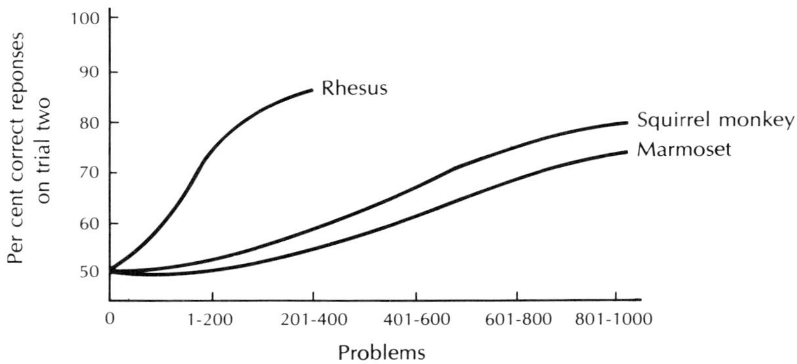

Fig. 27.1 Learning set performance of marmosets, squirrel monkeys, and rhesus monkeys. (After Miles and Meyer, 1956.)

attention and relatively more manipulation. Others gave numerous short bouts of attention and frequent manipulation. Jolly points out how such differences may be related to the species ecology. "Animals whose food is likely to fly away must watch, bide their time, and pounce with precision. Animals whose food waits dangling from a branch will haul in the branch, rather than touch the food and perhaps cause it to drop. Differences in attention span, orientation and manipulation thus relate directly to food habits." A similar point arises from a comparison of squirrel monkeys (*Saimiri sciureus*) and fox squirrels (*Sciurus niger*) in a delayed response task. The fox squirrels were superior, but this could be attributed to their use of gross bodily orientations toward the baited food well during the delay (King et al., 1968).

Inter-phyletic comparisons must, of course, take into account many other characteristics of behaviour besides learning. Sensory and perceptual abilities, motivational mechanisms, and moter skills must all be analysed, related to structure and compared among a wide range of species. Some examples of species and phyletic differences in sensory/perceptual mechanisms were discussed in Chaps. 4, 5, and 6. Inter-species differences in motivational mechanisms have been related to structure by a number of workers: for instance, Beach (e.g., 1948) has studied the role of the forebrain and of previous sexual experience in the mating behaviour of mammals, and Wünschmann (1963) has studied the relation between the complexity of central nervous organization and the development of exploratory tendencies. The relations between motor skills and brain structure are discussed by Paillard (1960): in herbivorous mammals the buccal organs of prehension have a relatively large cortical representation and the forelimbs a small one, while in the cat, and even more in the monkey, the cortical area concerned with the forelimb is enormously increased. Clearly such studies must be integrated with those of learning ability before a real understanding of evolutionary differences in the complexity of behaviour can be achieved.

27.2 THE USE OF BEHAVIOURAL CHARACTERS IN TAXONOMY

Comparison of the behaviour of distantly related animals can give us a picture of the broad sweep of its evolution, but tells us little about its detailed course. How can the small changes in behaviour, which must have accompanied the gradual evolution of species, be investigated? Here the student of behaviour is in a different position from the morphologist, for palaeontological data are not available, and ontogenetic evidence has so far been of little value. He must thus rely on the comparative method. This involves, first, the formation of hypotheses about the phylogenetic relationships between the taxonomic units (species, genera, or higher categories) on the basis of a comparison of a large number of characters. These hypotheses are more likely to be reliable if the characters used are numerous and diverse, so that it is most profitable if morphological, physiological, biochemical, and behavioural studies can be combined.[2] When the evolutionary relationships between the taxonomic units have been established with reasonable certainty, hypotheses about the course of evolution of the characters themselves can be formulated.

When the student of behaviour is concerned with groups whose rela-

[2] Such a comparative study shows the degree of resemblance between contemporary forms. It can be taken as indicating their phylogenetic relationships only with additional assumptions about similarities in the rate of divergence of different species. This assumption is known to be incorrect in some cases where a fossil record is available (Simpson, 1961), but is the only possible one in most cases.

tionships are already well known from other evidence, the first step is unnecessary, and comparison of behavioural characteristics can yield directly hypotheses about their evolution. But if the results of the behavioural study are not consistent with the previously accepted classification of the group, a reassessment of all characters may lead to its revision. We must thus consider first the use of behavioural characters for assessing phylogenetic relationships (Lorenz, 1935, 1950).

The first requirement is to select characters which are reasonably constant within groups but show moderate diversity between them. For instance all species of the genus *Parus* (most tits and chickadees) nest in holes and use moss for nest construction, differing in these characters from the other genera usually included in the family Paridae. These characters could thus be used to characterize the genus, but would be useless for elucidating relationships within it. For assessing relationships between species the most useful characters are fixed action patterns (pages 18–23). They are directly observable, they can be recorded on film, and their development is usually stable with respect to environmental influences. Indeed, fixed action patterns may be as characteristic of a species as any of its morphological characters. Among the detailed studies which exploited this were those of Heinroth (1911) and Lorenz (1941) on ducks and geese; Whitman (1919) on pigeons; Tinbergen (1959) on gulls; Crane (1949, 1952) on spiders and mantids; Baerends and Baerends-van Roon (1950) on cichlid fishes, and Jacobs (1953) and Faber (e.g., 1953) on grasshoppers.

Whatever the characters chosen, it is necessary to ensure that differences between the species (or other taxonomic groups concerned) are not environmentally determined. In principle this can be controlled by rearing them in the same environment: if the differences persist, then they must ultimately be of genetic origin (see page 431). In practice such a procedure is not often possible, and circumstantial evidence must suffice. As discussed in Chap. 18, the conclusion that the differences are ultimately genetic does not mean that the development of the characters is uninfluenced by the environment.

It is also necessary to assess the extent to which similarities between characters could be the result of independent evolution for a common function ("convergent evolution"), and the extent to which dissimilarities could be due to selection for divergence as such. Some examples of such characters, which could give a false indication of phylogenetic relationship, are mentioned in Chap. 28.

Having selected the characters and established that the intergroup differences in them are genetically determined, it may then be possible to use them directly as taxonomic characters on a presence or absence basis. An example is provided by Lorenz's (1941) study of the ducks and geese. In Fig. 27.2 the vertical lines represent species, while the horizontal lines represent characters. Some characters, such as the monosyllabic piping of a lost chick, *EPV*, are common to all members of the group. Others, such as displacement shaking given as a form of display, *Js*, are common to nearly all ducks but not to the geese, while yet others are present only in particular groups of ducks. In some cases the unexpected presence or absence of characters in a particular species has led to a reassessment of its taxonomic position (see also Johnsgard, 1961).

When reasonably closely related species are compared, presence or absence may be a relative matter. The pivoting movement (Fig. 16.6), so characteristic of goldfinch (*Carduelis carduelis*) courtship, is present to varying degrees in other members of the family, in some being represented only by occasional incipient movements (Hinde, 1955/1956).

In other cases variations in the characteristics of a movement provide profitable material for comparison between the taxonomic units. When passerine birds move through foliage they flick their tails in a characteristic fashion. These tail flicks vary in lateral displacement, spread, and vertical amplitude. Although one of the types of flicking has been developed several times by insectivores feeding on the ground, the form of the tail flicks has proved, on the whole, to be rather conservative within families, and provides useful evidence both for assigning genera to families and for assessing relationships between families. This is shown in Table 27.1 (Andrew, 1956d; see also Immelmann, 1962). Wickler (1960, 1961a, b) has used a similar method extensively in his studies of the fin movements of fishes, and of locomotory and other patterns in birds; these cover a wide range of species and relate movements to anatomy (see also Brereton and Immelmann, 1962; Simmons, 1961).

TABLE 27.1. Characters of the tail flicks of certain passerine groups useful in defining the groups and assessing the relationships between them. In D-U tail flicks a downstroke is followed by an upstroke, and in U-D the opposite is the case (*after Andrew, 1956d*).

	Vertical Amplitude	*Spread*	*Internal Displacement*	*Type*
Richmondeninae	Large	Large	?	D-U (variable)
Emberizinae	Large	Large	Small	U-D
Ploceidae (excluding Estrildinae)	Small	Small	Small	Variable due to small size
Estrildinae	Small	Small	Large	Variable due to small size
Carduelinae	Large	Small	Small	D-U

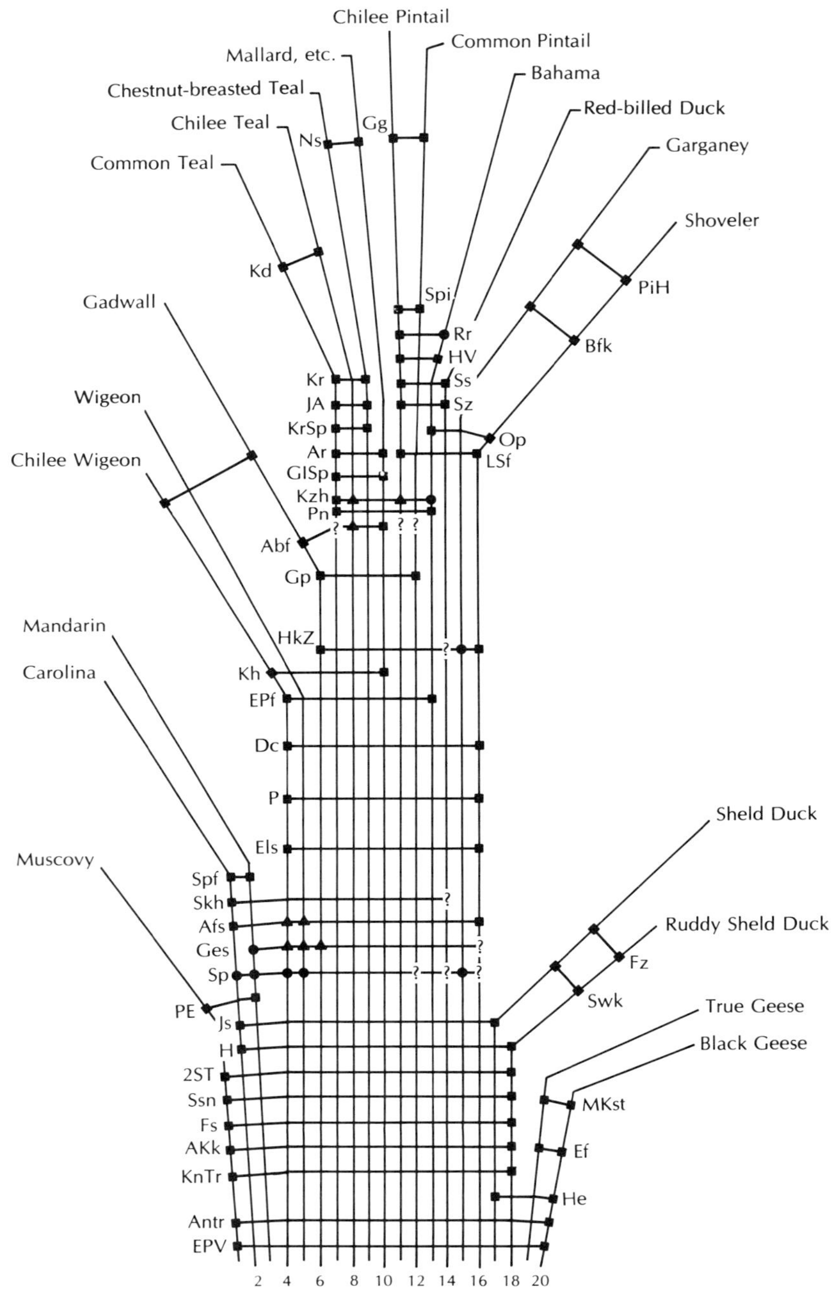
Chilee Pintail
Common Pintail
Mallard, etc.
Bahama
Chestnut-breasted Teal
Red-billed Duck
Chilee Teal
Garganey
Common Teal
Shoveler
Gadwall
Wigeon
Chilee Wigeon
Mandarin
Carolina
Muscovy
Sheld Duck
Ruddy Sheld Duck
True Geese
Black Geese
Ns
Gg
Kd
Spi
PiH
Rr
Bfk
HV
Kr
Ss
JA
Sz
KrSp
Op
Ar
LSf
GlSp
Kzh
Pn
Abf
Gp
HkZ
Kh
EPf
Dc
P
Els
Spf
Skh
Afs
Ges
Sp
PE
Js
H
2ST
Ssn
Fs
AKk
KnTr
Antr
EPV
Fz
Swk
MKst
Ef
He
2 4 6 8 10 12 14 16 18 20

Although the most useful behavioural characters for studies of systematics have been fixed action patterns, other aspects of behaviour are sometimes valuable; for instance Schutz (1956) found that the fright reaction, elicited in fishes by a substance from the injured skin of another individual, is not perfectly species-specific: the effectiveness of the substance produced by one species on the behaviour of another species varies with the distance of their phylogenetic relationship. Schutz was able to use the response to provide evidence on the systematic position of some species of Cyprinidae (see also Pfeiffer, 1960).

In practice, of course, taxonomic relationships are never assessed on the basis of one character alone. Comparisons of many different characters —morphological, behavioural, physiological—are essential. When evidence from different sources conflicts, it is necessary to assess their relative value (e.g., Fiedler, 1964). But characters of behaviour have been of proven value in many groups (see, for example, review by Cullen, 1959; Lorenz, 1961), and in some cases have given the first clue to taxonomic distinctiveness (Tinbergen, 1963).

27.3 THE MICROEVOLUTION OF BEHAVIOUR ILLUSTRATED BY THE EVOLUTION OF AVIAN DISPLAYS

Once the phylogenetic relationships between the taxonomic groups have been reasonably firmly established, comparison of behavioural characters can provide evidence about the course of their evolution. Once again, the study of fixed action patterns has proved especially fertile.

The argument involves first the recognition of formal similarities between the FAPs of different but closely related taxonomic units (usually species). If the species are believed to be closely related on other grounds, and convergence (see page 682) can reasonably be ruled out, such similarities suggest that the movements have a common evolutionary origin.

Having established basic similarities between the movements of the different species, the next stage is to describe the slight differences which occur. If these are genetically determined, it is possible to erect hypotheses as to which variants are phylogenetically older. For this additional informa-

---

Fig. 27.2 The use of behavioural characters in taxonomy, as exemplified by Lorenz's study of the Anatinae. (After Lorenz, 1941.)

Vertical and oblique lines, species and genera; *each horizontal line,* a behavioural character present in the species whose line it crosses; *a cross,* the character is absent in the species in question; *a circle,* exceptional differentiation of a character.

---

tion may be used, such as knowledge about the distribution of the variants between species, and about their causation and function. Once some idea about which variant is more primitive has been gained, comparison of that variant with other movements of the same or other species permits further hypotheses about its probable evolutionary history. It will be apparent that there is a strong possibility of circular argument; behavioural characters can be used to assess relationships between species, and known phylogenetic relationships between species can be used as a basis for hypotheses about the evolution of behavioural characters. This danger can be avoided by the use of as many independent sources of evidence about the relationships between the species studied as possible.

The stereotyped movements used in the threat and courtship movements of birds and fishes have proved a rich source of material for studying the evolution of behaviour, the extent of inter-species diversity making them especially suitable for comparative study. These displays have already been discussed in some detail in Chaps. 16 and 17, where it was shown that they are usually associated with some degree of ambivalence. Courtship displays, for instance, are often associated with tendencies to attack, flee from, and behave sexually toward the mate. The relations between these tendencies and the components of the displays are qualitatively similar among closely related species (e.g., Stokes, 1962b).

Comparative study of display movements has led to the recognition of three fundamental evolutionary sources of displays (Daanje, 1950; Lorenz, 1935; Tinbergen, 1952, 1959 and refs. cited, 1962; Hinde and Tinbergen, 1958).

1. *INTENTION MOVEMENTS*

These are the incomplete or preparatory movements which often appear at the beginning of an activity (pages 400–401). Many avian displays are derived from intention movements of locomotion: for instance the displays shown in Fig. 27.3 are all derived from the take-off leap (see Fig. 17.3), though the relative extent and coordination of the components have altered in ways to be discussed later. Intention movements of biting or striking are a common source of the components of threat movements: the upright threat posture of the herring gull (see Fig. 16.8 and page 376) provides several examples. In other cases intention movements of preening, nesting, self-protection, copulation, and many other types of behaviour have given rise to display movements (e.g., Kunkel, 1959; Immelmann, 1962; Andrew, 1961e). As discussed in Chap. 17, in conflict situations movements elicited by one object are sometimes redirected on to another. The intention movements of such redirected activities may also give rise to displays (e.g., Moynihan, 1955).

2. *DISPLACEMENT ACTIVITIES*

As discussed earlier (pages 406–418), the category of displacement activities is a causally heterogeneous one; nevertheless it is useful in the study of the evolution of display movements. Comparative study shows that many display movements have been derived, not from patterns associated with the major conflicting tendencies, but from apparently irrelevant activities such as preening, incubating, and so on. Since the movements are incomplete, this category overlaps with the previous one. Among birds, the commonest examples are bill-wiping and preening movements, which in some species have become elaborated into complex displays (e.g., cranes, Lorenz, 1935). Similarly, in some fishes the movement used for shaking off ectoparasites has become part of male courtship (Fiedler, 1954).

3. *AUTONOMIC RESPONSES*

Many displays are derived directly or indirectly from autonomic responses; for instance those which are derived from movements of the hair or feathers, movements of urination and defaecation, or changes in skin coloration due to changes in the surface vessels, belong in this category. In addition, many somatic movements, such as those of grooming and preening, are in part responses to stimuli resulting from autonomic activity and could thus also be included here. A number of examples in birds are reviewed by Andrew (1956a, c) and Morris (1956).

In addition, some displays have evolved from other displays: for instance the lateral and horizontal displays used in finch courtship develop during the course of individual courtship, and presumably also in evolution, from the head-forward horizontal threat posture (see Figs. 16.3, 16.4, and pages 365–370).

Comparative studies of the evolution of signal movements in mammals are few, but recent studies have included the Equidae (Trumler, 1959) and primates (e.g., van Hooff, 1962, 1967; Hinde, 1966), and indicate that many of them are derived from similar sources. As indicated in Chap. 16, however, such a generalisation must not be taken to imply that *all* signal movements depend on ambivalence (Andrew, 1963b, c).

When the evolutionary origin of a movement has been identified, the changes which it has undergone in the course of evolution can be described. These are usually such as to improve its adaptive value in the functional context in which it appears. In the case of display movements, the principal function is communication with other individuals, and they become modified to this end, though the extent and nature of the changes may be limited by other factors, such as the need for crypticity.

The changes involved are usually described collectively as "ritualiza-

(a)
(b)
(c)

tion" (Tinbergen, 1952; Blest, 1961; Manning, 1965; Ewing and Manning, 1967). They include changes in the display movement itself, and the development of conspicuous structures by which it is enhanced. These two processes usually proceed in parallel, so that there is correlation between the evolutionary development of the structure and that of the movement. For instance, among tits of the genus *Parus* the "head-up" threat posture is elaborated most in those species which have a conspicuous throat and breast coloration (see Fig. 16.3*a*). Furthermore, the blue tit (*Parus caeruleus*), in which the head-up posture is relatively inconspicuous, makes much use of a head-forward posture in fighting over territories: unlike many other species, it can raise the cheek feathers in a special way to make this conspicuous (see Fig. 16.3*b*); Tinbergen, 1937; Hinde, 1952).

In addition to changes in the form and speed of the movement and its sequential relations to other movements, there may also be changes in the relationship with the factors which elicited it. One of these is the development of "typical intensity" (pages 402–403); the form or frequency of the movement remains constant over a wide range of intensities of the eliciting factors. This also no doubt enhances the signal value of the movement by making it more distinctive. To be effective, "typical intensity" must be characteristic of the species, though it may involve a considerably greater reduction in the variability of the displays of individuals than occurs in the species as a whole.

Other types of change in the relations between the movement and its causal factors also occur during ritualization. Sometimes, for instance, homologous displays in related species depend on different strengths of the associated tendencies. Baerends and Blokzijl (1963), comparing the agonistic and sexual displays of two closely related cichlid fish (Genus *Tilapia*), found that each display was associated with a particular range of values of the ratio (tendency to attack)/(tendency to flee). The value of this quotient appeared to be higher in *T. nilotica* than that for the homologous display in *T. mossambica*. They thus suggest that the primary difference between the two species lies in a lower threshold for aggressive responses in *T. nilotica*. Similar differences in the courtship behaviour of two species of *Drosophila* have been analysed by Manning (1959). Again, some differences in the dis-

---

Fig. 27.3 Some examples of avian displays derived from the take-off leap (*cf. Fig. 17.3*). (*a*, After Stresemann, 1927–1934; *b*, after Bernhardt, 1940; *c*, after Daanje, 1950; *d*, *e*, after Verwey, 1930; *f*, *g*, after Kortlandt, 1940.)

a, Huddled, and *b*, stretching phase of display of male goldeneye (*Bucephala clangula*); *c*, display of turkey cock; *d* and *e*, two phases of display of blue heron (*Ardea cinerea*); *f* and *g*, rhythmic wing-flapping of the cormorant (*Phalacrocorax carbo*).

Fig. 27.3 Continued.

plays between male and female of the same species can be understood in terms of threshold differences (see also McKinney, 1961).

Sometimes threshold changes result in homologous displays of closely related species differing in relative frequency (e.g., Hörmann-Heck, 1957; Crook, 1963). If such changes proceed far enough, they may result in a display disappearing from the repertoire of one species, though still present in closely related ones. In some cases the potentiality for a display may be revealed only by genetic studies. For instance the "down-up" display, found in many species of duck, is absent in the pintail (*Dafila acuta*) and yellow-billed duck (*Nettion flavirostra*), but reappears in inter-species hybrids between them (Lorenz, 1958). Similarly, observations on the swordtail *Xiphophorus maculatus* and related species suggest that certain movements in its courtship are reduced elements which were originally integrated into one courtship movement (Franck, 1968).

Another type of evidence showing that loss of a behaviour pattern may not be all-or-none comes from Haas's (1962) study of bumblebees. In stressful situations some bumblebees abandon the species-characteristic patterns and show those normally found only in related species. Similarly cats, while hunting, catching, and killing prey, sometimes show phylogenetically older patterns if the species-characteristic ones are obstructed (Leyhausen, 1965).

It has also been suggested that in evolution there may be more marked motivational changes such that the movement comes to be governed by causal factors quite different from those which governed it originally. A well-known example of this is courtship feeding in birds: around the time of egg-laying and incubation the females of many species beg from their males with a posture used otherwise only by young begging food from their parents. Since courtship feeding is confined to a rather limited part of the reproductive cycle, it must depend on a fairly specific internal state which is presumably different from that of the begging juvenile. As another example, the Characid fish *Glandulocauda inequalis* has a sound-producing movement which appears to have evolved from "nipping at the surface", a movement used in feeding (Nelson, 1965; see also page 528).

Lorenz (1951, 1960, 1961) has stressed that this sort of change occurs in many displays; for instance in many ducks displacement drinking and preening have become associated into a stereotyped sequence in which one always follows the other. The precise extent of the "emancipation" that has occurred in any one case is, however, difficult to establish: there may be hidden causal relationships between patterns of behaviour which appear to be functionally quite unrelated. Thus in many fishes there are interactions between certain feeding movements and attacking behaviour: in the cichlid *Pelmatochromis subocellatus* sifting and digging movements often occur during attacking bouts, and the probability of attacking behaviour is reduced both by biting other fish and by digging (Heiligenberg, 1965b; see also Albrecht, 1966).

Although these generalisations have been concerned primarily with movements used for communication by birds and fishes, the comparative approach has proved its value in elucidating the probable course of evolution of behaviour in a wide variety of contexts (for recent reviews, see, e.g., Manning, 1965; Ewing and Manning, 1967; Wickler, 1967b). Although most work has been concerned with display movements, the evolution of maintenance activities has also been studied in this way (Baerends, 1966)—for example, the grooming movements used by insects (Jander, 1966). Although mammalian behaviour is often regarded as more plastic than that of lower forms, the study of the evolution of their fixed action patterns is no less profitable. For example, amongst their communicatory movements, van Hooff (1967) has studied a wide variety of facial expressive movements of primates: the principles involved in their evolution and ritualization are similar to those discussed above for lower vertebrates (see also Andrew, 1963b; Hinde, 1966). Using the slightly different approach of focussing on a category of displays defined in terms of their origin from a similar functional category, Wickler (1967a) has called attention to the numerous mammalian displays which have evolved from urine-marking, including the penile displays found in some subhuman primates and primitive man. In mammals, just as in lower vertebrates, the effectiveness of display movements may be augmented by the development of conspicuous structures, such as horns (Geist, 1966) or coloured skin or fur. The evolution of mammalian maintenance activities can also be studied in this way: for instance Zannier-Tanner (1965) took two simple movements (settling into and rising from the resting position) and traced their evolution through a wide range of families of the Artiodactyla (pigs, cows, deer, etc.) and Perissodactyla (horses, etc.).

Although the methods have wide applicability, in this field of study more almost than any other, it must be admitted that the diversity of nature defeats practically any generalisation. A final example illustrates this. The males of a mouth-breeding cichlid fish (*Haplochromis*) have conspicuous orange or yellow spots near the base of the anal fin. These spots resemble the eggs of the species. During courtship, they elicit the response of snapping up the eggs, and make the male more attractive to the female. In spawning, the female snaps up the eggs before the male can fertilize them, but her response to the male's spots ensures that she also snaps up some sperm, and fertilization occurs in her mouth (Wickler, 1962).

Wherever ritualization occurs, there may also be corresponding changes in responsiveness such that the recipient of the signal responds to it optimally. As yet, however, these have been little studied experimentally.

27.4 SUMMARY

1. Animals at different phyletic levels differ markedly in their behavioural capacities, but the task of defining the differences is one of considerable difficulty. Some examples from the study of learning abilities are given.
2. Behavioural characters can be used to provide evidence concerning the taxonomy of species.
3. Once the phylogenetic relationship between species is reasonably well known, hypotheses can be formed about the evolution of behavioural characters. The evolutionary origins of avian display movements are discussed. In the course cf their evolution these displays have undergone evolution to adapt them for their signal function.

28

THE ADAPTEDNESS OF BEHAVIOUR AND ITS ROLE IN SPECIATION

28.1 THE ADAPTEDNESS OF DIFFERENCES IN BEHAVIOUR

The behaviour we observe is only one point in a nexus of cause-effect relations. So far we have been occupied in tracing these relations backwards in time by studying either the immediate causation of the behaviour, or the development of the behaviour and the mechanisms which control it in the life of the individual. We have also discussed evidence concerning the evolution of the behaviour, thereby relating it to the even more remote past. However, to gain a full understanding of behaviour we must treat it as a cause as well as an effect, for the behaviour itself has consequences. Some of these consequences may be of little interest to us, and we can treat them as incidental by-products. Others will affect the individual's behaviour if it returns to the same, or a similar, situation (e.g., Chaps. 13 and 24–26). Yet others will influence the animal's chances of survival or reproduction, and will thus provide material for natural selection and affect later evolution. Thus, if one species is social while a relative is not, if one bird is a hole-nester while another is not, then the consequences of these differences are likely to be significant with respect to natural selection. We must thus ask precisely what the consequences of sociability or hole-nesting are, which of them are advantageous, and exactly why. As with morphological characters, the most significant questions are those which concern differences between characters rather than the characters themselves. To ask what is the selective significance of this animal's legs or pecking response, leads to less interesting answers than to ask why it has three toes and eats insects when most of its relations have four and eat seeds. Indeed, for the most part questions about adaptedness imply comparison even when it is not stated: "Why is it advantageous for this species to nest socially?" implies "When that species does not". Of even more interest, in that they lead to study of evolutionary changes actually in progress, are questions about the consequences of differences within a species: "Are the individuals which nest socially better off than those which do not, and why?" There is, of course, no reason why we should not ask similar questions about the selective significance of differences between individuals in learning ability or powers of perception, though such problems have not yet been tackled experimentally.

Questions about differences involve also questions about similarities. If we ask why this group differs from that, we are asking also why the members within the groups are similar. But when questions are posed by similarities in behaviour between distantly related species (convergence), it is the differences each shows from its own close relatives which pin-point the problem.

Progress in the study of the adaptedness of behaviour has been slow largely because it is only recently that the problems have been seen clearly. The clarification is largely due to the work of Tinbergen and his colleagues, who recognized problems which had previously been missed altogether. This is demonstrated by their work on egg-shell removal by the black-headed gull (*Larus ridibundus*). After their eggs hatch the adults of some species of birds leave the fragments of egg-shell in the nest, but others, such as the black-headed gull, remove them and drop them at a distance. This response takes about half a minute a year, and appears insignificant, yet the very fact that some species do it and others do not suggests that there is a problem in adaptedness. The further observation that in species nesting on open ground which do not remove the egg-shells, the young usually leave the nest very soon after hatching, and vice versa, also favours the view that the difference is adaptive and related to the life history of the species. In addition, the behaviour carries some disadvantages, for even the temporary absence of the parent from the nest required for egg-shell removal may expose the chick to a marauding predator. If the behaviour is potentially disadvantageous, there must surely be counterbalancing advantages.

Among the gulls and terns, those species with camouflaged young tend to show egg-shell removal, while those with conspicuous young do not. Since the broken egg-shells are themselves conspicuous, Tinbergen suggested that their removal was advantageous in that it reduced the conspicuousness of the nest site.

This suggestion was open to experimental testing. By laying out differently coloured eggs in a gull colony, it was shown that white eggs were taken by predators more rapidly than were those coloured naturally. The natural camouflage of the gull's egg is thus advantageous. Tinbergen also showed that this advantageous effect of camouflage could be reduced by the proximity of a conspicuous broken egg shell. Eggs were laid out in the colony, half of them with a broken shell nearby and half without. The former were taken by predators much more rapidly than the latter. There is thus strong reason for thinking that egg-shell removal has the advantageous consequence of reducing the conspicuousness of the nest.

This, however, turned out to be only the beginning of the problem. Comparisons with other species showed that black-headed gulls, which often do not remove the egg-shells for an hour or two after hatching, are more tardy than many other ground-nesting birds. Here, again, the finding of a difference has indicated a problem in adaptedness. Tinbergen has suggested that the difference is related to the fact that black-headed gulls themselves prey on newly hatched wet chicks, but not on dry ones. There may therefore be an advantage in delaying egg-shell removal until the parents' absence will not expose the chicks to intra-specific predation. This suggestion awaits experimental support, but does show that it is not only the occurrence of egg-shell removal which poses a problem in adaptedness, but also all its properties (Tinbergen, Broekhuysen et al., 1962; Tinbergen, Kruuk et al., 1962; Tinbergen, 1967).

Another example of the way in which an experimental approach to problems of function can produce rapid progress is provided by territorial behaviour. This is found in diverse species, invertebrate (e.g., Moore, 1952) as well as vertebrate, but there has been much disagreement as to whether it does in fact limit density (e.g., Lack, 1966), and the precise function it serves (e.g., Hinde, 1956b; Tinbergen, 1957). While it is recognized that the function certainly differs between species, field observation and comparative study have proved unable to provide answers. Experiment is, however, possible. In the field, a crude approach is to remove all the birds from a given area: the finding that it is quickly repopulated implies that the territorial behaviour of the occupants had been limiting density. A more sophisticated approach has formed part of the studies of Tinbergen and his colleagues on gulls. The black-headed gull nests in colonies, but within each colony the nests are spaced out. By comparing the behaviour of intruders on territories in which the owners were present but immobilized by a stupefying drug, with that of intruders on territories in which the owners were displaying normally, it was shown that the displays have a deterring effect. The territorial behaviour thus promotes spacing-out of nests. This in turn was shown to have two functions: it reduced predation on pipped eggs and newly hatched chicks by other gulls, and, together with the camouflage of the eggs, reduced predation by predators of other species (Patterson, 1965; Tinbergen, 1967).

Van den Assem (1967) has also used an experimental approach to this problem of the function of territorialism, but in this case in the laboratory.

By breeding sticklebacks, singly and in groups, in tanks of varied sizes, he showed that territorial behaviour could limit density. He also showed that the possession of an adequate territory had diverse effects on reproductive success, for males unable to claim territories could not build nests, and owners of large territories were more successful than owners of small ones both in courtship and in rearing the eggs laid in their nests. As Tinbergen (1968) points out, the effects of higher densities demonstrated here are all intra-specific, and really amount to demonstrating how the behaviour of the species is adjusted to lead to success with this type of territory: to assess fully the function of territoriality it would be necessary to manipulate such variables as food supply and predator pressure.

Only rarely is it possible to obtain evidence about qualitative changes in functional significance in the course of a movement's evolutionary history, but signal movements provide some examples. One case is the protective displays given by Saturnioid and Sphingid moths when disturbed. After studying respectively thirty-five and six species of these groups, Blest (1957) divided the responses to disturbance into the following types:

1. Generalized escape behaviour.
2. Rhythmic displays, involving coordinated movements of wings, legs, and abdomen repeated many times.
3. Sustained static displays, in which a posture may be maintained for several minutes.
4. Mixed displays, in which the response to mild disturbance is a static posture and that to more intense disturbance is rhythmic.
5. Vestigial displays, involving probably functionless remnants of any of the preceding types.
6. Cryptic behaviour.

These displays are not distributed according to the systematic divisions of these groups of moths, but are correlated with colouration: for instance, species showing generalized escape behaviour are cryptic, but those showing displays usually have conspicuous patterns consisting either of eyespots or mimetic patterns resembling harmful insects such as wasps. Experimental evidence shows that the displays are effective in eliciting escape responses from male passerines which prey on the moths. They thus have a functional significance.

The movements appear to have evolved from the intention movements of flight made by moths too cold to fly, and to have evolved in the order: inadequate flight movements, rhythmic displays, mixed displays, static displays. The efficiency of the displays in the various present-day species in scaring off predators increases in the same order. Thus the gradual evolution of these displays has been accompanied by a corresponding increase in efficiency. A comparably detailed study of defensive behaviour in mantids is given by Crane (1952).

The diversity of the signal movements shown by animals pose further problems in adaptedness. As discussed in Chap. 27, these movements have been elaborated in evolution in a manner which enhances their signal function. That they are in fact effective as signals is suggested in the first place by quite simple observations made under natural conditions. These show that the territorial song of a robin will cause an intruder to fly away (Lack, 1939), or that the threat posture of a great tit is likely to cause a rival to flee. Examination of the efficacy of the components of display movements requires more sophisticated techniques, but leads to a similar conclusion. Stokes (1962a, b) assessed the extent to which certain components of the displays given by tits (*Parus* sp.) at a winter feeding station were followed by the displaying bird attacking, fleeing, or remaining where it was. Having established that certain components and combinations of components were associated with one or other type of subsequent behaviour (page 375), he then assessed the responses of other birds to these components, and found them to be appropriate. For instance, birds with erect body feathers usually escape and only rarely attack. The rivals of such birds are more likely to attack and less likely to escape than would be expected on a chance basis. Similarly, a bird which faces its rival is more likely to attack than a bird which does not: the rival is more likely to flee from birds in the former condition than the latter.

Using a similar approach, Dunham et al. (1968) have demonstrated the efficacy of threat and submissive postures in reducing attack in a cyprinid fish. Hazlett and Bossert (1965), studying aggressive interactions between hermit crabs, compared the distribution of acts by one crab which followed the execution of each act by a second crab. Most of the behaviour patterns were found to elicit some change in behaviour of the other animal (see also Altmann, 1965, for a similar approach to communication in rhesus monkeys).

Additional evidence for the efficacy of these displays in communication is provided by dummy experiments which demonstrate that the structures used or postures given in a particular social context are, in fact, effective in eliciting the response appropriate to that context. For instance, Perdeck (1958) has shown that the species song is more effective in eliciting sexual approaches from certain grasshoppers than is that of related species, and Thielcke (1962) has shown that the species song is more effective in eliciting a reply from certain tree creepers (*Certhia*) than is the song of closely related species: this is of particular interest since the song probably promotes reproductive isolation. Some other examples were given in Chap. 4 (see also Lorenz, 1935).

For some signals, experiments with dummies can be misleading, because their communicatory significance depends on the context in which they are given. The "head toss" of gulls is given both by females and by large young soliciting food, in precopulatory displays, and in hostile encounters (Moynihan, 1955; Tinbergen, 1959). The common factor seems to be a tendency to escape, but this may conflict with one of several incompatible tendencies. Interpretation by another individual must depend on other signals which accompany the display, or on the context (Smith, 1965). The importance of context is particularly obvious when the signal has the effect of calling the attention of other individuals to something in the environment, as with certain vocalizations of monkeys (Rowell and Hinde, 1962) and lemurs (Jolly, 1966).

Signal movements have been subject to selection for their efficiency: examination shows that the sensory modality, and the detailed characteristics of the signal, are suited to the context in which they are used (e.g., Marler, 1957b, 1965; see also below). Nevertheless, there is no apparent trend towards some sort of "ideal" signal within each sensory modality. Nor could there be, for each species has a repertoire of signals which must be distinguishable from each other. Although Lorenz earlier regarded the form of each signal as a matter of mere convention, implying that it was more or less a matter of chance that, for instance, one species had one sort of greeting ceremony and another a different one, later study has shown that each signal is adapted to the particular context in which it is used.

In the first place, the different signal movements of any one species are distinct from each other, and are thus rarely confused. For instance, the submissive postures of passerine birds, which involve raising the feathers, a rather hunched position of the body and a generally rounded appearance, are in many respects the opposite of the horizontal sleeked threat posture (e.g., Hinde, 1952, 1953a; Sparks, 1965). Further, submissive postures frequently involve a turning away from the more dominant animal. This may be regarded as the ritualization of an intention movement of fleeing, which is efficacious in part also because it removes the sign stimuli for aggression (Tinbergen, 1959; Baeumer, 1959). In primates, and also in many other species, a direct frontal gaze is a stimulus for attack, and submissive postures are characterized by looking away from the opponent. (It has also been suggested by Chance (1962) that such postures "cut off" the incoming aggressive stimuli and thus reduce the tendency to retreat in the posturing individual. Grant and Mackintosh (1963) go so far as to regard catalepsy as a form of "cut off".) Functionally comparable submissive postures are known also in invertebrates (e.g., Crane, 1958; Reese, 1962). This necessity for displays to be distinctive may lead to "competition" between displays and to the extinction of some of them (Dilger, 1960): cases are known in which interspecies hybrids show display movements which are absent in both parents but present in various relatives (Lorenz, 1960; see page 673).

In some cases displays serve not only as intra-specific signals, but also as individual recognition marks. This has been studied especially in the vocal signals of birds, where inter-individual differences within a characteristic species pattern are sometimes considerable (e.g., Hutchison et al., 1968b), and have been shown experimentally to produce differential responses (Falls, 1969).

In addition to individual and intra-species diversity, selection has acted to regulate the amount of inter-species diversity. The evidence here is mainly not experimental, but depends on the establishment of cross-species correlations between particular characteristics of the signals and the natural situations in which they occur. Marler (1957b) has examined the circumstances in which selection will favour specific distinctiveness, or the lack of it, in signal movements and their associated structures. Where the signal plays a role in maintaining reproductive isolation from closely related forms, divergence is likely to occur. Thus in birds some aspects of the signals used in pair formation are usually markedly different between closely related species—in the Cardueline finches the displays used in the early stages of pair formation diverge to a greater extent than those which immediately precede copulation, which occurs after pair formation is accomplished (Hinde, 1955/1956). The manner in which the form of bird songs is adapted to their function of mate selection has also been discussed by Marler (1960), and the occurrence and adaptive significance of inter-species, inter-population, local, and individual differences in bird vocalizations have been reviewed by Marler and Hamilton (1966) and Thielcke (1969). Similarly the grasshoppers *Chorthippus brunneus* and *C. biguttulus* have distinctive calling songs, but if females are attracted to the vicinity of males of another species by tape-recorded songs of their own species, the later stages of courtship proceed normally and sterile hybrids are produced (Perdeck, 1958). In other species, however, the converse may hold: in dragonflies the initial stage of mating depends on relatively unspecific stimuli, though at closer range more specific ones operate (Buchholtz, 1956). But even here the very final stages, after mounting has occurred, may depend on relatively unspecific stimuli.

On the other hand, selection for convergence will occur in many displays given to predators—the prey species living in a given area will be better protected if they respond to each others' alarm signals. Marler (1959) has shown that the convergence shown by the alarm calls given by many passerine birds to a flying predator has an additional significance. The calls (see Fig. 28.1) all have just those properties which make them difficult to locate by the predator. They lack discontinuities and begin and end gradually, making binaural comparisons of time of arrival of the sound impossible; the frequency is too high for phase differences between the ears to be used for location, and too low for intensity differences to be appreciable.

This raises the further issue of the conflicting values of conspicuousness and crypticity. To be effective a signal must be conspicuous, but conspicuousness may bring danger from predators. A compromise must be effected, and the relative values of conspicuousness and crypticity will vary with the circumstances. Often this compromise is assisted by close adapta-

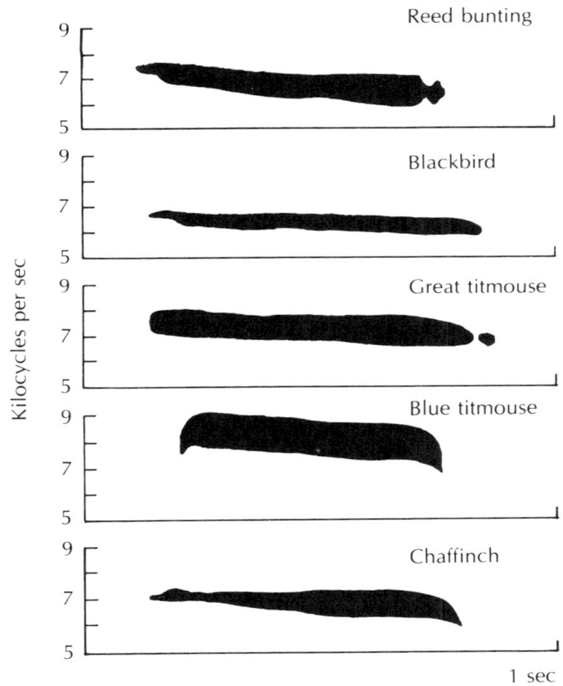

Fig. 28.1 Alarm calls of various passerine species, showing the similarities in their form. (After Marler, 1959.)

tion of the signal to its functional context: thus the calls used by birds at close range need not be loud, and hence conspicuousness is reduced (Marler, *loc. cit.*). The nature of the environment in which the species lives also has a crucial influence on the nature of the display given: discussion of the variations with habitat in the "distraction displays" given by nesting birds to potential predators is given by Armstrong (1954).

The nature of the display movements used in agonistic, social, and sexual encounters thus suggests that they have been influenced by natural selection in adaptation to the circumstances in which they are used. That the nature of the signal movements can be influenced by selection for crypticity is an aspect of the important principle that all the characters of a species form part of an adaptive complex suiting it to its habitat and way of life. Since most aspects of the behaviour of a species are adaptively interrelated, change in one may lead to other changes ramifying through the whole of the system. A classic example is given by the work of E. Cullen (1957) on the kittiwake (*Rissa tridactyla*). This is the only gull species which nests on steep cliffs, a trait which protects it from both avian and mammalian predators. It differs in behaviour from other gulls in many respects, but many of the differences from other gulls, including the nature of their intra-specific fighting behaviour, nesting, copulatory and parental behaviour, and various aspects of the behaviour of the young, can be regarded as evolutionary consequences of cliff-nesting. For instance the kittiwake builds a more elaborate nest than do ground-nesting species, and the young do not run away when attacked. Certain other species of gulls (*Larus thayeri, glaucoides,* and *hyperboreus*), which nest either on cliff ledges or on level ground, are intermediate between the kittiwake and typical ground-nesting gulls, such as the herring gull (*L. argentatus*) (N. G. Smith, 1966a, b). The gannet, also a cliff-nester but quite unrelated to the gulls, resembles the kittiwake in many of the characters in which the latter differs from its ground-nesting relatives (Nelson, 1967). The black-billed gull (*Larus bulleri*) shows quite a different set of adaptations, related to its habit of nesting in river beds subject to flooding (Beer, 1966b).

Comparable data are available also for many other species (e.g., Crook, 1960c, 1962, 1963, 1964a; Haartman, 1957; J. M. Cullen, 1960; Hoogland et al., 1956/1957; Immelmann, 1962; Wickler, 1958, 1961a; Brown, 1965; Nelson, 1964b). That apparently diverse characters may prove on examination to form part of an adaptive complex must be borne in mind when behavioural characters are used to assess systematic relations between species: if, during the course of evolution, selection for one character inevitably involves selection for several others, the group of characters will be of little more significance for systematics than will any one of them.

28.2 THE ROLE OF BEHAVIOUR IN SPECIATION

The final group of problems to be mentioned here, the role of behaviour in evolution, is closely related to the others. Every question about the adaptedness of behaviour is also one about the role of behaviour in evolution. Here, however, discussion will be confined to the role of behaviour in speciation. When two species or incipient species inhabit the same area, their continued independent existence depends on the maintenance of some degree of reproductive isolation between them, and on the reduction of competition for biological essentials (e.g., Dobzhansky, 1941; Mayr, 1942, 1963; Huxley, 1942; Lack, 1947; Sibley, 1961). Inter-species differences in the behaviour involved in mate selection, food selection, and in the selection of habitat, nest site, and so on, are thus likely to be crucial. Some evidence for this view, relating principally to birds, will be reviewed briefly; more detailed references are given in Hinde (1959a), and a recent review of the literature on *Drosophila* is given by Manning (1965).

1. *MATE SELECTION*

That inter-species differences in behaviour are in fact efficacious in ensuring reproductive isolation in nature is indicated by several types of evidence.

(*a*) Interspecies hybrids which can be bred readily in the laboratory are relatively very rare in nature. This is because in captivity, when male and female are confined together in the absence of conspecific partners, the behavioural mechanisms normally involved in pair formation are unable to operate. Thus in captivity matings between goldfinch (*Carduelis carduelis*) and greenfinch (*Chloris chloris*) are almost as sucessful as those between pure species pairs, but wild hybrids are very seldom found.

(*b*) Closely related sympatric species usually differ markedly in some aspect of behaviour or structure which plays a part in courtship or pair formation. To cite but one example, two sympatric species of the warbler genus *Phylloscopus* are closely similar morphologically but have quite different songs. That such differences are of adaptive value in the maintenance of reproductive isolation is indicated by several types of evidence:
   (1) Divergence is most marked in characteristics which are important in pair formation.
   (2) Divergence is often most marked where mating is otherwise possible; for instance, in the overlap zone of partially sympatric species (see Thielcke and Linsenmaier, 1963, for data on the *Phylloscopus* spp. cited above).
   (3) The species characteristics in question often disappear on islands where no closely related species are present and thus no selection against hybridization operates.
   (4) Divergence is usually more marked in males than in females. Since gamete loss is more serious for females than for males it is primarily female preference that will be selected for.

Some examples of the means whereby pair formation and mating are effected in birds and fishes have already been discussed in Chaps. 16, 17, and 27. Clearly it is the differences between the signals used by closely related species that are likely to be important in the maintenance of reproductive isolation. These can be classified as follows:

(*a*) Differences in the strengths and relative strengths of the tendencies to attack, flee from, behave sexually towards, or behave in other ways in the presence of, the mate.
(*b*) Differences in the extent of sexual dimorphism in behaviour.
(*c*) Differences in display movements. Many of these are related to changes of the type referred to under (*a*).
(*d*) Differences in the structures or colours shown off by the display

Selection for reproductive isolation between two populations will occur only if the populations have already diverged sufficiently for hybrid matings to be disadvantageous. The initial divergence is likely to occur when the populations are isolated from each other in some other way; for instance, by a geographical barrier. However, some divergence in the behaviour involved in courtship and pair formation is likely to occur even in the absence of selection for divergence as such. As we have seen (pages 684), the characters of a species must be regarded as an adaptive complex suiting it to the particular ecological niche which it inhabits. If the areas inhabited by the two populations differ ecologically, each will tend to become adapted to local conditions: the changes involved may have ramifying consequences through the adaptive complex, and thus affect the displays. Among song birds, for instance, local conditions will affect the extent of selection for territorial behaviour, and this may affect male song, the divergence between male and female behaviour and colour patterns, and the suppression of male aggressiveness in courtship: the courtship behaviour will thus be affected (Mayr et al., 1956; see also Smith, 1966).

Another example is provided by the relationship between nest-form and courtship behaviour in the weaver-birds (Ploceinae). The primitive nest-form was apparently globular, with the entrance near the top: species with such nests perform a nest-invitation display while perched immediately below the entrance. Other species have added perches, tunnels, etc. to the entrance so that the hole is now underneath: this presumably provides better protection against weather or predators. In species with nests of this type the males still display in the same position with respect to the entrance, so that they now hang upside down (Crook, 1964a).

Changes in courtship display can be of value in promoting reproductive isolation only if individuals of the new stock respond more strongly to the new pattern than to the old. It seems probable that there is a certain degree of plasticity in the stimuli eliciting sexual behaviour, and that learned tendencies to pair with individuals similar to those with which the individual was reared (see Chap. 21) may play an important role here. For instance, in the polymorphic goose *Anser caerulescens* assortative mating occurs: Cooke and Cooch (1968) suggest that this is due to selection by the males of mates which have plumage patterns similar to one of their parents. In Poeciliid fishes the role of experience is rather different. In laboratory experiments males reared with females only of their own species courted females also of related species with only a slight preference for their own. By contrast, males reared with females of other species showed much greater selectivity in their courtship. This was probably acquired as a consequence of their acceptance or rejection by the females they courted (Liley, quoted by Tinbergen, 1965).

2. *FOOD SELECTION*

When two closely related species inhabit the same area, they are likely to compete for biological essentials. Their continued coexistence will depend on the elimination of this competition. In this, behavioural mechanisms play an essential role.

Among birds, each species has a repertoire of fixed action patterns which are used in hunting for and preparing food (Chap. 3). The stimuli which elicit these movements, however, are often at first extremely generalized; for instance, young passerines will peck at almost any spot which contrasts with its background, and only slowly come to restrict their pecking to food items (Chap. 21). This involves both positive learning and habituation.

The food items an individual learns to take will be those which it can find and prepare most efficiently. Among the seed-eating finches (Fringillidae and Carduelidae), for instance, the size of seeds taken by adults is related to the size of their bills: each species has a tendency to take seeds it can dehusk easily to obtain the maximum kernel weight in a given time spent feeding, though, of course, there are also other factors that influence the choice of seed. Young birds take more of the smaller seeds than do older ones, and only gradually learn to take the larger varieties. The course of this learning depends on which seeds can be opened most efficiently, which will depend, in turn, on the size of the beak and the behaviour patterns available (Kear, 1962).

Similar mechanisms may serve to increase the efficiency with which sexually dimorphic species utilize food supplies (Selander, 1966). Furthermore the food a species takes is influenced by the precise locations in which it feeds, and this in turn by the behaviour it uses in feeding. For instance, a finch which readily hangs upside down will be able to exploit food sources not available to species which cannot. Such differences could be due to a genetically determined difference in tendency to hang, or to individually learned differences consequent upon structural differences. Since there is a correlation between body weight, leg proportions, and ability to hang, the latter seems likely. Of course not all inter-species differences in feeding locality depend on experience in this way. For instance, *Fringilla* (e.g., chaffinch) never seems to pull out thistle fluff to get the seeds, as *Carduelis* (e.g., goldfinch) does, although this simple action would seem to require no structural specialization (Newton, 1967).

Thus what a bird learns to eat will be determined in part by the structures and behaviour patterns available to it for feeding. These, however, become adapted in evolution to the food items which are locally abundant. Thus if two populations are isolated from each other, they will acquire methods of feeding and structures suitable to local conditions. If they subsequently come together in the same area, even though they do not differ genetically in responsiveness to food items, the individuals of the two groups may learn to feed on different items because of their differences in structure.

3. *HABITAT SELECTION*

Differences in habitat preference between incipient species may be important in reducing both interbreeding and competition. The evidence for the latter is especially strong:

(*a*) Sympatric closely related species often occupy different habitats.
(*b*) Where they do not occupy different habitats, they differ in feeding habits sufficiently for competition to be avoided.
(*c*) When one of two closely related species which occupy different habitats within the same general area is locally absent, the other may spread to occupy both habitats.
(*d*) Individuals attempting to breed outside their normal habitats are rarely successful.

Habitat selection among birds usually depends on characters of the habitat which are not essential to survival, but which are visibly prominent characteristics of the landscape and which thus serve as sign stimuli (Chap. 4). Learning plays some role in habitat selection in many species (e.g., Klopfer, 1963, 1967; Wecker, 1963). Selection of an abnormal habitat by some individuals may thus become perpetuated, providing scope for evolutionary adaptation to the new conditions.

It will be apparent from this brief summary of some of the issues raised by consideration of the adaptedness of behaviour and the role of behaviour in speciation that many of the problems discussed in the earlier sections of this book are of fundamental importance for the study of evolution. Conversely, a knowledge of the processes involved in speciation adds considerably to our appreciation and understanding of the inter-species differences in behaviour found in ethology and comparative psychology.

28.3 SUMMARY

1. Behaviour produces consequences which may or may not be of significance for the survival or reproduction of the individual. Consequences which are of such significance provide material for natural selection. Some examples of the manner in which the adaptive significance of behavioural characters can be studied are given.
2. All the characters of a species, behavioural and morphological, form an adaptive complex. Evolutionary change in any one may have ramifying consequences.
3. When two species or incipient species inhabit the same area, their continued independent existence depends on the maintenance of some degree of reproductive isolation and on reduction of competition for biological essentials. Some behavioural mechanisms involved in mate, food, and habitat selection are discussed.

REFERENCES

Adams, D. B. (1968) Cells related to fighting behaviour recorded from midbrain central gray neuropil of cat. *Science,* **159,** 894–896.

Ader, R. (1966) Frequency of stimulation during early life and subsequent emotionality in the rat. *Psychol. Rep.,* **18,** 695–701.

Ader, R., Friedman, S. B., and Grota, L. J. (1967) 'Emotionality' and adrenal cortical function: effects of strain, test and the 24-hour corticosterone rhythm. *Anim. Behav.,* **15,** 37–44.

Adler, H. E. (1963) Sensory factors in migration. *Anim. Behav.,* **11,** 566–577.

Adler, N., and Bermant, G. (1966) Sexual behaviour of male rats: effects of reduced sensory feedback. *J. comp. physiol. Psychol.,* **61,** 240–243.

Adler, N., and Hogan, J. A. (1963) Classical conditioning and punishment of an instinctive response in *Betta splendens. Anim. Behav.,* **11,** 351–354.

Adolph, E. F. (1947) Urges to eat and drink in rats. *Amer. J. Physiol.,* **151,** 110–125.

Adolph, E. F., Barker, J. P., and Hoy, P. A. (1954) Multiple factors in thirst. *Amer. J. Physiol.,* **178,** 538–562.

Adrian, E. D. (1928) The Basis of Sensation. Christophers, London.

Adrian, E. D. (1931) Potential changes in the isolated nervous system of *Dytiscus marginalis. J. Physiol,* **72,** 132–151.

Adrian, E. D. (1954) The physiological basis of perception. In "Brain Mechanisms and Consciousness", Ed. J. F. Delafresnaye, Blackwell, Oxford.

Adrian, E. D., and Zotterman, Y. (1926) The impulses produced by sensory nerve endings. Pt. 3, Impulses set up by touch and pressure. *J. Physiol.,* **61,** 465–483.

Ahrens, R. (1954) Beitrage zur Entwicklung des Physiognomie und Mimikerkennes. *Z. exp. und ang. Psychol.,* **2,** 402–454, 599–633.

Ainsworth, M. D., and Wittig, B. A. (1968) In Determinants of Infant Behaviour, **4,** Ed. B. Foss, Methuen, London.

Akerman, B. (1966a, b) Behavioural effects of electrical stimulation in the forebrain of the pigeon, I and II. *Behaviour,* **26,** 323–338 and 339–350.

Akerman, B., Andersson, B., Fabricius, E., and Svensson, L. (1960) Observations on central regulation of body temperature and of food and water intake in the pigeon (*Columba livia*). *Acta physiol. scand.,* **50,** 328–336.

Akert, K. (1961) Diencephalon. In Sheer (1961).

Akimoto, H., Saito, Y., and Nakamura, Y. (1961) Effects of arousal stimuli on evoked neuronal activities in cat's visual cortex. In "The Visual System", Ed. R. Jung and H. Kornhuber, Springer, Berlin.

Albrecht, H. (1966) Zur Stammesgeschichte einiger Bewegungsweisen bei Fischen, untersucht am Verhalten von *Haplochromis* (Pisces, Cichlidae). *Z. Tierpsychol,* **23,** 270–302.

Alexander, G., and Williams, D. (1966) Teat-seeking activity in lambs during the first hours of life. *Anim. Behav.,* **14,** 166–176.

Allee, W. C. (1938) The Social Life of Animals. Norton, New York.

Allport, G. W. (1937) Personality: a Psychological Interpretation. Holt, New York.

Alpern, H. P., and McGaugh, J. L. (1968) Retrograde amnesia as a function of duration of electroshock stimulation. *J. comp. physiol. Psychol.,* **65,** 265–269.

Altman, J., Das, G. D., and Anderson, W. J. (1968) Effects of infantile handling on morphological development of the rat brain: an explanatory study. *Developmental Psychobiology,* **1,** 10–20.

Altmann, S. A. (1962) Social behaviour of anthropoid primates: analysis of recent concepts. In Bliss (1962).

Altmann, S. A. (1965) Sociobiology of rhesus monkeys. II. Stochastics of social communication. *J. Theoret. Biol.,* **8,** 490–522.

Ambrose, J. A. (1960) The smiling response in early human infancy. Ph.D. thesis, London.

Ambrose, J. A. (1961) The development of the smiling response in early infancy. In "Determinants of Infant Behaviour, I", Ed. B. M. Foss, Methuen, London.

Ammons, R. B. (1947) Acquisition of motor skill: II. Rotary pursuit performance with continuous practice before and after a single rest. *J. exp. Psychol.,* **37,** 393–411.

Amsel, A. (1949) Selective association and the anticipatory goal response mechanism as explanatory concepts in learning theory. *J. exp. Psychol.,* **39,** 785–799.

Amsel, A. (1950) The effect upon level of consummatory response of the addition of anxiety to a motivational complex. *J. exp. Psychol.,* **40,** 709–715.

Amsel, A. (1958) The role of frustrative non reward in noncontinuous reward situations. *Psychol. Bull.,* **55,** 102–119.

Amsel A., and Maltzman, I. (1950) The effect upon generalized drive strength of emotionality as inferred from the level of consummatory response. *J. exp. Psychol.,* **40,** 563–569.

Amsel, A., and Work, M. S. (1961) The role of learned factors in "spontaneous" activity. *J. comp. physiol. Psychol.*, **54**, 527–532.

Anand, B. K. (1961) Nervous regulation of food intake. *Physiol. Rev.*, **41**, 677–708.

Anand, B. K., and Brobeck, J. R. (1951) Hypothalamic control of food intake in rats and cats. *Yale J. Biol. Med.*, **24**, 123–140.

Anand, B. K., Chhina, G. S., and Singh, B. (1962) Effect of glucose on the activity of hypothalamic "feeding centers". *Science*, **138**, 597–598.

Anderson, E. E. (1937/1938) The interrelationship of drives in the male albino rat: II. *Comp. Psychol. Monogr.*, **14**, No. 72.

Anderson, E. E. (1941) The externalization of Drive: I. Theoretical considerations. *Psychol. Rev.*, **48**, 204–224.

Andersson, B. (1953) The effect of injections of hypertonic NaCl solutions into different parts of the hypothalamus of goats. *Acta physiol. scand.*, **28**, 188–201.

Andersson, B. (1966) The physiology of thirst. In "Progress in Physiological Psychology: I", Ed. E. Stellar and J. M. Sprague, 191–207.

Andersson, B., and Larsson, B. (1961) Influence of local temperature changes in the preoptic area and rostral hypothalamus on the regulation of food and water intake. *Acta physiol. scand.*, **52**, 75–89.

Andersson, B., and McCann, S. M. (1955) A further study of polydipsia evoked by hypothalamic stimulation in the goat. *Acta physiol. scand.*, **33**, 333–346.

Andersson, B., and Wyrwicka, W. (1957) The elicitation of a drinking motor conditioned reaction by electrical stimulation of the hypothalamic "drinking area" in the goat. *Acta physiol. scand.*, **41**, 194–198.

Andrew, R. J. (1956a) Some remarks on behaviour in conflict situations, with special reference to *Emberiza* spp. *Brit. J. Anim. Behav.*, **4**, 41–45.

Andrew, R. J. (1956b) Normal and irrelevant toilet behaviour in *Emberiza* spp. *Brit. J. Anim. Behav.*, **4**, 85–91.

Andrew, R. J. (1956c) Fear responses in *Emberiza* spp. *Brit. J. Anim. Behav.*, **4**, 125–132.

Andrew, R. J. (1956d) Intention movements of flight in certain passerines, and their use in systematics. *Behaviour*, **10**, 179–204.

Andrew, R. J. (1957a) The aggressive and courtship behaviour of certain Emberizines. *Behaviour*, **10**, 255–308.

Andrew, R. J. (1957b) Influence of hunger on aggressive behavior in certain buntings of the genus *Emberiza*. *Physiol. Zool.*, **30**, 177–185.

Andrew, R. J. (1961a–d) The motivational organisation controlling the mobbing calls of the blackbird (*Turdus merula*): I, II, III, and IV. *Behaviour*, **17**, 224–246; **17**, 288–321; **18**, 25–43; **18**, 161–176.

Andrew, R. J. (1961e) The displays given by passerines in courtship and reproductive fighting: a review. *Ibis*, **103**a, 315–348.

Andrew, R. J. (1962) The situations that evoke vocalization in primates. *Ann. N.Y. Acad. Sci.*, **102**, 296–315.

Andrew, R. J. (1963a) Effect of testosterone on the behavior of the domestic chick. *J. comp. physiol. Psychol.*, **56,** 933–940.

Andrew, R. J. (1963b) The origin and evolution of the calls and facial expressions of the primates. *Behaviour,* **20,** 1–109.

Andrew, R. J. (1963c) Trends apparent in the evolution of vocalization in the old world monkeys and apes. *Symp. zool. Soc. Lond.,* **10,** 89–101.

Andrew, R. J. (1964a) Vocalization in chicks, and the concept of "stimulus contrast". *Anim. Behav.,* **12,** 64–76.

Andrew, R. J. (1964b) The development of adult responses from responses given during imprinting by the domestic chick. *Anim. Behav.,* **12,** 542–548.

Andrew, R. J. (1966) Precocious adult behaviour in the young chick. *Anim. Behav.,* **14,** 485–500.

Andrew, R. J. (1967) Intracranial self-stimulation in the chick. *Nature,* **213,** 847–848.

Andrew, R. J. (1969) The effects of testosterone on avian vocalizations. In Hinde (1969).

Angermeier, W. F. (1960) Some basic aspects of social reinforcements in albino rats. *J. comp. physiol. Psychol.,* **53,** 364–367.

Angermeier, W. F., Locke, D., and Harris, A. (1964) Direct stomach feeding in the pigeon. *Psychol. Rep.,* **15,** 771–774.

Anthoney, T. R. (1968) The ontogeny of greeting, grooming, and sexual motor patterns in captive baboons (Superspecies *Papio cynocephalus*). *Behaviour,* **31,** 358–372.

Arden, G. B. (1963) Complex receptive fields and responses to moving objects in cells of the rabbit's lateral geniculate body. *J. Physiol.,* **166,** 468–488.

Arlinsky, M. B. (1967) On the failure of fowl to adapt to prism induced displacement: some theoretical analyses. *Psychon. Sci.,* **7,** 237–238.

Armstrong, E. A. (1947) Courtship and Display amongst Birds. Lindsay Drummond, London.

Armstrong, E. A. (1950) The nature and function of displacement activities. *Sym. Soc. exp. Biol.,* **4,** 361–384.

Armstrong, E. A. (1954) The ecology of distraction display. *Brit. J. Anim. Behav.,* **2,** 121–135.

Aronson, L. R. (1949) An analysis of reproductive behavior in the mouth-breeding Cichlid fish, *Tilapia macrocephala* (Bleeker). *Zoologica,* **34,** 133–158.

Aronson, L. R. (1965) Environmental stimuli altering the physiological condition of the individual among lower vertebrates. In Beach (1965).

Aronson, L. R., and Cooper, M. (1966) The appearance of seasonal cycles in sexual behaviour in domestic male cats following desensitization of the glans penis. *Proc. 18th Int. Cong. Psychol. Moscow,* **1,** 190–197.

Aronson, L. R., and Noble, G. K. (1945) The sexual behavior of Anura: 2 Neural mechanisms controlling mating in the male leopard frog, *Rana pipiens*. *Bull. Amer. Mus. Nat. Hist.,* **86,** 87–139.

Aschoff, J. (1955) Exogene und endogene Komponente der 24 Stunden-Periodik bei Tier und Mensch. *Naturwiss.*, **42**, 569–575. Springer, Berlin.

Aschoff, J. (1958) Tierische Periodik unter dem Einfluss von Zeitgebern. *Z. Tierpsychol.*, **15**, 1–30.

Aschoff, J. (1960) Exogenous and endogenous components in circadian rhythms. *Cold Spring Harbor Symp. quant. Biol.*, **25**, 11–28.

Aschoff, J. (1963) Diurnal rhythms. *Ann. Rev. Physiol.*, **25**, 581–600.

Aschoff, J. (1964) Tagesperiodik licht- und dunkelaktiver Tiere. *Rev. suisse Zool.*, **71**, 528–558.

Aschoff, J. (1967) Circadian rhythms in birds. *Proc. 14th Int. Orn. Cong.*, 1966, 81–105.

Aschoff, J., and Wever, R. (1962a) Aktivitätsmenge und $\alpha:\rho$ Verhältnis als Messgrössen der Tagesperiodik. *Z. vergl. Physiol.*, **46**, 88–101, Springer, Berlin.

Aschoff, J., and Wever, R. (1962b) Uber Phasenbeziehungen zwischen biologischer Tagesperiodik und Zeitgeberperiodik. *Z. vergl. Physiol.*, **46**, 115–128.

Aschoff, J., and Wever, R. (1966) Circadian period and phase-angle difference in chaffinches (*Fringilla coelebs*). *Comp. Biochem. Physiol.*, **18**, 397–404.

Ashmole, N. P., and Ashmole, M. J. (1967) Comparative feeding ecology of sea birds of a tropical oceanic island. *Peabody Mus. Nat. Hist. Bull.*, **24**, New Haven.

Assem, J. van den (1967) Territory in the three-spined stickleback, *Gasterosteus aculeatus*. *Behaviour Suppl.* **16**, 1–164.

Autrum, H., and Gallwitz, U. (1951) Zur Analyse der Belichtungspotentiale des Insektenauges. *Z. vergl. Physiol.*, **33**, 407–435.

Autrum, H., and Stumpf, H. (1950) Das Bienenauge als Analysator für polarisiertes Licht. *Z. Naturf.*, **56**, 116–122.

Averbach, E., and Sperling, G. (1961) Short-term storage of information in vision. In "Information Theory", Ed. C. Cherry, Butterworths, London.

Ax, A. F. (1953) The physiological differentiation between fear and anger in humans. *Psychosom. Med.*, **15**, 433–442.

Azrin, N. H., Hake, D. F., and Hutchinson, R. R. (1965a) Elicitation of aggression by a physical blow. *J. exp. Anal. Behav.*, **8**, 55–57.

Azrin, N. H., and Hutchinson, R. R. (1967) Conditioning of the aggressive behavior of pigeons by a fixed-interval schedule of reinforcement. *J. exp. Anal. Behav.*, **10**, 395–402.

Azrin, N. H., Hutchinson, R. R., and Hake, D. F. (1966) Extinction-induced aggression. *J. exp. Anal. Behav.*, **9**, 191–204.

Azrin, N. H., Hutchinson, R. R., and McLaughlin, R. (1965b) The opportunity for aggression as an operant reinforcer during aversive stimulation. *J. exp. Anal. Behav.*, **8**, 171–180.

Azrin, N. H., Hutchinson, R. R., and Sallery, R. D. (1964a) Pain-aggression toward inanimate objects. *J. exp. Anal. Behav.*, **7**, 223–228.

Azrin, N. H., Ulrich, R. E., Hutchinson, R. R., and Norman, D. G. (1964b) Effect of shock duration on shock-induced fighting. *J. exp. Anal. Behav.*, **7**, 9–11.

Bacon, W. E., and Bindra, D. (1967) The generality of the incentive-motivational effects of classically conditioned stimuli in instrumental learning. *Acta Biol. Exper.*, **27**, 185–197.

Baenninger, R. (1966) Waning of aggressive motivation in *Betta splendens*. *Psychon. Sci.*, **4**, 241–242.

Baenninger, R. (1967) Contrasting effects of fear and pain on mouse killing by rats. *J. comp. physiol. Psychol.*, **63**, 298–303.

Baerends, G. P. (1941) Fortpflanzungsverhalten und Orientierung der Grabwespe *Ammophilia campestris* Jur. *Tijdschr Ent.*, **84**, 68–275.

Baerends, G. P. (1950) Specializations in organs and movements with a releasing function. *Symp. Soc. exp. Biol.*, **4**, 337–360.

Baerends, G. P. (1957a) The ethological analysis of fish behavior. In "The Physiology of Fishes", Ed. M. E. Brown, Academic Press, New York.

Baerends, G. P. (1957b) The ethological concept "releasing mechanism" illustrated by a study of the stimuli eliciting egg-retrieving in the herring gull. *Anat. Rec.*, **128**, 518–519.

Baerends, G. P. (1957b) The Ethological studies of insect behavior. *Ann. Rev. Ent.*, **4**, 207–234.

Baerends, G. P. (1959b) The value of the concept "releasing mechanism". *Proc. XVth Int. Cong. Zool., Lond.*

Baerends, G. P. (1964) La reconnaissance de l'oeuf par le Goéland argenté. *Bull. Soc. Scient. de Bretagne*, **37**, 193–208.

Baerends, G. P. (1966) Uber einen möglichen Einfluss von Triebkonflikten auf die Evolution von Verhaltensweisen ohne Mitteilungsfunktion. *Z. Tierpsychol.*, **23**, 385–394.

Baerends, G. P., and Baerends-van Roon, J. M. (1950) An introduction to the study of the ethology of cichlid fishes. *Behaviour Suppl.* **1**, 1–242.

Baerends, G. P., and Blokzijl, G. J. (1963) Gedanken über das Entstehen von Formdivergenzen zwischen homologen Signalhandlungen verwandter Arten. *Z. Tierpsychol.*, **20**, 517–528.

Baerends, G. P., Bril, K. A., and Bult, P. (1965) Versuche zur Analyse einer erlernten Reizsituation bei einem Schweinsaffen (*Macaca nemestrina*). *Z. Tierpsychol.*, **22**, 394–411.

Baerends, G. P., Brouwer, R., and Waterbolk, H. Tj. (1955) Ethological studies on *Lebistes reticulatus* (Peters): I. An analysis of the male courtship pattern. *Behaviour*, **8**, 249–334.

Baerends, G. P., and Cingel, N. A. van der (1962) On the phylogenetic origin of the snap display in the common heron (*Ardea cinerea L.*). *Symp. zool. Soc. Lond.*, **8**, 7–24.

Baeumer, E. (1955) Lebensart des Haushuhns. *Z. Tierpsychol.*, **12**, 387–401.

Baeumer, E. (1959) Verhaltensstudie über das Haushuhn,-dessen Lebensart: 2. *Z. Tierpsychol.*, **16**, 284–296.

Baeumer, E. (1962) Lebensart des Haushuhns, dritter Teil-über seine Laute und allgemeine Ergänzungen. *Z. Tierpsychol.*, **19,** 394–416.

Baggerman, B. (1958) An experimental study on the timing of breeding and migration in the three-spined stickleback (*Gasterosteus aculeatus* L.). *Archs, néerl. Zool.*, **12,** 105–317.

Baggerman, B., Baerends, G. P., Heikens, H. S., and Mook, J. H. (1956) Observations on the behaviour of the black tern, *Chlidonias n. niger* (L.), in the breeding area. *Ardea*, **44,** 1–71.

Baillie, P., and Morrison, S. D. (1963) The nature of the suppression of food intake by lateral hypothalamic lesions in rats. *J. Physiol.*, **165,** 227–245.

Bainbridge, R., and Waterman, T. H. (1957) Polarized light and the orientation of two marine Crustacea. *J. exp. Biol.*, **34,** 342–364.

Baldwin, B. A., and Ingram, D. L. (1967) The effect of heating and cooling the hypothalamus on behavioural thermoregulation in the pig. *J. Physiol.*, **191,** 375–392.

Ball, J. (1934) Sex behavior of the rat after removal of the uterus and vagina. *J. comp. Psychol.*, **18,** 419–422.

Bambridge, R. (1962) Early experience and sexual behaviour in the domestic chicken. *Science*, **136,** 259–260.

Bangert, H. (1960) Untersuchungen zur Koordination der Kopf-und Beinbewegungen beim Haushuhn. *Z. Tierpsychol.*, **17,** 143–164.

Banks, E. M. (1962) A time and motion study of pre-fighting behavior in mice. *J. genet. Psychol.*, **101,** 165–183.

Barcroft, J., and Barron, D. H. (1939) Movement in the mammalian foetus. *Ergeb. Physiol.*, **42,** 107–152.

Bard, P. (1935) The effects of denervation of the genitalia on the oestrual behavior of cats. *Amer. J. Physiol.*, **113,** 5–6.

Bard, P., and Mountcastle, V. B. (1948) Some forebrain mechanisms involved in expression of rage with special reference to suppression of angry behavior. *Res. Publ. Ass. nerv. ment. Dis.*, **27,** 362–402.

Barfield, R. J. (1964) Induction of copulatory behaviour by intracranial placement of androgen in capons. *Amer. Zool.*, **4,** 133.

Barfield, R. J., and Sachs, B. D. (1968) Sexual behaviour: stimulation by painful electrical shock to skin in male rats. *Science*, **161,** 392–395.

Barlow, G. W. (1962a) Ethology of the Asian Teleost, *Badis badis:* III. Aggressive behavior. *Z. Tierpsychol.* **19,** 29–55.

Barlow, G. W. (1962b) Ethology of the Asian Teleost, *Badis badis:* IV. Sexual behavior. *Copeia*, **2,** 346–360.

Barlow, G. W. (1963) Ethology of the Asian Teleost, *Badis badis:* II. Motivation and signal value of the colour patterns. *Anim. Behav.*, **11,** 97–105.

Barlow, G. W. (1968) Ethological units of behavior. In "Central Nervous System and Fish Behaviour", Ed. D. Ingle, Chicago Univer. Press.

Barlow, H. B. (1953) Summation and inhibition in the frog's retina. *J. Physiol.*, **119,** 68–88.

Barlow, H. B. (1961) The coding of sensory messages. In Thorpe and Zangwill (1961).

Barlow, H. B. (1963) Slippage of contact lenses and other artefacts in relation to fading and regeneration of supposedly stable retinal images. *Q. J. exp. Psychol.*, **15,** 36–51.

Barlow, H. B., FitzHugh, R., and Kuffler, S. W. (1957) Change of organization in the receptive fields of the cat's retina during dark adaptation. *J. Physiol.*, **137,** 338–354.

Barlow, H. B., and Hill, R. M. (1963) Selective sensitivity to direction of movement in ganglion cells of the rabbit retina. *Science*, **139,** 412–414.

Barlow, H. B., Hill, R. M., and Levick, W. R. (1964) Retinal ganglion cells responding selectively to direction and speed of image motion in the rabbit. *J. Physiol.*, **173,** 377–407.

Barlow, H. B., and Levick, W. R. (1965) The mechanism of directionally selective units in rabbit's retina. *J. Physiol.*, **178,** 477–504.

Barlow, W. (1959) Anxiety and muscle tension pain. *Brit. J. Clin. Practice*, **13,** 339–350.

Barnes, G. W., and Baron, A. (1961) Stimulus complexity and sensory reinforcement. *J. comp. physiol. Psychol.*, **54,** 466–469.

Barnes, G. W., and Kish, G. B. (1961) Reinforcing properties of the onset of auditory stimulation. *J. exp. Psychol.*, **62,** 164–170.

Barnett, S. A. (1958) Physiological effects of "social stress" in wild rats: I. Adrenal cortex. *J. psychosom. Res.*, **3,** 1–11.

Barnett, S. A. (1963) A Study in Behaviour. Methuen, London.

Barnett, S. A., and Burn, J. (1967) Early stimulation and maternal behaviour. *Nature*, **213,** 150–152.

Barnett, S. A., Evans, C. S., and Stoddart, R. C. (1968) Influence of females on conflict among wild rats. *J. Zool.*, **154,** 391–396.

Baron, A. (1964) Suppression of exploratory behaviour by aversive stimulation. *J. comp. physiol. Psychol.*, **57,** 299–301.

Baron, A., and Kish, G. B. (1960) Early social isolation as a determinant of aggregative behavior in the domestic chicken. *J. comp. physiol. Psychol.*, **53,** 459–463.

Baron, A., and Kish, G. B. (1962) Low-intensity auditory and visual stimuli as reinforcers for the mouse. *J. comp. physiol. Psychol.*, **55,** 1011–1013.

Barondes, S. H., and Cohen, H. D. (1966) Puromycin effect on successive phases of memory storage. *Science*, **151,** 594–595.

Barraclough, C. A., and Cross, B. A. (1963) Unit activity in the hypothalamus of the cyclic female rat: effect of genital stimuli and progesterone. *J. Endocrin.*, **26,** 339–359.

Barrass, R. (1961) A quantitative study of the behaviour of the male *Mormoniella vitripennis* (Walker) (Hymenoptera, Pteromalidae) towards two constant stimulus-situations. *Behaviour*, **18,** 288–312.

Barraud, E. M. (1955) Notes on the territorial behaviour of captive ten-spined sticklebacks (*Pygosteus pungitius*). *Brit. J. Anim. Behav.*, **3**, 134–136.

Barraud, E. M. (1961) The development of behaviour in some young passerines. *Bird Study*, **8**, 111–118.

Barron, D. H. (1950) Genetic neurology and the behaviour problem. In "Genetic Neurology", Ed. P. Weiss, Chicago.

Barry, H., and Symmes, D. (1963) Reinforcing effects of illumination change in different phases of the rat's diurnal cycle. *J. comp. physiol. Psychol.*, **56**, 117–119.

Barth, R. H. (1962) The endocrine control of mating behaviour in the cockroach *Byrsotria fumigata* (Guerin). *Gen. comp. Endocrin.*, **2**, 53–69.

Bartholomew, G. A., and Collias, N. E. (1962) The role of vocalization in the social behaviour of the northern elephant seal. *Anim. Behav.*, **10**, 7–14.

Bartlett, Sir Frederic (1953) Psychological criteria of fatigue. In Floyd and Welford (1953).

Bartoshuk, A. K. (1962a) Response decrement with repeated elicitation of human neonatal cardiac acceleration to sound. *J. comp. physiol. Psychol.*, **55**, 9–13.

Bartoshuk, A. K. (1962b) Human neonatal cardiac acceleration to sound: habituation and dishabituation. *Percept. mot. Skills*, **15**, 15–27.

Bash, K. W. (1939a) An investigation into a possible organic basis for the hunger drive. *J. comp. Psychol.*, **28**, 109–134.

Bash, K. W. (1939b) Contribution to a theory of the hunger drive. *J. comp. Psychol.*, **28**, 137–160.

Bastock, M. (1956) A gene mutation which changes a behavior pattern. *Evolution*, **10**, 421–439.

Bastock, M., and Blest, A. D. (1958) An analysis of behaviour sequences in *Automeris aurantiaca* Weym (Lepidoptera). *Behaviour*, **12**, 243–284.

Bastock, M., and Manning, A. (1955) The courtship of *Drosophila melanogaster*. *Behaviour*, **8**, 85–111.

Bastock, M., Morris, D., and Moynihan, M. (1953) Some comments on conflict and thwarting in animals. *Behaviour*, **6**, 66–84.

Bateson, P. P. G. (1964a) Effect of similarity between rearing and testing conditions on chicks' following and avoidance responses. *J. comp. physiol. Psychol.*, **57**, 100–103.

Bateson, P. P. G. (1964b) An effect of imprinting on the perceptual development of domestic chicks. *Nature*, **202**, 421–422.

Bateson, P. P. G. (1964c) Relation between conspicuousness of stimuli and their effectiveness in the imprinting situation. *J. comp. physiol. Psychol.*, **58**, 407–411.

Bateson, P. P. G. (1964d) Changes in the activity of isolated chicks over the first week after hatching. *Anim. Behav.*, **12**, 490–492.

Bateson, P. P. G. (1966) The characteristics and context of imprinting. *Biol. Rev.*, **41**, 177–220.

Bateson, P. P. G., and Reese, E. P. (1968) Reinforcing properties of conspicuous objects before imprinting has occurred. *Psychon. Sci.*, **10,** 379–380.

Bateson, P. P. G., and Reese, E. P. (1969) The reinforcing properties of conspicuous stimuli in the imprinting situation. *Anim. Behav.*

Batham, E. J., and Pantin, C. F. A. (1950) Phases of activity in the sea-anemone, *Metridium senile* (L.), and their relation to external stimuli. *J. exp. Biol.*, **27,** 377–399.

Bauer, J. (1968) Vergleichende Untersuchungen zum Kontaktverhalten verschiedener Arten der Gattung *Tilapia* (Cichlidae, Pisces) und ihrer Bastarde. *Z. Tierpsychol.*, **25,** 22–70.

Baumeister, A., Hawkins, W. F., and Cromwell, R. L. (1964) Need states and activity level. *Psychol. Bull.*, **61,** 438–453.

Baumgarten, R. von, and Nakayama, S. (1964) Spontane und reizbedingte Änderungen der antidromen Erregbarkeit von bulbären respiratorischen Nervenzellen der Katze. *Pflügers Arch. ges. Physiol.*, **281,** 245–258.

Baumgartner, G., Brown, J. L., and Schulz, A. (1965) Responses of single units of the cat visual system to rectangular stimulus patterns. *J. Neurophysiol.*, **28,** 1–18.

Beach, F. A. (1940) Effects of cortical lesions upon the copulatory behavior of male rats. *J. comp. Psychol.*, **29,** 193–244.

Beach, F. A. (1942a) Comparison of copulatory behavior of male rats raised in isolation, cohabitation, and segregation. *J. genet. Psychol.*, **60,** 121–136.

Beach, F. A. (1942b) Analysis of factors involved in the arousal, maintenance and manifestation of sexual excitement in male animals. *Psychosom. Med.*, **4,** 173–198.

Beach, F. A. (1944) Relative effects of androgen upon the mating behavior of male rats subjected to forebrain injury or castration. *J. exp. Zool.*, **97,** 249–295.

Beach, F. A. (1947) A review of physiological and psychological studies of sexual behavior in mammals. *Physiol. Rev.*, **27,** 240–307.

Beach, F. A. (1948) Hormones and Behavior. Hoeber, New York.

Beach, F. A. (1951a) Instinctive behavior: reproductive activities. In Stevens (1951).

Beach, F. A. (1951b) Effects of forebrain injury upon mating behavior in male pigeons. *Behaviour*, **4,** 36–59.

Beach, F. A. (1955) The descent of instinct. *Psychol. Rev.*, **62,** 401–410.

Beach, F. A. (1958a) Normal sexual behavior in male rats isolated at fourteen days of age. *J. comp. physiol. Psychol.*, **51,** 37–38.

Beach, F. A. (1958b) Neural and chemical regulation of behavior. In "Biological and Biochemical Bases of Behavior", Ed. H. F. Harlow and C. N. Woolsey, Univ. of Wisconsin, Madison.

Beach, F. A. (1965) Sex and Behavior. Wiley, New York.

Beach, F. A. (1966) Ontogeny of "coitus-related" reflexes in the female guinea pig. *Proc. Nat. Acad. Sci.*, **56,** 526–533.

Beach, F. A. (1967) Cerebral and hormonal control of reflexive mechanisms involved in copulatory behaviour. *Physiol. Rev.*, **47**, 289–316.

Beach, F. A. (1968) Coital behaviour in dogs. III. Effects of early isolation on mating in males. *Behaviour*, **30**, 218–238.

Beach, F. A., and Fowler, H. (1959a) Individual differences in the response of male rats to androgen. *J. comp. physiol. Psychol.*, **52**, 50–52

Beach, F. A., and Fowler, H. (1959b) Effects of "situational anxiety" on sexual behavior in male rats. *J. comp. physiol. Psychol.*, **52**, 245–248.

Beach, F. A., Goldstein, A. C., and Jacoby, G. A. (1955) Effects of electro-convulsive shock on sexual behavior in male rats. *J. comp. physiol. Psychol.*, **48**, 173–179.

Beach, F. A., and Holz, A. M. (1946) Mating behavior in male rats castrated at various ages and injected with androgen. *J. exp. Zool.*, **101**, 91–142.

Beach, F. A., and Holz-Tucker, A. M. (1949) Effects of different concentrations of androgen upon sexual behavior in castrated male rats. *J. comp. physiol. Psychol.*, **42**, 433–453.

Beach, F. A., and Jaynes, J. (1954) Effects of early experience upon the behavior of animals. *Psychol. Bull.*, **51**, 239–263.

Beach, F. A., and Jaynes, J. (1956a) Studies of maternal retrieving in rats: I. Recognition of young. *J. Mammalogy*, **37**, 177–180.

Beach, F. A., and Jaynes, J. (1956b) Studies of maternal retrieving in rats: III. Sensory cues involved in the lactating female's response to her young. *Behaviour*, **10**, 104–125.

Beach, F. A., and Jordan, L. (1956) Sexual exhaustion and recovery in the male rat. *Q. J. exp. Psychol.*, **8**, 121–133.

Beach, F. A., and LeBoeuf, B. J. (1967) Coital behaviour in dogs. 1. Preferential mating in the bitch. *Anim. Behav.*, **15**, 546–558.

Beach, F. A., and Levinson, G. (1950) Effects of androgen on the glans penis and mating behavior of castrated male rats. *J. exp. Zool.*, **114**, 159–168.

Beach, F. A., and Rabedeau, R. G. (1959) Sexual exhaustion and recovery in the male hamster. *J. comp. physiol. Psychol.*, **52**, 56–61.

Beach, F. A., and Ransom, T. W. (1967) Effects of environmental variation on ejaculatory frequency in male rats. *J. comp. physiol. Psychol.*, **64**, 384–387.

Beach, F. A., Westbrook, W. H., and Clemens, L. (1966) Comparisons of the ejaculatory response in men and animals. *Psychosom. Med.*, **28**, 749–763.

Beach, F. A., and Whalen, R. E. (1959a) Effects of ejaculation on sexual behavior in the male rat. *J. comp. physiol. Psychol.*, **52**, 249–254.

Beach, F. A., and Whalen, R. E. (1959b) Effects of intromission without ejaculation upon sexual behavior in male rats. *J. comp. physiol. Psychol.*, **52**, 476–481.

Beach, F. A., and Wilson, J. R. (1963) Mating behavior in male rats after removal of the seminal vesicles. *Proc. Nat. Acad. Sci.*, **49**, 624–626.

Beatty, W. W., and Schwartzbaum, J. S. (1968) Consummatory behavior for sucrose following septal lesions in the rat. *J. comp. physiol. Psychol.*, **65**, 93–102.

Beck, C. H., Warren, J. M., and Sterner, R. (1966) Overtraining and reversal learning by cats and rhesus monkeys. *J. comp. physiol. Psychol.*, **62**, 332–335.

Beebe-Center, J. G., Block, P., Hoffman, A. C., and Wade, M. (1948) Relative per diem consumption as a measure of preference in the rat. *J. comp. physiol. Psychol.*, **41**, 239–251.

Beer, C. G. (1961) Incubation and nestbuilding behaviour of black-headed gulls: I. Incubation behaviour in the incubation period. *Behaviour*, **18**, 62–106.

Beer, C. G. (1963a) Incubation and nestbuilding behaviour of black-headed gulls: III. The pre-laying period. *Behaviour*, **21**, 13–77.

Beer, C. G. (1963b) Incubation and nestbuilding behaviour of black-headed gulls: IV. Nest-building in the laying and incubation periods. *Behaviour*, **21**, 155–176.

Beer, C. G. (1966a) Incubation and nestbuilding behaviour of black-headed gulls: V. The post-hatching period. *Behaviour*, **26**, 189–214.

Beer, C. G. (1966b) Adaptations to nesting habitat in the reproductive behaviour of the black-billed gull *Larus bulleri*. *Ibis*, **108**, 394–410.

Beidler, L. M. (1961) The chemical senses. *Ann. Rev. Psychol.*, **12**, 363–388.

Beidler, L. M., Fishman, I. Y., and Hardiman, C. W. (1955) Species differences in taste responses. *Amer. J. Physiol.*, **181**, 235–239.

Bélanger, D., and Feldman, S. M. (1962) Effects of water deprivation upon heart rate and instrumental activity in the rat. *J. comp. physiol. Psychol.*, **55**, 220–225.

Bell, R. Q. (1960) Relations between behavior manifestations in the human neonate. *Child Development*, **31**, 463–477.

Bell, R. Q. (1965) Developmental Psychology. *Ann. Rev. Psychol.*, **16**, 1–38.

Bell, R. W., and Denenberg, V. H. (1963) The interrelationships of shock and critical periods in infancy as they affect adult learning and activity. *Anim. Behav.*, **11**, 21–27.

Bellows, R. T. (1939) Time factors in water drinking in dogs. *Amer. J. Physiol.*, **125**, 87–97.

Beniest-Noirot, E. (1958) Analyse du comportement dit maternel chez la souris. *Monogr. Françaises de Psychol.*, **1**, C.N.R.S., Paris.

Bennet-Clark, H. C. (1963) The control of meal size in the bloodsucking bug, *Rhodnius prolixus*. *J. exp. Biol.*, **40**, 741–750.

Berkowitz, L. (1963) Aggression. McGraw-Hill, New York.

Berkson, G., and Davenport, R. K. (1962) Stereotyped movements of mental defectives: I. Initial survey. *Amer. J. mental Deficiency*, **66**, 849–852.

Berkson, G., and Mason, W. A. (1964a) Stereotyped behaviors of chimpanzees: relation to general arousal and alternative activities. *Percept. and Mot. Skills*, **19**, 635–652.

Berkson, G., and Mason, W. A. (1964b) Stereotyped movements of mental defectives: IV. The effects of toys and the character of the acts. *Amer. J. mental Deficiency*, **68,** 511–524.

Berkson, G., Mason, W. A., and Saxon, S. V. (1963) Situation and stimulus effects on stereotyped behaviour of chimpanzees. *J. comp. physiol. Psychol.*, **56,** 786–792.

Berkun, M. M., Kessen, M. L., and Miller, N. E. (1952) Hunger-reducing effects of food by stomach fistula versus food by mouth measured by a consummatory response. *J. comp. physiol. Psychol.*, **45,** 550–554.

Berlyne, D. E. (1960) Conflict, Arousal and Curiosity. McGraw-Hill, New York.

Berlyne, D. E. (1967) Arousal and reinforcement. Nebraska Symposium on Motivation, 1–110.

Berlyne, D. E., Koenig, I. D. V., and Hirota, T. (1966) Novelty, arousal, and the reinforcement of diversive exploration in the rat. *J. comp. physiol. Psychol.*, **62,** 222–226.

Berlyne, D. E., and Slater, J. (1957) Perceptual curiosity, exploratory behavior and maze learning. *J. comp. physiol. Psychol.*, **50,** 228–232.

Bermant, G. (1961) Response latencies of female rats during sexual intercourse. *Science*, **133,** 1771–1773.

Bermant, G. (1964) Effects of single and multiple enforced intercopulatory intervals on the sexual behavior of male rats. *J. comp. physiol. Psychol.*, **57,** 398–403.

Bermant, G., Lott, D. F., and Anderson, L. (1968) Temporal characteristics of the Coolidge effect in male rat copulatory behavior. *J. comp. physiol. Psychol.*, **65,** 447–452.

Bernhardt, P. (1940) Beitrag zur Biologie der Schellente (*Bucephala clangula*). *J. Ornith.*, **88,** 488–497.

Bernstein, L. (1952) A note on Christie's "Experimental naïveté and experiential naïveté". *Psychol. Bull.*, **49,** 38–40.

Bernstein, N. (1967) The Coordination and Regulation of Movements. Pergamon, London.

Bertram, B. C. R. (1969) Ph.D. thesis, Cambridge.

Bethe, A. (1931) Plastizität und Zentren lehre. *Handb. norm. path. Physiol.*, **15,** 1175–1220.

Beukema, J. T. (1968) Predation by the three-spined stickleback (*Gasterosteus aculeatus*): the influence of hunger and experience. *Behaviour*, **31,** 1–126.

Beusekom, G. van (1948) Some experiments on the optical orientation in *Philanthus triangulum* Fabr. *Behaviour*, **1,** 195–225.

Bevan, W. (1961) Perceptual learning: an overview. *J. Gen. Psychol.*, **64,** 69–99.

Bevan, W., Daves, W. F., and Levy, G. W. (1960) The relation of castration, androgen therapy and pre-test fighting experience to competitive aggression in male C57 BL/10 mice. *Anim. Behav.*, **8,** 6–12.

Bexton, W. H., Heron, W., and Scott, T. H. (1954) Effects of decreased variation in the sensory environment. *Canad. J. Psychol.*, **8,** 70–76.

Bignami, G., and Beach, F. A. (1968) Mating behaviour in the chinchilla. *Anim. Behav.*, **16,** 45–53.

Bilodeau, E. A. (1966) Acquisition of Skill. Academic Press, New York.

Bindra, D. (1959) Motivation: A Systematic Reinterpretation. Ronald Press, New York.

Bindra, D. (1961) Components of general activity and the analysis of behavior. *Psychol. Rev.*, **68,** 205–215.

Bindra, D. (1968) Neuropsychological interpretation of the effects of drive and incentive-motivation on general activity and instrumental behavior. *Psychol. Rev.*, **75,** 1–22.

Bindra, D. (1969) *Ann. N.Y. Acad. Sci.*, in press.

Bindra, D., and Palfai, T. (1967) Nature of positive and negative incentive-motivational effects on general activity. *J. comp. physiol. Psychol.*, **63,** 288–297.

Bindra, D., and Spinner, N. (1958) Response to varying degrees of novelty: the incidence of various activities. *J. exp. Anal. Behav.*, **1,** 341–350.

Bingham, W. E., and Griffiths, W. J. (1952) The effect of different environments during infancy on adult behavior in the rat. *J. comp. physiol. Psychol.*, **45,** 307–312.

Birch, D., Burnstein, E., and Clark, R. A. (1958) Response strength as a function of hours of food deprivation under a controlled maintenance schedule. *J. comp. physiol. Psychol.*, **51,** 350–354.

Birch, H. G. (1956) Sources of order in the maternal behavior of animals. *Amer. J. Orthopsychiatr.*, **26,** 279–284.

Birch, H. G., and Clark, G. (1950) Hormonal modification of social behavior: IV. The mechanism of estrogen-induced dominance in chimpanzees. *J. comp. physiol. Psychol.*, **43,** 181–193.

Birukow, G. (1951) Ermüdung und Umstimmung bei Gleichgewichtsreaktionen der Amphibien. *Verh. d. zool. Ges. Wilhelmshaven*, 1951, 144–150.

Birukow, G. (1954) Photo-geomenotaxis bei *Geotrupes silvaticus* Panz. und ihre zentralnervöse Koordination. *Z. vergl. Physiol.*, **36,** 176–211.

Birukow, G. (1958) Zur Funktion der Antennen beim Mistkäfer (*Geotrupes silvaticus* Panz.). *Z. Tierpsychol.*, **15,** 265–276.

Birukow, G. (1964) Aktivitäts und Orientierungsrhythmik beim Kornkäfer (*Calandra granaria* L.). *Z. Tierpsychol.*, **21,** 279–301.

Birukow, G., and Oberdorfer, H. (1959) Schwerkraftorientierung beim Wasserläufer *Velia currens* F. (Heteroptera) am Tage und zur Nachtzeit. *Z. Tierpsychol.*, **16,** 693–705.

Bishop, L. G., Keehn, D. G., and McCann, G. D. (1968) Motion detection by interneurons of optic lobes of the flies (*Calliphora phaenicea* and *Musca domestica*. *J. Neurophysiol.*, **31,** 509–525.

Bitterman, M. E. (1960) Toward a comparative psychology of learning. *Amer. Psychologist*, **15,** 704–712.

Blanchard, R. J., and Blanchard, D. C. (1966) Food deprivation and reactivity to shock. *Psychon. Sci.*, **4,** 317–318.

Blase, B. (1960) Die Lautäusserungen des Neuntöters (*Lanius c. collurio* L.), Freilandbeobachtungen und Kasper-Hauser-Versuche. *Z. Tierpsychol.*, **17**, 293–344.

Blauvelt, H. (1956) Neonate-mother relationship in goat and man. In "Group Processes", *Trans. of 2nd Conf.* Ed. B. Schaffner, Josiah Macy Jr., New York.

Blest, A. D. (1957) The evolution of protective displays in the Saturnioidea and Sphingidae (Lepidoptera). *Behaviour*, **11**, 257–309.

Blest, A. D. (1958) Interaction between consecutive responses in a Hemileucid moth, and the evolution of insect communication. *Nature*, **181**, 1077–1078.

Blest, A. D. (1959) Central control of interactions between behaviour patterns in a Hemileucine moth. *Nature*, **184**, 1164–1165.

Blest, A. D. (1961) The concept of ritualization. In Thorpe and Zangwill (1961).

Blest, A. D., and Collett, T. S. (1965a, b) Micro-electrode studies of the medial protocerebrum of some Lepidoptera. I and II. *J. insect Physiol.*, **11**, 1079–1103 and 1289–1306.

Bligh, J. (1966) The thermosensitivity of the hypothalamus and thermoregulation in mammals. *Biol. Rev.*, **41**, 317–367.

Bliss, E. L. (1962) Roots of Behavior. Hoeber, New York.

Blodgett, H. C., and McCutchan, K. (1948) Relative strength of place and response learning in the T maze. *J. comp. physiol. Psychol.*, **41**, 17–24.

Blough, D. S. (1955) Method for tracing dark adaptation in the pigeon. *Science*, **121**, 703–704.

Blough, D. S. (1961) Experiments in animal psychophysics. *Sci. Amer.*, **205**, 113–122.

Blough, D. S. (1966) The study of animal sensory processes by operant methods. In Honig (1966).

Blough, D. S., and Millward, R. B. (1965) Learning: operant conditioning and verbal learning. *Ann. Rev. Psychol.*, **16**, 63–94.

Blurton-Jones, N. G. (1958/1959) Experiments on the causation of the threat postures of Canada geese. *Rep. Wildfowl Trust*, 1960, 46–52.

Blurton-Jones, N. G. (1968) Observations and experiments on causation of threat displays of the great tit (*Parus major*). *Anim. Behav. Monogr.*, **1**, 2.

Bolles, R. C. (1958) A replication and further analysis of a study on position reversal learning in hungry and thirsty rats. *J. comp. physiol. Psychol.*, **51**, 349.

Bolles, R. C. (1959) Group and individual performance as a function of intensity and kind of deprivation. *J. comp. physiol. Psychol.*, **52**, 579–585.

Bolles, R. C. (1960) Grooming behavior in the rat. *J. comp. physiol. Psychol.*, **53**, 306–310.

Bolles, R. C. (1963) Effect of food deprivation upon the rat's behavior in its home cage. *J. comp. physiol. Psychol.*, **56**, 456–460.

Bolles, R. C. (1965) Effects of deprivation conditions upon the rat's home cage behavior. *J. comp. physiol. Psychol.*, **60**, 244–248.

Bolles, R. C. (1967) Theory of Motivation. Harper & Row, New York.

Bolles, R. C., and de Lorge, J. (1962) Effect of hunger on exploration in a familiar locale. *Psychol. Rep.*, **10**, 54.

Bolles, R. C., and Morlock, H. (1960) Some asymmetrical drive summation phenomena. *Psychol. Rep.*, **6**, 373–378.

Bolles, R. C., Rapp, H. M., and White, G. C. (1968) Failure of sexual activity to reinforce female rats. *J. comp. physiol. Psychol.*, **65**, 311–313.

Bolles, R. C., and Stokes, L. W. (1965) Rat's anticipation of diurnal and a-diurnal feeding. *J. comp. physiol. Psychol.*, **60**, 290–294.

Bolles, R. C., and Woods, P. J. (1964) The ontogeny of behaviour in the albino rat. *Anim. Behav.*, **12**, 427–441.

Booth, D. A., and Quartermain, D. (1965) Taste sensitivity of eating elicited by chemical stimulation of the rat hypothalamus. *Psychon. Sci.*, **3**, 525–526.

Borg, G., Diamant, H., Ström, L., and Zotterman, Y. (1967) The relation between neural and perceptual intensity: a comparative study on the neural and psychophysical response to taste stimuli. *J. Physiol.*, **192**, 13–20.

Boring, E. G. (1935) The relation of the attributes of sensation to the dimensions of the stimulus. *Philos. Sci.*, **2**, 236–245.

Boshka, S. C., Weisman, H. M., and Thor, D. H. (1966) A technique for inducing aggression in rats utilizing morphine withdrawal. *Psychol. Rec.*, **16**, 541–543.

Bovard, E. W. (1958) The effects of early handling on the viability of the albino rat. *Psychol. Rev.*, **65**, 257–271.

Bovet, J. (1960) Experimentelle Untersuchungen über das Heimfindevermögen von Mäusen. *Z. Tierpsychol.*, **17**, 728–755.

Bovet, J. (1962) Influence d'un effet directionnel sur le retour au gîte des Mulots fauve et sylvestre (*Apodemus flavicollis* Melch. et *A sylvaticus* L.) et du Campagnol roux (*Clethrionomys glareolus* Schr.) *Z. Tierpsychol.*, **19**, 472–488.

Bower, G. H., and Miller, N. E. (1958) Rewarding and punishing effects from stimulating the same place in the rat's brain. *J. comp. physiol. Psychol.*, **51**, 669–674.

Bower, T. G. R. (1966) Heterogeneous summation in human infants. *Anim. Behav.*, **14**, 395–398.

Bowlby, J. (1952) Critical phases in the development of social responses in man and other animals. In "Prospects in Psychiatric Research", Ed. J. M. Tanner, Oxford Univ. Press.

Bowlby, J. (1960) Separation anxiety. *Int. J. Psycho-anal.*, **41**, 89–113.

Boyd, H., and Fabricius, E. (1965) Observations on the incidence of following of visual and auditory stimuli in naive mallard ducklings (*Anas platyrhynchos*). *Behaviour*, **25**, 1–15.

Bradley, W. E., and Wright, F. S. (1966) Visceral reflex activity: development in postnatal rabbit. *Science*, **152**, 216–217.

Brady, J. (1967a, b) Control of the circadian rhythm of activity in the cockroach. 1. The role of the corpora cardiaca, brain and stress. 2. The role of the suboesophageal ganglion and ventral nerve cord. *J. exp. Biol.*, **47**, 153–164 and 165–178.

Brady, J. V. (1960) Emotional behavior. In "Handbook of Physiology", Sec. I, Vol. III, Ed. J. Field, H. W. Magoun, and V. E. Hall, Amer. Physiol. Soc., Washington.

Brady, J. V. (1961) Motivational-emotional factors and intracranial self-stimulation. In Sheer, 1961.

Brady, J. V., Boren, J. J., Conrad, D., and Sidman, M. (1957) The effect of food and water deprivation upon intracranial self-stimulation. *J. comp. physiol. Psychol.*, **50**, 134–137.

Brady, J. V., Kelly, D., and Plumlee, L. (1969) Experimental approaches to the study of emotional behaviour. *Ann. N.Y. Acad. Sci.*, in press.

Braemer, W. (1957) Verhaltensphysiologische Untersuchungen am optischen Apparat bei Fischen. *Z. vergl. Physiol.*, **39**, 374–398.

Braemer, W. (1960) A critical review of the sun-azimuth hypothesis. *Cold Spring Harbor Symp. quant. Biol.*, **25**, 413–427.

Braemer, W., and Braemer, H. (1958) Zur Gleichgewichtsorientierung schrägstehender Fische. *Z. vergl. Physiol.*, **40**, 529–542.

Braemer, W., and Schwassmann, H. O. (1960) Von Rhythmus der Sonnenorientierung am Aquator (bei Fischen). *Ergeb. Biol.*, **26**, 182–201.

Brattgard, S. (1952) The importance of adequate stimulation for the chemical composition of retinal ganglion cells during early post-natal development. *Acta Radio. Suppl.*, 96.

Braun, E. (1963) Die ersten Beutefanghandlungen junger Blaufelchen (*Coregonus wartmanni* Bloch) und Hechte (*Esox lucius* L.). *Z. Tierpsychol.*, **20**, 257–266.

Breed, F. S. (1911) The development of certain instincts and habits in chicks. *Behaviour Monographs*, **1**, 1–78.

Brennan, W. M., Ames, E. W., and Moore, R. W. (1966) Age differences in infants' attention to patterns of different complexities. *Science*, **151**, 354–356.

Brereton, J. le G., and Immelmann, K. (1962) Head-scratching in the Psittaciformes. *Ibis*, **104**, 169–175.

Bridger, W. H. (1960) Signalling systems in the development of cognitive functions. In "The Central Nervous System and Behaviour", Ed. M. A. B. Brazier, Macy Foundation, New York.

Bridger, W. H. (1962) Ethological concepts and human development. In "Recent Advances in Biological Psychiatry", **4**, 95–107, Plenum Press, New York.

Brindley, G. S., and Merton, P. A. (1960) The absence of position sense in human eye. *J. Physiol.*, **153**, 127–130.

Broadbent, D. E. (1952) Listening to one of two synchronous messages. *J. exp. Psychol.*, **44**, 51–55.

Broadbent, D. E. (1958) Perception and Communication. Pergamon Press, London.

Broadbent, D. E. (1966) Recent analyses of short-term memory. *Proc. 18th Int. Cong. Psychol. Moscow*, **21**, 18–25.

Broadbent, D. E., and Gregory, M. (1963a) Vigilance considered as a statistical decision. *Brit. J. Psychol.*, **54,** 309–323.

Broadbent, D. E., and Gregory, M. (1963b) Division of attention and the decision theory of signal detection. *Proc. Roy. Soc. B,* **158,** 222–231.

Broadhurst, P. L. (1959) The interaction of task difficulty and motivation. *Acta Psychol.*, **16,** 321–338.

Broadhurst, P. L. (1961) Analysis of maternal effects in the inheritance of behaviour. *Anim. Behav.*, **9,** 129–141.

Broadhurst, P. L., and Jinks, J. L. (1963) The inheritance of mammalian behaviour re-examined. *J. Hered.*, **54,** 170–176.

Brobeck, J. R. (1955) Energy exchange. In "Textbook of Physiology" (17th ed.), Ed. J. F. Fulton, Saunders, Philadelphia.

Brobeck, J. R. (1960) Regulation of feeding and drinking. In "Handbook of Physiology, Sec. 1, Vol. II", Ed. J. Field, H. W. Magoun, and V. E. Hall, Amer. Physiol. Soc., Washington.

Brobeck, J. R., Larsson, J., and Reyes, E. A. (1956) A study of the electrical activity of the hypothalamic feeding mechanism. *J. Physiol.*, **132,** 358–364.

Brockway, B. F. (1963) Ethological studies of the budgerigar (*Melopsittacus undalatus*): non-reproductive behavior. *Behaviour,* **22,** 193–222.

Brogden, W. J. (1939) Sensory pre-conditioning. *J. exp. Psychol.*, **25,** 323–332.

Bromley, R. B., and Bard, P. (1940) A study of the effect of estrin on the responses to genital stimulation shown by decapitate and decerebrate female cats. *Amer. J. Physiol.*, **129,** P318–P319.

Bronfenbrenner, U. (1966) Early deprivation in animals and man. In "Early Experience and Behavior", Ed. G. Newton, Thomas, Springfield, Ill.

Bronson, F. H., and Desjardins, C. (1968) Aggression in adult mice: modification by neonatal injections of gonadal hormones. *Science,* **161,** 705–706.

Bronson, G. W. (1968a) The development of fear in man and other animals. *Child Devel.*, **39,** 409–430.

Bronson, G. W. (1968b) The fear of novelty. *Psychol. Bull.*, **69,** 350–358.

Broom, D. M. (1968) Specific habituation by chicks. *Nature,* **217,** 880–881.

Brower, J. van Z., and Brower, L. P. (1965) Experimental studies of mimicry: 8. further investigations of honey bees (*Apis mellifora*) and their dronefly mimics (*Eristalis* spp.). *Amer. Nat.*, **99,** 173–188.

Brower, L. P., Brower, J. van Z., and Cranston, F. P. (1965) Courtship behavior of the queen butterfly *Danaus gilippus berenice* (Cramer). *Zoologica,* **50,** 1–39.

Brower, L. P., Brower, J. van Z., and Westcott, P. (1960) Experimental studies of mimicry: 5. the reactions of toads (*Bufo terrestris*) to bumble bees (*Bombus americanorum*) and their robberfly mimics (*Mallophora bomboides*), with a discussion of aggressive mimicry. *Amer. Nat.*, **44,** 343–355.

Brown, F. A. (1959) Living clocks. *Science,* **130,** 1535–1544.

Brown, F .A. (1960) Response to pervasive geophysical factors and the biological clock problem. *Cold Spring Harbor Symp. quant. Biol.*, **25,** 57–71.

Brown, F. A., and Webb, H. M. (1948) Temperature relations of an endogenous daily rhythmicity in the fiddler crab *Uca*. *Physiol. Zool.*, **21**, 371–381.

Brown, J. L. (1964) Goals and terminology in ethological motivation research. *Anim. Behav.*, **12**, 538–541.

Brown, J. L. (1969) The control of avian vocalization by the central nervous system. In Hinde (1969).

Brown, J. L., and Hunsperger, R. W. (1963) Neuroethology and the motivation of agonistic behaviour. *Anim. Behav.*, **11**, 439–448.

Brown, J. S. (1953) Problems presented by the concept of acquired drives. In "Current Theory and Research in Motivation", University of Nebraska Press, Lincoln, Neb.

Brown, J. S. (1961) The Motivation of Behavior. McGraw-Hill, New York.

Brown, J. S., and Farber, I. E. (1951) Emotions conceptualised as intervening variables—with suggestions toward a theory of frustration. *Psychol. Bull.*, **48**, 465–495.

Brown, J. S., and Farber, I. E. (1968) Secondary motivational systems. *Ann. Rev. Psychol.*, **19**, 99–134.

Brown, J. S., and Jacobs, A. (1949) The role of fear in the motivation and acquisition of responses. *J. exp. Psychol.*, **39**, 747–759.

Brown, R. G. B. (1965) Courtship behaviour in the *Drosophila obscura* group. Part II. Comparative studies. *Behaviour*, **25**, 281–323.

Brown, R. G. B., Blurton-Jones, N. G., and Hussell, D. J. T. (1967) The breeding behaviour of Sabines' gull, *Xema sabini*. *Behaviour*, **28**, 110–140.

Brown, R. T. (1968) Early experience and problem-solving ability. *J. comp. physiol. Psychol.*, **65**, 433–440.

Bruce, V. G. (1960) Environmental entrainment of circadian rhythms. *Cold Spring Harbor Symp. quant. Biol.*, **25**, 29–48.

Brückner, G. H. (1933) Untersuchungen zur Tiersoziologie, insbesondere zur Auflösung der Familie. *Z. Psychol.*, **128**, 1–120.

Bruner, J., and Tauc, L. (1966a) Habituation at the synaptic level in *Aplysia*. *Nature*, **210**, 37–39.

Bruner, J., and Tauc, L. (1966b) Long-lasting phenomena in the molluscan nervous system. *Symp. Soc. exp. Biol.*, **20**, 457–475.

Buchholtz, C. (1956) Eine Analyse des Paarungsverhaltens und der dabei wirkenden Auslöser bei den Libellen *Platycnemis pennipes* Pull und *Pl. dealbata* Klug. *Z. Tierpsychol.*, **13**, 13–25.

Bullock, T. H. (1958) Parameters of integrative action of the nervous system at the neuronal level. *Exper. Cell. Res. Suppl.*, **5**, 323–337.

Bullock, T. H. (1959a) Neuron doctrine and electrophysiology. *Science*, **129**, 997–1002.

Bullock, T. H. (1959b) Initiation of nerve impulses in receptor and central neurons. *Revs. Mod. Physics*, **31**, 504–514.

Bullock, T. H. (1961) The origins of patterned nervous discharge. *Behaviour*, **17**, 48–59.

Bullock, T. H. (1966) Simple systems for the study of learning mechanisms. *Neurosciences Res. Prog. Bull.*, **4**, 2.

Bullock, T. H., and Terzuolo, C. A. (1957) Diverse forms of activity in the somata of spontaneous and integrating ganglion cells. *J. Physiol.*, **138**, 341–364.

Bullough, W. S. (1942) The reproductive cycles of the British and Continental races of the starling (*Sturnus vulgaris* L.). *Phil. Trans. Roy. Soc. Lond. B*, **21**, 165–246.

Bunnell, B. N., and Flesher, C. K. (1965) Copulatory behavior of male hamsters as a function of time since androgen withdrawal. *Psychon. Sci.*, **3**, 181–182.

Bunnell, B. N., and Kimmel, M. E. (1965) Some effects of copulatory experience on postcastration mating behavior in the male hamster. *Psychon. Sci.*, **3**, 179–180.

Burghardt, G. (1966) Stimulus control of the prey attack response in naive garter snakes. *Psychon. Sci.*, **4**, 37–38.

Burghardt, G. M. (1967) Chemical-cue preferences of inexperienced snakes: comparative aspects. *Science*, **157**, 718–721.

Burnstein, D. D., and Wolff, P. C. (1967) Vocal conditioning in the guinea pig. *Psychon. Sci.*, **8**, 39–40.

Bush, B. M. H. (1962) Peripheral reflex inhibition in the claw of the crab, *Carcinus maenas* (L.). *J. exp. Biol.*, **39**, 71–88.

Bush, B. M. H., Wiersma, C. A. G., and Waterman, T. H. (1964) Efferent mechanoreceptive responses in the optic nerve of the crab *Podophthalmus*. *J. cell. comp. Physiol.*, **64**, 327–345.

Buss, A. H. (1966) Instrumentality of aggression, feedback and frustration as determinants of physical aggression. *J. Pers. soc. Psychol.*, **3**, 153–162.

Butenandt, A. (1955) Uber Wirkstoffe des Insektenreiches: II. Zur Kenntnis der Sexual-Lockstoffe. *Naturw. Rdsch.*, **12**, 457–464.

Butler, R. A. (1953) Discrimination learning by rhesus monkeys to visual-exploration motivation. *J. comp. physiol. Psychol.*, **46**, 95–98.

Butler, R. A. (1957a) Discrimination learning by rhesus monkeys to auditory incentives. *J. comp. physiol. Psychol.*, **50**, 239–241.

Butler, R. A. (1957b) The effect of deprivation of visual incentives on visual exploration motivation in monkeys. *J. comp. physiol. Psychol.*, **50**, 177–179.

Butler, R. A. (1964) Incentive conditions which influence visual exploration. *J. exp. Psychol.*, **48**, 19–23.

Butler, R. A., and Harlow, H. F. (1954) Persistence of visual exploration in monkeys. *J. comp. physiol. Psychol.*, **47**, 257–263.

Butler, R. A., and Woolpy, J. H. (1963) Visual attention in the rhesus monkey. *J. comp. physiol. Psychol.*, **56**, 324–328.

Butz-Kuenzer, E. (1957) Optische und labyrinthäre Auslösung der Lagereaktionen bei Amphibien. *Z. Tierpsychol.*, **14**, 429–447.

Caggiula, A. R. (1967) Specificity of copulation-reward systems in the posterior hypothalamus. *Proc. 75th Ann. Conv. Amer. Psychol. Ass.*, 125–126.

Caïn, J. (1959) Le Problème des Névroses Expérimentales. Brouwer, Bruges.

Cairns, R. B. (1966a) Development, maintenance and extinction of social attachment behavior in sheep. *J. comp. physiol. Psychol.*, **62,** 298–306.

Cairns, R. B. (1966b) Attachment behavior of mammals. *Psychol. Rev.*, **73,** 409–426.

Campbell, B. A. (1960) Effects of water deprivation on random activity. *J. comp. physiol. Psychol.*, **53,** 240–241.

Campbell, B. A. (1967) Developmental studies of learning and motivation in infra-primate mammals. In Stevenson *et al.* (1967).

Campbell, B. A., and Campbell, E. H. (1962) Retention and extinction of learned fear in infant and adult rats. *J. comp. physiol. Psychol.*, **55,** 1–8.

Campbell, B. A., and Pickleman, J. R. (1961) The imprinting object as a reinforcing stimulus. *J. comp. physiol. Psychol.*, **54,** 592–596.

Campbell, B. A., and Riccio, D. C. (1966) Cold-induced stress in rats as a function of age. *J. comp. physiol. Psychol.*, **61,** 234–239.

Campbell, B. A., and Sheffield, F. D. (1953) Relation of random activity to food deprivation. *J. comp. physiol. Psychol.*, **46,** 320–322.

Campbell, B. A., Smith, N. F., Misanin, J. R., and Jaynes, J. (1966) Species differences in activity during hunger and thirst. *J. comp. physiol. Psychol.*, **61,** 123–127.

Campbell, F. W., and Robson, J. G. (1961) A fresh approach to stabilized retinal images. *J. Physiol.*, **158,** 11P.

Cane, V. (1961) Some ways of describing behaviour. In Thorpe and Zangwill (1961).

Cane, V. R., and Vince, M. A. (1968) Age and learning in quail. *Brit. J. Psychol.*, **59,** 37–46.

Capretta, P. J. (1962) Saccharin consumption under varied conditions of hunger drive. *J. comp. physiol. Psychol.*, **55,** 656–660.

Capretta, P. J., and Rea, R. (1967) Discrimination reversal learning in the crayfish. *Anim. Behav.*, **15,** 6–7.

Carbaugh, B. T., Schein, M. W., and Hale, E. B. (1962) Effects of morphological variations of chicken models on sexual responses of cocks. *Anim. Behav.*, **10,** 235–238.

Carlsson, S. G., and Larsson, K. (1964) Mating in male rats after local anesthetization of the glans penis. *Z. Tierpsychol.*, **21,** 854–856.

Carmichael, L. (1926) The development of behavior in vertebrates experimentally removed from the influence of external stimulation. *Psychol. Rev.*, **33,** 51–58.

Carmichael, L. (1927) A further study of the development of behavior in vertebrates experimentally removed from the influence of external stimulation. *Psychol. Rev.*, **34,** 34–47.

Carmichael, L. (1934) An experimental study in the pre-natal guinea-pig of the origin and development of reflexes and patterns of behaviour in relation to the stimulation of specific receptor areas during the period of active fetal life. *Genet. Psychol. Monogr.*, **16**, 337–491.

Carr, W. J. (1952) The effect of adrenalectomy upon the NaCl taste threshold in rat. *J. comp. physiol. Psychol.*, **45**, 377–380.

Carr, W. J., and Caul, W. F. (1962) The effect of castration in rat upon the discrimination of sex odours. *Anim. Behav.*, **10**, 20–27.

Carr, W. J., Loeb, L. S., and Dissinger, M. L. (1965) Responses of rats to sex odours. *J. comp. physiol. Psychol.*, **59**, 370–377.

Carr, W. J., Loeb, L. S., and Wylie, N. R. (1966) Responses to feminine odors in normal and castrated male rats. *J. comp. physiol. Psychol.*, **62**, 336–338.

Carr, W. J., and McGuigan, D. I. (1965) The stimulus basis and modification of visual cliff performance in the rat. *Anim. Behav.*, **13**, 25–29.

Carr, W. J., Solberg, B., and Pfaffmann, C. (1962) The olfactory threshold for estrous female urine in normal and castrated male rats. *J. comp. physiol. Psychol.*, **55**, 415–417.

Carthy, J. D. (1951) Instinct. *New Biology*, **10**, 95–105.

Caryl, P. G. (1969) Social control of fighting behaviour in the zebra finch, *Taeniopygia guttata* (Vieillot). Ph.D. thesis, Cambridge.

Caspari, E. (1958) Genetic basis of behavior. In "Behavior and Evolution", Ed. A. Roe and G. G. Simpson, Yale University Press, New Haven.

Catania, A. C. (1966) Concurrent operants. In Honig (1966).

Cate, J. ten (1965) Automatic activity of the locomotor centres of the lumbar cord in lizards. *J. exp. Biol.*, **43**, 181–184.

Chambers, R. M. (1956) Effects of intravenous glucose injections on learning, general activity and hunger drive. *J. comp. physiol. Psychol.*, **49**, 558–564.

Chance, M. R. A. (1962) An interpretation of some agonistic postures; the role of "cut-off" acts and postures. *Symp. zool. Soc. Lond.*, **8**, 71–89.

Chapman, R. M., and Bragdon, H. R. (1964) Evoked responses to numerical and non-numerical visual stimuli while problem solving. *Nature*, **203**, 1155–1157.

Chapman, R. M., and Levy, N. (1957) Hunger drive and reinforcing effect of novel stimuli. *J. comp. physiol. Psychol.*, **50**, 233–238.

Charlesworth, W. R., and Thompson, W. R. (1957) Effect of lack of visual stimulus variation on exploratory behavior in the adult white rat. *Psychol. Rep.*, **3**, 509–512.

Chase, R. A., Harvey, S., Standfast, S., Rapin, I., and Sutton, S. (1961) Studies on sensory feedback: I. *Q. J. exp. Psychol.*, **13**, 141–152.

Chernetski, K. E. (1964) Facilitation of a somatic reflex by sound in *Rana clamitans:* effects of sympathectomy and decerebration. *Z. Tierpsychol.*, **21**, 813–821.

Cherry, E. C. (1953) Some experiments on the recognition of speech, with one and with two ears. *J. acoust. Soc. Amer.*, **25**, 975–979.

Chiang, M. (1967) Use of tools by wild macaque monkeys in Singapore. *Nature*, **214**, 1258–1259.

Chitty, D. (1957) Self-regulation of numbers through changes in viability. *Cold Spring Harbor Symp. quant. Biol.*, **22,** 277–280.

Chorover, S. L., and Schiller, P. H. (1965) Short-term retrograde amnesia in rats. *J. comp. physiol. Psychol.*, **59,** 73–78.

Chorover, S. L., and Schiller, P. H. (1966) Re-examination of prolonged retrograde amnesia in one-trial learning. *J. comp. physiol. Psychol.*, **61,** 34–41.

Chow, K. L., and Nissen, H. W. (1955) Interocular transfer of learning in visually naive and experienced infant chimpanzees. *J. comp. physiol. Psychol.*, **48,** 229–237.

Christophersen, E. R., and Wagman, W. (1965) Maternal behavior in the albino rat as a function of self-licking deprivation. *J. comp. physiol. Psychol.*, **60,** 142–144.

Cicala, G. A. (1961) Running speed in rats as a function of drive level and presence or absence of competing response trials. *J. exp. Psychol.*, **62,** 329–334.

Clark, R. B. (1960) Habituation of the polychaete *Nereis* to sudden stimuli: 1. General properties of the habituation process. *Anim. Behav.*, **8,** 82–91.

Clark, R., Schuster, C. R., and Brady, J. V. (1961) Instrumental conditioning of jugular self-infusion in the rhesus monkey. *Science,* **133,** 1829–1830.

Clarke, J. R. (1953) The effect of fighting on the adrenals, thymus and spleen of the vole (*Microtus agrestis*). *J. Endocr.*, **9,** 114–126.

Claus, H. J., and Bindra, D. (1960) Reactions to novelty and stimulus-change induced response decrement. *Canad. J. Psychol.*, **14,** 101–110

Clayton, F. L. (1958) Light reinforcement as a function of water deprivation. *Psychol. Rep.*, **4,** 63–66.

Clayton, F. L., and Hinde, R. A. (1968) The habituation and recovery of aggressive display in *Betta splendens. Behaviour,* **30,** 96–106.

Clemens, L. G. (1967) Effect of stimulus female variation on sexual performance of the male deermouse, *Peromyscus maniculatus. Proc. 75th Ann. Conv. Amer. Psychol. Ass.*, 119–120.

Coghill, G. E. (1929) Anatomy and the Problem of Behaviour. Cambridge University Press.

Cohen, H. B. (1961) The effect of contralateral visual stimulation on visibility with stabilized retinal images. *Canad. J. Psychol.*, **15,** 212–219.

Cohen, M. J. (1955) The function of receptors in the statocyst of the lobster *Homarus americanus. J. Physiol.*, **130,** 9–34.

Cohen, M. J. (1960) The response patterns of single receptors in the crustacean statocyst. *Proc. Roy. Soc. Lond. B,* **152,** 30–49.

Cohen, M. J., Hagiwara, S., and Zotterman, Y. (1955) The response spectrum of taste fibres in the cat: a single fibre analysis. *Acta physiol. scand.*, **33,** 316–332.

Cole, D. D., and Shafer, J. N. (1966) A study of social dominance in cats. *Behaviour,* **27,** 39–53.

Collett, T. S., and Blest, A. D. (1966) Binocular, directionally selective neurones, possibly involved in the optomotor response of insects. *Nature*, **212**, 1330–1333.

Collias, N. E. (1952) The development of social behaviour in birds. *Auk*, **69**, 127–159.

Collias, N. E. (1956) The analysis of socialization in sheep and goats. *Ecology*, **37**, 228–239.

Collias, N. E., and Collias, E. C. (1962) An experimental study of the mechanisms of nest building in a weaverbird. *Auk*, **79**, 568–595.

Collias, N. E., and Collias E. C. (1964) Evolution of nest-building in the weaverbirds (*Ploceidae*). *Univ. Calif. Publ. Zool.*, **73**, 1–162.

Collias, N., and Joos, M. (1953) The spectrographic analysis of sound signals of the domestic fowl. *Behaviour*, **5**, 175–188.

Collier, G., Levitsky, D., and Squibb, R. L. (1967) Instrumental performance as a function of the energy content of the diet. *J. comp. physiol. Psychol.*, **64**, **68–72.**

Collier, G., and Willis, F. N. (1961) Deprivation and reinforcement. *J. exp. Psychol.*, **62**, 377–384.

Collins, T. B. (1965) Strength of the following response in the chick in relation to degree of "parent" contact. *J. comp. physiol. Psychol.*, **60**, 192–195.

Cooke, F., and Cooch, F. G. (1968) The genetics of polymorphism in the goose *Anser caerulescens*. *Evolution*, **22**, 289–300.

Coons, E. E., and Cruce, J. A. F. (1968) Lateral hypothalamus: food current intensity in maintaining self-stimulation of hunger. *Science*, **159**, 1117–1119.

Coons, E. E., and Miller, N. E. (1960) Conflict versus consolidation of memory traces to explain "retrograde amnesia" produced by E. C. S. *J. comp. physiol. Psychol.*, **53**, 524–531.

Cooper, M., and Aronson, L. (1962) Effect of a sensory deprivation on the sexual behavior of experienced adult male cats. *Amer. Zool.*, **2**, 114 (abstract).

Coppock, H. W., and Chambers, R. M. (1954) Reinforcement of position preference by automatic intravenous injections of glucose. *J. comp. physiol. Psychol.*, **47**, 355–357.

Corbit, J. D. (1965a) Effect of intravenous sodium chloride on drinking in the rat. *J. comp. physiol. Psychol.*, **60**, 397–406.

Corbit, J. D. (1965b) Hyperphagic hyperreactivity to adulteration of drinking water with quinine HCl. *J. comp. physiol. Psychol.*, **60**, 123–124.

Corbit, J. D., and Stellar, E. (1964) Palatability, food intake, and obesity in normal and hyperphagic rats. *J. comp. physiol. Psychol.*, **58**, 63–67.

Cott, H. B. (1940) Adaptive Coloration in Animals. Methuen, London.

Cotton, J. W. (1953) Running time as a function of amount of food deprivation. *J. exp. Psychol.*, **46**, 188–198.

Cowey, A. (1968) Discrimination. In Weiskrantz (1968a).

Cowles, J. T., and Nissen, H. W. (1937) Reward-expectancy in delayed responses of chimpanzees. *J. comp. physiol. Psychol.*, **24**, 345–348.

Cragg, B. G. (1967) Changes in visual cortex on first exposure of rats to light. *Nature*, **215**, 251–252.

Craik, K. J. W. (1943) The Nature of Explanation. Cambridge University Press, London.

Crane, J. (1949) Comparative biology of Salticid spiders at Rancho Grande, Venezuela: IV. An analysis of display. *Zoologica*, **34**, 159–214.

Crane, J. (1952) A comparative study of innate defensive behavior in Trinidad mantids (Orthoptera, Mantoidea). *Zoologica*, **37**, 259–294.

Crane, J. (1955) Imaginal behavior of a Trinidad butterfly, *Heliconius erato hydara* Hewitson, with special reference to the social use of color. *Zoologica*, **40**, 167–196.

Crane, J. (1957) Basic patterns of display in fiddler crabs (Ocypodidae, Genus *Uca*). *Zoologica*, **42**, 69–82.

Crane, J. (1958) Aspects of social behaviour in fiddler crabs, with special reference to *Uca maracoani*. *Zoologica*, **43**, 113–130.

Creer, T. L., Hitzing, E. W., and Schaeffer, R. W. (1966) Classical conditioning of reflexive fighting. *Psychon. Sci.*, **4**, 89–90.

Crespi, L. P. (1944) Amount of reinforcement and level of performance. *Psychol Rev.*, **51**, 341–357.

Crook, J. H. (1960a) Studies on the reproductive behaviour of the Baya weaver (*Ploceus philippinus*). *J. Bombay Nat. Hist. Soc.*, **57**, 1–44.

Crook, J. H. (1960b) Studies on the social behaviour of *Quelea q. quelea* (Linn.) in French West Africa. *Behaviour*, **16**, 1–55.

Crook, J. H. (1960c) Nest form and construction in certain West African weaver-birds. *Ibis*, **102**, 1–25.

Crook, J. H. (1962) The adaptive significance of pair formation types in weaver birds. *Symp. zool. Soc. Lond.*, **8**, 57–70.

Crook, J. H. (1963) Comparative studies on the reproductive behaviour of two closely related weaver bird species (*Ploceus cucullatus* and *Ploceus nigerrimus*) and their races. *Behaviour*, **21**, 177–232.

Crook, J. H. (1964a) The evolution of social organisation and visual communication in the weaver birds (Ploceinae). *Behaviour Suppl.* 10, 1–178.

Crook, J. H. (1964b) Field experiments on the nest construction and repair behaviour of certain weaver birds. *Proc. Zool. Soc. Lond.*, **142**, 217–255.

Cross, B. A. (1964) The hypothalamus in mammalian homeostasis. *Symp. Soc. exp. Biol.*, **18**, 157–194.

Cross, B. A., and Silver, I. A. (1966) Electrophysiological studies on the hypothalamus. *Brit. med. Bull.*, **22**, 254–260.

Cruze, W. W. (1935) Maturation and learning in chicks. *J. comp. Psychol.*, **19**, 371–409.

Cullen, E. (1957) Adaptations in the kittiwake to cliff-nesting. *Ibis*, **99**, 275–302.

Cullen, E. (1960) Experiment on the effect of social isolation on reproductive behaviour in the three-spined stickleback. *Anim. Behav.*, **8**, 235.

Cullen, J. M. (1959) Behaviour as a help in taxonomy. *Systematics Assn. Publ. No.* 3, 131–140.

Cullen, J. M. (1960) Some adaptations in the nesting behaviour of terns. *Proc. XIIth Int. Orn. Cong. Helsinki*, 1958, 153–157.

Cullen, J. M., and Ashmole, N. P. (1963) The Black Noddy (*Anous tenuirostris*) on Acension Island, Pt 2, Behaviour. *Ibis*, **103**b, 423–446.

Curio, E. (1959) Verhaltensstudien am Trauerschnäpper. *Z. Tierpsychol. Suppl.* **3,** 1–110.

Curio, E. (1960) Ontogenese und Phylogenese einiger Triebäusserungen von Fliegenschnäppern. *J. Orn.*, **101,** 291–309.

Curtius, A. (1954) Uber angeborene Verhaltensweisen bei Vögeln, insbesondere bei Hühnerkücken. *Z. Tierpsychol.*, **11,** 94–109.

Daanje, A. (1950) On locomotory movements in birds and the intention movements derived from them. *Behaviour*, **3,** 48–98.

Dachowski, L. (1964) Irrelevant thirst drive and light aversion. *Psych. Rep.*, **14,** 899–904.

D'Amato, M. R., and Jagoda, H. (1962) Effect of early exposure to photic stimulation on brightness discrimination and exploratory behavior. *J. Genet. Psychol.*, **101,** 267–271.

Dane, B., and Kloot, W. G. van der (1964) An analysis of the display of the Goldeneye duck (*Bucephala clangula* L.). *Behaviour*, **22,** 283–328.

Dane, B., Walcott, C., and Drury, W. H. (1959) The form and duration of the display actions of the goldeneye (*Bucephala clangula*). *Behaviour*, **14,** 265–281.

Darwin, C. (1872) The Expression of the Emotions in Man and the Animals. John Murray, London.

Datel, W. E., and Seward, J. P. (1952) On the persistence of an ear-scratching response in the rat. *J. abnorm. soc. Psychol.*, **47,** 58–61.

Davenport, D., Camougis, G., and Hickok, J. F. (1960) Analyses of the behaviour of commensals in host-factor: 1. *Anim. Behav.*, **8,** 209–218.

Davenport, R. K., and Menzel, E. W. (1963) Stereotyped behaviour of the infant chimpanzee. *Archiv. gen. Psychiatr.*, **8,** 99–104.

Davenport, R. K., Menzel, E. W., and Rogers, C. M. (1966) Effects of severe isolation on "normal" juvenile chimpanzees. *Archiv. gen. Psychiatr.*, **14,** 134–138.

Davidson, J. M. (1966a) Characteristics of sex behaviour in male rats following castration. *Anim. Behav.*, **14,** 266–272.

Davidson, J. M. (1966b) Activation of the male rat's sexual behavior by intracerebral implantation of androgen. *Endocrinology*, **79,** 783–794.

Davies, S. J. J. F. (1961) The orientation of pecking in very young magpie geese *Anseranas semipalmata*. *Ibis*, **103**a, 277–283.

Davies, S. J. J. F. (1965) The behaviour of *Streptopelia* doves and their hybrids. Ph.D. thesis, Cambridge.

Davis, H., and Donenfeld, I. (1967) Extinction induced social interaction in rats. *Psychon. Sci.*, **7,** 85–86.

Davis, J. (1957) Comparative foraging behavior of the spotted and brown towhees. *Auk*, **74**, 129–166.

Davis, J. D. (1958) The reinforcing effect of weak-light onset as a function of amount of food deprivation. *J. comp. physiol. Psychol.*, **51**, 496–498.

Davis, J. D. (1965) Food intake following blood mixing of hungry and satiated rats. *Psychon. Sci.*, **3**, 177–178.

Davis, J. D., Gallagher, R. L., and Ladove, R. (1967) Food intake controlled by a blood factor. *Science*, **156**, 1247–1248.

Dawkins, R. (1968) The ontogeny of a pecking preference in domestic chicks. *Z. Tierpsychol.*, **25**, 170–186.

Dawson, B. V., and Foss, B. M. (1965) Observational learning in budgerigars. *Anim. Behav.*, **13**, 470–474.

Decker, J. D., and Hamburger, V. (1967) The influence of different brain regions on periodic motility of the chick embryo. *J. exp. Zool.*, **165**, 371–384.

De Coursey, P. (1960) Phase control of activity in a rodent. *Cold Spring Harbor Symp. quant. Biol.*, **25**, 49–55.

De Coursey, P. (1961) Effect of light on the circadian activity rhythm of the flying squirrel, *Glaucomys volans*. *Z. vergl. Physiol.*, **44**, 331–354.

Deese, J. (1951) The extinction of a discrimination without performance of the choice response. *J. comp. physiol. Psychol.*, **44**, 362–366

DeFries, J. C., Thomas, E. A., Hegmann, J. P., and Weir, M. W. (1967a) Open-field behavior in mice: analysis of maternal effects by means of ovarian transplantation. *Psychon. Sci.*, **8**, 207–208.

DeFries, J. C., Weir, M. W., and Hegmann, J. P. (1967b) Differential effects of prenatal maternal stress on offspring behavior in mice as a function of genotype and stress. *J. comp. physiol. Psychol.*, **63**, 332–334.

Delgado, J. M. R. (1962) Brain centers and control of behavior—animals. *First Hahnemann Symp. on Psychosom. Med.*, 221–227, Lea and Febiger.

Delgado, J. M. R. (1963) Effect of brain stimulation on task-free situations. *Electroenceph. clin. Neurophysiol. Suppl.* **24**, 260–281.

Delgado, J. M. R. (1964) Free behaviour and brain stimulation. *Int. Rev. Neurobiology*, **6**, 349–449.

Delgado, J. M. R. (1966) Aggressive behavior evoked by radio-stimulation in monkey colonies. *Amer. Zool.*, **6**, 669–681.

Delgado, J. M. R. (1967) Social rank and radio-stimulated aggressiveness in monkeys. *J. nerv. ment. Dis.*, **144**, 383–390.

Delgado, J. M. R., and Anand, B. K. (1953) Increase of food intake induced by electrical stimulation of the lateral hypothalamus. *Amer. J. Physiol.*, **172**, 162–168.

Delius, J. D. (1963) Das Verhalten der Feldlerche. *Z. Tierpsychol.*, **20**, 297–348.

Delius, J. D. (1967) Displacement activities and arousal. *Nature*, **214**, 1259–1260.

Dell, P. C. (1958) Some basic mechanisms of the translation of bodily needs into behaviour. In Ciba Foundation Symposium on the "Neurological Basis of Behaviour", Ed. G. E. W. Wolstenholme and C. M. O'Connor, Boston.

Dember, W. N. (1956) Response by the rat to environmental change. *J. comp. physiol. Psychol.*, **49**, 93–95.

Dember, W. N., and Fowler, H. (1958) Spontaneous alternation behavior. *Psychol. Bull.*, **55**, 412–428.

Dempsey, E. W., and Rioch, D. McK. (1939) The localization in the brain stem of the oestrous responses of the female guineapig. *J. Neurophysiol.*, **2**, 9–18.

DeNelsky, G. Y., and Denenberg, V. H. (1967a) Infantile stimulation and adult exploratory behaviour in the rat: effects of handling upon visual variation-seeking. *Anim. Behav.*, **15**, 568–573.

DeNelsky, G. Y., and Denenberg, V. H. (1967b) Infantile stimulation and adult exploratory behavior. *J. comp. physiol. Psychol.*, **63**, 309–312

Denenberg, V. H. (1962) An attempt to isolate critical periods of development in the rat. *J. comp. physiol. Psychol.*, **55**, 813–815.

Denenberg, V. H. (1964) Critical periods, stimulus input, and emotional reactivity: a theory of infantile stimulation. *Psychol. Rev.*, **71**, 335–351.

Denenberg, V. H. (1966) Animal studies on developmental determinants of behavioral adaptability. In Experience, Structure, and Adaptability, Ed. O. J. Harvey, Springer, New York.

Denenberg, V. H. (1968) A consideration of the usefulness of the critical period hypothesis as applied to the stimulation of rodents in infancy. In "Early Experience and Behavior", Ed. G. Newton and S. Levine, Thomas, Springfield, Ill.

Denenberg, V. H., and Bell, R. W. (1960) Critical periods for the effects of infantile experience on adult learning. *Science*, **131**, 227–228.

Denenberg, V. H., Brumaghim, J. T., Haltmeyer, G. C., and Zarrow, M. X. (1967) Increased adrenocortical activity in the neonatal rat following handling. *Endocrin.*, **81**, 1047–1052.

Denenberg, V. H., and Haltmeyer, G. C. (1967) Test of the monotonicity hypothesis concerning infantile stimulation and emotional reactivity. *J. comp. physiol. Psychol.*, **63**, 394–396.

Denenberg, V. H., Hudgens, G. A., and Zarrow, M. X. (1966) Mice reared with rats. *Psychol. Rep.*, **18**, 451–456.

Denenberg, V. H., and Karas, G. G. (1959) Effects of differential infantile handling upon weight gain and mortality in the rat and mouse. *Science*, **130**, 629–630.

Denenberg, V. H., and Kline, N. J. (1958) The relationship between age and avoidance learning in the hooded rat. *J. comp. physiol. Psychol.*, **51**, 488–491.

Denenberg, V. H., and Morton, J. R. C. (1962a) Effects of environmental complexity and social groupings upon modification of emotional behavior. *J. comp. physiol. Psychol.*, **55**, 242–246.

Denenberg, V. H., and Morton, J. R. C. (1962b) Effects of preweaning and postweaning manipulations upon problem-solving behavior. *J. comp. physiol. Psychol.*, **55**, 1096–1098.

Denenberg, V. H., Paschke, R. E., and Zarrow, M. X. (1968) Killing of mice by rats prevented by early interaction between the two species. *Psychon. Sci.*, **11,** 39.

Denenberg, V. H., and Rosenberg, K. M. (1967) Non-genetic transmission of information. *Nature*, **216,** 549–550.

Denenberg, V. H., Schell, S. F., Karas, G. G., and Haltmeyer, G. C. (1966) Comparison of background stimulation and handling as forms of infantile stimulation. *Psych. Rep.*, **19,** 943–948.

Denenberg, V. H., and Smith, S. A. (1963) Effects of infantile stimulation and age upon behavior. *J. comp. physiol. Psychol.*, **56,** 307–312.

Denenberg, V. H., and Whimbey, A. E. (1963a) Infantile stimulation and animal husbandry: a methodological study. *J. comp. physiol. Psychol.*, **56,** 877–878.

Denenberg, V. H., and Whimbbey, A. E. (1963b) Behavior of adult rats is modified by the experiences their mothers had as infants. *Science*, **142,** 1192–1193.

Dennis, W. (1934) Congenital cataract and unlearned behavior. *J. genet. Psychol.*, **44,** 340–351.

Desmedt, J. E. (1960) Neurophysiological mechanisms controlling acoustic input. In "Neural Mechanisms of the Auditory and Vestibular Systems", Ed. G. L. Rasmussen and W. Windle, Thomas, Springfield, Ill.

Desmedt, J. E. (1962) Auditory-evoked potentials from cochlea to cortex as influenced by activation of the efferent olivo-cochlear bundle. *J. Acoustical Soc. Amer.*, **34,** 1478–1496.

Desmedt, J. E., and Delwaide, P. J. (1965) Functional properties of the efferent cochlear bundle of the pigeon revealed by stereotaxic stimulation. *Exp. Neurol.*, **11,** 1–26.

Desmedt, J .E., and Mechelse, K. (1958) Suppression of acoustic input by thalamic stimulant. *Proc. Soc. exp. Biol. Med.*, **99,** 772–775.

Desmedt, J. E., and Monaco, P. (1961) Mode of action of the efferent olivo-cochlear bundle on the inner ear. *Nature*, **192,** 1263–1265.

Dethier, V. G. (1943) The dioptric apparatus of lateral ocelli: II. Visual capacities of the ocellus. *J. cell. comp. Physiol.*, **22,** 115–126.

Dethier, V. G. (1963) The Physiology of Insect Senses. Methuen, London.

Dethier, V. G. (1966) Insects and the concept of motivation. Nebraska Symp. on Motivation, 1966, 105–136.

Dethier, V. G., and Bodenstein, D. (1958) Hunger in the blowfly. *Z. Tierpsychol.*, **15,** 129–140.

Dethier, V. G., and Gelperin, A. (1967) Hyperphagia in the blowfly. *J. exp. Biol.*, **47,** 191–200.

Dethier, V. G., Solomon, R. L., and Turner, L. H. (1965) Sensory input and central excitation and inhibition in the blowfly. *J. comp. physiol. Psychol.*, **60,** 303–313.

Dethier, V. G., Solomon, R. L., and Turner, L. H. (1968) Central inhibition in the blowfly. *J. comp. physiol. Psychol.*, **66,** 144–150.

Deutsch, J. A. (1960) The Structural Basis of Behaviour. Cambridge Univ. Press, London and University of Chicago.

Deutsch, J. A. (1962) A system for shape recognition. *Psychol. Rev.*, **69**, 492–500.

Deutsch, J. A., and Deutsch, D. (1966) Physiological Psychology. Dorsey Press, Homewood, Ill.

De Vito, J. L., and Smith, O. A. (1959) Effects of temperature and food deprivation on the random activity of *Macaca mulatta*. *J. comp. physiol. Psychol.*, **52**, 29–32.

De Vore, I. (1965) Primate Behaviour. Holt, Rinehart and Winston, New York.

Dews, P. B. (1962) The effect of multiple S^{Δ} periods on responding on a fixed-interval schedule. *J. exp. Anal. Behav.*, **5**, 369–374.

Dewsbury, D. A. (1967a) A quantitative description of the behavior of rats during copulation. *Behaviour*, **29**, 154–179.

Dewsbury, D. A. (1967b) Changes in inter-intromission interval during uninterrupted copulation in rats. *Psychon. Sci.*, **7**, 177–178.

Dewsbury, D. A. (1968) Copulatory behavior in rats: changes as satiety is approached. *Psychol. Rep.*, **22**, 937–943.

Dewson, J. H. (1967) Efferent olivocochlear bundle: some relationships to noise masking and to stimulus attenuation. *J. Neurophysiol.*, **30**, 817–832.

Dewson, J. H. (1968) Efferent olivocochlear bundle: some relationships to stimulus discrimination in noise. *J. Neurophysiol.*, **31**, 122–130.

Diamond, M. C., Law, F., Rhodes, H., Lindner, B., Rosenzweig, M. R., Krech, D., and Bennett, E. L. (1966) Increases in cortical depth and glia numbers in rats subjected to enriched environment. *J. comp. Neurol.*, **128**, 117–126.

Di Cara, L. V., and Miller, N. E. (1968) Instrumental learning of vasomotor responses by rats: learning to respond differentially in the two ears. *Science*, **159**, 1485–1486.

Dieterlen, F. (1959) Das Verhalten des syrischen Goldhamsters (*Mesocricetus auratus* Waterhouse). *Z. Tierpsychol.*, **16**, 47–103.

Dilger, W. C. (1960) The comparative ethology of the African parrot genus *Agapornis*. *Z. Tierpsychol.*, **17**, 649–685.

Dilger, W. C. (1962) The behavior of lovebirds. *Sci. Amer.*, **206**, 88–98.

Dimond, S. J. (1968) Effects of photic stimulation before hatching on the development of fear in chicks. *J. comp. physiol. Psychol.*, **65**, 320–324.

Doane, B. K., Mahatoo, W., Heron, W., and Scott, T. H. (1959) Changes in perceptual function after isolation. *Canad. J. Psychol.*, **13**, 210–219.

Dobrzecka, C., Szwejkowska, G., and Konorski, J. (1966) Qualitative versus directional cues in two forms of differentiation. *Science*, **153**, 87–89.

Dobzhansky, Th. (1941) Genetics and the origin of species. Columbia Univ. Press, New York.

Dobzhansky, Th., Ehrman, L., and Kastritsis, P. A. (1968) Ethological isolation between sympatric and allopatric species of the *obscura* group of *Drosphila*. *Anim. Behav.*, **16**, 79–87.

Dodwell, P. C. (1961) Coding and learning in shape discrimination. *Psychol. Rev.*, **68,** 373–382.

Dodwell, P. C. (1962) A test of two theories of shape discrimination. *Q. J. exp. Psychol.*, **14,** 65–70.

Döhl, J. (1966) Manipulierfähigkeit und "einsichtiges" Verhalten eines Schimpansen bei komplizierten Handlungsketten. *Z. Tierpsychol.*, **23,** 77–113.

Dollard, J., Doob, L. W., Miller, N. E., Mowrer, O. H., and Sears, R. R. (1939) Frustration and Aggression. Yale Univ. Press, New Haven.

Dollard, J., and Miller, N. E. (1950) Personality and Psychotherapy. McGraw-Hill, New York.

Dolley, W. L., and Wierda, J. L. (1929) Relative sensitivity to light of different parts of the compound eye in *Eristalis tenax*. *J. exp. Zool.*, **53,** 129–139.

Doty, B. A., Jones, C. N., and Doty, L. A. (1967) Learning-set formation by mink, ferrets, skunks and cats. *Science*, **155,** 1579–1580.

Doty, R. W. (1951) Influence of stimulus pattern on reflex deglutition. *Amer. J. Physiol.*, **166,** 142–158.

Doty, R. W. (1967) In "Comprehensive Textbook of Psychiatry", Ed. A. M. Freedman and H. I. Kaplan, Williams and Wilkins, Baltimore.

Doty, R. W. (1968) Neural organization of deglutition. In "Handbook of Physiology—Alimentary Canal, IV", Ed. W. Heidel, American Physiological Society, Washington.

Doty, R. W., and Bosma, J. F. (1956) An electromyographic analysis of reflex deglutition. *J. Neurophysiol.*, **19,** 44–60.

Doty, R. W., and Giurgea, C. (1961) Conditioned reflexes established by coupling electrical excitation of two cortical areas. In "Brain Mechanisms and Learning", Ed. A. Fessard, R. W. Gerard, and J. Konorski, Thomas, Springfield, Ill.

Doty, R. W., Richmond, W. H., and Storey, A. T. (1967) Effect of medullary lesions on coordination of deglutition. *Exp. Neurol.*, **17,** 91–106.

Douglas, R. J. (1966) Cues for spontaneous alternation. *J. comp. physiol. Psychol.*, **62,** 171–183.

Dowling, J. E., and Boycott, B. B. (1966) Organization of the primate retina: electron microscopy. *Proc. Roy. Soc. Ser. B.*, **166,** 80–111.

Doyle, G., and Pratt Yule, E. (1959) Grooming activities and freezing behavior in relation to emotionality in albino rats. *Anim. Behav.*, **7,** 18–22.

Draper, W. A. (1967) A behavioural study of the home-cage activity of the white rat. *Behaviour*, **28,** 280–306.

Draper, W. A., and Bernstein, I. S. (1963) Stereotyped behavior and cage size. *Percept. and Mot. Skills*, **16,** 231–234.

Drees, O. (1952) Untersuchungen über die angeborenen Verhaltensweisen bei Springspinnen (*Salticidae*). *Z. Tierpsychol.*, **9,** 169–207.

Drori, D., and Folman, Y. (1967) The sexual behaviour of male rats unmated to 16 months of age. *Anim. Behav.*, **15,** 20–24.

Duffy, E. (1951) The concept of energy mobilization. *Psychol. Rev.*, **58,** 30–40.

Duffy, E. (1962) Activation and Behavior. Wiley, New York.

Dufort, R. H. (1964) The rat's adjustment to 23-, 47-, and 71-hour food-deprivation schedules. *Psychol. Rep.*, **14**, 663–669.

Duncan, C. J., and Sheppard, P. M. (1963) Continuous and quantal theories of sensory discrimination. *Proc. Roy. Soc. B.*, **158**, 343–363.

Duncan, C. P. (1949) The retroactive effect of electroshock on learning. *J. comp. physiol. Psychol.*, **42**, 32–44.

Duncan, I. J. H., and Wood-Gush, D. (in prep.).

Dunham, D. W. (1966) Agonistic behavior in captive rose-breasted grosbeaks, *Pheucticus ludovicianus* (L.). *Behaviour*, **27**, 160–173.

Dunham, D. W., Kortmulder, K., and van Iersel, J. J. A. (1968) Threat and appeasement in *Barbus stoliczkanus* (Cyprinidae). *Behaviour*, **30**, 15–26.

Dürrwächter, G. (1957) Untersuchungen über Phototaxis und Geotaxis einiger *Drosophila*–Mutanten nach Aufzucht in verschiedenen Lichtbedingungen. *Z. Tierpsychol.*, **14**, 1–28.

Dzimirski, I. (1959) Untersuchungen über Bewegungssehen und Optomotorik bei Springspinnen (Salticidae). *Z. Tierpsychol.*, **16**, 385–402.

Eacker, J. N. (1967) Behaviorally produced illumination change. *J. comp. physiol. Psychol.*, **64**, 140–145.

Earl, R. W. (1957) Motivation, performance and extinction. *J. comp. physiol. Psychol.*, **50**, 248–251.

Eayrs, J. T. (1954) Spontaneous activity in the rat. *Brit. J. Anim. Behav.*, **2**, 25–30.

Edwards, D. A. (1968) Mice: fighting by neonatally androgenized females. *Science*, **161**, 1027–1028.

Efstathiou, A., Bauer, J., Greene, M., and Held, R. (1967) Altered reaching following adaptation to optical displacement of the hand. *J. exp. Psychol.*, **73**, 113–120.

Egger, M. D., and Flynn, J. P. (1963) Effects of electrical stimulation of the amygdala on hypothalamically elicited attack behavior in cats. *J. Neurophysiol.*, **26**, 705–720.

Ehrlich, A. (1963) Effects of tegmental lesions on motivated behavior in rats. *J. comp. physiol. Psychol.*, **56**, 390–396.

Eibl-Eibesfeldt, I. (1951a) Beobachtungen zur Fortpflanzungsbiologie und Jungendentwicklung des Eichhörnchens. *Z. Tierpsychol.*, **8**, 370–400.

Eibl-Eibesfeldt, I. (1951b) Nahrungserwerb und Beuteschema der Erdkröte (*Bufo bufo* L.). *Behaviour*, **4**, 1–35.

Eibl-Eibesfeldt, I. (1951c) Gefangenschaftsbeobachtungen an der persischen Wüstenmaus (*Meriones persicus persicus* Blanford): Ein Beitrag zur vergleichenden Ethologie der Nager. *Z. Tierpsychol.*, **8**, 400–423.

Eibl-Eibesfeldt, I. (1953) Zur Ethologie des Hamsters (*Cricetus cricetus* L.). *Zeits Tierpsychol.*, **10**, 204–254.

Eibl-Eibesfeldt, I. (1955a) Angeborenes und Erworbenes im Nestbauverhalten der Wanderratte. *Naturwiss.*, **42**, 633–634.

Eibl-Eibesfeldt, I. (1955b) Der Kommentkampf der Meerechse (*Amblyrhynchus cristatus* Bell) nebst einigen Notizen zur Biologie dieser Art. *Z. Tierpsychol.*, **12**, 49–62.

Eibl-Eibesfeldt, I. (1958) Das Verhalten der Nagetiere. Handb. Zool. Berlin, 8, 12, 10.

Eibl-Eibersfeldt, I. (1961) The interactions of unlearned behaviour patterns and learning in mammals. In "Brain Mechanisms and Learning", Ed. J. F. Delafresnaye, C.I.O.M.S. Symp., Blackwell, Oxford.

Eibl-Eibesfeldt, I. (1962) Die Verhaltensentwicklung des Krallenfrosches (*Xenopus laevis*) und des Scheibenzünglers (*Discoglossus pictus*) unter besonderer Berücksichtigung der Beutefanghandlungen. *Z. Tierpsychol.*, **19**, 385–393.

Eibl-Ebesfeldt, I. (1963) Angeborenes und Erworbenes im Verhalten einiger Säuger. *Z. Tierpsychol.*, **20**, 705–754.

Eibl-Eibesfeldt, I. (1967) Concepts of ethology and their significance in the study of human behavior. In Stevenson *et al.* (1967).

Eibl-Eibesfeldt, I., and Kramer, S. (1958) Ethology, the comparative study of animal behavior. *Quart. Rev. Biol.*, **33**, 181–211.

Eichengreen, J. M., Coren, S., and Nachmias, J. (1966) Visual-cliff preference by infant rats: effects of rearing and test conditions. *Science*, **151**, 830–831.

Eikmanns, K. H. (1955) Verhaltensphysiologische Untersuchungen über den Beutefang und das Bewegungssehen der Erdkröte (*Bufo bufo* L.). *Z. Tierpsychol.*, **12**, 229–253.

Eisner, E. (1960) The relationship of hormones to the reproductive behaviour of birds, referring especially to parental behaviour: a review. *Anim. Behav.*, **8**, 155–179.

Elliott, O., and King, J. A. (1960) Effect of early food deprivation upon later consummatory behavior in puppies. *Psychol. Rep.*, **6**, 391–400.

Ellis, P. E. (1953) Social aggregation and gregarious behaviour in hoppers of *Locusta migratoria migratorioides* (R. & F.). *Behaviour*, **5**, 225–258.

Ellis, P. E. (1962) The behaviour of locusts in relation to phases and species. *Colloques internationaux du centre national de la recherche scientifique, Paris*, **114**, 123–143.

Ellis, P. E. (1963) Changes in the social aggregation of locust hoppers with changes in rearing conditions. *Anim. Behav.*, **11**, 152–160.

Ellis, P. E. (1964) Marching and colour in locust hoppers in relation to social factors. *Behaviour*, **23**, 177–192.

Ellison, G. D., and Konorski, J. (1964) Separation of the salivary and motor responses in instrumental conditioning. *Science*, **146**, 1071–1072.

Emlen, S. T. (1967) Migratory orientation in the indigo bunting, *Passerina cyanea*. I and II. *Auk*, **84**, 309–342, 463–489.

Engelmann, C. (1941) Versuche über den Geschmackssinn des Huhns. *Z. Tierpsychol.*, **4**, 204–218.

Engen, T., and Lipsitt, L. P. (1965) Decrement and recovery of responses to olfactory stimuli in the human neonate. *J. comp. physiol. Psychol.*, **59,** 312–316.

Epstein, A. N. (1960) Reciprocal changes in feeding behavior produced by intrahypothalamic chemical injections. *Amer. J. Physiol.*, **199,** 969–974.

Epstein, A. N. (1967) Oropharyngeal factors in feeding and drinking. *Handb. Physiol.*, **6,** I, 197–218.

Epstein, A. N., and Teitelbaum, P. (1962) Regulation of food intake in the absence of taste, smell, and other oropharyngeal sensations. *J. comp. physiol. Psychol.*, **55,** 753–759.

Epstein, A. N., and Teitelbaum, P. (1967) Specific loss of hypoglycemic control of feeding in recovered lateral rats. *Amer. J. Physiol.*, **213,** 1159–1167.

Erickson, C. J., Bruder, R. H., Komisaruk, B. R., and Lehrman, D. S. (1967) Selective inhibition by progesterone of androgen-induced behavior in male ring doves (*Streptopelia risoria*). *Endocrin.*, **81,** 39–44.

Erickson, R. P. (1963) Sensory neural patterns and gustation. In "Int. Symp. Olfaction and Taste", *Proc.* **1,** Pergamon, New York.

Estes, W. K. (1942) Discriminative conditioning: I. A discriminative property of conditioned anticipation. *J. exp. Psychol.*, **32,** 150–155.

Estes, W. K. (1948) Discriminative conditioning: II. Effects of a Pavlovian conditioned stimulus upon a subsequently established operant response. *J. exp. Psychol.*, **38,** 173–177.

Estes, W. K. (1959) The statistical approach to learning theory. In Koch (1959).

Estes, W. K., Koch, S., McCorquodale, K., Meehl, P. E., Mueller, C. G., Schoenfeld, W. N., and Verplanck, W. S. (1954) Modern Learning Theory. Appleton-Century-Crofts, New York.

Etienne, A. S. (1968) Die Beantwortung von Flimmerfrequenzen durch die Libellenlarve *Aeschna cyanea* M. *Z. vergl. Physiol.*, **61,** 34–40.

Evans, C. S., and Goy, R. W. (1968) Social behaviour and reproductive cycles in captive ring-tailed lemurs (*Lemur catta*). *J. Zool.*, **156,** 181–198.

Evans, D. R., and Browne, L. B. (1960) The physiology of hunger in the blowfly. *Amer. Midl. Nat.*, **64,** 282–300.

Evans, R. M. (1967) Early aggressive responses in domestic chicks. *Anim. Behav.*, **16,** 24–28.

Ewer, R. F. (1959) Suckling behaviour in kittens. *Behaviour,* **15,** 146–162.

Ewer, R. F. (1961) Further observations on suckling behaviour in kittens, together with some general considerations of the interrelations of innate and acquired responses. *Behaviour,* **17,** 247–260.

Ewert, J-P. (1967a) Untersuchungen über die Anteile zentralnervöser Aktionen an der taxisspezifischen Ermüdung beim Beutefang der Erdkröte (*Bufo bufo* L.). *Z. vergl. Physiol.*, **57,** 263–298.

Ewert, J-P. (1967b) Die Einfluss von Störreizen auf die Antwortbereitschaft bei der Richtbewegung der Erdkröte (*Bufo bufo* L.). *Z. Tierpsychol.*, **24,** 298–312.

Ewert, J-P., and Birukow, G. (1965) Uber den Einfluss des Zentralnervensystems auf die Ermüdbarkeit der Richtbewegung im Beuteschema der Erdkröte (*Bufo bufo* L.). *Naturwiss.*, **52,** 68–69.

Ewing, A. W., and Manning, A. (1967) The evolution and genetics of insect behaviour. *Ann. Rev. Entomol.*, **12,** 471–494.

Faber, A. (1953) Die Laut- und Gebärdensprache bei Insekten. Orthopetera I. *Mitt. Staatl. Mus. Naturk. Stuttgart*, **287,** 1–198.

Fabricius, E. (1951) Zur Ethologie junger Anatiden. *Acta Zool. Fenn.*, **68,** 1–178.

Fabricius, E. (1962) Some aspects of imprinting in birds. *Sym. Zool. Soc. Lond.*, **8,** 139–148.

Fabricius, E., and Boyd, H. (1954) Experiments on the following-reaction of ducklings. *Rep. Wildfowl Trust*, 1952–3, 84–89.

Fabricius, E. and Gustafson, K-J. (1954) Further aquarium observations on the spawning behavior of the Char, *Salmo alpinus* L. Inst. Freshwater Res. Drottningholm, Rep. No. 35, 58–104.

Fabry, K. E. (1966) The importance of the intraband manipulating activity of monkeys in their colonies. *Proc. 18th Int. Cong. Psychol.*, **1,** 209–214.

Falls, J. B. (1963) Properties of bird song eliciting responses from territorial males. *Proc. 13th Int. Orn. Cong.*, **1,** 259–271.

Falls, J. B. (1969) Functions of territorial song in the white-throated sparrow. In Hinde (1969).

Fantz, R. L. (1957) Form preferences in newly hatched chicks. *J. comp. physiol. Psychol.*, **50,** 422–430.

Fantz, R. L. (1958) Pattern vision in young infants. *Psychol. Rec.*, **8,** 43–47.

Fantz, R. L. (1965) Ontogeny of Perception. In Schrier *et al.* (1961).

Fantz, R. L. (1967) Visual perception and experience in early infancy. In Stevenson *et al.* (1967).

Farner, D. S. (1966) The control of avian reproductive cycles. In *Proc. XIV. Int. Ornith. Cong.*, 107–134, Ed. D. W. Snow, Blackwell, Oxford.

Farris, H. E. (1967) Classical conditioning of courting behavior in the Japanese quail, *Coturnix coturnix japonica. J. exp. Anal. Behav.*, **10,** 213–217.

Faure, J. C. (1932) The phases of locusts in South Africa. *Bull. ent. Res.*, **23,** 293–405.

Fehrer, E. (1956) The effects of hunger and familiarity of locale on exploration. *J. comp. physiol. Psychol.*, **49,** 549–552.

Feldberg, W., and Myers, R. D. (1965) Changes in temperature produced by micro-injections of amines into the anterior hypothalamus of cats. *J. Physiol.*, **177,** 239–245.

Feldman, S. M., and Waller, J. H. (1962) Dissociation of electro-cortical activation and behavioral arousal. *Nature*, **196,** 1320–1322.

Fentress, J. C. (1968) Interrupted ongoing behaviour in two species of vole (*Microtus agrestis* and *Clethrionomys britannicus*) I and II. *Anim. Behav.*, **16,** 135–153, 154–167.

Ferguson, D. E., and Landreth, H. F. (1966) Celestial orientation of Fowler's toad *Bufo fowleri*. *Behaviour*, **26**, 105–123.

Ferguson, D. E., Landreth, H. F., and McKeown, J. P. (1967) Sun compass orientation of the Northern cricket frog, *Acris crepitans*. *Anim. Behav.*, **15**, 45–53.

Fernández-Guardiola, A., Roldán, E., Fanjal, M. L., and Castells, C. (1961) Role of the pupillary mechanism in the process of habituation of the visual pathways. *Electroenceph. clin. Neurophysiol.*, **13**, 564–576.

Ferster, C. B., and Skinner, B. F. (1957) Schedules of Reinforcement. Appleton-Century-Crofts, New York.

Feshback, S. (1964) The function of aggression and the regulation of aggressive drive. *Psychol. Rev.*, **71**, 257–272.

Ficken, M. S., and Dilger, W. C. (1960) Comments on redirection with examples of avian copulations with substitute objects. *Anim. Behav.*, **8**, 219–222.

Fidura, F. G., and Gray, J. A. (1966) Visual discrimination of color, pattern, and form in the Japanese quail (*Coturnix coturnix japonica*). *Psychon. Sci.*, **5**, 427–428.

Fiedler, K. (1954) Vergleichende Verhaltensstudien an Seenadeln, Schlangennadeln und Seepferdchen. *Z. Tierpsychol.*, **11**, 358–416.

Fiedler, K. (1964) Verhaltensstudien an Lippfischen der Gattung *Crenilabrus* (Labridae, Perciformes). *Z. Tierpsychol.*, **21**, 521–591.

Finch, G. (1938) Salivary conditioning in atropinized dogs. *Amer. J. Physiol.*, **124**, 136–141.

Finger, F. W. (1965) Effect of food deprivation on running-wheel activity in naive rats. *Psychol. Rep.*, **16**, 753–757.

Fischer, G. J. (1966) Auditory stimuli in imprinting. *J. comp. physiol. Psychol.*, **61**, 271–273.

Fischer, G. J., and Campbell, G. L. (1964) The development of passive avoidance conditioning in Leghorn chicks. *Anim. Behav.*, **12**, 268–269.

Fischer, H. (1965) Das Triumphgeschrei der Graugans (*Anser anser*). *Z. Tierpsychol.*, **22**, 247–304.

Fischer, K. (1961) Untersuchungen zur Sonnenkompassorientierung und Laufaktivität von Smaragdeidechsen (*Lacerta viridis* Laur.). *Z. Tierpsychol.*, **18**, 450–470.

Fisher, A. E. (1956) Maternal and sexual behavior induced by intracranial chemical stimulation. *Science*, **124**, 228–229.

Fisher, A. E. (1962) Effects of stimulus variation and sexual satiation in the male rat. *J. comp. physiol. Psychol.*, **55**, 614–620.

Fisher, A. E., and Hale, E. B. (1956–7) Stimulus determinants of sexual and aggressive behaviour in male domestic fowl. *Behaviour*, **10**, 309–323.

Flanagan, J. (1967) Galvanic skin response: emotion or attention? *Proc. 75th Ann. Conv., Amer. Psychol. Ass.*, 7–8.

Floyd, N. F., and Welford, A. T. (1953) Ergonomics Symposium on Fatigue. H. K. Lewis, London.

Flynn, J. P., and Jerome, E. A. (1952) Learning in an automatic multiple-choice box with light as incentive. *J. comp. physiol. Psychol.*, **45**, 336–340.

Folman, Y., and Drori, D. (1965) Normal and aberrant copulatory behaviour in male rats (*R. norvegicus*) reared in isolation. *Anim. Behav.*, **13**, 427–429.

Fonberg, E. (1967) The motivational role of the hypothalamus in animal behaviour. *Acta Biol. Exper.*, **27**, 303–318.

Forgays, D. G., and Forgays, J. W. (1952) The nature of the effect of free-environmental experience in the rat. *J. comp. physiol. Psychol.*, **45**, 322–328.

Forgays, D. G., and Levin, H. (1958) Learning as a function of change of sensory stimulation in food-deprived and food-satiated animals. *J. comp. physiol. Psychol.*, **51**, 50–54.

Forgays, D. G., and Levin, H. (1959) Discrimination and reversal learning as a function of change of sensory stimulation. *J. comp. physiol. Psychol.*, **52**, 191–194.

Forgays, D. G., and Read, J. M. (1962) Crucial periods for free-environmental experience in the rat. *J. comp. physiol. Psychol.*, **55**, 816–818.

Forgus, R. H. (1955) Early visual and motor experience as determiners of complex maze-learning ability under rich and reduced stimulation. *J. comp. physiol. Psychol.*, **48**, 215–220.

Forgus, R. H. (1958) The interaction between form pre-exposure and test requirements in determining form discrimination. *J. comp. physiol. Psychol.*, **51**, 588–591.

Forselius, S. (1957) Studies of Anabantid fishes: I, II, and III. *Zool. bidrag fran Uppsala*, **32**, 97–597.

Fowler, H. (1963) Facilitation and inhibition of performance by punishment: the effects of shock intensity and distribution of trials. *J. comp. physiol. Psychol.*, **56**, 531–538.

Fowler, H. (1967) Satiation and curiosity. In "Psychology of Learning and Motivation", Ed. K. W. and J. T. Spence, Academic Press, New York.

Fowler, H., and Whalen, R. E. (1961) Variation in incentive stimulus and sexual behavior in the male rat. *J. comp. physiol. Psychol.*, **54**, 68–71.

Fox, M. W. (1962) Neuro-behavioral ontogeny. A synthesis of ethological and neurophysiological concepts. *Brain Research*, **2**, 3–20.

Fox, M. W., Inman, O., and Glisson, S. (1968) Age differences in central nervous effects of visual deprivation in the dog. *Developmental Psychobiology*, **1**, 48–54.

Fox, M. W., and Spencer, J. W. (1967) Development of the delayed response in the dog. *Anim. Behav.*, **15**, 162–168.

Fox, M. W., and Stelzner, D. (1966) Behavioural effects of differential early experience in the dog. *Anim. Behav.*, **14**, 273–281.

Fox, M. W., and Stelzner, D. (1967) The effects of early experience on the development of inter and intraspecies social relationships in the dog. *Anim. Behav.*, **15**, 377–386.

Fox, S. S. (1962) Self-maintained sensory input and sensory deprivation in monkeys: a behavioral and neuropharmacological study. *J. comp. physiol. Psychol.*, **55**, 438–444.

Fox, W. (1952) Behavioral and evolutionary significance of the abnormal growth of beaks of birds. *Condor*, **54**, 160–162.

Fraenkel, G. S., and Gunn, D. L. (1940) The Orientation of Animals. Clarendon Press, Oxford.

Franck, D. (1966) Möglichkeiten zur vergleichenden Analyse auslösender und richtender Reize mit Hilfe des Attrapenversuchs, ein Vergleich der Successiv- und Simultanmethode. *Behaviour*, **27**, 150–159.

Franck, D. (1968) Weitere Untersuchungen zur vergleichenden Ethologie der Gattung *Xiphophorus* (Pisces). *Behaviour*, **30**, 76–95.

Franken, R. E. (1967) Stimulus change, exploration, and latent learning. *J. comp. physiol. Psychol.*, **64**, 301–307.

Franzisket, L. (1953) Untersuchungen zur Spezifität und Kumulierung der Erregungsfähigkeit und zur Wirkung einer Ermüdung in der Afferenz bei Wischbewegungen des Rückenmarksfrosches. *Z. vergl. Physiol.*, **34**, 525–538.

Franzisket, L. (1959) Zentralnervöse Steuerung des peripheren Nervenwachstums. *Verh. Deutsch. Zool. Gesell, Münster*, **8**, 139–145.

Franzisket, L. (1963) Characteristics of instinctive behaviour and learning in reflex activity of the frog. *Anim. Behav.*, **11**, 318–324.

Fredericson, E. (1951) Competition: the effects of infantile experience upon adult behavior. *J. abnorm. soc. Psychol.*, **46**, 406–409.

Fredericson, E. (1952) Perceptual homeostasis and distress vocalization in puppies. *J. Personal.*, **20**, 472–477.

Free, J. B. (1956) A study of the stimuli which release the food-begging and offering responses of worker honeybees. *Brit. J. Anim. Behav.*, **4**, 94–101.

Free, J. B. (1961) The stimuli releasing the stinging response of honeybees. *Anim. Behav.*, **9**, 193–196.

Freedman, D. G. (1964) Smiling in blind infants and the issue of innate vs acquired. *J. Child Psychol. Psychiatr.*, **5**, 171–184.

Freedman, S. J. (1961) Sensory deprivation: facts in search of a theory. *J. nerv. ment. Dis.*, **132**, 17–21.

Freud, S. (1915) Instincts and Their Vicissitudes. Coll. Papers Vol. IV, Hogarth, London.

Frisch, K. von (1950) Bees, their Vision, Chemical Senses and Language. Ithaca, New York.

Frisch, K. von (1953) Aus dem Leben der Bienen. By permission from Karl von Frisch (1954), "The Dancing Bees", Methuen, London.

Frisch, O. von (1962) Zur Biologie des Zwergchamäleons (*Microsaurus punilus*). *Z. Tierpsychol.* **19**, 276–289.

Fromme, A. (1941) An experimental study of the factors of maturation and practice in the behavioural development of the embryo of the frog, *Rana pipiens*. *Genet. Psychol. Monogr.*, **24**, 219–256.

Fromme, H. G. (1961) Untersuchungen über das Orientierungsvermögen nächtlich ziehender Kleinvögel (*Erithacus rubecula, Sylvia communis*). *Z. Tierpsychol.*, **18**, 205–220.

Fuller, J. L., and Clark, L. D. (1966a) Genetic and treatment factors modifying the postisolation syndrome in dogs. *J. comp. physiol. Psychol.*, **61**, 251–257.

Fuller, J. L., and Clark, L. D. (1966b) Effects of rearing with specific stimuli upon postisolation behavior in dogs. *J. comp. physiol. Psychol.*, **61**, 258–263.

Fuller, J. L., and Thompson, R. W. (1960) Behavior Genetics. Wiley, New York.

Fuortes, M. G. F. (1958) Electrical activity of cells in the eye of Limulus. *Amer. J. Ophthal.*, **46**, 210.

Furchtgott, E., and Cureton, E. E. (1964) Factor analysis of emotionality and conditioning in mice. *Psychol. Rep.*, **15**, 787–794.

Fuster, J. M. (1958) Effects of stimulation of brain stem on tachistoscopic perception. *Science*, **127**, 150.

Galambos, R. (1956) Suppression of auditory nerve activity by stimulation of efferent fibres to cochlea. *J. Neurophysiol.*, **19**, 424–437.

Galambos, R. (1959) Electrical correlates of conditioned learning. In the "Central Nervous System and Behaviour, I", Ed. M. A. B. Brazier, Macy Foundation, New York.

Galambos, R. (1960) Studies of the auditory system with implanted electrodes. In "Neural Mechanisms of the Auditory and Vestibular Systems", Ed. G. L. Rasmussen and W. F. Windle, Thomas, Springfield, Ill.

Galambos, R., and Davis, H. (1944) Inhibition of activity in single auditory nerve fibers by acoustic stimulation. *J. Neurophysiol.*, **7**, 287–303.

Galambos, R., Sheatz, G., and Vernier, V. G. (1956) Electrophysiological correlates of a conditioned response in cats. *Science*, **123**, 376–377.

Gallistel, C. R. (1966) Motivating effects in self-stimulation. *J. comp. physiol. Psychol.*, **62**, 95–101.

Ganz, L., and Riesen, A. H. (1962) Stimulus generalization to hue in the dark-reared macaque. *J. comp. physiol. Psychol.*, **55**, 92–99.

Ganz, L., and Wilson, P. D. (1967) Innate generalization of a form discrimination without contouring eye movements. *J. comp. physiol. Psychol.*, **63**, 258–269.

Gard, C., Hard, E., Larsson, K., and Petersson, V-A. (1967) The relationship between sensory stimulation and gross motor behaviour during the postnatal development in the rat. *Anim. Behav.*, **15**, 563–567.

Gardner, B. T. (1964) Hunger and sequential responses in the hunting behavior of Salticid spiders. *J. comp. physiol. Psychol.*, **58**, 167–173.

Gastaut, H., Jus, A., Jus, C., Morrell, F., van Leeuwen, W. S. Dongier, S., Naquet, R., Regis, H., Roger, A., Bekkering, D., Kamp A., and Werre, J. (1957) Étude topographique des réactions électroencéphalographiques conditionnées chez l'homme. *Electroenceph. clin. Neurophysiol.*, **9**, 1–34.

Gauss, G. H. (1961) Ein Beitrag zur Kenntnis des Balzverhaltens einheimischer Molche. *Z. Tierpsychol.*, **18**, 60–66.

Gauthier-Pilters, H. (1959) Einige Beobachtungen zum Droh-, Angriffs- und

Kampfverhalten des Dromedarhengstes, sowie über Geburt und Verhaltensentwicklung des Jungtieres, in der nordwestlichen Sahara. *Z. Tierpsychol.*, **16,** 593–604.

Gaze, R. M., and Jacobson, M. (1963a) 'Convexity detectors' in the frog's visual system, *J. Physiol.*, **169,** 1P–3P.

Gaze, R. M., and Jacobson, M. (1963b) A study of the retinotectal projection during regeneration of the optic nerve in the frog. *Proc. Roy. Soc. B,* **157,** 420–448.

Gaze, R. M., Jacobson, M., and Székely, G. (1963) The retinotectal projection in *Xenopus* with compound eyes. *J. Physiol.*, **165,** 484–499.

Gaze, R. M., Jacobson, M., and Székely, G. (1965) On the formation of connexions by compound eyes in *Xenopus. J. Physiol.*, **176,** 409–417.

Gaze, R. M., and Waton, W. E. (1968) Cell division and migration in the brain after optic nerve lesions. In "Growth of the Nervous System", Ed. G. E. W. Wolstenholme and M. O'Connor, Churchill, London.

Geisler, M. (1961) Untersuchungen zur Tagesperiodik des Mistkäfers *Geotrupes silvaticus* Panz. *Z. Tierpsychol.*, **18,** 389–420.

Geist, V. (1963) On the behaviour of the North American Moose (*Alces alces andersoni* Peterson 1950) in British Columbia. *Behaviour,* **20,** 377–416.

Geist, V. (1966) The evolution of horn-like organs. *Behaviour,* **27,** 175–214.

Gelperin, A. (1966) Investigations of a foregut receptor essential to taste threshold regulation in the blowfly. *J. insect Physiol.*, **12,** 829–841.

Gelperin, A., and Dethier, V. G. (1967) Long-term regulation of sugar intake by the blowfly. *Physiol. Zool.*, **40,** 218–228.

Gerall, A. A. (1963) An exploratory study of the effect of social isolation variables on the sexual behaviour of male guinea pigs. *Anim. Behav.*, **11,** 274–282.

Gerall, A. A. (1966) Hormonal factors influencing masculine behavior of female guineapigs. *J. comp. physiol. Psychol.*, **62,** 365–369.

Gerall, A. A., Hendricks, S. E., Johnson, L. L., and Bounds, T. W. (1967) Effects of early castration in male rats on adult sexual behavior. *J. comp. physiol. Psychol.*, **64,** 206–212.

Gerall, A. A., and Ward, I. L. (1966) Effects of prenatal exogenous androgen on the sexual behavior of the female albino rat. *J. comp. physiol. Psychol.*, **62,** 370–375.

Gerall, H. D., Ward, I. L., and Gerall, A. A. (1967) Disruption of the male rat's sexual behaviour induced by social isolation. *Anim. Behav.*, **15,** 54–58.

Gerard, R. W. (1961) The fixation of experience. In "Brain Mechanisms and Learning", Ed. J. F. Delafresnaye, C.I.O.M.S. Symposium, Oxford Univ. Press.

Gerard, R. W., Marshall, W. H., and Saul, L. J. (1936) Electrical activity of the cat's brain. *A. M. A. Arch. Neurol. Psychiat.*, **36,** 675–738.

Gernandt, B. (1949) Response of mammalian vestibular neurons to horizontal rotation and caloric stimulation. *J. Neurophysiol.*, **12,** 173–184.

Gesell, A. (1954) The ontogenesis of infant behaviour. In "Manual of Child Psychology", Ed. L. Carmichael, New York.

Ghent, L. (1957) Some effects of deprivation on eating and drinking behavior. *J. comp. physiol. Psychol.*, **50**, 172–176.

Gibb, J. (1954) Feeding ecology of tits, with notes on treecreeper and goldcrest. *Ibis*, **96**, 513–543.

Gibb, J. A. (1962) L. Tinbergen's hypothesis of the role of specific search images. *Ibis*, **104**, 106–111.

Gibson, E. J., and Walk, R. D. (1956) The effect of prolonged exposure to visually presented patterns on learning to discriminate them. *J. comp. physiol. Psychol.*, **49**, 239–242.

Gibson, E. J., Walk, R. D., Pick, H. L., and Tighe, T. J. (1958) The effect of prolonged exposure to visual patterns on learning to discriminate similar and different patterns. *J. comp. physiol. Psychol.*, **51**, 584–587.

Gibson, E. J., Walk, R. D., and Tighe, T. J. (1959) Enhancement and deprivation of visual stimulation during rearing as factors in visual discrimination learning. *J. comp. physiol. Psychol.*, **52**, 74–81.

Gibson, J. J. (1950) The Perception of the Visual World. Houghton Mifflin, Cambridge, Mass.

Gibson, J. J. (1959) Perception as a function of stimulation. In Koch (1959).

Gibson, J. J., and Gibson, E. J. (1955) Perceptual learning; differentiation or enrichment? *Psychol. Rev.*, **62**, 32–41. *See also Psychol. Rev.*, **62**, 447–450.

Gideiri, Y. B. A. (1966) The behaviour and neuro-anatomy of some developing teleost fishes. *J. Zool. Lond.*, **149**, 215–241.

Giebel, H-D. (1958) Visuelles Lernvermögen bei Einhufern. *Zool. Jahrb.*, **67**, 487–520.

Ginsburg, B., and Allee, W. C. (1942) Some effects of conditioning on social dominance and subordination in inbred strains of mice. *Physiol. Zool.*, **15**, 485–506.

Ginsburg, N. (1960) Conditioned vocalization in the budgerigar. *J. comp. physiol. Psychol.*, **53**, 183–186.

Girdner, J. B. (1953) An experimental analysis of the behavioural effects of a perceptual consequence unrelated to organic drive states. *Amer. Psychol.*, **8**, 354–355.

Glanzer, M. (1953a) Stimulus satiation: an explanation of spontaneous alternation and related phenomena. *Psychol. Rev.*, **60**, 257–269.

Glanzer, M. (1953b) The role of stimulus satiation in spontaneous alternation. *J. exp. Psychol.*, **45**, 387–393.

Glanzer, M. (1958) Curiosity, exploratory drive, and stimulus satiation. *Psychol. Bull.*, **55**, 302–315.

Glickman, S. E., and Hartz, K. E. (1964) Exploratory behaviour in several species of rodents. *J. comp. physiol. Psychol.*, **58**, 101–104.

Glickman, S. E., and Schiff, B. (1967) A biological theory of reinforcement. *Psychol. Rev.*, **74**, 81–109.

Goethe, F. (1954) Vergleichende Beobachtungen zum Verhalten der Silbermöwe (*Larus argentatus*) und der Heringsmöwe (*Larus fuscus*). *Proc. XI. Int. Orn. Congr.*, 577–582.

Goldberg, J. M., and Greenwood, D. D. (1966) Response of neurons of the dorsal and posteroventral nuclei of the cat to acoustic stimuli of long duration. *J. Neurophysiol.*, **29**, 72–93.

Goldman, P. S. (1965) Conditioned emotionality in the rat as a function of stress in infancy. *Anim. Behav.*, **13**, 434–442.

Goldman, P. S. (1969) The relationship between amount of stimulus input in infancy and subsequent emotionality: monotonic or curvilinear? *Ann. N.Y. Acad. Sci.* (in press).

Goldstein, S. R. (1967) Mirror image as a reinforcer in Siamese fighting fish: a repetition with additional controls. *Psychon. Sci.*, **7**, 331–332.

Gonzalez, R. C., Behrend, E. R., and Bitterman, M. E. (1965) Partial reinforcement in the fish: experiments with spaced trials and partial delay. *Amer. J. Psychol.*, **78**, 198–207.

Gonzalez, R. C., and Bitterman, M. E. (1967) Partial reinforcement effect in the goldfish as a function of amount of reward. *J. comp. physiol. Psychol.*, **64**, 163–167.

Goodhart, C. B. (1960) The evolutionary significance of human hair patterns and skin colouring. *Advanc. Sci.*, **17**, 53–59.

Goodman, I. J., and Brown, J. L. (1966) Stimulation of positively and negatively reinforcing sites in the avian brain. *Life Sciences*, **5**, 693–704.

Goodman, L. J. (1965) The role of certain optomotor reactions in regulating stability in the rolling plane during flight in the desert locust, *Schistocerca gregaria*. *J. exp. Biol.*, **42**, 385–408.

Goodrich, C. L. (1966) Activity and exploration as a function of age and deprivation. *J. genet. Psychol.*, **108**, 239–252.

Goodrich, E. S. (1930) Studies on the Structure and Development of Vertebrates. Macmillan, London.

Gordon, G., and Jukes, M. G. M. (1962) Correlation of different excitatory and inhibitory influences on cells in the nucleus gracilis of the cat. *Nature*, **196**, 1183–1185.

Gordon, G., and Manson, J. R. (1967) Cutaneous receptive fields of single nerve cells in the thalamus of the cat. *Nature*, **215**, 597–599.

Górska, T., and Jankowska, E. (1961) The effect of deafferentation on instrumental (Type II) conditioned reflexes in dogs. *Acta Biol. Exp.*, **21**, 219–234.

Górska, T., Jankowska, E., and Kozak, W. (1961) The effect of deafferentation on instrumental (Type II) cleaning reflex in cats. *Acta Biol. Exp.*, **21**, 207–217.

Gossette, R. L. (1967) Successive discrimination reversal (SDR) performances of four avian species on a brightness discrimination task. *Psychon. Sci.*, **8**, 17–18.

Gossette, R. L., Gossette, M. F., and Riddell, W. (1966) Comparisons of successive discrimination reversal performances among closely and remotely related avian species. *Anim. Behav.*, **14**, 560–564.

Gottlieb, G. (1961) Developmental age as a baseline for determination of the critical period in imprinting. *J. comp. physiol. Psychol.*, **54**, 422–427.

Gottlieb, G. (1963) A naturalistic study of imprinting in wood ducklings (*Aix sponsa*). *J. comp. physiol. Psychol.*, **56,** 86–91.

Gottlieb, G. (1965) Imprinting in relation to parental and species identification by avian neonates. *J. comp. physiol. Psychol.*, **59,** 345–356.

Gottlieb, G. (1966) Species identification by avian neonates: contributory effect of perinatal auditory stimulation. *Anim. Behav.*, **14,** 282–290.

Gottlieb, G. (1968a) Prenatal behavior of birds. *Quart. Rev. Biol.*, **43,** 148–174.

Gottlieb, G. (1968b) Species recognition in ground-nesting and hole-nesting ducklings. *Ecology*, **49,** 87–95.

Gottlieb, G., and Klopfer, P. H. (1962) The relation of developmental age to auditory and visual imprinting. *J. comp. physiol. Psychol.*, **55,** 821–826.

Gottlieb, G., and Kuo, Z. Y. (1965) Development of behavior in the duck embryo. *J. comp. physiol. Psychol.*, **59,** 183–188.

Gottlieb, G., and Simner, M. L. (1969) Auditory versus visual flicker in the directing the approach response of domestic chicks. *J. comp. physiol. Psychol.*, **67,** 58–63.

Goy, R. W. (1968) Organizing effects of androgen on the behaviour of rhesus monkeys. In "Endocrinology and Human Behaviour", Ed. R. P. Michael, Oxford Univ. Press.

Goy, R. W., and Young, W. C. (1956–57) Strain differences in the behavioural responses of female guineapigs to alpha-estradiol benzoate and progesterone. *Behaviour*, **10,** 340–354.

Graff, H., and Stellar, E. (1962) Hyperphagia, obesity and finickiness. *J. comp. physiol. Psychol.*, **55,** 418–424.

Graham-Brown, T. (1911) The intrinsic factors in the act of progression in the mammal. *Proc. Roy. Soc. B*, **84,** 308–319.

Granit, R. (1947) The Sensory Mechanisms of the Retina. Oxford Univ. Press.

Granit, R. (1955a) Receptors and Sensory Perception. Yale Univ. Press, New Haven.

Granit, R. (1955b) Centrifugal and antidromic effects on ganglion cells of retina. *J. Neurophysiol.*, **18,** 388–411.

Granit, R., and Kaada, B. R. (1952) Influence of stimulation of central nervous structures on muscle spindles in cat. *Acta physiol. scand.*, **27,** 130–160.

Grant, E. C. (1963) An analysis of the social behaviour of the male laboratory rat. *Behaviour*, **21,** 260–281.

Grant, E. C., and Mackintosh, J. H. (1963) A comparison of the social postures of some common laboratory rodents. *Behaviour*, **21,** 256–259.

Graue, L. C. (1965) Experience effect on initial orientation in pigeon homing. *Anim. Behav.*, **13,** 149–153.

Gray, J. (1950) The role of peripheral sense organs during locomotion in the vertebrates. *Sym. Soc. exp. Biol.*, **4,** 112–126.

Gray, J., and Lissmann, H. W. (1946a) Further observations on the effect of deafferentation on the locomotory activity of amphibian limbs. *J. exp. Biol.*, **23,** 121–132.

Gray, J., and Lissmann, H. W. (1946b) The coordination of limb movements in the amphibia. *J. exp. Biol.*, **23,** 133–142.

Gray, J., and Lissmann, H. W. (1947) The effect of labyrinthectomy on the co-ordination of limb movements in the toad. *J. exp. Biol.*, **24,** 36–40.

Gray, J., and Sand, A. (1936a) The locomotory rhythm of the dogfish (*Scyllium canicula*). *J. exp. Biol.*, **13,** 200–209.

Gray, J., and Sand, A. (1936b) Spinal reflexes of the dogfish, *Scyllium canicula. J. exp. Biol.*, **13,** 210–218.

Gray, J. A., Levine, S., and Broadhurst, P. L. (1965) Gonadal hormone injections in infancy and adult emotional behaviour. *Anim. Behav.*, **13,** 33–45.

Gray, P. H. (1961) The releasers of imprinting: differential reactions to colors as a function of maturation. *J. comp. physiol. Psychol.*, **54,** 597–601.

Green, G. W. (1964a, b) The control of spontaneous locomotor activity in *Phormia regina* I and II. *J. insect Physiol.*, **10,** 711–726, 727–752.

Green, M., Green, R., and Carr, W. J. (1966) The hawk-goose phenomenon: a replication and an extension. *Psychon. Sci.*, **4,** 185–186.

Greenberg, B. (1963a) Parental behaviour and recognition of young in *Cichlasoma biocellatum. Anim. Behav.*, **11,** 578–582.

Greenberg, B. (1963b) Parental behaviour and imprinting in Cichlid fishes. *Behaviour*, **21,** 127–144.

Gregory, R. L. (1958) Eye movements and the stability of the visual world. *Nature*, **182,** 1214–1216.

Gregory, R. L., and Wallace, J. G. (1963) Recovery from early blindness. *Exp. Psychol. Soc. Monogr. No.* 2.

Grey Walter, W. (1959) Intrinsic rhythms of the brain. In "Handbook of Physiology", Sec. 1, Neurophysiology, Vol. 1, Ed. J. Field, H. W. Magoun and V. E. Hall, American Psysiol. Soc., Washington.

Griffin, D. R., and Hock, R. J. (1949) Airplane observations of homing birds. *Ecology*, **30,** 176–198.

Griffin, J. B., and Pearson, J. A. (1967) Habituation of flexor reflex in spinal rats and rats with frontal cortical lesions followed by spinal transection. *Brain Res.*, **6,** 777–780.

Griffiths, W. J. (1962) Effect of food and water deprivation on shock tolerance of albino rats. *Psychol. Rep.*, **11,** 163–166.

Grindley, G. C. (1932) The formation of a simple habit in guinea-pigs. *Brit. J. Psychol.*, **23,** 127–147.

Grohmann, J. (1939) Modifikation oder Funktionsreifung? Ein Beitrag zur Klärung der wechselseitigen Beziehungen zwischen Instinkthandlung und Erfahrung. *Z. Tierpsychol.*, **2,** 132–144.

Gross, C. G. (1968) General activity. In Weiskrantz (1968).

Grossman, S. P. (1960) Eating or drinking elicited by direct adrenergic or cholinergic stimulation of hypothalamus. *Science*, **132,** 301–302.

Grossman, S. P. (1964a) Effect of chemical stimulation of the septal area on motivation. *J. comp. physiol. Psychol.*, **58,** 194–200.

Grossman, S. P. (1964b) Behavioral effects of chemical stimulation of the ventral amygdala. *J. comp. physiol. Psychol.*, **57**, 29–36.

Grossman, S. P. (1967) Physiological Psychology. Wiley, New York.

Grossman, S. P., and Grossman, L. (1966) Effects of chemical stimulation of the midbrain reticular formation on appetitive behavior. *J. comp. physiol. Psychol.*, **61**, 333–338.

Grossman, S. P., Peters, R. H., Freedman, P. E., and Willer, H. I. (1965) Behavioral effects of cholinergic stimulation of the thalamic reticular formation. *J. comp. physiol. Psychol.*, **59**, 57–65.

Grunt, J. A., and Young, W. C. (1952) Psychological modification of fatigue following orgasm (ejaculation) in the male guineapig. *J. comp. physiol. Psychol.*, **45**, 508–510.

Grunt, J. A., and Young, W. C. (1953) Consistency of sexual behavior patterns in individual male guineapigs following castration and androgen therapy. *J. comp. physiol. Psychol.*, **46**, 138–144.

Grüsser, O. J., and Grüsser-Cornehls, U. (1968) Neurophysiologische Grundlagen visueller angeborener Auslösemechanismen beim Frosch. *Z. vergl. Physiol.*, **59**, 1–24.

Grüsser-Cornehls, U., Grüsser, O. J., and Bullock, T. H. (1963) Unit responses in the frog's tectum to moving and nonmoving visual stimuli. *Science*, **141**, 820–822.

Guhl, A. M. (1961) Gonadal hormones and social behavior in infrahuman vertebrates. In Young (1961a).

Guhl, A. M. (1962) The behaviour of chickens. In "The Behaviour of Domestic Animals", Ed. E. S. E. Hafez, Baillière, Tindall and Cox, London.

Guiton, P. (1959) Socialisation and imprinting in brown leghorn chicks. *Anim. Behav.*, **7**, 26–34.

Guiton, P. (1961) The influence of imprinting on the agonistic and courtship responses of the brown leghorn cock. *Anim. Behav.*, **9**, 167–177.

Guiton, P. (1962a) The development of sexual response in domestic fowl in relation to the concept of imprinting. *Anim. Behav.*, **10**, 184.

Guiton, P. (1962b) The development of sexual responses in the domestic fowl, in relation to the concept of imprinting. *Sym. Zool. Soc. Lond.*, **8**, 227–234.

Guiton, P. (1966) Early experience and sexual object-choice in the brown leghorn. *Anim. Behav.*, **14**, 534–538.

Gunther, M. (1955) Instinct and the nursing couple. *Lancet*, **1955**, 575–578.

Guthrie, E. R. (1952) The Psychology of Learning. Harper, New York.

Guttman, N. (1954) Equal-reinforcement values of sucrose and glucose solutions compared with equal-sweetness values. *J. comp. physiol. Psychol.*, **47**, 358–361.

Gwinner, E. (1965) Uber den Einfluss des Hungers und anderer Faktoren auf die Versteck-Aktivität des Kolkraben (*Corvus corax*). *Vogelwarte*, **23**, 1–4.

Haartman, L. von (1957) Adaptation in hole-nesting birds. *Evolution*, **11**, 339–347.

Haas, A. (1962) Phylogenetisch bedeutungsvolle Verhaltensänderungen bei Hummeln. *Z. Tierpsychol.*, **19**, 356–370.

Haber, R. N. (1958) Discrepancy from adaptation level as a source of affect. *J. exp. Psychol.*, **56**, 370–375.

Hagamen, W. D., Zitzmann, E. K., and Reeves, A. G. (1963) Sexual mounting of diverse objects in a group of randomly selected, unoperated male cats. *J. comp. physiol. Psychol.*, **56**, 298–302.

Hagbarth, K.-E., and Fex, J. (1959) Centrifugal influences on single unit activity in spinal sensory paths. *J. Neurophysiol.*, **22**, 321–338.

Hagbarth, K.-E., and Kerr, D. I. B. (1954) Central influences on spinal afferent conduction. *J. Neurophysiol.*, **17**, 295–307.

Hagiwara, S. (1961) Nervous activities of the heart in Crustacea. *Ergebn. Biol.*, **24**, 287–311, Springer, Berlin.

Hagiwara, S., and Bullock, T. H. (1957) Intracellular potentials in pacemaker and integrative neurons of the lobster cardiac ganglion. *J. cell. comp. Physiol.*, **50**, 25–48.

Hagiwara, S., and Watanabe, A. (1956) Discharges in motoneurons of cicada. *J. cell. comp. Physiol.*, **47**, 415–428.

Hailman, J. P. (1962) Pecking of laughing gull chicks at models of the parental head. *Auk*, **79**, 89–98.

Hailman, J. P. (1966) Mirror-image color-preferences for background and stimulus object in gull-chicks. *Experientia*, **22**, 257.

Hailman, J. P. (1967) The ontogeny of an instinct. *Behaviour Suppl.* **15.**

Hall, J. F. (1956) The relationship between external stimulation, food deprivation and activity. *J. comp. physiol. Psychol.*, **49**, 339–341.

Hall, K. R. L. (1963) Tool-using performances as indicators of behavioral adaptability. *Current Anthropology*, **4**, 479–494.

Hall-Craggs, J. (1962) The development of song in the blackbird *Turdus merula. Ibis*, **104**, 277–300.

Halliday, M. S. (1966) Exploration and fear in the rat. *Symp. Zool. Soc. Lond.*, **18**, 45–59.

Halliday, M. S. (1968) Exploratory behavior. In Weiskrantz (1968a).

Haltmeyer, G. C., Denenberg, V. H., and Zarrow, M. X. (1967) Modification of the plasma corticosterone response as a function of infantile stimulation and electric shock parameters. *Physiol. and Behav.*, **2**, 61–63.

Hamburger, V. (1963) Some aspects of the embryology of behavior. *Quart. Rev. Biol.*, **38**, 342–365.

Hamburger, V. (1964) Ontogeny of behaviour and its structural basis. *Proc. 5th Int. Neurochem. Symp., Austria*, 21–34.

Hamburger, V. (1968) The beginnings of coordinated movements in the chick embryo. In "Growth of the Nervous System", Ed. G. E. W. Wolstenholme and M. O'Connor, Churchill, London.

Hamburger, V., Balaban, M., Oppenheim, R., and Wenger, E. (1965) Periodic motility of normal and spinal chick embryos between 8 and 17 days incubation. *J. exp. Zool.*, **159,** 1–14.

Hamburger, V., and Oppenheim, R. (1967) Prehatching motility and hatching behavior in the chick. *J. exp. Zool.*, **166,** 171–204.

Hamburger, V., Wenger, E., and Oppenheim, R. (1966) Motility in the chick embryo in the absence of sensory input. *J. exp. Zool.*, **162,** 133–160.

Hamilton, C. L. (1963) Interactions of food intake and temperature regulation in the rat. *J. comp. physiol. Psychol.*, **56,** 476–488.

Hammel, H. T., Hardy, J. D., and Fusco, M. M. (1960) Thermoregulatory responses to hypothalamic cooling in unanaesthetized dogs. *Amer. J. Physiol.*, **198,** 481–486.

Hammer, L. R. (1968) Relationship of reinforcement value to consummatory behavior. *J. comp. physiol. Psychol.*, **66,** 667–672.

Hammond, P. H., Merton, P. A., and Sutton, G. G. (1956) Nervous gradation of muscular contraction. *Brit. Med. Bull.*, **12,** 214–218.

Hans, H., and Thorsteinson, A. J. (1961) The influence of physical factors and host plant odour on the induction and termination of dispersal flights in *Sitona cylindriollis* Fahr. *Ent. exp. Appl.*, **4,** 165–177.

Hansen, E. W. (1966) The development of maternal and infant behavior in the rhesus monkey. *Behaviour*, **27,** 107–149.

Hanson, N. R. (1955) Causal chains. *Mind*, **64,** 289–311.

Harker, J. E. (1964) The Physiology of Diurnal Rhythms. Cambridge Univ. Press.

Harlow, H. F. (1959a) The development of learning in the rhesus monkey. *Amer. Scientist*, **47,** 459–479.

Harlow, H. F. (1959b) Learning set and error factory theory. In Koch (1959).

Harlow, H. F. (1962) Development of the second and third affectional systems in macaque monkeys. In "Research Approaches to Psychiatric Problems", Ed. T. T. Tourlentes *et al.*, U.S.A.

Harlow, H. F. (1963) The maternal affectional system. In "Determinants of Infant Behavior, II", Ed. B. M. Foss, Methuen, London.

Harlow, H. F. (1965) Sexual behavior in rhesus monkeys. In Beach (1965).

Harlow, H. F., and Harlow, M. K. (1962) Principles of Primate Learning. In "Little Club Clinics in Developmental Medicine", No. 7, Heinemann, London.

Harlow, H. F., and Harlow, M. K. (1965) The affectional systems. In Schrier (1965).

Harlow, H. F., Harlow, M. K., and Meyer, D. R. (1950) Learning motivated by a manipulation drive. *J. exp. Psychol.*, **40,** 228–234.

Harlow, H. F., and Woolsey, C. N. (1958) Biological and Biochemical Bases of Behavior. Univ. Wisconsin Press, Madison.

Harlow, H. F., and Zimmermann, R. R. (1959) Affectional responses in the infant monkey. *Science*, **130,** 421–432.

Harnly, M. H. (1941) Flight capacity in relation to phenotypic and genotypic variations in the wings of *Drosophila melanogaster*. *J. exp. Zool.*, **88,** 263–273.

Harper, L. V. (1968) The effects of isolation from birth on the social behaviour of guineapigs in adulthood. *Anim. Behav.*, **16,** 58–64.

Harrington, G. M., and Linder, W. K. (1962) A positive reinforcing effect of electrical stimulation. *J. comp. physiol. Psychol.*, **55,** 1014–1015.

Harris, C. S. (1965) Perceptual adaptation to inverted, reversed, and displaced vision. *Psychol. Rev.*, **72,** 419–444.

Harris, G. W. (1955) Neural Control of the Pituitary Gland. Arnold, London.

Harris, G. W., and Levine, S. (1965) Sexual differentiation of the brain and its experimental control. *J. Physiol.*, **181,** 379–400.

Harris, G. W., Michael, R. P., and Scott, P. P. (1958) Neurological site of action of stilboestrol in eliciting sexual behaviour. Ciba Foundation Symp. on the "Neurological Basis of Behaviour", Churchill, London.

Harris, J. E., and Whiting, H. P. (1954) Control of rhythmical activity in the skeletal muscle of the embryonic dogfish. *J. Physiol.*, **124,** 63P.

Hart, B. L. (1967a) Testosterone regulation of sexual reflexes in spinal male rats. *Science*, **155,** 1283–1284.

Hart, B. L. (1967b) Sexual reflexes and mating behavior in the male dog. *J. comp. physiol. Psychol.*, **64,** 388–399.

Hart, B. L. (1968a) Alteration of quantitative aspects of sexual reflexes in spinal male dogs by testosterone. *J. comp. physiol. Psychol.*, **66,** 726–730.

Hart, B. L. (1968b) Role of prior experience in the effects of castration on sexual behaviour of male dogs. *J. comp. physiol. Psychol.*, **66,** 719–725.

Hart, B. L. (1968c) Neonatal castration: influence on neural organization of sexual reflexes in male rats. *Science*, **160,** 1135–1136.

Hart, B. L. (1968d) Sexual reflexes and mating behavior in the male rat. *J. comp. physiol. Psychol.*, **65,** 453–460.

Hartline, D. K. (1967) Impulse identification and axon mapping of the nine neurons in the cardiac ganglion of the lobster *Homarus americanus*. *J. exp. Biol.*, **47,** 327–340.

Hartline, H. K. (1938) The response of single optic nerve fibers of the vertebrate eye to illumination of the retina. *Amer. J. Physiol.*, **121,** 400–415.

Hartline, H. K. (1949) Inhibition of activity of visual receptors by illuminating nearby retinal areas in the *Limulus* eye. *Fed. Proc.*, **8,** 69.

Hartline, H. K., and Graham, C. H. (1932) Nerve impulses from single receptors in the eye. *J. cell. comp. Physiol.*, **1,** 277–295.

Hartline, H. K., and Ratliff, F. (1956) Inhibitory interaction of receptor units in the eye of *Limulus*. *J. gen. Physiol.*, **40,** 357–376.

Hartline, H. K., Wagner, H. G., and Ratliff, F. (1956) Inhibition in the eye of *Limulus*. *J. gen. Physiol.*, **39,** 651–673.

Harwood, D., and Vowles, D. M. (1966) Forebrain stimulation and feeding behavior in the ring dove (*Streptopelia risoria*). *J. comp. physiol. Psychol.*, **62,** 388–396.

Harwood, D., and Vowles, D. M. (1967) Defensive behaviour and the after effects of brain stimulation in the ring dove (*Streptopelia risoria*). *Neuropsychologia*, **5,** 345–366.

Haskell, P. T. (1956) Hearing in certain Orthoptera: II. The nature of the response of certain receptors to natural and imitation stridulation. *J. exp. Biol.*, **33,** 767–776.

Haskell, P. T. (1957) Stridulation and associated behaviour in certain Orthoptera: 1. Analysis of the stridulation of, and behaviour between, males. *Brit. J. Anim. Behav.*, **5,** 139–148.

Haskell, P. T. (1960) Stridulation and associated behaviour in certain Orthoptera: 3. The influence of the gonads. *Anim. Behav.*, **8,** 76–81.

Hasler, A. D. (1960) Guideposts of migrating fishes. *Science*, **132,** 785–792.

Hassenstein, B. (1959) Optokinetische Wirksamkeit bewegter periodischer Muster. *Z. Naturf.*, **14,** 659–674.

Hassenstein, B. (1961) Wie sehen Insekten Bewegungen? *Naturwiss.*, **48,** 207–214.

Hassenstein, B., and Reichardt, W. (1959) Wie sehen Insekten Bewegungen? *Umschau*, **10,** 302–305.

Haude, R. H., and Ray, O. S. (1967) Visual exploration in monkeys as a function of visual incentive duration and sensory derivation. *J. comp. physiol. Psychol.*, **64,** 332–336.

Hayes, K. J., and Hayes, C. (1952) Imitation in a home-raised chimpanzee. *J. comp. physiol. Psychol.*, **45,** 450–459.

Hayes, W. N., Hertzler, D. R., and Hogberg, D. K. (1968) Visual responsiveness and habituation in the turtle. *J. comp. physiol. Psychol.*, **65,** 331–335.

Hayes, W. N., and Saiff, E. I. (1967) Visual alarm reactions in turtles. *Anim. Behav.*, **15,** 102–106.

Hazlett, B. A. (1966) Factors affecting the aggressive behavior of the hermit crab (*Calcinus tibicen*). *Z. Tierpsychol.*, **23,** 655–671.

Hazlett, B. A., and Bossert, W. H. (1965) A statistical analysis of the aggressive communications system of some hermit crabs. *Anim. Behav.*, **13,** 357–373.

Head, H. (1920) Studies in Neurology. Oxford Univ. Press.

Hearst, E. (1962) Concurrent generalization gradients for food-controlled and shock-controlled behavior. *J. exp. Anal. Behav.*, **5,** 19–31.

Hearst, E. (1967) Oscillatory behavior during approach-avoidance conflict. *J. exp. Anal. Behav.*, **10,** 75–84.

Heart, E., and Koresko, M. B. (1964) Self-presentation and self-termination of a conflict-producing stimulus. *Science*, **146,** 415–416.

Hebb, D. O. (1937a) The innate organization of visual activity: I. Perception of figures by rats reared in total darkness. *J. genet. Psychol.*, **51,** 101–126.

Hebb, D. O. (1937b) The innate organization of visual activity: II. Transfer of response in the discrimination of brightness and size by rats reared in total darkness. *J. comp. Psychol.*, **24,** 277–299.

Hebb, D. O. (1946) On the nature of fear. *Psychol. Rev.*, **53,** 259–276.

Hebb, D. O. (1949) The Organization of Behaviour. Wiley, New York.

Hebb, D. O. (1953) Heredity and environment in mammalian behaviour. *Brit. J. Anim. Behav.*, **1**, 43–47.

Hebb, D. O. (1955) Drives and the C.N.S. (conceptual nervous system). *Psychol. Rev.*, **62**, 243–254.

Hebb, D. O. (1963) The semiautonomous process. *Amer. Psychol.*, **18**, 16–27.

Hebb, D. O., and Thompson, W. A. (1954) The social significance of animal studies. In "Handbook of Social Psychology", Ed. G. Lindzey, Addison-Wesley.

Hediger, H. (1950) Wild Animals in Captivity. Butterworths, London.

Heiligenberg, W. (1963) Ursachen für das Auftreten von Instinktbewegungen bei einem Fische (*Pelmatochromis subocellatus kribensis* Boul., Cichlidae). *Z. vergl. Physiol.*, **47**, 339–380.

Heiligenberg, W. (1965a) The suppression of behavioral activities by frightening stimuli. *Z. vergl. Physiol.*, **50**, 660–672.

Heiligenberg, W. (1965b) A quantitative analysis of digging movements and their relationship to aggressive behaviour in Cichlids. *Anim. Behav.*, **13**, 163–170.

Heiligenberg, W. (1965c) The effect of external stimuli on the attack readiness of a Cichlid fish. *Z. vergl. Physiol.*, **49**, 459–464.

Heiligenberg, W. (1966) The stimulation of territorial singing in house crickets (*Acheta domesticus*). *Z. vergl. Physiol.*, **53**, 114–129, Springer, Berlin.

Hein, A. (1968) Exposure history in early perceptual-motor development. In "Percep. Motor Foundations", Ed. Muriel Sloan, A.A.A., Washington.

Hein, A., and Held, R. (1962) A neural model for labile sensori-motor coordinations. From "Biological Prototypes and Synthetic Systems", 1, Plenum Press, New York.

Hein, A., and Held, R. (1967) Dissociation of the visual placing response into elicited and guided components. *Science*, **158**, 390–392.

Heinemann, E. G., Chase, S., and Mandell, C. (1968) Discriminative control of attention. *Science*, **160**, 553–554.

Heinemann, E. G., and Rudolph, R. L. (1963) The effect of discriminative training on the gradient of stimulus generalization. *Amer. J. Psychol.*, **76**, 653–658.

Heinroth, O. (1911) Beiträge zur Biologie, namentlich Ethologie und Psychologie der Anatiden. *Verh. 5 Int. Orn. Kong.*, 589–702.

Heinroth, O. (1924) Lautäusserungen der Vögel. *J. Ornith.*, **72**, 223–244.

Held, R. (1961) Exposure-history as a factor in maintaining stability of perception and coordination. *J. nerv. ment. Dis.*, **132**, 26–32.

Held, R. (1967/1968) Dissociation of visual functions by deprivation and rearrangement. *Psychol. Forschung.*, **31**, 338–348.

Held, R., and Bauer, J. (1967) Visually guided reaching in infant monkeys after restricted rearing. *Science*, **155**, 718–720.

Held, R., and Bossom, J. (1961) Neonatal deprivation and adult rearrangement: complementary techniques for analyzing plastic sensory-motor coordinations. *J. comp. physiol. Psychol.*, **54**, 33–37.

Held, R., Efstathiou, A., and Greene, M. (1966) Adaptation to displaced and delayed visual feedback from hand. *J. exp. Psychol.*, **72,** 887–891.

Held, R., and Freedman, S. J. (1963) Plasticity in human sensorimotor control. *Science,* **142,** 455–462.

Held, R., and Hein, A. V. (1958) Adaptation of disarranged hand-eye coordination contingent upon re-afferent stimulation. *Percept. Mot. Skills,* **8,** 87–90.

Held, R., and Hein, A. (1963) Movement-produced stimulation in the development of visually guided behavior. *J. comp. physiol. Psychol.,* **56,** 872–876.

Held, R., and Hein, A. (1967) On the modifiability of form perception. In "Models for Perception of Speech and Visual Forms", Ed. W. Wathen-Dunn, M.I.T. Press.

Heller, D. P. (1968) Absence of size constancy in visually deprived rats. *J. comp. physiol. Psychol.,* **65,** 336–339.

Hellon, R. F. (1967) Hypothalamic neurones responding to temperature in conscious rabbits. *J. Physiol.,* **191,** 37 P.

Helmholtz, H. von (1867) Handbook der physiologischen Optik, Voss, Leipzig. English translation by J. P. C. Southall (1925), Optical Society of America, Wisconsin.

Helson, H. (1948) Adaptation-level as a basis for a quantitative theory of frames of reference. *Psychol. Rev.,* **55,** 297–313.

Henderson, N. D. (1964) Behavioral effects of manipulation during different stages in the development of mice. *J. comp. physiol. Psychol.,* **57,** 284–289.

Henderson, N. D. (1965) Acquisition and retention of conditioned fear during different stages in the development of mice. *J. comp. physiol. Psychol.,* **59,** 439–442.

Henderson, N.D. (1966) Effects of intensity and spacing of prior stimulation on later emotional behavior. *J. comp. physiol. Psychol.,* **62,** 441–448.

Hendry, D. P., and Rasche, R. H. (1961) Analysis of a new nonnutritive positive reinforcer based on thirst. *J. comp. physiol. Psychol.,* **54,** 477–483.

Henke, K. (1930) Die Lichtorientierung und die Bedingungen der Lichtstimmung bei der Rollassel *Armadillidium cinereum* Zenker. *Z. vergl. Physiol.,* **13,** 534–625.

Henriques, J. (in prep.) Ph.D. thesis, Cambridge.

Hensel, H., and Zotterman, Y. (1951) The response of mechanoreceptors to thermal stimulation. *J. Physiol.,* **115,** 16–24.

Herberg, J., and Blundell, J. E. (1967) Lateral hypothalamus: hoarding behaviour elicited by electrical stimulation. *Science,* **155,** 349–350.

Herberg, L. J. (1963) Seminal ejaculation following positively reinforcing electrical stimulation of the rat hypothalamus. *J. comp. physiol. Psychol.,* **56,** 679–685.

Herbert, J. (1968) Sexual preference in the rhesus monkey *Macaca mulatta* in the laboratory. *Anim. Behav.,* **16,** 120–128.

Herbert, J., and Trimble, M. R. (1967) Effect of oestradiol and testosterone on the sexual receptivity and attractiveness of the female rhesus monkey. *Nature,* **216,** 165–166.

Hernández-Peón, R., Guzman-Flores, C., Alcares, M., and Fernández-Guardiola, A. (1957a) Sensory transmission in the visual pathway during "attention" in unanaesthetised cats. *Acta Neur. Lat.-Amer.,* **3,** 1–8.

Hernández-Peón, R., Jouvet, M., and Scherrer, H. (1957b) Auditory potentials at cochlear nucleus during acoustic habituation. *Acta Neur. Lat.-Amer.,* **3,** 114–116.

Hernández-Peón, R., Scherrer, H., and Jouvet, M. (1956) Modification of electric activity in cochlear nucleus during "attention" in unanaesthetized cats. *Science,* **123,** 331–332.

Herren, R. Y. (1933) The effect of high and low female sex hormone concentration on the two-point threshold of pain and touch and upon tactile sensitivity. *J. exp. Psychol.,* **16,** 324–327.

Herrnstein, R. J. (1966) Superstition: a corollary of the principles of operant conditioning. In Honig (1966).

Herrnstein, R. J., and Loveland, D. H. (1964) Complex visual concept in the pigeon. *Science,* **146,** 549–551.

Herrnstein, R. J., and van Sommers, P. (1962) Method for sensory scaling with animals. *Science,* **135,** 40–41.

Hersher, L., Moore, A. U., and Richmond, J. B. (1958) Effect of post partum separation of mother and kid on maternal care in the domestic goat. *Science,* **128,** 1342–1343.

Hersher, L., Richmond, J. B., and Moore, A. U. (1963) Modifiability of the critical period for the development of maternal behavior in sheep and goats. *Behaviour,* **20,** 311–320.

Herter, K. (1927) Reizphysiologische Untersuchungen an der Karpfenlaus (*Argulus foliaceus* L.). *Z. vergl. Physiol.,* **5,** 283–370.

Hertz, M. (1929, 1930, 1931) Die Organisation des optischen Feldes bei der Biene: I, II, and III. *Z. vergl. Physiol.,* **8,** 693–748; **11,** 107–145; **14,** 629–674.

Hertz, M. (1933) Uber figurale Intensitäten und Qualitäten in der optischen Wahrnehmung der Biene. *Biol. Zbl.,* **53,** 10–40.

Hertz, M. (1935) Die Untersuchungen über den Formensinn der Honigbiene. *Naturwiss.,* **23,** 618–624.

Hess, E. H. (1956a) Natural preferences of chicks and ducklings for objects of different colours. *Psych. Rep.,* **2,** 477–483.

Hess, E. H. (1956b) Space perception in the chick. *Sci. Amer.,* **195,** 71–80.

Hess, E. H. (1959a) Imprinting. *Science,* **130,** 133–141.

Hess, E. H. (1959b) Two conditions limiting critical age for imprinting. *J. comp. physiol. Psychol.,* **52,** 515–518.

Hess, E. H. (1964) Imprinting in birds. *Science,* **146,** 1128–1139.

Hess, W. R. (1943) Das Zwischenhirn als Koordinationsorgan. *Helv. Physiol. Acta,* **1,** 549–565.

Hess, W. R. (1954) Diencephalon: Autonomic and Extrapyramidal Functions. Grune, New York.

Heusser, H. (1958) Uber die Berziehungen der Erdkröte (*Bufo bufo* L.) zu ihrem Laichplatz: I. *Behaviour,* **12,** 208–232.

Hildreth, P. E. (1962) Quantitative aspects of mating behavior in *Drosophila. Behaviour,* **19,** 57–73.

Hilgard, E. R. (1948, 1956) Theories of Learning. Appleton-Century-Crofts, New York.

Hilgard, E. R., and Marquis, D. G. (1961) Conditioning and Learning. Revised G. A. Kimble, Methuen, London.

Hill, W. F. (1958) The effect of varying periods of confinement on activity in tilt cages. *J. comp. physiol. Psychol.,* **51,** 570–574.

Hinde, R. A. (1952) The behaviour of the great tit (*Parus major*) and some other related species. *Behaviour Suppl.* **2,** 1–201.

Hinde, R. A. (1953a) The conflict between drives in the courtship and copulation of the Chaffinch. *Behaviour,* **5,** 1–31.

Hinde, R. A. (1953b) A possible explanation of paper-tearing behaviour in birds. *Brit. Birds,* **46,** 21–23.

Hinde, R. A. (1954a, b) Factors governing the changes in strength of a partially inborn response, as shown by the mobbing behaviour of the chaffinch (*Fringilla coelebs*): I. The nature of the response, and an examination of its course. *Proc. Roy. Soc. B,* **142,** 306–331. II. The waning of the response. *Proc. Roy. Soc. B,* **142,** 331–358.

Hinde, R. A. (1954c) The courtship and copulation of the greenfinch (*Chloris chloris*). *Behaviour,* **7,** 207–232.

Hinde, R. A. (1954d) Changes in responsiveness to a constant stimulus. *Brit. J. Anim. Behav.,* **2,** 41–55.

Hinde, R. A. (1955) The following response of moorhens and coots. *Brit. J. Anim. Behav.,* **3,** 121–122.

Hinde, R. A. (1955/1956) A comparative study of the courtship of certain finches (*Fringillidae*). *Ibis,* **97,** 706–745; **98,** 1–23.

Hinde, R. A. (1956a) Ethological models and the concept of drive. *Brit. J. Philos. Sci.,* **6,** 321.

Hinde, R. A. (1956b) The biological significance of the territories of birds. *Ibis,* **98,** 340–369.

Hinde, R. A. (1956c) The behaviour of certain cardueline F, interspecies hybrids. *Behaviour,* **9,** 202–213.

Hinde, R. A. (1958a) The nest-building behaviour of domesticated canaries. *Proc. Zool. Soc. Lond.,* **131,** 1–48.

Hinde, R. A. (1958b) Alternative motor patterns in chaffinch song. *Anim. Behav.,* **6,** 211–218.

Hinde, R. A. (1959a) Behaviour and speciation in birds and lower vertebrates. *Biol. Rev.,* **34,** 85–128.

Hinde, R. A. (1959b) Some recent trends in ethology. In Koch (1959).

Hinde, R. A. (1959c) Some factors influencing sexual and aggressive behaviour in male Chaffinches. *Bird Study*, **6**, 112–122.

Hinde, R. A. (1959d) Unitary drives. *Anim. Behav.*, **7**, 130–141.

Hinde, R. A. (1960a) Energy models of motivation. *Sym. Soc. exp. Biol.*, **14**, 199–213.

Hinde, R. A. (1960b) Factors governing the changes in strength of a partially inborn response, as shown by the mobbing behaviour of the chaffinch (*Fringilla coelebs*): III. The interaction of short-term and long-term incremental and decremental effects. *Proc. Roy. Soc. Ser. B*, **153**, 398–420.

Hinde, R. A. (1962a) Sensitive periods and the development of behaviour. From "Lessons from Animal Behaviour for the Clinician", Little Club Clinic in Developmental Medicine, No. 7, The National Spastics Society Study Group, 1962.

Hinde, R. A. (1962b) Some aspects of the imprinting problem. *Sym. Zool. Soc. Lond.*, **8**, 129–138.

Hinde, R. A. (1965) Interaction of internal and external factors in integration of canary reproduction. In Beach (1965).

Hinde, R. A. (1966) Ritualization and social communication in rhesus monkeys. *Phil. Trans. Roy. Soc. B*, **251**, 285–294.

Hinde, R. A. (1967a) Aspects of the control of avian reproductive development within the breeding season. In *Proc. XIV Int. Ornith. Congr.*, 135–153.

Hinde, R. A. (1968) Dichotomies in the study of development. In "Genetic and Environmental Influences on Behavior", Ed. J. M. Thoday and A. S. Parkes, Oliver and Boyd, Edinburgh.

Hinde, R. A. (1969) Bird Vocalizations in Relation to Current Problems in Biology and Psychology. Cambridge Univ. Press.

Hinde, R. A. (in press) Social development. In "Behavior of Nonhuman Primates", **III**, Ed. A. M. Schrier, H. F. Harlow, F. Stollnitz, Academic Press, New York.

Hinde, R. A., Bell, R. Q., and Steel, E. A. (1963) Changes in sensitivity of the canary brood patch during the natural breeding season. *Anim. Behav.*, **11**, 553–560.

Hinde, R. A., and Fisher, J. (1951) Further observations on the opening of milk bottles by birds. *Brit. Birds*, **44**, 393–396.

Hinde, R. A., and Rowell, T. E. (1962) Communication by postures and facial expressions in the rhesus monkey (*Macaca mulatta*). *Proc. Zool. Soc. Lond.*, **138**, 1–21.

Hinde, R. A., Rowell, T. E., and Spencer-Booth, Y. (1964) Behaviour of socially living rhesus monkeys in their first six months. *Proc. Zool. Soc. Lond.*, **143**, 609–649.

Hinde, R. A., and Spencer-Booth, Y. (1968) The study of mother-infant interaction in captive group-living rhesus monkeys. *Proc. Roy. Soc. Ser. B*, **169**, 177–201.

Hinde, R. A., and Steel, E. A. (1962) Selection of nest material by female canaries. *Anim. Behav.*, **10**, 67–75.

Hinde, R. A., and Steel, E. A. (1964) Effect of exogenous hormones on the tactile sensitivity of the canary brood patch. *J. Endocrin.*, **30**, 355–359.

Hinde, R. A., and Steel, E. A. (1966) Integration of the reproductive behaviour of female canaries. *Symp. Soc. exp. Biol.*, **20**, 401–426.

Hinde, R. A., and Stevenson, J. G. (1969a) Goals and response control. In "Development and Evolution of Behaviour", **1**, Ed. L. R. Aronson, E. Tobach, J. S. Rosenblatt, D. S. Lehrman, Freeman, New York.

Hinde, R. A., and Stevenson, J. G. (1969b) Integration of response sequences. In "Advances in the Study of Behaviour", **2**, Ed. D. S. Lehrman, R. A. Hinde, E. Shaw, Academic Press, New York.

Hinde, R. A., Thorpe, W. H., and Vince, M. A. (1956) The following response of young coots and moorhens. *Behaviour*, **9**, 2–3, 214–242.

Hinde, R. A., and Tinbergen, N. (1958) The comparative study of species-specific behavior. From "Behavior and Evolution", Ed. A. Roe and G. G. Simpson, Yale Univ. Press, New Haven.

Hines, M. (1942) The development and regression of reflexes, postures and progression in the young macaque. *Contr. Embryol. Carnegie Instn., Wash.*, **30**, 153–209.

Hirayoshi, I., and Warren, J. M. (1967) Overtraining and reversal learning by experimentally naive kittens. *J. comp. physiol. Psychol.*, **64**, 507–510.

Hirsch, J. (1967a) Book review. *Anim. Behav.*, **15**, 394–395.

Hirsch, J. (1967b) Behavior—Genetic Analysis. McGraw-Hill, New York.

Hitchcock, L., Michels, K. M., and Brown, D. R. (1963) Discrimination learning: squirrels vs. raccoons. *Percept. and Mot. Skills*, **16**, 405–414.

Hoagland, H. (1933a) Electrical responses from the lateral-line nerves of catfish: I. *J. gen. Physiol.*, **16**, 695–714.

Hoagland, H. (1933b) Quantitative analysis of responses from lateral-line nerves of fishes: II. *J. gen. Physiol.*, **16**, 715–732.

Hochberg, J., and Held, R. (1964) Nonvisual components in visual form perception. *Percept. and Mot. Skills*, **18**, 559–560.

Hockman, C. H. (1961) Prenatal maternal stress in the rat: its effects on emotional behavior in the offspring. *J. comp. physiol. Psychol.*, **54**, 679–684.

Hodgson, E. S. (1957) Electrophysiological studies of arthropod chemoreception: II. Responses of labellar chemoreceptors of the blowfly to stimulation by carbohydrates. *J. insect Physiol.*, **1**, 240–247.

Hodgson, E. S. (1965) The chemical senses and changing viewpoints in sensory physiology. In "Viewpoints in Biology", **4**, Ed. J. D. Carthy and C. L. Duddington, Butterworths, London.

Hodgson, E. S., and Roeder, K. D. (1956) Electrophysiological studies of arthropod chemoreception: I. General properties of the labellar chemoreceptors of Diptera. *J. cell. comp. Physiol.*, **48**, 51–76.

Hodos, W. (1965) Motivational properties of long durations of rewarding brain stimulation. *J. comp. physiol. Psychol.*, **59**, 219–224.

Hoebel, B. G., and Teitelbaum, P. (1962) Hypothalamic control of feeding and self-stimulation. *Science,* **135,** 375–377.

Hoebel, B. G., and Teitelbaum, P. (1966) Weight regulation in normal and hypothalamic hyperphagic rats. *J. comp. physiol. Psychol.,* **61,** 189–193.

Hof, M. W. van, Hof-van Duin, J. van, Mark, F. van der, and Rietveld, W. J. (1962) The effect of image formation and that of flash-counting on the occipitocortical response to lightflashes. *Acta physiol. pharm. néerl.,* **11,** 485–493.

Hoffman, H. S., and Kozma, F., Jr. (1967) Behavioral control by an imprinted stimulus: long-term effects. *J. exp. Anal. Behav.,* **10,** 495–501.

Hoffman, H. S., Searle, J. L., Toffey, S., and Kozma, F., Jr. (1966) Behavioral control by an imprinted stimulus. *J. exp. Anal. Behav.,* **9,** 177–189.

Hoffmann, K. (1953) Experimentelle Aenderung des Richtungsfindens beim Star durch Beeinflussung der "inneren Uhr". *Naturwiss.,* **40,** 148.

Hoffmann, K. (1959) Die Aktivitätsperiodik von im 18- und 36-Studen-tag erbrüteten Eidechsen. *Z. vergl. Physiol.,* **42,** 422–432.

Hoffmann, K. (1965) Overt circadian frequencies and circadian rule. In "Circadian Clocks", Ed. J. Aschoff, North-Holland, Amsterdam.

Hogan, J. A. (1964) Operant control of preening in pigeons. *J. exp. anal. Behav.,* **7,** 351–352.

Hogan, J. A. (1965) An experimental study of conflict and fear: an analysis of behaviour of young chicks toward a mealworm. I. The behaviour of chicks which do not eat the mealworm. *Behaviour,* **25,** 45–97.

Hogan, J. A. (1966) *Ibid.* II. The behaviour of chicks which eat the mealworm. *Behaviour,* **27,** 273–289.

Hogan, J. A. (1967) Fighting and reinforcement in the Siamese fighting fish (*Betta splendens*). *J. comp. physiol. Psychol.,* **64,** 356–359.

Holling, C. S. (1958) Sensory stimuli involved in the location and selection of sawfly cocoons by small mammals. *Canad. J. Zool.,* **36,** 633–653.

Holling, C. S. (1959) The components of predation as revealed by a study of the small mammal predation of the European pine sawfly. *Canad. Ent.,* **91,** 293–320.

Holling, C. S. (1961) Principles of insect predation. *Ann. Rev. Entom.,* **6,** 163–182.

Holling, C. S. (1965) The functional response of predators to prey density and its role in mimicry and population regulation. *Memoirs Entom. Soc. Canada,* **45,** 1–60.

Holmgren, B., and Merton, P. A. (1954) Local feedback control of motoneurones. *J. Physiol.,* **123,** 47P.

Holst, E. von (1935a) Erregungsbildung und Erregungsleitung im Fischrückenmark. *Pflüg. Arch. ges. Physiol.,* **235,** 345–359.

Holst, E. von (1935b) Uber den Prozess der zentralnervösen Koordination. *Pflüg. Arch. ges. Physiol.,* **236,** 149–158.

Holst, E. von (1935c) Alles oder Nichts, Block, Alternans, Bigemini und verwandte Phänomene als Eigenschaften des Rückenmarks. *Pflüg. arch. ges. Physiol.,* **236,** 515–532.

Holst, E. von (1935d) Uber den Lichtrückenreflex bei Fischen. *Publ. Staz. Zool. Napoli*, **15**, 143–158.

Holst, E. von (1936) Versuche zur Theorie der relativen Koordination. *Pflüg. arch. ges. Physiol.*, **237**, 93–121.

Holst, E. von (1939) Entwurf eines Systems der Lokomotorischen Periodenbildungen bei Fischen. *Z. vergl. Physiol.*, **26**, 481–528.

Holst, E. von (1948) Von der Mathematik der nervösen Ordnungsleistung. *Experientia*, **4**, 374–381.

Holst, E. von (1950a) Quantitative Messung von Stimmungen im Verhalten der Fische. *Sym. Soc. Exp. Biol.*, **IV**, 143–172.

Holst, E. von (1950b) Die Arbeitsweise des Statolithenapparates bei Fischen. *Z. vergl. Physiol.*, **32**, 60–120.

Holst, E. von (1954) Relations between the central nervous system and the peripheral organs. *Brit. J. Anim. Behav.*, **2**, 89–94.

Holst, E. von, and Mittelstaedt, H. (1950) Das Reafferenzprinzip. *Naturwiss.*, **37**, 464–476.

Holst, E. von, and Saint Paul, U. von (1963) On the functional organisation of drives. *Anim. Behav.*, **11**, 1–20, translated from *Naturwiss.*, **18**, 409–422.

Holst, E. von, and Schoen, L. (1954) Der Einfluss mechanisch veränderter Augenstellungen auf die Richtungslokalisation bei Fischen. *Z. vergl. Physiol.*, **36**, 433–442.

Holzapfel, M. (1939a) Uber Bewegungstereotypien bei gehaltenen Säugern: Ill. *Der Zoologische Garten*, **10**, 184–193.

Holzapfel, M. (1939b) Analyse des Sperrens und Pickens in der Entwicklung des Stars. *J. f. Ornith.*, **87**, 525–553.

Holzapfel, M. (1949) Die Beziehungen zwischen den Trieben junger und erwachsener Tiere. *Schweiz. Z. Psych.*, **8**, 32–60.

Holzapfel, M. (1956a) Das Spiel bei Säugetieren. In "Handbuch der Zoologie", **8**, Ed. Kükenthal.

Holzapfel, M. (1956b) Uber die Bereitschaft zu Spiel-und Instinkthandlungen. *Z. Tierpsychol.*, **13**, 442–462.

Honig, W. K. (1966) Operant Behavior. Appleton-Century-Crofts, New York.

Hooff, J. A. R. A. M. van (1962) Facial expressions in higher primates. *Symp. Zool. Soc. Lond.*, **8**, 97–125.

Hooff, J. A. R. A. M. van (1967) The facial displays of the Catarrhine monkeys and apes. In Morris (1967).

Hoogland, R., Morris, D., and Tinbergen, N. (1956/1957) The spines of sticklebacks (*Gasterosteus* and *Pygosteus*) as means of defence against predators (*Perca* and *Esox*). *Behaviour*, **10**, 205–236.

Hooker, T., and Hooker-Lade, B. (1969) Duetting. In Hinde (1969).

Hörmann-Heck, S. von (1957) Untersuchungen über den Erbgang einiger Verhaltensweisen bei Grillenbastarden. *Z. Tierpsychol.*, **14**, 137–183.

Horn, G. (1952) The neurological basis of thought. *Mermaid*, **18**, 17–25.

Horn, G. (1960) Electrical activity of the cerebral cortex of the unanaesthetized cat during attentive behaviour. *Brain,* **83,** 57–76.

Horn, G. (1962) Some neural correlates of perception. "Viewpoints in Biology", I, Ed. J. Carthy, Butterworths, London.

Horn, G. (1963) The response of single units in the striate cortex of unrestrained cats to photic and somaesthetic stimuli. *J. Physiol.,* **165,** 80–81P.

Horn, G. (1965) Physiological and psychological aspects of selective perception. In Lehrman *et al.* (1965).

Horn, G. (1967) Neuronal mechanisms of habituation. *Nature,* **215,** 707–711.

Horn, G. (1969) A neural basis for the novelty response. In "Single Unit Mechanisms of the Orientation Response", Ed. E. N. Sokolov and O. Vinogradova, Moscow Univ. Press.

Horn, G., and Blundell, J. (1959) Evoked potentials in visual cortex of the unanaesthetised cat. *Nature,* **184,** 173–174.

Horn, G., and Hill, R. M. (1966) Responsiveness to sensory stimulation of units in the superior colliculus and subjacent tectotegmental regions of the rabbit. *Exper. Neurol.,* **14,** 199–223.

Horn, G., and Hill, R. M. (1969) Modification of sensory field occurring spontaneously and with bodily tilt. *Nature,* **221,** 186–188.

Horn, G., and Hinde, R. A. (in press) Short-term changes in neural activity and behaviour. Cambridge Univ. Press.

Horn, G., and Rowell, C. H. F. (1968) Medium and long-term changes in the behaviour of visual neurones in the tritocerebrum of locusts. *J. exp. Biol.,* **49,** 143–170.

Horridge, G. A. (1959a) Analysis of the rapid responses of *Nereis* and *Harmothöe* (Annelida). *Proc. Roy. Soc. B,* **150,** 245–262.

Horridge, G. A. (1959b) The nerves and muscles of medusae: VI. The rhythm. *J. exp. Biol.,* **36,** 72–91.

Horridge, G. A. (1961) The centrally determined sequence of impulses initiated from a ganglion of the clam *Mya. J. Physiol.,* **155,** 320–336.

Horridge, G. A. (1962) Learning of leg position by the ventral nerve cord in headless insects. *Proc. Roy. Soc. B,* **157,** 33–52.

Horridge, G. A. (1966a) Study of a system, as illustrated by the optokinetic response. *Symp. Soc. exp. Biol.,* **20,** 179–198.

Horridge, G. A. (1966b) Perception of edges versus areas by the crab *Carcinus. J. exp. Biol.,* **44,** 247–254.

Horridge, G. A. (1966c) Optokinetic memory in the locust. *J. exp. Biol.,* **44,** 255–262.

Horridge, G. A. (1966d) Direct response of the crab *Carcinus* to the movement of the sun. *J. exp. Biol.,* **44,** 275–284.

Horridge, G. A. (1968a) Interneurones. Freeman, London.

Horridge, G. A. (1968b) Affinity of neurones in regeneration. *Nature,* **219,** 737–740.

Horridge, G. A., and Burrows, M. (1968) Efferent copy and voluntary eye cup movement in the crab *Carcinus*. *J. exp. Biol.*, **49**, 315–324.

Horridge, G. A., and Sandeman, D. C. (1964) Nervous control of optokinetic responses in the crab, *Carcinus*. *Proc. Roy. Soc. B*, **161**, 216–246.

Howard, E. (1935) The Nature of a Bird's World. Cambridge Univ. Press, London.

Howarth, C. I., and Ellis, K. (1961) The relative intelligibility threshold for one's own name compared with other names. *Q. J. exp. Psychol.*, **13**, 236–239.

Howell, T. R., and Bartholomew, G. A. (1952) Experiments on the mating behavior of the brewer blackbird. *Condor*, **54**, 140–151.

Hoyle, G. (1965) Neurophysiological studies on "learning" in headless insects. In "Physiology of the Insect Central Nervous System", Ed. J. Treherne, Academic Press, New York.

Hsiao, S. (1965) Effect of female variation on sexual satiation in the male rat. *J. comp. physiol. Psychol.*, **60**, 467–469.

Hsiao, S. (1969) The Coolidge effect in male rat copulatory behaviour: failure to replicate Fisher's results. *Psychon. Sci.*, **14**, 1.

Hubel, D. H., and Wiesel, T. N. (1959) Receptive field of single neurones in the cat's striate cortex. *J. Physiol.*, **148**, 574–591.

Hubel, D. H., and Wiesel, T. N. (1960) Receptive fields of optic nerve fibres in the spider monkey. *J. Physiol.*, **154**, 572–580.

Hubel, D. H., and Wiesel, T. N. (1961) Integrative action in the cat's lateral geniculate body. *J. Physiol.*, **155**, 385–393.

Hubel, D. H., and Wiesel, T. N. (1962) Receptive fields, binocular interaction and functional architecture in the cat's visual cortex. *J. Physiol.*, **160**, 106–154.

Hubel, D. H., and Wiesel, T. N. (1963) Receptive fields of cells in striate cortex of very young, visually inexperienced kittens. *J. Neurophysiol.*, **26**, 994–1002.

Hubel, D. H., and Wiesel, T. N. (1964) Responses of monkey geniculate calls to monochromatic and white spots of light. *Physiologist*, **7**, 162–163.

Hubel, D. H., and Wiesel, T. N. (1965a) Receptive fields and functional architecture in two nonstriate visual areas (18 and 19) of the cat. *J. Neurophysiol.*, **28**, 229–289.

Hubel, D. H., and Wiesel, T. N. (1965b) Binocular interaction in striate cortex of kittens reared with artificial squint. *J. Neurophysiol.*, **28**, 1041–1059.

Hubel, D. H., and Wiesel, T. N. (1968) Receptive fields and functional architecture of monkey striate cortex. *J. Physiol.*, **195**, 215–243.

Huber, F. (1955) Sitz und Bedeutung nervöser Zentren für Instinkthandlungen beim Männchen von *Gryllus campestris* L. *Z. Tierpsychol.*, **12**, 12–48.

Huber, F. (1960) Untersuchungen über die Funktion des Zentralnervensystems und insbesondere des Gehirnes bei der Fortbewegung und der Lauterzeugung der Grillen. *Z. vergl. Physiol.*, **44**, 60–132.

Huber, F. (1964) The role of the central nervous system in Orthoptera during the co-ordination and control of stridulation. In "Acoustic Behaviour of Animals", Ed. R. G. Busnel, Elsevier, Amsterdam.

Hudgens, G. A., Denenberg, V. H., and Zarrow, M. X. (1967) Mice reared with rats: relations between mother's activity level and offspring's behavior. *J. comp. physiol. Psychol.*, **63**, 304–308.

Hudgens, G. A., Denenberg, V. H., and Zarrow, M. X. (1968) Mice reared with rats: effects of preweaning and postweaning social interactions upon adult behaviour. *Behaviour*, **30**, 259–274.

Hudson, B. B. (1950) One trial learning in the domestic rat. *Genet. Psychol. Monogr.*, **41**, 99–145.

Hughes, A. (1965) The development of behaviour in *Eleutherodactylus martinicensis* (Amphibia, Anura). *Proc. Zool. Soc. Lond.*, **144**, 153–161.

Hughes, A., Bryant, S. V., and Bellairs, A. d'A. (1967) Embryonic behaviour in the lizard *Lacerta vivipara*. *J. Zool.*, **153**, 139–152.

Hughes, G. M. (1957) The coordination of insect movements: II. The effect of limb amputation and the cutting of commissures in the cockroach (*Blatta orientalis*). *J. exp. Biol.*, **34**, 306–333.

Hughes, G. M. (1958) The coordination of insect movements: III. Swimming in *Dytiscus, Hydrophilus,* and a dragonfly nymph. *J. exp. Biol.*, **35**, 567–583.

Hughes, G. M., and Kerkut, G. A. (1956) Electrical activity in a slug ganglion in relation to the concentration of Locke solution. *J. exp. Biol.*, **33**, 282–294.

Hughes, G. M., and Wiersma, C. A. G. (1960) The co-ordination of swimmeret movements in the crayfish, *Procambarus clarkii* (Girard). *J. exp. Biol.*, **37**, 657–670.

Hull, C. L. (1933) Differential habituation to internal stimuli in the albino rat. *J. comp. Psychol.*, **16**, 255–273.

Hull, C. L. (1943) Principles of Behavior. Appleton-Century-Crofts, New York.

Hull, C. L. (1952) A Behavior System. Yale Univ. Press, New Haven.

Hulse, S. H. (1966) Stimulus intensity and the magnitude of the licking reflex in rats. *Psychon. Sci.*, **6**, 33–34.

Hulse, S. H. (1967) Licking behavior of rats in relation to saccharin concentration and shifts in fixed-ratio reinforcement. *J. comp. physiol. Psychol.*, **64**, 478–484.

Hulse, S. H., and Suter, S. (1968) One-drop licking in rats. *J. comp. physiol. Psychol.*, **66**, 536–539.

Humphrey, G. (1933) The Nature of Learning. Kegan Paul, New York.

Hunt, G. L., and Smith, W. J. (1967) Pecking and initial drinking responses in young domestic fowl. *J. comp. physiol. Psychol.*, **64**, 230–236.

Hunt, H. H., and Otis, L. S. (1955) Restricted experience and "timidity" in the rat. *Amer. Psychologist*, **10**, 432.

Hunt, J. McV. (1941) The effects of infant feeding-frustration upon adult hoarding in the albino rat. *J. abnorm. soc. Psychol.*, **36**, 338–360.

Hunt, J. McV., Schlosberg, H., Solomon, R. L., and Stellar, E. (1947) Studies of the effects of infantile experience on adult behavior in rats: I. *J. comp. physiol. Psychol.*, **40**, 291–304.

Hurwitz, H. M. B. (1956) Conditioned responses in rats reinforced by light. *Brit. J. Anim. Behav.*, **4**, 31–33.

Hurwitz, H. M. B. (1957) Periodicity of response in operant extinction. *Q. J. exp. Psychol.*, **9**, 177–184.

Hurwitz, H. M. B., and De, S. C. (1958) Studies in light reinforced behavior. *Psychol. Rep.*, **4**, 71–77.

Hutchings, D. E. (1963) Early "experience" and its effects on later behavioral processes in rats: III. Effects of infantile handling and body temperature reduction on later emotionality. *Trans. N.Y. Acad. Sci.*, **25**, 890–901.

Hutchinson, R. R., Azrin, N. H., and Hunt, G. M. (1968b) Attack produced by intermittent reinforcement of a concurrent operant response. *J. exp. Anal. Behav.*, **11**, 489–495.

Hutchinson, R. R., Azrin, N. H., and Renfrew, J. W. (1968a) Effects of shock intensity and duration on the frequency of biting attack by squirrel monkeys. *J. exp. Anal. Behav.*, **11**, 83–88.

Hutchinson, R. R., and Renfrew, J. W. (1966) Stalking attack and eating behaviors elicited from the same sites in the hypothalamus. *J. comp. physiol. Psychol.*, **61**, 360–367.

Hutchinson, R. R., and Renfrew, J. W. (1967) Modification of eating and drinking. *J. comp. physiol. Psychol.*, **63**, 408–416.

Hutchinson, R. R., Ulrich, R. E., and Azrin, N. H. (1965) Effects of age and related factors on the pain-aggression reaction. *J. comp. physiol. Psychol.*, **59**, 365–369.

Hutchison, J. B. (1964) Investigation on the neural control of clasping and feeding in *Xenopus laevis* (Daudin). *Behaviour*, **24**, 47–65.

Hutchison, J. B. (1967a) A study of the neural control of sexual clasping behaviour in *Rana angolensis* Bocage and *Bufo regularis* Reuss with a consideration of self-regulatory hind-brain systems in the Anura. *Behaviour*, **28**, 1–57.

Hutchison, J. B. (1967b) Initiation of courtship by hypothalamic implants of testosterone propionate in castrated doves (*Streptopelia risoria*). *Nature*, **216**, 591–592.

Hutchison, J. B. (1968) Neuro-endocrine systems underlying male courtship in *Streptopelia risoria*. Ph.D. thesis, Cambridge.

Hutchison, J. B. (1969) Changes in hypothalamic responsiveness to testosterone in male Barbary doves (*Streptopelia risoria*). *Nature*, **222**, 176.

Hutchison, J. B., and Poynton, J. C. (1963) A neurological study of the clasp reflex in *Xenopus laevis* (Daudin). *Behaviour*, **22**, 41–63.

Hutchison, R. E., Hinde, R. A., and Bendon, B. (1968a) Oviduct development and its relation to other aspects of reproduction in domesticated canaries. *J. Zool.*, **155**, 87–102.

Hutchison, R. E., Hinde, R. A., and Steel, E. A. (1967) The effects of oestrogen, progesterone and prolactin on brood patch formation in ovariectomised canaries. *J. Endocrin.*, **39**, 379–385.

Hutchison, R. E., Stevenson, J. G., and Thorpe, W. H. (1968b) The basis for individual recognition by voice in the sandwich tern (*Sterna sandvicensis*). *Behaviour*, **32**, 150–157.

Hutt, C. (1966) Exploration and play in children. *Symp. Zool. Soc. Lond.*, **18**, 61–81.

Hutt, C., and Hutt, S. J. (1965) Effects of environmental complexity on stereotyped behaviours of children. *Anim. Behav.*, **13**, 1–4.

Hutt, C., Hutt, S. J., and Ounsted, C. (1965) The behaviour of children with and without upper C.N.S. lesions. *Behaviour*, **24**, 246–268.

Hutt, S. J., and Hutt, C. (1968) Stereotypy, arousal and autism. *Human Development*, **11**, 277–286.

Huxley, J. S. (1942) Evolution, the Modern Synthesis. Allen and Unwin, London.

Hymovitch, B. (1952) The effects of experimental variations on problem solving in the rat. *J. comp. physiol. Psychol.*, **45**, 313–321.

Ibbotson, A. (1966) Aggregation in *Aphis fabae* Scop. II. Visual components. *Anim. Behav.*, **14**, 251–260.

Iersel, J. J. A. van (1953) An analysis of the parental behaviour of the male three-spined stickleback (*Gasterosteus aculeatus* L.) *Behaviour Suppl.* **3**, 1–159.

Iersel, J. J. A. van (1958) Some aspects of territorial behaviour of the male three-spined stickleback (*Gasterosteus aculeatus* L.). *Behaviour Suppl.* **3**, 1–159.

Iersel, J. J. A. van, and Assem, J. van den (1965) Aspects of orientation in the diggerwasp *Bembix rostrata*. In "Learning and Associated Phenomena in Invertebrates", *Anim. Behav. Suppl.* **1**.

Iersel, J. J. A. van, and Bol, A. C. A. (1958) Preening of two tern species. A study on displacement activities. *Behaviour*, **13**, 1–88.

Igel, G. J., and Calvin, A. D. (1960) The development of affectional responses in infant dogs. *J. comp. physiol. Psychol.*, **53**, 302–305.

Ikeda, K., and Wiersma, C. A. G. (1964) Autogenic rhythmicity in the abdominal ganglia of the crayfish: the control of swimmeret movements. *Comp. Biochem. Physiol.*, **12**, 107–115.

Immelmann, K. (1962) Beiträge zu einer vergleichenden Biologie australischer Prachtfinken (Spermestidae). *Zool. Jb. Syst. Bd.*, **90**, 1–196.

Immelmann, K. (1965) Objektfixierung geschlechtlicher Triebhandlungen bei Prachtfinken. *Naturwiss.*, **52**, 169.

Immelmann, K. (1969) Song development in the zebra finch and other Estrildid finches. In Hinde (1969).

Ingle, D. (1967/1968) Two visual mechanisms underlying the behavior of fish. *Psychol. Forschung.*, **31**, 44–51.

Inhelder, E. (1955) Zur Psychologie einiger Verhaltensweisen-besonders des Spiels von Zootieren. *Z. Tierpsychol.*, **12**, 88–144.

Isaac, D., and Marler, P. (1963) Ordering of sequences of singing behaviour of Mistle Thrushes in relation to timing. *Anim. Behav.*, **11**, 179–188.

Isaac, W. (1960) Arousal and reaction times in cats. *J. comp. physiol. Psychol.*, **53**, 234–236.

Ishikawa, S. (1966) Electrical response and function of a bitter substance receptor associated with the maxillary sensilla of the larva of the silkworm, *Bombyx mori* L. *J. cell. comp. Physiol.*, **67**, 1–11.

Ishiko, N., and Loewenstein, W. R. (1959) Spontaneous fluctuations in generator potential in a receptor membrane. *Nature,* **183,** 1725–1726.

Jacobs, H. L. (1964) Evaluation of the osmotic effects of glucose loads in food satiation. *J. comp. physiol. Psychol.,* **57,** 309–310.

Jacobs, W. (1953) Verhaltensbiologische Studien an Feldheuschrecken. *Z. Tierpsychol., Beit.* **1,** 1–228.

Jacobson, M., and Gaze, R. M. (1964) Types of visual response from single units in the optic tectum and optic nerve of the goldfish. *Q. J. exp. Physiol.,* **49,** 199–209.

Jacobson, M., and Gaze, R. M. (1965) Selection of appropriate tectal connections by regenerating optic nerve fibers in adult goldfish. *Exper. Neurol.,* **13,** 418–430.

Jahn, T. (1960) Optische Gleichgewichtsregelung und zentrale Kompensation bei Amphibien, insbesondere bei der Erdkröte (*Bufo bufo* L.). *Z. vergl. Physiol.,* **43,** 119–140.

James, H. (1959) Flicker: an unconditioned stimulus for imprinting. *Canad. J. Psychol.,* **13,** 59–67.

James, H. (1960a) Imprinting with visual flicker: evidence for a critical period. *Canad. J. Psychol.,* **14,** 13–20.

James, H. (1960b) Social inhibition of the domestic chick's response to visual flicker. *Anim. Behav.,* **8,** 223–224.

James, W. (1892) Textbook of Psychology. Macmillan, London.

James, W. T. (1953) Social facilitation of eating behavior in puppies after satiation. *J. comp. physiol. Psychol.,* **46,** 427–428.

James, W. T. (1957) The effect of satiation on the sucking response in puppies. *J. comp. physiol. Psychol.,* **50,** 375–378.

James, W. T. (1963a) An analysis of esophageal feeding as a form of operant reinforcement in the dog. *Psychol. Rep.,* **12,** 31–39.

James, W. T. (1963b) Suppression of sucking by food injection in puppies. *Psychol. Rep.,* **13,** 862.

James, W. T., and Gilbert, T. F. (1956) The effect of social facilitation on food intake of puppies fed separately and together for the first 90 days of life. *Brit. J. Anim. Behav.,* **3,** 131–133.

James, W. T., and Rollins, J. (1965) Effect of various degrees of stomach loading on the sucking response in puppies. *Psychol. Rep.,* **17,** 844–846.

Jander, R. (1957) Die optische Richtungsorientierung der roten Waldameise (*Formica rufa* L.). *Z. vergl. Physiol.,* **40,** 162–238.

Jander, R. (1963) Insect orientation. *Ann. Rev. Entomol.,* **8,** 95–114.

Jander, R. (1964) Die Detektortheorie optischer Auslösemechanismen von Insekten. *Z. Tierpsychol.,* **21,** 302–307.

Jander, R. (1965) Die Phylogenie von Orientierungsmechanismen der Arthropoden. *Verh. Deutschen Zool. Gesell. Jena,* 1965, 266–306.

Jander, R., and Barry, C. K. (1968) Die phototaktische Gegenkopplung von

Stirnocellen und Facettenaugen in der Phototropotaxis der Heuschrecken und Grillen. *Z. vergl. Physiol.*, **57**, 432–458.

Jander, R., and Waterman, T. H. (1960) Sensory discrimination between polarized light and light intensity patterns by arthropods. *J. cell. comp. Physiol.*, **56**, 137–159.

Jander, U. (1966) Untersuchungen zur Stammesgeschichte von Putzbewegungen von Tracheaten. *Z. Tierpsychol.*, **23**, 799–844.

Janowitz, H. D., and Grossman, M. I. (1949) Some factors affecting the food intake of normal dogs and dogs with esophagostomy and gastric fistula. *Amer. J. Physiol.*, **159**, 143–148.

Janowitz, H. D., Hanson, M. E., and Grossman, M. I. (1949) Effects of intravenously administered glucose on food intake in the dog. *Amer. J. Physiol.*, **156**, 87–91.

Jarmon, H., and Gerall, A. A. (1961) The effect of food deprivation upon the sexual performance of male guinea pigs. *J. comp. physiol. Psychol.*, **54**, 306–309.

Jarvik, M. E., and Kopp, R. (1967) Transcorneal electroconvulsive shock and retrograde amnesia in mice. *J. comp. physiol. Psychol.*, **64**, 431–433.

Jasper, H. H. (1949) Diffuse projection systems: the integrative action of the thalamic system. *Electroenceph. clin. Neurophysiol.*, **1**, 405–420.

Jasper, H. H. (1961) Thalamic reticular system. In Sheer (1961).

Jasper, H. H., Proctor, L. D., Kaighton, R. S., Noshay, W. C., and Costello, R. T. (1958) Reticular Formation of the Brain. Little, Brown, Boston.

Jaynes, J. (1956) Imprinting: the interaction of learned and innate behavior: I. Development and generalization. *J. comp. physiol. Psychol.*, **49**, 201–206.

Jaynes, J. (1957) Imprinting: the interaction of learned and innate behavior: II. The critical period. *J. comp. physiol. Psychol.*, **50**, 6–10.

Jaynes, J. (1958a) Imprinting: the interaction of learned and innate behavior: III. Practice effects on performance, retention and fear. *J. comp. physiol. Psychol.*, **51**, 234–237.

Jaynes, J. (1958b) Imprinting: the interaction of learned and innate behavior: IV. Generalization and emergent discrimination. *J. comp. physiol. Psychol.*, **51**, 238–242.

Jenkins, H. M., and Harrison, R. H. (1960) Effect of discrimination training on auditory generalization. *J. exp. Psychol.*, **59**, 246–253.

Jenkins, J. J., and Hanratty, J. A. (1949) Drive intensity discrimination in the albino rat. *J. comp. physiol. Psychol.*, **42**, 228–232.

Jennings, H. S. Behavior of the Lower Organisms. Columbia Univ. Press, New York.

Jensen, D. D. (1961) Operationism and the question "Is this behavior learned or innate?". *Behaviour*, **17**, 1–8.

Johnsgard, P. A. (1961) The taxonomy of the Anatidae—a behavioural analysis. *Ibis*, **103**a, 71–85.

Johnson, D. L. (1967) Honey bees: do they use the direction information contained in their dance maneuver? *Science*, **155**, 844–847.

Johst, V. (1967) Vergleichende Untersuchung des agonistischen Verhaltens einiger Arten von *Clethrionomys*. *Z. Tierpsychol.*, **24**, 558–579.

Jolly, A. (1964a) Choice of cue in Prosimian learning. *Anim. Behav.*, **12**, 571–577.

Jolly, A. (1964b) Prosimians' manipulation of simple object problems. *Anim. Behav.*, **12**, 560–570.

Jolly, A. (1966) Lemur Behavior. Univ. of Chicago.

Joly, J. (1966) Sur l'éthologie sexuelle de *Salamandra salamandra* (L.). *Z. Tierpsychol.*, **23**, 8–27.

Jones, F. R. H. (1955) Photo-kinesis in the ammocoete larva of the brook lamprey. *J. exp. Biol.*, **32**, 492–503.

Joslin, J., Fletcher, H., and Emlen, J. (1964) A comparison of the responses to snakes of lab- and wild-reared rhesus macaques. *Anim. Behav.*, **12**, 348–352.

Jouvet, M., and Desmedt, J. E. (1956) Contrôle central des messages acoustiques afférents. *C. R. Acad. Sci.*, **243**, 1916–1917.

Jouvet, M., and Lapras, C. (1959) Variations des réponses électriques somesthésiques au niveau du thalamus chez l'homme au cours de l'attention. *C. R. Soc. Biol.*, **153**, 98–101.

Jung, R., and Hassler, R. (1960) The extrapyramidal motor system. "Handbook of Physiology", Sec. 1, Vol. 2, Ed. J. Field, H. W. Magoun, and V. E. Hall, Amer. Physiol. Soc., Washington.

Kaess, D. W., and Wilson, J. P. (1964) Modification of the rat's avoidance of visual depth. *J. comp. physiol. Psychol.*, **58**, 151–152.

Kagan, J. (1955) Differential reward value of incomplete and complete sexual behavior. *J. comp. physiol. Psychol.*, **48**, 59–64.

Kagan, J., and Beach, F. A. (1953) Effects of early experience on mating behavior in male rats. *J. comp. physiol. Psychol.*, **46**, 204–208.

Kagan, J., and Berkun, M. (1954) The reward value of running activity. *J. comp. physiol. Psychol.*, **47**, 108.

Kahl, M. P. (1966) Comparative ethology of the Ciconiidae. Part I. The Marabou stork *Leptoptilos crumeniferus*. *Behaviour*, **27**, 76–106.

Kahn, M. W. (1951) The effect of severe defeat at various age levels on the aggressive behavior of mice. *J. genet. Psychol.*, **79**, 117–130.

Kalish, H. I. (1954) Strength of fear as a function of the number of acquisition and extinction trials. *J. exp. Psychol.*, **47**, 1–9.

Kalmus, H. (1934) Uber die Natur des Zeitgedächtnisses der Bienen. *Z. vergl. Physiol.*, **20**, 405–419.

Kalmus, H. (1966) Regulation and Control in Living Systems. Wiley, New York.

Kalter, J., and Warkang, J. (1959) Experimental production of congenital malformations in mammals by metabolic procedure. *Physiol. Rev.*, **39**, 69–115.

Kamback, M. C. (1966) Effects of food deprivation on object and non-object directed behavior in the rat. *Psychon. Sci.*, **5**, 107–108.

Kammer, A. E. (1967) Muscle activity during flight in some large Lepidoptera. *J. exp. Biol.*, **47**, 277–295.

Kammer, A. E. (1968) Motor patterns during flight and warm-up in Lepidoptera. *J. exp. Biol.*, **48**, 89–109.

Kandel, E. R., and Spencer, W. A. (1968) Cellular neurophysiological approaches in the study of learning. *Physiol. Rev.*, **48**, 65–134.

Karlson, P., and Butenandt, A. (1959) Pheromones (ectohormones) in insects. *Ann. Rev. Ent.*, **4**, 39–58.

Kästle, W. (1964) Verhaltensstudien an Taggeckonen der Gattungen *Lygodactylus* und *Phelsuma*. *Z. Tierpsychol.*, **21**, 486–507.

Katz, D. (1937) Animals and Men. Longmans Green, New York.

Kaufman, I. C., and Hinde, R. A. (1961) Factors influencing distress calling in chicks, with special reference to temperature changes and social isolation. *Anim. Behav.*, **9**, 197–204.

Kaufman, I. C., and Rosenblum, L. A. (1967) Depression in infant monkeys separated from their mothers. *Science*, **155**, 1030–1031.

Kaufmann, H. (1965) Definitions and methodology in the study of aggression. *Psychol. Bull.*, **64**, 351–364.

Kavanau, J. L. (1962) Twilight transitions and biological rhythmicity. *Nature*, **194**, 1293–1295.

Kavanau, J. L. (1963) Compulsory regime and control of environment in animal behavior: I. Wheel running. *Behaviour*, **20**, 251–281.

Kavanau, J. L. (1966) Wheel running preferences of mice. *Z. Tierpsychol.*, **23**, 858–866.

Kavanau, J. L. (1967) Behavior of captive white-footed mice. *Science*, **155**, 1623–1639.

Kavanau, J. L. (1968) Activity and orientational responses of white-footed mice to light. *Nature*, **218**, 245–252.

Kear; J. (1962) Food selection in finches with special reference to interspecific differences. *Proc. Zool. Soc. Lond.*, **138**, 163–204.

Kear, J. (1967) Wildfowl Trust 18th Annual Report, 122–124.

Keeler, C. E., and King, H. D. (1942) Multiple effects of coat color genes in the Norway rat, with special reference to temperament and domestication. *J. comp. Psychol.*, **34**, 241–250.

Keenleyside, M. H. A. (1955) Some aspects of the schooling behaviour of fish. *Behaviour*, **8**, 183–248.

Keenleyside, M. H. A., and Yamamoto, F. T. (1962) Territorial behaviour of juvenile Atlantic salmon (*Salmo salar* L.). *Behaviour*, **19**, 139–169.

Keesey, R. E. (1964) Duration of stimulation and the reward properties of hypothalamic stimulation. *J. comp. physiol. Psychol.*, **58**, 201–207.

Keesey, R. E. (1966) Hypothalamic stimulation as a reinforcer of discrimination learning. *J. comp. physiol. Psychol.*, **62**, 231–236.

Kelleher, R. T. (1966) Chaining and conditioned reinforcement. In Honig (1966).

Keller, F. S. (1941) Light-aversion in the white rat. *Psychol. Rec.*, **4,** 235–250.

Keller, F. S., and Hill, L. M. (1936) Another "insight" experiment. *J. genet. Psychol.*, **48,** 484–489.

Kellogg, W. N. (1968) Communication and language in the home-raised chimpanzee. *Science,* **162,** 423–427.

Kendig, J. J. (1968) Motor neurone coupling in locust flight. *J. exp. Biol.*, **48,** 389–404.

Kendler, H. H. (1946) The influence of simultaneous hunger and thirst drives upon the learning of two opposed spatial responses of the white rat. *J. exp. Psychol.*, **36,** 212–220.

Kendler, H. H., and Gasser, W. P. (1948) Variables in spatial learning: I. Number of reinforcements during training. *J. comp. physiol. Psychol.*, **41,** 178–187.

Kennedy, D., and Takeda, K. (1965) Reflex control of abdominal flexor muscles in the crayfish. II. The tonic system. *J. exp. Biol.*, **43,** 229–246.

Kennedy, G. C. (1953) The role of depot fat in the hypothalamic control of food intake in the rat. *Proc. Roy. Soc. B,* **140,** 578–592.

Kennedy, J. S. (1945) Classification and nomenclature of animal behaviour. *Nature,* **156,** 754.

Kennedy, J. S. (1954) Is modern ethology objective? *Brit. J. Anim. Behav.*, **2,** 12–19.

Kennedy, J. S. (1958) The experimental analysis of aphid behaviour and its bearing on current theories of instinct. *Proc. 10th Int. Cong. Entom., Montreal,* **2,** 397–404.

Kennedy, J. S. (1965) Coordination of successive activities in an aphid. Reciprocal effects of settling on flight. *J. exp. Biol.*, **43,** 489–509.

Kennedy, J. S. (1966) The balance between antagonistic induction and depression of flight activity in *Aphis fabae* Scopoli. *J. exp. Biol.*, **45,** 215–228.

Kennedy, J. S., and Booth, C. O. (1963) Co-ordination of successive activities in an aphid. The effect of flight on the settling responses. *J. exp. Biol.*, **40,** 351–369.

Kennedy, J., and Booth, C. O. (1964) Coordination of successive activities in an aphid. Depression of settling after flight. *J. exp. Biol.*, **41,** 805–824.

Kent, G. C., and Liberman, R. J. (1949) Induction of psychic estrus in the hamster with progesterone administered via the lateral brain ventricle. *Endocrinology,* **45,** 29–32.

Kerfoot, W. B. (1967) Dorsal light receptors. *Nature,* **215,** 305–306.

Kerkut, G. A., and Taylor, B. J. R. (1958) The effect of temperature changes on the activity of poikilotherms. *Behaviour,* **13,** 259–279.

Kerpelman, L. C. (1965) Preexposure to visually presented forms and nondifferential reinforcement in perceptual learning. *J. exp. Psychol.*, **69,** 257–262.

Kessen, W. (1967) Sucking and looking: two organized congenital patterns of behavior in the human newborn. In Stevenson *et al.* (1967).

Khavari, K. A., and Russell, R. W. (1966) Acquisition, retention, and extinction under conditions of water deprivation and of central cholinergic stimulation. *J. comp. physiol. Psychol.*, **61**, 339–345.

Kiernan, C. C. (1965) Effect of food deprivation and unconditioned operant pretests on bar pressing for light in the albino rat. *J. comp. physiol. Psychol.*, **60**, 268–271.

Kimble, D. P., and Ray, R. S. (1965) Reflex habituation and potentiation in *Rana pipiens*. *Anim. Behav.*, **13**, 530–533.

Kimble, G. A. (1966) Classical and instrumental conditioning: one process or two? *Proc. 18th Int. Cong. Psychol., Moscow*, **1**, 110.

Kincaid, J. P. (1968) Different temporal gradients of retrograde amnesia produced by metrazol or electroconvulsive shock. *Psychon. Sci.*, **11**, 329–330.

Kinder, E. F. (1927) A study of the nest-building activity of the albino rat. *J. exp. Zool.*, **47**, 117–161.

King, D. L. (1966) A review and interpretation of some aspects of the infant-mother relationship in mammals and birds. *Psychol. Bull.*, **65**, 143–155.

King, J. A., and Eleftheriou, B. E. (1959) Effects of early handling upon adult behavior in two subspecies of deermice, *Peromyscus maniculatus*. *J. comp. physiol. Psychol.*, **52**, 82–88.

King, J. A., and Gurney, N. L. (1954) Effect of early social experience on adult aggressive behavior in C57BL/10 mice. *J. exp. physiol. Psychol.*, **47**, 326–330.

King, J. A., and Weisman, R. G. (1964) Sand digging contingent upon bar pressing in deermice (*Peromyscus*). *Anim. Behav.*, **12**, 446–450.

King, J. E., Flaningam, M. R., and Rees, W. W. (1968) Delayed response with different delay conditions by squirrel monkeys and fox squirrels. *Anim. Behav.*, **16**, 271–275.

Kinosita, H., and Okajima, A. (1968) Analysis of shell-searching behavior of the land hermit-crab, *Coenobita rugosus*, H. Milne Edwards. *J. Faculty of Science*, Sec. IV, **II**, 293–358.

Kirby, R. H. (1963) Acquisition, extinction and retention of an avoidance response in rats as a function of age. *J. comp. physiol. Psychol.*, **56**, 158–162.

Kirkman, F. B. (1937) Bird Behaviour. Nelson, London.

Kish, G. B. (1955) Learning when the onset of illumination is used as reinforcing stimulus. *J. comp. physiol. Psychol.*, **48**, 261–264.

Kish, G. B. (1966) Studies of sensory reinforcement. In Honig (1966).

Kish, G. B., and Barnes, G. W. (1961) Reinforcing effects of manipulation in mice. *J. comp. physiol. Psychol.*, **54**, 713–715.

Kish, G. B., and Baron, A. (1962) Satiation of sensory reinforcement. *J. comp. physiol. Psychol.*, **55**, 1007–1010.

Kislak, J. W., and Beach, F. A. (1955) Inhibition of aggressiveness by ovarian hormones. *Endocrinology*, **56**, 684–692.

Klausewitz, W. (1953) Die Korrelation von Verhaltensphysiologie und Farbphysiologie bei *Agama cyanogaster atricollis*. *Z. Tierpsychol.*, **10**, 169–180.

Kling, J. W., Horowitz, L., and Delhagen, J. E. (1956) Light as a positive reinforcer for rat responding. *Psychol. Rep.*, **2**, 337–340.

Klinghammer, E. (1967) Factors influencing choice of mate in altricial birds. In Stevenson *et al.* (1967).

Klinghammer, E., and Hess, E. H. (1964) Parental feeding in ring doves (*Streptopelia roseogrisea*): innate or learned? *Z. Tierpsychol.*, **21**, 338–347.

Klopfer, P. H. (1957) An experiment on empathic learning in ducks. *Amer. Naturalist*, **91**, 61–63.

Klopfer, P. H. (1959a) An analysis of learning in young Anatidae. *Ecology*, **40**, 90–102.

Klopfer, P. H. (1959b) Social interactions in discrimination learning with special reference to feeding behavior in birds. *Behaviour*, **14**, 282–299.

Klopfer, P. H. (1962) Behavioral Aspects of Ecology. Prentice-Hall, Englewood Cliffs, N.J.

Klopfer, P. H. (1963) Behavioral aspects of habitat selection: the role of early experience. *Wilson Bull.*, **75**, 15–22.

Klopfer, P. H. (1967) Behavioral stereotypy in birds. *Wilson Bull.*, **79**, 290–300.

Klopfer, P. H., and Gamble, J. (1966) Maternal "imprinting" in goats: the role of chemical senses. *Z. Tierpsychol.*, **23**, 588–592.

Klopfer, P. H., and Hailman, J. P. (1964) Perceptual preferences and imprinting in chicks. *Science*, **145**, 1333–1334.

Klopfer, P. H., and Klopfer, M. S. (1968) Maternal "imprinting" in goats: fostering of alien young. *Z. Tierpsychol.*, **25**, 862–866.

Knoll, M. (1953) Uber das Tages—und Dämmerungssehen des Grasfrosches (*Rana temporaria* L.) nach Aufzucht in veränderten Lichtbedingungen. *Z. vergl. Physiol.*, **35**, 42–70.

Koch, S. (1959) Psychology, a Study of a Science: Study I, Vols. 1 and 2, McGraw-Hill, New York.

Koepke, J. E., and Pribram, K. H. (1966) Habituation of GSR as a function of stimulus duration and spontaneous activity. *J. comp. physiol. Psychol.*, **61**, 442–448.

Koepke, J. E., and Pribram, K. H. (1967) Effect of food reward on the maintenance of sucking behavior during infancy. *Proc. 75th Ann. Conv. Amer. Psychol. Assoc.*, 111–112.

Koffka, K. (1935) Principles of Gestalt Psychology. Harcourt Brace, New York.

Köhler, I. (1951) Ueber Aufbau und Wandlungen der Wahrnehmungswelt, Sitzungsber. *Oest. Akad. Wiss.*, **227**, 1–118.

Köhler, O. (1955) "Zählende" Voegel und vergleichende Verhaltensforschung. *Acta XI Congr. Int. Orn.*, 1954, 588–598.

Köhler, W. (1925) The Mentality of Apes. Harcourt Brace, New York.

Kohn, M. (1951) Satiation of hunger from food injected directly into the stomach versus food ingested by mouth. *J. comp. physiol. Psychol.*, **44**, 412–422.

Kolb, E. (1955) Untersuchungen über zentrale Kompensation und Kompensationsbewegungen einseitig enstateter Frösche. *Z. vergl. Physiol.*, **37**, 136–160.

Komisaruk, B. R. (1967) Effects of local brain implants of progesterone on reproductive behavior in ring doves. *J. comp. physiol. Psychol.*, **64**, 219–224.

Komisaruk, B. R., McDonald, P. G., Whitmoyer, D. I., and Sawyer, C. H. (1967) Effects of progesterone and sensory stimulation on EEG and neuronal activity in the rat. *Exp. Neurol.*, **19**, 494–507.

Konishi, M. (1963) The role of auditory feedback in the vocal behavior of the domestic fowl. *Z. Tierpsychol.*, **20**, 349–367.

Konishi, M. (1964) Effects of deafening on song development in two species of Juncos. *Condor*, **66**, 85–102.

Konishi, M. (1965a) Effects of deafening on song development in American robins and black-headed grosbeaks. *Z. Tierpsychol.*, **22**, 584–599.

Konishi, M. (1965b) The rôle of auditory feedback in the control of vocalization in the white-crowned sparrow. *Z. Tierpsychol.*, **22**, 770–783.

Konishi, M. (1966) The attributes of instinct. *Behaviour*, **27**, 316–328.

Konishi, M., and Nottebohm, F. (1969) Experimental studies in the ontogeny of avian vocalizations. In Hinde (1969).

Konorski, J. (1948) Conditioned Reflexes and Neuron Organization. Cambridge Univ. Press.

Konorski, J. (1962) The role of central factors in differentiation. *Proc. Int. Union Physiol. Sci.*, **3**, 318–329.

Koronakos, C., and Arnold, W. J. (1957) The formation of learning sets in rats. *J. comp. physiol. Psychol.*, **50**, 11–14.

Kortlandt, A. (1940) Eine Ubersicht der angeborenen Verhaltensweisen des Mitteleuropäischen Kormorans (*Phalacrocorax carbo sinensis*). *Arch. néerl. Zool.*, **14**, 401–442.

Kortlandt, A. (1965) How do chimpanzees use weapons when fighting leopards? *Amer. Philos. Soc. Year Book*, 1965, 327–332.

Kovach, J .K. (1964) Effects of autonomic drugs on imprinting. *J. comp. physiol. Psychol.*, **57**, 183–187.

Kovach, J. K., Fabricius, E., and Fält, L. (1966) Relationships between imprinting and perceptual learning. *J. comp. physiol. Psychol.*, **61**, 449–454.

Kovach, J. K., and Kling, A. (1967) Mechanisms of neonate sucking behaviour in the kitten. *Anim. Behav.*, **15**, 91–101.

Kovach, J. K., Paden, P., and Wilson, G. (1968) Stimulus variables in the elicitation and short-range reversibility of early approach and following responses. *J. comp. physiol. Psychol.*, **66**, 175–178.

Kozak, W., Rodieck, R. W., and Bishop, P. O. (1965) Responses of single units in lateral geniculate nucleus of cat to moving visual patterns. *J. Neurophysiol.*, **28**, 19–47.

Kramer, G. (1952) Experiments on bird orientation. *Ibis*, **94**, 265–285.

Kramer, G. (1957) Experiments on bird orientation and their interpretation. *Ibis*, **99**, 196–227.

Kramer, G. (1959) Recent experiments on bird orientation. *Ibis*, **101**, 399–416.

Kramer, G. (1961) Long distance orientation. In "Biology and Comparative Physiology of Birds", Ed. A. J. Marshall, Academic Press, New York.

Kramer, G., and St. Paul, U. von (1951) Über angeborenes und erworbenes Feinderkennen beim Gimpel (*Pyrrhula pyrrhula* L.). *Behaviour,* **3,** 243–251.

Krames, L., and Carr, W. J. (1968) The effect of previous visual experience upon the response to depth in domestic chickens. *Psychon. Sci.,* **10,** 249–250.

Krames, L., Costanzo, D. J., and Carr, W. J. (1967) Responses of rats to odors from novel versus original sex partners. *Proc. 75th Ann. Conv. Amer. Psychol. Ass.,* 117–118.

Krasne, F. B. (1965) Escape from recurring tactile stimulation in *Branchiomma vesiculosum. J. exp. Biol.,* **42,** 307–322.

Krasne, F. B. (1966) Decreased tolerance of hypothalamic hyperphagics to quinine in drinking water. *Psychon. Sci.,* **4,** 313–314.

Krasne, F. B., and Roberts, A. (1967) Habituation of the crayfish escape response during release from inhibition induced by picrotoxin. *Nature,* **215,** 769–770.

Kravitz, J. H., and Wallach, H. (1966) Adaptation to displaced vision contingent upon vibrating stimulation. *Psychonomic Sci.,* **6,** 465–466.

Krechevsky, I. (1937) Brain mechanisms and variability I, II and III. *J. comp. Psychol.,* **23,** 121–138, 139–164, 351–364.

Krieckhaus, E. E., and Wolf, G. (1968) Acquisition of sodium by rats. *J. comp. physiol. Psychol.,* **65,** 197–201.

Kruijt, J. P. (1962) Imprinting in relation to drive interactions in Burmese red junglefowl. *Symp. Zool. Soc. Lond.,* **8,** 219–226.

Kruijt, J. P. (1964) Ontogeny of social behaviour in Burmese red junglefowl (*Gallus gallus spadiceus* Bonnaterre). *Behaviour Suppl.* **12.**

Kuczka, H. (1956) Verhaltensphysiologische Untersuchungen über die Wischhandlung der Erdkröte (*Bufo bufo* L.). *Z. Tierpsychol.,* **13,** 185–207.

Kuehn, R. E., and Beach, F. A. (1963) Quantitative measurement of sexual receptivity in female rats. *Behaviour,* **21,** 282–299.

Kuenzer, E., and Kuenzer, P. (1962) Untersuchungen zur Brutpflege der Zwergcichliden *Apistogramma reitzigi* und *A. borellii. Z. Tierpsychol.,* **19,** 56–83.

Kuenzer, P. (1958) Verhaltensphysiologische Untersuchungen über das Zucken des Regenwurms. *Z. Tierpsychol.,* **15,** 31–49.

Kuenzer, P. (1968) Die Auslösung der Nachfolgereaktion bei erfahrungslosen Jungfischen von *Nannacara anomala* (Cichlidae). *Z. Tierpsychol.,* **25,** 257–314.

Kuffler, S. W. (1953) Discharge patterns and functional organization of mammalian retina. *J. Neurophysiol.,* **16,** 37–68.

Kuffler, S. W., FitzHugh, R., and Barlow, H. B. (1957) Maintained activity in the cat's retina in light and darkness. *J. gen. Physiol.,* **40,** 683–702.

Kühme, W. (1961) Beobachtungen am afrikanischen Elefanten (*Loxodonta africana* Blumenbach) in Gefangenschaft. *Z. Tierpsychol.,* **18,** 285–296.

Kühme, W. (1962) Das Schwarmverhalten elterngeführter Jungcichliden (Pisces). *Z. Tierpsychol.,* **19,** 513–538.

Kühme, W. (1963) Ergänzende Beobachtungen an afrikanischen Elefanten (*Loxodonta africana* Blumenbach 1797) im Freigehege. *Z. Tierpsychol.,* **20,** 66–79.

Kühn, A. (1919) Die Orientierung der Tiere im Raum. Jena.

Kunkel, P. (1959) Zum Verhalten einiger Prachtfinken (Estrilidinae). *Z. Tierpsychol.*, **16**, 302–350.

Kunkel, P. (1962) Bewegungsformen, Sozialverhalten, Balz und Nestbau des Gangesbrillenvogels (*Zosterops palpebrosa* Temm.). *Z. Tierpsychol.*, **19**, 559–576.

Kunkel, P. (1967) Displays facilitating sociability in waxbills of the genera *Estrilda* and *Lagonosticta* (Fam. Estrilidinae). *Behaviour*, **29**, 237–261.

Kuo, Z. Y. (1932) Ontogeny of embryonic behavior in Aves: IV. The influence of embryonic movements upon the behavior after hatching. *J. comp. Psychol.*, **14**, 109–122.

Kuo, Z. Y. (1967) The Dynamics of Behaviour Development. Random House, New York.

Kurtz, K. H., and Siegel, A. (1966) Conditioned fear and magnitude of startle response. *J. comp. physiol. Psychol.*, **62**, 8–14.

Kutscher, C. (1967) Inhibition of drinking in thirsty rats by food deprivation: measures of potential internal stimuli. *Proc. 75th Ann. Conv. Amer. Psychol. Assoc.*, 113–114.

Lacey, J. I. (1966) Somatic response patterning and stress: some revisions of activation theory. Paper delivered at a symposium on "Issues in Stress", York University, Toronto, Canada, May 10–12, 1965.

Lacey, J. I., Kagan, J., Lacey, B. C., and Moss, H. A. (1963) The visceral level: situational determinants and behavioral correlates of autonomic response patterns. In "Expression of the Emotions in Man", Ed. P. H. Knapp, International Univ. Press, New York.

Lack, D. (1939) The behaviour of the robin: I and II. *Proc. Zool. Soc. Lond. A*, **109**, 169–178.

Lack, D. (1947) Darwin's Finches. Cambridge Univ. Press.

Lack, D. (1966) Population Studies of Birds. Oxford Univ. Press.

Lade, B. I., and Thorpe, W. H. (1964) Dove songs as innately coded patterns of specific behaviour. *Nature*, **202**, 366–368.

Lagerspetz, K. (1964) Studies on the aggressive behaviour of mice. *Suomalaisen Tiedeakatemian Toimituksia. Ann. Acad. Scient. Fenn.*, **131**, 1–131.

Lal, H. (1967) Operant control of vocal responding in rats. *Psychonomic Sci.*, **8**, 35–36.

Landreth, H. F., and Ferguson, D. E. (1967) Newts: sun-compass orientation. *Science*, **158**, 1459–1461.

Landreth, H. F., and Ferguson, D. E. (1968) The sun compass of Fowler's toad *Bufo woodhousei* Fowleri. *Behaviour*, **30**, 27–43.

Langworthy, O. R. (1929) A correlated study of the development of reflex action in fetal and young kittens and the myelinization of tracts in the nervous system. *Contr. Embryol. Carneg. Inst.*, **20**, 127–171.

Lannoy, J. de (1967) Zur Prägung von Instinkthandlungen. *Z. Tierpsychol.*, **24**, 162–200.

Lanyon, W. E. (1957) The comparative biology of the meadowlarks in Wisconsin. *Pub. Nuttall Ornith. Club*, **1**, Cambridge, Mass.

Lanyon, W. E. (1960) The ontogeny of vocalisations in birds. In "Animal Sounds and Communication", *Amer. Inst. Biol. Sciences*, **7**.

Larimer, J. L., and Tindel, J. R. (1966) Sensory modifications of heart rate in crayfish. *Anim. Behav.*, **14**, 239–245.

Larsson, K. (1956) Conditioning and Sexual Behavior in the Male Albino Rat. Almquist and Wiksell, Stockholm.

Larsson, K. (1958) Aftereffects of copulatory activity of the male rat: II. *J. comp. physiol. Psychol.*, **51**, 417–420.

Larsson, K. (1959a) Experience and maturation in the development of sexual behaviour in male puberty rat. *Behaviour*, **14**, 101–107.

Larsson, K. (1959b) The effect of restraint upon copulatory behaviour in the rat. *Anim. Behav.*, **7**, 23–25.

Larsson, K. (1962) Spreading cortical depression and the mating behaviour in male and female rats. *Z. Tierpsychol.*, **19**, 321–331.

Larsson, K. (1963) Non-specific stimulation and sexual behaviour in the male rat. *Behaviour*, **20**, 110–114.

Larsson, K. (1964) Mating behavior in male rats after cerebral cortex ablation. *J. exp. Zool.*, **155**, 203–213.

Larsson, K. (1967a) Effects of neonatal castration and androgen replacement therapy upon the development of the mating behavior of the male rat. *Z. Tierpsychol.*, **24**, 471–475.

Larsson, K. (1967b) Testicular hormone and developmental changes in mating behaviour of the male rat. *J. comp. physiol. Psychol.*, **63**, 223–230.

Larsson, S. (1954) On the hypothalamic organisation of the nervous mechanism regulating food intake. *Acta physiol. scand.*, **32**, *Suppl.* 115, 1–40.

Lashley, K. S. (1917) The accuracy of movement in the absence of excitation from the moving organ. *Amer. J. Physiol.*, **43**, 169–194.

Lashley, K. S. (1930) The mechanism of vision: I. A method for rapid analysis of pattern-vision in the rat. *J. genet. Psychol.*, **37**, 453–460.

Lashley, K. S. (1938a) Experimental analysis of instinctive behavior. *Psychol. Rev.*, **45**, 445–471.

Lashley, K. S. (1938b) The mechanism of vision: XV. Preliminary studies of the rat's capacity for detail vision. *J. gen. Psychol.*, **18**, 123–193.

Lashley, K. S. (1951) The problem of serial order in behavior. In "Cerebral Mechanisms in Behavior", Ed. L. A. Jeffress, Wiley, New York.

Lashley, K. S. (1952) Functional interpretation of anatomic patterns. *Proc. Ass. Res. Nerv. Ment. Dis.*, **30**, 529–547.

Lashley, K. S., and Ball, J. (1929) Spinal conduction and kinesthetic sensitivity in the maze habit. *J. comp. Psychol.*, **9**, 71–105.

Lashley, K. S., and Russell, J. T. (1934) The mechanism of vision: XI. A preliminary test of innate organisation. *J. genet. Psychol.*, **45**, 136–144

Lashley, K. S., and Wade, M. (1946) The Pavlovian theory of generalization. *Psychol. Rev.*, **53,** 72–87.

Lassek, A. M., and Mayer, E. K. (1953) An ontogenetic study of motor deficits following dorsal brachial rhizotomy. *J. Neurophysiol.*, **16,** 247–251.

Laszlo, J. I. (1967) Training of fast tapping with reduction of kinaesthetic, tactile, visual and auditory sensations. *Q. J. exp. Psychol.*, **19,** 344–349.

Lát, J. (1963) Factors affecting the basic non-specific tonus of the central nervous system. In "Central and Peripheral Mechanisms of Motor Functions", Ed. A. Gutman and P. Hník, *Proc. Conf. Czech. Acad. Sci. Prague*, 225–263.

Lát, J. (1967a) Self selection of dietary components. 'Handbook of Physiology **6,** 1', Ed. C. F. Code, 367–386, Amer. Psychol. Soc.

Lát, J. (1967b) Nutrition, learning and adaptive capacity. In "The Chemical Senses and Nutrition", Ed. M. R. Kare and O. Maller, Johns Hopkins, Baltimore.

Lát, J., and Gollová-Hémon, E. (1969) Permanent effects of nutrition and endocrinological intervention in early ontogeny on the level of non-specific excitability and on lability (emotionality). *Ann. N.Y. Acad. Sci.* (in press).

Lát, J., Widdowson, E. M., and McCance, R. A. (1960) Some effects of accelerating growth. III. Behaviour and nervous activity. *Proc. Roy. Soc. B*, **153,** 347–356.

Lawick-Goodall, J. van (1967) Mother-offspring relationship in free-ranging chimpanzees. In Morris (1967).

Lawick-Goodall, J. van (1968) Behaviour of free-living chimpanzees of the Gombe Stream area. *Anim. Behav. Monogr.*, **3.**

Lawick-Goodall, J. van, and Lawick H. van (1966) Use of tools by the Egyptian vulture, *Neophron percnopterus. Nature*, **212,** 1468–1469.

Lawrence, D. H. (1949) Acquired distinctiveness of cues: I. Transfer between discriminations on the basis of familiarity with the stimulus. *J. exp. Psychol.*, **39,** 770–784.

Lawrence, D. H. (1950) Acquired distinctiveness of cues: II. Selective association in a constant stimulus situation. *J. exp. Psychol.*, **40,** 175–188.

Lawrence, D. H., and Mason, W. A. (1955) Food intake in the rat as a function of deprivation intervals and feeding rhythms. *J. comp. physiol. Psychol.*, **48,** 267–271.

Leblond, C. P., and Nelson, W. O. (1937) Maternal behavior in hypophysectomized male and female mice. *Amer. J. Physiol.*, **120,** 160–172.

Lee-Teng, E. (1967) Retrograde amnesia in relation to current intensity and seizure pattern in chicks. *Proc. 75th Ann. Conv. Amer. Psychol. Assoc.*, 87–88.

Lehr, E. (1967) Experimentelle Untersuchungen an Affen und Halbaffen über Generalisation von Insekten- und Blütenabbildungen. *Z. Tierpsychol.*, **24,** 208–244.

Lehrman, D. S. (1953) A critique of Konrad Lorenz's theory of instinctive behaviour. *Quart. Rev. Biol.*, **28,** 337–363.

Lehrman, D. S. (1955) The physiological basis of parental feeding behaviour in the Ring Dove (*Streptopelia risoria*). *Behaviour*, **7**, 241–286.

Lehrman, D. S. (1961) Gonadal hormones and parental behavior in birds and infra-human mammals. In Young (1961a).

Lehrman, D. S. (1962) Interaction of hormonal and experiential influences on development of behavior. In Bliss (1962).

Lehrman, D. S. (1965) Interaction between internal and external environments in the regulation of the reproductive cycle of the ring dove. In Beach (1965).

Lehrman, D. S., Hinde, R. A., and Shaw, E. (1965) Advances in the Study of Behavior, Vol. 1, Academic Press, New York.

Lehrman, D. S., and Wortis, R. P. (1960) Previous breeding experience and hormone-induced incubation behavior in the ring dove. *Science*, **132**, 1667–1668.

Le Magnen, J. (1952) Les phénomènes olfacto-sexuels chez le rat blanc. *Arch. Sci. Physiol.*, **6**, 295–331.

Le Magnen, J. (1953) L'olfaction. *J. physiol. path. gén.*, **45**, 285–326.

Lemon, R. E. (1967) The response of cardinals to songs of different dialects. *Anim. Behav.*, **15**, 538–545.

Lenneberg, E. H. (1967) Biological Foundations of Language. Wiley, New York.

Lester, D. (1968) Two tests of a fear motivated theory of exploration. *Psychon. Sci.*, **10**, 385–386.

Lester, D., and Cheses, K. T. (1968) Effects of deprivation upon aggression in rats. *Psychol. Rep.*, **22**, 1129–1133.

Lettvin, J. Y., Maturana, H. R., McCulloch, W. S., and Pitts, W. H. (1959) What the frog's eye tells the frog's brain. *Proc. Instit. Radio Engrs.*, **47**, 1940–1951.

Lettvin, J. Y., Maturana, H. R., Pitts, W. H., and McCulloch, W. S. (1961) Two remarks on the visual system of the frog. In Rosenblith (1961)

Leuba, C. (1955) Toward some integration of learning theories: the concept of optimal stimulation. *Psychol. Rep.*, **1**, 27–33.

Levick, W. R. (1967) Receptive fields and trigger features of ganglion cells in the visual streak of the rabbit's retina. *J. Physiol.*, **188**, 285–307.

Levin, H., and Forgays, D. G. (1959) Learning as a function of sensory stimulation of various intensities. *J. comp. physiol. Psychol.*, **52**, 195–201.

Levine, L., Barsel, G. E., and Diakow, C. A. (1965a) Interaction of aggressive and sexual behavior in male mice. *Behaviour*, **25**, 272–280.

Levine, L., Diakow, C. A., and Barsel, G. E. (1965b) Interstrain fighting in male mice. *Anim. Behav.*, **13**, 52–58.

Levine, S. (1953) The role of irrelevant drive stimuli in learning. *J. exp. Psychol.*, **45**, 410–416.

Levine, S. (1957) Infantile experience and resistance to physiological stress. *Science*, **126**, 405.

Levine, S. (1962a) The effects of infantile experience on adult behavior. In "Experimental Foundations of Clinical Psychology", Ed. A. J. Bachrach, Basic Books, New York.

Levine, S. (1962b) "Some effects of stimulation in infancy", Little Club Clinics in Developmental Medicine, No. 7, Heinemann, London.

Levine, S. (1962c) Plasma-free corticosteroid response to electric shock in rats stimulated in infancy. *Science,* **135,** 795–796.

Levine, S. (1962d) The psychophysiological effect of early stimulation. In Bliss (1962).

Levine, S. (1967) Maternal and environmental influences on the adrenocortical response to stress in weanling rats. *Science,* **156,** 258–260.

Levine, S., and Alpert, M. (1959) Differential maturation of the central nervous system as a function of early experience. *Arch. gen. Psychiat.,* **1,** 403–405.

Levine, S., Chevalier, J. A., and Korchin, S. J. (1956) The effects of early shock and handling on later avoidance learning. *J. Pers.,* **24,** 475–493.

Levine, S., and Mullins, R. F. (1966) Hormonal influence on brain organization in infant rats. *Science,* **152,** 1585–1592.

Levison, C. A., Levison, P. K., and Norton, H. P. (1968) Effects of early visual conditions on stimulation-seeking behavior in infant rhesus monkeys. *Psychon. Sci.,* **11,** 101–102.

Levison, P. K., and Flynn, J. P. (1965) The objects attacked by cats during stimulation of the hypothalamus. *Anim. Behav.,* **13,** 217–220.

Lewis, D. J., and Adams, H. E. (1963) Retrograde amnesia from conditioned competing responses. *Science,* **141,** 516–517.

Lewis, D. J., and Maher, B. A. (1966) Neural consolidation and electroconvulsive shock. *Psychol. Rev.,* **72,** 225–239.

Lewis, D. J., Miller, R. R., and Misanin, J. R. (1968) Control of retrograde amnesia. *J. comp. physiol. Psychol.,* **66,** 48–52.

Lewis, M. M. (1963) Language, Thought and Personality. Harrap, London.

Leyhausen, P. (1956) Verhaltensstudien bei Katzen. *Z. Tierpsychol., Beiheft,* **2.**

Leyhausen, P. (1965) Uber die Funktion der relativen Stimmungshierarchie. *Z. Tierpsychol.,* **22,** 412–494.

Liesenfeld, F. J. (1956) Untersuchungen am Netz und über den Erschütterungssinn von *Zygiella x-notata* (Cl.) (Araneidae). *Z. vergl. Physiol.,* **38,** 563–592.

Lincoln, D. W., and Cross, B. A. (1967) Effect of oestrogen on the responsiveness of neurones in the hypothalamus, septum and pre-optic area of rats with light-induced persistent oestrous. *J. Endrocin.,* **37,** 191–203.

Lind, H. (1959) The activation of an instinct caused by a "transitional action". *Behaviour,* **14,** 123–135.

Lind, H. (1961) Studies on the behaviour of the black-tailed godwit (*Limosa limosa* L.). *Naturfredningsradets reservatudvalg nr* 66, *Copenhagen.*

Lind, H. (1962) Zur Analyse des sexuellen Verhaltens der Kolbenente, *Netta rufina* (Pallas). *Z. Tierpsychol.,* **19,** 607–625.

Lindauer, M. (1961) Communication among Social Bees. Harvard Univ. Press.

Lindauer, M. (1964) Allgemeine Sinnesphysiologie: Orientierung im Raum. *Fortschr. der. Zool.*, **16**, 58–140.

Lindenlaub, E. (1965) Ueber das Heimfindevermögen von Säugetieren: II. Versuche an Maüsen. *Z. Tierpsychol.*, **12**, 452–458.

Lindenlaub, E. (1960) Neue Befunde über die Anfangsorientierung von Maüsen. *Z. Tierpsychol.*, **17**, 555–578.

Lindsley, D. B. (1956) Physiological psychology. *Ann. Rev. Psychol.*, **7**, 323–348.

Lindsley, D. B. (1960) Attention, consciousness, sleep and wakefulness. In "Handbook of Physiology", Sec. 1, Vol. III, Ed. J. Field, H. W. Magoun, and V. E. Hall, Amer. Physiol. Soc., Washington.

Lindsley, D. B. (1961) The reticular activating system and perceptual integration. In Sheer (1961).

Lindzey, G., Winston, H. D., and Manosevitz, M. (1963) Early experience, genotype and temperament in *Mus musculus*. *J. comp. physiol. Psychol.*, **56**, 622–629.

Linsenmair, K. E. (1967) Konstruktion und Signalfunktion der Sandpyramide der Reiterkrabbe *Ocypode saratan* Forsk. (Decapoda Brachyura Ocypodidae). *Z. Tierpsychol.*, **24**, 403–456.

Lipsitt, L. P. (1967) Learning in the human infant. In Stevenson *et al.* (1967).

Lipsitt, L. P., and Kaye, H. (1965) Change in neonatal response to optimizing and non-optimizing sucking stimulation. *Psychon. Sci.*, **2**, 221–222.

Lisk, R. D. (1962a) Diencephalic placement of estradiol and sexual receptivity in the female rat. *Amer. J. Physiol.*, **203**, 493–496.

Lisk, R. D. (1962b) Testosterone-sensitive centers in the hypothalamus of the rat. *Acta Endocrin.*, **41**, 195–204.

Lisk, R. D. (1966a) Inhibitory centers in sexual behavior in the male rat. *Science*, **152**, 669–670.

Lisk, R. D. (1966b) Increased sexual behavior in the male rat following lesions in the mamillary region. *J. exp. Zool.*, **161**, 129–136.

Lisk, R. D. (1966c) The hypothalamus and hormone sensing systems. *Exc. Med. Int. Cong. Ser.* **132**, 925–957.

Lisk, R. D. (1966d) Hormonal implants in the central nervous system and behavioral receptivity in the female rat. In "Brain and Behaviour", III, Ed. R. A. Gorski and R. H. Whalen, Univ. Calif. Press.

Lisk, R. D. (1967a) Neural localization for androgen activation of copulatory behavior in the male rat. *Endocrinology*, **80**, 754–761.

Lisk, R. D. (1967b) Neural control of gonad size by hormone feedback in the desert iguana *Dipsosaurus d. dorsalis*. *Gen. comp. Endocrin.*, **8**, 258–266.

Lisk, R. D., and Newlon, M. (1963) Estradiol: evidence for its direct effect on hypothalamic neurons. *Science*, **139**, 223–224.

Lissavoy, V. de (1963) Head banging in early childhood: a suggested cause. *J. genet. Psychol.*, **102**, 109–114.

Lissmann, H. W. (1932) Die Umwelt des Kampffisches (*Betta splendens* Regan). *Z. vergl. Physiol.*, **18,** 65–111.

Lissmann, H. W. (1946a) The neurological basis of the locomotory rhythm in the spinal dogfish (*Scyllium canicula, Acanthias vulgaris*): I. Reflex behaviour. *J. exp. Biol.*, **23,** 143–161.

Lissmann, H. W. (1946b) The neurological basis of the locomotory rhythm in the spinal dogfish (*Scyllium canicula, Acanthias vulgaris*): II. The effect of de-afferentation. *J. exp. Biol.*, **23,** 162–176.

Lissmann, H. W. (1958) On the function and evolution of electric organs in fish. *J. exp. Biol.*, **34,** 156–191.

Lissmann, H. W., and Machin, K. E. (1958) The mechanism of object location in *Gymnarchus niloticus* and similar fish. *J. exp. Biol.*, **35,** 451–486.

Littlejohn, M. J., and Michaud, T. J. (1959) Mating call discrimination by females of Strecker's chorus frog (*Pseudacris streckeri*). *Texas J. Sci.*, **11,** 86–92.

Livingston, R. B. (1959) Central control of receptors and sensory transmission systems. In "Handbook of Physiology", Sec. 1, Vol. 1, Ed. J. Field, H. W. Magoun, and V. E. Hall, Amer. Physiol. Soc., Washington.

Livingston, W. K., Haugen, F. P., and Brookhart, J. M. (1954) Functional organization of the central nervous system. *Neurology*, **4,** 485–496.

Lockard, R. B. (1963a) Self-regulated exposure to light by albino rats as a function of rearing luminance and test luminance. *J. comp. physiol. Psychol.*, **56,** 558–564.

Lockard, R. B. (1963b) Some effects of light upon the behavior of rodents. *Psychol. Bull.*, **60,** 509–529.

Lockard, R. B. (1966) Several tests of stimulus-change and preference theory in relation to light-controlled behavior of rats. *J. comp. physiol. Psychol.*, **62,** 415–426.

Loewenstein, W. R. (1956) Modulation of cutaneous mechanoreceptors by sympathetic stimulation. *J. Physiol.*, **132,** 40–60.

Loher, W., and Huber, F. (1966) Nervous and endocrine control of sexual behaviour in a grasshopper (*Gomptocerus rufus*, L. Acridinae). *Symp. Soc. exp. Biol.*, **20,** 381–400.

Löhrl, H. (1959) Zur Frage des Zeitpunktes einer Prägung auf die Heimatregion beim Halsbandschnäpper (*Ficedula albicollis*). *J. Ornith.*, **100,** 132–140.

Loizos, C. (1966) Play in mammals. *Symp. Zool. Soc. Lond.*, **18,** 1–10.

Loizos, C. (1967) Play behaviour in higher primates: a review. In Morris (1967).

London, I. D. (1954) Research on sensory interaction in the Soviet Union. *Psych. Bull.*, **51,** 531–568.

Lorber, S. H., Komarov, S. A., and Shay, H. (1950) Effect of sham feeding on gastric motor activity of the dog. *Amer. J. Physiol.*, **162,** 447–451.

Lorenz, K. (1935) Der Kumpan in der Umwelt des Vogels. *J. f. Ornith.*, **83,** 137–213, 289–413.

Lorenz, K. (1937) Uber die Bildung des Instinktbegriffes. *Naturwiss.*, **25**, 289–300, 307–318, 324–331.

Lorenz, K. (1939) Vergleichende Verhaltensforschung. *Zoo. Anz. Suppl. Bd.*, **12**, 69–102.

Lorenz, K. (1941) Vergleichende Bewegungsstudien an Anatinen. *Suppl. J. Ornith.*, **89**, 194–294.

Lorenz, K. (1950) The comparative method in studying innate behaviour patterns. *Sym. Soc. exp. Biol.*, **4**, 221–268.

Lorenz, K. (1951) Uber die Entstehung auslösender "Zeremonien". *Vogelwarte*, **16**, 9–13.

Lorenz, K. (1956) Play and vacuum activity in animals. In "Symposium. L'Instinct dans le comportement des animaux et de l'homme", 633–645, Masson, Paris.

Lorenz, K. (1958) The evolution of behavior. *Sci. Amer.*, **199**, 67–78.

Lorenz, K. (1960) Prinzipien der vergleichenden Verhaltensforschung. *Fortschr. Zool.*, **12**, 265–294.

Lorenz, K. (1961) Phylogenetische Anpassung und adaptive Modifikation des Verhaltens. *Z. Tierpsychol.*, **18**, 139–187.

Lorenz, K. (1965) Evolution and Modification of Behavior. Univ. of Chicago.

Lorenz, K. (1966) On Aggression. Methuen, London.

Lorenz, K., and Tinbergen, N. (1939) Taxis und Instinkthandlung in der Eirollbewegung der Graugans: I. *Z. Tierpsychol.*, **2**, 1–29.

Lovejoy, E. (1966) Analysis of the overlearning reversal effect. *Psychol. Rev.*, **73**, 87–103.

Löwenstein, O., and Sand, A. (1940) The mechanism of the semicircular canal: a study of the responses of single fibre preparations to angular accelerations and to rotation at constant speed. *Proc. Roy. Soc. B*, **129**, 256–275.

Ludwig, J. (1965) Beobachtungen über das Spiel bei Boxern. *Z. Tierpsychol.*, **22**, 813–838.

Lukaszewska, I. (1968) Some further failures to find the visual overlearning-reversal effect in rats. *J. comp. physiol. Psychol.*, **65**, 359–361.

McBride, G. (1963) The "teat order" and communication in young pigs. *Anim. Behav.*, **11**, 53–56.

McCall, R. B. (1965) Stimulus change in light-contingent bar-pressing. *J. comp. physiol. Psychol.*, **59**, 258–262.

McCall, R. B. (1966) Initial-consequent change surface in light-contingent bar-pressing. *J. comp. physiol. Psychol.*, **62**, 35–42.

McCleary, R. A. (1953) Taste and post-ingestion factors in specific-hunger behavior. *J. comp. physiol. Psychol.*, **46**, 411–421.

McClelland, D. C., and Atkinson, J. W. (1948) The projective expression of needs: I. *J. Psychol.*, **25**, 205–222.

MacCorquodale, K., and Meehl, P. E. (1954) Edward C. Tolman. In "Modern Learning Theory", Ed. Estes *et al.*, Appleton-Century-Crofts, New York.

MacDonald, G. E. (1968) Imprinting: drug produced isolation and the sensitive period. *Nature,* **217,** 1158–1159.

MacDonald, G. E., and Solandt, A. (1966) Imprinting: effects of drug induced immobilization. *Psychon., Sci.,* **5,** 95–96.

MacDonnell, M. F., and Flynn, J. P. (1964) Attack elicited by stimulation of the thalamus of cats. *Science,* **144,** 1249–1250.

MacDonnell, M. F., and Flynn, J. P. (1966a) Sensory control of hypothalamic attack. *Anim. Behav.,* **14,** 399–405.

MacDonnell, M. F., and Flynn, J. P. (1966b) Control of sensory fields by stimulation of hypothalamus. *Science,* **152,** 1406–1408.

MacDonnell, M. F., and Flynn, J. P. (1968) Attack elicited by stimulation of thalamus and adjacent structures of cats. *Behaviour,* **31,** 185–202.

McDougall, W. (1923) An Outline of Psychology. Methuen, London.

McFarland, D. J. (1965a) The effect of hunger on thirst motivated behaviour in the Barbary dove. *Anim. Behav.,* **13,** 286–292.

McFarland, D. J. (1965b) Control theory applied to the control of drinking in the Barbary dove. *Anim. Behav.,* **13,** 478–492.

McFarland, D. J. (1965c) Hunger, thirst and displacement pecking in the Barbary dove. *Anim. Behav.,* **13,** 293–300.

McFarland, D. J. (1966) The role of attention in the disinhibition of displacement activities. *Q. J. exp. Psychol.,* **18,** 19–30.

McFarland, D. J., and L'Angellier, A. B. (1966) Disinhibition of drinking during satiation of feeding behaviour in the Barbary dove. *Anim. Behav.,* **14,** 463–467.

McFarland, J. H., Werner, H., and Wapner, S. (1962) The effect of postural factors on the distribution of tactual sensitivity and the organization of tactual kinaesthetic space. *J. exp. Psychol.,* **63,** 148–154.

McGaugh, J. L. (1965) Facilitation and impairment of memory storage processes. In "The Anatomy of Memory", Vol. I, Ed. D. P. Kimble, Science and Behavior Books, Palo Alto.

McGill, T. E. (1962a) Reduction in "head-mounts" in the sexual behavior of the mouse as a function of experience. *Psych. Rep.,* **10,** 284.

McGill, T. E. (1962b) Sexual behavior in three inbred strains of mice. *Behaviour,* **19,** 341–350.

McGill, T. E. (1965) Studies of the sexual behavior of male laboratory mice: effects of genotype, recovery of sex drive, and theory. In Beach (1965).

McGill, T. E., and Tucker, G. R. (1964) Genotype and sex drive in intact and in castrated male mice. *Science,* **145,** 514–515.

McGinty, D., Epstein, A. N., and Teitelbaum, P. (1965) The contribution of oropharyngeal sensations to hypothalamic hyperphagia. *Anim. Behav.,* **13,** 413–418.

Machin, K. E. (1964) Feedback theory and its application to biological systems. *Symp. Soc. exp. Biol.,* **18,** 421–446.

MacKay, D. M. (1962) In "Aspects of the Theory of Artificial Intelligence", Ed. C. A. Muses, Plenum Press, London.

McKay, J. (1968) Some aspects of the physiology of the Tettigonid ear. Ph.D. thesis, Makerere, Univ. of E. Africa.

McKelvey, R. K., and Marx, M. H. (1951) Effects of infantile food and water deprivation on adult hoarding in the rat. *J. comp. physiol. Psychol.*, **44**, 423–430.

McKinney, F. (1961) An analysis of the displays of the European eider *Somateria mollissima mollissima* (Linnaeus) and the Pacific eider *Somateria mollissima v. nigra* Bonaparte. *Behaviour Suppl.* **7**.

McKinney, F. (1965) The comfort movements of Anatidae. *Behaviour*, **25**, 120–220.

Mackintosh, J. H., and Grant, E. C. (1966) The effect of olfactory stimuli on the agonistic behaviour of laboratory mice. *Z. Tierpsychol.*, **23**, 584–587.

Mackintosh, J., and Sutherland, N. S. (1963) Visual discrimination by the goldfish: the orientation of rectangles. *Anim. Behav.*, **11**, 135–141.

Mackintosh, N. J. (1963) Extinction of a discrimination habit as a function of over-training. *J. comp. physiol. Psychol.*, **56**, 842–847.

Mackintosh, N. J. (1964a) Overtraining and transfer within and between dimensions in the rat. *Q. J. exp. Psychol.*, **16**, 250–256.

Mackintosh, N. J. (1964b) Discrimination learning in the Octopus. *Anim. Behav. Suppl.* **1**, 129–134.

Mackintosh, N. J. (1965a) Selective attention in animal discrimination learning. *Psychol. Bull.*, **64**, 124–150.

Mackintosh, N. J. (1965b) Overtraining, reversal and extinction in rats and chicks. *J. comp. physiol. Psychol.*, **59**, 31–36.

Mackintosh, N. J. (1965c) Overtraining, transfer to proprioceptive control and position reversal. *Q. J. exp. Psychol.*, **17**, 26–36.

Mackintosh, N. J. (1969) Further analysis of the overtraining reversal effect. *J. comp. physiol. Psychol.*, **67**, Monograph 1–18.

Mackintosh, N. J., and Mackintosh, J. (1963) Reversal learning in *Octopus vulgaris* Lamarck, with and without irrelevant cues. *Q. J. exp. Psychol.*, **15**, 236–242.

Mackintosh, N. J., and Mackintosh, J. (1964) Performance of *Octopus* over a series of reversals of a simultaneous discrimination. *Anim. Behav.*, **12**, 321–324.

Mackintosh, N. J., Mackintosh, J., Safriel-Jorne, O., and Sutherland, N. S. (1966) Overtraining, reversal and extinction in the goldfish. *Anim. Behav.*, **14**, 314–318.

Mackintosh, N. J., Mackintosh, J., and Sutherland, N. S. (1963) The relative importance of horizontal and vertical extents in shape discrimination by *Octopus. Anim. Behav.*, **11**, 355–358.

Mackworth, J. F. (1962) The visual image and the memory trace. *Canad. J. Psychol.*, **16**, 55–59.

McMichael, R. E. (1961) The effects of preweaning shock and gentling on later resistance to stress. *J. comp. physiol. Psychol.*, **54**, 416–421.

McMichael, R. E. (1966) Early-experience effects as a function of infant treatment and other experimental conditions. *J. comp. physiol. Psychol.*, **62**, 433–436.

McNiven, M. A. (1960) "Social-releaser mechanisms" in birds—a controlled replication of Tinbergen's study. *Psychol. Rec.*, **10**, 259–265.

MacPhail, E. M. (1967) Positive and negative reinforcement from intracranial stimulation in pigeons. *Nature*, **213**, 947–948.

Magni, F. (1967) Central and peripheral mechanisms in the modulation of flashing in the firefly *Luciola italica*, L. *Arch. ital. Biol.*, **105**, 339–360.

Magnus, D. (1955) Zum Problem der überoptimalen Schlüsselreize. (Versuche am Kaisermantel *Argynnis paphia*.) *Verh. Dtsch. zool. Ges. Tubingen*, 1954 (*Zool. Anz.*) *Suppl.* **18**, 317–325.

Magnus, D. (1958) Experimentelle Untersuchungen zur Bionomie und Ethologie des Kaisermantels *Argynnis paphia* L. (Lep. Nymph): I Uber optische Auslöser von Anfliegereaktionen und ihre Bedeutung für das Sichfinden der Geschlechter. *Z. Tierpsychol.*, **15**, 397–426.

Maier, N. R. F., and Schneirla, T. C. (1935) Principles of Animal Psychology, McGraw-Hill, New York.

Mainardi, D. (1965) Some observations on the interaction between inverted sexual behaviour and social status in female mice. *Atti della Società Italiana di Scienze Naturali e del Museo Civico di Storia Naturale di Milano*, **104**, 238–242.

Mainardi, D., Marsan, M., and Pasquali, A. (1965) Causation of sexual preferences of the house mouse. *Atti della Società Italiana di Scienze Naturali e del Museo Civico di Storia Naturale di Milano*, **104**, 325–338.

Majendie, M. (1960) The para-visual detector. *J. Inst. Navigation*, **13**, 4.

Maldonado, H. (1964) The control of attack by octopus. *Z. vergl. Physiol.*, **47**, 656.

Malmo, R. B. (1959) Activation: a neurophysiological dimension. *Psychol. Rev.*, **66**, 367–386.

Malmo, R. B., Shagass, C., and Davis, J. F. (1951) Electromyographic studies of muscular tension in psychiatric patients under stress. *J. clin. exp. Psychopath.*, **12**, 45–66.

Malott, M. K. (1968) Stimulus control in stimulus deprived chickens. *J. comp. physiol. Psychol.*, **66**, 276–282.

Mandler, J. M. (1958) Effect of early food deprivation on adult behavior in the rat. *J. comp. physiol. Psychol.*, **51**, 513–517.

Manning, A. (1956) The effect of honey-guides. *Behaviour*, **9**, 114–139.

Manning, A. (1959) The sexual behavior of two sibling *Drosophila* species. *Behaviour*, **15**, 123–145.

Manning, A. (1965) *Drosophila* and the evolution of behaviour. In "Viewpoints in Biology, 4", Ed. J. D. Carthy and C. L. Duddington, Butterworths, London.

Manning, A. (1967a) "Pre-imaginal conditioning" in *Drosophila*. *Nature*, **216**, 338–340.

Manning, A. (1967b) The control of female receptivity in female *Drosophila*. *Anim. Behav.*, **15**, 239–250.

Manton, S. M. (1953) Locomotory habits and the evolution of the larger arthropodan groups. *Symp. soc. exp. Biol.*, **7**, 339–376.

Mark, R. F., Campenhausen, G. von, and Lischinsky, D. J. (1967) Nerve-muscle relations in the salamander: possible relation to nerve regeneration and muscle specificity. *Exp. Neurol.*, **16**, 438–449.

Marler, P. (1955) Studies of fighting in chaffinches: (2) The effect on dominance relations of disguising females as males. *Brit. J. Anim. Behav.*, **3**, 137–146.

Marler, P. (1956a) The voice of the chaffinch and its function as a language. *Ibis*, **98**, 231–261.

Marler, P. (1956b) Behaviour of the chaffinch (*Fringilla coelebs*). *Behaviour Suppl.* **5**, 1–184.

Marler, P. (1956c) Studies of fighting in chaffinches: (3) Proximity as a cause of aggression. *Brit. J. Anim. Behav.*, **4**, 23–30.

Marler, P. (1957a) Studies of fighting in chaffinches: (4) Appetitive and consummatory behaviour. *Brit. J. Anim. Behav.*, **5**, 29–37.

Marler, P. (1957b) Specific distinctiveness in the communication signals of birds. *Behaviour*, **11**, 13–39.

Marler, P. (1959) Developments in the study of animal communication. In "Darwin's Biological Work", Ed. P. R. Bell, Cambridge Univ. Press.

Marler, P. (1960) Bird songs and mate selection. In "Animal Sounds and Communication", *Amer. Inst. Biol. Sci.*, **7**, 348–367.

Marler, P. (1961) The filtering of external stimuli during instinctive behaviour. In Thorpe and Zangwill (1961).

Marler, P. (1965) Communication in monkeys and apes. In De Vore (1965).

Marler, P. (1967) Comparative study of song development in sparrows. *Proc. 14th Int. Ornith. Cong. Oxford*, 231–244, Blackwell, Oxford.

Marler, P., and Hamilton, W. J. (1966) Mechanisms of Animal Behavior. Wiley, New York.

Marler, P., Kreith, M., and Tamura, M. (1962) Song development in hand-raised Oregon juncos. *Auk*, **79**, 12–30.

Marler, P., and Tamura, M. (1964) Culturally transmitted patterns of vocal behavior in sparrows. *Science*, **146**, 1483–1486.

Marley, E., and Morse, W. H. (1966) Operant conditioning in the newly hatched chicken. *J. exp. Anal. Behav.*, **9**, 95–104.

Marrone, R. L. (1965) Effects of food deprivation on the fighting response of *Betta splendens*. *Psychol. Rep.*, **17**, 632.

Marsh, G., and Walz, G. (1967) Selective attention in imprinting. *Psychonomic Sci.*, **9**, 373–374.

Marshall, S. L. A. (1947) Men against Fire. Morrow, New York.

Marx, M. H. (1952) Infantile deprivation and adult behavior in the rat: retention of increased rate of eating. *J. comp. physiol. Psychol.*, **45**, 43–49.

Marx, M. H., Henderson, R. L., and Roberts, C. L. (1955) Positive reinforcement of the bar-pressing response by a light stimulus following dark operant pretests with no aftereffect. *J. comp. physiol. Psychol.*, **48**, 73–76.

Marx, M. H., and Knarr, F. A. (1963) Long-term development of reinforcing properties of a stimulus as a function of temporal relationship to food reinforcement. *J. comp. physiol. Psychol.*, **56**, 546–550.

Marzocco, F. N. (1951) Doctoral dissertation, Iowa, cited by Brown (1961).

Mason, W. A. (1965a) Determinants of social behavior in young chimpanzees. In Schrier *et al.* (1965).

Mason, W. A. (1965b) The social behavior of monkeys and apes. In De Vore (1965).

Mason, W. A. (1967) Motivational aspects of social responsiveness in young chimpanzees. In Stevenson *et al.* (1967).

Mason, W. A., and Berkson, G. (1962) Conditions influencing vocal responsiveness of infant chimpanzees. *Science*, **137**, 127–128.

Mason, W. A., and Harlow, H. F. (1958) Formation of conditioned responses in infant monkeys. *J. comp. physiol. Psychol.*, **51**, 68–70.

Mason, W. A., and Sponholz, R. R. (1963) Behavior of rhesus monkeys raised in isolation. *J. Psychiatric Res.*, **1**, 299–306.

Mast, S. O. (1911) Light and the Behavior of Organisms. Wiley, London.

Mast, S. O. (1912) Behavior of fireflies, with special reference to the problem of orientation. *J. anim. Behav.*, **2**, 256–272.

Masters, W. H., and Johnson, V. E. (1965) The sexual response cycles of the human male and female: comparative anatomy and physiology. In Beach (1965).

Matthews, B. H. C. (1931) The response of a single end organ. *J. Physiol.*, **71**, 64–110.

Matthews, G. V. T. (1961) "Nonsense" orientation in Mallard *Anas platyrhynchos* and its relation to experiments on bird navigation. *Ibis*, **103**a, 211–230.

Matthews, G. V. T. (1968) Bird Navigation. Cambridge Univ. Press.

Matthews, P. B. C. (1964) Muscle spindles and their motor control. *Physiol. Rev.*, **44**, 219–288.

Matthews, W. A. (1964) Shape discrimination in tropical fish. *Anim. Behav.*, **12**, 111–115.

Matthews, W. A., and Hemmings, G. (1963) A theory concerning imprinting. *Nature*, **198**, 1183–1184.

Maturana, H. R. (1964) Functional organization of the pigeon retina. In "Information Processing in the Nervous System", Ed. R. W. Gerard, Excerpta Medica Foundation, Amsterdam.

Maturana, H. R., and Frank, S. (1963) Directional movement and horizontal edge detectors in the pigeon retina. *Science*, **142**, 977–979.

May, J. G., and Dorr, D. (1968) Initiative pecking in chicks as a function of early social experience. *Psychon. Sci.*, **11**, 175–176.

Maynard, D. M. (1955) Activity in a crustacean ganglion: II. Pattern and interaction in burst formation. *Biol. Bull.*, **109**, 420–436.

Mayr, E. (1942) Systematics and the Origin of Species. Columbia Univ. Press, New York.

Mayr, E. (1963) Animal Species and Evolution. Harvard Univ. Press, Cambridge, Mass.

Mayr, E., Andrew, R. J., and Hinde, R. A. (1956) Die systematische Stellung der Gattung *Fringilla*. *J. f. Ornith.*, **97**, 258–273.

Meier, G. W. (1961) Infantile handling and development in Siamese kittens. *J. comp. physiol. Psychol.*, **54**, 284–286.

Meier, G. W. (1965) Other data on the effects of social isolation during rearing upon adult reproductive behaviour in the rhesus monkey (*Macaca mulatta*). *Anim. Behav.*, **13**, 228–231.

Meier, G. W., and McGee, R. K. (1959) A re-evaluation of the effect of early perceptual experience on discrimination performance during adulthood. *J. comp. physiol. Psychol.*, **52**, 390–395.

Mello, N. K., and Peterson, N. J. (1964) Behavioral evidence for color discrimination in cat. *J. Neurophysiol.*, **27**, 323–333.

Melton, A. W. (1963) Implications of short-term memory for a general theory of memory. *J. Verbal Learning and Verbal Behav.*, **2**, 1–21.

Melzack, R. (1952) Irrational fears in the dog. *Canad. J. Psychol.*, **6**, 141–147.

Melzack, R. (1961) On the survival of mallard ducks after "habituation" to the hawk-shaped figure. *Behaviour*, **17**, 9–16.

Melzack, R. (1965) Effects of early experience on behavior: experimental and conceptual considerations. In "Psychopathology of Perception", Grune and Stratton, New York.

Melzack, R., Penick, E., and Beckett, A. (1959). The problem of "innate fear" of the hawk shape: an experimental study with mallard ducks. *J. comp. physiol. Psychol.*, **52**, 694–698.

Melzack, R., and Scott, T. H. (1957) The effects of early experience on the response to pain. *J. comp. physiol. Psychol.*, **50**, 155–161.

Mendelson, J. (1966) Role of hunger in T-maze learning for food by rats. *J. comp. physiol. Psychol.*, **62**, 341–349.

Meng, M. (1957) Untersuchungen zum Farben- und Formensehen der Erdkröte (*Bufo bufo* L.). *Zool. Beitr.*, **3**, 313–364.

Menzel, E. W. (1963) The effects of cumulative experience on responses to novel objects in young isolation-reared chimpanzees. *Behaviour*, **21**, 1–12.

Menzel, E. W. (1964a) Responsiveness to object-movement in young chimpanzees. *Behaviour*, **24**, 147–160.

Menzel, E. W. (1964b) Patterns of responsiveness in chimpanzees reared through infancy under conditions of environmental restriction. *Psychol. Forsch.*, **27**, 337–365.

Menzel, E. W., Davenport, R. K., and Rogers, C. M. (1963) The effects of environmental restriction upon the chimpanzee's responsiveness to objects. *J. comp. physiol. Psychol.*, **56,** 78–85.

Merton, P. A. (1964) Human position sense and sense of effort. *Symp. soc. exp. Biol.*, **18,** 387–400.

Messenger, J. B. (1968) The visual attack of the cuttlefish, *Sepia officinalis. Anim. Behav.*, **16,** 342–357.

Messmer, E., and Messmer, I. (1956) Die Entwicklung der Lautäusserungen und einiger Verhaltensweisen der Amsel (*Turdus merula merula* L.) unter natürlichen Bedingungen und nach Einzelaufzucht in schalldichten Räumen. *Z. Tierpsychol.*, **13,** 341–441.

Meyer, D. R. (1952) The stability of human gustatory sensitivity during changes in time of food deprivation. *J. comp. physiol. Psychol.*, **45,** 373–376.

Meyer, M. E. (1964) Discriminative basis for astronavigation in birds. *J. comp. physiol. Psychol.*, **58,** 403–406.

Meyer, M. E. (1966) The internal clock hypothesis for astronavigation in homing pigeons. *Psychon. Sci.*, **5,** 259–260.

Meyers, B. (1964) Discrimination of visual movement in perceptually deprived cats. *J. comp. physiol. Psychol.*, **57,** 152–153.

Meyers, B., and McCleary, R. (1964) Interocular transfer of a pattern discrimination in pattern deprived cats. *J. comp. physiol. Psychol.*, **57,** 16–21.

Meyers, W. J. (1962) Critical period for the facilitation of exploratory behavior by infantile experience. *J. comp. physiol. Psychol.*, **55,** 1099–1101.

Michael, C. R. (1968) Receptive fields of single optic nerve fibers in a mammal with an all-cone retina. II. *J. Neurophysiology*, **31,** 257–267.

Michael, R. P. (1961a) An investigation of the sensitivity of circumscribed neurological areas to hormonal stimulation. In "Regional Neurochemistry", Ed. S. S. Katy and J. Elkes, 465–480, Pergamon Press, Oxford.

Michael, R. P. (1961b) Observations upon the sexual behaviour of the domestic cat (*Felis catus* L.) under laboratory conditions. *Behaviour*, **18,** 1–21.

Michael, R. P., and Herbert, J. (1963) Menstrual cycle influences grooming behavior and sexual activity in the rhesus monkey. *Science*, **140,** 500–501.

Michael, R. P., and Keverne, E. B. (1968) Pheromones in the communication of sexual status in primates. *Nature*, **218,** 746–749.

Michael, R. P., and Saayman, G. S. (1967) Sexual performance index of male rhesus monkeys. *Nature*, **214,** 425.

Michael, R. P., and Saayman, G. S. (1968) Differential effects on behaviour of the subcutaneous and intravaginal administration of oestrogen in the rhesus monkey (*Macaca mulatta*). *J. Endocrin.*, **41,** 231–246.

Michael, R. P., Saayman, G. S., and Zumpe, D. (1968) The suppression of mounting behaviour and ejaculation in male rhesus monkeys (*Macaca mulatta*) by administration of progesterone to their female partners. *J. Endocrin.*, **41,** 421–431.

Michels, K. M., Bevan, W., and Strasel, H. C. (1958) Discrimination learning and interdimensional transfer under conditions of systematically controlled visual experience. *J. comp. physiol. Psychol.*, **51**, 778–781.

Michels, K. M., Pittman, G. G., Hitchcock, L., and Brown, D. R. (1962) Visual discrimination: tree squirrels and quantified stimulus dimensions. *Percept. Mot. Skills*, **15**, 443–450.

Michelsen, A. (1966) Pitch discrimination in the locust ear: observations on single sense cells. *J. insect Physiol.*, **12**, 1119–1131.

Michelsen, A. (1968) Frequency discrimination in the locust ear by means of four groups of receptor cells. *Nature*, **220**, 585–586.

Michener, M. C., and Walcott, C. (1967) Homing of single pigeons—analysis of tracks. *J. exp. Biol.*, **47**, 99–131.

Middlebush, C. W., Mountjoy, P. T., and Riley, V. (1968) Unconditioned aggression in the Japanese quail *Coturnix coturnix japonica*. *Psychon. Sci.*, **11**, 113–114.

Miles, R. C., and Meyer, D. R. (1956) Learning sets in marmosets. *J. comp. physiol. Psychol.*, **49**, 219–222.

Mill, P. J., and Hughes, G. M. (1966) The nervous control of ventilation in dragonfly larvae. *J. exp. Biol.*, **44**, 297–316.

Millenson, J. R. (1962) Acquired counting behaviour in mice maintained under two reinforced procedures. *Anim. Behav.*, **10**, 171–173.

Miller, A. J. (1968) Variations in retrograde amnesia with parameters of electroconvulsive shock and time of testing. *J. comp. physiol. Psychol.*, **66**, 40–47.

Miller, G. A. (1964) Kenneth Craik lecture delivered in Cambridge.

Miller, G. A., Galanter, E., and Pribram, K. H. (1960) Plans and the Structure of Behavior. Holt, Rinehart and Winston, New York.

Miller, J. D., and Murray, F. S. (1966) Guinea pig's immobility response to sound. *J. comp. physiol. Psychol.*, **61**, 227–233.

Miller, N .E. (1944) Experimental studies of conflict. In "Personality and Behavior Disorders", Ed. J. McV. Hunt, Ronald Press, New York.

Miller, N. E. (1948a) Theory and experiment relating psychoanalytic displacement to stimulus-response generalization. *J. abnorm. soc. Psychol.*, **43**, 155–178.

Miller, N. E. (1948b) Studies of fear as an acquirable drive: I. Fear as motivation and fear-reduction as reinforcement in the learning of new responses. *J. exp. Psychol.*, **38**, 89–101.

Miller, N. E. (1956) Effects of drugs on motivation: the value of using a variety of measures. *Ann. N.Y. Acad. Sci.*, **65**, 318–333.

Miller, N. E. (1957) Experiments on motivation. *Science*, **126**, 1271–1278.

Miller, N. E. (1959) Liberalization of basic S-R concepts. In Koch (1959).

Miller, N. E. (1960) Motivational effects of brain stimulation and drugs. *Fed. Proc.*, **19**, 846–854.

Miller, N. E. (1961a) Learning and performance motivated by direct stimulation of the brain. In Sheer (1961).

Miller, N. E. (1961b) Implications for theories of reinforcement. In Sheer (1961).

Miller, N. E. (1961c) Integration of neurophysiological and behavioural research. *Ann. N.Y. Acad. Sci.*, **92,** 830–839.

Miller, N. E. (1969) Psychosomatic effects of specific types of training. In "Experimental Approaches to the Study of Emotional Behaviour", Ed. E. Tobach, *Ann. N.Y. Acad. Sci.* (in press).

Miller, N. E., and Carmona, A. (1967) Modification of a visceral response, salivation in thirsty dogs, by instrumental training with water reward. *J. comp. physiol. Psychol.*, **63,** 1–6.

Miller, N. E., and DiCara, L. (1967) Instrumental learning of heart rate changes in curarized rats. *J. comp. physiol. Psychol.*, **63,** 12–19.

Miller, N. E., and Dollard, J. (1941) Social Learning and Imitation. Yale Univ. Press, New Haven.

Miller, N. E., and Kessen, M. L. (1952) Reward effects of food via stomach fistula compared with those of food via mouth. *J. comp. physiol. Psychol.*, **45,** 555–564.

Miller, N. E., Sampliner, R. I., and Woodrow, P. (1957) Thirst-reducing effects of water by stomach fistula vs. water by mouth measured by both a consummatory and an instrumental response. *J. comp. physiol. Psychol.*, **50,** 1–5.

Miller, N. E., and Stevenson, S. S. (1936) Agitated behavior of rats during experimental extinction and a curve of spontaneous recovery. *J. comp. Psychol.*, **21,** 205–231.

Miller, P. L. (1960) Respiration in the desert locust. *J. exp. Biol.*, **37,** 224–278.

Miller, P. L. (1964) The central nervous control of respiratory movements. *Proc. XIIth Int. Cong. Entomol. Lond.*, 1964, 198.

Mittelstaedt, H. (1949) Telotaxis und Optomotorik von *Eristalis* bei Augeninversion. *Naturwiss.*, **36,** 90–91.

Mittelstaedt, H. (1960) The analysis of behavior in terms of control systems. *5th Cong. on Group Processes*, Macy Foundation, New York.

Mittelstaedt, H. (1961) Die Regelungstheorie als methodisches Werkzeug der Verhaltensanalyse. *Naturwiss.*, **48,** 246–254.

Mittelstaedt, H. (1962) Control systems of orientation in insects. *Ann. Rev. Entomol.*, **7,** 177–198.

Mittelstaedt, H. (1963) Bikomponenten—Theorie der Orientierung. *Ergeb. Biol.*, **26,** 253.

Mittelstaedt, H. (1964) Basic control patterns of orientational homeostasis. *Symp. soc. exp. Biol.*, **18,** 365–386.

Moll, R. P. (1964) Drive and maturation effects in the development of consummatory behavior. *Psychol. Rep.*, **15,** 295–302.

Moltz, H. (1960) Imprinting: empirical basis and theoretical significance. *Psychol. Bull.*, **57,** 291–314.

Moltz, H. (1963) Imprinting: an epigenetic approach. *Psychol. Rev.*, **70,** 123–138.

Moltz, H. (1965) Contemporary instinct theory and the fixed action pattern. *Psychol. Rev.*, **72,** 27–47.

Moltz, H., Geller, D., and Levin, R. (1967) Maternal behavior in the totally mammectomized rat. *J. comp. physiol. Psychol.*, **64**, 225–229.

Moltz, H., and Stettner, L. J. (1961) The influence of patterned-light deprivation on the critical period for imprinting. *J. comp. physiol. Psychol.*, **54**, 279–283.

Moltz, H., and Stettner, L. J. (1962) Interocular mediation of the following response after patterned-light deprivation. *J. comp. physiol. Psychol.*, **55**, 626–632.

Monteith, L. G. (1962) Apparent continual changes in the host preferences of *Drino bohemica* Mesn (Diptera: Tachinidae), and their relation to the concept of host-conditioning. *Anim. Behav.*, **10**, 292–299.

Montgomery, A. V., and Holmes, J. H. (1955) Gastric inhibition of the drinking response. *Amer. J. Physiol.*, **182**, 227–231.

Montgomery, K. C. (1952) A test of two explanations of spontaneous alternation. *J. comp. physiol. Psychol.*, **45**, 287–293.

Montgomery, K. C. (1954) The role of the exploratory drive in learning. *J. comp. physiol. Psychol.*, **47**, 60–64.

Montgomery, K. C. (1955) The relation between fear induced by novel stimulation and exploratory behavior. *J. comp. physiol. Psychol.*, **48**, 254–260.

Montgomery, K. C., and Monkman, J. A. (1955) The relation between fear and exploratory behaviour. *J. comp. physiol. Psychol.*, **48**, 132–136.

Montgomery, K. C., and Segall, M. (1955) Discrimination learning based upon the exploratory drive. *J. comp. physiol. Psychol.*, **48**, 225–228.

Mook, D. G. (1963) Oral and postingestional determinants of the intake of various solutions in rats with esophageal fistulas. *J. comp. physiol. Psychol.*, **56**, 645–659.

Mook, D. G., and Blass, E. M. (1968) Quinine-aversion thresholds and "finickiness" in hyperphagic rats. *J. comp. physiol. Psychol.*, **65**, 202–207.

Mook, J. H., Mook, L. J., and Heikens, H. S. (1960) Further evidence for the role of "searching images" in the hunting behaviour of titmice. *Arch. Néerl. Zool.*, **13**, 448–465.

Moore, N. W. (1952) On the so-called "territories" of dragonflies (*Odonata-Anisoptera*). *Behaviour*, **4**, 85–100.

Moray, N., and Jordan, A. (1967) Vision in chicks with distorted visual fields. *Psychon. Sci.*, **9**, 303.

Morgan, C. T. (1943) Physiological Psychology. McGraw-Hill, New York.

Morgan, C. T. (1959) Physiological theory of drive. In Koch (1959).

Morgan, C. T., and Morgan, J. D. (1940a, b) Studies in hunger: I and II. *J. genet. Psychol.*, **56**, 137–147; **57**, 153–163.

Morgane, P. J. (1961) Medial forebrain bundle and "feeding centers" of the hypothalamus. *J. comp. Neurol.*, **117**, 1–26.

Morris, D. (1954a) The reproductive behaviour of the zebra finch (*Poephila guttata*), with special reference to pseudofemale behaviour and displacement activities. *Behaviour*, **6**, 271–322.

Morris, D. (1954b) The reproductive behaviour of the river bullhead (*Cottus gobi* L.) with special reference to the fanning activity. *Behaviour*, **7**, 1–32.

Morris, D. (1955) The causation of pseudofemale and pseudomale behaviour: a further comment. *Behaviour*, **8**, 46–56.

Morris, D. (1956) The feather postures of birds and the problem of the origin of social signals. *Behaviour*, **9**, 75–113.

Morris, D. (1957a) "Typical intensity" and its relation to the problem of ritualisation. *Behaviour*, **11**, 1–12.

Morris, D. (1957b) The reproductive behaviour of the bronze mannikin (*Lonchura cucullata*). *Behaviour*, **11**, 156–201.

Morris, D. (1958a) The reproductive behaviour of the Ten-spined Stickleback (*Pygosteus pungitius* L.). *Behaviour Suppl.* **61**, 1–154.

Morris, D. (1958b) The function and causation of courtship ceremonies. In "L'instinct dans le Comportement des Animaux et de L'homme", Fondation Singer Polignac, Paris.

Morris, D. (1958c) The comparative ethology of grassfinches (*Erythrurae*) and Mannikins (*Amadinae*). *Proc. Zool. Soc. Lond.*, **131**, 389–439.

Morris, D. (Ed.) (1967) Primate Ethology. Weidenfeld and Nicolson, London.

Morrison, B. J., and Hill, W. F. (1967) Socially facilitated reduction of the fear response in rats raised in groups or in isolation. *J. comp. physiol. Psychol.*, **63**, 71–76.

Morrison, S. D., and Mayer, J. (1957a) Adipsia and aphagia in rats after lateral subthalamic lesions. *Amer. J. Physiol.*, **191**, 248–254.

Morrison, S. D., and Mayer, J. (1957b) Effect of sham operations in the hypothalamus on food and water intake of the rat. *Amer. J. Physiol.*, **191**, 255–258.

Morse, W. H. (1966) Intermittent reinforcement. In Honig (1966).

Moruzzi, G., and Magoun, H. W. (1949) Brain-stem reticular formation and activation of the EEG. *Electroenceph. clin. Neurophysiol.*, **1**, 455–473.

Mote, F. A., and Finger, F. W. (1942) Exploratory drive and secondary reinforcement in the acquisition and extinction of a simple running response. *J. exp. Psychol.*, **31**, 57–68.

Mott, F. W., and Sherrington, C. S. (1895) Experiments upon the influence of sensory nerves upon movement and nutrition of the limbs. Preliminary communication. *Proc. Roy. Soc.*, **57**, 481–488.

Mountcastle, V. B. (1957) Modality and topographic properties of single neurons in the cat's somatic sensory cortex. *J. Neurophysiol.*, **20**, 408–434.

Mowrer, O. H. (1936) "Maturation" v. "Learning" in the development of vestibular and optokinetic nystagmus. *J. genet. Psychol.*, **48**, 383–404.

Mowrer, O. H. (1939) A stimulus-response analysis of anxiety and its role as a reinforcing agent. *Psychol. Rev.*, **46**, 553–565.

Moyer, K. E. (1963) Startle response: habituation over trials and days, and sex and strain differences. *J. comp. physiol. Psychol.*, **56**, 863–865.

Moyer, K. E., and Bunnell, B. N. (1962) Effect of stomach distention caused by water on food and water consumption in the rat. *J. comp. physiol. Psychol.*, **55**, 652–655.

Moynihan, M. (1955) Some aspects of reproductive behavior in the black-headed gull (*Larus ridibundus ridibundus* L.) and related species. *Behaviour Suppl.* **4**, 1–201.

Moynihan, M. (1956) Notes on the behavior of some North American gulls: I. Aerial hostile behavior. *Behaviour*, **10**, 126–178.

Moynihan, M. (1958a) Notes on the behavior of some North American gulls: II. Non-aerial hostile behavior of adults. *Behaviour*, **12**, 95–182.

Moynihan, M. (1958b) Notes on the behavior of some North American gulls: III. Pairing behavior. *Behaviour*, **13**, 112–130.

Moynihan, M. (1959) Notes on the behavior of some North American gulls: IV. The ontogeny of hostile behavior and display patterns. *Behaviour*, **14**, 214–239.

Moynihan, M. (1962) Hostile and sexual behavior patterns of South American and Pacific Laridae. *Behaviour Suppl.* **8**, 1–365.

Moynihan, M., and Hall, M. F. (1954) Hostile, sexual and other social behaviour patterns of the spice finch (*Lonchura punctulata*) in captivity. *Behaviour*, **7**, 33–76.

Mrosovsky, N. (1964) The performance of dormice and other hibernators on tests of hunger motivation. *Anim. Behav.*, **12**, 454–469.

Mueller, H. C. (1965) Homing and distance orientation in bats. *Z. Tierpsychol.*, **23**, 403–421.

Muenzinger, K. F., Brown, W. O., Crow, W. J., and Powlowski, R. F. (1952) Motivation in learning: XI. An analysis of electric shock for correct responses into its avoidance and accelerating components. *J. exp. Psychol.*, **43**, 115–119.

Müller, A. (1925) Uber Lichtreaktionen von Landasseln. *Z. vergl. Physiol.*, **3**, 113–144.

Müller, D. (1961) Quantitative Luftfeind-Attrapenversuche bei Auer-und Birkhühnern. *Z. Naturf.*, **16**, 551–553.

Mulligan, J. A. (1966) Singing behavior and its development in the song sparrow, *Melospiza melodia. Univ. Calif. Publ. Zool.*, **81**, 1–76.

Munn, N. L. (1950) Handbook of Psychological Research on the Rat. Houghton Mifflin, New York.

Murphey, R. K., and Phillips, R. E. (1967) Central patterning of a vocalization in fowl. *Nature*, **216**, 1125–1126.

Myer, J. S. (1964) Stimulus control of mouse-killing rats. *J. comp. physiol. Psychol.*, **58**, 112–117.

Myer, J. S., and Baenninger, R. (1966) Some effects of punishment and stress on mouse killing by rats. *J. comp. physiol. Psychol.*, **62**, 292–297.

Myer, J. S., and White, R. T. (1965) Aggressive motivation in the rat. *Anim. Behav.*, **13**, 430–433.

Myers, R. D. (1964) Modification of drinking patterns by chronic intracranial chemical infusion. *1st Int. Symp. on Thirst*, Florida State Univ., 533–549, Pergamon Press, Oxford.

Myers, R. D. (1966) Release of chemical factors from the diencephalic region of the unanaesthetized monkey during changes in body temperature. *Proc. Physiol. Soc.*, **188,** 50–51.

Myers, R. D., and Sharpe, L. G. (1968) Chemical activation of ingestive and other hypothalamic mechanisms. *Physiol. Behav.*, **3** (in press).

Myers, R. D., and Yaksh, T. L. (1969) Feeding and temperature responses in the unrestrained rat after injections of cholinergic and aminergic substances into the cerebral ventricles. *Physiol. Behav.* (in press).

Myers, S. A., Horel, J. A., and Pennypacker, H. S. (1965) Operant control of vocal behavior in the monkey *Cebus albifrons Psychon. Sci.*, **3,** 389–390.

Myrberg, A. A. (1964) An analysis of the preferential care of eggs and young by adult Cichlid fishes. *Z. Tierpsychol.*, **21,** 53–98.

Myrberg, A. A. (1966) Parental recognition of young in Cichlid fishes. *Anim. Behav.*, **14,** 565–571.

Myrberg, A., Kramer, E., and Heinecke, P. (1965) Sound production by Cichlid fishes. *Science*, **149,** 555–558.

Nachman, M. (1967) Hypothalamic hyperphagia, finickiness, and taste preferences in rats. *Proc. 75th Ann. Conv. Amer. Psychol. Assoc.*, 127–128.

Nachman, M., and Pfaffmann, C. (1963) Gustatory nerve discharge in normal and sodium-deficient rats. *J. comp. physiol. Psychol.*, **56,** 1007–1011.

Nachman, M., and Valentino, D. A. (1966) Roles of taste and postingestional factors in the satiation of sodium appetite in rats. *J. comp. physiol. Psychol.*, **62,** 280–283.

Nachtigall, W., and Wilson, D. M. (1967) Neuro-muscular control of Dipteran flight. *J. exp. Biol.*, **47,** 77–97.

Nadler, R. D. (1968) Masculinization of female rats by intracranial implantation of androgen in infancy. *J. comp. physiol. Psychol.*, **66,** 157–167.

Nakayama, T., Hammel, H. T., Hardy, J. D., and Eisenman, J. S. (1963) Thermal stimulation of electrical activity of single units of the preoptic region. *Amer. J. Physiol.*, **204,** 1122–1126.

Naquet, R., Regis, H., Fischer-Williams, M., and Fernandez-Guardiola, A. (1960) Variations in the responses evoked by light along the specific pathways. *Brain*, **83,** 52–56.

Nauta, W. J. H. (1946) Hypothalamic regulation of sleep in rats: an experimental study. *J. Neurophysiol.*, **9,** 285–316.

Neil, E. H. (1964) An analysis of color changes and social behavior of *Tilapia mossambica. Univ. Calif. Publ. Zool.*, **75,** 1–58.

Nelson, J. B. (1967) Colonial and cliff nesting in the gannet. *Ardea*, **55,** 60–90.

Nelson, J. E. (1964) Vocal communication in Australian flying foxes (*Pteropodidae: Megachiroptera*). *Z. Tierpsychol.*, **21,** 857–870.

Nelson, K. (1964a) The temporal patterning of courtship behaviour in the glandulocaudine fishes (*Ostariophysi, Characidae*). *Behaviour,* **24,** 90–146.

Nelson, K. (1964b) Behavior and morphology in the glandulocaudine fishes (*Ostariophysi, Characidae*). *Univ. Calif. Publ. Zool.,* **75,** 59–152.

Nelson, K. (1965) The evolution of a pattern of sound production associated with courtship in the characid fish *Glandulocauda inequalis. Evolution,* **18,** 526–540.

Neumann, G. H. (1961) Die visuelle Lernfähigkeit primitiver Säugetiere. *Z. Tierpsychol.,* **18,** 71–83.

Newell, T. G. (1967) Effects of maternal deprivation on later behavior in two inbred strains of mice. *Psychon. Sci.,* **9,** 119–120.

Newton, I. (1967) The adaptive radiation and feeding ecology of some British finches. *Ibis,* **109,** 33–98.

Nice, M. M. (1943) Studies in the life-history of the song sparrow: II. *Trans. Linn. Soc. N.Y.,* **6,** 1–328.

Nicolai, J. (1956) Zur Biologie und Ethologie des Gimpels (*Pyrrhula pyrrhula* L.). *Z. Tierpsychol.,* **13,** 93–132.

Nicolai, J. (1964) Der Brutparasitismus der Viduinae als ethologisches Problem. *Z. Tierpsychol.,* **21,** 129–204.

Nissen, H. W. (1930) A study of exploratory behaviour in the white rat by means of the obstruction method. *J. genet. Psychol.,* **37,** 361–376.

Nissen, H. W. (1954) Cited in Young (1961a).

Nissen, H. W. (1956) Individuality in the behaviour of chimpanzees. *Amer. Anthropologist,* **58,** 407–413.

Nissen, H. W., Chow, K. L., and Semmes, J. (1951) Effects of restricted opportunity for tactual, kinesthetic and manipulative experience on the behavior of a chimpanzee. *Amer. J. Psychol.,* **64,** 485–507.

Noble, G. K. (1936) Courtship and sexual selection of the flicker (*Colaptes auratus luteus*). *Auk,* **53,** 269–282.

Noble, G. K., and Curtis, B. (1935–36) Sexual selection in fishes. *Anat. Rec.,* **64,** 84–5.

Noble, G. K., and Curtis, B. (1939) The social behavior of the jewel fish, *Hemichromis bimaculatus* Gill. *Bull. Amer. Mus. Nat. Hist.,* **76,** 1–46.

Noble, G. K., and Vogt, W. (1935) An experimental study of sex recognition in birds. *Auk,* **52,** 278–286.

Noble, G. K., and Zitrin, A. (1942) Induction of mating behaviour in male and female chicks following injections of sex hormones. *Endocrinology,* **30,** 327–334.

Noirot, E. (1964a) Changes in responsiveness to young in the adult mouse: I. The problematical effect of hormones. *Anim. Behav.,* **12,** 52–58.

Noirot, E. (1964b) Changes in responsiveness to young in the adult mouse: II. The effect of external stimuli. *J. comp. physiol. Psychol.,* **57,** 97–99.

Noirot, E. (1964c) Changes in responsiveness to young in the adult mouse: IV. The effect of an initial contact with a strong stimulus. *Anim. Behav.*, **12,** 442–445.

Noirot, E. (1965) Changes in responsiveness to young in the adult mouse: III. The effect of immediately preceding performances. *Behaviour,* **24,** 318–325.

Noirot, E., and Richards, M. (1966) Maternal behaviour in virgin female golden hamsters: changes consequent upon initial contact with pups. *Anim. Behav.,* **14,** 7–10.

Nottebohm, F. (1967) The role of sensory feedback in the development of avian vocalizations. *Proc. 14th Int. Ornith. Cong. Oxford,* 265–280, Blackwell, Oxford.

Novin, D., Hutchins, C., and Mundy, P. (1966) Effects of intravenous injections of water on the consummatory behavior of rats. *J. comp. physiol. Psychol.,* **61,** 473–474.

Nyman, A. J. (1967) Problem solving in rats as a function of experience at different ages. *J. genet. Psychol.,* **110,** 31–39.

Oberholzer, R. J. H., and Tofani, W. O. (1960) The neural control of respiration. In "Handbook of Physiology", Ed. J. Field, Sec. 1, Vol. 2, 1111–1129, Amer. Physiol. Soc., Washington.

Oehlert, B. (1958) Kampf und Paarbildung einiger Cichliden. *Z. Tierpsychol.,* **15,** 141–174.

O'Kelly, L. (1954) The effect of preloads of water and sodium chloride on voluntary water intake of thirsty rats. *J. comp. physiol. Psychol.,* **47,** 7–13.

O'Kelly, L. I., Hatton, G. I., Tucker, L., and Westall, D'A. (1965) Water regulation in the rat. *J. comp. physiol. Psychol.,* **59,** 159–165.

Oldfield, R. C., and Zangwill, O. L. (1942a, b) Head's concept of the schema and its application in contemporary British psychology: I and II. *Brit. J. Psychol.,* **32,** 267–286; **33,** 58–64.

Olds, J. (1958) Selective effects of drives and drugs on "reward" systems of the brain. In "Neurological Basis of Behaviour", Eds. G. E. W. Wolstenholme and C. M. O'Connor, Churchill, London.

Olds, J. (1961) Differential effects of drives and drugs on self stimulation at different brain sites. In Sheer (1961).

Olds, J. (1962) Hypothalamic substrates of reward. *Physiol. Rev.,* **42,** 554–604.

Olds, J., and Milner, P. (1954) Positive reinforcement produced by electrical stimulation of septal area and other regions of rat brain. *J. comp. physiol. Psychol.,* **47,** 419–427.

Olds, J., and Peretz, B. (1960) A motivational analysis of the reticular activating system. *Electroenceph. and clinical Neurophysiol.,* **12,** 445–454.

Oppenheim, R. W. (1968) Light responsivity in chick and duck embryos just prior to hatching. *Anim. Behav.,* **16,** 276–280.

Orr, D. W., and Windle, W. F. (1934) The development of behavior in chick embryos: the appearance of somatic movements. *J. comp. Neurol.,* **60,** 271–285.

Osgood, C. E. (1953) Method and Theory in Experimental Psychology. Oxford Univ. Press, New York.

Otter, C. J. den, and Starre, H. van der (1967) Responses of tarsal hairs of the bluebottle, *Calliphora erythrocephala* Meig., to sugar and water. *J. insect Physiol.*, **13,** 1177–1188.

Ottinger, D. R., and Simmons, J. E. (1964) Behavior of human neonates and prenatal maternal anxiety. *Psychol. Rep.*, **14,** 391–394.

Overall, J. E. (1964) Note on the scientific status of factors. *Psychol. Bull.*, **61,** 270–276.

Padilla, S. G. (1935) Further studies on the delayed pecking of chicks. *J. comp. Psychol.*, **20,** 413–443.

Paillard, J. (1960) The patterning of skilled movements. In "Handbook of Physiology", Sec. 1, Vol. III, Eds. J. Field, H. W. Magoun, and V. E. Hall, Amer. Physiol. Soc., Washington.

Palen, G. F. (1965) Focusing cues in the visual cliff behavior of day-old chicks. *J. comp. physiol. Psychol.*, **59,** 452–454.

Palka, Y. S., Ramirez, V. D., and Sawyer, C. H. (1966) Distribution and biological effects of tritiated estradiol implanted in the hypothalamo-hypophysal region of female rats. *Endocrinology*, **78,** 487–499.

Palka, Y. S., and Sawyer, C. H. (1966a) The effects of hypothalamic implants of ovarian steroids on oestrous behaviour in rabbits. *J. Physiol.*, **185,** 251–269.

Palka, Y. S., and Sawyer, C. H. (1966b) Induction of oestrous behavior in rabbits by hypothalamic implants of testosterone. *Amer. J. Physiol.*, **211,** 225–228.

Pantin, C. F. A. (1964) Learning, world-models and pre-adaptation. *Anim. Behav. Suppl.* **1,** 1–8.

Paolino, R. M., Quartermain, D., and Miller, N. E. (1966) Different temporal gradients of retrograde amnesia produced by carbon dioxide anaesthesia and electroconvulsive shock. *J. comp. physiol. Psychol.*, **62,** 270–274.

Pardi, L. (1960) Innate components in the solar orientation of littoral amphipods. *Cold Spring Harbor Symp. quant. Biol.*, **25,** 395–401.

Paré, W. P. (1964) Relationship of various behaviors in the open-field test of emotionality. *Psychol. Rep.*, **14,** 19–22.

Parks, E. R. (1968) The orientation reaction as a mediator of sensory preconditioning. *Psychon. Sci.*, **11,** 11–12.

Parriss, J. R. (1963) Retention of shape discrimination after regeneration of the optic nerves in the toad. *Q. J. exp. Psychol.*, **15,** 22–26.

Parriss, J. R., and Young, J. Z. (1962) The limits of transfer of a learned discrimination to figures of larger and smaller sizes. *Z. vergl. Physiol.*, **45,** 618–635.

Pastore, N. (1958) Form perception and size constancy in the duckling. *J. Psychol.*, **45,** 259–261.

Pastore, N. (1959) Perceptual functioning in the duckling. *J. genet. Psychol.*, **95,** 157–169.

Pastore, N. (1962) Perceptual functioning in the duckling. *J. Psychol.*, **54,** 293–298.

Patlak, C. S. (1953) A mathematical contribution to the study of orientation of organisms. *Bull. Math. Biophys.*, **15,** 431–476.

Patterson, I. J. (1965) Timing and spacing of broods in the black-headed gull, *Larus ridibundus. Ibis,* **107,** 433–459.

Pavlov, I. P. (1955) Selected Works. Central Books, London.

Payne, R. S., Roeder, K. D., and Wallman, J. (1966) Directional sensitivity of the ears of noctuid moths. *J. insect Physiol.*, **44,** 17–31.

Peirce, J. T., and Nuttall, R. L. (1961a) Duration of sexual contacts in the rat. *J. comp. physiol. Psychol.*, **54,** 585–587.

Peirce, J. T., and Nuttall, R. L. (1961b) Self-paced sexual behavior in the female rat. *J. comp. physiol. Psychol.*, **54,** 310–313.

Pelkwijk, J. J. ter, and Tinbergen, N. (1937) Eine reizbiologische Analyse einiger Verhaltensweisen von *Gasterosteus aculeatus* L. *Z. Tierpsychol.*, **1,** 193–200.

Penney, R. L., and Emlen, J. T. (1967) Further experiments on distance navigation in the Adelie penguin *Pygoscelis adeliae. Ibis,* **109,** 99–109.

Pennycuick, C. J. (1960) The physical basis of astro-navigation in birds; theoretical considerations. *J. exp. Biol.*, **37,** 573–593.

Pepelko, W. E., and Clegg, M. T. (1965) Studies of mating behaviour and some factors influencing the sexual response in the male sheep *Ovis aries. Anim. Behav.*, **13,** 249–258.

Perdeck, A. C. (1956) Vogeltrekstation (Texel). *Jaarsverslag*, 1956, 66–74.

Perdeck, A. C. (1958) The isolating value of specific song patterns in two sibling species of grasshoppers (*Chorthippus brunneus* Thunb. and *C. biguttulus* L.). *Behaviour,* **12,** 1–75.

Perdeck, A. C. (1963) Does navigation without visual clues exist in Robins? *Ardea,* **51,** 91–104.

Peters, J. J., and Isaacson, R. L. (1963) Acquisition of active and passive responses in two breeds of chickens. *J. comp. physiol. Psychol.*, **56,** 793–796.

Peters, J. J., Vonderahe, A. R., and Powers, T. H. (1958) Electrical studies of functional development of the eye and optic lobes in the chick embryo. *J. exp. Zool.*, **139,** 459–468.

Peters, J. J., Vonderahe, A., and Schmid, D. (1965) Onset of cerebral electrical activity associated with behavioral sleep and attention in the developing chick. *J. exp. Zool.*, **160,** 255–262.

Peters, R. S. (1958) The Concept of Motivation. Routledge and Kegan-Paul, London.

Petersen, B., Lundgren, L., and Wilson, L. (1956–57) The development of flight capacity in a butterfly. *Behaviour,* **10,** 324–339.

Peterson, N. (1960) Control of behaviour by presentation of an imprinted stimulus. *Science,* **132,** 1395–1396.

Peterson, N. (1962) Effect of monochromatic rearing on the control of responding by wavelength. *Science,* **136,** 774–775.

Pfaffmann, C. (1963) Taste stimulation and preference behavior. In "Olfaction and Taste", Ed. Y. Zotterman, Vol. 1, 257–273, Macmillan, New York.

Pfaffmann, C. (1964) Taste, its sensory and motivating properties. *Amer. Scientist*, **52,** 187–206.

Pfaffmann, C., and Bare, J. K. (1950) Gustatory nerve discharges in normal and adrenalectomized rats. *J. comp. physiol. Psychol.*, **43,** 320–324.

Pfeiffer, W. (1960) Uber die Schreckreaktion bei Fischen und die Herkunft des Schreckstoffes. *Z. vergl. Physiol.*, **43,** 578–614.

Phillips, C. G. (1964) Changing concepts of the precentral motor area. In "Brain and Conscious Experience", Ed. J. C. Eccles, Springer-Verlag, New York.

Phillips, R. E., and Siegel, P. B. (1966) Development of fear in chicks of two closely related genetic lines. *Anim. Behav.*, **14,** 84–88.

Phoenix, C. H. (1961) Hypothalamic regulation of sexual behaviour in male guinea pigs. *J. comp. physiol. Psychol.*, **54,** 72–77.

Piaget, J. (1952) The Origins of Intelligence in Children. Int. Univ. Press, New York.

Piaget, J. (1955) The Child's Construction of Reality. Routledge and Kegan-Paul, London.

Pierrel, R., and Sherman, J. G. (1963) Barnabus, the rat with college training. *The Brown Alumni Monthly*, Brown University.

Pittendrigh, C. S. (1960) Circadian rhythms and the circadian organization of living systems. *Cold Spring Harbor Symp. quant. Biol.*, **25,** 159–184.

Pitz, G. F., and Ross, R. B. (1961) Imprinting as a function of arousal. *J. comp. physiol. Psychol.*, **54,** 602–604.

Platt, J. J., Sutker, L. W., and James, W. T. (1968) Social facilitation of eating behavior in young opossums: II. The effects of isolation. *Psychon. Sci.*, **10,** 267–268.

Plotnik, R. J., and Tallarico, R. B. (1966) Object-quality learning-set formation in the young chicken. *Psychon. Sci.*, **5,** 195–196.

Pollack, L. J., and Davis, L. (1931) Studies in decerebration: VI. The effect of deafferentation upon decerebrate rigidity. *Amer. J. Physiol.*, **98,** 47–49.

Pomeroy, D. E. (1962) Birds with abnormal bills. *Brit. Birds*, **55,** 49–72.

Poole, T. B. (1966) Aggressive play in polecats. *Symp. zool. Soc. Lond.*, **18,** 23–44.

Poole, T. B. (1967) Aspects of aggressive behaviour in polecats. *Z. Tierpsychol.*, **24,** 351–369.

Portmann, A. (1961) Sensory organs: skin, taste and olfaction. In "Biology and Comparative Physiology of Birds", Ed. A. J. Marshall, Academic Press, New York.

Poschel, B. P. H. (1966) Comparison of reinforcing effects yielded by lateral versus medial hypothalamic stimulation. *J. comp. physiol. Psychol.*, **61,** 346–352.

Postman, L. (1955) Association theory and perceptual learning. *Psychol. Rev.*, **62,** 438–446.

Poulsen, H. (1954) On the song of the linnet (*Carduelis cannabina*). *Dansk orn. Foren. Tidsskr.*, **48,** 32–37.

Poulsen, H. (1959) Song learning in the domestic canary. *Z. Tierpsychol.*, **16,** 173–178.

Poulton, E. C. (1952) Perceptual anticipation in tracking with two-pointer and one-pointer displays. *Brit. J. Psychol.*, **43,** 222–229.

Poulton, E. C. (1956) Listening to overlapping calls. *J. exp. Psychol.*, **52,** 334–339.

Precht, H., and Freytag, G. (1958) Uber Ermüdung und Hemmung angeborener Verhaltensweisen bei Springspinnen (Salticidae). Zugleich ein Beitrag zum Triebproblem. *Behaviour*, **13,** 143–211.

Prechtl, H. F. R. (1952) Angeborene Bewegungsweisen junger Katzen. *Experientia*, **8,** 220.

Prechtl, H. F. R. (1953) Zur Physiologie der angeborenen auslösenden Mechanismen: I. Quantitative Untersuchungen über die Sperrbewegung junger Singvögel. *Behaviour*, **5,** 32–50.

Prechtl, H. F. R. (1958) The directed head turning response and allied movements of the human baby. *Behaviour*, **13,** 212–242.

Prechtl, H. F. R. (1965) Problems of behavioural studies in the newborn infant. In Lehrman *et al.* (1965).

Premack, D. (1959) Toward empirical behavior laws: I. Positive reinforcement. *Psychol. Rev.*, **66,** 219–233.

Premack, D., Collier, G., and Roberts, C. L. (1957) Frequency of light contingent bar pressing as a function of the amount of deprivation for light. *Amer. Psychol.*, **12,** 411.

Prescott, R. G. W. (1964) Brain-stem stimulation and reinforcement in the albino rat. Ph.D. thesis, Cambridge.

Prescott, R. G. W. (1966) Estrous cycle in the rat: effects on self-stimulation behavior. *Science*, **152,** 796–797.

Prescott, R. G. W. (1967) Diurnal activity cycles and intracranial self-stimulation in the rat. *J. comp. physiol. Psychol.*, **64,** 346–349.

Pribram, K. H. (1958) Neocortical function in behaviour. In Harlow and Woolsey (1958).

Pribram, K. H. (1961) Limbic system. In Sheer (1961).

Pribram, K. H. (1963) Reinforcement revisited: a structural view. Nebraska symposium on motivation, 1963.

Pringle, J. W. S. (1951) On the parallel between learning and evolution. *Behaviour*, **3,** 174–215.

Pringle, J. W. S. (1965) Locomotion: flight. In "The Physiology of Insects", Ed. M. Rockstein, Vol. II, Academic Press, New York.

Pritchard, R. M., Heron, W., and Hebb, D. O. (1960) Visual perception approached by the method of stabilized images. *Canad. J. Psychol.*, **14,** 67–77.

Pritchatt, D. (1968) Avoidance of electric shock by the cockroach *Periplaneta americana*. *Anim. Behav.*, **16,** 178–185.

Pumphrey, R. J. (1940) Hearing in insects. *Biol. Rev.*, **15,** 107–132.

Pumphrey, R. J. (1948) The sense organs of birds. *Ibis*, **90,** 171–199.

Quartermain, D., and Miller, N. E. (1966) Sensory feedback in time-response of drinking elicited by carbachol in preoptic area of rat. *J. comp. physiol. Psychol.*, **62,** 350–353.

Quine, D. A., and Cullen, J. M. (1964) The pecking response of young Arctic terns (*Sterna macrura*) and the adaptiveness of the releasing mechanism". *Ibis*, **106**, 145–173.

Rabb, G. B., and M. S. (1963) On the behavior and breeding biology of the African pipid frog *Hymenochirus boettgeri*. *Z. Tierpsychol.*, **20**, 215–241.

Räber, H. (1948) Analyse des Balzverhaltens eines domestizierten Truthahns (*Meleagris*). *Behaviour*, **1**, 237–266.

Rajalakshmi, R., and Jeeves, M. A. (1965) The relative difficulty of reversal learning (reversal index) as a basis of behavioural comparisons. *Anim. Behav.*, **13**, 203–211.

Ralls, K. (1967) Auditory sensitivity in mice: *Peromyscus* and *Mus musculus*. *Anim. Behav.*, **15**, 123–128.

Ranson, S. W., Fisher, C., and Ingram, W. R. (1937) Hypothalamic regulation of temperature in the monkey. *Arch. neurol. Psychiat.*, **38**, 445–466.

Rasmussen, E. W. (1939) Wildness in rats. *Acta Psychol.*, **4**, 295–304.

Ratner, S. C. (1965) Comparisons between behaviour development of normal and isolated domestic fowl. *Anim. Behav.*, **13**, 497–503.

Ratner, S. C., Gawronski, J. J., and Rice, F. E. (1964) The variable of concurrent action in the language of children: effects of delayed speech feedback. *Psychol. Rec.*, **14**, 47–56.

Ratner, S. C., and Thompson, R. W. (1960) Immobility reactions (fear) of domestic fowl as a function of age and prior experience. *Anim. Behav.*, **8**, 186–191.

Razran, G. H. S. (1933) Conditioned responses in animals other than dogs. *Psychol. Bull.*, **30**, 261–324.

Razran, G. H. S. (1935) Conditioned responses: an experimental study and a theoretical analysis. *Arch. Psychol., N.Y.*, 191.

Razran, G. (1961) The observable unconscious and the inferable conscious in current Soviet psychophysiology: interoceptive conditioning, semantic conditioning and the orienting reflex. *Psychol. Rev.*, **68**, 81–147.

Reading, A. J. (1966) Effect of maternal environment on the behavior of inbred mice *J. comp. physiol. Psychol.*, **62**, 437–440.

Reese, E. S. (1962) Submissive posture as an adaptation to aggressive behaviour in hermit crabs. *Z. Tierpsychol.*, **19**, 645–651.

Reese, E. S. (1963) The behavioral mechanisms underlying shell selection by hermit crabs. *Behaviour*, **21**, 78–126.

Reichardt, W. (1961) Autocorrelation: a principle for the evaluation of sensory information by the central nervous system. In Rosenblith (1961).

Reid, L. S., and Finger, F. W. (1955) The rat's adjustment to 23-hour food-deprivation cycles. *J. comp. physiol. Psychol.*, **48**, 110–113.

Renner, K. E. (1967) Temporal integration: modification of the incentive value of a food reward by early experience with deprivation. *J. exp. Psychol.*, **75**, 400–407.

Renner, M. (1957) Neue Versuche über den Zeitsinn der Honigbiene. *Z. vergl. Physiol.*, **40**, 85–118.

Rensch, B. (1961) Malversuche mit Affen. *Z. Tierpsychol.*, **18,** 347–364.

Rensch, B., and Dücker, G. (1959) Die Spiele von Mungo und Ichneumon. *Behaviour*, **14,** 185–213.

Rescorla, R. A., and Solomon, R. L. (1967) Two-process learning theory. *Psychol. Rev.*, **74,** 151–182.

Ressler, R. H. (1963) Genotype-correlated parental influences in two strains of mice. *J. comp. physiol. Psychol.*, **56,** 882–886.

Ressler, R. H. (1966) Inherited environmental influences on the operant behavior of mice. *J. comp. physiol. Psychol.*, **61,** 264–267.

Reyniers, J. A. (1953) Germfree life. *Lancet*, 933–934.

Reynolds, G. S. (1961a) Attention in the pigeon. *J. exp. Anal. Behav.*, **4,** 203–208.

Reynolds, G. S. (1961b) Contrast, generalisation and the process of discrimination. *J. exp. Anal. Behav.*, **4,** 289–294.

Reynolds, G. S., Catania, A. C., and Skinner, B. F. (1963) Conditioned and unconditioned aggression in pigeons. *J. exp. Anal. Behav.*, **1,** 73–75.

Reynolds, R. W. (1958) The relationship between stimulation voltage and rate of hypothalamic self-stimulation in the rat. *J. comp. physiol. Psychol.*, **51,** 193–198.

Rheingold, H. L., and Hess, E. H. (1957) The chick's "preference" for some visual properties of water. *J. comp. physiol. Psychol.*, **50,** 417–421.

Riccio, D. G., Hamilton, D. M., and Triechler, F. R. (1967) Effects of age and ingestive experience on air-drinking behavior in the rat. *Psychon. Sci.*, **7,** 295–296.

Richards, M. P. M. (1966a) Maternal behaviour in the golden hamster: responsiveness to young in virgin, pregnant, and lactating females. *Anim. Behav.*, **14,** 310–313.

Richards, M. P. M. (1966b) Activity measured by running wheels and observation during the oestrous cycle, pregnancy and pseudopregnancy in the golden hamster. *Anim. Behav.*, **14,** 450–458.

Richter, C. P. (1927) Animal behavior and internal drives. *Quart. Rev. Biol.*, **2,** 307–343.

Richter, C. P. (1937) Hypophyseal control of behavior. *Cold Spring Harbor Symp. quant. Biol.*, **5,** 258–268.

Richter, C. P., and Hawkes, C. D. (1939) Increased spontaneous activity and food intake produced in rats by removal of the frontal poles of the brain. *J. Neurol. Psychiatr.*, **2,** 231–242.

Riesen, A. H. (1958) Plasticity of behavior: psychological aspects. In Harlow and Woolsey (1958).

Riesen, A. H. (1961) Studying perceptual development using the technique of sensory deprivation. *J. nerv. ment. Dis.*, **132,** 21–25.

Riesen, A. H. (1966) Sensory deprivation. In "Progress in Physiological Psychology", Ed. E. Stellar and J. M. Sprague, **1,** 117–147, Academic Press, New York.

Riesen, A. H., and Aarons, L. (1959) Visual movement and intensity discrimination in cats after early deprivation of pattern vision. *J. comp. physiol. Psychol.*, **52,** 142–149.

Riesen, A. H., and Mellinger, J. C. (1956) Interocular transfer of habits in cats after alternating monocular visual experience. *J. comp. physiol. Psychol.*, **49,** 516–520.

Riess, B. F. (1950) The isolation of factors of learning and native behavior in field and laboratory studies. *Ann. N.Y. Acad. Sci.*, **51,** 1093–1102.

Rigby, W. K. (1954) Approach and avoidance gradients and conflict behavior in a predominantly temporal situation. *J. comp. physiol. Psychol.*, **47,** 83–89.

Riggs, L. A., and Schick, A. M. L. (1968) Accuracy of retinal image stabilization achieved with a plane mirror on a tightly fitting contact lens. *Vision Research*, **8,** 159–169.

Riggs, L. A., and Tulunay, S. U. (1959) Visual effects of varying the extent of compensation for eye movements. *J. Opt. Soc. Amer.*, **9,** 741–745.

Rilling, S., Mittelstaedt, H., and Roeder, K. D. (1959) Prey recognition in the praying mantis. *Behaviour*, **14,** 164–184.

Riss, W., Valenstein, E. S., Sinks, J., and Young, W. C. (1955) Development of sexual behavior in male guinea pigs from genetically different stocks under controlled conditions of androgen treatment and caging. *Endocrinology*, **57,** 139–146.

Ritchie, B. F. (1948) Studies in spatial learning: VI. Place orientation and direction orientation. *J. exp. Psychol.*, **38,** 659–669.

Roberts, B. L. (1969a, b) Spontaneous rhythms in the motoneurons of spinal dogfish (*Scyliorhinus canicula*). *J. Marine Biol. Ass.*

Roberts, C. L., Marx, M. H., and Collier, G. (1958) Light onset and light offset as reinforcers for the albino rat. *J. comp. physiol. Psychol.*, **51,** 575–579.

Roberts, M. B. V. (1962) The giant fibre reflex of the earthworm, *Lumbricus terrestris* L: II. Fatigue. *J. exp. Biol.*, **39,** 229–237.

Roberts, M. B. V. (1966) Facilitation in the rapid response of the earthworm, *Lumbricus terrestris* L. *J. exp. Biol.*, **45,** 141–150.

Roberts, S. K. de F. (1965) Photoreception and entrainment of cockroach activity rhythms. *Science*, **148,** 958–959.

Roberts, S. K. de F. (1966) Circadian activity rhythms in cockroaches: III. *J. cell. Physiol.*, **67,** 473–486.

Roberts, W. W. (1958) Both rewarding and punishing effects from stimulation of posterior hypothalamus of cat with same electrode at same intensity. *J. comp. physiol. Psychol.*, **51,** 400–407.

Roberts, W. W., and Carey, R. J. (1965) Rewarding effect of performance of gnawing aroused by hypothalamic stimulation in the rat. *J. comp. physiol. Psychol.*, **59,** 317–324.

Roberts, W. W., and Kiess, H. O. (1964) Motivational properties of hypothalamic aggression in cats. *J. comp. physiol. Psychol.*, **58,** 187–193.

Roberts, W. W., Steinberg, M. L., and Means, L. W. (1967) Hypothalamic mechanisms for sexual, aggressive, and other motivational behaviors in the opossum, *Didelphis virginiana. J. comp. physiol. Psychol.*, **64,** 1–15.

Robinson, J. S. (1961) The reinforcing effects of response-contingent light increment and decrement in hooded rats. *J. comp. physiol. Psychol.*, **54,** 470–473.

Rodgers, W. L., Epstein, A. N., and Teitelbaum, P. (1965) Lateral hypothalamic aphagia: motor failure or motivational deficit? *Amer. J. Physiol.*, **208,** 334–342.

Rodgers, W. L., Melzack, R., and Segal, J. R. (1963) "Tail flip response" in goldfish. *J. comp. physiol. Psychol.*, **56,** 917–923.

Rodieck, R. W. (1967) Receptive fields in the cat retina: a new type. *Science*, **157,** 90–92.

Rodieck, R. W., and Stone, J. (1965) Response of cat retinal ganglion cells to moving visual patterns. *J. Neurophysiol.*, **28,** 819–832.

Roeder, K. D. (1962a) The behaviour of free-flying moths in the presence of artificial ultrasonic pulses. *Anim. Behav.*, **10,** 300–304.

Roeder, K. D. (1962b) Neural mechanisms of animal behavior. *Am. Zoologist*, **2,** 105–115.

Roeder, K. D. (1964) Aspects of noctuid tympanic nerve response having significance in the avoidance of bats. *J. insect Physiol.*, **10,** 529–546.

Roeder, K. D. (1966a) Auditory system of noctuid moths. *Science*, **154,** 1515–1521.

Roeder, K. D. (1966b) Interneurons of the thoracic nerve cord activated by tympanic nerve fibres in noctuid moths. *J. insect Physiol.*, **12,** 1227–1244.

Roeder, K. D. (1967) Nerve cells and insect behavior. Harvard Univ. Press.

Roeder, K. D., Tozian, L., and Weiant, E. A. (1960) Endogenous nerve activity and behaviour in the mantis and cockroach. *J. insect Physiol.*, **4,** 45–62.

Ropartz, P. (1968) The relation between olfactory stimulation and aggressive behaviour in mice. *Anim. Behav.*, **16,** 97–100.

Rosen, J., and Hart, F. M. (1963) Effects of early social isolation upon adult timidity and dominance in *Peromyscus. Psychol. Rep.*, **13,** 47–50.

Rosenblatt, J. S. (1950) Hoarding or Hauling? M.Sc. thesis, New York Univ. Library.

Rosenblatt, J. S. (1965a) Effects of experience on sexual behavior in male cats. In Beach (1965).

Rosenblatt, J. S. (1965b) The basis of synchrony in the behavioral interaction between the mother and her offspring in the laboratory rat. In "Determinants of Infant Behaviour", **3,** Ed. B. Foss, Methuen, London.

Rosenblatt, J. S. (1967) Nonhormonal basis of maternal behavior in the rat. *Science*, **156,** 1512–1513.

Rosenblatt, J. S., and Aronson, L. R. (1958a) The influence of experience on the behavioural effects of androgen in prepuberally castrated male cats. *Anim. Behav.*, **6,** 171–182.

Rosenblatt, J. S., and Aronson, L. R. (1958b) The decline of sexual behavior in male cats after castration with special reference to the role of prior sexual experience. *Behaviour*, **12**, 285–338.

Rosenblatt, J. S., and Lehrman, D. S. (1963) Maternal behavior of the laboratory rat. In "Maternal Behavior in Mammals", Ed. H. Rheingold, Wiley, New York.

Rosenblatt, J. S., Turkewitz, G., and Schneirla, T. C. (1961) Early socialization in the domestic cat. In "Determinants of Infant Behaviour", **2**, Ed. B. Foss, Methuen, London.

Rosenblith, W. A. (Ed.) (1961) Sensory Communication. Wiley, New York.

Rosenblueth, A., Wiener, N., and Bigelow, J. (1943) Behavior, purpose and teleology. *Philosophy of Science*, **10**, 18–24.

Rosenblum, L. A. (1961) The development of social behavior in the rhesus monkey. Ph.D. thesis, Univ. of Wisconsin.

Rosenblum, L. A., and Cross, H. A. (1963) Performance of neonatal monkeys in the visual-cliff situation. *Amer. J. Psychol.*, **76**, 318–320.

Rosenzweig, M. R., Krech, D., Bennett, E. L., and Diamond, M. C. (1962) Effects of environmental complexity and training on brain chemistry and anatomy: a replication and extension. *J. comp. physiol. Psychol.*, **55**, 429–437.

Rosenzweig, M. R., and Leiman, A. L. (1968) Brain functions. *Ann. Rev. Psychol.*, **19**, 55–98.

Ross, S. (1950) Sucking behavior in neonate dogs. *J. abn. soc. Psychol.*, **46**, 142–149.

Ross, S., Goldstein, I., and Kappel, S. (1962) Perceptual factors in eating behavior in chicks. *J. comp. physiol. Psychol.*, **55**, 240–241.

Roth, L. L., and Rosenblatt, J. S. (1966) Mammary glands of pregnant rats: development stimulated by licking. *Science*, **151**, 1403–1404.

Roth, L. M., and Barth, R. H. (1967) The sense organs employed by cockroaches in mating behaviour. *Behaviour*, **28**, 58–94.

Rowell, C. H. F. (1961) Displacement grooming in the chaffinch. *Anim. Behav.*, **9**, 38–63.

Rowell, C. H. F. (1963) A method for chronically implanting stimulating electrodes in the brain of locusts, and some results of stimulation. *J. exp. Biol.*, **40**, 271–284.

Rowell, C. H. F. (1964) Comments on a recent discussion of some ethological terms. *Anim. Behav.*, **12**, 535–537.

Rowell, C. H. F. (1965) The control of reflex responsiveness and the integration of behavior. In "Physiology of the Insect Central Nervous System", Ed. J. E. Treherne and J. W. L. Beament, 157–164, Academic Press, New York.

Rowell, C. H. F. (1966) Activity of interneurones in the arm of *Octopus* in response to tactile stimulation. *J. exp. Biol.*, **44**, 589–605.

Rowell, C. H. F. (1967) Experiments on aggregations of *Phymateus purpurascens* (Orthoptera, Acrididae, Pyrgomorphinae). *J. Zool. Lond.*, **152**, 179–193.

Rowell, C. H. F., and Horn, G. (1968) Dishabituation and arousal in the response of single nerve cells in an insect brain. *J. exp. Biol.*, **49**, 171–183.

Rowell, T. E. (1961) Maternal behaviour in non-maternal golden hamsters (*Mesocricetus auratus*). *Anim. Behav.*, **9**, 11–15.

Rowell, T. E. (1962) Agonistic noises of the rhesus monkey (*Macaca mulatta*). *Symp. Zool. Soc. Lond.*, **8**, 91–96.

Rowell, T. E. (1963) Behaviour and female reproductive cycles of rhesus macaques. *J. Reprod. Fertil.*, **6**, 193–203.

Rowell, T. E. (1966) Hierarchy in the organization of a captive baboon group. *Anim. Behav.*, **14**, 430–443.

Rowell, T. E., and Hinde, R. A. (1962) Vocal communication by the rhesus monkey (*Macaca mulatta*). *Proc. Zool. Soc. Lond.*, **138**, 279–294.

Rowell, T. E., Hinde, R. A., and Spencer-Booth, Y. (1964) "Aunt"-infant interaction in captive rhesus monkeys. *Anim. Behav.*, **12**, 219–226.

Rowland, V. (1957) Differential electroencephalographic response to conditioned auditory stimuli in arousal from sleep. *EEG Clin. Neurophysiol.*, **9**, 585–594.

Rubin, H. B., and Azrin, N. H. (1967) Temporal patterns of sexual behavior in rabbits as determined by an automatic recording technique. *J. exp. Anal. Behav.*, **10**, 219–231.

Ruch, T. C. (1951) Motor systems. In "Handbook of Experimental Psychology", Ed. S. S. Stevens, Wiley, New York.

Rudel, R. G., and Teuber, H. L. (1963) Discrimination of direction of line in children. *J. comp. physiol. Psychol.*, **56**, 892–898.

Rudolph, R. L., Honig, W. K., and Gerry, J. E. (1969) Effects of monochromatic rearing on the acquisition of stimulus control. *J. comp. physiol. Psychol.*, **67**, 50–57.

Ruiter, L. de (1952) Some experiments on the camouflage of stick caterpillars. *Behaviour*, **4**, 222–232.

Ruiter, L. de (1956) Countershading in caterpillars. *Arch. néerl. Zool.*, **11**, 285–341.

Ruiter, L. de (1963) The physiology of vertebrate feeding behaviour. *Z. Tierpsychol.*, **20**, 498–516.

Ruiter, L. de (1967a) Critic's comments. In "The Chemical Senses and Nutrition", Ed. M. R. Kare and O. Maller, Johns Hopkins, Baltimore.

Ruiter, L. de (1967b) Feeding behaviour of vertebrates in the natural environment. In "Handbook of Physiology", Ed. C. F. Code, 97–116, Amer. Physiol. Soc.

Ruiter, L. de, and Beukema, J. J. (1963) Foraging and feeding behaviour in the three-spined stickleback. *Anim. Behav.*, **11**, 605.

Ruiter, L. de, Wiepkema, P. R., and Reddingius, J. (1968) Ethological and neurological aspects of the regulation of food intake. *Ann. N.Y. Acad. Sci.*

Rüppell, G. (1962) Vorläufige Beiträge zur Kenntnis des Drohverhaltens beim Rotschenkel *Tringa totanus totanus* Linné. *Z. Tierpsychol.*, **19**, 465–471.

Russek, M. (1963) Participation of hepatic glucoreceptors in the control of intake of food. *Nature*, **197**, 79–80.

Russell, E. M. (1967a) Changes in the behaviour of *Lebistes reticulatus* upon a repeated shadow stimulus. *Anim. Behav.*, **15**, 574–585.

Russell, E. M. (1967b) The effect of experience of surroundings on the response of *Lebistes reticulatus* to a strange object. *Anim. Behav.*, **15**, 586–594.

Russell, E. S. (1943) Perceptual and sensory signs in instinctive behaviour. *Proc. Linn. Soc. Lond.*, **154**, 195–216.

Russell, E. S. (1945) The Directiveness of Organic Activities. Cambridge Univ. Press, London.

Russell, W. M. S. (1954) Experimental studies of the reproductive behaviour of *Xenopus laevis*: I. The control mechanisms for clasping and unclasping, and the specificity of hormone action. *Behaviour*, **7**, 113–188.

Sachs, B. D. (1965) Sexual behavior of male rats after one to nine days without food. *J. comp. physiol. Psychol.*, **60**, 144–146.

Sackett, G. P. (1963) A neural mechanism underlying unlearned, critical period and developmental aspects of visually controlled behavior. *Psychol. Rec.*, **70**, 40–50.

Sackett, G. P. (1965) Effects of sensory deprivation level, visual complexity, and age upon light contingent responses during rearing. *Anim. Behav.*, **13**, 393–399.

Sackett, G. P. (1966) Development of preference for differentially complex patterns by infant monkeys. *Psychon. Sci.*, **6**, 441–442.

Sackett, G. P. (1967) Response to stimulus novelty and complexity as a function of rats' early rearing experiences. *J. comp. physiol. Psychol.*, **63**, 369–375.

Salt, G. (1937) The sense used by *Trichogramma* to distinguish between parasitized and unparasitized hosts. *Proc. Roy. Soc. Lond. B.*, **122**, 57–75.

Salzen, E. A. (1962) Imprinting and fear. *Symp. Zool. Soc. Lond.*, **8**, 199–217.

Salzen, E. A. (1966) The interaction of experience, stimulus characteristics and exogenous androgen in the behaviour of domestic chicks. *Behaviour*, **26**, 286–322.

Sand, A. (1937) The mechanism of the lateral sense organs of fishes. *Proc. Roy. Soc. B*, **123**, 472–495.

Sargent, T. D. (1965) The role of experience in the nest-building of the zebra finch. *Auk*, **82**, 48–61.

Sargent, T. D., and Keiper, R. R. (1967) Stereotypies in caged canaries. *Anim. Behav.*, **15**, 62–66.

Satinoff, E., and Stanley, W. C. (1963) Effect of stomach loading on sucking behaviour in neonatal puppies. *J. comp. physiol. Psychol.*, **56**, 66–68.

Sauer, F. (1954) Die Entwicklung der Lautäusserungen vom Ei ab schalldicht gehaltener Dorngrasmücken (*Sylvia c. communis*, Latham) im Vergleich mit später isolierten und mit wildlebenden Artgenossen. *Z. Tierpsychol.*, **11**, 10–93.

Sauer, F. (1955) Uber Variationen der Artgesänge bei Grasmücken. *J. f. Ornith.*, **96**, 129–146.

Sauer, F. (1957) Die Sternenorientierung nächtlich ziehender Grasmücken (*Sylvia atricapilla, bovin und curruca*). *Z. Tierpsychol.*, **14,** 29–70.

Saunders, F. J. (1968) Effects of sex steroids and related compounds on pregnancy and on development of the young. *Physiol. Rev.*, **48,** 601–643.

Saxena, A. (1960) Lernkapazität, Gedächtnis und Transpositionsvermögen bei Forellen. *Zool. Jahrb.*, **69,** 63–94.

Schaefer, H. H., and Hess, E. H. (1959) Color preferences in imprinting objects. *Z. Tierpsychol.*, **16,** 161–172.

Schaefer, T. (1963) Early "experience" and its effects on later behavioral processes in rats: II. A critical factor in the early handling phenomenon. *Trans. N.Y. Acad. Sci.*, **25,** 871–889.

Schaller, G. B., and Emlen, J. T. (1961) The development of visual discrimination patterns in the crouching reactions of nestling grackles. *Auk*, **78,** 125–137.

Schaller, G. B., and Emlen, J. T. (1962) The ontogeny of avoidance behaviour in some precocial birds. *Anim. Behav.*, **10,** 370–381.

Schein, M. W. (1963) On the irreversibility of imprinting. *Z. Tierpsychol.*, **20,** 462–467.

Schein, M. W., and Hale, E. B. (1959) The effect of early social experience on male sexual behaviour of androgen injected turkeys. *Anim. Behav.*, **7,** 189–200.

Schein, M. W., and Hale, E. B. (1965) Stimuli eliciting sexual behavior. In Beach (1965).

Schenkel, R. (1966) Play, exploration and territoriality in the wild lion. *Symp. Zool. Soc. Lond.*, **18,** 11–22.

Schiff, W. (1965) Perception of impending collision. *Psychol. Monogr.*, **79,** No. 64.

Schiffman, H. R. (1968) Physical support with and without optical support: reaction to apparent depth by chicks and rats. *Science*, **159,** 892–894.

Schiffman, H. R., and Walk, R. D. (1963) Behavior on the visual cliff of monocular as compared to binocular chicks. *J. comp. physiol. Psychol.*, **56,** 1064–1068.

Schiller, P. H. (1957) Manipulative patterns in the chimpanzee. In "Instinctive Behavior", Ed. P. H. Schiller, Methuen, London.

Schjelderup-Ebbe, T. (1922) Beiträge zur Sozialpsychologie des Haushuhns. *Z. Psychol.*, **88,** 225–252.

Schleidt, M. (1954) Untersuchungen über die Auslösung des Kollerns beim Truthahn (*Meleagris gallopavo*). *Z. Tierpsychol.*, **11,** 417–435.

Schleidt, W. M. (1961) Reaktionen von Truthühnern auf fliegende Raubvögel und Versuche zur Analyse ihrer AAM's. *Z. Tierpsychol.*, **18,** 534–560.

Schleidt, W. (1962) Die historische Entwicklung der Begriffe "Angeborenes auslösendes Schema" und "Angeborener Auslösemechanismus" in der Ethologie. *Z. Tierpsychol.*, **19,** 697–722.

Schleidt, W. M. (1964) Uber die Spontaneität von Erbkoordinationen. *Z. Tierpsychol.*, **21,** 235–256.

Schleidt, W. M., Schleidt, M., and Magg, M. (1960) Störung der Mutter-Kind-Beziehung bei Truthuhern durch Gehörverlust. *Behaviour*, **16,** 254–260.

Schlosberg, H. (1937) The relationship between success and the laws of conditioning. *Psychol. Rev.*, **44**, 379–394.

Schlosberg, H. (1954) Three dimensions of emotion. *Psychol. Rev.*, **61**, 81–88.

Schmidt, H.-D. (1956) Das Verhalten von Haushunden in Konfliktsituationen. *Z. Psychol.*, **159**, 161–245.

Schmidt-Koenig, K. (1961) Die Sonne als Kompass im Heim-Orientierungssystem der Brieftauben. *Z. Tierpsychol.*, **18**, 221–244.

Schmidt-Koenig, K. (1964) Uber die Orientierung der Vogel: Experimente und Probleme. *Naturwiss.*, **18**, 423–431.

Schmidt-Koenig, K. (1965) Current problems in bird orientation. In Lehrman *et al.* (1965).

Schneider, D. (1957) Elektrophysiologische Untersuchungen von Chemo- und Mechanorezeptoren der Antenne des Seidenspinners *Bombyx mori* L. *Z. vergl. Physiol.*, **40**, 4–41.

Schneider, D. (1962) Electrophysiological investigation on the olfactory specificity of sexual attracting substances in different species of moths. *J. insect Physiol.*, **8**, 15–30.

Schneider, G. E. (1967/1968) Contrasting visuomotor functions of tectum and cortex in the golden hamster. *Psychol. Forschung*, **31**, 52–62.

Schneider, G. E., and Gross, C. G. (1965) Curiosity in the hamster. *J. comp. physiol. Psychol.*, **59**, 150–152.

Schneirla, T. C. (1933) Some important features of ant learning. *Z. vergl. Physiol.*, **19**, 439–452.

Schneirla, T. C. (1946) Problems in the biopsychology of social organisation. *J. abn. soc. Psychol.*, **41**, 385–402.

Schneirla, T. C. (1949) Levels in the psychological capacities of animals. In "Philosophy for the Future", Ed. R. W. Sellars *et al.* Macmillan, New York.

Schneirla, T. C. (1951) A consideration of some problems in the ontogeny of family life and social adjustments in various infra-human animals. In "Problems of Infancy and Childhood", Ed. M. J. E. Senn, Macy, New York.

Schneirla, T. C. (1952) A consideration of some conceptual trends in comparative psychology. *Psych. Bull.*, **49**, 559–597.

Schneirla, T. C. (1956) Interrelationships of the "innate" and the "acquired" in instinctive behavior. In "L'Instinct dans le Comportement des Animaux et de L'homme", Fondation Singer Polignac, Paris.

Schneirla, T. C. (1959) An evolutionary and developmental theory of biphasic processes underlying approach and withdrawal. In "Nebraska Symposium on Motivation", Lincoln.

Schneirla, T. C. (1965) Aspects of stimulation and organisation in approach/withdrawal processes underlying vertebrate behavioural development. In Lehrman *et al.* (1965).

Schneirla, T. C. (1966) Behavioural development and comparative psychology. *Quart. Rev. Biol.*, **41**, 283–302.

Schneirla, T. C., and Rosenblatt, J. S. (1961) Behavioural organisation and genesis of the social bond in insects and mammals. *Amer. J. Orthopsych.*, **31**, 223–253.

Schneirla, T. C., and Rosenblatt, J. S. (1963) "Critical periods" in the development of behavior. *Science*, **139**, 1110–1115.

Schneirla, T. C., Rosenblatt, J. S., and Tobach, E. (1963) Maternal behavior in the cat. In "Maternal Behavior in Mammals", Ed. H. Rheingold, Wiley, New York.

Schnetter, B. (1968) Visuelle Formunterscheidung der Honigbiene im Bereich von Vier- und Sechsstrahlsternen. *Z. vergl. Physiol.*, **59**, 90–109.

Schoen, L. (1950) Quantitative Untersuchungen über die zentrale Kompensation nach einseitiger Utriculusausschaltung bei Fischen. *Z. vergl. Physiol.*, **32**, 121–150.

Schoen, L. (1951) Das Zusammenspiel beider Augen als Gleichgewichtsorgane der Fische. *Verh. dt. zool. Ges.*, **45**, 191–195.

Schoenfeld, W. N., Antonitis, J. J., and Bersh, P. J. (1950) A preliminary study of training conditions necessary for secondary reinforcement. *J. exp. Psychol.*, **40**, 40–45.

Schöne, H. (1951a) Die Lichtorientierung der Larven von *Acilius sulcatus* L. und *Dytiscus marginalis* L. *Z. vergl. Physiol.*, **33**, 63–98.

Schöne, H. (1951b) Die statische Gleichgewichtsorientierung dekapoder Crustaceen. *Verh. dt. zool. Ges.*, **16**, 157–162.

Schöne, H. (1962) Optisch gesteuerte Lageänderungen (Versuche an Dytiscidenlarven zur Vertikalorientierung). *Z. vergl. Physiol.*, **45**, 590–604.

Schöne, H. (1965) Release and orientation of behaviour and the role of learning as demonstrated in Crustacea. *Anim. Behav. Suppl.* **1**, 135–143.

Schoonhoven, L. M. (1967) Chemoreception of mustard oil glucocides in larvae of *Pieris brassicae*. *Proc. Koninkl. Nederl. Akad. Wetensch, Amsterdam, Series C*, **70**, 5, 556–568.

Schoonhoven, L. M., and Dethier, V. G. (1966) Sensory aspects of host-plant discrimination by lepidopterous larvae. *Arch. néerl. Zool.*, **16**, 497–530.

Schrier, A. M., Harlow, H. F., and Stollnitz, F. (Ed.) (1965) Behavior of Nonhuman primates. I and II. Academic Press, New York.

Schuck, J. R. (1960) Pattern discrimination and visual sampling by the monkey. *J. comp. physiol. Psychol.*, **53**, 251–255.

Schultz, D. P. (1965) Sensory Restriction. Academic Press, New York.

Schutz, F. (1956) Vergleichende Untersuchungen über die Schreckreaktion bei Fischen und deren Verbreitung. *Z. vergl. Physiol.*, **38**, 84–135.

Schutz, F. (1965a) Sexuelle Prägung bei Anatiden. *Z. Tierpsychol.*, **22**, 50–103.

Schutz, F. (1965b) Homosexualität und Prägung. *Psychol. Forschung*, **28**, 439–463.

Schüz, E. (1949) Die Spät-Auflassung ostpreussischer Jungstörche in West-Deutschland durch die Vogelwarte Rossitten 1933. *Vogelwarte*, **15**, 63–78.

Schwartz, M., and Beach, F. A. (1954) Effects of adrenalectomy upon mating behavior in castrated male dogs. *Amer. Psychol.*, **9**, 467–468.

Schwartz, S. (1964) Effect of neonatal cortical lesions and early environmental factors on adult rat behavior. *J. comp. physiol. Psychol.*, **57**, 72–77.

Schwassmann, H. O. (1960) Environmental cues in the orientation rhythms of fish. *Cold Spring Harbor Symp. quant. Biol.*, **25**, 443–450.

Scott, J. P. (1945) Social behavior, organisation and leadership in a small flock of domestic sheep. *Comp. Psychol. Monogr.*, **18**, No. 4.

Scott, J. P. (1954) The process of socialization in higher animals. *Child Family Digest*, 1954, 69–84.

Scott, J. P. (1958) Aggression. Univ. Chicago Press.

Scott, J. P. (1962) Critical periods in behavioral development. *Science*, **138**, 949–958.

Scott, J. P., and Fredericson, E. (1951) The causes of fighting in mice and rats. *Physiol. Zool.*, **24**, 273–309.

Scott, J. P., and Marston, M. V. (1950) Critical periods affecting the development of normal and maladjustive social behavior of puppies. *J. genet. Psychol.*, **77**, 25–60.

Scott, J. W. (1967) Brain stimulation reinforcement with distributed practice. *J. comp. physiol. Psychol.*, **63**, 175–183.

Searle, L. V. (1949) The organization of hereditary maze-brightness and maze-dullness. *Genet. Psychol. Monogr.*, **39**, 279–325.

Seay, B., and Harlow, H. F. (1965) Maternal separation in the rhesus monkey. *J. nerv. ment. Dis.*, **140**, 434–441.

Sechzer, J. A. (1964) Successful interocular transfer of pattern discrimination in "split-brain" cats with shock-avoidance motivation. *J. comp. physiol. Psychol.*, **58**, 76–83.

Sechzer, J. A., and Brown, J. L. (1964) Color discrimination in the cat. *Science*, **144**, 427–429.

Seidel, R. J. (1959) A review of sensory preconditioning. *Psych. Bull.*, **56**, 58–73.

Seitz, A. (1940/1941) Die Paarbildung bei einigen Cichliden: I. *Z. Tierpsychol.*, **4**, 40–84.

Seitz, A. (1943) Die Paarbildung bei einigen Cichliden: II. *Z. Tierpsychol.*, **5**, 74–101.

Seitz, P. E. D. (1954) The effects of infantile experiences upon adult behavior in animal subjects: 1. Effects of litter size during infancy upon adult behavior in the rat. *Amer. J. Psychiatr.*, **110**, 916–927.

Selander, R. K. (1966) Sexual dimorphism and differential niche utilization in birds. *Condor*, **68**, 113–151.

Senden, M. von (1932) Raum- und Gestaltauffassung bei operierten Blindgeborenen vor und nach der Operation. Barth, Leipzig.

Sevenster, P. (1961) A causal analysis of a displacement activity (Fanning in *Gasterosteus aculeatus* L.). *Behaviour Suppl.* **9**, 1–170.

Sevenster, P. (1968) Motivation and learning in sticklebacks. In "The Central Nervous System and Fish Behavior", Ed. D. Ingle, Univ. Chicago Press.

Sevenster-Bol, A. C. A. (1962) On the causation of drive reduction after a consummatory act. *Arch. néerl. Zool.*, **15**, 175–236.

Sexton, O. J., and Fitch, J. (1967) A test of Klopfer's empathic learning hypothesis. *Psychonomic Sci.*, **7**, 181–182.

Sharma, K. N. (1967) Receptor mechanisms in the alimentary tract: their excitation and functions. "Handbook of Physiol.", Sec. 6, Vol. I. Food and Water Intake, Amer. Physiol. Soc., Baltimore.

Sharma, K. N., Anand, B. K., Dua, S., and Singh, B. (1961) Role of stomach in regulation of activities of hypothalamic feeding centers. *Amer. J. Physiol.*, **201**, 593–598.

Sharpe, R. S., and Johnsgard, P. A. (1966) Inheritance of behavioral characters in F_2 mallard x pintail (*Anas platyrhynchos* L. x *Anas acuta* L.) hybrids. *Behaviour*, **27**, 259–272.

Sharpless, S. K. (1961) Effects of intravenous injections of epinephrine and norepinephrine in a choice situation. *J. comp. physiol. Psychol.*, **54**, 103–108.

Sharpless, S., and Jasper, H. (1956) Habituation of the arousal reaction. *Brain*, **79**, 655–680.

Shaw, E. (1960) The development of schooling behavior in fishes. *Physiol. Zool.*, **33**, 79–86.

Shaw, E. (1962) Environmental conditions and the appearance of sexual behavior in the platyfish. In Bliss (1951).

Sheer, D. E. (1961) Electrical Stimulation of the Brain. Hogg Foundation for Mental Health, Univ. of Texas.

Sheffield, F. D., and Campbell, B. A. (1954) The role of experience in the "spontaneous" activity of hungry rats. *J. comp. physiol. Psychol.*, **47**, 97–100.

Sheffield, F. D., and Roby, T. B. (1950) Reward value of a non-nutritive sweet taste. *J. comp. physiol. Psychol.*, **43**, 471–481.

Sheffield, F. D., Wulff, J. J., and Backer, R. (1951) Reward value of copulation without sex drive reduction. *J. comp. physiol. Psychol.*, **44**, 3–8.

Sheldon, M. H. (1968) Learning. In Weiskrantz (1968).

Sheridan, J. A., Cimbalo, R. S., Sills, J. A., and Alluisi, E. A. (1966) Effects of darkness, constant illumination and synchronised photic stimulation on auditory sensitivity to pulsed tones. *Psychon. Sci.*, **5**, 311–312.

Sherrington, C. S. (1906) Integrative Action of the Nervous System. Cambridge Univ. Press.

Sherrington, C. S. (1910) Flexion-reflex of the limb, crossed extension reflex, and reflex stepping and standing. *J. Physiol.*, **40**, 28–121.

Sherrington, C. S. (1913) Further observations on the production of reflex stepping by combination of reflex excitation with reflex inhibition. *J. Physiol.*, **47**, 196–214.

Sherrington, C. S. (1917) Reflexes elicitable in the cat from pinna vibrissae and jaws. *J. Physiol.*, **51**, 404–431.

Sherrington, C. S. (1931) Quantitative management of contraction in lowest level coordination. *Brain*, **54,** 1–28.

Shinkman, P. G. (1962) Visual depth discrimination in animals. *Psychol. Bull.*, **59,** 489–501.

Shinkman, P. G. (1963) Visual depth-discrimination in day-old chicks. *J. comp. physiol. Psychol.*, **56,** 410–414.

Shipley, W. U. (1963) The demonstration in the domestic guinea pig of a process resembling classical imprinting. *Anim. Behav.*, **11,** 470–474.

Shreck, P. K., Sterritt, G. M., Smith, M. P., and Stilson, D. W. (1963) Environmental factors in the development of eating in chicks. *Anim. Behav.*, **11,** 306–309.

Sibley, C. G. (1961) Hybridization and isolating mechanisms. In "Vertebrate Speciation", Ed. G. W. F. Blair, 69–88, Texas Univ. Press.

Siegel, A. I. (1953a) Deprivation of visual form definition in the ring dove: I. Discriminatory learning. *J. comp. physiol. Psychol.*, **46,** 115–119.

Siegel, A. I. (1953b) Deprivation of visual form definition in the ring dove: II. Perceptual-motor transfer. *J. comp. physiol. Psychol.*, **46,** 249–252.

Siegel, P. B., Phillips, R. E., and Folsom, E. F. (1965) Genetic variation in the crow of adult chickens. *Behaviour*, **24,** 229–235.

Siegel, P. S., and Brantley, J. J. (1951) The relationship of emotionality to the consummatory response of eating. *J. exp. Psychol.*, **42,** 304–306.

Siegel, P. S., and Dorman, L. B. (1954) Food intake of the rat following the intragastric administration of "hungry" and "satiated" blood. *J. comp. physiol Psychol.*, **47,** 227–229.

Siegel, P. S., and Siegel, H. S. (1949) The effect of emotionality on the water intake of the rat. *J. comp. physiol. Psychol.*, **42,** 12–16.

Siegel, S. (1967) Overtraining and transfer processes. *J. comp. physiol. Psychol.*, **64,** 471–477.

Sigg, E. B., Day, C., and Colombo, C. (1966) Endocrine factors in isolation-induced aggressiveness in rodents. *Endocrinology*, **78,** 679–684.

Simmons, K. E. L. (1955) The nature of the predator-reactions of waders towards humans; with special reference to the role of the aggressive-, escape- and brooding-drives. *Behaviour*, **8,** 130–173.

Simmons, K. E. L. (1961) Problems of head-scratching in birds. *Ibis*, **103a,** 37–49.

Simner, M. L. (1967) Age changes in preference for visual patterned stimulation in the newly hatched chick. *Proc. 75th Ann. Convention, Amer. Psychol. Ass.*, 107–108.

Simons, L. A., Dunlop, C. W., Webster, W. R., and Aitkin, L. M. (1966) Acoustic habituation in cats as a function of stimulus rate and the role of temporal conditioning of the middle ear muscles. *Electroenceph. Clin. Neurophysiol.*, **20,** 485–493.

Simpson, G. G. (1961) Principles of Animal Taxonomy. Columbia Univ. Press, New York.

Simpson, M. J. A. (1968) The display of the Siamese fighting fish, *Betta splendens. Anim. Behav. Monogr.*, 1, No. **1.**

Skinner, B. F. (1938) The Behavior of Organisms; an Experimental Analysis. Appleton-Century-Crofts, New York.

Skinner, B. F. (1948) "Superstition" in the pigeon. *J. exp. Psychol.*, **38,** 168–172.

Skinner, B. F. (1966) Operant behaviour. In Honig (1966).

Sladen, W. J. L. (1955) The pygoscelid penguins. *F.I.D.S., Sci. Rep.*, **17,** London.

Slotnick, B. M. (1967) Intercorrelations of maternal activities in the rat. *Anim. Behav.*, **15,** 267–269.

Sluckin, W. (1962) Perceptual and associative learning. *Symp. zool. Soc. Lond.*, **8,** 193–198.

Sluckin, W., and Salzen, E. A. (1961) Imprinting and perceptual learning. *Q. J. exp. Psychol.*, **13,** 65–77.

Smith, C. J. (1959) Mass action and early environment in the rat. *J. comp. physiol. Psychol.*, **52,** 154–56.

Smith, F. V. (1962) Perceptual aspects of imprinting. *Symp. zool. Soc. Lond.*, **8,** 171–191.

Smith, F. V., and Bird, M. W. (1964a) The approach response of chicks in groups in relation to the strength of the stimulus. *Anim. Behav.*, **12,** 252–258.

Smith, F. V., and Bird, M. W. (1964b) The correlation of responsiveness to visual and auditory stimuli in the domestic chick. *Anim. Behav.*, **12,** 259–263.

Smith, K. U. (1962) Delayed Sensory Feedback and Behavior. W. B. Saunders, Philadelphia and London.

Smith, M. H. (1966) Effects of intravenous injections on eating. *J. comp. physiol. Psychol.*, **61,** 11–14.

Smith, M., and Duffy, M. (1955) The effects of intragastric injection of various substances on subsequent bar-pressing. *J. comp. physiol. Psychol.*, **48,** 387–391.

Smith, M., and Duffy, M. (1957) Some physiological factors that regulate eating behavior. *J. comp. physiol. Psychol.*, **50,** 601–608.

Smith, M. H., Pool, R., and Weinberg, H. (1962) The role of bulk in the control of eating. *J. comp. physiol. Psychol.*, **55,** 115–120.

Smith, H. H., Salisbury, R., and Weinberg, H. (1961) The reaction of hypothalamic-hyperphagic rats to stomach preloads. *J. comp. physiol. Psychol.*, **54,** 660–664.

Smith, N. G. (1966a) Evolution of some arctic gulls (*Larus*): an experimental study of isolating mechanisms. Amer. Ornith. Union, *Ornithol. Monogr.*, **4,** 1–99.

Smith, N. G. (1966b) Adaptations to cliff-nesting in some arctic gulls (*Larus*). *Ibis*, **108,** 68–83.

Smith, O. A. (1961) Food intake and hypothalamic stimulation. In Sheer (1961).

Smith, W. J. (1965) Message, meaning and context in ethology. *Amer. Nat.*, **99,** 405–409.

Snowdon, C. T., Bell, D. D., and Henderson, N. D. (1964) Relationships between heart rate and open field behavior. *J. comp. physiol. Psychol.*, **58,** 423–426.

Sokolov, E. N. (1960) Neuronal models and the orienting reflex. In "The Central Nervous System and Behavior", Ed. M. A. B. Brazier, Macy Foundation, New York.

Sokolov, E. N. (1966) Neuronal mechanisms of the orienting reflex. *Proc. 18th Int. Cong. Psychol., Moscow,* 31–33.

Solomon, P., *et al.* (1961) Sensory Deprivation. Harvard Univ. Press.

Solomon, R. L., and Turner, L. H. (1962) Discriminative classical conditioning in dogs paralyzed by curare can later control discriminative avoidance responses in the normal state. *Psychol. Rev.*, **69,** 202–219.

Southwick, C. H., Beg, M. A., and Siddiqi, M. R. (1965) Rhesus monkeys in North India. In de Vore (1965).

Spalding, D. A. (1873) Instinct with original observations on young animals. *MacMillan's Magazine,* **27,** 282–293, reprinted *Brit. J. anim. Behav.*, **2,** 2–11.

Sparks, J. H. (1965) On the role of allopreening invitation behaviour in reducing aggression among red avadavats. *Proc. zool. Soc. Lond.*, **145,** 387–404.

Spence, J. T., and Maher, B. A. (1962) Handling and noxious stimulation of the albino rat: I. Effects on subsequent emotionality. *J. comp. physiol. Psychol.*, **55,** 247–251.

Spence, K. W. (1956) Behavior Theory and Conditioning. Yale Univ. Press, New Haven.

Spencer, W. A., Thompson, R. F., and Neilson, D. R. (1966) Decrement of ventral root electrotonus and intracellularly recorded PSP's produced by iterated cutaneous afferent volleys. *J. Neurophysiol.*, **29,** 253–274.

Spencer-Booth, Y. (1968) The behaviour of group companions towards rhesus monkey infants. *Anim. Behav.*, **16,** 541–557.

Spencer-Booth, Y., and Hinde, R. A. (1967) The effects of separating rhesus monkey infants from their mothers for six days. *J. Child. Psychol. Psychiatr.*, **7,** 179–197.

Spencer-Booth, Y., Hinde, R. A., and Bruce, M. (1965) Social companions and the mother-infant relationship in rhesus monkeys. *Nature,* **208,** 301.

Sperry, R. W. (1941) The effect of crossing nerves to antagonistic muscles in the hind limb of the rat. *J. comp. Neurol.*, **75,** 1–19.

Sperry, R. W. (1943) Effect of 180 degrees rotation of the retinal field on visuomotor coordination. *J. exp. Zool.*, **92,** 263–279.

Sperry, R .W. (1947) Effect of crossing nerves in antagonistic limb muscles in the monkey. *Arch. Neurol. Psychiatr., Lond.*, **58,** 452–473.

Sperry, R. W. (1950) Neural basis of the spontaneous optokinetic response produced by visual inversion. *J. comp. physiol. Psychol.*, **43,** 482–489.

Sperry, R. W. (1951) Mechanisms of neural maturation. In "Handbook of Experimental Psychology", Ed. S. S. Stevens, Wiley, New York.

Sperry, R. W. (1958) Physiological plasticity and brain circuit theory. In Harlow and Woolsey (1958).

Sperry, R. W. (1963) Chemoaffinity in the orderly growth of nerve fibre patterns and connections. *Proc. Nat. Acad. Sci.*, **50**, 703–710.

Sperry, R. W. (1964) Problems outstanding in the evolution of brain function. James Arthur Lecture, *Amer. Mus. Nat. Hist.*, New York.

Sperry, R. W., and Deupree, N. (1956) Functional recovery following alterations in nerve-muscle connections of fishes. *J. comp. Neurol.*, **106**, 143–158.

Sperry, R. W., and Hibbard, E. (1968) Regulative factors in the orderly growth of retino-tectal connexions. In "Growth of the Nervous System", Ed. G. E. W. Wolstenholme and M. O'Connor, Churchill, London.

Spies, G. (1965) Food versus intracranial self-stimulation reinforcement in food-deprived rats. *J. comp. physiol. Psychol.*, **60**, 153–157.

Spinelli, D. N. (1966) Visual receptive fields in the cat's retina: complications. *Science*, **152**, 1768–1769.

Stachnik, T. J., Ulrich, R. E., and Mabry, J. H. (1966) Reinforcement of aggression through intracranial stimulation. *Psychon. Sci.*, **5**, 101–102.

Stamm, R. A. (1962) Aspekte des Paarverhaltens von *Agapornis personata* Reichenow (Aves, Psittacidae, Loriini). *Behaviour*, **19**, 1–56.

Stanley, W. C., and Bacon, W. E. (1963) Suppression of sucking behaviour in non-deprived puppies. *Psychol. Rep.*, **13**, 175–178.

Steel, E. A., and Hinde, R. A. (1964) Effect of exogenous oestrogen on brood patch development of intact and ovariectomized canaries. *Nature*, **202**, 718–719.

Steffens, A. B. (1967) Blood glucose levels and food intake in normal and hypothalamic hyperphagic rats. *Acta physiol. pharmacol. Neurol.*, **14**, 524–526.

Stein, L. (1958) Secondary reinforcement established with subcortical stimulation. *Science*, **127**, 466–467.

Steinberg, J., and Bindra, D. (1962) Effects of pregnancy and salt-intake on genital licking. *J. comp. physiol. Psychol.*, **55**, 103–106.

Stellar, E. (1954) The physiology of motivation. *Psychol. Rev.*, **61**, 5–22.

Stellar, E. (1960) Drive and Motivation. In "Handbook of Physiology", Sec. 1, Vol. III, Ed. J. Field, H. W. Magoun, and V. E. Hall, Amer. Physiol. Soc., Washington.

Stellar, E., and Hill, J. H. (1952) The rat's rate of drinking as a function of water deprivation. *J. comp. physiol. Psychol.*, **45**, 96–102.

Stellar, E., Hyman, R., and Samet, S. (1954) Gastric factors controlling water- and salt-solution-drinking. *J. comp. physiol. Psychol.*, **47**, 220–226.

Sterritt, G. M. (1962) Inhibition and facilitation of eating by electric shock. *J. comp. physiol. Psychol.*, **55**, 226–229.

Sterritt, G. M. (1965) Inhibition and facilitation of eating by electric shock. III. *Psychon. Sci.*, **2**, 319–320.

Sterritt, G. M., and Smith, M. P. (1965) Reinforcement effects of specific components of feeding in young leghorn chicks. *J. comp. physiol. Psychol.*, **59,** 171–175.

Stettner, L. J., Schultz, W. J., and Levy, A. (1967) Successive reversal learning in the bob-white quail (*Colurius virginianus*). *Anim. Behav.*, **15,** 1–5.

Stettner, L. J., and Tilds, B. N. (1966) Effect of presence of an imprinted object on response of ducklings in an open field and when exposed to a fear stimulus. *Psychon. Sci.*, **4,** 107–108.

Stevens, S. S. (1951) Handbook of Experimental Psychology. Wiley, New York.

Stevens, S. S. (1958) Measurement and man. *Science,* **127,** 383–389.

Stevens, S. S. (1961) To honor Fechner and repeal his law. *Science,* **133,** 80–86.

Stevenson, H. W., Hess, E. H., and Rheingold, H. L. (1967) Early behavior. Wiley, New York.

Stevenson, J. G. (1967) Reinforcing effects of chaffinch song. *Anim. Behav.*, **15,** 427–432.

Stevenson, J. G. (1969) Song as a reinforcer. In Hinde (1969).

Stewart, J., and Palfai, T. (1967) Castration, androgen, and dominance status in the rat. *Psychonomic Sci.*, **7,** 1–2.

Stichmann, W. (1962) Transpositionsversuche mit Haushuhnrassen stark verschiedener Körpergrösse. *Z. Tierpsychol.*, **19,** 290–320.

Stokes, A. W. (1962a) Agonistic behaviour among blue tits at a winter feeding station. *Behaviour,* **19,** 118–138.

Stokes, A. W. (1962b) The comparative ethology of great, blue, marsh and coal tits at a winter feeding station. *Behaviour,* **19,** 208–218.

Stone, J., and Fabian, M. (1966) Specialized receptive fields of the cat's retina. *Science,* **152,** 1277–1279.

Stone, L. S. (1948) Functional polarisation in developing and regenerating retinae of transplanted eyes. *Ann. N.Y. Acad. Sci.*, **49,** 856–865.

Stout, J. F. (1963) The significance of sound production during the reproductive behaviour of *Notropis analostanus* (Family Cyprinidae). *Anim. Behav.*, **11,** 83–92.

Stratton, G. (1897) Vision without inversion of the retinal image. *Psychol. Rev.*, **4,** 341–360.

Stresemann, E. (1927–1934) Aves in Kükenthal-Krumbach, Handbüch der Zoologie. Vol. VIII. Berlin and Leipzig.

Stretch, R. G. A. (1960) Exploratory behaviour in the rat. *Nature,* **186,** 454–456.

Strong, P. N., and Hedges, M. (1966) Comparative studies in simple oddity learning: 1. Cats, raccoons, monkeys and chimpanzees. *Psychonomic Sci.*, **5,** 13–14.

Strongman, K. T. (1965) The effect of anxiety on food intake in the rat. *Q. J. exp. Psychol.*, **17,** 255–260.

Surtees, G. (1964a, b) Laboratory studies on dispersion behaviour of adult beetles in grain. VIII and IX. *Anim. Behav.*, **12,** 374–377, 378–381.

Sutherland, N. S. (1957a) Visual discrimination of orientation and shape by the *Octopus*. *Nature*, **179**, 11–13.

Sutherland, N. S. (1957b) Spontaneous alternation and stimulus avoidance. *J. comp. physiol. Psychol.*, **50**, 358–362.

Sutherland, N. S. (1959) Stimulus analysing mechanisms. In Proc. Sym. Mechanization of Thought Processes. Her Majesty's Stationery Office, London, 575–609.

Sutherland, N. S. (1961) The methods and findings of experiments on the visual discrimination of shape by animals. *Exp. Psychol. Soc. Monogr.* **1.**

Sutherland, N. S. (1962) Visual discrimination of shape by *Octopus*: squares and crosses. *J. comp. physiol. Psychol.*, **55**, 939–943.

Sutherland, N. S. (1963a) Cat's ability to discriminate oblique rectangles. *Science*, **139**, 209–210.

Sutherland, N. S. (1963b) Shape discrimination and receptive fields. *Nature*, **197**, 118–122.

Sutherland, N. S. (1964) The learning of discrimination by animals. *Endeavour*, **23**, 140–152.

Sutherland, N. S. (1966) Partial reinforcement and breadth of learning. *Q. J. exp. Psychol.*, **18**, 289–301.

Sutherland, N. S. (1969) Shape discrimination in rat, octopus and goldfish. *J. comp. physiol. Psychol..*, **67**, 160–176.

Sutherland, N. S., and Andelman, L. (1967) Learning with one and two cues. *Psychon. Sci.*, **7**, 107–108.

Sutherland, N. S., and Holgate, V. (1966) Two-cue discrimination learning in rats. *J. comp. physiol. Psychol.*, **67**, 160–176.

Sutherland, N. S., and Mackintosh, J. (1964) Discrimination learning: non-additivity of cues. *Nature*, **201**, 528–530.

Sutherland, N. S., Mackintosh, N. J., and Mackintosh, J. (1963) Simultaneous discrimination training of *Octopus* and transfer of discrimination along a continuum. *J. comp. physiol. Psychol.*, **56**, 150–156.

Swanson, H. H. (1967) Alteration of sex-typical behaviour of hamsters in open field and emergence tests by neonatal administration of androgen or oestrogen. *Anim. Behav.*, **15**, 209–216.

Symons, P. E. K. (1966) Analysis of spine-raising in the male three-spined stickleback. *Behaviour*, **26**, 1–75.

Székely, G. (1968) Development of limb movements: embryological, physiological and model studies. In "Growth of the Nervous System", Ed. G. E. W. Wolstenholme and M. O'Connor, Churchill, London.

Szlep, R. (1952) On the plasticity of instinct of a garden spider (*Aranea diadema* L.) construction of a cobweb. *Acta Biol. Exp.*, **16**, 5–22.

Tallarico, R. B., and Farrell, W. M. (1964) Studies of visual depth perception: an effect of early experience on chicks on a visual cliff. *J. comp. physiol. Psychol.*, **57**, 94–96.

Tapp, J. T., and Markowitz, H. (1963) Infant handling: effects on avoidance learning, brain weight and cholinesterase activity. *Science*, **140**, 486–487.

Tapp, J. T., and Simpson, L. L. (1966) Motivational and response factors as determinants of the reinforcing value of light onset. *J. comp. physiol. Psychol.*, **62,** 143–146.

Taub, E., Bacon, R. C., and Berman, A. J. (1965) Acquisition of a trace-conditioned avoidance response after deafferentation of the responding limb. *J. comp. physiol. Psychol.*, **59,** 275–279.

Taub, E., Ellman, S. J., and Berman, A. J. (1966a) Deafferentation in monkeys: effect on conditioned grasp response. *Science*, **151,** 593–594.

Taub, E., Teodoru, D., Ellman, S. J., Bloom, R. F., and Berman, A. J. (1966b) Deafferentation in monkeys. *Psychon. Sci.*, **4,** 323–324.

Taylor, J. G. (1962) The Behavioural Basis of Perception. Yale Univ. Press.

Tees, R. C. (1967) Effects of early auditory restriction in the rat on adult pattern discrimination. *J. comp. physiol. Psychol.*, **63,** 389–393.

Teghtsoonian, R., and Campbell, B. A. (1960) Random activity of the rat during food deprivation as a function of environmental conditions. *J. comp. physiol. Psychol.*, **53,** 242–244.

Teitelbaum, P. (1955) Sensory control of hypothalamic hyperphagia. *J. comp. physiol. Psychol.*, **48,** 156–163.

Teitelbaum, P. (1957) Random and food-directed activity in hyperphagic and normal rats. *J. comp. physiol. Psychol.*, **50,** 486–490.

Teitelbaum, P. (1961) Disturbances in feeding and drinking behaviour after hypothalamic lesions. *Nebr. Symp. Motiv.*, **9,** 39–65.

Teitelbaum, P. (1966) The use of operant methods in the assessment and control of motivational states. In Honig (1966).

Teitelbaum, P., and Campbell, B. A. (1958) Ingestion patterns in hyperphagic and normal rats. *J. comp. physiol. Psychol.*, **51,** 135–141.

Tembrock, G. (1958) Spielverhalten beim Rotfuchs. *Zool. Beitr.*, **3,** 423–496.

Tembrock, G. (1962a) Versuch einer Analyse des Imponierverhaltens beim Rotfuchs, *Vulpes vulpes* (L.). *Z. Tierpsychol.*, **19,** 577–585.

Tembrock, G. (1962b) Zur Strukturanalyse des Kampfverhaltens bei *Vulpes*. *Behaviour*, **19,** 261–282.

Tenen, S. S., and Miller, N. E. (1964) Strength of electrical stimulation of lateral hypothalamus, food deprivation, and tolerance for quinine in food. *J. comp. physiol. Psychol.*, **58,** 55–62.

Terrace, H. S. (1963) Errorless transfer of a discrimination across two continua. *J. exp. Anal. Behav.*, **6,** 223–232.

Terrace, H. S. (1966) Stimulus control. In Honig (1966).

Teuber, H.-L. (1960) Perception. In "Handbook of Physiology", Sec. I, Vol. III, Ed. J. Field, H. W. Magoun, and V. E. Hall, Amer. Physiol. Soc., Washington.

Teuber, H.-L. (1961) Sensory deprivation, sensory suppression and agnosia: notes for a neurologic theory. *J. nerv. ment. Dis.*, **132,** 32–40.

Thielcke, G. (1962) Versuche mit Klangattrapen zur Klärung der Verwandtschaft der Baumläufer *Certhia familiaris* L., *C. brachydactyla* Brehm, and *C. americana* Bonaparte. *J. Orn.*, **103,** 266–271.

Thielcke, G. (1969) Geographic variation in bird vocalizations. In Hinde (1969).

Thielcke, G., and Linsenmaier, K. E. (1963) Zur geographischen Variation des Gesanges des Zilpzalps *Phylloscopus collybita,* in Mittel- und Südwesteuropa mit einem Vergleich des Gesanges des Fitis *Phylloscopus trochilus. J. Orn.,* **104,** 372–402.

Thielcke-Poltz, H., and Thielcke, G. (1960) Akustisches Lernen verschieden alter schallisolierter Amseln (*Turdus merula* L.) und die Entwicklung erlernter Motive ohne und mit künstlichen Einfluss von Testosteron. *Z. Tierpsychol.,* **17,** 211–244.

Thines, G., and Heuts, B. (1968) The effect of submissive experiences on dominance and aggressive behaviour of *Xiphophorus* (Pisces, Poeciliidae). *Z. Tierpsychol.,* **25,** 139–154.

Thistlethwaite, D. (1951) A critical review of latent learning and related experiments. *Psychol. Bull.,* **48,** 97–129.

Thoman, E., Wetzel, A., and Levine, S. (1968) Learning in the neonatal rat. *Anim. Behav.,* **16,** 54–57.

Thomas, D. W., and Mayer, J. (1968) Meal taking and regulation of food intake by normal and hypothalamic hyperphagic rats. *J. comp. physiol. Psychol.,* **66,** 642–653.

Thompson, R. F. (1967) Foundations of Physiological Psychology. Harper & Row, New York.

Thompson, R. F., and Spencer, W. A. (1966) Habituation. *Psychol. Rev.,* **73,** 16–43.

Thompson, R. W., Koenigsberg, L. A., and Tennison, J. C. (1965) Effects of age on learning and retention of an avoidance response in rats. *J. comp. physiol. Psychol.,* **60,** 457–459.

Thompson, T. I. (1963) Visual reinforcement in Siamese fighting fish. *Science,* **141,** 55–57.

Thompson, T. I. (1964) Visual reinforcement in fighting cocks. *J. exp. Anal. Behav.,* **7,** 45–49.

Thompson, T. I. (1969a) Aggressive behavior of Siamese fighting fish: analysis and synthesis of conditioned and unconditioned components. In "Biology of Aggression", Ed. S. Garattin and E. Sigg, Excerpta Medica Foundation, Amsterdam.

Thompson, T. I. (1969b) Conditioned avoidance of the mobbing call by chaffinches. *Anim. Behav.* (in press).

Thompson, T. I., and Bloom, W. (1966) Aggressive behaviour and extinction-induced response rate increase. *Psychonomic Sci.,* **5,** 335–336.

Thompson, T. I., and Sturm, T. (1965a) Visual-reinforcer color, and operant behavior in Siamese fighting fish. *J. exp. Anal. Behav.,* **8,** 341–344.

Thompson, T. I., and Sturm, T. (1965b) Classical conditioning of aggressive display in Siamese fighting fish. *J. exp. Anal. Behav.,* **8,** 397–403.

Thompson, W. R. (1953) The inheritance of behaviour; behavioural differences in fifteen mouse strains. *Canad. J. Psychol.,* **7,** 145–153.

Thompson, W. R. (1957) Influence of prenatal maternal anxiety on emotionality in young rats. *Science,* **125,** 698–699.

Thompson, W. R., and Heron, W. (1954) The effects of restricting early experience on the problem-solving capacity of dogs. *Canad. J. Psychol.*, **8**, 17–31.

Thompson, W. R., Watson, J., and Charlesworth, W. R. (1962) The effects of prenatal maternal stress on offspring behavior in rats. *Psychol. Monogr.*, **76**, 1–26.

Thor, D. H., and Hoats, D. L. (1968) A circadian variable in self-exposure to light by the rat. *Psychon. Sci.*, **12**, 1–2.

Thorpe, W. H. (1945) The evolutionary significance of habitat selection. *J. Anim. Ecol.*, **14**, 67–70.

Thorpe, W. H. (1954) The process of song-learning in the chaffinch as studied by means of the sound spectrograph. *Nature*, **173**, 465–469.

Thorpe, W. H. (1958) The learning of song patterns by birds, with especial reference to the song of the chaffinch, *Fringilla coelebs*. *Ibis*, **100**, 535–570.

Thorpe, W. H. (1961a) Sensitive periods in the learning of animals and men. In Thorpe and Zangwill (1961).

Thorpe, W. H. (1961b) Bird-song. Cambridge Univ. Press.

Thorpe, W. H. (1963a) Learning and Instinct in Animals. (1st ed., 1956) Methuen, London.

Thorpe, W. H. (1963b) Ethology and the coding problem in germ cell and brain. *Z. Tierpsychol.*, **20**, 529–551.

Thorpe, W. H., and Jones, F. G. W. (1937) Olfactory conditioning and its relation to the problem of host selection. *Proc. Roy. Soc. Ser. B*, **124**, 56–81.

Thorpe, W. H., and North, M. E. W. (1965) Origin and significance of the power of vocal imitation: with special reference to the antiphonal singing of birds. *Nature*, **208**, 219–222.

Thorpe, W. H., and Zangwill, O. L. (1961) Current Problems in Animal Behavior. Cambridge Univ. Press.

Thorson, J. (1966) Small-signal analysis of a visual reflex in the locust. I and II. *Kybernetik*, **3**, 41–53 and 53–66.

Thorsteinson, A. J. (1960) Host selection in phytophagous insects. *An. Rev. Ent.*, **5**, 193–218.

Tinbergen, L. (1960) The natural control of insects in pinewoods: I. Factors influencing the intensity of predation by song birds. *Arch. néerl. Zool.*, **13**, 265–343.

Tinbergen, N. (1935) Uber die Orientierung des Bienenwolfes (*Philanthus triangulum* Fabr.). *Z. vergl. Physiol.*, **21**, 699–716.

Tinbergen, N. (1937) Uber das Verhalten kämpfender Kohlmeisen (*Parus m. major L.) Ardea*, **26**, 222–223.

Tinbergen, N. (1940–41) Die Ubersprungbewegung. *Z. Tierpsychol.*, **4**, 1–10.

Tinbergen, N. (1942) An objectivistic study of the innate behaviour of animals. *Biblioth. Biother.*, **1**, 39–98.

Tinbergen, N. (1948) Social releasers and the experimental method required for their study. *Wilson Bull.*, **60**, 6–51.

Tinbergen, N. (1951) The Study of Instinct. Clarendon Press, Oxford.

Tinbergen, N. (1952) Derived activities: their causation, biological significance, origin and emancipation during evolution. *Quart. Rev. Biol.*, **27**, 1–32.

Tinbergen, N. (1953) The Herring Gull's World. Collins, London.

Tinbergen, N. (1955) In "Trans. First Conf. on Group Processes (1954)", Ed. B. Schaffner, Macy Foundation, New York.

Tinbergen, N. (1957) The functions of territory. *Bird Study*, **4**, 14–27.

Tinbergen, N. (1959) Comparative studies of the behaviour of gulls (*Laridae*): a progress report. *Behaviour*, **15**, 1–70.

Tinbergen, N. (1962) The evolution of animal communication—a critical examination of methods. *Symp. zool. Soc. Lond.*, **8**, 1–6.

Tinbergen, N. (1963) On aims and methods of ethology. *Z. Tierpsychol.*, **20**, 410–433.

Tinbergen, N. (1965a) Behavior and natural selection. In "Ideas in Modern Biology", Ed. J. A. Moore, *Proc. XVI Int. Zool. Cong., Washington*, 521–542.

Tinbergen, N. (1965b) Some recent studies of the evolution of sexual behavior. In Beach (1965).

Tinbergen, N. (1967) Adaptive features of the black-headed gull *Larus ridibundus* L. *Proc. XIV Int. Orn. Cong.*, 43–59.

Tinbergen, N. (1968) Book review. *Anim. Behav.*, **16**, 398–399.

Tinbergen, N., Broekhuysen, G. J., Feekes, F., Houghton, J. C. W., Kruuk, H., and Szulc, E. (1962) Egg shell removal by the black-headed gull, *Larus ridibundus*, L.; a behaviour component of camouflage. *Behaviour*, **19**, 74–117.

Tinbergen, N., Impekoven, M., and Franck, D. (1967) An experiment on spacing-out as a defence against predation. *Behaviour*, **28**, 307–321.

Tinbergen, N., Kruuk, H., and Paillette, M. (1962) Egg shell removal by the black-headed gull, *Larus r. ridibundus*: II. Bird Study, **9**, 123–131.

Tinbergen, N., and Kruyt, W. (1938) Uber die Orientierung des Bienenwolfes (*Philanthus triangulum* Fabr.): III. Die Bevorzugung bestimmter Wegmarken. *Z. vergl. Physiol.*, **25**, 292–334.

Tinbergen, N., and Kuenen, D. J. (1939) Uber die auslösenden und die richtunggebenden Reizsituationen der Sperrbewegung von jungen Drosseln (*Turdus m. merula* L. und *T. e. ericetorum* Turton). *Z. Tierpsychol.*, **3**, 37–60.

Tinbergen, N., Meeuse, B. J. D., Boerema, L. K., and Varossieau, W. W. (1943) Die Balz des Samtfalters, *Eumenis* (= *Satyrus*) *semele* (L.). *Z. Tierpsychol.*, **5**, 182–226.

Tinbergen, N., and Perdeck, A. C. (1950) On the stimulus situation releasing the begging response in the newly hatched herring gull chick (*Larus argentatus argentatus* Pont.). *Behaviour*, **3**, 1–39.

Tinkelpaugh, O. L. (1928) An experimental study of representative factors in monkeys. *J. comp. Psychol.*, **8**, 197–236.

Tirala, L. G. (1923) Die Form als Reiz. *Zool. Jb. Allg. Zool. Physiol.*, **39**, 395–442.

Tobach, E., and Schneirla, T. C. (1962) Eliminative responses in mice and rats and the problem of "emotionality". In Bliss (1962).

Tolman, C. W. (1964) Social facilitation of feeding behaviour in the domestic chick. *Anim. Behav.*, **12**, 245–251.

Tolman, C. W. (1967a) The effects of tapping sounds on feeding behaviour of domestic chicks. *Anim. Behav.*, **15**, 145–148.

Tolman, C. W. (1967b) The feeding behaviour of domestic chicks as a function of rate of pecking by a surrogate companion. *Behaviour*, **29**, 57–62.

Tolman, C. W. (1968) The varieties of social stimulation in the feeding behaviour of domestic chicks. *Behaviour*, **30**, 275–286.

Tolman, E. C. (1932) Purposive Behavior in Animals and Men. Century, New York.

Tolman, E. C. (1959) Principles of purposive behavior. In Koch (1959).

Tolman, E. C., Ritchie, B. F., and Kalish, D. (1946/1947) Studies in spatial learning: I and V. *J. exp. Psychol.*, **36**, 221–229; **37**, 285–292.

Toulmin, S. E. (1953) The Philosophy of Science. Hutchinson, London.

Towbin, E. J. (1949) Gastric distension as a factor in the satiation of thirst in esophagostomized dogs. *Amer. J. Physiol.*, **159**, 533–541.

Towbin, E. J. (1964) The role of the gastrointestinal tract in the regulation of water intake. In "Thirst", Ed. M. J. Wayner, Pergamon, New York.

Tracy, H. C. (1926) The development of motility and behavior reactions in the toadfish (*Opsanus tau*). *J. comp. Neurol.*, **40**, 253–269.

Treichler, F. R., and Hamilton, D. M. (1967) Relationships between deprivation and air-drinking behavior. *J. comp. physiol. Psychol.*, **63**, 541–544.

Tretzel, E. (1967) Imitation und Transposition menschlicher Pfiffe durch Amseln *Turdus m. merula* L. *Z. Tierpsychol.*, **24**, 137–161.

Trevarthen, C. B. (1967–68) Two mechanisms of vision in primates. *Psychol. Forschung*, **31**, 299–337.

Tribukait, B. (1956) Die Aktivitätsperiodik der weissen Maus im Kunsttag von 16–29 Stunden Länge. *Z. vergl. Physiol.*, **38**, 479–490.

Trimble, M. R., and Herbert, J. (1968) The effect of testosterone or oestradiol upon the sexual and associated behaviour of the adult female rhesus monkey. *J. Endocrin.*, **42**, 171–185.

Trowill, J. A. (1967) Instrumental conditioning of the heart rate in the curarized rat. *J. comp. physiol. Psychol.*, **63**, 7–11.

Trumler, E. (1959) Das "Rossigkeitsgesicht" und ähnliches Ausdrucksverhalten bei Einhufern. *Z. Tierpsychol.*, **16**, 478–488.

Tsang, Y. C. (1937) Maze learning in rats hemidecorticated in infancy. *J. comp. Psychol.*, **24**, 221–254.

Tsang, Y. C. (1938) Hunger motivation in gastrectomized rats. *J. comp. Psychol.*, **26**, 1–17.

Tschanz, B. (1959) Zur Brutbiologie der Trottellumme (*Uria aalge aalge* Pont.). *Behaviour*, **14**, 1–100.

Tschanz, B. (1968) Trottellummen. *Z. Tierpsychol.*, Beiheft, **4.**

Tucker, D., and Beidler, L. M. (1956) Autonomic nervous system influence on olfactory receptors. *Amer. J. Physiol.*, **187,** 637.

Tugendhat, B. (1960a) The normal feeding behavior of the three-spined stickleback (*Gasterosteus aculeatus*). *Behaviour*, **15,** 284–318.

Tugendhat, B. (1960b) The disturbed feeding behavior of the three-spined stickleback: I. Electric shock is administered in the food area. *Behaviour*, **16,** 159–187.

Tugendhat Gardner, B. (1964) Hunger and sequential responses in the hunting behavior of Salticid spiders. *J. comp. physiol. Psychol.*, **58,** 167–173.

Turner, E. R. A. (1964) Social feeding in birds. *Behaviour*, **24,** 1–46.

Tyhurst, J. S. (1951) Individual reactions to community disaster. *Amer. J. Psychiatr.*, **107,** 764–769.

Tyler, S. (1969) Ph.D. thesis, Cambridge.

Uexküll, J. von (1934) Streifzüge durch die Umwelten von Tieren und Menschen. Springer Berlin. Translated in "Instinctive Behaviour" (1957), Ed. C. H. Schiller, Methuen, London.

Ullyott, P. (1936) The behaviour of *Dendrocoelum lacteum:* I and II. *J. exp. Biol.*, **13,** 253–264, 265–278.

Ulrich, R. E. (1966) Pain as a cause of aggression. *Amer. Zool.*, **6,** 643–662.

Ulrich, R. E., Johnston, M., Richardson, J., and Wolff, P. C. (1963) The operant conditioning of fighting behavior in rats. *Psychol. Rec.*, **13,** 465–470.

Ulrich, R. E., Stachnik, T. J., Brierton, G. R., and Mabry, J. H. (1966) Fighting and avoidance in response to aversive stimulation. *Behaviour*, **26,** 124–129.

Ulrich, R. E., Wolff, P. C., and Azrin, N. H. (1964) Shock as an elicitor of intra- and inter-species fighting behaviour. *Anim. Behav.*, **12,** 14–15.

Uttley, A. M. (1954) The classification of signals in the nervous system. *EEG clin. Neurophysiol.*, **6,** 479–494.

Uyeno, E. T., and White, M. (1967) Social isolation and dominance behavior. *J. comp. physiol. Psychol.*, **63,** 157–159.

Valenstein, E. S. (1964) Problems of measurement and interpretation with reinforcing brain stimulation. *Psychol. Rev.*, **71,** 415–438.

Valenstein, E. S. (1968) Biology of drives. *Neurosci. Res. Bull.*, **6.**

Valenstein, E. S., Cox, V. C., and Kakolewski, J. W. (1968) Modification of motivated behavior elicited by electrical stimulation of the hypothalamus. *Science*, **159,** 1119–1120.

Valenstein, E. S., and Goy, R. W. (1957) Further studies of the organization and display of sexual behavior in male guinea pigs. *J. comp. physiol. Psychol.*, **50,** 115–119.

Valenstein, E. S., Riss, W., and Young, W. C. (1955) Experiential and genetic factors in the organization of sexual behavior in male guinea pigs. *J. comp. physiol. Psychol.*, **48,** 397–403.

Varley, M., and Symmes, D. (1966) The hierarchy of dominance in a group of macaques. *Behaviour,* **27,** 54–75.

Vaughan, E., and Fisher, A. E. (1962) Male sexual behavior induced by intracranial electrical stimulation. *Science,* **137,** 758–760.

Verney, E. B. (1947) The antidiuretic hormone and the factors which determine its release. *Proc. Roy Soc. B,* **135,** 25–106.

Vernon, M. D. (1952) A Further Study of Visual Perception. Cambridge Univ. Press.

Vernon, W., and Ulrich, R. (1966) Classical conditioning of pain-elicited aggression. *Science,* **152,** 668.

Verplanck, W. S. (1954) Burrhus F. Skinner. In "Modern Learning Theory", Ed. Estes *et al.,* Appleton-Century-Crofts, New York.

Verplanck, W. S., and Hayes, J. R. (1953) Eating and drinking as a function of maintenance schedule. *J. comp. physiol. Psychol.,* **46,** 327–333.

Verwey, J. (1930) Die Paarungsbiologie des Fischreihers. *Zool. Jb., Abt. allg. Zool. Physiol.,* **48,** 1–120.

Vestal, B. M., and King, J. A. (1968) Relationship of age at eye opening to first optokinetic response in deermice (*Peromyscus*). *Developmental Psychobiology,* **1,** 30–34.

Vilks, E. K. (1966) Experimental study of instinctive behaviour of birds under natural conditions. *Proc. 18th Int. Cong. Psychol. Moscow,* **1,** 82–83.

Vince, M. A. (1960) Developmental changes in responsiveness in the great tit (*Parus major*). *Behaviour,* **15,** 219–243.

Vince, M. A. (1961a) Developmental changes in learning capacity. In Thorpe and Zangwill (1961).

Vince, M. A. (1961b) "String-pulling" in birds: III. The successful response in greenfinches and canaries. *Behaviour,* **17,** 103–129.

Vince, M. A. (1964a) Use of the feet in feeding by the great tit *Parus major. Ibis,* **106,** 508–529.

Vince, M. A. (1964b) Social facilitation of hatching in the bobwhite quail. *Anim. Behav.,* **12,** 531–534.

Vince, M. A. (1969) Embryonic communication, respiration and the synchronisation of hatching. In Hinde (1969).

Vince, M. A., and Warren, R. P. (1963) Individual differences in taste discrimination in the great tit (*Parus major*). *Anim. Behav.,* **11,** 548–552.

Vinogradova, O. S. (1966) Dynamic classification of the reactions of hippocampal neurons to sensory stimuli. *Fed. Proc. Trans. Suppl.,* **25,** T397–T403.

Volkonsky, M. (1939) Sur la photo-akinèse des acridiens. *Arch. Inst. Pasteur Algér.,* **17,** 194–220.

Voronin, L. G. (1962) Some results of comparative-physiological investigations of higher nervous activity. *Psychol. Bull.,* **59,** 161–195.

Voss, C. (1967) Uber das Formensehen der roten Waldameise (*Formica rufa* Gruppe). *Z. vergl. Physiol.,* **55,** 225–254.

Vowles, D. M. (1954a) The orientation of ants: I. The substitution of stimuli. *J. exp. Biol.*, **31**, 341–355.

Vowles, D. M. (1954b) The orientation of ants: II. Orientation to light, gravity and polarized light. *J. exp. Biol.*, **31**, 356–375.

Vowles, D. M., and Harwood, D. (1966) The effect of exogenous hormones on aggressive and defensive behaviour in the ring dove *(Streptopelia risoria)*. *J. Endocrin.*, **36**, 35–51.

Wagner, A. R., Thomas, E., and Norton, T. (1967) Conditioning with electrical stimulation of the motor cortex. *J. comp. physiol. Psychol.*, **64**, 191–199.

Wahlström, G. (1964) The circadian rhythm of the canary studied by self-selection of light and darkness. *Acta Soc. Med. Upsal.*, **69**, 241–271.

Walden, A. M. (1968) Studies of exploratory behavior in the albino rat. *Psychol. Rep.*, **22**, 483–493.

Waldron, I. (1967a) Mechanisms for the production of the motor output pattern in flying locusts. *J. exp. Biol.*, **47**, 201–212.

Waldron, I. (1967b) Neural mechanism by which controlling inputs influence motor output in the flying locust. *J. exp. Biol.*, **47**, 213–228.

Waldron, I. (1968) The mechanism of coupling of the locust flight oscillator to oscillatory inputs. *Z. Physiol.*, **57**, 331–347.

Walk, R. D. (1958) "Visual" and "visual-motor" experience: a replication. *J. comp. physiol. Psychol.*, **51**, 785–787.

Walk, R. D. (1965) The study of visual depth and distance perception in animals. In Lehrman *et al.* (1965).

Walk, R. D., and Gibson, E. J. (1961) A comparative and analytical study of visual depth perception. *Psychol. Monogr.*, **75**, 44 pp.

Walk, R. D., Gibson, E. J., and Tighe, T. J. (1957) Behavior of light- and dark-reared rats on a visual cliff. *Science*, **126**, 80–81.

Walker, E. L., Dember, W. N., Earl, R. W., Fawl, C. L., and Karoly, A. J. (1955) Choice alternation: III. Response intensity vs. response discriminability. *J. comp. physiol. Psychol.*, **48**, 80–85.

Wallace, G. K. (1958) Some experiments on form perception in the nymphs of the desert locust *Schistocerca gregaria* Forskal. *J. exp. Biol.*, **35**, 765–775.

Wallace, G. K. (1959) Visual scanning in the desert locust *Schistocerca gregaria* Forskal. *J. exp. Biol.*, **36**, 512–545.

Wallace, G. K. (1962) Experiments on visually controlled orientation in the desert locust, *Schistocerca gregaria* (Forskal). *Anim. Behav.*, **10**, 361–369.

Wallis, D. I. (1962) Aggressive behaviour in the ant, *Formica fusca*. *Anim. Behav.*, **10**, 267–274.

Wallraff, H. G. (1960) Können Grasmücken mit Hilfe des Sternenhimmels navigieren? *Z. Tierpsychol.*, **17**, 165–177.

Wallraff, H. G. (1967) The present status of our knowledge about pigeon homing. *Proc. 14th Int. Ornith. Cong. Oxford*, 331–358, Blackwell, Oxford.

Walther, F. (1958) Zum Kampf- und Paarungsverhalten einiger Antilopen. *Z. Tierpsychol.*, **15**, 340–380.

Walther F. (1960) "Antilopenhafte" Verhaltensweisen im Paarungszeremoniell des Okapi (*Okapia johnstoni* Sclater, 1901). *Z. Tierpsychol.*, **17**, 188–210.

Walther, F. R. (1965) Verhaltensstudien an der Grantgazelle (*Gazella granti* Brooke, 1872) im Ngorongoro-Krater. *Z. Tierpsychol.*, **22**, 167–208.

Ward, J. A., and Barlow, G. W. (1967) The maturation and regulation of glancing off the parents by young orange chromides (*Etroplus maculatus*: Pisces—Cichlidae). *Behaviour*, **29**, 1–56.

Warden, C. J. (1931) Animal Motivation: Experimental Studies on the Albino Rat. Columbia Univ. Press, New York.

Ware, R. (1968) Development of differential reinforcing values of sexual responses in the male albino rat. *J. comp. physiol. Psychol.*, **65**, 461–465.

Warren, J. M. (1953) Additivity of cues in a visual pattern discrimination by monkeys. *J. comp. physiol. Psychol.*, **46**, 484–486.

Warren, J. M. (1957) The phylogeny of maze learning: (i) Theoretical orientation. *Brit. J. Anim. Behav.*, **5**, 90–93.

Warren, J. M. (1960) Reversal learning by paradise fish (*Macropodus opercularis*). *J. comp. physiol. Psychol.*, **53**, 376–378.

Warren, J. M. (1965a) Primate learning in comparative perspective. In Schrier *et al.* (1965).

Warren, J. M. (1965b) The comparative psychology of learning. *Ann. Rev. Psychol.*, **16**, 95–118.

Warren, J. M. (1967) An assessment of the reversal index. *Anim. Behav.*, **15**, 493–498.

Warren, R. P. (1963) Preference aversion in mice to bitter substance. *Science*, **140**, 808–809.

Warren, R. P., and Aronson, L. R. (1957) Sexual behavior in adult male hamsters castrated-adrenalectomized prior to puberty. *J. comp. physiol. Psychol.*, **50**, 475–480.

Warren, R. P., and Hinde, R. A. (1959) The effect of oestrogen and progesterone on the nest-building of domesticated canaries. *Anim. Behav.*, **7**, 209–213.

Warriner, C. C., Lemmon, W. B., and Ray, T. S. (1963) Early experience as a variable in mate selection. *Anim. Behav.*, **11**, 221–224.

Wasman, M., and Flynn, J. P. (1962) Directed attack elicited from hypothalamus. *Arch. Neurol.*, **6**, 220–227.

Waterman, T. H., Wiersma, C. A. G., and Bush, B. M. H. (1964) Afferent visual responses in the optic nerve of the crab *Podophthalamus*. *J. cell. comp. Physiol.*, **63**, 135–155.

Watson, A. J. (1961) The place of reinforcement in the explanation of behaviour. In Thorpe and Zangwill (1961).

Watson, J. B. (1919) Psychology from the Standpoint of a Behaviorist. Lippincott, Philadelphia.

Weasner, M. H., Finger, F. W., and Reid, L. S. (1960) Activity changes under food deprivation as a function of recording device. *J. comp. physiol. Psychol.*, **53**, 470–474.

Webb, W. B. (1949) The motivational aspect of an irrelevant drive in the behavior of the white rat. *J. exp. Psychol.*, **39**, 1–14.

Wecker, S. C. (1963) The role of early experience in habitat selection by the prairie deermouse, *Peromyscus maniculatus bairdi. Ecol. Monogr.*, **33**, 307–325.

Weddell, G. (1955) Somesthesis and the chemical senses. *Ann. Rev. Psychol.*, **6**, 119–136.

Weddell, G., Taylor, D. A., and Williams, C. M. (1955) Studies on the innervation of skin: III. The patterned arrangement of the spinal sensory nerves to the rabbit ear. *J. Anat. Lond.*, **89**, 317–342.

Weene, P., and Held, R. (1966) Changes in perceived size of angle as a function of orientation in the frontal plane. *J. exp. Psychol.*, **71**, 55–59.

Wehmer, F. (1965) Effects of prior experience with objects on maternal behaviours in the rat. *J. comp. physiol. Psychol.*, **60**, 294–296.

Wehner, R. (1967) Zur Physiologie des Formensehens bei der Honigbiere. *Z. vergl. Physiol.*, **55**, 145–166.

Weidmann, R., and Weidmann, U. (1958) An analysis of the stimulus situation releasing food-begging in the black-headed gull. *Anim. Behav.*, **6**, 114.

Weidmann, U. (1958) Verhaltensstudien an der Stockente (*Anas platyrhynchos* L.): II. *Z. Tierpsychol.*, **15**, 277–300.

Weiner, I. H., and Stellar, E. (1951) Salt preference of the rat determined by a single-stimulus method. *J. comp. physiol. Psychol.*, **44**, 394–401.

Weingarten, M., and Spinelli, D. N. (1966) Retinal receptive field changes produced by auditory and somatic stimulation. *Exptl. Neurol.*, **15**, 363–376.

Weinstein, S., Sersen, E. A., Fisher, L., and Weisinger, M. (1964) Is reafference necessary for visual adaptation? *Percept. mot. Skills*, **18**, 641–648.

Weir, M. W., and De Fries, J. C. (1964) Prenatal maternal influence on behavior in mice. *J. comp. physiol. Psychol.*, **58**, 412–417.

Weis-Fogh, T. (1956) Biology and physics of locust flight: II. Flight performance of the desert locust (*Schistocerca gregaria*). *Phil. Trans. Roy. Soc. B*, **239**, 459–510.

Weis-Fogh, T. (1964) Control of basic movements in flying insects. *Symp. Soc. exp. Biol.*, **18**, 343–361.

Weiskrantz, L. (1958) Sensory deprivation and the cat's optic nervous system. *Nature*, **181**, 1047–1050.

Weiskrantz, L. (1964) Neurological studies and animal behaviour. *Brit. Med. Bull.*, **20**, 49–53.

Weiskrantz, L. (Ed.) (1968a) Analysis of Behavioral Change. Harper & Row, New York.

Weiskrantz, L. (1968b) Memory. In Weiskrantz (1968a).

Weiskrantz, L. (1968c) Emotion. In Weiskrantz (1968a).

Weiskrantz, L., and Cowey, A. (1963) The aetiology of food reward in monkeys. *Anim. Behav.*, **11**, 225–234.

Weiss, B. A., and Schneirla, T. C. (1967) Inter-situational transfer in the ant

Formica schanfussi as tested in a two-phase single choice-point maze. *Behaviour*, **28**, 269–279.

Weiss, J. M. (1968) Effects of coping responses on stress. *J. comp. physiol. Psychol.*, **65**, 251–260.

Weiss, P. (1941a) Does sensory control play a constructive role in the development of motor coordination? *Schweiz. Med. Wschr.*, **71**, 591–595.

Weiss, P. (1941b) Self-differentiation of the basic patterns of coordination. *Comp. Psychol. Monogr.*, **17**.

Weiss, P. (1950) Experimental analysis of coordination by the disarrangement of central-peripheral relations. *Symp. Soc. exp. Biol.*, **4**, 92–111.

Welford, A. T. (1959) Evidence of a single channel decision mechanism limiting performance in a serial reaction task. *Q. J. exp. Psychol.*, **11**, 193–210.

Welford, A. T. (1962) Experimental psychology and the study of social behaviour. In "Society: Problems and Methods of Study", Ed. A. T. Welford *et al.*, Routledge and Kegan-Paul, London.

Welford, A. T., Brown, R. A., and Gabb, J. E. (1950) Two experiments on fatigue as affecting skilled performance in civilian air crew. *Brit. J. Psychol.*, **40**, 195–211.

Welker, W. I. (1956) Effects of age and experience on play and exploration of young chimpanzees. *J. comp. physiol. Psychol.*, **49**, 223–226.

Welker, W. I. (1964) Analysis of sniffing of the albino rat. *Behaviour*, **22**, 223–244.

Welker, W. I., and King, W. A. (1962) Effects of stimulus novelty on gnawing and eating by rats. *J. comp. physiol. Psychol.*, **55**, 838–842.

Wells, G. P. (1950) Spontaneous activity cycles in polychaete worms. *Symp. Soc. exp. Biol.*, **4**, 127–142.

Wells, G. P. (1955) The sources of animal behaviour. Inaugural lecture, Univ. Coll. London.

Wells, M. J. (1958) Factors affecting reactions to *Mysis* by newly hatched *Sepia*. *Behaviour*, **13**, 96–111.

Wells, M. J. (1959) Functional evidence for neurone fields representing the individual arms within the central nervous system of *Octopus*. *J. exp. Biol.*, **36**, 501–511.

Wells, M. J. (1961) Weight discrimination by *Octopus*. *J. exp. Biol.*, **38**, 127–133.

Wells, M. J. (1962a) Early learning in Sepia. *Symp. zool. Soc. Lond.*, **8**, 149–169.

Wells, M. J. (1962b) Brain and Behaviour in Cephalopods. Heinemann, London: Stanford Univ. Press, Stanford.

Wells, M. J. (1963) The orientation of *Octopus*. *Ergeb. Biol.*, **26**, 40–54.

Wells, M. J. (1964a) Tactile discrimination of shape by *Octopus*. *Q. J. exp. Psychol.*, **16**, 156–162.

Wells, M. J. (1964b) Tactile discrimination of surface curvature and shape by the octopus. *J. exp. Biol.*, **41**, 433–445.

Wells, M. J., and Wells, J. (1957) The function of the brain of *Octopus* in tactile discrimination. *J. exp. Biol.*, **34,** 131–142.

Wenner, A. M. (1962) Sound production during the waggle dance of the honey bee. *Anim. Behav.*, **10,** 79–95.

Wenner, A. M. (1967) Honey bees: do they use the distance information contained in their dance maneuver? *Science*, **155,** 847–849.

Werner, G., and Mountcastle, V. B. (1963) The variability of central nervous activity in a sensory system, and its implications for the central reflection of sensory events. *J. Neurophysiol.*, **26,** 958–977.

Whalen, R. E. (1961) Effects of mounting without intromission and intromission without ejaculation on sexual behaviour and maze learning. *J. comp. physiol. Psychol.*, **54,** 409–415.

Whalen, R. E. (1963a) The initiation of mating in naive female cats. *Anim. Behav.*, **11,** 461–463.

Whalen, R. E. (1963b) Sexual behavior of cats. *Behaviour*, **20,** 321–342.

Whalen, R. E., Beach, F. A., and Kuehn, R. E. (1961) Effects of exogenous androgen on sexually responsive and unresponsive male rats. *Endocrinology*, **69,** 373–380.

Whalen, R. E., and Robertson, R. T. (1968) Sexual exhaustion and recovery of masculine copulatory behavior in virilized female rats. *Psychon. Sci.*, 11, 319–320.

Whimbey, A. E., and Denenberg, V. H. (1967) Two independent behavioral dimensions in open-field performance. *J. comp. physiol. Psychol.*, **63,** 500–504.

White, B. L., Castle, P., and Held, R. (1964) Observations on the development of visually directed reaching. *Child Developm.*, **35,** 349–364.

White, B. L., and Held, R. (1966) Plasticity of sensorimotor development in the human infant. In "Causes of Behavior", Ed. J. F. Rosenblith and W. Allinsmith, Allyn and Bacon, Boston.

Whitfield, I. C. (1967) Coding in the auditory nervous system. *Nature*, **213,** 756–760.

Whiting, H. P. (1955) Functional development in the nervous system. *Proc. 1st Int. Neurochem. Symp.*, Academic Press, New York.

Whitman, C. O. (1919) The behavior of pigeons. *Publ. Carneg. Inst.*, **257,** 1–161.

Wickelgren, B. G. (1967a) Habituation of spinal motoneurons. *J. Neurophysiol.*, **30,** 1404–1423.

Wickelgren, B. G. (1967b) Habituation of spinal interneurons. *J. Neurophysiol.*, **30,** 1424–1438.

Wickens, D. D., Hall, J., and Reid, L. S. (1949) Associative and retro-active inhibition as a function of the drive stimulus. *J. comp. physiol. Psychol.*, **42,** 398–403.

Wickler, W. (1958) Vergleichende Verhaltensstudien an Grundfischen: II. *Z. Tierpsychol.*, **15,** 427–446.

Wickler, W. (1960) Die Stammesgeschichte typischer Bewegungsformen der Fisch-Brustflosse. *Z. Tierpsychol.*, **17**, 31–66.

Wickler, W. (1961a) Okologie und Stammesgeschichte von Verhaltensweisen. *Fortschr. Zool.*, **13**, 303–365.

Wickler, W. (1961b) Uber die Stammesgeschichte und den taxonomischen Wert einiger Verhaltensweisen der Vögel. *Z. Tierpsychol.*, **18**, 320–342.

Wickler, W. (1962) Ei-Attrappen und Maulbrüten bei afrikanischen Cichliden. *Z. Tierpsychol.*, **19**, 129–164.

Wickler, W. (1967a) Socio-sexual signals and their intraspecific imitation among primates. In Morris (1967).

Wickler, W. (1967b) Vergleichende Verhaltensforschung und Phylogenetik. In "Die Evolution der Organismen", 3, 1, Ed. G. Heberer, G. Fischer, Stuttgart.

Wickler, W. (1968) Mimicry. Weidenfeld and Nicolson, London.

Wiemer, F.-K. (1955) Mittelhirnfunktion bei Urodelen nach Regeneration und Transplantation. *Arch. Entw. Mech. Org.*, **147**, 560–633.

Wiepkema, P. R. (1961) An ethological analysis of the reproductive behaviour of the bitterling. *Archs. néerl. Zool.*, **14**, 103–199.

Wiepkema, P. R., de Ruiter, L., and Reddingius, J. (1966) Circadian rhythms in the feeding behaviour of CBA mice. *Nature*, **209**, 935–936.

Wiersma, C. A. G. (1931) An experiment on the "Resonance theory" of muscular activity. *Archs. néerl. Physiol.*, **16**, 337–345.

Wiersma, C. A. G. (1962) The organization of the arthropod central nervous system. *Amer. Zoologist*, **2**, 67–78.

Wiersma, C. A. G., Bush, B. M. H., and Waterman, T. H. (1964) Efferent visual responses of contralateral origin in the optic nerve of the crab *Podophthalamus*. *J. cell. comp. Physiol.*, **64**, 309–326.

Wiersma, C. A. G., and Ikeda, K. (1964) Interneurons commanding swimmeret movements in the crayfish *Procambarus clarkii* (Girard). *Comp. Biochem. Physiol.*, **12**, 509–525.

Wiersma, C. A. G., Waterman, T. H., and Bush, B. M. H. (1961) Impulse traffic in the optic nerve of decapod Crustacea. *Science*, **134**, 1435.

Wiersma, C. A. G., and Yamaguchi, T. (1967) Integration of visual stimuli by the crayfish central nervous system. *J. exp. Biol.*, **47**, 409–431.

Wiesel, T. N. (1960) Receptive fields of ganglion cells in the cat's retina. *J. Physiol.*, **153**, 583–594.

Wiesel, T. N., and Hubel, D. H. (1963a) Single-cell responses in striate cortex of kittens deprived of vision in one eye. *J. Neurophysiol.*, **26**, 1003–1017.

Wiesel, T. N., and Hubel, D. H. (1963b) Effects of visual deprivation on morphology and physiology of cells in the cat's lateral geniculate body. *J. Neurophysiol.*, **26**, 978–993.

Wiesel, T. N., and Hubel, D. H. (1965a) Extent of recovery from the effects of visual deprivation in kittens. *J. Neurophysiol.*, **28**, 1060–1072.

Wiesel, T. N., and Hubel, D. H. (1965b) Comparison of the effects of unilateral and bilateral eye closure on cortical unit responses in kittens. *J. Neurophysiol.*, **28,** 1029–1040.

Wiesel, T. N., and Hubel, D. H. (1966) Spatial and chromatic interactions in the lateral geniculate body of the rhesus monkey. *J. Neurophysiol.*, **29,** 1115–1156.

Wiesner, B. P., and Sheard, N. M. (1933) Maternal Behaviour in the Rat. Oliver and Boyd, Edinburgh.

Wigglesworth, V. B. (1941) The sensory physiology of the human louse *Pediculus humanus corporis de Beer* (Anoplura). *Parasitology,* **33,** 67–109.

Wigglesworth, V. B. (1949) The utilisation of reserve substances in *Drosophila* during flight. *J. exp. Biol.*, **28,** 150–163.

Wikler, A. (1952) Pharmacologic dissociation of behavior and EEG "sleep patterns" in dogs: Morphine, N-allylnormorphine, and atropine. *Proc. Soc. exp. Biol. Med.*, **79,** 261–265.

Wilcott, R.C. (1967) Effects of cerebral lesions in young and mature rats on later maze performance. *Proc. 75th Ann. Conv. Amer. Psychol. Ass.*, 75–76.

Wilkins, B. R. (1966a) Regulation and control in engineering. In Kalmus (1966).

Wilkins, B. R. (1966b) Basic mathematics of control. In Kalmus (1966).

Williams, A. O. D. (1967) Behavioral acts elicited by stimulation of single, identifiable brain cells. *Science,* **157,** 570–574.

Williams, C. D., Carr, R. M., and Peterson, H. W. (1966) Maze exploration in rats of four ages. *J. genet. Psychol.*, **109,** 241–247.

Williams, D. I. (1967a) The overtraining reversal effect in the pigeon. *Psychonomic Sci.*, **7,** 261–262.

Williams, D. I. (1967b) Constant irrelevant cue learning in the pigeon. *Anim. Behav.*, **15,** 229–230.

Williams, E., and Scott, J. P. (1953) The development of social behavior patterns in the mouse, in relation to natural periods. *Behaviour,* **6,** 35–65.

Willis, E. O. (1967) The behavior of bicolored ant birds. *Univ. Calif. Publ. Zool.*, **79,** 1–127.

Wilson, D. M. (1961) The central nervous control of flight in a locust. *J. exp. Biol.*, **38,** 471–490.

Wilson, D. M. (1964) Relative refractoriness and patterned discharge of locust flight motor neurones. *J. exp. Biol.*, **41,** 191–205.

Wilson, D. M. (1965) Proprioceptive leg reflexes in cockroaches. *J. exp. Biol.*, **43,** 397–409.

Wilson, D. M. (1966a) Central nervous mechanisms for the generation of rhythmic behaviour in arthropods. *Symp. soc. exp. Biol.*, **20,** 199–228.

Wilson, D. M. (1966b) Insect walking. *Ann. Rev. Ent.*, **11,** 103–122.

Wilson, D. M. (1967) Stepping patterns in Tarantula spiders. *J. exp. Biol.*, **47,** 133–151.

Wilson, D. M. (1968) Inherent asymmetry and reflex modulation of the locust flight pattern. *J. exp. Biol.*, **48,** 631–641.

Wilson, D. M., and Davis, W. J. (1965) Nerve impulse patterns and reflex control in the motor system of the crayfish claw. *J. exp. Biol.*, **43**, 193–210.

Wilson, D. M., and Gettrup, E. (1963) A stretch reflex controlling wingbeat frequency in grasshoppers. *J. exp. Biol.*, **40**, 171–185.

Wilson, D. M., and Hoy, R. R. (1968) Optomotor reaction, locomotory bias, and reactive inhibition in the milkweed bug *Oncopeltus* and the beetle *Zophobas*. *Z. vergl. Physiol.*, **58**, 136–152.

Wilson, D. M., and Weis-Fogh, T. (1963) Patterned activity of coordinated motor units, studied in flying locusts. *J. exp. Biol.*, **39**, 643–667.

Wilson, D. M., and Wyman, R. J. (1963) Phasically unpatterned nervous control of dipteran flight. *J. insect Physiol.*, **9**, 859–865.

Wilson, D. M., and Wyman, R. J. (1965) Motor output patterns during random and rhythmic stimulation of locust thoracic ganglia. *Biophys. Journ.*, **5**, 121–143.

Wilson, E. O. (1962) Chemical communication among workers of the fire ant *Solenopsis saevissima* (Fr. Smith): 3. The experimental induction of social responses. *Anim. Behav.*, **10**, 159–164.

Wilson, G. T., and Radloff, W. P. (1967) Degree of arousal and performance: effects of reticular stimulation on an operant task. *Psychonomic Sci.*, **7**, 13–14.

Wilson, J. R., Adler, N., and Le Boeuf, B. (1965) The effects of intromission frequency on successful pregnancy in the female rat. *Proc. Nat. Acad. Sci.*, **53**, 1392–1395.

Wilson, J. R., Kuehn, R. E., and Beach, F. A. (1963) Modification in the sexual behavior of male rats produced by changing the stimulus female. *J. comp. physiol. Psychol.*, **56**, 636–644.

Wilson, P. D., and Riesen, A. H. (1966) Visual deprivation in rhesus monkeys neonatally deprived of patterned light. *J. comp. physiol. Psychol.*, **61**, 87–95.

Wilson, R. S., and Bartels, B. L. (1968) Cardiac responses. *J. comp. physiol. Psychol.*, **65**, 132–139.

Wiltschko, W. (1968) Uber den Einfluss statischer Magnetfelder auf die Zugorientierung der Rotkehlchen (*Erithacus rubecula*). *Z. Tierpsychol.*, **25**, 537–558.

Wilz, K. J. (1967) The organization of courtship behaviour in sticklebacks. D. Phil. thesis, Oxford.

Windle, W. F. (1940) Physiology of the Foetus: Origin and Extent of Function in Prenatal Life. W. B. Saunders, Philadelphia.

Windle, W. F. (1944) Genesis of somatic motor function in mammalian embryos: a synthesizing article. *Physiol. Zool.*, **17**, 247–260.

Winn, H. E. (1957) Egg site selection by three species of darters (*Pisces, Percidae*). *Brit. J. Anim. Behav.*, **5**, 25–28.

Winn, H. E., Salmon, M., and Roberts, N. (1964) Sun-compass orientation by parrot fishes. *Z. Tierpsychol.*, **21**, 798–812.

Wolda, H. (1961) Response decrement in the prey-catching activity of *Notonecta glauca* L. (Hemiptera). *Archs. néerl. Zool.*, **14,** 61–89.

Wolf, E., and Zerrahn-Wolf, G. (1937) Flicker and the reactions of bees to flowers. *J. genet. Physiol.*, **20,** 511–518.

Wolfe, J. B. (1936) Effectiveness of token-records for chimpanzees. *Comp. Psychol. Monogr.*, **12,** 5.

Wolfe, J. B. (1939) An exploratory study of food-storing in rats. *J. comp. Psychol.*, **28,** 97–108.

Wood-Gush, D. G. M. (1956) The agonistic and courtship behaviour of the brown leghorn cock. *Brit. J. Anim. Behav.*, **4,** 133–142.

Woodruff, A. B., and Slövak, M. L. (1965) The effects of severely restricted visual experience on the perception of "identity" *Psychon. Sci.*, **2,** 41–42.

Woods, P. J. (1959) The effects of free and restricted environmental experience on problem-solving behavior in the rat. *J. comp. physiol. Psychol.*, **52,** 399–402.

Woods, P. J. (1962) Behavior in a novel situation as influenced by the immediately preceding environment. *J. exp. Anal. Behav.*, **5,** 185–190.

Woodworth, R. S. (1918) Dynamic Psychology. Columbia Univ. Press, New York.

Worden, F. G. (1966) Attention and auditory electrophysiology. In "Progress in Physiological Psychology", Ed. E. Stellar and J. M. Sprague, **1,** 45–116.

Worden, F. G., and Livingston, R. B. (1961) Brain-stem reticular formation. In Sheer (1961).

Worden, F. G., and Marsh, J. T. (1963) Amplitude changes of auditory potentials evoked at cochlear nucleus during acoustic habituation. *Electroenceph. clin. Neurophysiol.*, **15,** 866–881.

Wright, J. H., Gescheider, G. A., and Johnson, M. L. (1966) Energizing effects of combined food and water deprivation upon general activity. *Psychon. Sci.*, **5,** 415–416.

Wünschmann, A. (1963) Quantitative Untersuchungen zum Neugierverhalten von Wirbeltieren. *Z. Tierpsychol.*, **20,** 80–109.

Wüstehube, C. (1960) Beiträge zur Kenntnis besonders des Spiel- und Beuteverhaltens einheimischer Musteliden. *Z. Tierpsychol.*, **17,** 579–613.

Wyckoff, L. B. (1954) The role of observing responses in discrimination learning: Part I. *Psychol. Rev.*, **59,** 431–442.

Wynne, L. C., and Solomon, R. L. (1955) Traumatic avoidance learning: acquisition and extinction in dogs deprived of normal peripheral autonomic function. *Genet. Psychol. Monogr.*, **52,** 241–284.

Wyrwicka, W., and Dobrzecka, C. (1960) Relationship between feeding and satiation centers of the hypothalamus. *Science*, **132,** 805–806.

Wyrwicka, W., Dobrzecka, C., and Tarnocki, A. (1960) The effect of electrical stimulation of the hypothalamic feeding centre in satiated goats on alimentary conditioned reflexes, Type II. *Acta Biol. Exp.*, **20,** 121–136.

Yaschine, T., Mena, F., and Beyer, C. (1967) Gonadal hormones and mounting behavior in the female rabbit. *Amer. J. Physiol.*, **213,** 867–872.

Yates, A. J. (1963) Delayed auditory feedback. *Psychol. Bull.*, **60**, 213–232.

Yehle, A., Dauth, G., and Schneiderman, N. (1967) Correlates of heart-rate classical conditioning in curarized rabbits. *J. comp. physiol. Psychol.*, **64**, 98–104.

Young, J. Z. (1960) Regularities in the retina and optic lobes of *Octopus* in relation to form discrimination. *Nature*, **186**, 836–839.

Young, J. Z. (1961) Learning and discrimination in the octopus. *Biol. Rev.*, **36**, 32–96.

Young, J. Z. (1964) Paired centres for the control of attack by *Octopus*. *Proc. Roy. Soc. B*, **159**, 565–588.

Young, J. Z. (1965) The organization of a memory system. *Proc. Roy. Soc. B*, **163**, 285–320.

Young, P. T. (1949) Food-seeking drive, affective process and learning. *Psychol. Rev.*, **56**, 98–121.

Young, W. C. (1961a) Sex and Internal Secretion. Williams and Wilkins, Baltimore.

Young, W. C. (1961b) The hormones and mating behaviour. In Young (1961a).

Young, W. C. (1965) The organization of sexual behaviour by hormonal action during the prenatal and larval periods in vertebrates. In Beach (1965).

Young, W. C., Goy, R. W., and Phoenix, C. H. (1964) Hormones and sexual behaviour. *Science*, **143**, 212–218.

Youngstrom, K. A. (1938) Studies of the developing behavior of Anura. *J. comp. Neurol.*, **68**, 351–379.

Zannier-Tanner, E. (1965) Vergleichende Verhaltensuntersuchungen über das Hinlegen und Aufstehen bei Huftieren. *Z. Tierpsychol.*, **22**, 696–723.

Zeigler, H. P. (1961) Learning-set formation in pigeons. *J. comp. physiol. Psychol.*, **54**, 252–254.

Zeigler, H. P. (1964) Displacement activity and motivational theory. A case study in the history of ethology. *Psychol. Bull.*, **61**, 362–376.

Zeigler, H. P., Karten, H., and Green, H. L. (1967) Neural mechanisms in the feeding behavior of the pigeon. Paper delivered at Eastern Psychological Association, April, 1967. *Amer. Psychol.*, **22**, 748.

Zeigler, H. P., and Schmerler, S. (1965) Visual discrimination of orientation by pigeons. *Anim. Behav.*, **13**, 475–477.

Zeigler, H. P., and Witkowsky, P. (1968) The main sensory trigeminal nucleus of the pigeon: a single-unit analysis. *J. Comp. Neurol.*

Zeigler, H. P., and Wyckoff, L. B. (1961) Observing responses and discrimination learning. *Q. J. exp. Psychol.*, **13**, 129–140.

Zener, K. (1937) The significance of behavior accompanying conditioned salivary secretion for theories of the conditioned response. *Amer. J. Psychol.*, **50**, 384–403.

Zerrahn, G. (1933) Formdressur und Formunterscheidung bei der Honigbiene. *Z. vergl. Physiol.*, **20**, 117–161.

Zimbardo, P. G. (1958) The effects of early avoidance training and rearing conditions upon the sexual behaviour of the male rat. *J. comp. physiol. Psychol.*, **51,** 764–769.

Zimbardo, P. G., and Montgomery, K. C. (1957) Effects of "free environment" rearing upon exploratory behavior. *Psychol. Rep.*, **3,** 589–594.

Zimmermann, R. R. (1962) Form generalization in the infant monkey. *J. comp. physiol. Psychol.*, **55,** 918–923.

Zimmermann, R. R., and Torrey, C. C. (1965) Ontogeny of learning. In Schreier *et al.* (1965).

Zippelius, H. M., and Schleidt, W. M. (1956) Ultraschallaute bei jungen Mäusen. *Naturwiss.*, **21,** 502.

Zitrin, A., Jaynes, J., and Beach, F. A. (1956) Neural mediation of mating in male cats. III. *J. comp. Neurol.*, **105,** 111–125.

Zolman, J. F., and Becker, D. (1968) Spatial discrimination learning in young chicks. *Psychon. Sci.*, **10,** 361–362.

Zolman, J. F., and Martin, R. C. (1967) Instrumental aversive conditioning in newly hatched domestic chicks. *Psychon. Sci.*, **8,** 183–184.

Zucker, I. (1966) Facilitatory and inhibitory effects of progesterone on sexual responses of spayed guinea pigs. *J. comp. physiol. Psychol.*, **62,** 376–381.

Zucker, I. (1968) Biphasic effects of progesterone on sexual receptivity in the female guinea pig. *J. comp. physiol. Psychol.*, **65,** 472–478.

INDEXES

NAME INDEX

SUBJECT INDEX